Name Reactions

Jie Jack Li

Name Reactions

A Collection of Detailed Mechanisms and Synthetic Applications

Sixth Edition

 Springer

Jie Jack Li, Ph.D
Discovery Chemistry
ChemPartner
San Francisco, CA, USA

ISBN 978-3-030-50867-8   ISBN 978-3-030-50865-4   (eBook)
https://doi.org/10.1007/978-3-030-50865-4

This Springer imprint is published by the registered company Springer Nature Switzerland AG.
The registered company address is: Gewerbestrasse 11, 6330 Cham, Switzerland

Dedicated to Prof. David R. Williams

Kurt Alder
1902–1958
Nobel Prize, 1950

Eduard Buchner
1860–1917
Nobel Prize, 1907

Adolf von Baeyer
1835–1917
Nobel Prize, 1905

Elias James Corey
1928–
Nobel Prize, 1990

Derek H. R. Barton
1918–1999
Nobel Prize, 1969

Otto Paul Herman Diels
1876–1954
Nobel Prize, 1950

Emil Fisher
1852–1919
Nobel Prize, 1902

Victor Grignard
1871–1935
Nobel Prize, 1912

Hermann Staudinger
1881–1965
Nobel Prize, 1953

Robert Robinson
1886–1975
Nobel Prize, 1947

Georg Wittig
1897–1987
Nobel Prize, 1979

Otto Wallach
1847–1931
Nobel Prize, 1910

Karl Ziegler
1898–1973
Nobel Prize, 1963

Preface

Five years have elapsed since the fifth edition was published. Much has happened since then. The author has migrated from academia back to industry. I have taken out some name reactions from the fifth edition because the book was physically getting too heavy and unwieldy. This change allows more space to expand and update the more popular name reactions. All references have been updated to 2020 when available.

As in previous editions, each reaction is delineated by detailed, step-by-step, electron-pushing mechanism, supplemented with the original and the latest references, especially review articles. Now, with addition of many synthetic applications, it is not only an indispensable resource for senior undergraduate and graduate students to learn mechanisms and synthetic utility of name reactions and to prepare for their exams, but also a good reference book for all organic chemists in both industry and academia.

As always, I welcome your critique. Please send your comments to this email address: lijiejackli@hotmail.com.

March 1, 2020
San Mateo, California

Jie Jack Li

Table of Contents

Abbreviations and Acronyms

⬤—	Polymer support
1,10-phen	1,10-Phenanthroline
3CC	Three-component condensation
3CR	Three-component reaction
4CC	Four-component condensation
9-BBN	9-Borabicyclo[3.3.1]nonane
A	Adenosine
Ac	Acetyl
acac	Acetylacetonate
ACC	Acetyl-CoA carboxylase
ADDP	1,1′-(Azodicarbonyl)dipiperidine
AIBN	2,2′-Azobisisobutyronitrile
Alpine-borane®	B-Isopinocampheyl-9-borabicyclo[3.3.1]-nonane
AOM	p-Anisyloxymethyl = p-MeOC$_6$H$_4$OCH$_2$-
Ar	Aryl
ARA	Asymmetric reductive amination
ATH	Asymmetric transfer hydrogenation
ATPH	Tris(2,6-diphenyl)phenoxyaluminane
B:	Generic base
BBEDA	Bis-benzylidene ethylenediamine
bmim	1-Butyl-3-methylimidazolium
BINAP	2,2′-Bis(diphenylphosphino)-1,1′-binaphthyl
BINOL	1,1′-Bi-2-naphthaol
Bn	Benzyl
Boc	*tert*-Butyloxycarbonyl
BQ	Benzoquinone
BPR	Back pressure regulator
BT	Benzothiazole
Bz	Benzoyl

CAN	Cerium ammonium nitrate
CBS	Corey–Bakshi–Shibata reaction
Cbz	Benzyloxycarbonyl
CCB	Calcium channel blockers
CD4	Cluster of differentiation 4
CDK	Cyclin-dependent kinase
CFC	Continuous flow centrifugation
cod	1,5-Cyclooctadiene
COPC	Carbonyl–olefin [2 + 2] photocycloaddition
Cp	Cyclopentyl
CPME	Cyclopentyl methyl ether
CSA	Camphorsulfonic acid
CuTC	Copper thiophene-2-carboxylate
Cy	Cyclohexyl
DABCO	1,4-Diazabicyclo[2.2.2]octane
dba	Dibenzylideneacetone
DBU	1,8-Diazabicyclo[5.4.0]undec-7-ene
o-DCB	*ortho*-Dichlorobenzene
DCC	1,3-Dicyclohexylcarbodiimide
DCE	Dichloroethane
DDQ	2,3-Dichloro-5,6-dicyano-1,4-benzoquinone
de	Diastereoselctive excess
DEAD	Diethyl azodicarboxylate
DEL	DNA-encoded library
DET	Diethyl tartrate
Δ	Reaction heated under reflux
(DHQ)$_2$-PHAL	1,4-Bis(9-*O*-dihydroquinine)-phthalazine
(DHQD)$_2$-PHAL	1,4-Bis(9-*O*-dihydroquinidine)-phthalazine
DIAD	Diisopropyl azodidicarboxylate
DIBAL	Diisobutylaluminum hydride
DIC	*N*,*N*′-Diisopropylcarbodimide
DIPT	Diisopropyl tartrate
DIPEA	Diisopropylethylamine
DKR	Dynamic kinetic resolution
DLP	Dilauroyl peroxide
DMA	*N*,*N*-dimethylacetamide
DMAP	4-*N*,*N*-Dimethylaminopyridine
DME	1,2-Dimethoxyethane
DMF	*N*,*N*-Dimethylformamide
DMFDMA	*N*,*N*-Dimethylformamide dimethyl acetal
DMP	Dess–Martin periodinane
DMPU	*N*,*N*′-Dimethylpropyleneurea
DMS	Dimethylsulfide
DMSO	Dimethylsulfoxide
DMSY	Dimethylsulfoxonium methylide
DMT	Dimethoxytrityl

DPP-4	Dipeptidyl peptidase IV
DPPA	Diphenylphosphoryl azide
dppb	1,4-Bis(diphenylphosphino)butane
dppe	1,2-Bis(diphenylphosphino)ethane
dppf	1,1'-Bis(diphenylphosphino)ferrocene
dppp	1,3-Bis(diphenylphosphino)propane
dr	Diastereoselective ratio
DTBAD	Di-*tert*-butylazodicarbonate
DTBMP	2,6-Di-*tert*-butyl-4-methylpyridine
DTBP	Di-*tert*-butyl peroxide
E1	Unimolecular elimination
E1cB	2-Step, base-induced β-elimination *via* carbanion
E2	Bimolecular elimination
EAN	Ethylammonium nitrate
EDCI	1-Ethyl-3-(3-dimethylaminopropyl)carbodiimide
EDDA	Ethylenediamine diacetate
EDG	Electron-donating group
EDTA	Ethylenediaminetetraacetic acid
ee	Enantiomeric excess
Ei	Two groups leave at about the same time and bond to each other as they are doing so.
EMC	Meerwein–Eschenmoser–Claisen
ERK	Extracellular signal-regulated kinase
Eq	Equivalent
Equiv	Equivalent
Et	Ethyl
EtOAc	Ethyl acetate
EWG	Electron-withdrawing group
FEP	Fluorinated ethylene propene
Fmoc	Fluorenylmethyloxycarbonyl protecting group
fod	1,1,1,2,2,3,3-heptafluoro-7,7-dimethyl-4,6-octanedionate = Siever's reagent
FVP	Flash vacuum pyrolysis
HCV	Hepatitis virus C
HFIP	Hexafluoroisopropanol
HKR	Hydrolytic kinetic resolution
HMDS	Hexamethyldisilazane
HMPA	Hexamethylphosphoramide
HMTA	Hexamethylenetetramine
HMTTA	1,1,4,7,10,10-Hexamethyltriethylenetetramine
HOMO	Highest occupied molecular orbital
IBDA	Iodosobenzene diacetate, also known as PIDA
IBX	*o*-Iodoxybenzoic acid
IDH1	Isocitrate dehydrogenase 1
IEDDA	Inverse-electron-demand Diels–Alder
Imd	Imidazole

IMDA	Intramolecular Diels–Alder reaction
IPA	Isopropyl alcohol (Indian pale ale)
IPB	Insoluble polymer bound
IPr	Diidopropyl-phenylimidazolium derivative
JAK	Janus kinase
KHMDS	Potassium hexamethyldisilazide
LAH	Lithium aluminum hydride
LDA	Lithium diisopropylamide
LED	Light-emitting diode
LHMDS	Lithium hexamethyldisilazide
LUMO	Lowest unoccupied molecular orbital
LTMP	Lithium 2,2,6,6-tetramethylpiperidide
M	Metal
MBI	Mechanism-based inhibitors
m-CPBA	*m*-Chloroperoxybenzoic acid
MCRs	Multicomponent reactions
Mes	Mesityl
Mincle	Macrophage-inducible C-type lectin
MLCT	Metal to ligand charge transfer
MOM ether	Methoxymethyl ether
MPL	Medium pressure lamp
MPM	Methyl phenylmethyl
MPS	Morpholine-polysulfide
Ms	Methanesulfonyl (mesyl)
MS	Molecular sieves
MWI	Microwave irradiation
MTBE	Methyl tertiary butyl ether
MVK	Methyl vinyl ketone
NaDA	Sodium diisopropylamide
NBE	Norbornene
NBS	*N*-Bromosuccinimide
NCL	Native chemical ligation
NCS	*N*-Chlorosuccinimide
nbd	2,5-Norbornadiene
NBE	Norbornene
Nf	Nonafluorobutanesulfonyl
NFSI	*N*-Fluorobenzenesulfonimide
NHC	*N*-Heterocyclic carbene
NIS	*N*-Iodosuccinimide
NMM	*N*-Methyl morpholine
NMO	*N*-Methylmorpholine N-oxide (NMMO)
NMP	1-Methyl-2-pyrrolidinone
Nos	Nosylate = 4-nitrobenzenesulfonyl = Ns
NRI	Noradrenaline reuptake inhibitor
N-PSP	*N*-Phenylselenophthalimide
N-PSS	*N*-Phenylselenosuccinimide

Nu	Nucleophile
Nuc	Nucleophile
Ns	Nosylate
PAR-1	Protease activated receptor-1
PARP	Poly(ADP-ribosyl) polymerase
PCC	Pyridinium chlorochromate
PDC	Pyridinium dichromate
PDI	Phosphinyl dipeptide isostere
PE	Premature ejaculation
PEG	Polyethylene glycol
PEPPSI	Pyridine-enhanced pre-catalyst preparation, stabilization, and initiation
phen	1,10-Phenanthroline
PIDA	Phenyliodine diacetate (same as IBDA)
Pin	Pinacol
Piv	Pivaloyl
PNB	p-Nitrobenzyl
PMB	*para*-Methoxybenzyl
PPA	Polyphosphoric acid
PPSE	Trimethylsilyl polyphosphate
PPTS	Pyridinium p-toluenesulfonate
PT	Phenyltetrazolyl
PTADS	Tetrakis[(R)-(+)-N-(p-dodecylphenylsulfonyl)prolinato]
PTSA	p-Toluenesulfonic acid
PyPh$_2$P	Diphenyl 2-pyridylphosphine
Pyr	Pyridine
rac	Racemic
Red-Al	Sodium bis(methoxy-ethoxy)aluminum hydride (SMEAH)
rr	*r*egioisomeric *r*atio
Salen	*N*,*N*′-Disalicylidene-ethylenediamine
SET	Single-electron transfer
SIBX	Stabilized IBX
SM	Starting material
SMC	Sodium methyl carbonate
SMEAH	Sodium bis(methoxy-ethoxy)aluminum hydride: trade name Red-Al
S$_N$1	Unimolecular nucleophilic substitution
S$_N$2	Bimolecular nucleophilic substitution
S$_N$Ar	Nucleophilic substitution on an aromatic ring
SSRI	Selective serotonin reuptake inhibitor
T3P	Propylphosphonic anhydride
TBABB	Tetra-*n*-butylammonium bibenzoate
TBAF	Tetra-*n*-butylammonium fluoride
TBAI	Tetra-*n*-butylammonium iodide
TBAO	1,3,3-Trimethyl-6-azabicyclo[3.2.1]octane

TBDMS	*tert*-Butyldimethylsilyl
TBDPS	*tert*-Butyldiphenylsilyl
TBHP	*tert*-Butyl hydroperoxide
TBS	*tert*-Butyldimethylsilyl
t-Bu	*tert*-Butyl
TDI	Thiophosphinyl dipeptide isostere
TDS	Thexyldimethylsilyl
TEA	Triethylamine
TEMPO	2,2,6,6-Tetramethylpiperidinyloxy
TEOC	Trimethysilylethoxycarbonyl
TES	Triethylsilyl
Tf	Trifluoromethanesulfonyl (triflate)
TFA	Trifluoroacetic acid
TFAA	Trifluoroacetic anhydride
TFE	Trifluoroethanol
TFEA	Trifluoroethyl trifluoroacetate
THF	Tetrahydrofuran
TFP	Tri-2-furylphosphine
TFPAA	Trifluoroperacetic acid
TIPS	Triisopropylsilyl
TMEDA	*N,N,N′,N′*-Tetramethylethylenediamine
TMG	1,1,3,3-Tetramethylguanidine
TMOF	Trimethyl orthoformate
TMP	Tetramethylpiperidine
TMS	Trimethylsilyl
TMSCl	Trimethylsilyl chloride
TMSCN	Trimethylsilyl cyanide
TMSI	Trimethylsilyl iodide
TMSOTf	Trimethylsilyl triflate
TMU	Tetramethylurea
Tol	Toluene or tolyl
Tol-BINAP	2,2′-Bis(di-*p*-tolylphosphino)-1,1′-binaphthyl
TosMIC	(*p*-Tolylsulfonyl)methyl isocyanide
TPPO	Triphenylphosphine oxide
TrxR	Thioredoxin reductase
Ts	Tosyl
TsO	Tosylate
TTBP	2,4,6-Tri-*tert*-butylpyrimidine
UHP	Urea-hydrogen peroxide
VAPOL	2,2′-Diphenyl-(4-biphen-anthrol)
VMR	Vinylogous Mannich reaction
WERSA	Water extract of rice straw ash

Alder Ene Reaction

The Alder ene reaction, also known as the hydro-allyl addition, is addition of an enophile to an alkene (ene) *via* allylic transposition. The four-electron system including an alkene π-bond and an allylic C–H σ-bond can participate in a pericyclic reaction in which the double bond shifts and new C–H and C–C σ-bonds are formed.

X=Y: C=C, C≡C, C=O, C=N, N=N, N=O, S=O, *etc.*

Example 1[5]

Example 2, Here the "ene" is a carbonyl of the aldehyde[7]

© Springer Nature Switzerland AG 2021
J. J. Li, *Name Reactions*, https://doi.org/10.1007/978-3-030-50865-4_1

Example 3, Intramolecular Alder ene reaction[8]

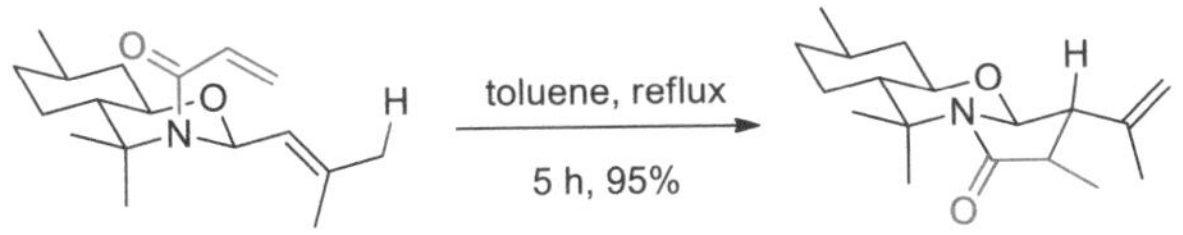
H
toluene, reflux
5 h, 95%
H

Example 4, Cobalt-catalyzed Alder ene reaction[9]

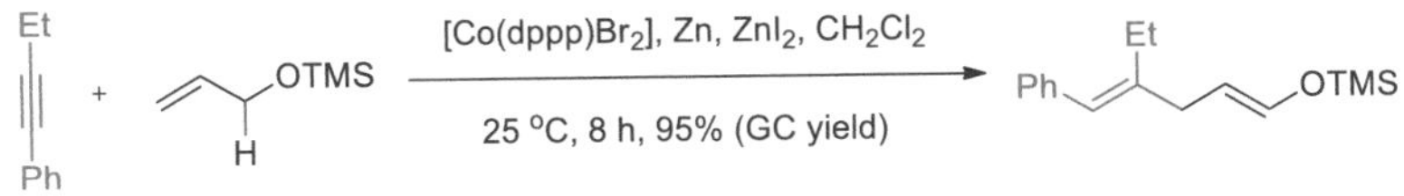
Et
Ph
OTMS
H
[Co(dppp)Br2], Zn, ZnI2, CH2Cl2
25 °C, 8 h, 95% (GC yield)
Et
Ph
OTMS

Example 5, Nitrile-migrating Alder ene reaction[10]

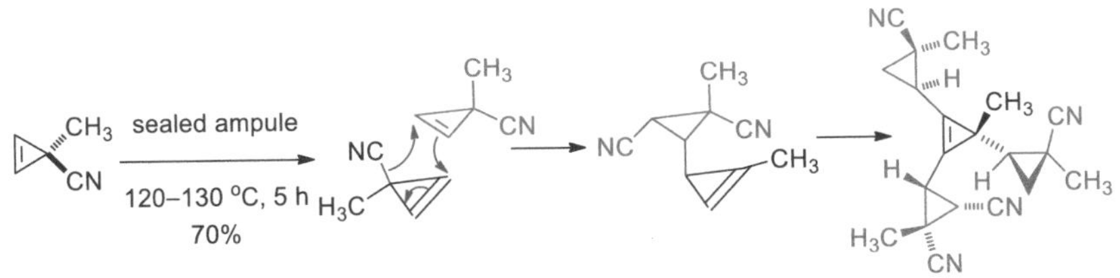
CH3
CN
sealed ampule
120–130 °C, 5 h
70%
NC
H3C
CH3
CN
NC
CH3
CN
CH3
NC
CH3
H
CH3
CN
H
H
CN
CH3
H3C
CN

Example 6[11]

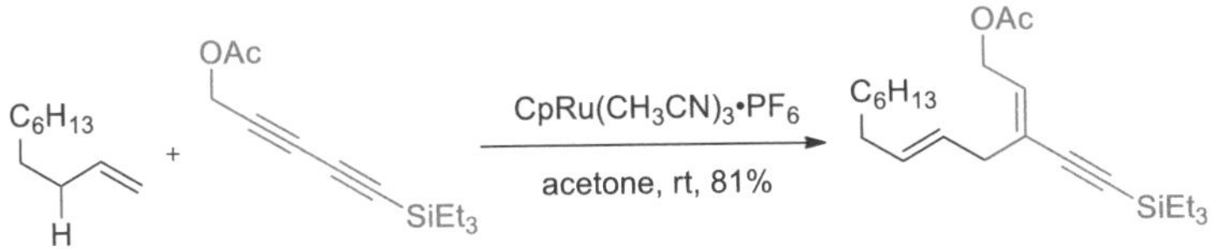
C6H13
H
OAc
SiEt3
CpRu(CH3CN)3•PF6
acetone, rt, 81%
C6H13
OAc
SiEt3

Example 7[13]

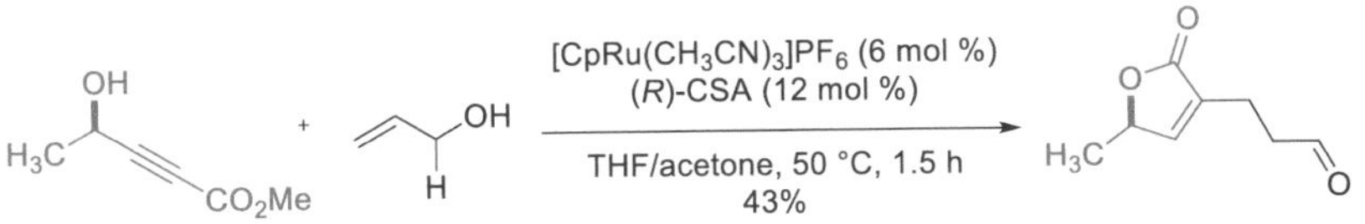
OH
H3C
CO2Me
OH
H
[CpRu(CH3CN)3]PF6 (6 mol %)
(R)-CSA (12 mol %)
THF/acetone, 50 °C, 1.5 h
43%
O
O
H3C
O

Example 8, Pd-catalyzed intramolecular Alder ene (BBEDA = bis-benzylidene ethylenediamine)[14]

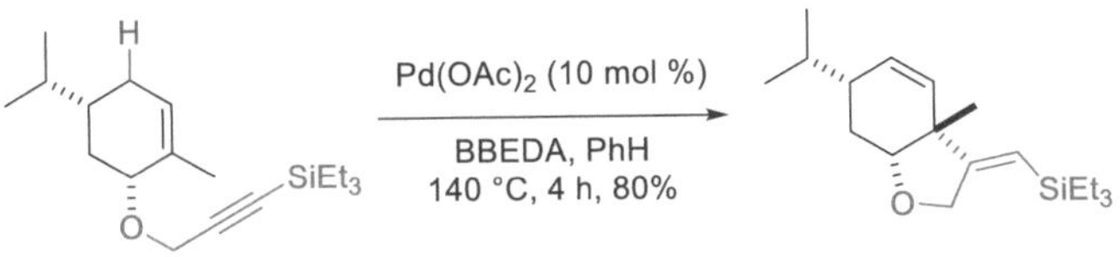
H
O
SiEt3
Pd(OAc)2 (10 mol %)
BBEDA, PhH
140 °C, 4 h, 80%
O
SiEt3

Example 9, Alder ene driven by high steric strain and bond angle distortion[15]

References

1. Alder, K.; Pascher, F.; Schmitz, A. *Ber.* **1943**, *76*, 27–53. Kurt Alder (Germany, 1902–1958) shared the Nobel Prize in Chemistry in 1950 with his teacher Otto Diels (Germany, 1876–1954) for the development of the diene synthesis.
2. Oppolzer, W. *Pure Appl. Chem.* **1981**, *53*, 1181–1201. (Review).
3. Johnson, J. S.; Evans, D. A. *Acc. Chem. Res.* **2000**, *33*, 325–335. (Review).
4. Mikami, K.; Nakai, T. In *Catalytic Asymmetric Synthesis;* 2nd edn.; Ojima, I., ed.; Wiley–VCH: New York, **2000**, 543–568. (Review).
5. Sulikowski, G. A.; Sulikowski, M. M. *e-EROS Encyclopedia of Reagents for Organic Synthesis* **2001**, Wiley: Chichester, UK.
6. Brummond, K. M.; McCabe, J. M. *The Rhodium(I)-Catalyzed Alder ene Reaction.* In *Modern Rhodium-Catalyzed Organic Reactions* **2005**, 151–172. (Review).
7. Miles, W. H.; Dethoff, E. A.; Tuson, H. H.; Ulas, G. *J. Org. Chem.* **2005**, *70*, 2862–2865.
8. Pedrosa, R.; Andres, C.; Martin, L.; Nieto, J.; Roson, C. *J. Org. Chem.* **2005**, *70*, 4332–4337.
9. Hilt, G.; Treutwein, J. *Angew. Chem. Int. Ed.* **2007**, *46*, 8500–8502.
10. Ashirov, R. V.; Shamov, G. A.; Lodochnikova, O. A.; Litvynov, I. A.; Appolonova, S. A.; Plemenkov, V. V. *J. Org. Chem.* **2008**, *73*, 5985–5988.
11. Cho, E. J.; Lee, D. *Org. Lett.* **2008**, *10*, 257–259.
12. Curran, T. T. *Alder Ene Reaction.* In *Name Reactions for Homologations-Part II*; Li, J. J., Ed.; Wiley: Hoboken, NJ, **2009**, pp 2–32. (Review).
13. Trost, B. M.; Quintard, A. *Org. Lett.* **2012**, *14*, 4698–4670.
14. Nugent, J.; Matousova, E.; Banwell, M. G.; Willis, A. C. *J. Org. Chem.* **2017**, *82*, 12569–12589.
15. Gupta, S.; Lin, Y.; Xia, Y.; Wink, D. J.; Lee, D. *Chem. Sci.* **2019**, *10*, 2212–2217.
16. Imino-ene reaction: Hou, L.; Kang, T.; Yang, L.; Cao, W.; Feng, X. *Org. Lett.* **2020**, *22*, 1390–1395.

Aldol Condensation

The aldol condensation is the coupling of an enolate ion with a carbonyl compound to form a β-hydroxycarbonyl, and sometimes, followed by dehydration to give a conjugated enone. A simple case is addition of an enolate to an **ald**ehyde to afford an alcohol, thus the name **aldol**.

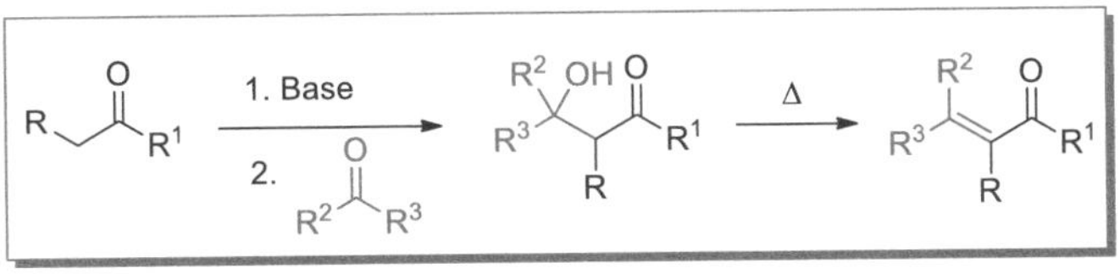

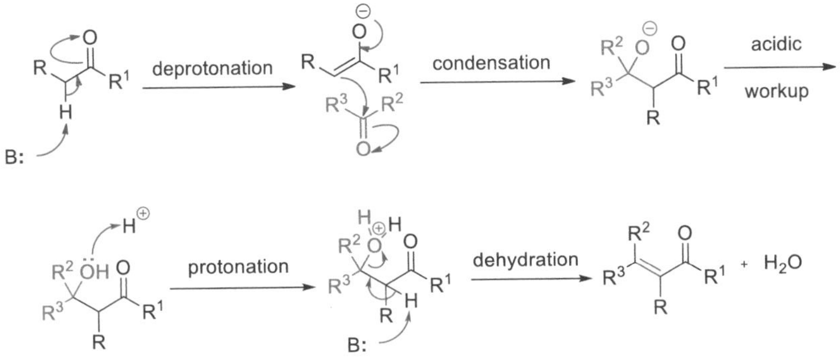

Example 1[3]

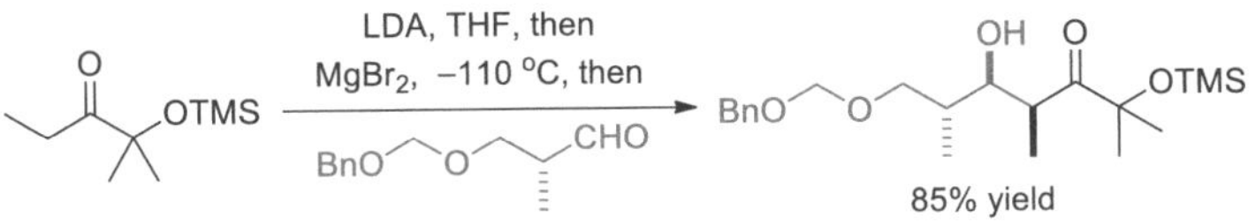

85% yield

Example 2[8]

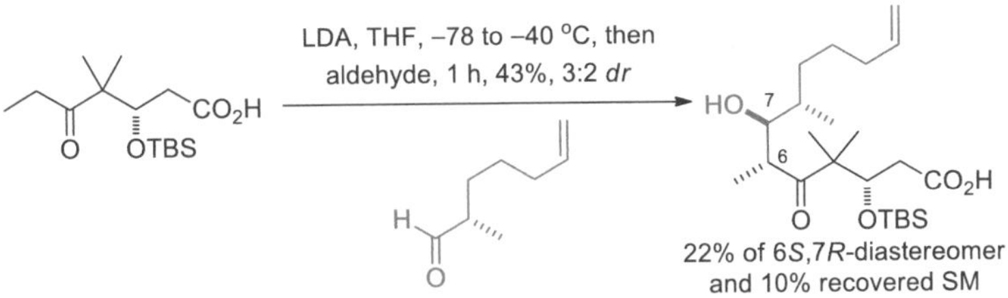

22% of 6S,7R-diastereomer
and 10% recovered SM

J. J. Li, *Name Reactions*, https://doi.org/10.1007/978-3-030-50865-4_2

Example 3, Enantioselective Mukaiyama aldol reaction[10]

Example 4, Intermolecular aldol reaction using organocatalyst[12]

Example 5, Intramolecular aldol reaction[13]

Example 6, Intramolecular vinylogous aldol reaction [ATPH = tris(2,6-diphenyl)phenoxyaluminane][14]

Example 7, A rare stereospecific retroaldol reaction[15]

Example 8, A rare intermolecular vinylogous aldol reaction, TFE = trifluoroethanol[16]

References

1. Wurtz, C. A. *Bull. Soc. Chim. Fr.* **1872**, *17*, 436–442. Charles Adolphe Wurtz (1817–1884) was born in Strasbourg, France. After his doctoral training, he spent a year under Liebig in 1843. In 1874, Wurtz became the Chair of Organic Chemistry at the Sorbonne, where he educated many illustrous chemists such as Crafts, Fittig, Friedel, and van't Hoff. The Wurtz reaction, where two alkyl halides are treated with sodium to form a new carbon–carbon bond, is no longer considered synthetically useful, although *the aldol reaction* that Wurtz discovered in 1872 has become a staple in organic synthesis. Alexander P. Borodin is also credited with the discovery of the aldol reaction together with Wurtz. In 1872 he announced to the Russian Chemical Society the discovery of a new by-product in aldehyde reactions with properties like that of an alcohol, and he noted similarities with compounds already discussed in publications by Wurtz from the same year.

2. Nielsen, A. T.; Houlihan, W. J. *Org. React.* **1968**, *16*, 1–438. (Review).

3. Still, W. C.; McDonald, J. H., III. *Tetrahedron Lett.* **1980**, *21*, 1031–1034.

4. Mukaiyama, T. *Org. React.* **1982**, *28*, 203–331. (Review).

5. Mukaiyama, T.; Kobayashi, S. *Org. React.* **1994**, *46*, 1–103. (Review on tin(II) enolates).

6. Johnson, J. S.; Evans, D. A. *Acc. Chem. Res.* **2000**, *33*, 325–335. (Review).

7. Denmark, S. E.; Stavenger, R. A. *Acc. Chem. Res.* **2000**, *33*, 432–440. (Review).

8. Yang, Z.; He, Y.; Vourloumis, D.; Vallberg, H.; Nicolaou, K. C. *Angew. Chem. Int. Ed.* **1997**, *36*, 166–168.

9. Mahrwald, R. (ed.) *Modern Aldol Reactions,* Wiley–VCH: Weinheim, Germany, **2004**. (Book).

10. Desimoni, G.; Faita, G.; Piccinini, F. *Eur. J. Org. Chem.* **2006**, 5228–5230.

11. Guillena, G.; Najera, C.; Ramon, D. J. *Tetrahedron: Asymmetry* **2007**, *18*, 2249–2293. (Review on enantioselective direct aldol reaction using organocatalysis.)

12. Doherty, S.; Knight, J. G.; McRae, A.; Harrington, R. W.; Clegg, W. *Eur. J. Org. Chem.* **2008**, 1759–1766.

13. O'Brien, E. M.; Morgan, B. J.; Kozlowski, M. C. *Angew. Chem. Int. Ed.* **2008**, *47*, 6877–6880.

14. Gazaille, J. A.; Abramite, J. A.; Sammakia, T. *Org. Lett.* **2012**, *14*, 178–181.

15. Wang, J.; Deng, Z.-X.; Wang, C.-M.; Xia, P.-J.; Xiao, J.-A.; Xiang, H.-Y.; Chen, X.-Q.; Yang, H. *Org. Lett.* **2018**, *20*, 7535–7538.

16. Kutwal, M. S.; Dev, S.; Appayee, C. *Org. Lett.* **2019**, *21*, 2509–2513.

17. Vojackova, P.; Michalska, L.; Necas, M.; Shcherbakov, D.; Bottger, E. C.; Sponer, J.; Sponer, J. E.; Svenda, J. *J. Am. Chem. Soc.* **2020**, *142*, 7306–7311.

Arndt–Eistert Homologation

One-carbon homologation of carboxylic acids using diazomethane.

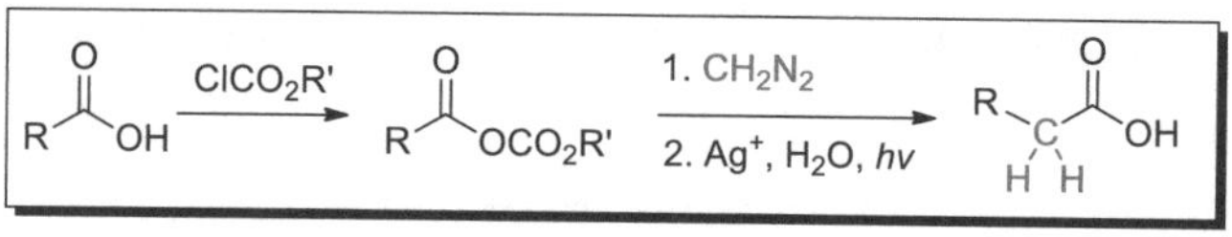

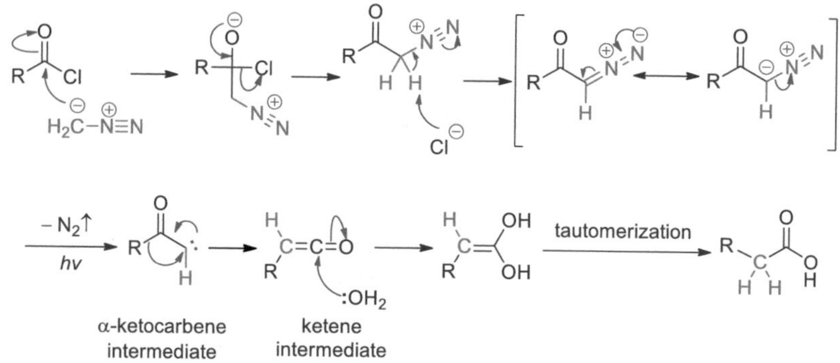

α-ketocarbene
intermediate

ketene
intermediate

Example 1, Homologation of an amino acid[7]

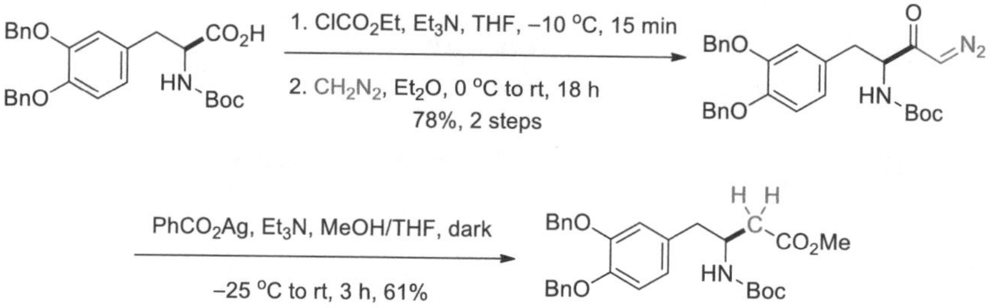

Example 2, An interesting variation[9]

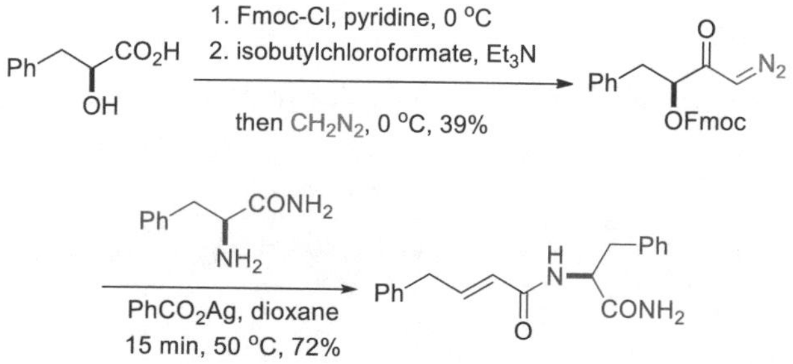

© Springer Nature Switzerland AG 2021
J. J. Li, *Name Reactions*, https://doi.org/10.1007/978-3-030-50865-4_3

Example 3[10]

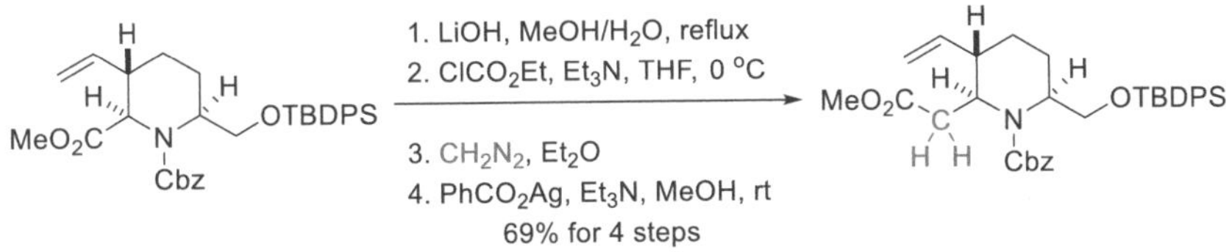
1. LiOH, MeOH/H2O, reflux
2. ClCO2Et, Et3N, THF, 0 °C
3. CH2N2, Et2O
4. PhCO2Ag, Et3N, MeOH, rt
69% for 4 steps
OTBDPS
MeO2C
Cbz

Example 4[10]

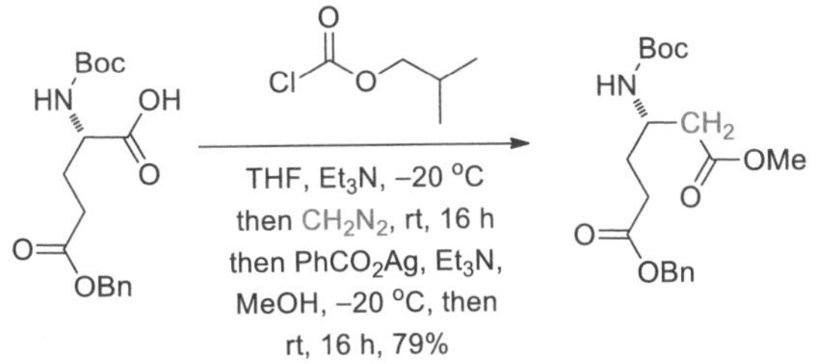
THF, Et3N, –20 °C
then CH2N2, rt, 16 h
then PhCO2Ag, Et3N,
MeOH, –20 °C, then
rt, 16 h, 79%
Boc
OBn
OMe

Example 5, Continuous flow silver-catalyzed Arndt–Eistert reaction/Wolff rearrangement[12]

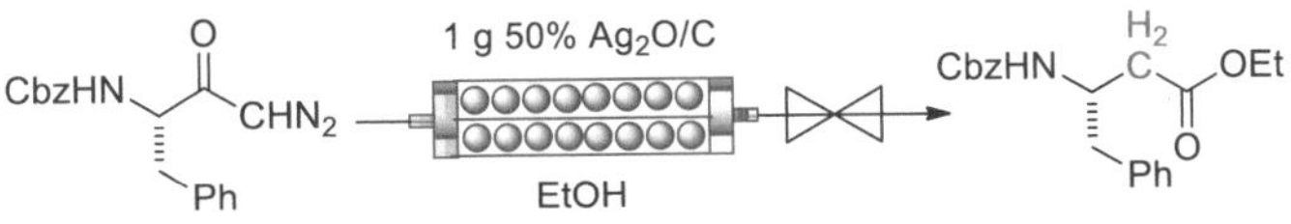
1 g 50% Ag2O/C
EtOH
CbzHN
CHN2
OEt
Ph

Example 6, Arndt–Eistert reaction/Wolff rearrangement sequence[13]

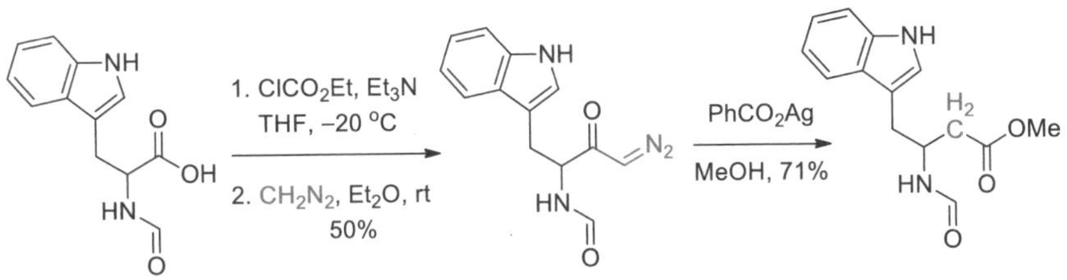
1. ClCO2Et, Et3N
THF, –20 °C
2. CH2N2, Et2O, rt
50%
PhCO2Ag
MeOH, 71%
OMe

Example 7, α-Arylamino diazoketones: reaction in the presence of diazoketones[14]

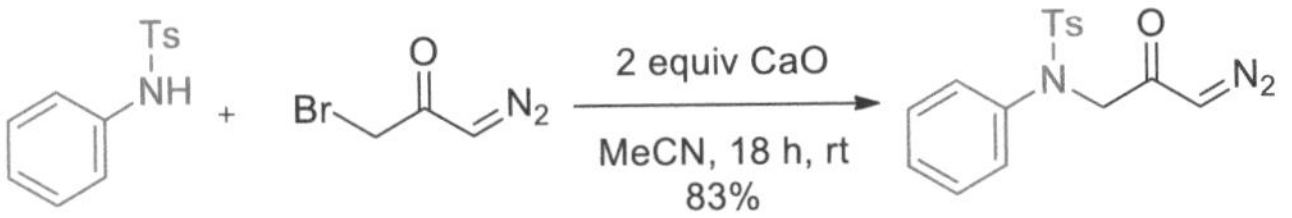
2 equiv CaO
MeCN, 18 h, rt
83%
Ts
Br
N2

References

1. Arndt, F.; Eistert, B. *Ber.* **1935**, *68*, 200–208. Fritz Arndt (1885–1969) was born in Hamburg, Germany. He discovered the Arndt–Eistert homologation at the University of Breslau where he extensively investigated the synthesis of diazomethane and its reactions with aldehydes, ketones, and acid chlorides. Fritz Arndt's chain-smoking of cigars ensured that his presence in the laboratories was always well advertised. Bernd Eistert (1902–1978), born in Ohlau, Silesia, was Arndt's Ph.D. student. Eistert later joined I. G. Farbenindustrie, which became BASF after the Allies broke up the conglomerate following WWII.

2. Podlech, J.; Seebach, D. *Angew. Chem. Int. Ed.* **1995**, *34*, 471–472.

3. Matthews, J. L.; Braun, C.; Guibourdenche, C.; Overhand, M.; Seebach, D. In *Enantioselective Synthesis of β-Amino Acids* Juaristi, E. ed.; Wiley-VCH: Weinheim, Germany, 1996, pp 105–126. (Review).

4. Katritzky, A. R.; Zhang, S.; Fang, Y. *Org. Lett.* **2000**, *2*, 3789–3791.

5. Vasanthakumar, G.-R.; Babu, V. V. S. *Synth. Commun.* **2002**, *32*, 651–657.

6. Chakravarty, P. K.; Shih, T. L.; Colletti, S. L.; Ayer, M. B.; Snedden, C.; Kuo, H.; Tyagarajan, S.; Gregory, L.; Zakson-Aiken, M.; Shoop, W. L.; Schmatz, D. M.; Wyvratt, M. J.; Fisher, M. H.; Meinke, P. T. *Bioorg. Med. Chem. Lett.* **2003**, *13*, 147–150.

7. Gaucher, A.; Dutot, L.; Barbeau, O.; Hamchaoui, W.; Wakselman, M.; Mazaleyrat, J.-P. *Tetrahedron: Asymmetry* **2005**, *16*, 857–864.

8. Podlech, J. In *Enantioselective Synthesis of β-Amino Acids (2nd Ed.)* Wiley: Hoboken, NJ, **2005**, pp 93–106. (Review).

9. Spengler, J.; Ruiz-Rodriguez, J.; Burger, K.; Albericio, F. *Tetrahedron Lett.* **2006**, *47*, 4557–4560.

10. Toyooka, N.; Kobayashi, S.; Zhou, D.; Tsuneki, H.; Wada, T.; Sakai, H.; Nemoto, H.; Sasaoka, T.; Garraffo, H. M.; Spande, T. F.; Daly, J. W. *Bioorg. Med. Chem. Lett.* **2007**, *17*, 5872–5875.

11. Fuchter, M. J. *Arndt–Eistert Homologation.* In *Name Reactions for Homologations-Part I*; Li, J. J., Ed.; Wiley: Hoboken, NJ, **2009**, pp 336–349. (Review).

12. Pinho, V. D.; Gutmann, B.; Kappe, C. O. *RSC Adv.* **2014**, *4*, 37419–37422.

13. Zarezin, D. P.; Shmatova, O. I.; Nenajdenko, V. G. *Org. Biomol. Chem.* **2018**, *16*, 5987–5998.

14. Castoldi, L.; Ielo, L.; Holzer, W.; Giester, G.; Roller, A.; Pace, V. *J. Org. Chem.* **2018**, *83*, 4336–4347.

Baeyer–Villiger Oxidation

General scheme:

The most electron-rich alkyl group (more substituted carbon) migrates first.

The general migration order: tertiary alkyl > cyclohexyl > secondary alkyl > benzyl > phenyl > primary alkyl > methyl >> H.

For substituted aryls: *p*-MeO-Ar > *p*-Me-Ar > *p*-Cl-Ar > *p*-Br-Ar > *p*-O$_2$N-Ar

Example 1, UHP = urea hydrogen peroxide[4]

J. J. Li, *Name Reactions*, https://doi.org/10.1007/978-3-030-50865-4_4

Example 2, Chemoselective over lactam[5]

Example 3, Chemoselective over lactone[6]

Example 4, Chemoselective over ester[8]

Example 5, A trifluoroperacetic acid (TFPAA)-mediated tandem reaction[11]

Example 6, A process route toward entecavir[12]

Example 7, One-pot Baeyer–Villiger oxidation/allylic oxidation[13]

(+)-salimabromide

References

1. v. Baeyer, A.; Villiger, V. *Ber.* **1899**, *32*, 3625–3633. Adolf von Baeyer (1835–1917) was one of the most illustrious organic chemists in history. He contributed to many areas of the field. The Baeyer–Drewson indigo synthesis made possible the commercialization of synthetic indigo. Another one of Baeyer's claim of fame is his synthesis of barbituric acid, named after his then girlfriend, Barbara. Baeyer's real joy was in his laboratory and he deplored any outside work that took him away from his bench. When a visitor expressed envy that fortune had blessed so much of Baeyer's work with success, Baeyer retorted dryly: "Herr Kollege, I experiment more than you." As a scientist, Baeyer was free of vanity. Unlike other scholastic masters of his time (Liebig for instance), he was always ready to acknowledge ungrudgingly the merits of others. Baeyer's famous greenish-black hat was a part of his perpetual wardrobe and he had a ritual of tipping his hat when he admired novel compounds. Adolf von Baeyer received the Nobel Prize in Chemistry in 1905 at age seventy. His apprentice, Emil Fischer, won it in 1902 when he was fifty, three years before his teacher. Victor Villiger (1868–1934), born in Switzerland, went to Munich and worked with Adolf von Baeyer for eleven years.

2. Krow, G. R. *Org. React.* **1993**, *43*, 251–798. (Review).

3. Renz, M.; Meunier, B. *Eur. J. Org. Chem.* **1999**, 737–750. (Review).

4. Wantanabe, A.; Uchida, T.; Ito, K.; Katsuki, T. *Tetrahedron Lett.* **2002**, *43*, 4481–4485.

5. Laurent, M.; Ceresiat, M.; Marchand-Brynaert, J. *J. Org. Chem.* **2004**, *69*, 3194–3197.

6. Brady, T. P.; Kim, S. H.; Wen, K.; Kim, C.; Theodorakis, E. A. *Chem. Eur. J.* **2005**, *11*, 7175–7190.

7. Curran, T. T. *Baeyer–Villiger Oxidation.* In *Name Reactions for Functional Group Transformations*; Li, J. J., Ed.; Wiley: Hoboken, NJ, **2007**, pp 160–182. (Review).

8. Demir, A. S.; Aybey, A. *Tetrahedron* **2008**, *64*, 11256–11261.

9. Zhou, L.; Liu, X.; Ji, J.; Zhang, Y.; Hu, X.; Lin, L.; Feng, X. *J. Am. Chem. Soc.* **2012**, *134*, 17023–17026. (Desymmetrization and Kinetic Resolution).

10. Uyanik, M.; Ishihara, K. *ACS Catal.* **2013**, *3*, 513–520. (Review).

11. Wang, B.-L.; Gao, H.-T.; Li, W.-D. Z. *J. Org. Chem.* **2015**, *80*, 5296–5301.

12. Xu, H.; Wang, F.; Xue, W.; Zheng, Y.; Wang, Q.; Qiu, F. G.; Jin, Y. *Org. Process Res. Dev.* **2018**, *22*, 377–384.

13. Palm, A.; Knopf, C.; Schmalzbauer, B.; Menche, D. *Org. Lett.* **2019**, *21*, 1939–1942.

14. Ma, X.; Liu, Y.; Du, L.; Zhou, J.; Marko, I. E. *Nat. Commun.* **2020**, *1*, 914.

Baker–Venkataraman Rearrangement

Base-catalyzed acyl transfer reaction that converts α-acyloxyketones to β-diketones, which are substrates for making flavones (flavonoids).

Example 1, Carbamoyl Baker–Venkataraman rearrangement[5]

© Springer Nature Switzerland AG 2021
J. J. Li, *Name Reactions*, https://doi.org/10.1007/978-3-030-50865-4_5

Example 2, Carbamoyl Baker–Venkataraman rearrangement, followed by cyclization[6]

Example 3, Baker–Venkataraman rearrangement[9]

Example 4, Baker–Venkataraman rearrangement[10]

Example 5, In the presence of the *C*-aryl glycoside[11]

Example 6, Soft-enolization Baker–Venkataraman rearrangement[12]

References

1.   Baker, W. *J. Chem. Soc.* **1933**, 1381–1389. Wilson Baker (1900–2002) was born in Runcorn, England. He studied chemistry at Manchester under Arthur Lapworth and at Oxford under Robinson. In 1943, Baker was the first to confirm that penicillin contained sulfur, of which Robinson commented: "This is a feather in your cap, Baker." Baker began his independent academic career at University of Bristol. He retired in 1965 as the Head of the School of Chemistry. Baker was a well-known chemist centenarian, spending 47 years in retirement!

2.   (a) Chadha, T. C.; Mahal, H. S.; Venkataraman, K. *Curr. Sci.* **1933**, *2*, 214–215. (b) Mahal, H. S.; Venkataraman, K. *J. Chem. Soc.* **1934**, 1767–1771. K. Venkataraman studied under Robert Robinson at Manchester. He returned to India and later arose to be the Director of the National Chemical Laboratory at Poona. He is also known as the "Father of Dyestuff Industry" in India.

3.   Kraus, G. A.; Fulton, B. S.; Wood, S. H. *J. Org. Chem.* **1984**, *49*, 3212–3214.

4.   Reddy, B. P.; Krupadanam, G. L. D. *J. Heterocycl. Chem.* **1996**, *33*, 1561–1565.

5.   Kalinin, A. V.; da Silva, A. J. M.; Lopes, C. C.; Lopes, R. S. C.; Snieckus, V. *Tetrahedron Lett.* **1998**, *39*, 4995–4998.

6.   Kalinin, A. V.; Snieckus, V. *Tetrahedron Lett.* **1998**, *39*, 4999–5002.

7.   Thasana, N.; Ruchirawat, S. *Tetrahedron Lett.* **2002**, *43*, 4515–4517.

8.   Santos, C. M. M.; Silva, A. M. S.; Cavaleiro, J. A. S. *Eur. J. Org. Chem.* **2003**, 4575–4585.

9.   Krohn, K.; Vidal, A.; Vitz, J.; Westermann, B.; Abbas, M.; Green, I. *Tetrahedron: Asymmetry* **2006**, *17*, 3051–3057.

10. Yu, Y.; Hu, Y.; Shao, W.; Huang, J.; Zuo, Y.; Huo, Y.; An, L.; Du, J.; Bu, X. *E. J. Org. Chem.* **2011**, 4551–4563.

11. Yao, C.-H.; Tsai, C.-H.; Lee, J.-C. *J. Nat. Prod.* **2016**, 1719–1723.

12. St-Gelais, A.; Alsarraf, J.; Legault, J.; Gauthier, C.; Pichette, A. *Org. Lett.* **2018**, *20*, 7424–7428.

13. Kshatriya, R.; Jejurkar, V. P.; Saha, S. *Tetrahedron* **2018**, *74*, 811–833. (Review).

14. Liu, Q.; Mu, Y.; An, Q.; Xun, J.; Ma, J.; Wu, W.; Xu, M.; Xu, J.; Han, L.; Huang, X. *Bioorg. Chem.* **2020**, *94*, 103420.

Bamford–Stevens Reaction

The Bamford–Stevens reaction and the Shapiro reaction share a similar mechanistic pathway. The former uses a base such as Na, NaOMe, LiH, NaH, NaNH$_2$, heat, *etc.*, whereas the latter employs bases such as alkyllithiums and Grignard reagents. As a result, the Bamford–Stevens reaction furnishes more-substituted olefins as the thermodynamic products, while the Shapiro reaction generally affords less-substituted olefins as the kinetic products.

In protic solvent (S–H):

In aprotic solvent:

© Springer Nature Switzerland AG 2021
J. J. Li, *Name Reactions*, https://doi.org/10.1007/978-3-030-50865-4_6

Example 1, Tandem Bamford–Stevens/thermal aliphatic Claisen rearrangement sequence[6]

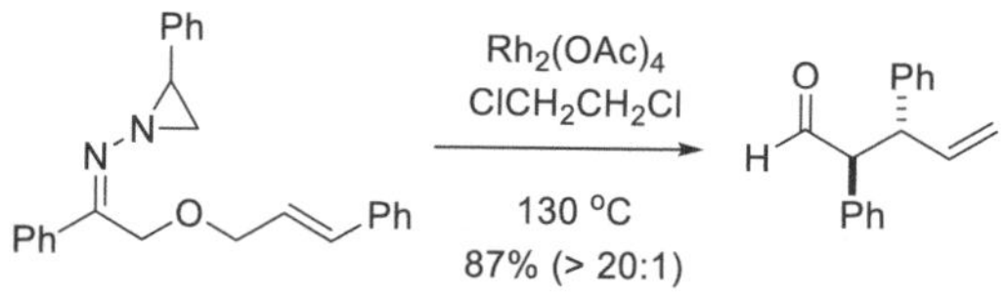

The starting material *N*-aziridinyl imine is also known as Eschenmoser hydrazone.

Example 2, Thermal Bamford–Stevens[6]

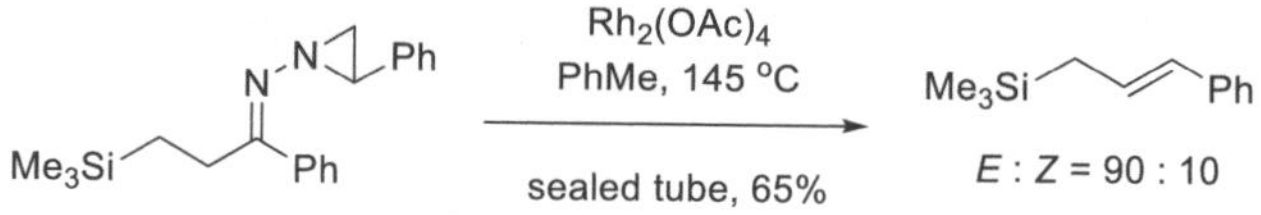

E : *Z* = 90 : 10

Example 3[7]

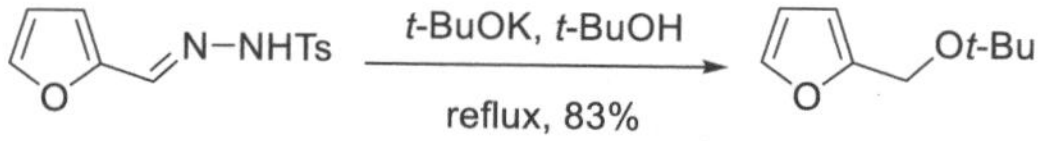

Example 4[8]

Example 5, Diazoesters from arylsulfonylhydrazones by means of in-flow Bamford–Stevens reactions[11]

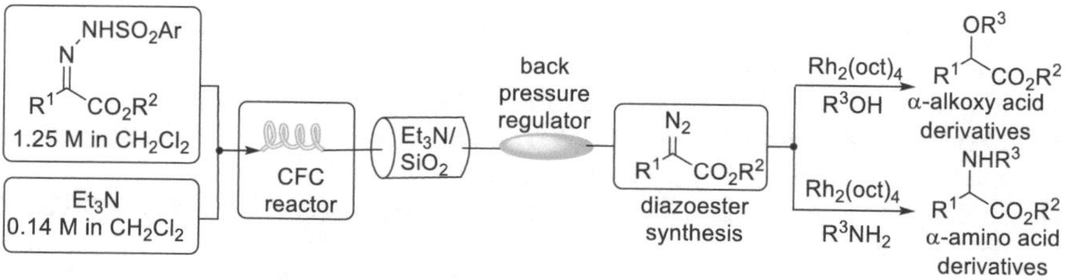

CFC = Continuous-Flow Centrifugation

Example 6, Synthesis of carbene precursor[12]

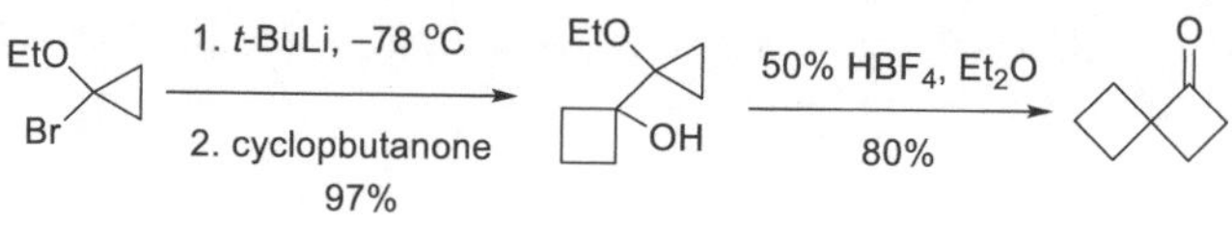

Example 7, Microwave-mediated synthesis of fullerene acceptors for organic photovoltaics[12]

References

1. Bamford, W. R.; Stevens, T. S. M. *J. Chem. Soc.* **1952**, 4735–4740. Thomas Stevens (1900–2000), another chemist centenarian, was born in Renfrew, Scotland. He and his student W. R. Bamford published this paper at the University of Sheffield, UK. Stevens also contributed to another name reaction, the McFadyen–Stevens reaction.

2. Felix, D.; Müller, R. K.; Horn, U.; Joos, R.; Schreiber, J.; Eschenmoser, A. *Helv. Chim. Acta* **1972**, *55*, 1276–1319.

3. Shapiro, R. H. *Org. React.* **1976**, *23*, 405–507. (Review).

4. Adlington, R. M.; Barrett, A. G. M. *Acc. Chem. Res.* **1983**, *16*, 55–59. (Review on the Shapiro reaction).

5. Chamberlin, A. R.; Bloom, S. H. *Org. React.* **1990**, *39*, 1–83. (Review).

6. Sarkar, T. K.; Ghorai, B. K. *J. Chem. Soc., Chem. Commun.* **1992**, *17,* 1184–1185.

7. Chandrasekhar, S.; Rajaiah, G.; Chandraiah, L.; Swamy, D. N. *Synlett* **2001**, 1779–1780.

8. Aggarwal, V. K.; Alonso, E.; Hynd, G.; Lydon, K. M.; Palmer, M. J.; Porcelloni, M.; Studley, J. R. *Angew. Chem. Int. Ed.* **2001**, *40*, 1430–1433.

9. May, J. A.; Stoltz, B. M. *J. Am. Chem. Soc.* **2002**, *124*, 12426–12427.

10. Humphries, P. *Bamford–Stevens Reaction*. In *Name Reactions for Homologations-Part II*; Li, J. J., Ed.; Wiley: Hoboken, NJ, **2009**, pp 642–652. (Review).

11. Bartrum, H. E.; Blakemore, D. C.; Moody, C. J.; Hayes, C. J. *Chem. Eur. J.* **2011**, *17*, 9586–9589.

12. Rosenberg, M.; Schrievers, T.; Brinker, U. H. *J. Org. Chem.* **2016**, *81*, 12388–12400.

13. Campisciano, V.; Riela, S.; Noto, R.; Gruttadauria, M.; Giacalone, F. *RSC Adv.* **2014**, *108*, 63200–63207.

14. Meichsner, E.; Nierengarten, I.; Holler, M.; Chesse, M.; Nierengarten, J.-F. *Helv. Chim. Acta* **2018**, *101*, e180059.

15. Jana, S.; Li, F.; Empel, C.; Verspeek, D.; Aseeva, P.; Koenigs, R. M. *Chem. Eur. J.* **2020**, *26*, 2586–2591.

Barbier Reaction

The Barbier reaction is an organic reaction between an alkyl halide and a carbonyl group as an electrophilic substrate in the presence of magnesium, aluminium, zinc, indium, tin or its salts. The reaction product is a primary, secondary or tertiary alcohol. *Cf.* Grignard reaction.

According to conventional wisdom,[3] the organometallic intermediate (M = Mg, Li, Sm, Zn, La, *etc.*) is generated *in situ*, which is intermediately trapped by the carbonyl compound. However, recent experimental and theoretical studies seem to suggest that the Barbier coupling reaction goes through a single electron transfer (SET) pathway.

Generation of the organometallic intermediate *in situ*:

SET = single electron transfer

Ionic mechanism,

Single electron transfer (SET) mechanism:

Example 1[6]

J. J. Li, *Name Reactions*, https://doi.org/10.1007/978-3-030-50865-4_7

Example 2[9]

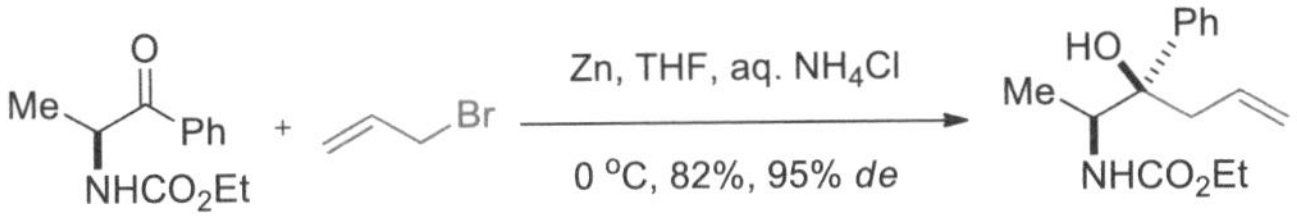

Example 3[10]

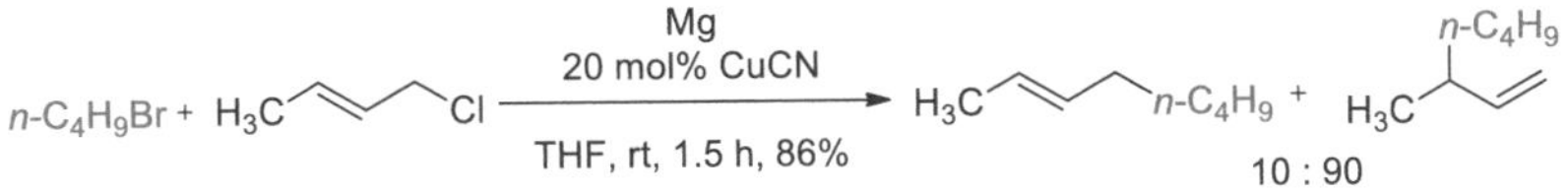

Example 4, Intramolecular Barbier reaction[11]

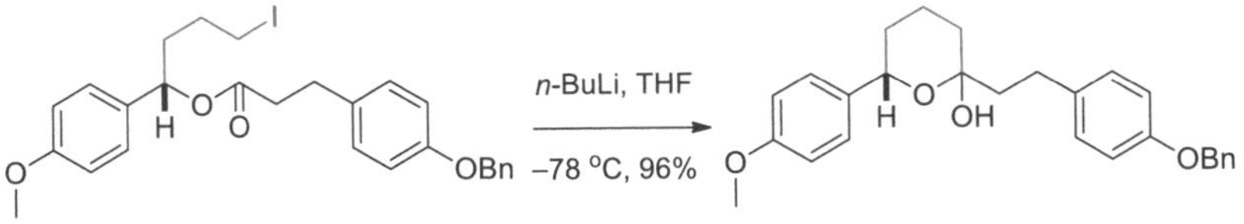

Example 5, The following whole sequence of 5 steps can also be carried out in one-pot[12]

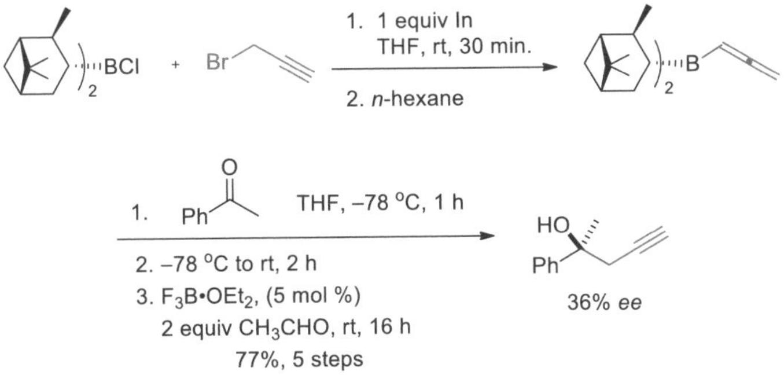

Example 6, A cuprate Barbier protocol to overcome strain and sterical hindrance[13]

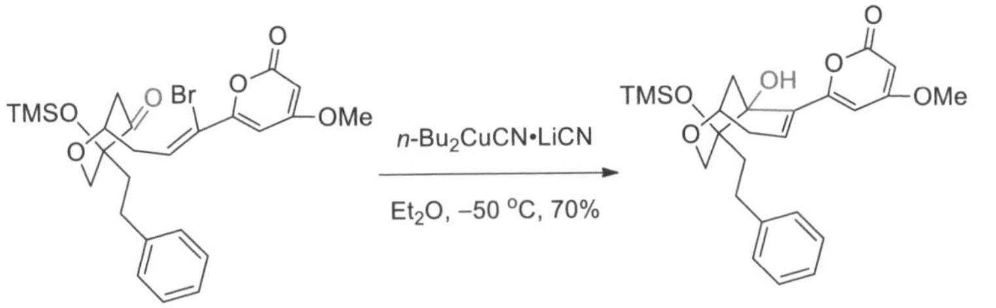

Example 7, CpTiCl$_2$ as an improved titanocene(III) catalyst[14]

Manganese dust reduces CpTiCl$_3$ to CpTiCl$_2$

Example 8, Co(I)-catalyzed Barbier reaction of an aromatic halide with an aromatic aldehyde or an imine[15]

References

1. Barbier, P. *C. R. Hebd. Séances Acad. Sci.* **1899**, *128*, 110–111. Phillippe Barbier (1848–1922) was born in Luzy, Nièvre, France. He studied terpenoids using zinc and magnesium. Barbier suggested the use of magnesium to his student, Victor Grignard, who later discovered the Grignard reagent and won the Nobel Prize in 1912.
2. Grignard, V. *C. R. Hebd. Séances Acad. Sci.* **1900**, *130*, 1322–1324.
3. Moyano, A.; Pericás, M. A.; Riera, A.; Luche, J.-L. *Tetrahedron Lett.* **1990**, *31*, 7619–7622. (Theoretical study).
4. Alonso, F.; Yus, M. *Rec. Res. Dev. Org. Chem.* **1997**, *1*, 397–436. (Review).
5. Russo, D. A. *Chem. Ind.* **1996**, *64*, 405–409. (Review).
6. Basu, M. K.; Banik, B. *Tetrahedron Lett.* **2001**, *42*, 187–189.
7. Sinha, P.; Roy, S. *Chem. Commun.* **2001**, 1798–1799.
8. Lombardo, M.; Gianotti, K.; Licciulli, S.; Trombini, C. *Tetrahedron* **2004**, *60*, 11725–11732.
9. Resende, G. O.; Aguiar, L. C. S.; Antunes, O. A. C. *Synlett* **2005**, 119–120.
10. Erdik, E.; Kocoglu, M. *Tetrahedron Lett.* **2007**, *48*, 4211–4214.
11. Takeuchi, T.; Matsuhashi, M.; Nakata, T. *Tetrahedron Lett.* **2008**, *49*, 6462–6465.
12. Hirayama, L. C.; Haddad, T. D.; Oliver, A. G.; Singaram, B. *J. Org. Chem.* **2012**, *77*, 4342–4353.
13. Rizzo, A.; Tauner, D. *Org. Lett.* **2018**, *20*, 1841–1844.
14. Roldan-Molina, E.; Padial, N. M.; Lezama, L.; Oltra, J. E. *Eur. J. Org. Chem.* **2018**, 5997–6001.
15. Presset, M.; Paul, J.; Cherif, G. N.; Ratnam, N.; Laloi, N.; Leonel, E.; Gosmini, C.; Le Gall, E. *Chem. Eur. J.* **2019**, *25*, 4491–4495.
16. Beaver, M. G.; Shi, Xi.; Riedel, J.; Patel, P.; Zeng, A.; Corbett, M. T.; Robinson, J. A.; Parsons, A. T.; Cui, S.; Baucom, K.; et al. *Org. Process Res. Dev.* **2020**, *24*, 490–499.

Barton–McCombie Deoxygenation

Deoxygenation of alcohols by means of radical scission of their corresponding thiocarbonyl derivatives.

AIBN = 2,2′-azobisisobutyronitrile

Example 1[2]

Example 2[5]

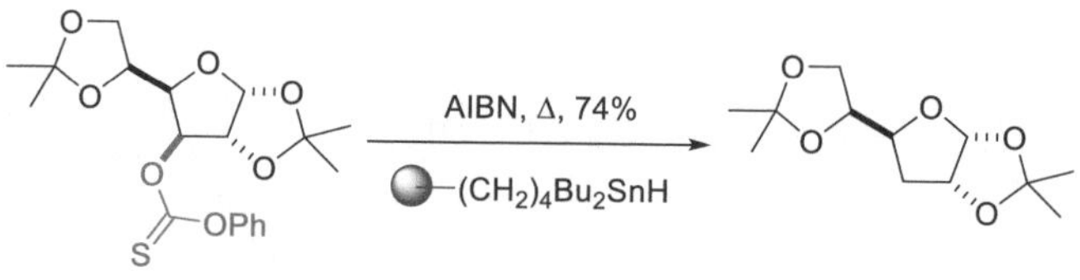

AIBN, Δ, 74%

⬤—(CH₂)₄Bu₂SnH

Example 3[6]

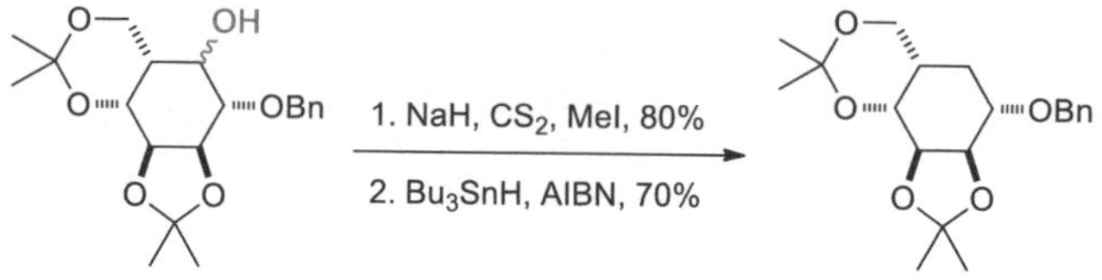

1. NaH, CS₂, MeI, 80%

2. Bu₃SnH, AIBN, 70%

Example 4[7]

NaH, CS₂

MeI, THF
rt

Bu₃SnH, AIBN

toluene, reflux
72%, 2 steps

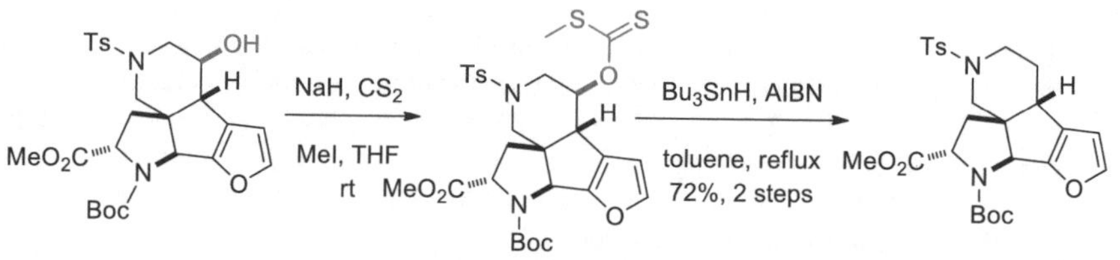

Example 5[8]

1. NaH, imidazole

2. CS₂, MeI
12%, 2 steps

Bu₃SnH, AIBN

toluene, 43%

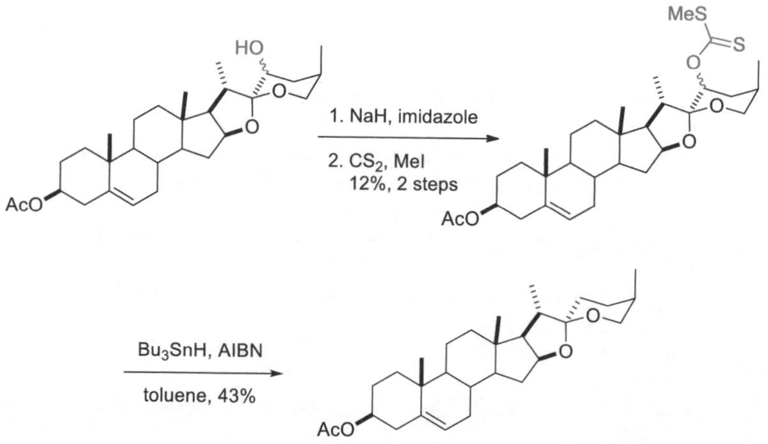

Example 6, CPME = cyclopentyl methyl ether[10]

Example 7[11]

References

1. Barton, D. H. R.; McCombie, S. W. *J. Chem. Soc., Perkin Trans. 1* **1975**, 1574–1585. Stuart McCombie, a Barton student, borrowed both substrate and tri-*n*-butylstannane from other group members to carry out the first Barton–McCombie deoxygenation re-cation. He worked at Schering–Plough for many years, but is now retired after the company was bought by Merck.
2. Gimisis, T.; Ballestri, M.; Ferreri, C.; Chatgilialoglu, C.; Boukherroub, R.; Manuel, G. *Tetrahedron Lett.* **1995**, *36*, 3897–3900.
3. Zard, S. Z. *Angew. Chem. Int. Ed.* **1997**, *36*, 673–685.
4. Lopez, R. M.; Hays, D. S.; Fu, G. C. *J. Am. Chem. Soc.* **1997**, *119*, 6949–6950.
5. Boussaguet, P.; Delmond, B.; Dumartin, G.; Pereyre, M. *Tetrahedron Lett.* **2000**, *41*, 3377–3380.
6. Gómez, A. M.; Moreno, E.; Valverde, S.; López, J. C. *Eur. J. Org. Chem.* **2004**, 1830–1840.
7. Deng, H.; Yang, X.; Tong, Z.; Li, Z.; Zhai, H. *Org. Lett.* **2008**, *10*, 1791–1793.
8. Mancuso, J. *Barton–McCombie deoxygenation*. In *Name Reactions for Homologa-tions-Part I*; Li, J. J., Ed.; Wiley: Hoboken, NJ, **2009**, pp 614–632. (Review).
9. McCombie, S. W.; Motherwell, W. B.; Tozer, M. J. *The Barton–McCombie Reaction*, In *Org. React.* **2012**, *77*, pp 161–591. (Review).
10. Sulake, R. S.; Lin, H.-H.; Hsu, C.-Y.; Weng, C.-F.; Chen, C. *J. Org. Chem.* **2015**, *80*, 6044–6051.
11. Satyanarayana, V.; Chaithanya Kumar, G.; Muralikrishna, K.; Singh Yadav, J. *Tetra-hedron Lett.* **20018**, *59*, 2828–2830.
12. McCombie, S. W.; Quiclet-Sire, B.; Zard, S. Z. *Tetrahedron* **2018**, *74*, 4969–4979. (Review of mechanism).
13. Wu, J.; Baer, R. M.; Guo, L.; Noble, A.; Aggarwal, V. K. *Angew. Chem. Int. Ed.* **2019**, *58*, 18830–18834.

Beckmann Rearrangement

Acid-mediated isomerization of oximes to amides.
In protic acid:

the substituent *trans* to the leaving group migrates

With PCl5:

Again, the substituent *trans* to the leaving group migrates

Example 1, Microwave (MW) reaction[3]

J. J. Li, *Name Reactions*, https://doi.org/10.1007/978-3-030-50865-4_9

Example 2[4]

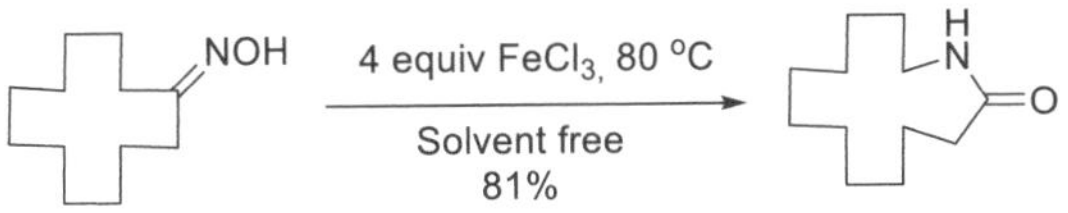
NOH
4 equiv FeCl3, 80 °C
Solvent free
81%
H
N
O

Example 3[6]

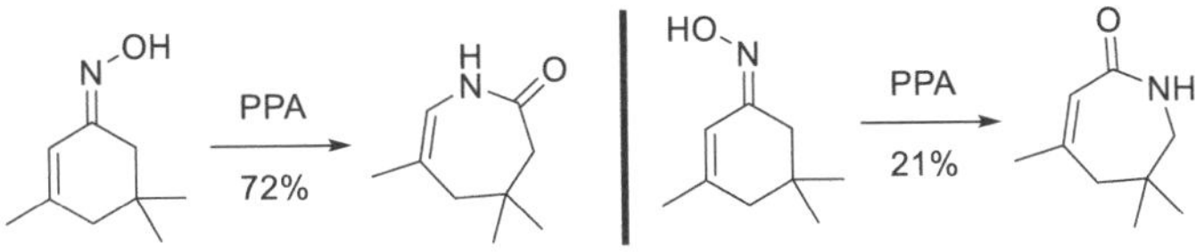
N OH
PPA
72%
H
N
O
HO
N
PPA
21%
O
NH

PPA = polyphosphoric acid

Example 4[8]

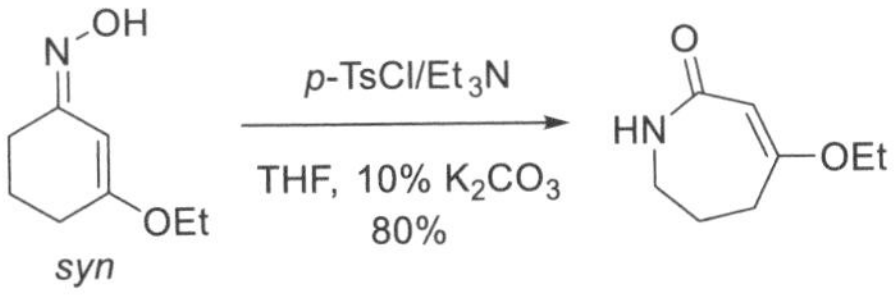
N OH
OEt
syn
p-TsCl/Et3N
THF, 10% K2CO3
80%
O
HN
OEt

Example 5, Radical Beckmann rearrangement[11]

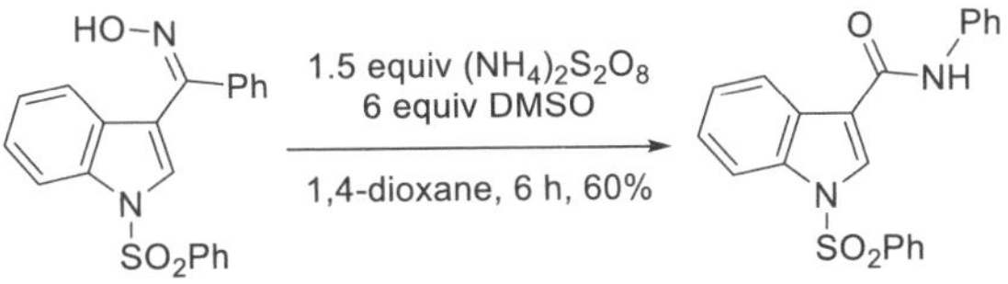
HO–N
Ph
N
SO2Ph
1.5 equiv (NH4)2S2O8
6 equiv DMSO
1,4-dioxane, 6 h, 60%
O
Ph
NH
N
SO2Ph

Example 6, Organocatalytic Beckmann rearrangement with a boronic acid/perfluoropinacol system under ambient conditions[142]

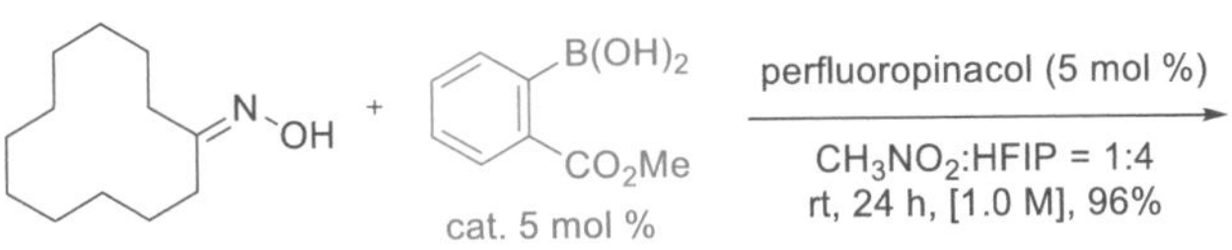
N OH
+
B(OH)2
CO2Me
cat. 5 mol %
perfluoropinacol (5 mol %)
CH3NO2:HFIP = 1:4
rt, 24 h, [1.0 M], 96%

HFIP = hexafluoroisopropanol

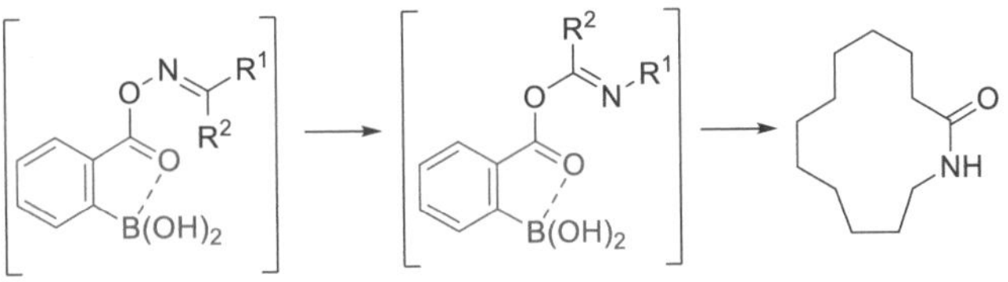
O
N
R1
R2
O
B(OH)2
R2
O
N
R1
O
B(OH)2
O
NH

Example 7, Novel androhrapholide Beckmann rearrangement derivatives[13]

TsCl, Et$_3$N, DMAP

CH$_2$Cl$_2$, rt, 3 h, 47%

References

1. Beckmann, E. *Chem. Ber.* **1886,** *89*, 988. Ernst Otto Beckmann (1853–1923) was born in Solingen, Germany. He studied chemistry and pharmacy at Leipzig. In addition to the Beckmann rearrangement of oximes to amides, his name is associated with the Beckmann thermometer, used to measure freezing and boiling point depressions to determine molecular weights.
2. Gawley, R. E. *Org. React.* **1988,** *35*, 1–420. (Review).
3. Thakur, A. J.; Boruah, A.; Prajapati, D.; Sandhu, J. S. *Synth. Commun.* **2000,** *30*, 2105–2011.
4. Khodaei, M. M.; Meybodi, F. A.; Rezai, N.; Salehi, P. *Synth. Commun.* **2001,** *31*, 2047–2050.
5. Torisawa, Y.; Nishi, T.; Minamikawa, J.-i. *Bioorg. Med. Chem. Lett.* **2002,** *12*, 387–390.
6. Hilmey, D. G.; Paquette, L. A. *Org. Lett.* **2005,** *7*, 2067–2069.
7. Fernández, A. B.; Boronat, M.; Blasco, T.; Corma, A. *Angew. Chem. Int. Ed.* **2005,** *44*, 2370–2373.
8. Collison, C. G.; Chen, J.; Walvoord, R. *Synthesis* **2006,** 2319–2322.
9. Kumar, R. R.; Vanitha, K. A.; Balasubramanian, M. *Beckmann Rearrangement.* In *Name Reactions for Homologations-Part II*; Li, J. J., Ed.; Wiley: Hoboken, NJ, **2009,** pp 274–292. (Review).
10. Faraldos, J. A.; Kariuki, B. M.; Coates, R. M. *Org. Lett.* **2011,** *13*, 836–839.
11. Mahajan, P. S.; Humne, V. T.; Tanpure, S. D.; Mhaske, S. B. *Org. Lett.* **2016,** *18*, 3450–3453.
12. Mo, X.; Morgan, T. D. R.; Ang, H. T.; Hall, D. G. *J. Am. Chem. Soc.* **2018,** *140*, 5264–5271.
13. Wang, W.; Wu, Y.; Yang, K.; Wu, C.; Tang, R.; Li, H.; Chen, L. *Eur. J. Med. Chem.* **2019,** *173*, 282–293.
14. Zhang, Y.; Shen, S.; Fang, H.; Xu, T. *Org. Lett.* **2020,** *22*, 1244–1248.

Abnormal Beckmann Rearrangement

Takes place when the migrating fragment (e.g., R^1) departs from the intermediate, leaving a nitrile as a stable product.

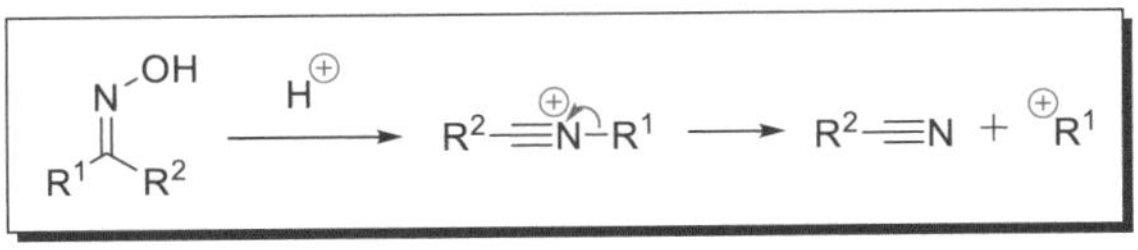

Example 1[9]

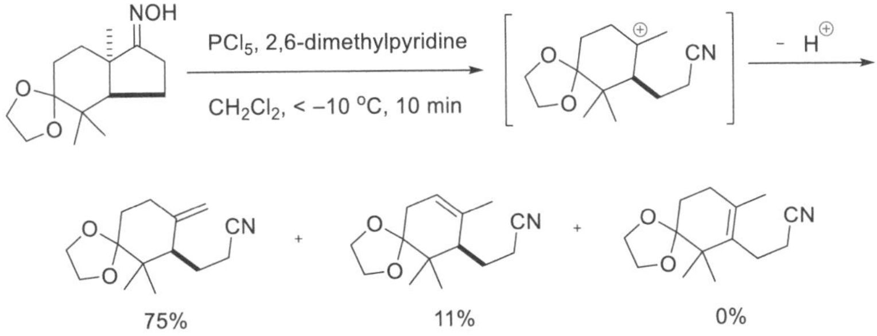

75%   11%   0%

Example 2[10]

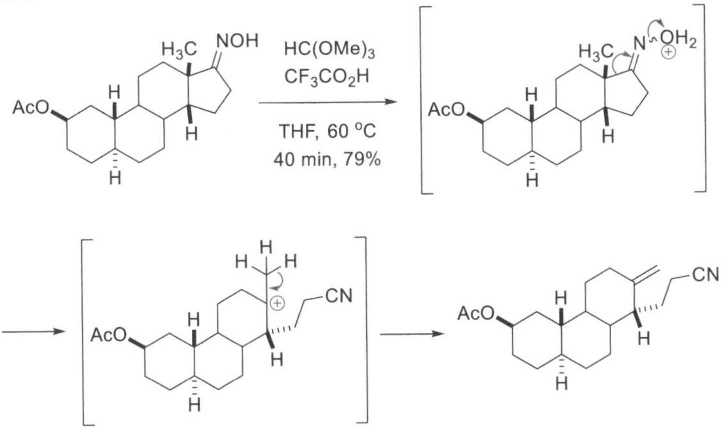

References

1. Cao, L.; Sun, J.; Wang, X.; Zhu, R. *Tetrahedron* **2007**, *63*, 5036–5041.
2. Wang, C.; Rath, N. P.; Covey, D. F. *Tetrahedron* **2007**, *63*, 7977–7984.
3. Gui, J.; Wang, Y.; Tian, H.; Gao, Y.; Tian, W. *Tetrahedron Lett.* **2014**, *55*, 4233–4235.
4. Alhifthi, A.; Harris, B. L.; Goerigk, L.; White, J. M.; Williams, S. J. *Org. Biomol. Chem.* **2017**, *15*, 0105–10115.

Benzilic Acid Rearrangement

Rearrangement of benzil to benzilic acid *via* aryl migration.

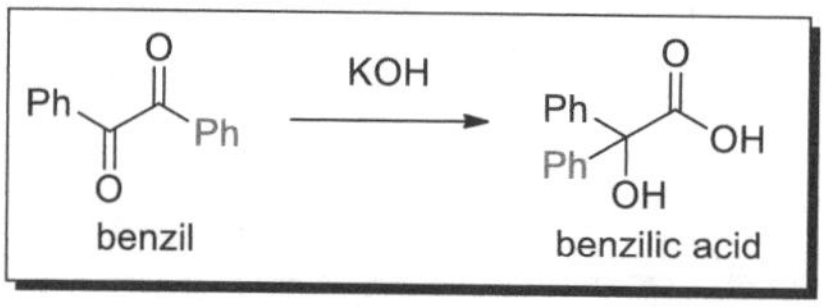
KOH
benzil
benzilic acid

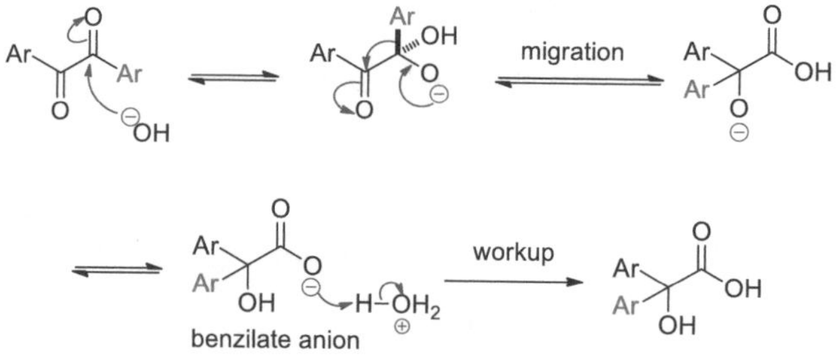
migration
workup
benzilate anion

Final deprotonation (before workup) of the carboxylate to afford the benzilate anion drives the reaction forward.

Example 1[3]

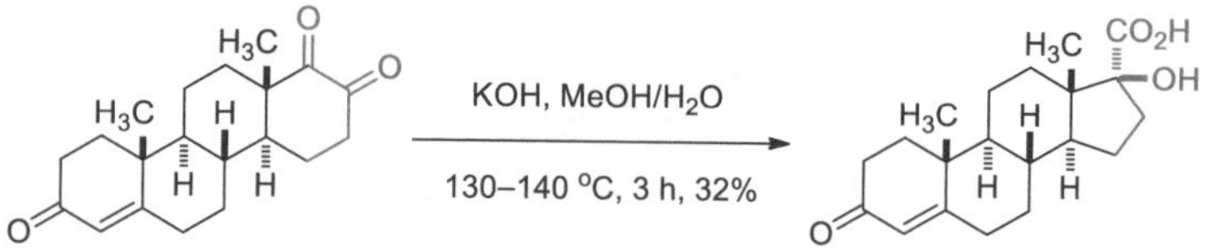
KOH, MeOH/H2O
130–140 °C, 3 h, 32%

Example 2[6]

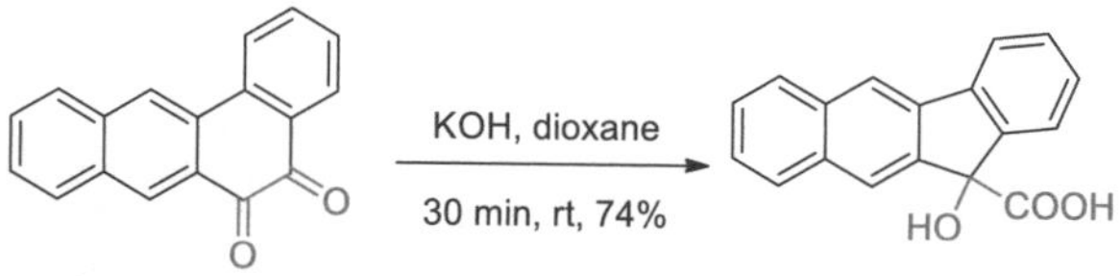
KOH, dioxane
30 min, rt, 74%

Example 3, Retro-benzilic acid rearrangement[7]

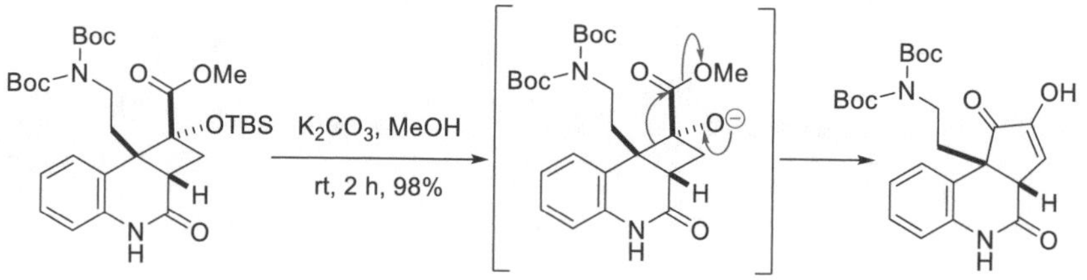
K2CO3, MeOH
rt, 2 h, 98%

J. J. Li, *Name Reactions*, https://doi.org/10.1007/978-3-030-50865-4_10

Example 4, Cyclobutane-1,2-diones (Computational Chemistry)[9]

Example 5, Biomimetic benzilic acid rearrangement[10]

Example 6, Benzoquinone ansamycin converted to cyclopentenone-containing ansamycin macrolactam via the benzilic acid rearrangement[11]

geldanamycin D   Mccrearmycin B

Example 7, Biomimetic benzilic acid rearrangement[12]

References

1. Liebig, J. *Justus Liebigs Ann. Chem.* **1838**, 27. Justus von Liebig (1803–1873) pursued his Ph.D. in organic chemistry in Paris under the tutelage of Joseph Louis Gay-Lussac (1778–1850). He was appointed the Chair of Chemistry at Giessen University, which incited a furious jealousy amongst several of the professors already working there because he was so young. Fortunately, time would prove the choice was a wise one for the department. Liebig would soon transform Giessen from a sleepy university to a mecca of organic chemistry in Europe. Liebig is now considered the father of organic chemistry. Many classic name reactions were published in the journal that still bears his name, *Justus Liebigs Annalen der Chemie.*[2]
2. Zinin, N. *Justus Liebigs Ann. Chem.* **1839**, *31*, 329.
3. Georgian, V.; Kundu, N. *Tetrahedron* **1963**, *19*, 1037–1049.
4. Robinson, J. M.; Flynn, E. T.; McMahan, T. L.; Simpson, S. L.; Trisler, J. C.; Conn, K. B. *J. Org. Chem.* **1991**, *56*, 6709–6712.
5. Fohlisch, B.; Radl, A.; Schwetzler-Raschke, R.; Henkel, S. *Eur. J. Org. Chem.* **2001**, 4357–4365.
6. Patra, A.; Ghorai, S. K.; De, S. R.; Mal, D. *Synthesis* **2006**, *15*, 2556–2562.
7. Selig, P.; Bach, T. *Angew. Chem. Int. Ed.* **2008**, *47*, 5082–5084.
8. Kumar, R. R.; Balasubramanian, M. *Benzilic Acid Rearrangement.* In *Name Reactions for Homologations-Part II*; Li, J. J., Ed.; Wiley: Hoboken, NJ, **2009**, pp 395–405. (Review).
9. Sultana, N.; Fabian, W. M. F. *Beilstein J. Org. Chem.* **2013**, *9*, 594–601.
10. Xiao, M.; Wu, W.; Wei, L.; Jin, X.; Yao, X.; Xie, Z. *Tetrahedron* **2015**, *71*, 3705–3714.
11. Wang, X.; Zhang, Y.; Ponomareva, L. V.; Qiu, Q.; Woodcock, R.; Elshahawi, S. I.; Chen, X.; Zhou, Z.; Hatcher, B. E.; Hower, J. C.; et al. *Angew. Chem. Int. Ed.* **2017**, *56*, 2994–2998.
12. Noack, F.; Hartmayer, B.; Heretsch, P. *Synthesis* **2018**, *50*, 809–820.
13. Novak, A. J. E.; Grigglestone, C. E.; Trauner, D. *J. Am. Chem. Soc.* **2019**, *141*, 15515–15518.

Benzoin Condensation

Cyanide-catalyzed condensation of aryl aldehyde to benzoin. Now cyanide is mostly replaced by thiazolium salts or *N*-heterocyclic carbenes. *Cf.* Stetter reaction.

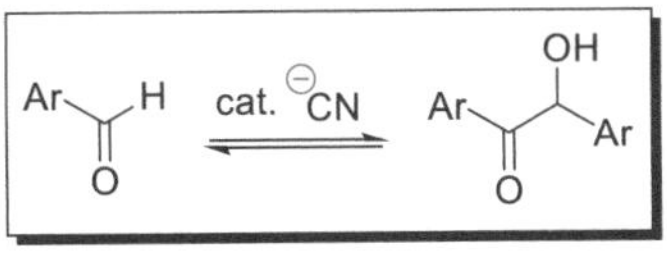

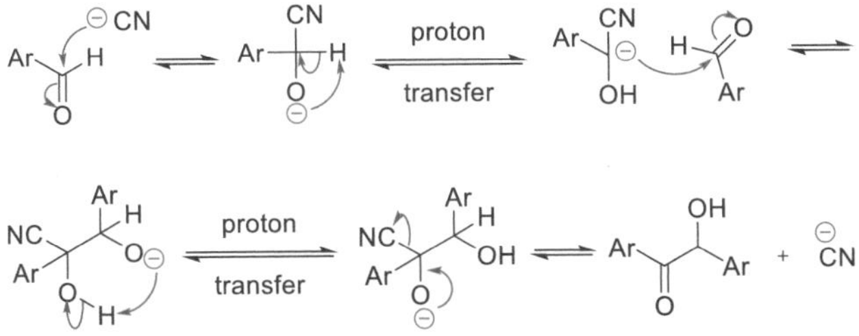

Example 1[2]

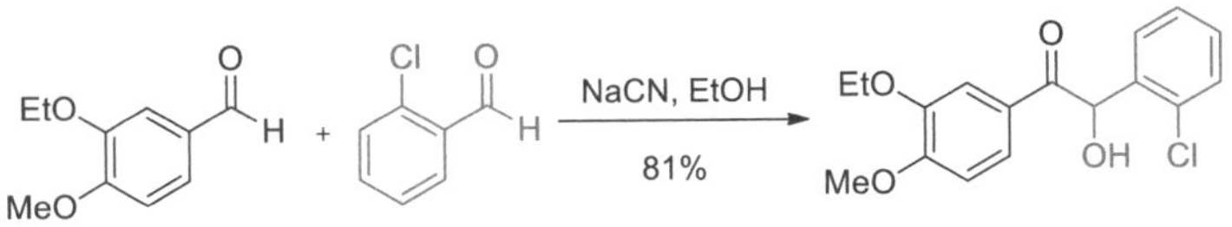

Example 2[7]

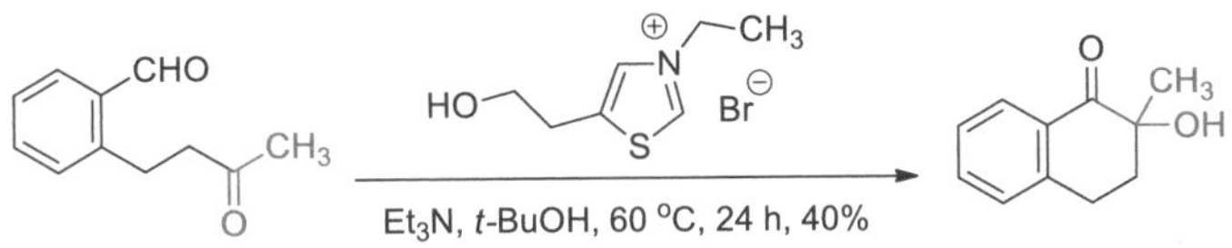

Example 3[7]

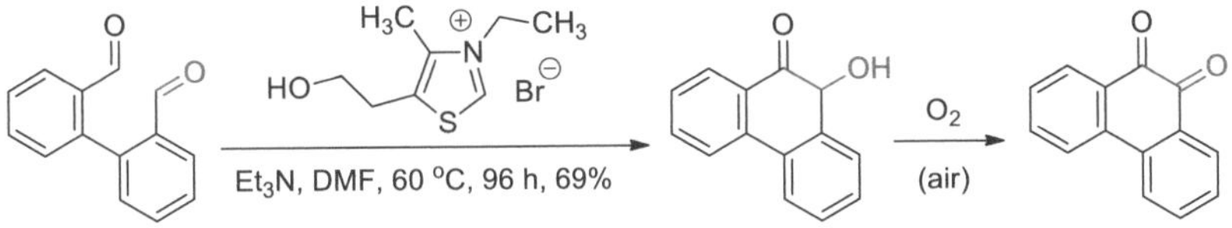

© Springer Nature Switzerland AG 2021
J. J. Li, *Name Reactions*, https://doi.org/10.1007/978-3-030-50865-4_11

Example 4, With Brook rearrangement[9]

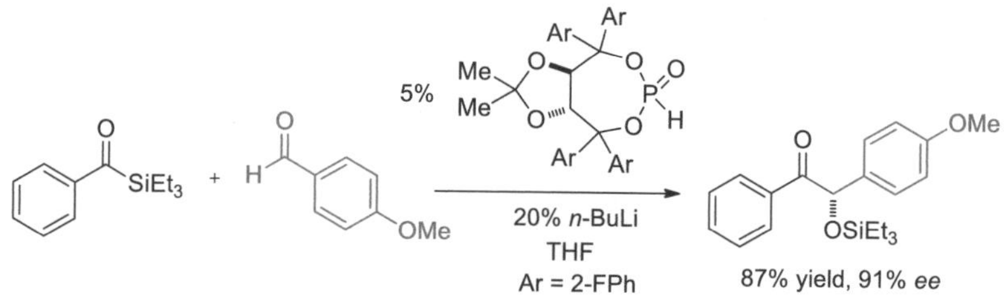
5%
20% n-BuLi
THF
Ar = 2-FPh
87% yield, 91% ee

Example 5[10]

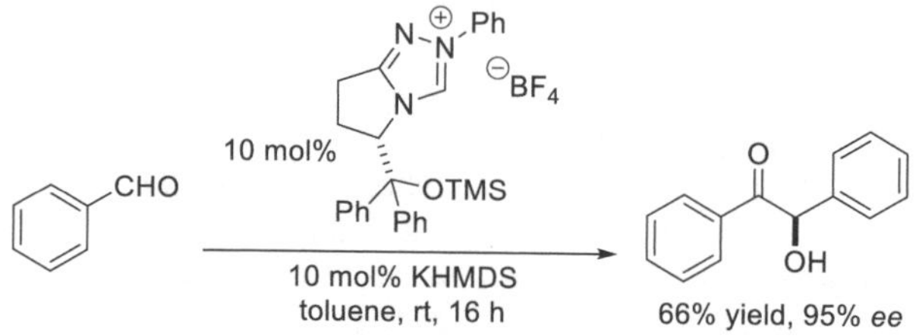
10 mol%
10 mol% KHMDS
toluene, rt, 16 h
66% yield, 95% ee

Example 6[12]

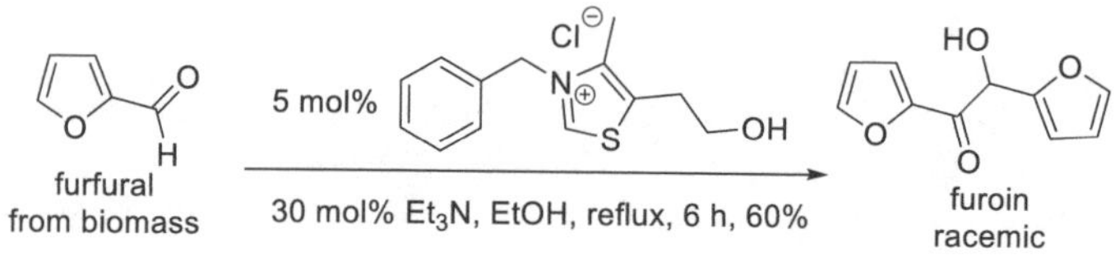
furfural
from biomass
5 mol%
30 mol% Et3N, EtOH, reflux, 6 h, 60%
furoin
racemic

Example 7, *N*-Heterocyclic carbene (NHC)-catalyzed cross benzoin reaction resulting in substrate-controlled diastereoselectvity reversal[13]

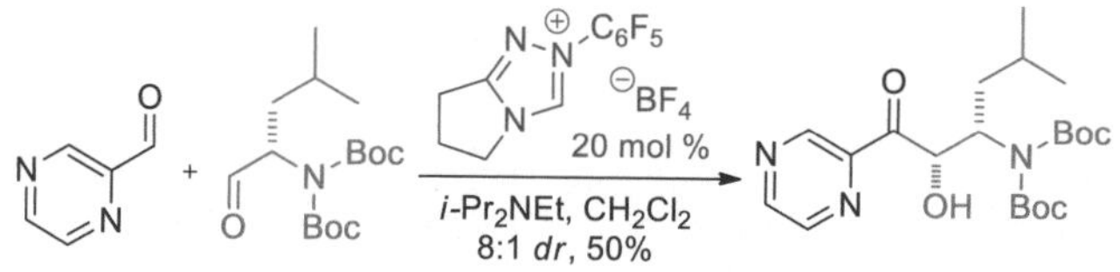
20 mol %
i-Pr2NEt, CH2Cl2
8:1 dr, 50%

Example 8, *N*-Heterocyclic carbene (NHC)-catalyzed asymmetric benzoin reaction in water[14]

References

1. Lapworth, A. J. *J. Chem. Soc.* **1903,** *83*, 995–1005. Arthur Lapworth (1872–1941) was born in Scotland. He was a figure in the development of the modern view of mechanisms of organic reactions. Lapworth investigated the benzoin condensation at the Chemical Department, The Goldsmiths' Institute, New Cross, UK.
2. Buck, J. S.; Ide, W. S. *J. Am. Chem. Soc.* **1932,** *54*, 3302–3309.
3. Ide, W. S.; Buck, J. S. *Org. React.* **1948,** *4*, 269–304. (Review).
4. Stetter, H.; Kuhlmann, H. *Org. React.* **1991,** *40*, 407–496. (Review).
5. White, M. J.; Leeper, F. J. *J. Org. Chem.* **2001,** *66*, 5124–5131.
6. Hachisu, Y.; Bode, J. W.; Suzuki, K. *J. Am. Chem. Soc.* **2003,** *125*, 8432–8433.
7. Enders, D.; Niemeier, O. *Synlett* **2004,** 2111–2114.
8. Johnson, J. S. *Angew. Chem. Int. Ed.* **2004,** *43*, 1326–1328. (Review).
9. Linghu, X.; Potnick, J. R.; Johnson, J. S. *J. Am. Chem. Soc.* **2004,** *126*, 3070–3071.
10. Enders, D.; Han, J. *Tetrahedron: Asymmetry* **2008,** *19*, 1367–1371.
11. Cee, V. J. *Benzoin Condensation.* In *Name Reactions for Homologations-Part I*; Li, J. J., Ed.; Wiley: Hoboken, NJ, **2009,** pp 381–392. (Review).
12. Kabro, A.; Escudero-Adan, E. C.; Grushin, V. V.; van Leeuwen, P. W. N. M. *Org. Lett.* **2012,** *14*, 4014–4017.
13. Duan, A.; Fell, J. S.; Yu, P.; Lam, C. Y.-h.; Gravel, M.; Houk, K. N. *J. Org. Chem.* **2019,** *84*, 13565–13571.

Bergman Cyclization

Formation of a substituted benzene through 1,4-benzenediyl diradical formation from enediyne *via* electrocyclization.

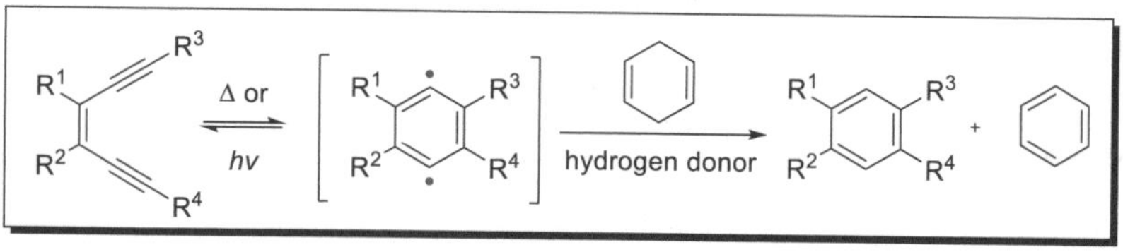

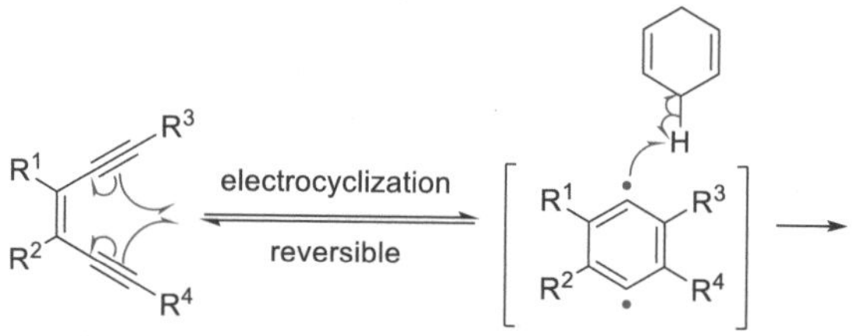

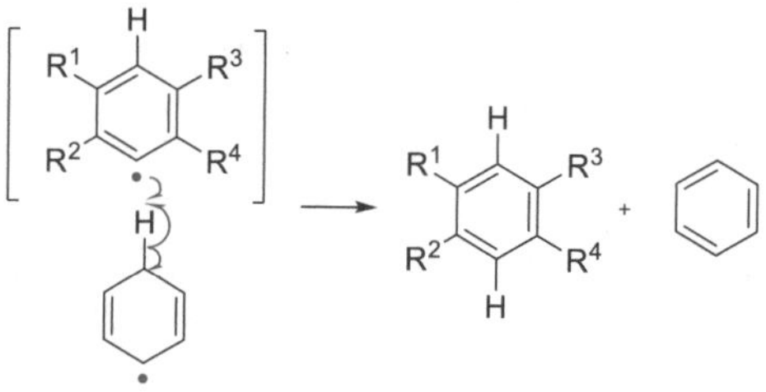

enediyne   1,4-benzenediyl diradical

Example 1[6]

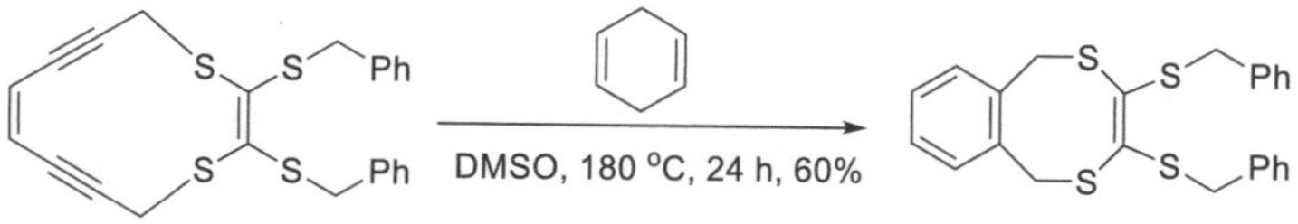

DMSO, 180 °C, 24 h, 60%

Example 2[7]

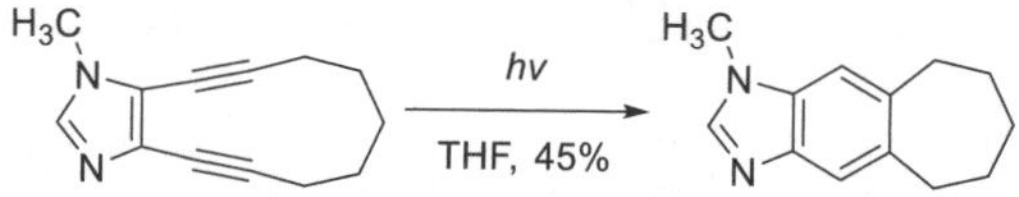

hv

THF, 45%

© Springer Nature Switzerland AG 2021
J. J. Li, *Name Reactions*, https://doi.org/10.1007/978-3-030-50865-4_12

Example 3, Wolff rearrangement followed by Bergman cyclization[8]

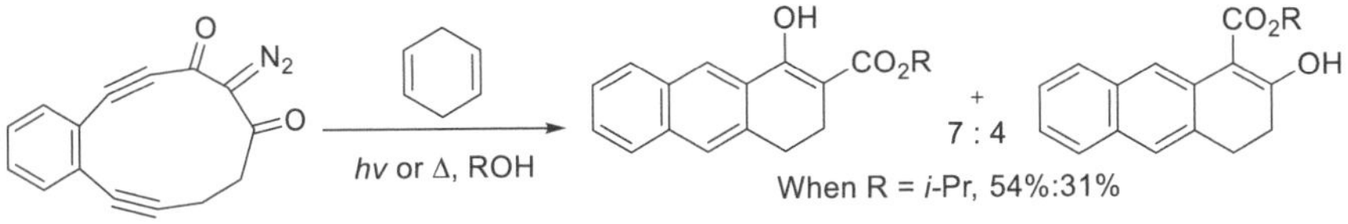

O
N2
O
hv or Δ, ROH
OH
CO2R
+
7 : 4
CO2R
OH
When R = i-Pr, 54%:31%

Example 4[10]

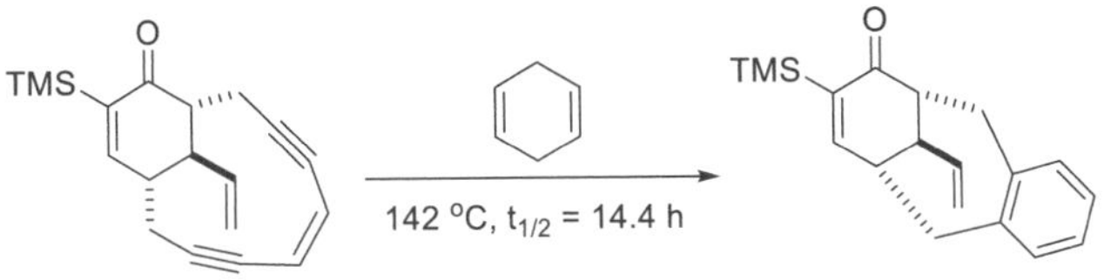

TMS
O
142 °C, t1/2 = 14.4 h
TMS
O

Example 5[12]

N
N
20% 1,4-cyclohexadiene
PhCl, 180 °C, 48 h, 65%
N
N
major product
N
N
minor product
major: minor
= 9:1

Example 5[13]

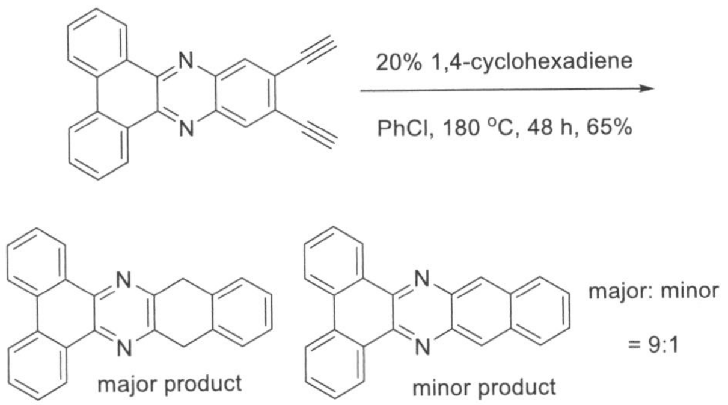

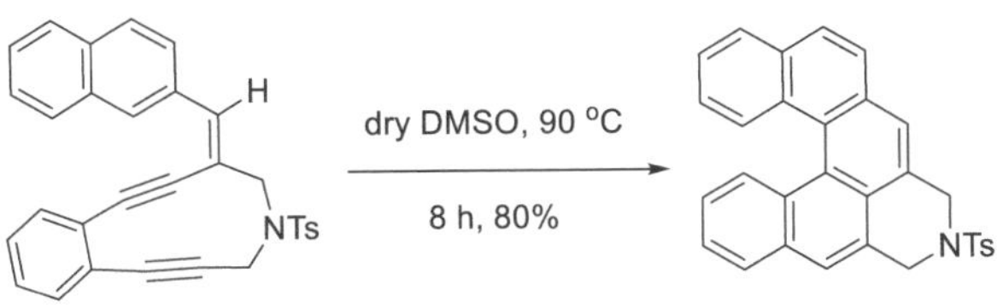

H
NTs
dry DMSO, 90 °C
8 h, 80%
NTs

Example 6, Barrierless nucleophilic addition to *p*-benzynes[15]

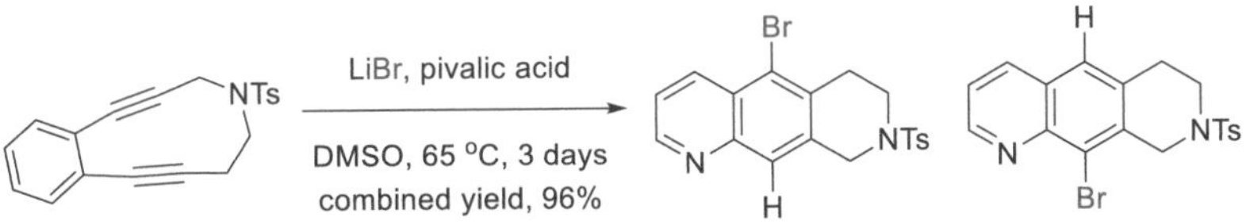

NTs
LiBr, pivalic acid
DMSO, 65 °C, 3 days
combined yield, 96%
Br
N
NTs
H
H
N
NTs
Br

Example 7, Barrierless nucleophilic addition to *p*-benzynes[14]

References

1. Jones, R. R.; Bergman, R. G. *J. Am. Chem. Soc.* **1972**, *94*, 660–661. Robert G. Bergman (1942–) is a professor at the University of California, Berkeley. His discovery of the Bergman cyclization was completed far in advance of the discovery of ene-diyne's anti-cancer properties.
2. Bergman, R. G. *Acc. Chem. Res.* **1973**, *6*, 25–31. (Review).
3. Myers, A. G.; Proteau, P. J.; Handel, T. M. *J. Am. Chem. Soc.* **1988**, *110*, 7212–7214.
4. Yus, M.; Foubelo, F. *Rec. Res. Dev. Org. Chem.* **2002**, *6*, 205–280. (Review).
5. Basak, A.; Mandal, S.; Bag, S. S. *Chem. Rev.* **2003**, *103*, 4077–4094. (Review).
6. Bhattacharyya, S.; Pink, M.; Baik, M.-H.; Zaleski, J. M. *Angew. Chem. Int. Ed.* **2005**, *44*, 592–595.
7. Zhao, Z.; Peacock, J. G.; Gubler, D. A.; Peterson, M. A. *Tetrahedron Lett.* **2005**, *46*, 1373–1375.
8. Karpov, G. V.; Popik, V. V. *J. Am. Chem. Soc.* **2007**, *129*, 3792–3793.
9. Kar, M.; Basak, A. *Chem. Rev.* **2007**, *107*, 2861–2890. (Review).
10. Lavy, S.; Pérez-Luna, A.; Kündig, E. P. *Synlett* **2008**, 2621–2624.
11. Pandithavidana, D. R.; Poloukhtine, A.; Popik, V. V. *J. Am. Chem. Soc.* **2009**, *131*, 351–356.
12. Spence, J. D.; Rios, A. C.; Frost, M. A.; et al. *J. Org. Chem.* **2012**, *77*, 10329–10339.
13. Das, E.; Basak, A. *Tetrahedron* **2013**, *69*, 2184–2192.
14. Williams, D. E.; Bottriell, H.; Davies, J.; Tietjen, I.; Brockman, M. A.; Andersen, R. J. *Org. Lett.* **2015**, *17*, 5304–5307.
15. Das, E.; Basak, S.; Anoop, A.; Chand, S.; Basak, A. *J. Org. Chem.* **2019**, *84*, 2911–2921.
16. Das, E.; Basak, A. *J. Org. Chem.* **2020**, *85*, 2697–2703.

Biginelli Reaction

Also known as Biginelli pyrimidone synthesis. One-pot condensation of an aromatic aldehyde, urea, and β-dicarbonyl compound in acidic ethanolic solution and expansion of such a condensation thereof. It belongs to a class of transformations called multicomponent reactions (MCRs).

Example 1[4]

© Springer Nature Switzerland AG 2021
J. J. Li, *Name Reactions*, https://doi.org/10.1007/978-3-030-50865-4_13

Example 2[5]

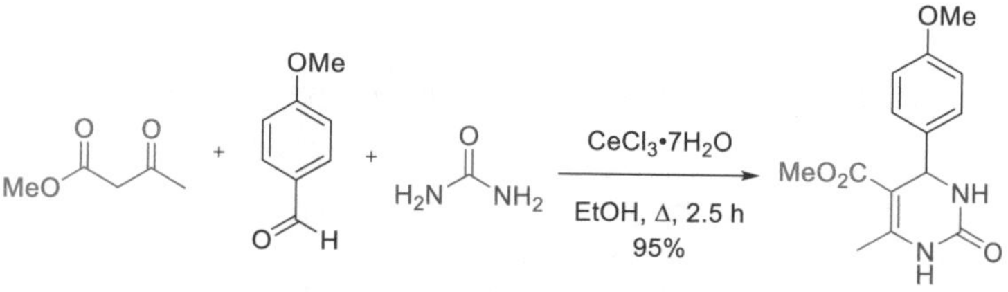

Example 3, Microwave (μW)-induced Biginelli condensation[9]

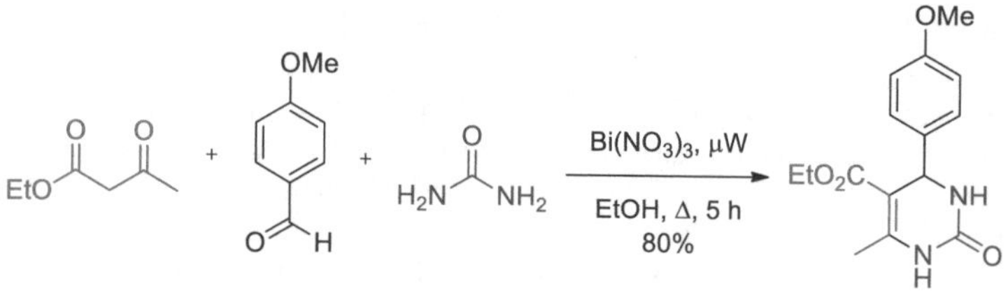

Example 3[10]

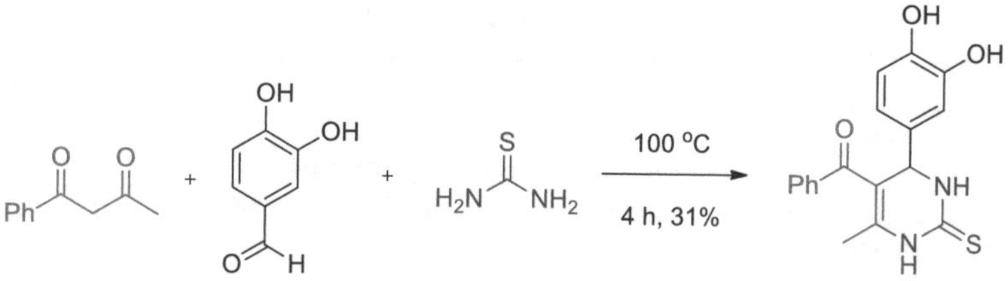

Example 4, Dual organocatalysis system[13]

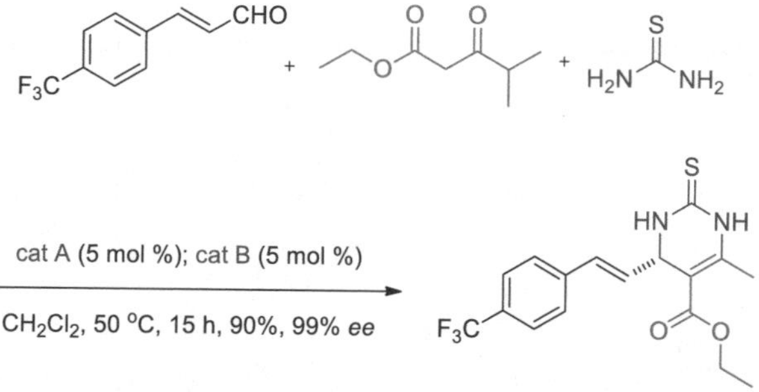

cat A (5 mol %); cat B (5 mol %)

CH₂Cl₂, 50 °C, 15 h, 90%, 99% ee

cat A = *trans*-4,5-methanoproline   cat B = quinidine urea

References

1. Biginelli, P. *Ber.* **1891**, *24*, 1317. Pietro Biginelli (1860–1937) published this paper at Instituto Superiore di Santitá (State Medicinal Institute) in Roma, Italy soon after joining the laboratory of Hugo Schiff (of the Schiff base fame) in Florence. As a student in Turin, Biginelli of Icilio Guareschi who discover the Guareschi reaction.
2. Kappe, C. O. *Tetrahedron* **1993**, *49*, 6937–6963. (Review).
3. Kappe, C. O. *Acc. Chem. Res.* **2000**, *33*, 879–888. (Review).
4. Kappe, C. O. *Eur. J. Med. Chem.* **2000**, *35*, 1043–1052. (Review).
5. Ghorab, M. M.; Abdel-Gawad, S. M.; El-Gaby, M. S. A. *Farmaco* **2000**, *55*, 249–255.
6. Bose, D. S.; Fatima, L.; Mereyala, H. B. *J. Org. Chem.* **2003**, *68*, 587–590.
7. Kappe, C. O.; Stadler, A. *Org. React.* **2004**, *68*, 1–116. (Review).
8. Limberakis, C. *Biginelli Pyrimidone Synthesis* In *Name Reactions in Heterocyclic Chemistry*; Li, J. J., Ed.; Wiley: Hoboken, NJ, **2005**, pp 509–520. (Review).
9. Banik, B. K.; Reddy, A. T.; Datta, A.; Mukhopadhyay, C. *Tetrahedron Lett.* **2007**, *48*, 7392–7394.
10. Wang, R.; Liu, Z.-Q. *J. Org. Chem.* **2012**, *77*, 3952–3958.
11. Nagarajaiah, H.; Mukhopadhyay, A.; Moorthy, J. N. *Tetrahedron Lett.* **2016**, *57*, 5135–5149.
12. Kaur, R.; Chaudhary, S.; Kumar, K.; Gupta, M. K.; Rawal, R. K. *Eur. J. Med. Chem.* **2017**, *132*, 108–134.
13. Yu, H.; Xu, P.; He, H.; Zhu, J.; Lin, H.; Han, S. *Tetrahedron: Asymmetry* **2017**, *28*, 257–265.
14. Yu, S.; Wu, J.; Lan, H.; Gao, L.; Qian, H.; Fan, K.; Yin, Z. *Org. Lett.* **2020**, *22*, 102–105.

Birch Reduction

The Birch reduction is the 1,4-reduction of aromatics to their corresponding cy-clohexadienes by alkali metals (Li, K, Na) dissolved in liquid ammonia in the presence of an alcohol.

Benzene ring bearing an electron-donating substituent:

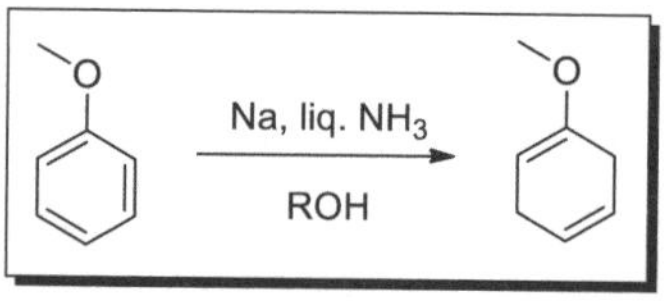

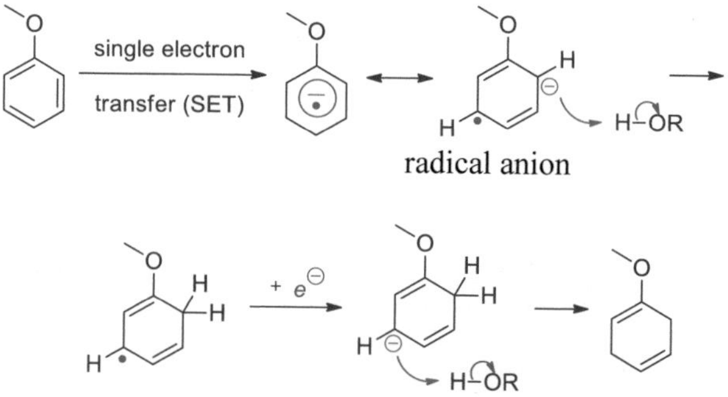

radical anion

Benzene ring with an electron-withdrawing substituent:

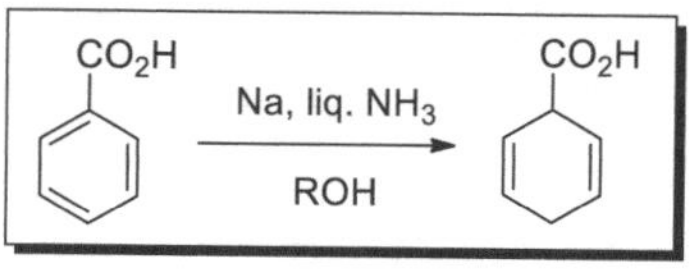

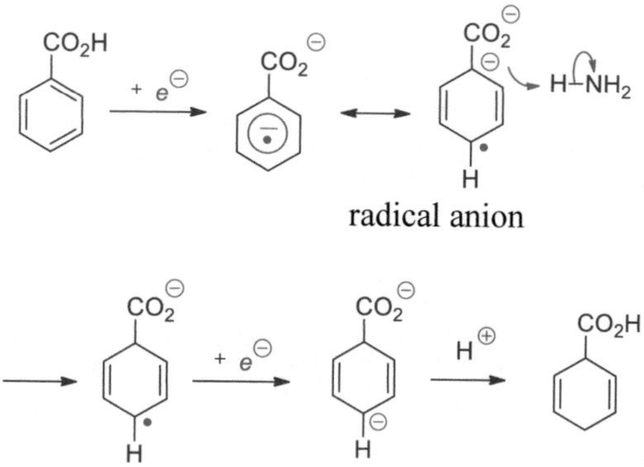

radical anion

© Springer Nature Switzerland AG 2021
J. J. Li, *Name Reactions*, https://doi.org/10.1007/978-3-030-50865-4_14

Example 1, Birch reductive alkylation[4]

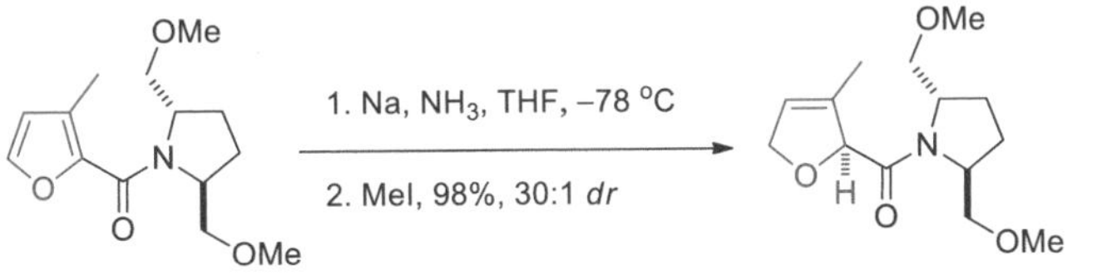
OMe
1. Na, NH_3, THF, –78 °C
2. MeI, 98%, 30:1 *dr*

Example 2[7]

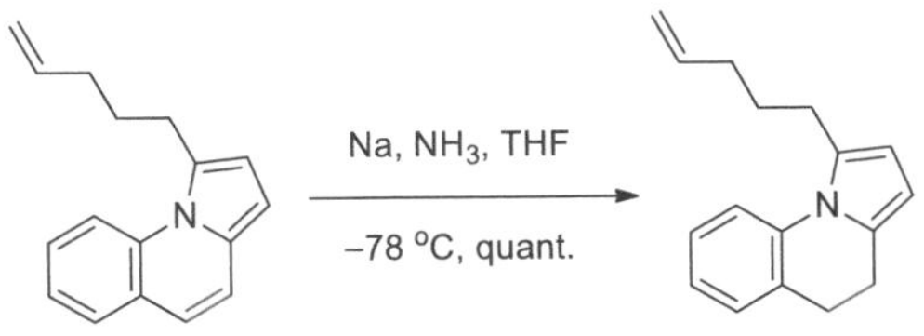
Na, NH_3, THF
–78 °C, quant.

Example 3, Fully reduced products[8]

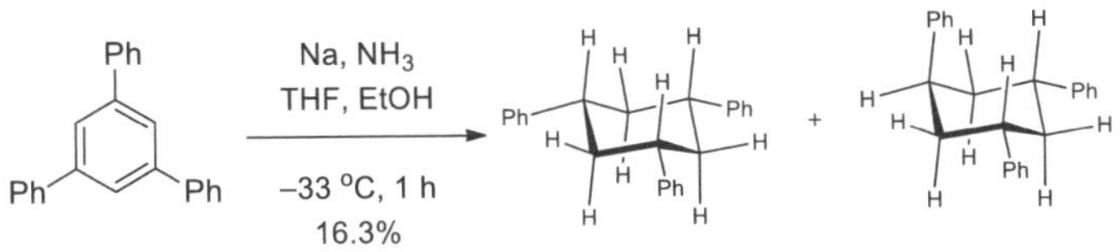
Ph
Na, NH_3
THF, EtOH
–33 °C, 1 h
16.3%

Example 4, Birch reductive alkylation[9]

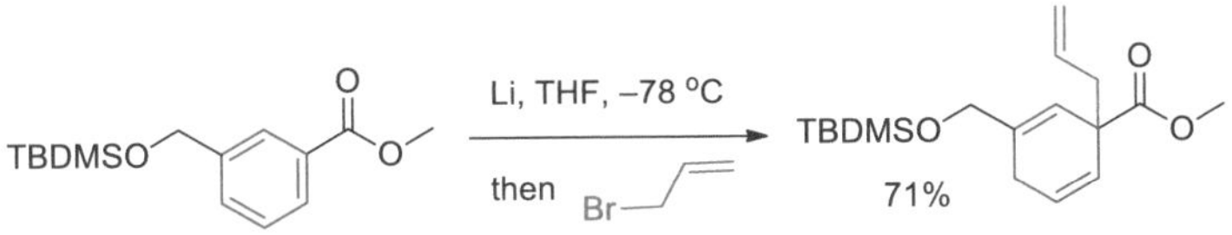
TBDMSO
Li, THF, –78 °C
then
Br
71%

Example 5[10]

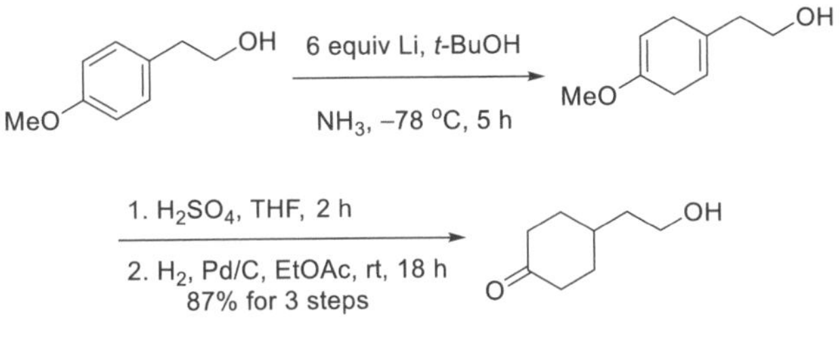
OH
MeO
6 equiv Li, *t*-BuOH
NH_3, –78 °C, 5 h
1. H_2SO_4, THF, 2 h
2. H_2, Pd/C, EtOAc, rt, 18 h
87% for 3 steps

Example 6, Chemoselective ammonia-free Birch reduction[11]

Example 7, Directed Birch reduction enabled by an intramolecular proton source[12]

The hydroxyl group serves as the intramolecular proton source.

Example 8, Birch reduction of α,β-unsaturated imide[13]

References

1. Birch, A. J. *J. Chem. Soc.* **1944,** 430–436. Arthur Birch (1915–1995), an Australian, developed the "Birch reduction" at Oxford University during WWII in Robert Robinson's laboratory. The Birch reduction was instrumental to the discovery of the birth control pills and many other drugs.
2. Rabideau, P. W.; Marcinow, Z. *Org. React.* **1992,** *42,* 1–334. (Review).
3. Birch, A. J. *Pure Appl. Chem.* **1996,** *68,* 553–556. (Review).
4. Donohoe, T. J.; Guillermin, J.-B. *Tetrahedron Lett.* **2001,** *42,* 5841–5844.
5. Pellissier, H.; Santelli, M. *Org. Prep. Proced. Int.* **2002,** *34,* 611–642. (Review).
6. Subba Rao, G. S. R. *Pure Appl. Chem.* **2003,** *75,* 1443–1451. (Review).
7. Kim, J. T.; Gevorgyan, V. *J. Org. Chem.* **2005,** *70,* 2054–2059.
8. Gealis, J. P.; Müller-Bunz, H.; Ortin, Y. *Chem. Eur. J.* **2008,** *14,* 1552–1560.
9. Fretz, S. J.; Hadad, C. M.; Hart, D. J.; Vyas, S.; Yang, D. *J. Org. Chem.* **2013,** *78,* 83–92.
10. Desrat, S.; Remeur, C.; Roussi, F. *Org. Biomol. Chem.* **2015,** *13,* 5520–5531.
11. Lei, P.; Ding, Y.; Zhang, X.; Adijiang, A.; Li, H.; Ling, Y.; An, J. *Org. Lett.* **2018,** *20,* 3439–3442.
12. Zhu, X.; McAtee, C. C.; Schindler, C. S. *J. Am. Chem. Soc.* **2019,** *141,* 3409–3413.
13. Sengupta, A.; Hosokawa, S. *Synlett* **2019,** *30,* 709–712.

Bischler–Napieralski Reaction

Dihydroisoquinolines from β-phenethylamides in refluxing phosphorus oxychloride.

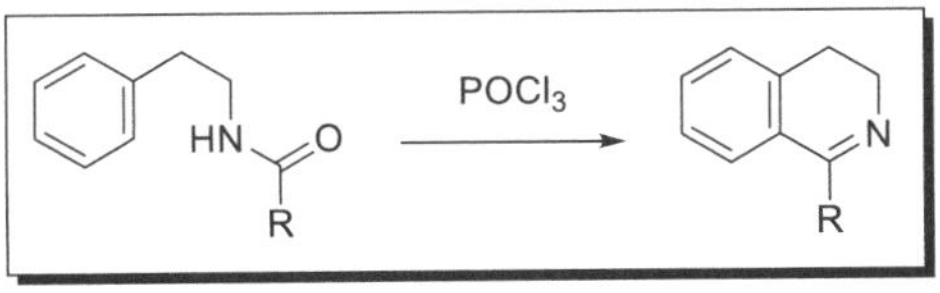

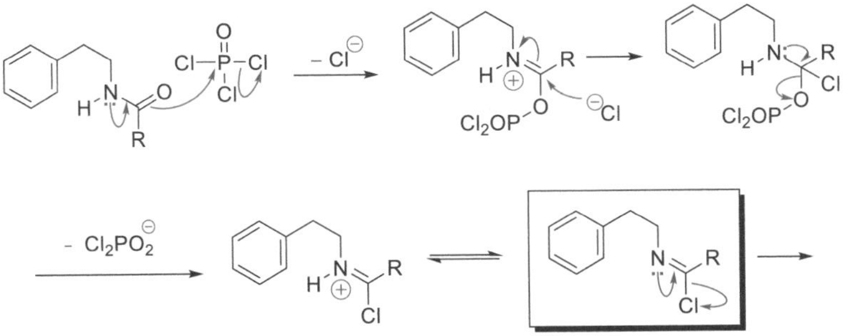

Imidoyl intermediate[2]

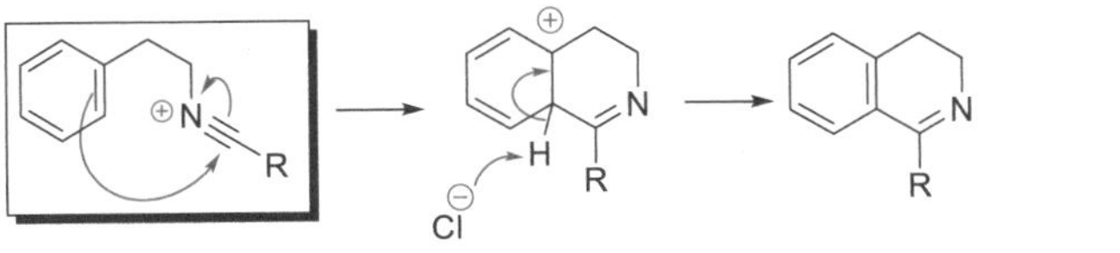

Nitrilium salt intermediate[2]

Example 1[3]

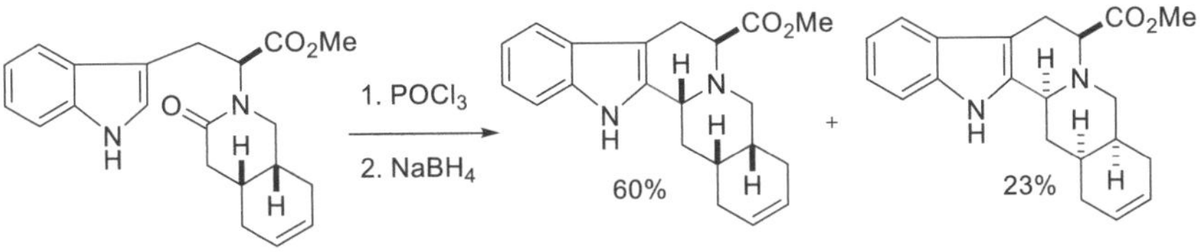

60%

23%

Example 2[5]

96%

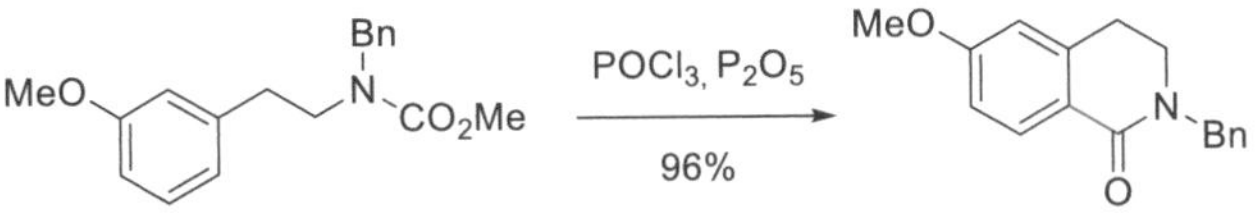

© Springer Nature Switzerland AG 2021
J. J. Li, *Name Reactions*, https://doi.org/10.1007/978-3-030-50865-4_15

Example 3[7]

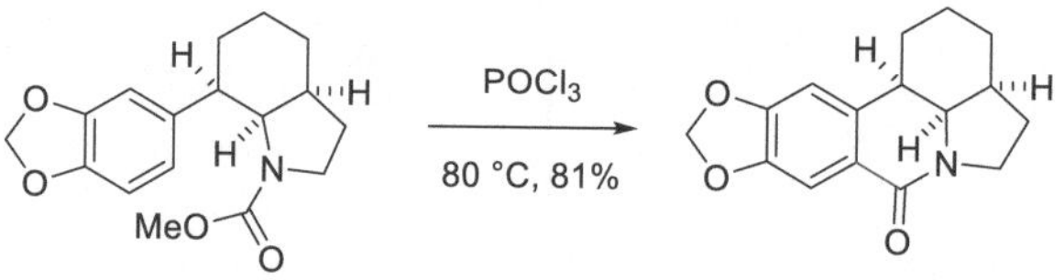

Example 4[8]

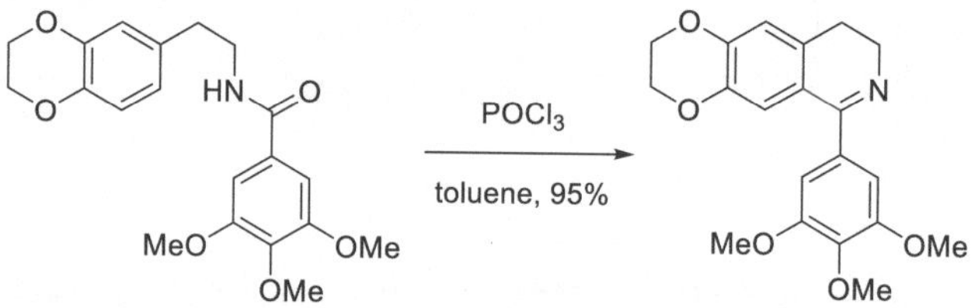

Example 5[10]

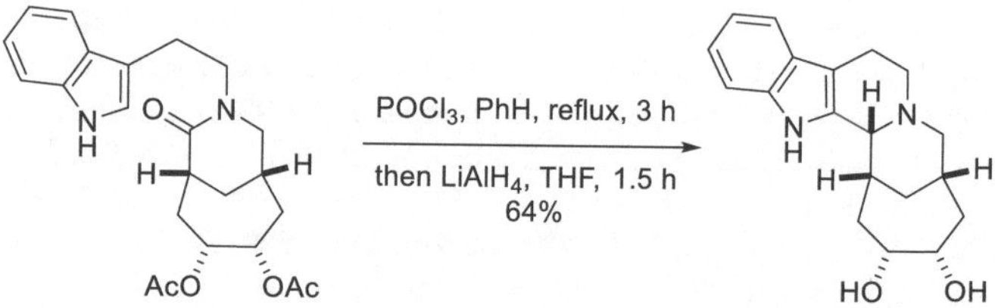

Example 6, An unprecedented Bischler–Napieralski (B–N) reaction[11]

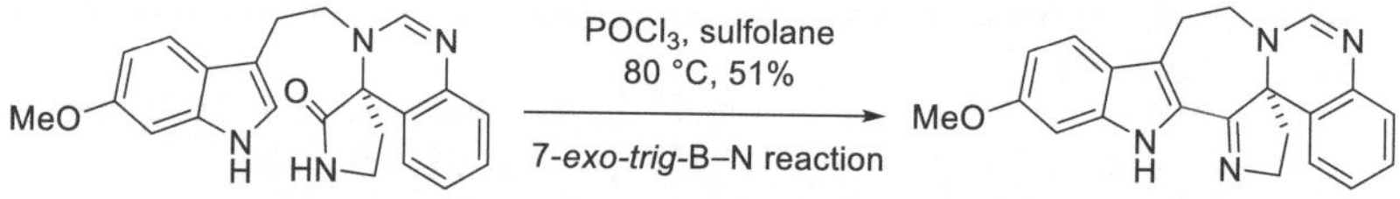

Example 7, Direct bservation of intermediates involved in the Bischler–Napieralski reaction[12]

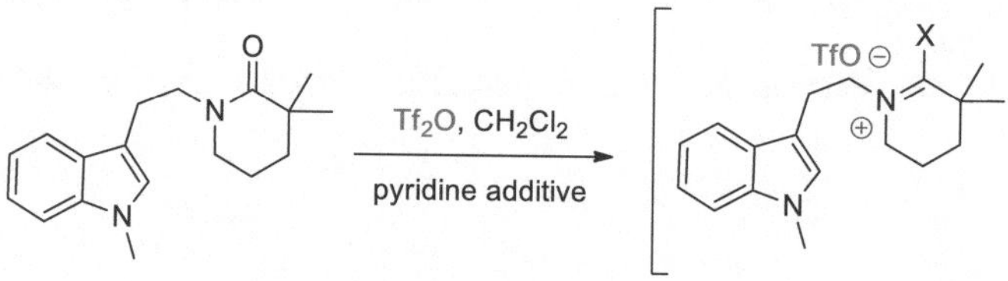

Example 8, A new Bischler–Napieralski reaction[13]

References

1. Bischler, A.; Napieralski, B. *Ber.* **1893**, *26*, 1903–1908. Augustus Bischler discovered the Bischler–Napieralski reaction while studying alkaloids at Basel Chemical Works, Switzerland with his coworker, B. Napieralski. Bernard Napieralski was affiliated with the University of Zurich.
2. Mechanistic studies: (a) Fodor, G.; Gal, J.; Phillips, B. A. *Angew. Chem. Int. Ed. Engl.* **1972**, *11*, 919–920. (b) Nagubandi, S.; Fodor, G. *J. Heterocycl. Chem.* **1980**, *17*, 1457–1463. (c) Fodor, G.; Nagubandi, S. *Tetrahedron* **1980**, *36*, 1279–1300.
3. Aubé, J.; Ghosh, S.; Tanol, M. *J. Am. Chem. Soc.* **1994**, *116*, 9009–9018.
4. Sotomayor, N.; Domínguez, E.; Lete, E. *J. Org. Chem.* **1996**, *61*, 4062–4072.
5. Wang, X.-j.; Tan, J.; Grozinger, K. *Tetrahedron Lett.* **1998**, *39*, 6609–6612.
6. Ishikawa, T.; Shimooka, K.; Narioka, T.; Noguchi, S.; Saito, T.; Ishikawa, A.; Yamazaki, E.; Harayama, T.; Seki, H.; Yamaguchi, K. *J. Org. Chem.* **2000**, *65*, 9143–9151.
7. Banwell, M. G.; Harvey, J. E.; Hockless, D. C. R., Wu, A. W. *J. Org. Chem.* **2000**, *65*, 4241–4250.
8. Capilla, A. S.; Romero, M.; Pujol, M. D.; Caignard, D. H.; Renard, P. *Tetrahedron* **2001**, *57*, 8297–8303.
9. Wolfe, J. P. *Bischler–Napieralski Reaction*. In *Name Reactions in Heterocyclic Chemistry*; Li, J. J., Ed.; Wiley: Hoboken, NJ, **2005**, pp 376–385. (Review).
10. Ho, T.-L.; Lin, Q.-x. *Tetrahedron* **2008**, *64*, 10401–10405.
11. Buyck, T.; Wang, Q.; Zhu, J. *Org. Lett.* **2012**, *14*, 1338–1341.
12. White, K. L.; Mewald, M.; Movassaghi, M. *J. Org. Chem.* **2015**, *80*, 7403–7411.
13. Xie, C.; Luo, J.; Zhang, Y.; Zhu, L.; Hong, R. *Org. Lett.* **2017**, *19*, 3592–3595.
14. Min, L.; Yang, W.; Weng, Y.; Zheng, W.; Wang, X.; Hu, Y. *Org. Lett.* **2019**, *21*, 2574–2577.
15. Amer, M. M.; Olaizola, O.; Carter, J.; Abas, H.; Clayden, J. *Org. Lett.* **2020**, *22*, 253–256.

Brook Rearrangement

Rearrangement of α-silyl oxyanions to α-silyloxy carbanions *via* a reversible process involving a pentacoordinate silicon intermediate is known as the [1,2]-Brook rearrangement, or [1,2]-silyl migration.

[1,2]-Brook rearrangement

pentacoordinate silicon intermediate

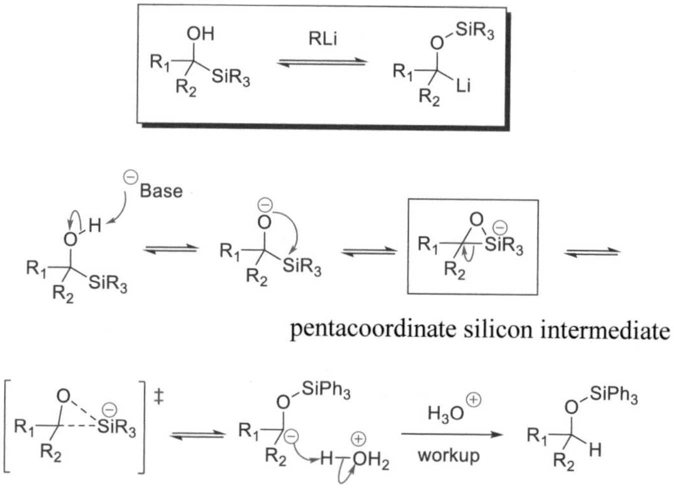

[1,3]-Brook rearrangement

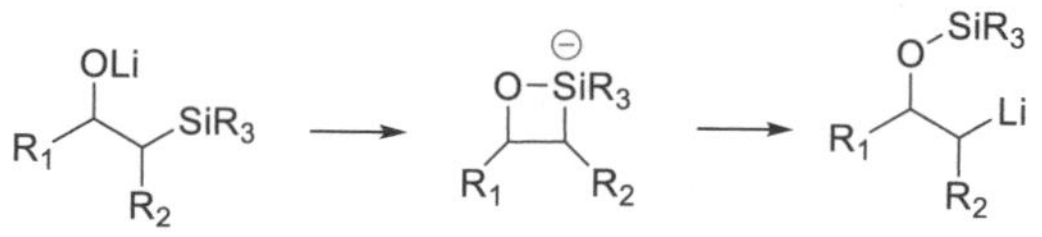

[1,4]-Brook rearrangement

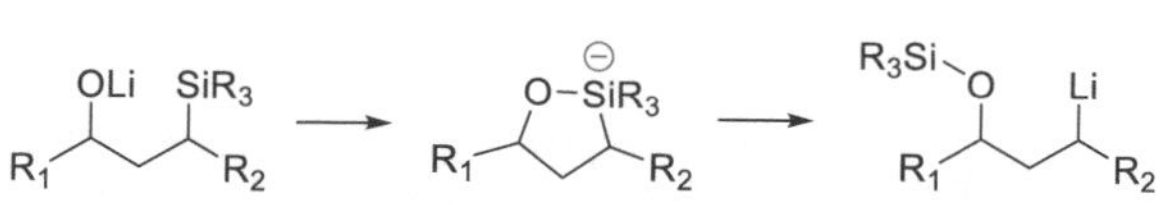

Example 1[6]

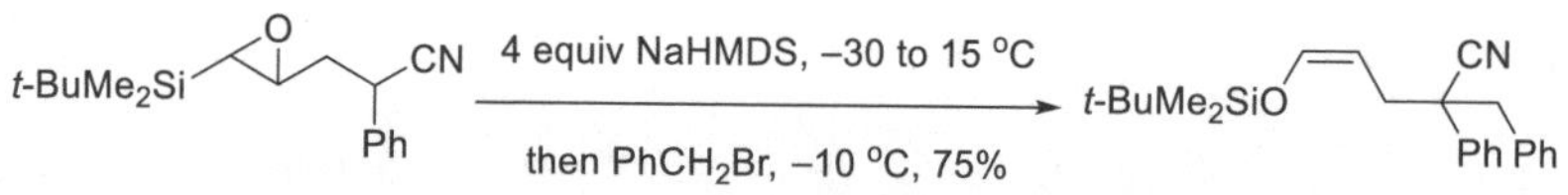

J. J. Li, *Name Reactions*, https://doi.org/10.1007/978-3-030-50865-4_16

Example 2, [1,2]-Brook rearrangement followed by a retro-[1,5]-Brook rearrangement[8]

Example 3, [1,5]-Brook rearrangement[9]

Example 4, Retro-[1,4]-Brook rearrangement[10]

Example 5, Retro-Brook rearrangement[12]

Example 6, Retro-Brook rearrangement[13]

Alternatively:

Example 7, Cyclopropanes from Brook rearrangement[14]

Alternatively:

References

1. Brook, A. G. *J. Am. Chem. Soc.* **1958,** *80,* 1886–1889. Adrian G. Brook (1924–2013) was born in Toronto, Canada. He was a professor in Lash Miller Chemical Laboratories, University of Toronto, Canada.
2. Brook, A. G. *Acc. Chem. Res.* **1974,** *7,* 77–84. (Review).
3. Bulman Page, P. C.; Klair, S. S.; Rosenthal, S. *Chem. Soc. Rev.* **1990,** *19,* 147–195. (Review).
4. Fleming, I.; Ghosh, U. *J. Chem. Soc., Perkin Trans. 1* **1994,** 257–262.
5. Moser, W. H. *Tetrahedron* **2001,** *57,* 2065–2084. (Review).
6. Okugawa, S.; Takeda, K. *Org. Lett.* **2004,** *6,* 2973–2975.
7. Matsumoto, T.; Masu, H.; Yamaguchi, K.; Takeda, K. *Org. Lett.* **2004,** *6,* 4367–4369.
8. Clayden, J.; Watson, D. W.; Chambers, M. *Tetrahedron* **2005,** *61,* 3195–3203.
9. Smith, A. B., III; Xian, M.; Kim, W.-S.; Kim, D.-S. *J. Am. Chem. Soc.* **2006,** *128,* 12368–12369.
10. Mori, Y.; Futamura, Y.; Horisaki, K. *Angew. Chem. Int. Ed.* **2008,** *47,* 1091–1093.
11. Greszler, S. N.; Johnson, J. S. *Org. Lett.* **2009,** *11,* 827–830.
12. He, Y.; Hu, H.; Xie, X.; She, X. *Tetrahedron* **2013,** *69,* 559–563.
13. Chari, J. V.; Ippoliti, F. M.; Garg, N. K. *J. Org. Chem.* **2019,** *84,* 3652–3655.
14. Tang, F; Ma, P.-J.; Yao, Y.; Xu, Y.-J.; Lu, C.-D. *Chem. Commun.* **2019,** *55,* 3777–3780.
15. Lee, N.; Tan, C.-H.; Leow, D. *Asian J. Org. Chem.* **2019,** *8,* 25–31. (Review).
16. Kondoh, A.; Aita, K.; Ishikawa, S.; Terada, M. *Org. Lett.* **2020,** *22,* 2105–2110.

Brown Hydroboration

Addition of boranes to olefins followed by alkalinic oxidation of the organoborane adducts to afford alcohols. Regiochemistry follows anti-Markovnikov's rule.

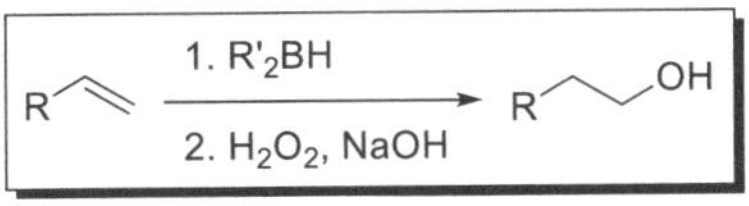

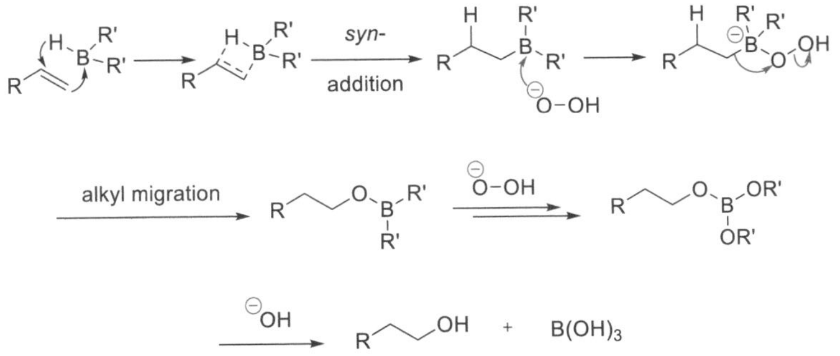

Example 1[2]

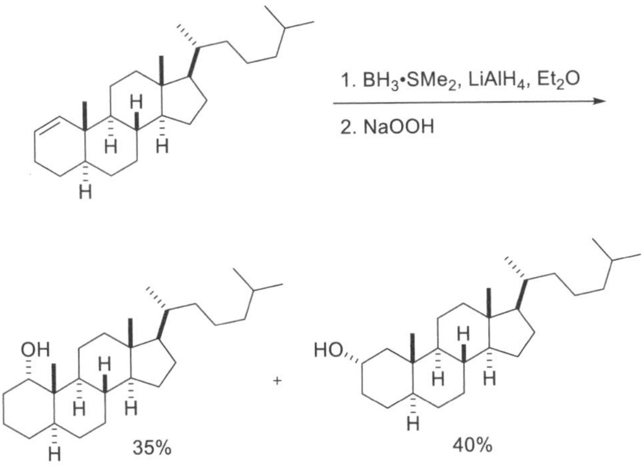

1. BH$_3$•SMe$_2$, LiAlH$_4$, Et$_2$O

2. NaOOH

35%   +   40%

Example 2[7]

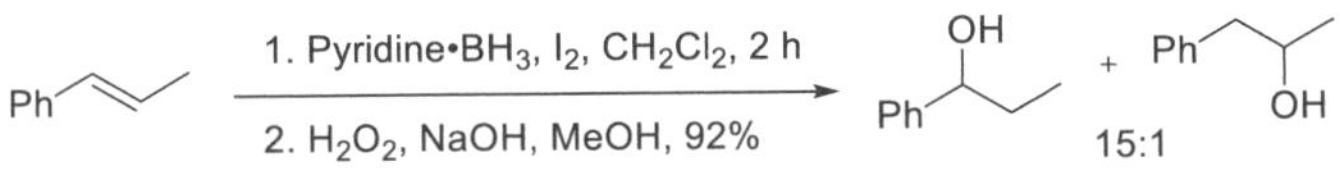

Ph

1. Pyridine•BH$_3$, I$_2$, CH$_2$Cl$_2$, 2 h

2. H$_2$O$_2$, NaOH, MeOH, 92%

OH
Ph   +   Ph

15:1   OH

© Springer Nature Switzerland AG 2021
J. J. Li, *Name Reactions*, https://doi.org/10.1007/978-3-030-50865-4_17

Example 3[8]

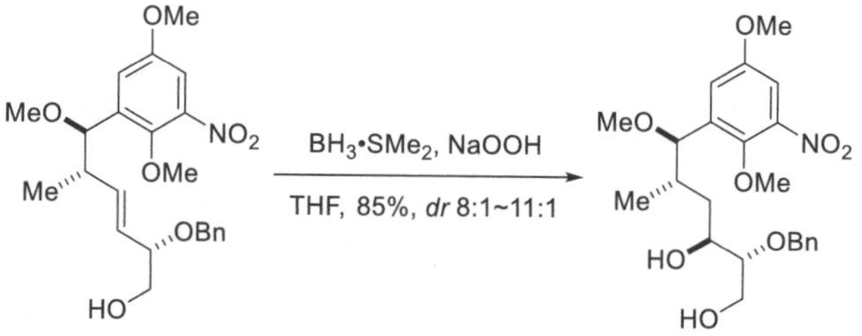

Example 4, Asymmetric hydroboration, nbd = 2,5-norbornadiene[10]

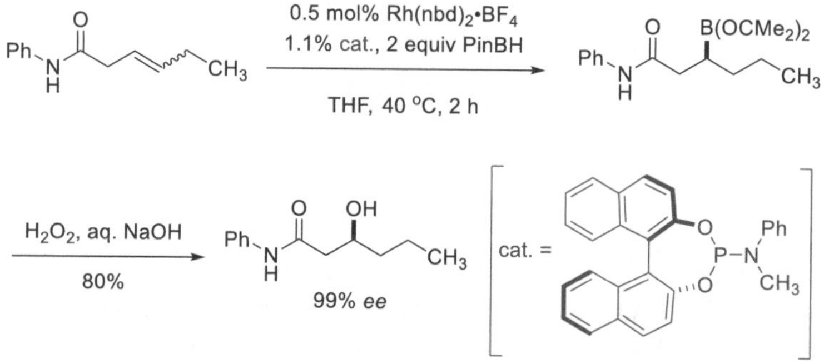

Example 5[11]

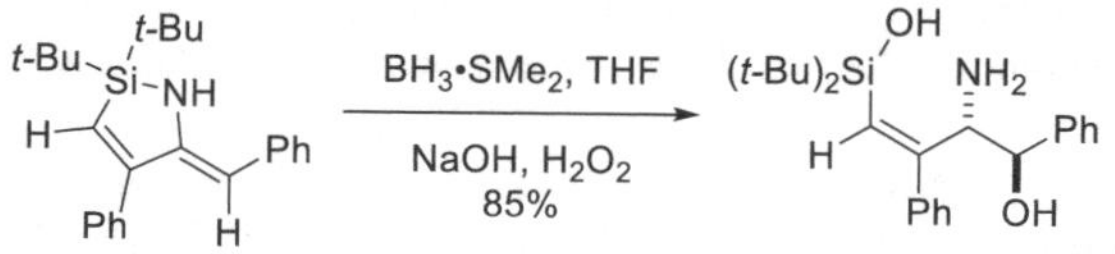

Example 6, Double whammy: One-pot CBS reduction/Brown hydroboration[12]

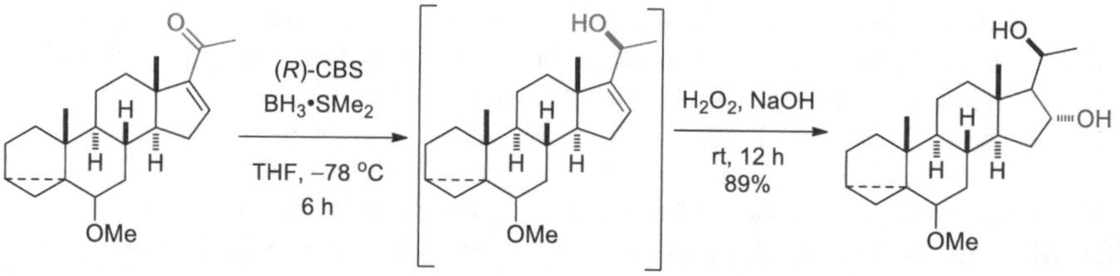

Example 7, The excellent diastereoselectivity of the Brown hydroboration likely resulted from complete shielding of the top face of the double bond by two piperidine rings[13]

Example 8, Simple, but get the job done[14]

References

1. Brown, H. C.; Tierney, P. A. *J. Am. Chem. Soc.* **1958**, *80*, 1552–1558. Herbert C. Brown (USA, 1912–2004) began his academic career at Wayne State University and moved on to Purdue University where he shared the Nobel Prize in Chemistry in 1981 with Georg Wittig (Germany, 1897–1987) for their development of organic boron and phosphorous compounds.
2. Nussim, M.; Mazur, Y.; Sondheimer, F. *J. Org. Chem.* **1964**, *29*, 1120–1131.
3. Pelter, A.; Smith, K.; Brown, H. C. *Borane Reagents,* Academic Press: New York, **1972**. (Book).
4. Brewster, J. H.; Negishi, E. *Science* **1980**, *207*, 44–46. (Review).
5. Fu, G. C.; Evans, D. A.; Muci, A. R. *Adv. Catal. Proc.* **1995**, *1*, 95–121. (Review).
6. Hayashi, T. *Comprehensive Asymmetric Catalysis I–III* **1995**, *1*, 351–364. (Review).
7. Clay, J. M.; Vedejs, E. *J. Am. Chem. Soc.* **2005**, *127*, 5766–5767.
8. Carter K. D.; Panek J. S. *Org. Lett.* **2004**, *6*, 55–57.
9. Clay, J. M. *Brown Hydroboration Reaction.* In *Name Reactions for Functional Group Transformations*; Li, J. J., Ed.; Wiley: Hoboken, NJ, **2007**, pp 183–188. (Review).
10. Smith, S. M.; Thacker, N. C.; Takacs, J. M. *J. Am. Chem. Soc.* **2008**, *130*, 3734–3735.
11. Anderson, L. L.; Woerpel, K. A. *Org. Lett.* **2009**, *11,* 425–428.
12. Cheng, S.-L.; Jiang, X.-L.; Shi, Y.; Tian, W.-S. *Org. Lett.* **2015**, *17*, 2346–2349.
13. Chen, Z.-T.; Xiao, T.; Tang, P.; Zhang, D.; Qin, Y. *Tetrahedron* **2018**, *74,* 1129–1134.
14. Reddy, M. S.; Manikanta, G.; Krishna, P. R. *Synthesis* **2019**, *51,* 1427–1434.
15. Srinivasu, K.; Nagaiah, K.; Yadav, J. S. *ChemistrySelect* **2020**, *5,* 2763–2766.

Bucherer–Bergs Reaction

Formation of hydantoins from carbonyl compounds with potassium cyanide (KCN) and ammonium carbonate [$(NH_4)_2CO_3$] or from cyanohydrins and ammonium carbonate. It belongs to the category of multiple component reactions (MCRs).

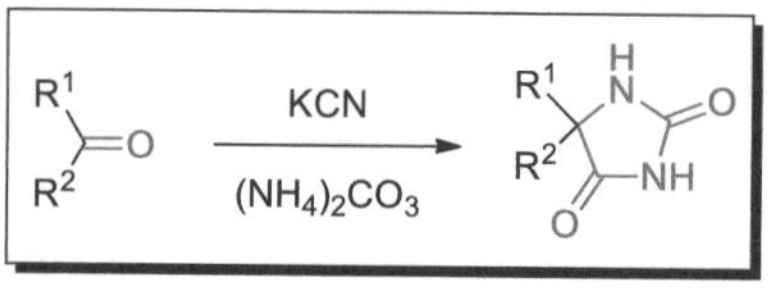

$$(NH_4)_2CO_3 = 2\ NH_3 + CO_2 + H_2O$$

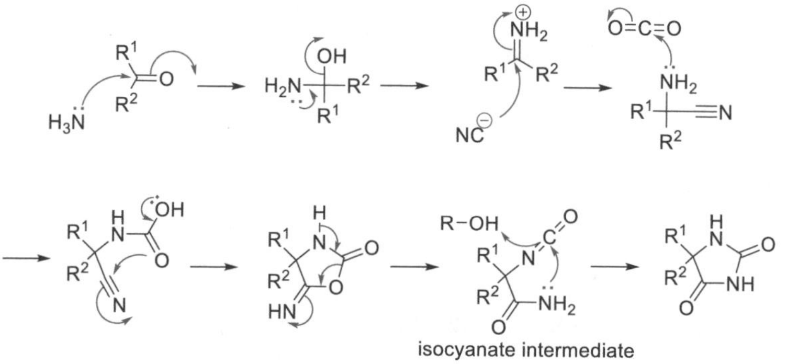

isocyanate intermediate

Example 1[5]

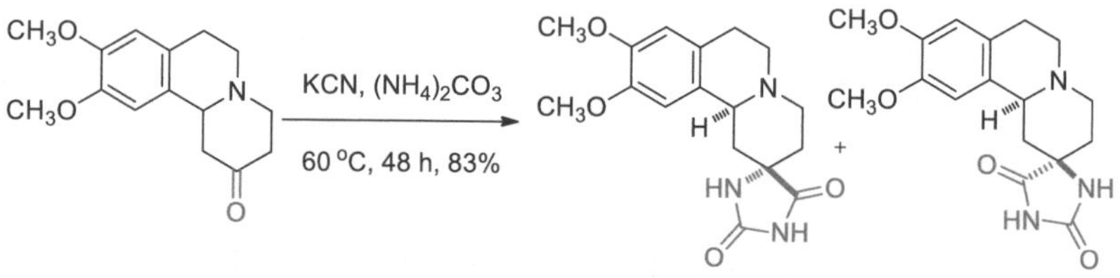

KCN, $(NH_4)_2CO_3$

60 °C, 48 h, 83%

Example 2[6]

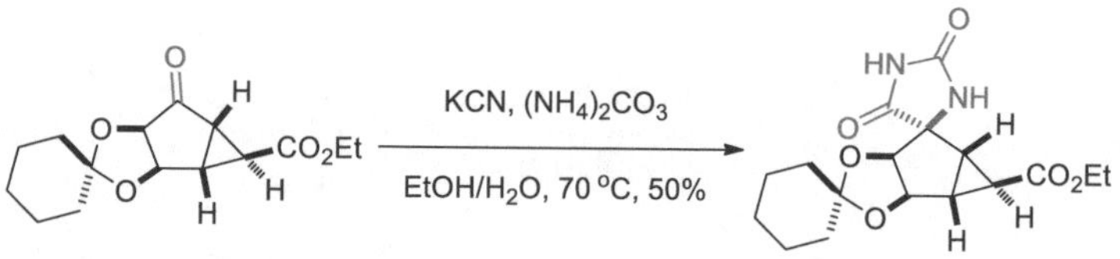

KCN, $(NH_4)_2CO_3$

EtOH/H_2O, 70 °C, 50%

© Springer Nature Switzerland AG 2021
J. J. Li, *Name Reactions*, https://doi.org/10.1007/978-3-030-50865-4_18

Example 3, In the presence of boronic acid[7]

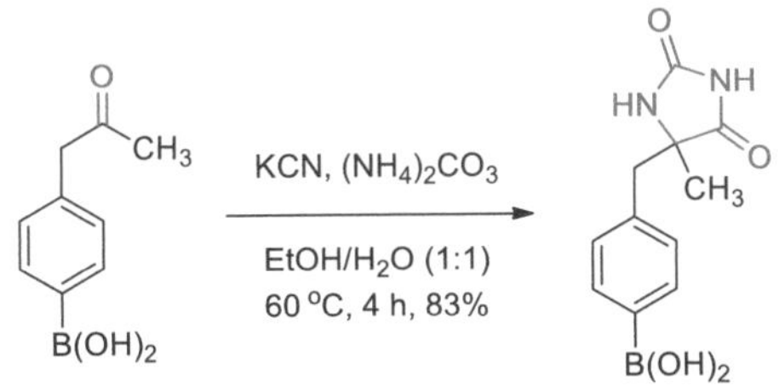
KCN, $(NH_4)_2CO_3$
EtOH/H_2O (1:1)
60 °C, 4 h, 83%

Example 4[9]

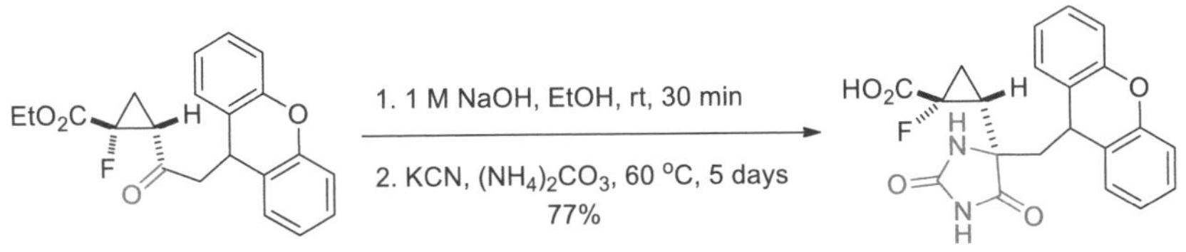
1. 1 M NaOH, EtOH, rt, 30 min
2. KCN, $(NH_4)_2CO_3$, 60 °C, 5 days
77%

Example 5[11]

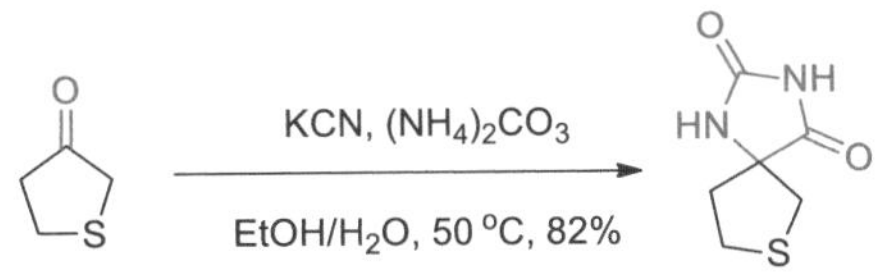
KCN, $(NH_4)_2CO_3$
EtOH/H_2O, 50 °C, 82%

Example 6, From β-keto-ester[12]

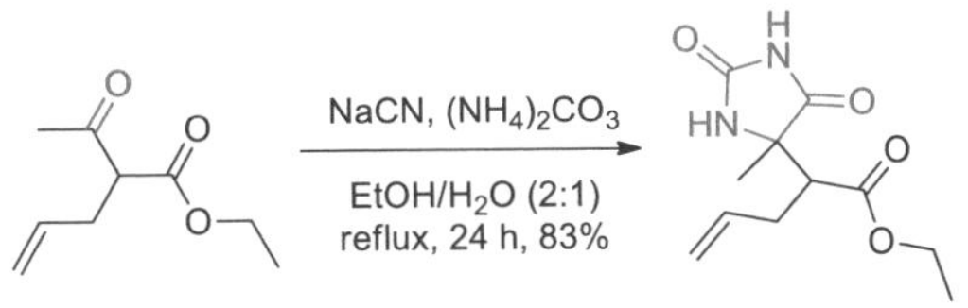
NaCN, $(NH_4)_2CO_3$
EtOH/H_2O (2:1)
reflux, 24 h, 83%

Example 7, Derivatization of natural product extracts via the Bucherer–Bergs reaction[13]

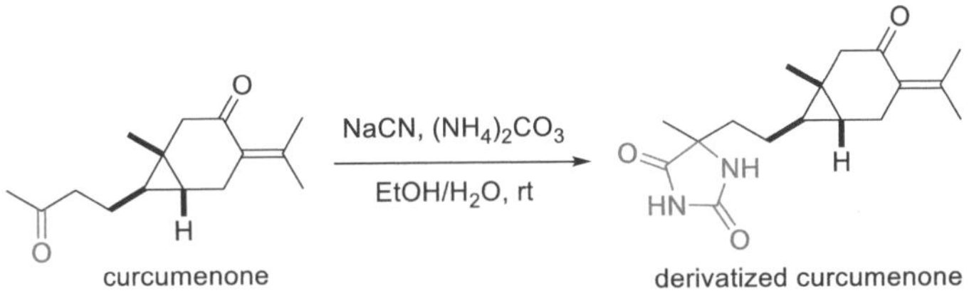
NaCN, $(NH_4)_2CO_3$
EtOH/H_2O, rt
curcumenone
derivatized curcumenone

Example 8, Ferrocenyl hydantoin[13]

References

1. Bergs, H. Ger. Pat. 566, 094, **1929**. Hermann Bergs worked at I. G. Farben in Germany.
2. Bucherer, H. T., Steiner, W. *J. Prakt. Chem.* **1934,** *140*, 291–316. (Mechanism).
3. Ware, E. *Chem. Rev.* **1950,** *46,* 403–470. (Review).
4. Wieland, H. In *Houben–Weyl's Methoden der organischen Chemie*, Vol. XI/2, **1958,** p 371. (Review).
5. Menéndez, J. C.; Díaz, M. P.; Bellver, C.; Söllhuber, M. M. *Eur. J. Med. Chem.* **1992,** *27*, 61–66.
6. Domínguez, C.; Ezquerra, A.; Prieto, L.; Espada, M.; Pedregal, C. *Tetrahedron: Asymmetry* **1997,** *8*, 511–514.
7. Zaidlewicz, M.; Cytarska, J.; Dzielendziak, A.; Ziegler-Borowska, M. *ARKIVOC* **2004,** *iii,* 11–21.
8. Li, J. J. *Bucherer–Bergs Reaction.* In *Name Reactions in Heterocyclic Chemistry*, Li, J. J., Ed.; Wiley: Hoboken, NJ, **2005,** pp 266–274. (Review).
9. Sakagami, K.; Yasuhara, A.; Chaki, S.; Yoshikawa, R.; Kawakita, Y.; Saito, A.; Taguchi, T.; Nakazato, A. *Bioorg. Med. Chem.* **2008,** *16*, 4359–4366.
10. Wuts, P. G. M.; Ashford, S. W.; Conway, B.; Havens, J. L.; Taylor, B.; Hritzko, B.; Xiang, Y.; Zakarias, P. S. *Org. Process Res. Dev.* **2009,** *13*, 331–335.
11. Oba, M.; Shimabukuro, A.; Ono, M.; Doi, M.; Tanaka, M. *Tetrahedron: Asymmetry* **2013,** *24*, 464–467.
12. Šmit, B. M.; Pavlović, R. Z. *Tetrahedron* **2015,** *71*, 1101–1108.
13. Tomohara, K.; Ito, T.; Furusawa, K.; Hasegawa, N.; Tsuge, K.; Kato, A.; Adachi, I. *Tetrahedron Lett.* **2017,** *58*, 3143–3147.
14. Bisello, A.; Cardena, R.; Rossi, S.; Crisma, M.; Formaggio, F.; Santi, S. *Organometal.* **2017,** *36*, 2190–2197.
15. Lamberth, C. *Bioorg. Med. Chem.* **2020,** *28*, 115471.

Büchner Ring Expansion

Reaction of a phenyl ring with a diazoacetic ester to give a cyclohepta-2,4,6-trienecarboxylic acid ester. Intramolecular Büchner reaction is more useful in synthesis. *Cf.* Pfau–Platter azulene synthesis.

rhodium carbenoid

[2 + 2]
cycloaddition

reductive
elimination

electrocyclic
ring opening

Example 1, Intramolecular Büchner reaction[7]

Rh$_2$(OAc)$_4$
CH$_2$Cl$_2$

Example 2, Intramolecular Büchner reaction[8]

cat. Rh$_2$(OCOt-Bu)$_4$, DMAP
Ac$_2$O, CH$_2$Cl$_2$, 73%

Example 3, Intramolecular Büchner reaction within the Grubbs' catalyst[9]

1 atm CO
CH$_2$Cl$_2$, 90%

Example 4, Intermolecular Büchner reaction[10]

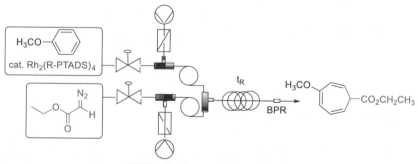

Example 5, Intramolecular Büchner ring expansion[12]

Example 6, regioselective and enantioselective intermolecular Büchner ring expansion in flow, PTADS = Tetrakis[(R)-(+)-N-(p-dodecylphenylsulfonyl)-prolinato][13]

BPR = back pressure regulator

Example 7, Intermolecular Büchner ring expansion in flow[14]

IPB Cu-BOX = insoluble polymer bound Cu-bis(oxazoline) ligand

55–83% *ee*

Example 8, Cyclopropanation by gold- or zinc-catalyzed retro-Büchner reaction at room temperature[15]

[(JohnPhos)Au(MeCN)]SbF$_6$ (5 mol %)
$\xrightarrow{\hspace{3cm}}$
EtOAc (0.1 M), 25 °C, 20 h, 39%

References

1. Büchner, E. *Ber.* **1896**, *29*, 106–109. Eduard Büchner (1860–1917) won Nobel Prize in 1907 for his work on fermentation. His name is imortalized with the Büchner funnels that we still use daily in organic laboratories.

2. von E. Doering, W.; Knox, L. H. *J. Am. Chem. Soc.* **1957**, *79*, 352–356.

3. Marchard, A. P.; Brockway, N. M. *Chem. Rev.* **1974**, *74*, 431–469. (Review).

4. Anciaux, A. J.; Demoncean, A.; Noels, A. F.; Hubert, A. J.; Warin, R.; Teyssié, P. *J. Org. Chem.* **1981**, *46*, 873–876.

5. Duddeck, H.; Ferguson, G.; Kaitner, B.; Kennedy, M.; McKervey, M. A.; Maguire, A. R. *J. Chem. Soc., Perkin Trans. 1* **1990**, 1055–1063.

6. Doyle, M. P.; Hu, W.; Timmons, D. J. *Org. Lett.* **2001**, *3*, 933–935.

7. Manitto, P.; Monti, D.; Speranza, G. *J. Org. Chem.* **1995**, *60*, 484–485.

8. Crombie, A. L; Kane, J. L., Jr.; Shea, K. M.; Danheiser, R. L. *J. Org. Chem.* **2004**, *69*, 8652–8667.

9. Galan, B. R.; Gembicky, M.; Dominiak, P. M.; Keister, J. B.; Diver, S. T. *J. Am. Chem. Soc.* **2005**, *127*, 15702–15703.

10. Panne, P.; Fox, J. M. *J. Am. Chem. Soc.* **2007**, *129*, 22–23.

11. Gomes, A. T. P. C.; Leão, R. A. C.; Alonso, C. M. A.; Neves, M. G. P. M. S.; Faustino, M. A. F.; Tomé, A. C.; Silva, A.M. S.; Pinheiro, S.; de Souza, M. C. B. V.; Ferreira, V. F.; Cavaleiro, J. A. S. *Helv. Chim. Acta* **2008**, *91*, 2270–2283.

12. Foley, D. A.; O'Leary, P.; Buckley, N. R.; Lawrence, S. E.; Maguire, A. R. *Tetrahedron* **2013**, *69*, 1778–1794.

13. Fleming, G. S.; Beeler, A. B. *Org. Lett.* **2017**, *19*, 5268–5271.

14. Crowley, D. C.; Lynch, D.; Maguire, A. R. *J. Org. Chem.* **2018**, *83*, 3794–3805.

15. Mato, M.; Herlé, B.; Echavarren, A. M. *Org. Lett.* **2018**, *20*, 4341–4345.

16. Hoshi, T.; Ota, E.; Inokuma, Y.; Yamaguchi, J. *Org. Lett.* **2019**, *21*, 10081–10084.

Buchwald–Hartwig Amination

The Buchwald–Hartwig amination is an exceedingly general method for generating an aromatic amine from an aryl halide or an aryl sulfonates. The key feature of this methodology is the use of catalytic palladium modulated by various electron-rich ligands. Strong bases, such as sodium *tert*-butoxide, are essential for catalyst turnover.

Mechanism:

The catalytic cycle is shown on the next page.

Example 1[3]

R[1] = EWG or EDG
amine = 2° cyclic or acyclic
amine = 1° aliphatic: low yield, unless R[1] *ortho*

© Springer Nature Switzerland AG 2021
J. J. Li, *Name Reactions*, https://doi.org/10.1007/978-3-030-50865-4_20

Catalytic cycle:

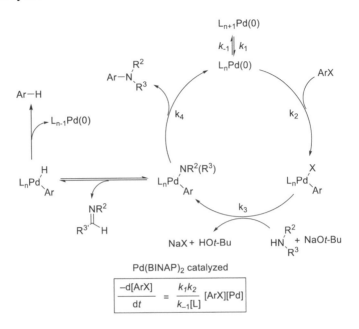

$$\frac{-d[ArX]}{dt} = \frac{k_1 k_2}{k_{-1}[L]} [ArX][Pd]$$

Example 2[4]

5 mol% (dppf)PdCl$_2$
15 mol% dppf

NaOt-Bu, THF
100 °C (sealed), 3 h
80–96% yield
(11 examples)

X = Br or I
R^1 = EWG or EDG
amine = 2° acyclic (one example)
amine = 1° aliphatic or aromatic

Example 3, Room temperature Buchwald–Hartwig amination[9]

1–2 mol% Pd(dba)$_2$
(t-Bu)$_3$P (P/Pd = 0.8/1)

NaOt-Bu, PhMe
22 °C, 1–6 h
81–99% yield

R^1 = EDG or EWG
amine = 2° cyclic or acyclic: aromatic, aliphatic, or azoles
amine = 1° anilines: no aliphatic

Example 4[10]

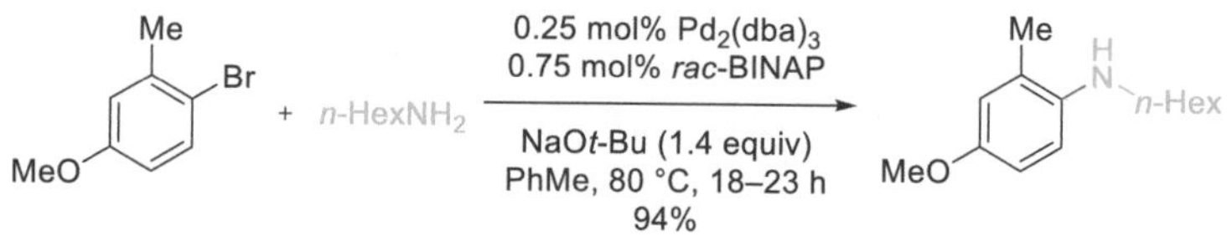
Me
Br
MeO
+ n-HexNH2
0.25 mol% Pd2(dba)3
0.75 mol% rac-BINAP
NaOt-Bu (1.4 equiv)
PhMe, 80 °C, 18–23 h
94%
Me
H
N
n-Hex
MeO

Example 5[11]

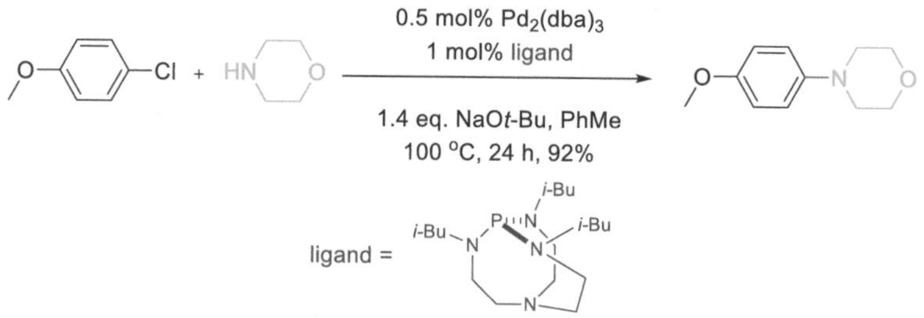
O
Cl
+
HN
O
0.5 mol% Pd2(dba)3
1 mol% ligand
1.4 eq. NaOt-Bu, PhMe
100 °C, 24 h, 92%
O
N
O
ligand =
i-Bu
i-Bu
i-Bu
P
N
N
N
N

Example 6[12]

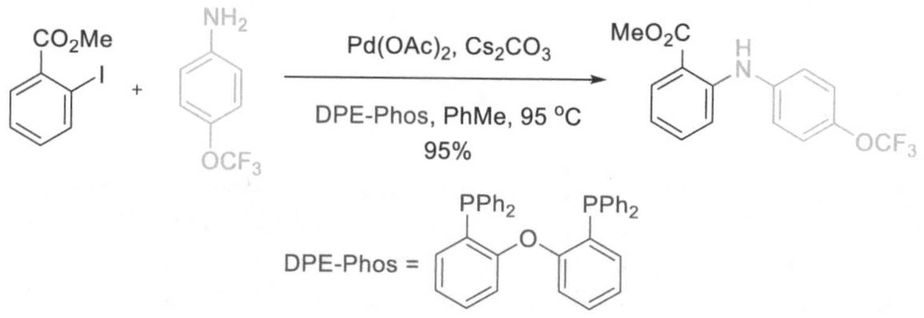
CO2Me
I
+
NH2
OCF3
Pd(OAc)2, Cs2CO3
DPE-Phos, PhMe, 95 °C
95%
MeO2C
H
N
OCF3
DPE-Phos =
PPh2
PPh2
O

Example 7, Amination of volatile amines[14]

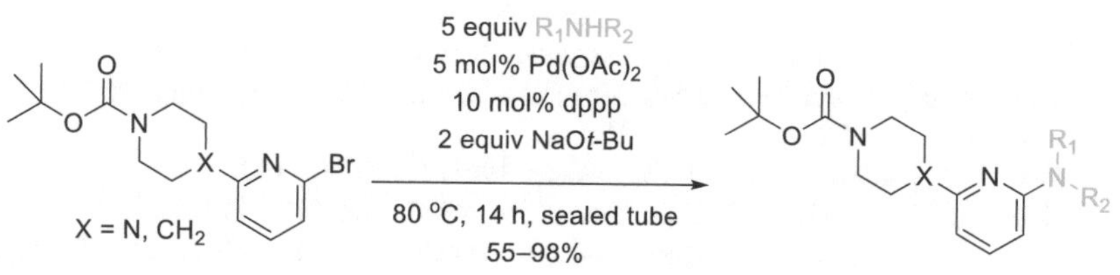
O
O
N
X
N
Br
X = N, CH2
5 equiv R1NHR2
5 mol% Pd(OAc)2
10 mol% dppp
2 equiv NaOt-Bu
80 °C, 14 h, sealed tube
55–98%
O
O
N
X
N
R1
N
R2

Example 8[15]

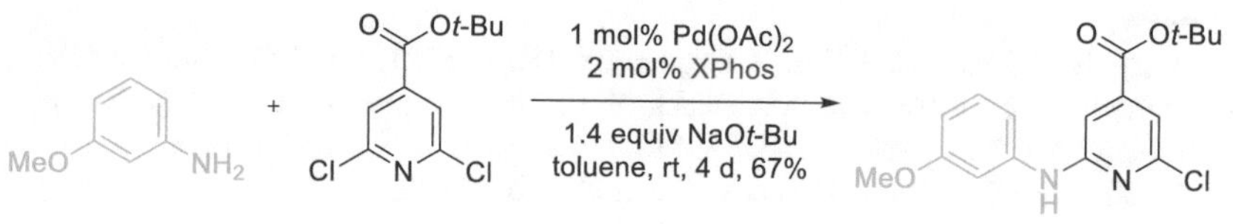
MeO
NH2
+
O
Ot-Bu
Cl
N
Cl
1 mol% Pd(OAc)2
2 mol% XPhos
1.4 equiv NaOt-Bu
toluene, rt, 4 d, 67%
O
Ot-Bu
MeO
N
H
N
Cl

XPhos =

Example 9, In the sysnthesis of HCV NS5A polymerase inhibitor pibrentasvir. Coupling between an arylchloride and an amide[17]

References

1. (a) Paul, F.; Patt, J.; Hartwig, J. F. *J. Am. Chem. Soc.* **1994**, *116*, 5969–5970. John Hartwig earned his Ph.D. at the University of California-Berkeley in 1990 under the guidance of Robert Bergman and Richard Anderson. He moved from Yale University to the University of Illinois at Urbana-Champaign in 2006 and moved from UI-UC to UC Berkeley in 2011. Hartwig and Buchwald independently discovered this chemistry. (b) Mann, G.; Hartwig, J. F. *J. Org. Chem.* **1997**, *62*, 5413–5418. (c) Mann, G.; Hartwig, J. F. *Tetrahedron Lett.* **1997**, *38*, 8005–8008.
2. (a) Guram, A. S.; Buchwald, S. L. *J. Am. Chem. Soc.* **1994**, *116*, 7901–7902. Stephen Buchwald received his Ph.D. in 1982 under Jeremy Knowles at Harvard University. He is currently a professor at MIT. (b) Palucki, M.; Wolfe, J. P.; Buchwald, S. L. *J. Am. Chem. Soc.* **1996**, *118*, 10333–10334.
3. Wolfe, J. P.; Buchwald, S. L. *J. Org. Chem.* **1996**, *61*, 1133–1135.
4. Driver, M. S.; Hartwig, J. F. *J. Am. Chem. Soc.* **1996**, *118*, 7217–7218.
5. Wolfe, J. P.; Wagaw, S.; Marcoux, J.-F.; Buchwald, S. L. *Acc. Chem. Res.* **1998**, *31*, 805–818. (Review).
6. Hartwig, J. F. *Acc. Chem. Res.* **1998**, *31*, 852–860. (Review).
7. Frost, C. G.; Mendonça, P. *J. Chem. Soc., Perkin Trans. 1* **1998**, 2615–2624. (Review).
8. Yang, B. H.; Buchwald, S. L. *J. Organomet. Chem.* **1999**, *576*, 125–146. (Review).
9. Hartwig, J. F.; Kawatsura, M.; Hauck, S. I.; Shaughnessy, K. H.; Alcazar-Roman, L. M. *J. Org. Chem.* **1999**, *64*, 5575–5580.
10. Wolfe, J. P.; Buchwald, S. L. *Org. Syn.* **2002**, *78*, 23–30.

11. Urgaonkar, S.; Verkade, J. G. *J. Org. Chem.* **2004,** *69,* 9135–9142.

12. Csuk, R.; Barthel, A.; Raschke, C. *Tetrahedron* **2004,** *60,* 5737–5750.

13. Janey, J. M. *Buchwald–Hartwig amination,* In *Name Reactions for Functional Group Transformations*; Li, J. J., Corey, E. J. Eds.; Wiley: Hoboken, NJ, **2007,** pp 564–609. (Review).

14. Li, J. J.; Wang, Z.; Mitchell, L. H. *J. Org. Chem.* **2007,** *72,* 3606–3607.

15. Lorimer, A. V.; O'Connor, P. D.; Brimble, M. A. *Synthesis* **2008,** 2764–2770.

16. Witt, A.; Teodorovic, P.; Linderberg, M.; Johansson, P.; Minidis, A. *Org. Process Res. Dev.* **2013,** *17,* 672–678.

17. (a) Rodgers, J. D.; Shepard, S.; Li, Y.-L.; Zhou, J.; Liu, P.; Meloni, D.; Xia, M. WO 2009114512 (2009); (b) Kobierski, M. E.; Kopach, M. E.; Martinelli, J. R.; Varie, D. L.; Wilson, T. M.; WO 2016205487 (2016).

18. Weber, P.; Biafora, A.; Doppiu, A.; Bongard, H.-J.; Kelm, H.; Goossen, L. J. *Org. Process Res. Dev.* **2019,** *23,* 1462–1470.

19. Kashani, S. K.; Jessiman, J. E.; Newman, S. G. *Org. Process Res. Dev.* **2020,** in press.

Burgess reagent

The Burgess reagent [methyl N-(triethylammoniumsulfonyl)carbamate], a neutral, white crystalline solid, is efficient at generating olefins from secondary and tertiary alcohols where the first-order thermolytic Ei (during the elimination—the two groups leave at about the same time and bond to each other concurrently) mechanism prevails.

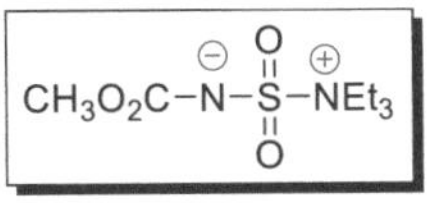

Preparation[2]

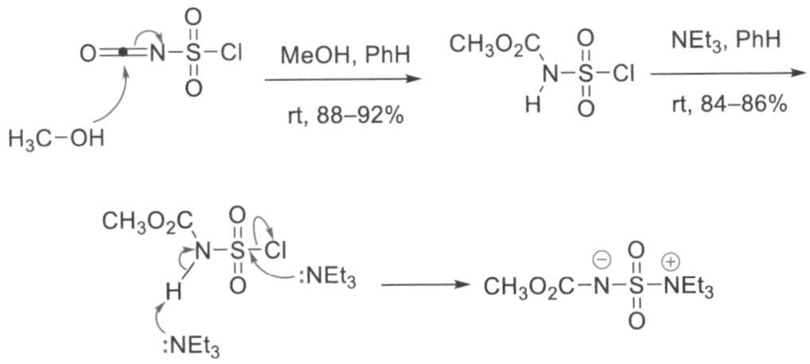

Mechanism of dehydration[5]

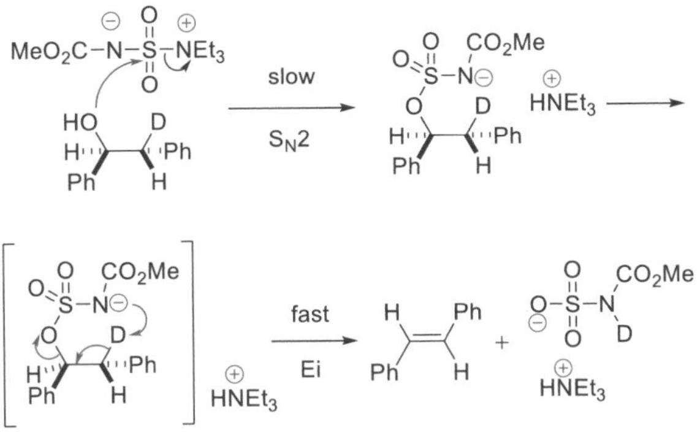

© Springer Nature Switzerland AG 2021
J. J. Li, *Name Reactions*, https://doi.org/10.1007/978-3-030-50865-4_21

Example 1, On primary alcohols, the hydroxyl group does not eliminate but rather undergoes substitution[3]

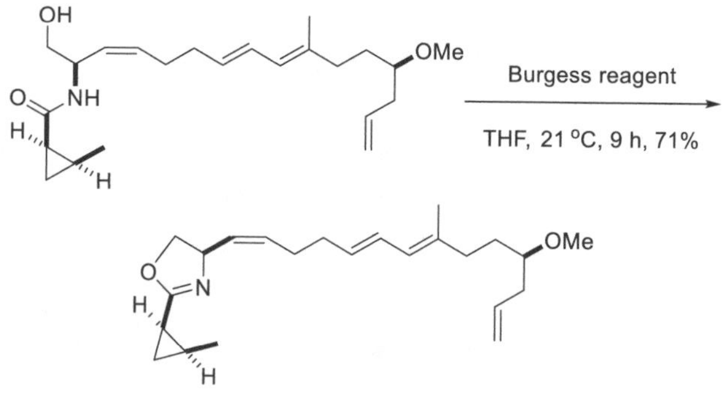

Example 2, Dehydration[6]

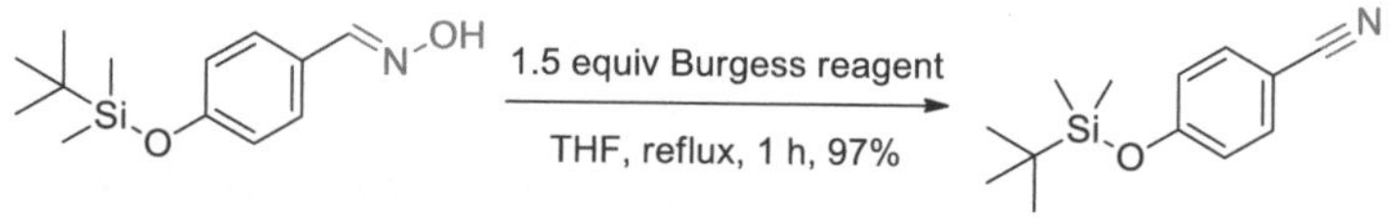

Example 3, Dehydration[7]

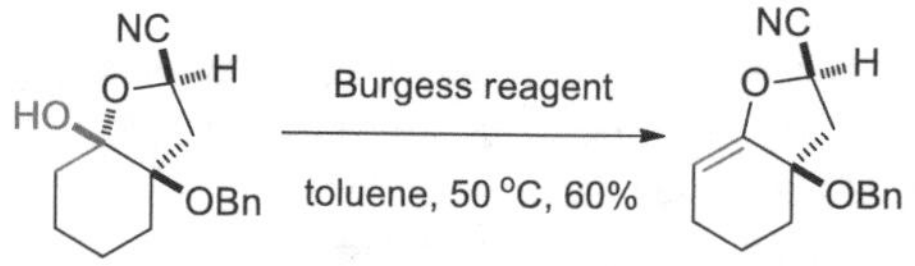

Example 4[8]

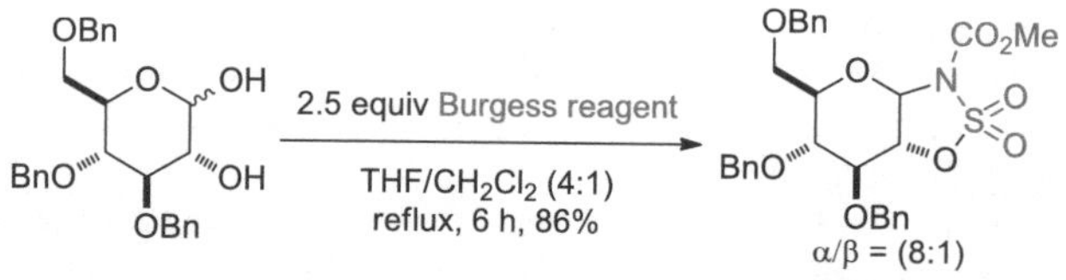

Example 5, Cyclodehydration followed by a novel carbamoylsulfonylation[10]

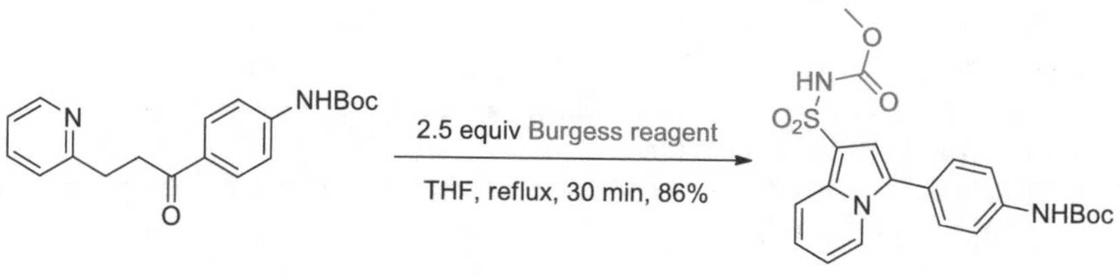

Example 6, Burgess reagent facilitates alcohol oxidation in DMSO[12]

Example 7, Burgess reagent-mediated ring expansion[13]

References

1   (a) Atkins, G. M., Jr.; Burgess, E. M. *J. Am. Chem. Soc.* **1968,** *90,* 4744–4745. (b) Burgess, E. M.; Penton, H. R., Jr.; Taylor, E. A., Jr. *J. Am. Chem. Soc.* **1970,** *92,* 5224–5226. (c) Atkins, G. M., Jr.; Burgess, E. M. *J. Am. Chem. Soc.* **1972,** *94,* 6135–6141. (d) Burgess, E. M.; Penton, H. R., Jr.; Taylor, E. A. *J. Org. Chem.* **1973,** *38,* 26–31.

2   (a) Burgess, E. M.; Penton, H. R., Jr.; Taylor, E. A.; Williams, W. M. *Org. Synth. Coll. Edn.* **1987,** *6,* 788–791. (b) Duncan, J. A.; Hendricks, R. T.; Kwong, K. S. *J. Am. Chem. Soc.* **1990,** *112,* 8433–8442.

3   Wipf, P.; Xu, W. *J. Org. Chem.* **1996,** *61,* 6556–6562.

4   Lamberth, C. *J. Prakt. Chem.* **2000,** *342,* 518–522. (Review).

5   Khapli, S.; Dey, S.; Mal, D. *J. Indian Inst. Sci.* **2001,** *81,* 461–476. (Review).

6   Miller, C. P.; Kaufman, D. H. *Synlett* **2000,** *8,* 1169–1171.

7   Keller, L.; Dumas, F.; D'Angelo, J. *Eur. J. Org. Chem.* **2003,** 2488–2497.

8   Nicolaou, K. C.; Snyder, S. A.; Longbottom, D. A.; Nalbandian, A. Z.; Huang, X. *Chem. Eur. J.* **2004,** *10,* 5581–5606.

9   Holsworth, D. D. *The Burgess Dehydrating Reagent.* In *Name Reactions for Functional Group Transformations*; Li, J. J., Ed.; Wiley: Hoboken, NJ, **2007,** pp 189–206. (Review).

10   Li, J. J.; Li, J. J.; Li, J.; et al. *Org. Lett.* **2008,** *10,* 2897–2900.

11   Werner, L.; Wernerova, M.; Hudlicky, T. et al. *Adv. Synth. Catal.* **2012,** *354,* 2706–2712.

12   Sultane, P. R.; Bielawski, C. W. *J. Org. Chem.* **2017,** *82,* 1046–1052.

13   Badarau, E.; Robert, F.; Massip, S.; Jakob, F.; Lucas, S.; Frormann, S.; Ghosez, L. *Tetrahedron* **2018,** *74,* 5119–5128.

14   Widlicka, D. W.; Gontcharov, A.; Mehta, R.; Pedro, D. J.; North, R. *Org. Process Res. Dev.* **2019,** *23,* 1970–1978.

Cadiot–Chodkiewicz Coupling

Bis-acetylene synthesis from alkynyl halides and alkynyl copper reagents.
Cf. Castro–Stephens reaction.

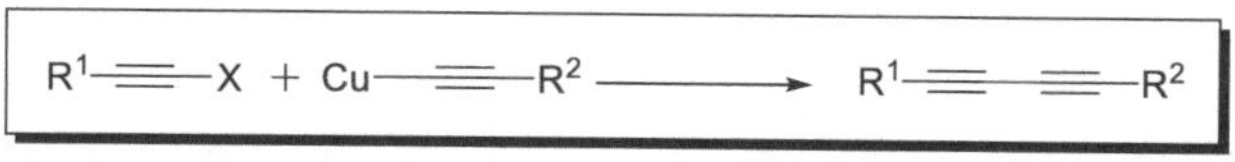

$$R^1 \!\!=\!\!=\!\!X + Cu\!\!=\!\!=\!\!R^2 \longrightarrow R^1\!\!=\!\!=\!\!=\!\!=\!\!R^2$$

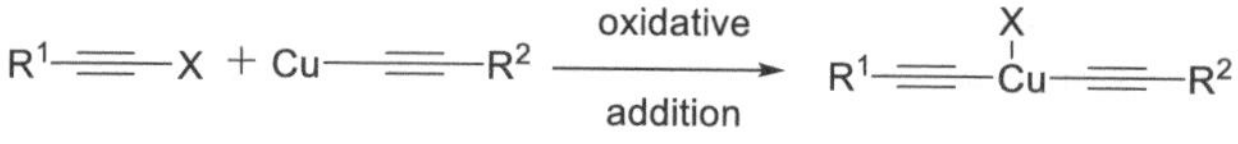

Cu(III) intermediate

Example 1[3]

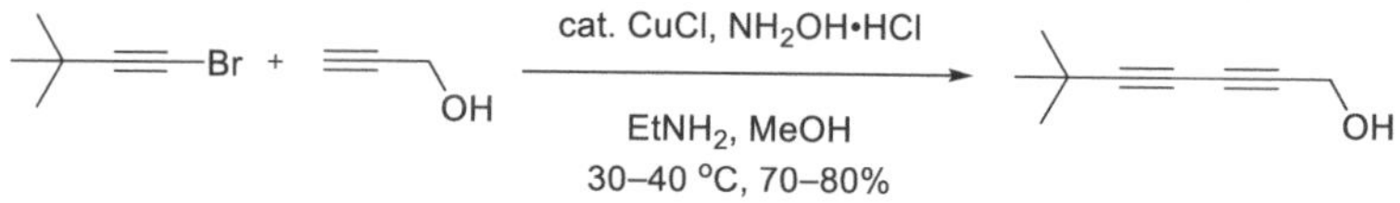

Example 2[7]

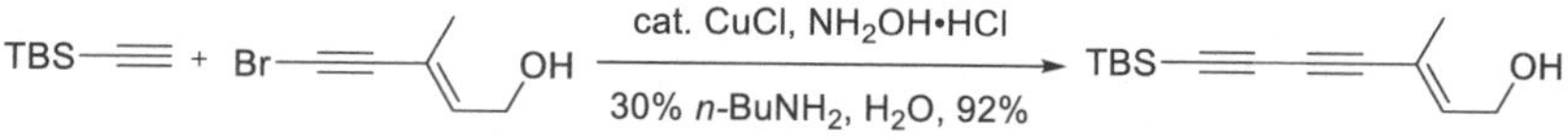

Example 3[9]

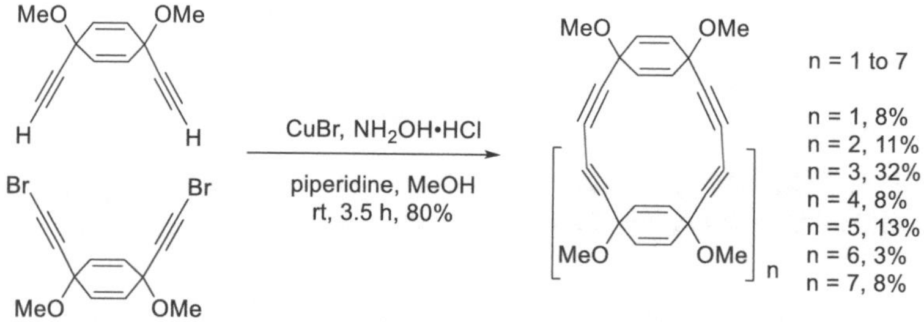

n = 1 to 7

n = 1, 8%
n = 2, 11%
n = 3, 32%
n = 4, 8%
n = 5, 13%
n = 6, 3%
n = 7, 8%

J. J. Li, *Name Reactions*, https://doi.org/10.1007/978-3-030-50865-4_22

Example 4, Cadiot–Chodkiewicz active template synthesis of rotaxanes and switchable molecular shuttles with weak intercomponent interactions[10]

Example 5, Gold-catalyzed Cadiot–Chodkiewicz cross-coupling of terminal alkynes with alkynyl hypervalent iodine reagents[13]

Ta-Au =   Ph₃P-Au-N   Phen = 1,10-phenanthroline =

Example 6, Cadiot–Chodkiewicz cross-coupling is superior to the Sonogashi coupling in this case:[14]

References

1. Chodkiewicz, W.; Cadiot, P. *C. R. Hebd. Seances Acad. Sci.* **1955**, *241*, 1055–1057. Both Paul Cadiot (1923–) and Wladyslav Chodkiewicz (1921–) were French chemists.
2. Cadiot, P.; Chodkiewicz, W. In *Chemistry of Acetylenes;* Viehe, H. G., ed.; Dekker: New York, **1969**, 597–647. (Review).
3. Gotteland, J.-P.; Brunel, I.; Gendre, F.; Désiré, J.; Delhon, A.; Junquéro, A.; Oms, P.; Halazy, S. *J. Med. Chem.* **1995**, *38*, 3207–3216.
4. Bartik, B.; Dembinski, R.; Bartik, T.; Arif, A. M.; Gladysz, J. A. *New J. Chem.* **1997**, *21*, 739–750.
5. Montierth, J. M.; DeMario, D. R.; Kurth, M. J.; Schore, N. E. *Tetrahedron* **1998**, *54*, 11741–11748.
6. Negishi, E.-i.; Hata, M.; Xu, C. *Org. Lett.* **2000**, *2*, 3687–3689.
7. Marino, J. P.; Nguyen, H. N. *J. Org. Chem.* **2002**, *67*, 6841–6844.
8. Utesch, N. F.; Diederich, F.; Boudon, C.; Gisselbrecht, J.-P.; Gross, M. *Helv. Chim. Acta* **2004**, *87*, 698–718.
9. Bandyopadhyay, A.; Varghese, B.; Sankararaman, S. *J. Org. Chem.* **2006**, *71*, 4544–4548–4548.
10. Berna, J.; Goldup, S. M.; Lee, A.-L.; Leigh, D. A.; Symes, M. D.; Teobaldi, G.; Zerbetto, F. *Angew. Chem. Int. Ed.* **2008**, *47*, 4392–4396.
11. Glen, P. E.; O'Neill, J. A. T.; Lee, A.-L. *Tetrahedron* **2013**, *69*, 57–68.
12. Sindhu, K. S.; Thankachan, A. P.; Sajitha, P. S.; Anilkumar, G. *Org. Biomol. Chem.* **2015**, *13*, 6891–6905. (Review).
13. Li, X.; Xie, X.; Sun, N.; Liu, Y. *Angew. Chem. Int. Ed.* **2017**, *56*, 6994–6998.
14. Kanikarapu, S.; Marumudi, K.; Kunwar, A. C.; Yadav, J. S.; Mohapatra, D. K. *Org. Lett.* **2017**, *19*, 4167–4170.
15. Geng, J.; Ren, Q.; Chang, C.; Xie, X.; Liu, J.; Du, Y. *RSC Adv.* **2019**, *9*, 10253–10263.
16. Radhika, S.; Harry, N. A.; Neetha, M.; Anilkumar, G. *Org. Biomol. Chem.* **2019**, *17*, 9081–9094. (Review).
17. Kaldhi, D.; Vodnala, N.; Gujjarappa, R.; Kabi, A. K.; Nayak, S.; Malakar, C. C. *Tetrahedron Lett.* **2020**, *61*, 151775.

Cannizzaro Reaction

Base-induced disproportionation between two aldehydes to produce an alcohol and a carboxylic acid. If the starting material is an α-ketoaldehyde, an intramolecular Cannizzaro disproportionation reaction can also occur (see Examples 1, 5 and 6). Aldehydes are aromatic aldehydes, formaldehyde or other aliphatic aldehydes without α-hydrogen.

Pathway A:

Final deprotonation of the carboxylic acid drives the reaction forward.

Pathway B:

Example 1, Intramolecular Cannizzaro disproportionation reaction[3]

© Springer Nature Switzerland AG 2021
J. J. Li, *Name Reactions*, https://doi.org/10.1007/978-3-030-50865-4_23

Example 2[4]

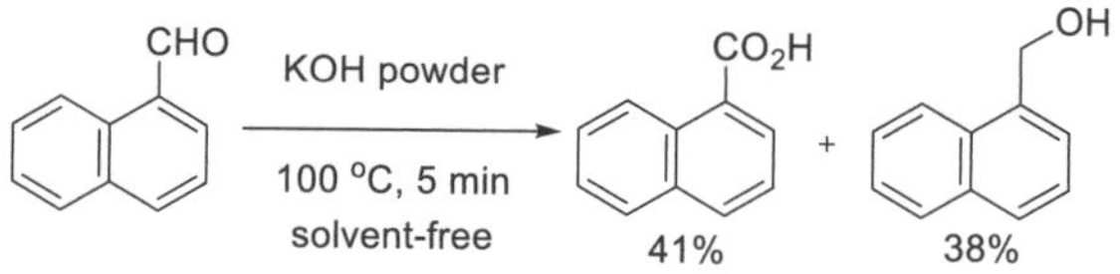

Example 3[6]

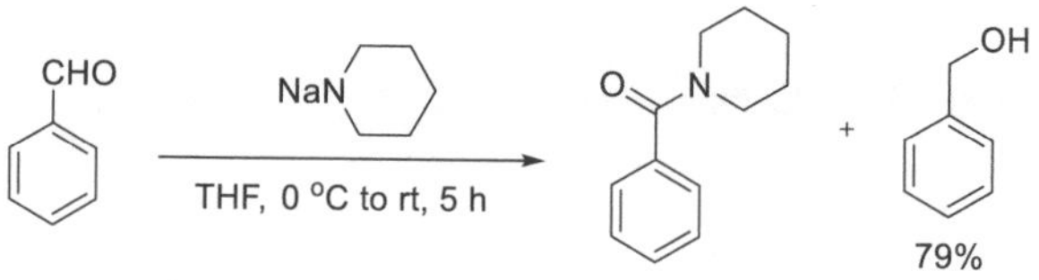

Example 4[8]

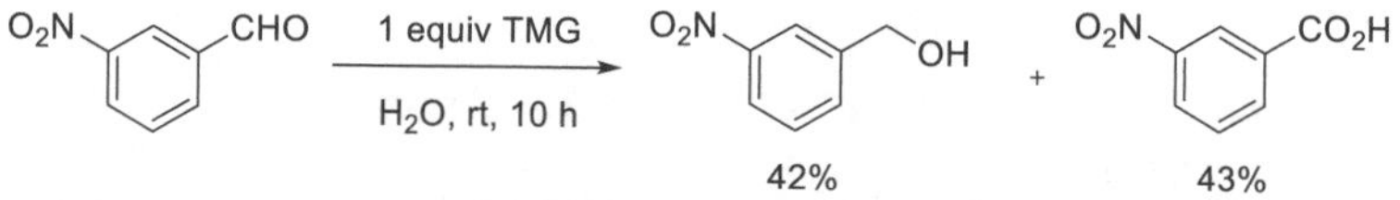

TMG = 1,1,3,3-tetramethylguanidine, an organic base

Example 5, Desymmetrization by intramolecular Cannizzaro reaction[9]

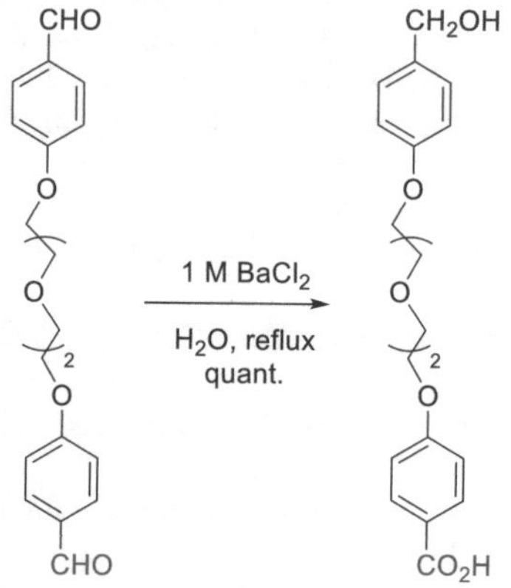

Example 6, Intramolecular aza-Cannizzaro reaction[11]

glycine   oxalic acid

Example 7, Ball mill Cannizzaro reaction[12]

References

1. Cannizzaro, S. *Ann.* **1853**, *88*, 129–130. Stanislao Cannizzaro (1826–1910) was born in Palermo, Sicily, Italy. In 1847, he had to escape to Paris for participating in the Sicilian Rebellion. Upon his return to Italy, he discovered the disproportionation reaction at the Collegio Nazionale di Alessandria (Piedmont) using bitter almond oil (benzaldehyde) and potash (potassium hydroxide) as the base. Political interests brought Cannizzaro to the Italian Senate and he later became its vice president.
2. Geissman, T. A. *Org. React.* **1944**, *1*, 94–113. (Review).
3. Russell, A. E.; Miller, S. P.; Morken, J. P. *J. Org. Chem.* **2000**, *65*, 8381–8383.
4. Yoshizawa, K.; Toyota, S.; Toda, F. *Tetrahedron Lett.* **2001**, *42*, 7983–7985.
5. Reddy, B. V. S.; Srinvas, R.; Yadav, J. S.; Ramalingam, T. *Synth. Commun.* **2002**, *32*, 219–223.
6. Ishihara, K.; Yano, T. *Org. Lett.* **2004**, *6*, 1983–1986.
7. Curini, M.; Epifano, F.; Genovese, S.; Marcotullio, M. C.; Rosati, O. *Org. Lett.* **2005**, *7*, 1331–1333.
8. Basavaiah, D.; Sharada, D. S.; Veerendhar, A. *Tetrahedron Lett.* **2006**, *47*, 5771–5774.
9. Ruiz-Sanchez, A. J.; Vida, Y.; Suau, R.; Perez-Inestrosa, E. *Tetrahedron* **2008**, *64*, 11661–11665.
10. Shen, M.-G.; Shang, S.-B.; Song, Z.-Q.; Wang, D.; Rao, X.-P.; Gao, H.; Liu, H. *J. Chem. Res.* **2013**, *37*, 51–52.
11. Sud, A.; Chaudhari, P. S.; Agarwal, I.; Mohammad, A. B.; Dahanukar, V. H.; Bandichhor, R. *Tetrahedron Lett.* **2017**, *58*, 1891–1894.
12. Chacon-Huete, F.; Messina, C.; Chen, F.; Cuccia, L.; Ottenwaelder, X.; Forgione, P. *Green Chem.* **2018**, *20*, 5261–5265.
13. Janczewski, L.; Walczak, M.; Fraczyk, J.; Kaminski, Z. J.; Kolesinska, B. *Synth. Commun.* **2019**, *49*, 3290–3300.

Catellani Reaction

Selective *ortho*-alkylation and -arylation of aryl iodides can be achieved by the cooperative catalytic action of palladium and norbornene.[1] The first reported case was the *ortho*-dialkylation of aryl iodides, followed by Heck reaction.[2] Here an aryl iodide with free *o*-positions reacts with an aliphatic iodide and a terminal olefin in the presence of palladium/norbornene (NBE) as catalyst and a base, to give a 2,6-substituted vinylarene. Analogously, an aryl iodide with one substituted *o*-position leads to a vinylarene containing two different *ortho* groups.[3]

Example 1, A three-component reaction allowing the construction of three adjacent C–C bonds through C–I and C–H activation.[2]

Mechanism for the reaction of an *o*-substitued aryl iodide: Pd(0), Pd(II) and Pd(IV) intermediates and catalytic role of palladium and norbornene.[1–3]

    The mechanism involves initial oxidative addition of an *o*-substituted aryl iodide to Pd(0) followed by a stereoselective norbornene (NBE) insertion leading to the *cis,exo* complex **2**. β-Hydrogen elimination is prevented by geometric constraints, and a five-membered palladacycle (**3**) readily forms through intramolecu-

© Springer Nature Switzerland AG 2021
J. J. Li, *Name Reactions*, https://doi.org/10.1007/978-3-030-50865-4_24

lar C–H activation. Oxidative addition of an alkyl iodide to **3** affords a Pd(IV) intermediate (**4**) which undergoes reductive elimination by selective migration of the alkyl moiety onto the aromatic ring to form **5**. Norbornene deinsertion occurs spontaneously at this point, likely due to steric hindrance, giving 2,6-disubstituted phenylpalladium(II) species (**6**) which finally react with the terminal olefin to liberate the organic product and Pd(0). Alternatively the sequence can be terminated by other well-known reactions of the aryl-Pd bond such as the Suzuki or Sonogashira couplings, hydrogenolysis, amination, or cyanation. The described methodology can also be extended to ring-forming reactions.[1e] Thus, the reaction is very versatile and offers countless possibilities for building up many types of functionalized aromatic compounds.

Example 2, The synthesis of fused aromatic compounds through final intramolecular Heck reaction was first reported by the Lautens group.[4,1e]

Example 3, The high tolerance to functional groups enabled a key step to the synthesis of a precursor of (+)-linoxepin by Lautens.[5]

ortho-Arylation of an aryl iodide leading to the construction of a biaryl moiety is also possible, provided that the starting aryl iodide bears an *ortho* substituent. The *o*-substituent in palladacycles of type **3** is essential for selectively directing the attack of an aryl halide onto the aromatic site (*ortho* effect).[1,6]

Example 4, Aryl-aryl coupling combined with Heck reaction.[7]

Example 5. The non symmetrical coupling of an aryl iodide bearing an *o*-electron-donating group, an aryl bromide containing an electron-withdrawing substituent, and a terminal olefin illustrates the importance of correctly tuning the electronic properties of the two aryl halides for selectivity control.[8]

Example 6, Internal chelation to Pd(IV)[9] can cancel the *ortho* effect.[10]

Example 7, Synthjesis of benzo[1,6]naphthyridinones[11]

Example 8, Iterative C–H bis-silylation[12]

Example 9, Borono-Catellani arylation for unsymmetrical biaryl synthesis[13,14]

References

1. (a) Tsuji, J. Palladium Reagents and Catalysts – New Perspective for the 21st Century, 2004, John Wiley & Sons, pp. 409–416. (b) Catellani, M. *Synlett* **2003**, 298–313. (c) Catellani, M. *Top. Organomet. Chem.* **2005**, *14*, 21–53. (d) Catellani, M.; Motti, E.; Della Ca', N. *Acc. Chem. Res.* **2008**, *41*, 1512–1522. (e) Martins, A.; Mariampillai, B.; Lautens, M. *Top Curr Chem* **2010**, *292*, 1–33. (f) Chiusoli, G. P.; Catellani, M.; Costa, M.; Motti, E.; Della Ca', N.; Maestri, G. *Coord. Chem. Rev.* **2010**, *254*, 456–469. Marta Catellani and coworkers at the University of Parma discovered an elegant entry into the synthesis of *o,o*-disubstituuted vinylarenes starting from aryl iodides. The reaction exploits a multicomponent protocol where, together with the reactants and catalyst, norbornene or another strained olefin is used. The latter is essential as it enters the complex catalytic cycle by activating three adjacent positions of the arene, being recycled at the end of the process.
2. (a) Catellani, M.; Frignani, F.; Rangoni, A. *Angew. Chem. Int. Ed. Engl.* **1997**, *36*, 119–122. (b) Catellani, M.; Fagnola, M. C. *Angew. Chem. Int. Ed. Engl.* **1994**, *33*, 2421–2422.
3. Catellani, M;. Cugini, F. *Tetrahedron*, **1999**, *55*, 6595–6602.
4. (a) Lautens, M.; Piguel, S.; Dahlmann, M. *Angew. Chem. Int. Ed. Engl.* **2000**, *39*, 1045–1046. (b) Lautens, M.; Paquin, J.-F.; Piguel, S. *J. Org. Chem.* **2001**, *66*, 8127–8134. (c) Lautens, M.; Paquin, J.-F.; Piguel, S. *J. Org. Chem.* **2002**, *67*, 3972–3974.
5. Weinstabl, H.; Suhartono, M.; Qureshi, Z.; Lautens, M. *Angew. Chem. Int. Ed.* **2013**, *125*, 5413–5416.
6. Maestri, G.; Motti, E.; Della Ca', N.; Malacria, M.; Derat, E.; Catellani, M. *J. Am. Chem. Soc.* **2011**, *133*, 8574–8585.
7. Motti, E.; Ippomei, G.; Deledda, S.; Catellani, M. *Synthesis* **2003**, 2671–2678.
8. Faccini, F.; Motti, E.; Catellani, M. *J. Am. Chem. Soc.* **2004**, *126*, 78–79.
9. Vicente, J.; Arcas, A.; Juliá-Hernández, F.; Bautista, D. *Angew. Chem. Int. Ed.* **2011**, *50*, 6896–6899.
10. Della Ca', N.; Maestri, G.; Malacria, M.; Derat, E.; Catellani, M. *Angew. Chem. Int. Ed.* **2011**, *50*, 12257–12261.
11. Elsayed, M. S. A.; Griggs, B.; Cushman, M. *Org. Lett.* **2018**, *20*, 5228–5232.
12. Lv, W.; Yu, J.; Ge, B.; Wen, S.; Cheng, G. *J. Org. Chem.* **2018**, *83*, 12683–12693.
13. Chen, S.; Liu, Z.-S.; Yang, T.; Hua, Y.; Zhou, Z.; Cheng, H.-G.; Zhou, Q. *Angew. Chem. Int. Ed.* **2018**, *57*, 7161–7165.
14. Wang, P.; Chen, S.; Zhou, Z.; Cheng, H.-G.; Zhou, Q. *Org. Lett.* **2019**, *21*, 323–3327.
15. Cheng, H.-G.; Chen, S.; Chen, R.; Zhou, Q. *Angew. Chem. Int. Ed.* **2019**, *58*, 5832–5844. (Review).

Chan–Lam C–X Coupling Reaction

Arylation, vinylation and alkylation of a wide range of NH/OH/SH substrates by oxidative cross-coupling with boronic acids in the presence of catalytic cupric acetate, weak base and in air (open-flask chemistry). The reaction works for amides, amines, amidines, anilines, azides, azoles, hydantoins, hydrazines, imides, imines, nitroso, pyrazinones, pyridines, purines, pyrimidines, sulfonamides, sulfinates, sulfoximines, ureas, alcohols, phenols, thiols, *etc.* The boronic acids can be replaced with siloxanes, stannanes or other organometalloids. The mild condition of this reaction is an advantage over Buchwald–Hartwig's Pd-catalyzed cross-coupling using halides, though boronic acids are more expensive than halides. The Chan–Lam C–X bond cross-coupling reaction has emerged as a powerful and popular methodogy similar to Suzuki–Miyaura's C–C bond cross-coupling reaction.

Proposed Mechanism:[4]

J. J. Li, *Name Reactions*, https://doi.org/10.1007/978-3-030-50865-4_25

Example 1[1a,d]

Example 2[5]

Example 3[6]

Example 4[14]

93% (α-ester assistance, acetal, lower yield)

Example 5[15]

Example 6, Chan–Lam coupling between aryl boroxines and enolates as sp^3-carbon nucleophiles[18]

Example 7, Using tertiary trifluoroborates[20]

References

1. (a) Chan, D. M. T.; Monaco, K. L.; Wang, R.-P.; Winters, M. P. *Tetrahedron Lett.* **1998**, *39*, 2933–2936. (b) Lam, P. Y. S.; Clark, C. G.; Saubern, S.; Adams, J.; Winters, M. P.; Chan, D. M. T.; Combs, A. *Tetrahedron Lett.* **1998**, *39*, 2941–2949. Dominic Chan is a chemist at DuPont Crop Protection, Wilmington, DE, USA. He did his PhD research with Prof. Barry Trost at the University of Wisconson, Madison. Patrick Lam is a research director at Bristol–Myers Squibb, Princeton, NJ, USA. He was formerly with DuPont Pharmaceuticals Company. He did his PhD research with Prof. Louis Friedrich in the Univeristy of Rochester and Post-doc research with Prof. Michael Jung and the late Prof. Donald Cram in UCLA. (c) Evans, D. A.; Katz, J. L.; West, T. R. *Tetrahedron Lett.* **1998**, *39*, 2937–2940. Prof. Evans' group found out about the discovery of this reaction on a National Organic Symposium poster and became interested in the *O*-arylation because of his long interest in vancomycin total synthesis. (d) Lam, P. Y. S.; Clark, C. G.; Saubern, S.; Adams, J.; Averill, K. M.; Chan, D. M. T.; Combs, A. *Synlett* **2000**, 674–676. (e) Lam, P. Y. S.; Bonne, D.; Vincent, G.; Clark, C. G.; Combs, A. P. *Tetrahedron Lett.* **2003**, *44*, 1691–1694.
2. Reviews: (a) Qiao, J. X.; Lam, P. Y. S. *Syn.* **2011**, 829–856; (b) Chan, D. M. T.; Lam, P. Y. S., Book chapter in *Boronic Acids* Hall, ed. **2005**, Wiley–VCH, 205–240. (c) Ley, S. V.; Thomas, A. W. *Angew. Chem., Int. Ed. Engl.* **2003**, *42*, 5400–5449.
3. Catalytic copper: (a) Lam, P. Y. S.; Vincent, G.; Clark, C. G.; Deudon, S.; Jadhav, P. K. *Tetrahedron Lett.* **2001**, *42*, 3415–3418. (b) Antilla, J. C.; Buchwald, S. L. *Org. Lett.* **2001**, *3*, 2077–2079. (c) Quach, T. D.; Batey, R. A. *Org. Lett.* **2003**, *5*, 4397–4400. (d) Collman, J. P.; Zhong, M. *Org. Lett.* **2000**, *2*, 1233–1236. (e) Lan, J.-B.; Zhang, G.-L.; Yu, X.-Q.; You, J.-S.; Chen, L.; Yan, M.; Xie, R.-G. *Synlett* **2004**, 1095–1097.

4. Mechanism (Part of the mechanistic work from Shannon's lab was funded and in collaboration with BMS: (a) Huffman, L. M.; Stahl, S. S. *J. Am. Chem. Soc.* **2008,** *130*, 9196–9197. (b) King, A. E.; Brunold, T. C.; Stahl, S. S. *J. Am. Chem. Soc.* **2009,** *131*, 5044. (c) King, A. E.; Huffman, L. M.; Casitas, A.; Costas, M.; Ribas, X.; Stahl, S. S. *J. Am. Chem. Soc.* **2010,** *132*, 12068–12073. (d) Casita, A.; King, A. E.; Prella, T.; Costas, M.; Stahl, S. S.; Ribas, X. *J. Chem. Sci.* **2010,** *1*, 326–330.

5. Vinyl boronic acids: Lam, P. Y. S.; Vincent, G.; Bonne, D.; Clark, C. G. *Tetrahe dron Lett.* **2003,** *44*, 4927–4931.

6. Intramolecular: Decicco, C. P.; Song, Y.; Evans, D.A. *Org. Lett.* **2001,** *3*, 1029–1032.

7. Solid phase: (a) Combs, A. P.; Saubern, S.; Rafalski, M.; Lam, P. Y. S. *Tetrahe dron Lett.* **1999,** *40*, 1623–1626. (b) Combs, A. P.; Tadesse, S.; Rafalski, M.; Haque, T. S.; Lam, P. Y. S. *J. Comb. Chem.* **2002,** *4*, 179–182.

8. Boronates/borates: (a) Chan, D. M. T.; Monaco, K. L.; Li, R.; Bonne, D.; Clark, C. G.; Lam, P. Y. S. *Tetrahedron Lett.* **2003,** *44*, 3863–3865. (b) Yu, X. Q.; Yamamoto, Y.; Miyuara, N. *Chem. Asian J.* **2008,** *3*, 1517–1522.

9. Siloxanes: (a) Lam, P. Y. S.; Deudon, S.; Averill, K. M.; Li, R.; He, M. Y.; DeShong, P.; Clark, C. G. *J. Am. Chem. Soc.* **2000,** *122*, 7600–7601. (b) Lam, P. Y. S.; Deudon, S.; Hauptman, E.; Clark, C. G. *Tetrahedron Lett.* **2001,** *42*, 2427–2429.

10. Stannanes: Lam, P. Y. S.; Vincent, G.; Bonne, D.; Clark, C. G. *Tetrahedron Lett.* **2002,** *43*, 3091–3094.

11. Thiols: (a) Herradura, P. S.; Pendora, K. A.; Guy, R. K. *Org. Lett.* **2000,** *2*, 2019–2022. (b) Savarin, C.; Srogl, J.; Liebeskind, L. S. . *Org. Lett.* **2002,** *4*, 4309–4312. (c) Xu, H.-J.; Zhao, Y.-Q.; Feng, T.; Feng, Y.-S. *J. Org. Chem.* **2012,** *77*, 2878–2884.

12. Sulfinates: (a) Beaulieu, C.; Guay. D.; Wang, C.; Evans, D. A. *Tetrahedron Lett.* **2004,** *45*, 3233–3236. (b) Huang, H.; Batey, R. A. *Tetrahedron.* **2007,** *63*, 7667–7672. (c) Kar, A.; Sayyed, L.A.; Lo, W.F.; Kaiser, H.M.; Beller, M.; Tse, M. K. *Org. Lett.* **2007,** *9*, 3405–3408.

13. Sulfoximines: Moessner, C.; Bolm, C. *Org. Lett.* **2005,** *7*, 2667–2669.

14. β-Lactam: Wang, W.; *et al.* *Bio. Med. Chem. Lett.* **2008,** *18*, 1939–1944.

15. Cyclopropyl boronic acid: Tsuritani, T.; Strotman, N. A.; Yamamoto, Y.; Kawa saki, M.; Yasuda, N.; Mase, T. *Org. Lett.* **2008,** *10*, 1653–1655.

16. Alcohols: Quach, T. D.; Batey, R. A. *Org. Lett.* **2003,** *5*, 1381–1384.

17. Fluorides: (a) Ye, Y.; Sanford, M. S. *J. Am. Chem. Soc.* **2013,** *135*, 4648–4651. (b) Fier, P. S.; Luo, J.; Hartwig, J. F. *J. Am. Chem. Soc.* **2013,** *135*, 2552–2559.

18. Moon, P. J.; Halperin, H. M.; Lundgren, R. J. *Angew. Chem., Int. Ed. Engl.* **2016,** *55*, 1894–1898.

19. Vantourout, J. C.; Law, R. P.; Isidro-Llobet, A.; Atkinson, S. J.; Watson, A. J. B. *J. Org. Chem.* **2016,** *81*, 3942–3950.

20. Harris, M. R.; Li, Q.; Lian, Y.; Xiao, J.; Londregan, A. T. *Org. Lett.* **2017,** *19*, 2450–2453.

21. Ando, S.; Hirota, Y.; Matsunaga, H.; Ishizuka, T. *Tetrahedron Lett.* **2019,** *60*, 1277–1280.

22. Clerc, A.; Beneteau, V.; Pale, P.; Chassaing, S. *ChemCatChem* **2020,** *12*, 2060–2065.

Chapman Rearrangement

Thermal aryl rearrangement of *O*-aryliminoethers to amides.

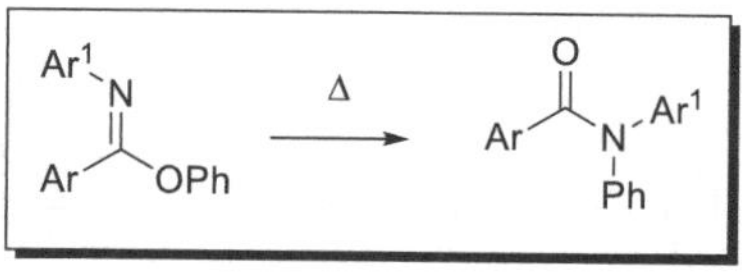

Mechanism:

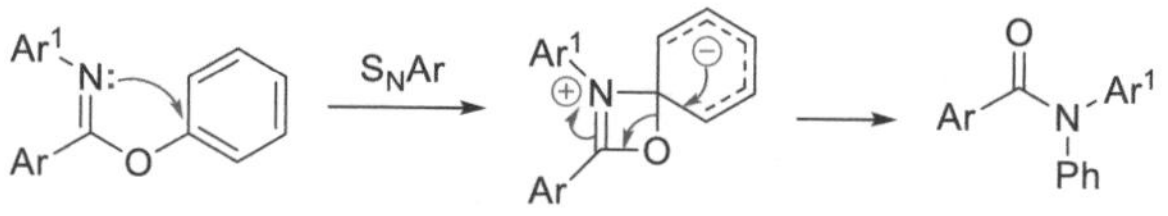

oxazete intermediate

Example 1[2]

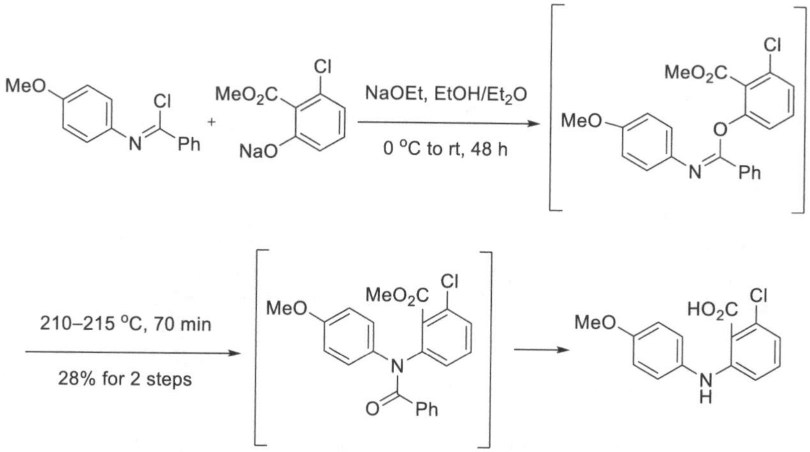

210–215 °C, 70 min

28% for 2 steps

Example 2[4]

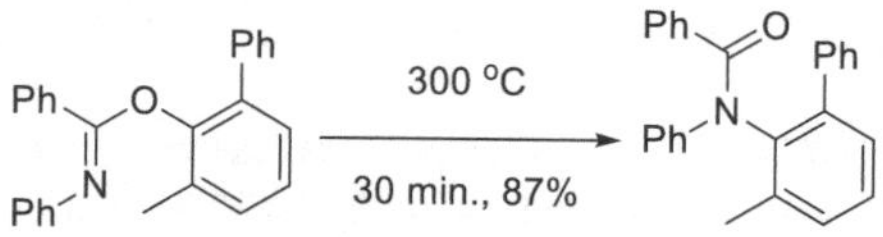

300 °C

30 min., 87%

© Springer Nature Switzerland AG 2021
J. J. Li, *Name Reactions*, https://doi.org/10.1007/978-3-030-50865-4_26

Example 3, Double Chapman rearrangement[9]

Example 4, Chapman-like thermal rearrangement[11]

Example 5, Chapman-like thermal rearrangement[12]

References

1. Chapman, A. W. *J. Chem. Soc.* **1925**, *127*, 1992–1998. Arthur William Chapman was born in 1898 in London, England. He was a Lecturer in Organic Chemistry and later became Registrar of the University of Sheffield from 1944 to 1963.
2. Dauben, W. G.; Hodgson, R. L. *J. Am. Chem. Soc.* **1950**, *72*, 3479–3480.
3. Schulenberg, J. W.; Archer, S. *Org. React.* **1965**, *14*, 1–51. (Review).
4. Relles, H. M. *J. Org. Chem.* **1968**, *33*, 2245–2253.
5. Shawali, A. S.; Hassaneen, H. M. *Tetrahedron* **1972**, *28*, 5903–5909.
6. Kimura, M.; Okabayashi, I.; Isogai, K. *J. Heterocycl. Chem.* **1988**, *25*, 315–320.
7. Farouz, F.; Miller, M. J. *Tetrahedron Lett.* **1991**, *32*, 3305–3308.
8. Dessolin, M.; Eisenstein, O.; Golfier, M.; Prange, T.; Sautet, P. *J. Chem. Soc., Chem. Commun.* **1992**, 132–134.
9. Marsh, A.; Nolen, E. G.; Gardinier, K. M.; Lehn, J. M. *Tetrahedron Lett.* **1994**, *35*, 397–400.
10. Almeida, R.; Gomez-Zavaglia, A.; Kaczor, A.; Cristiano, M. L. S.; Eusebio, M. E. S.; Maria, T. M. R.; Fausto, R. *Tetrahedron* **2008**, *64*, 3296–3305.
11. Noorizadeh, S.; Ozhand, A. *Chin. J. Chem.* **2010**, *28*, 1876–1884.
12. Patel, Sh. S.; Chandna, N.; Kumar, S.; Jain, N. *Org. Biomol. Chem.* **2016**, *14*, 56836–5689.
13. Fang, J.; Ke, M.; Huang, G.n; Tao, Y.; Cheng, D.; Chen, F.-E. *RSC Adv.* **2019**, *16*, 9270–9280.

Chichibabin Pyridine Synthesis

Also known as the Chichibabin reaction. Condensation of aldehydes with ammonia to afford pyridines.

$$3 \ R\text{—}CHO \ + \ NH_3 \ \rightleftharpoons$$

Example 1[3]

© Springer Nature Switzerland AG 2021

J. J. Li, *Name Reactions*, https://doi.org/10.1007/978-3-030-50865-4_27

Example 2[7]

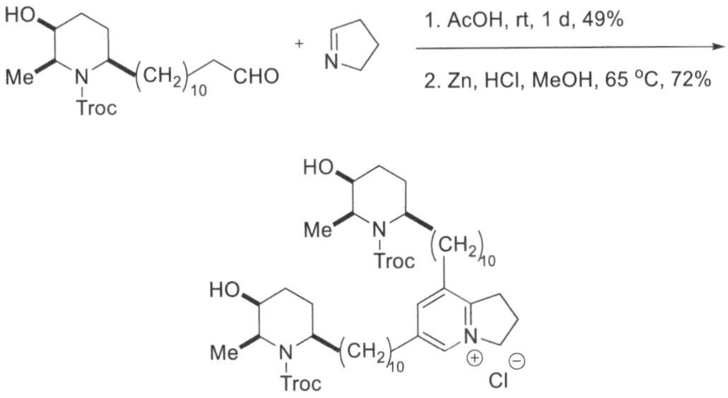

Example 3[8]

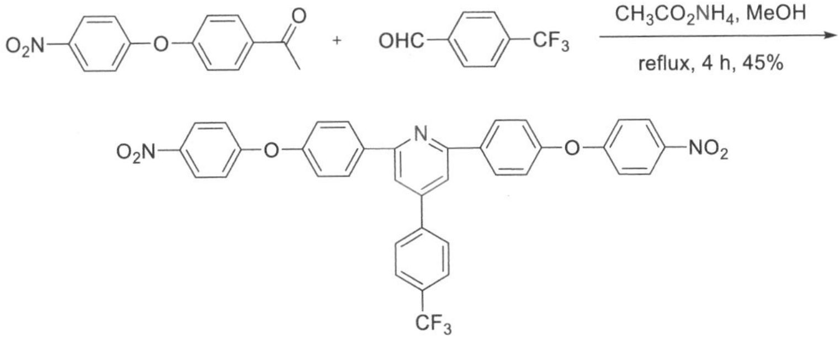

Example 4, An abnormal Chichibabin reaction[9]

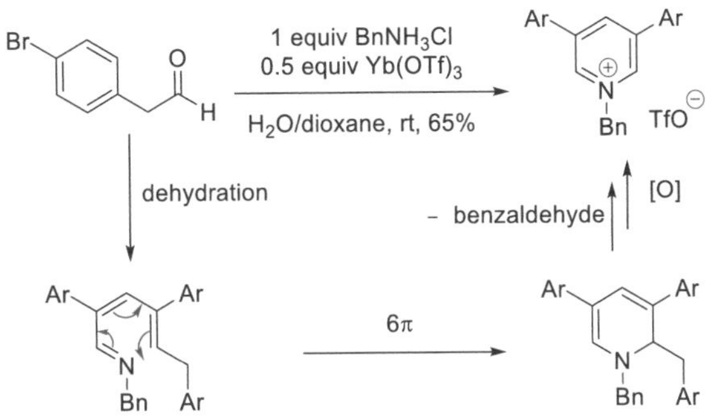

Example 5, Radiolabeled Chichibabin reaction[11]

Example 6, Radiolabeled Chichibabin reaction[13]

References

1. Chichibabin, A. E. *J. Russ. Phys. Chem. Soc.* **1906**, *37*, 1229. Alexei E. Chichibabin (1871–1945) was born in Kuzemino, Russia. He was Markovnikov's favorite student. Markovnikov's successor, Zelinsky (of Hell–Volhard–Zelinsky reaction fame) did not want to cooperate with the pupil and gave Chichibabin a negative judgment on his Ph.D. work, earning Chichibabin the nickname "the self-educated man."
2. Frank, R. L.; Riener, E. F. *J. Am. Chem. Soc.* **1950**, *72*, 4182–4183.
3. Weiss, M. *J. Am. Chem. Soc.* **1952**, *74*, 200–202.
4. Kessar, S. V.; Nadir, U. K.; Singh, M. *Indian J. Chem.* **1973**, *11*, 825–826.
5. Shimizu, S.; Abe, N.; Iguchi, A.; Dohba, M.; Sato, H.; Hirose, K.-I. *Microporous Mesoporous Materials* **1998**, *21*, 447–451.
6. Galatasis, P. *Chichibabin (Tschitschibabin) Pyridine Synthesis.* In *Name Reactions in Heterocyclic Chemistry*; Li, J. J., Ed.; Wiley: Hoboken, NJ, **2005**, pp 308–309. (Review).
7. Snider, B. B.; Neubert, B. J. *Org. Lett.* **2005**, *7*, 2715–2718.
8. Wang, X.-L.; Li, Y.-F.; Gong, C.-L.; Ma, T.; Yang, F.-C. *J. Fluorine Chem.* **2008**, *129*, 56–63.
9. Burns, N. Z.; Baran, P. S. *Angew. Chem. Int. Ed.* **2008**, *47*, 205–208.
10. Allais, C.; Grassot, J.-M.; Rodriguez, J.; Constantieux, T. *Chem. Rev.* **2014**, *114*, 10829–10868. (Review).
11. Tanigawa, T.; Komatsu, A.; Usuki, T. *Bioorg. Med. Chem. Lett.* **2015**, *25*, 2046–2049.
12. Khan, F. A. K.; Zaheer, Z.; Sangshetti, J. N.; Patil, R. H.; Farooqui, M. *Bioorg. Med. Chem. Lett.* **2017**, *27*, 567–573.
13. Fuse, W.; Imura, A.; Tanaka, N.; Usuki, T. *Tetrahedron Lett.* **2019**, *60*, 928–930.

Chugaev Elimination

Thermal elimination of xanthates to olefins.

Example 1[4]

Example 2[5]

© Springer Nature Switzerland AG 2021

J. J. Li, *Name Reactions*, https://doi.org/10.1007/978-3-030-50865-4_28

Example 3, Chugaev *syn*-elimination is followed by an intramolecular ene reaction[6]

Example 4, Aromatization via Chugaev elimination[10]

Example 5, Double Chugaev elimination[11]

References

1. Chugaev, L. *Ber.* **1899,** *32,* 3332. Lev A. Chugaev (1873–1922) was born in Moscow, Russia. He was a Professor of Chemistry at Petrograd, a position once held by Dimitri Mendeleyev and Paul Walden. In addition to terpenoids, Chugaev also investigated nickel and platinum chemistry. He completely devoted his life to science. The light in Chugaev's study would invariably burn until 4 or 5 am.

2. Harano, K.; Taguchi, T. *Chem. Pharm. Bull.* **1975,** *23,* 467–472.

3. Ho, T.-L.; Liu, S.-H. *J. Chem. Soc., Perkin Trans. 1* **1984,** 615–617.

4. Fu, X.; Cook, J. M. *Tetrahedron Lett.* **1990,** *31,* 3409–3412.

5. Meulemans, T. M.; Stork, G. A.; Macaev, F. Z.; Jansen, B. J. M.; de Groot, A. *J. Org. Chem.* **1999,** *64,* 9178–9188.

6. Nakagawa, H.; Sugahara, T.; Ogasawara, K. *Org. Lett.* **2000,** *2,* 3181–3183.

7. Fuchter, M. J. *Chugaev elimination.* In *Name Reactions for Functional Group Transformations*; Li, J. J., Ed.; Wiley: Hoboken, NJ, **2007,** pp 334–342. (Review).

8. Ahmed, S.; Baker, L. A.; Grainger, R. S.; Innocenti, P.; Quevedo, C. E. *J. Org. Chem.* **2008,** *73,* 8116–8119.

9. Tang, P.; Wang, L.; Chen, Q.-F.; Chen, Q.-H.; Jian, X.-X.; Wang, F.-P. *Tetrahedron* **2012,** *68,* 5031–5036.

10. He, S.; Hsung, R. P.; Presser, W. R.; Ma, Z.-X.; Haugen, B. J. *Org. Lett.* **2014,** *16,* 2180–2183.

11. Fukaya, K.; Kodama, K.; Tanaka, Y.; Yamazaki, H.; Sugai, T.; Yamaguchi, Y.; Watanabe, A.; Oishi, T.; Sato, T.; Chida, N. *Org. Lett.* **2015,** *17,* 2574–2577.

12. Burroughs, L.; Ritchie, J.; Woodward, S. *Tetrahedron* **2016,** *72,* 1686–1689.

13. He, W.; Ding, Y.; Tu, J.; Que, C.; Yang, Z.; Xu, J. *Org. Biomol. Chem.* **2018,** *16,* 1659–1666.

14. Langlais, M.; Coutelier, O.; Destarac, M. **2019,** *60,* 1522–1525

Claisen Condensation

Base-catalyzed condensation of esters to afford β-keto esters.

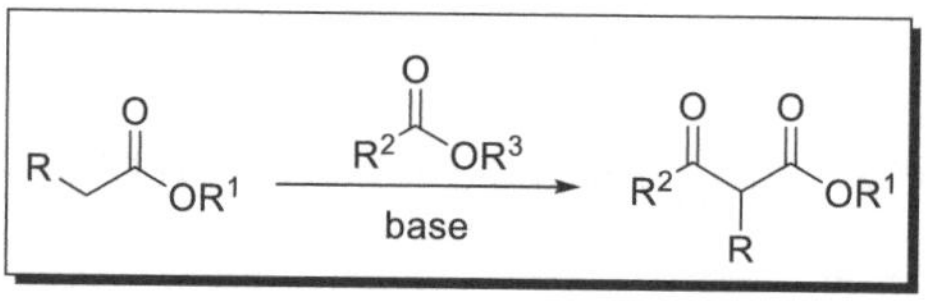

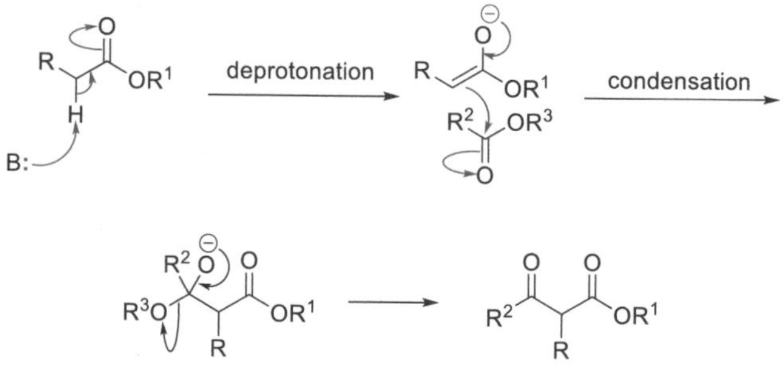

Example 1[4]

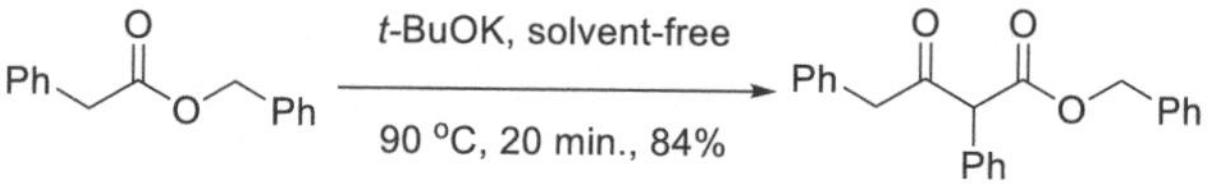

t-BuOK, solvent-free

90 °C, 20 min., 84%

Example 2[6]

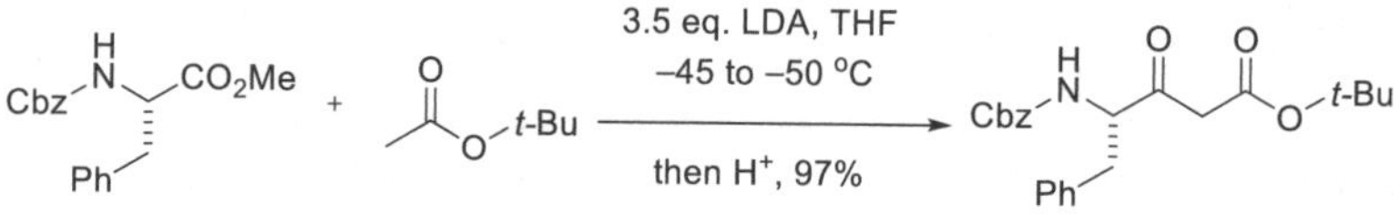

3.5 eq. LDA, THF
−45 to −50 °C

then H[+], 97%

Example 3, Retro-Claisen condensation[9]

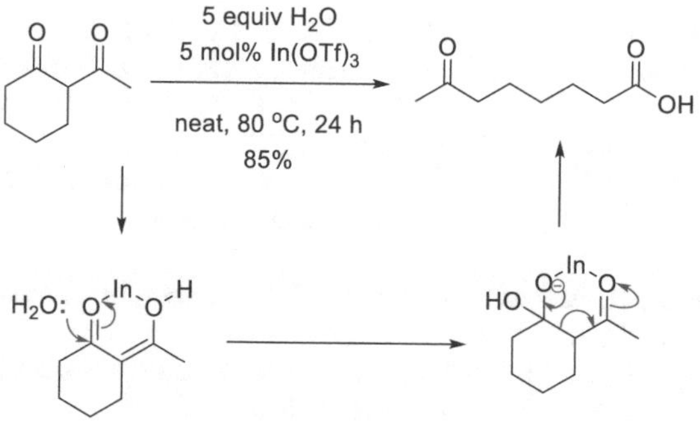

5 equiv H_2O
5 mol% In(OTf)$_3$

neat, 80 °C, 24 h
85%

© Springer Nature Switzerland AG 2021
J. J. Li, *Name Reactions*, https://doi.org/10.1007/978-3-030-50865-4_29

Example 4, Solvent-free Claisen condensation[10]

Example 5, Intramolecualar Claisen condensation (Dieckmann condensation)[11]

Example 6, Vinylogous Claisen condensation[12]

References

1   Claisen, R. L.; Lowman, O. *Ber.* **1887,** *20,* 651. Rainer Ludwig Claisen (1851–1930),
    born in Cologne, Germany, probably had the best pedigree in the history of organic
    chemistry. He apprenticed under Kekulé, Wöhler, von Baeyer, and Fischer before em-
    barking on his own independent research.

2   Hauser, C. R.; Hudson, B. E. *Org. React.* **1942,** *1,* 266–302. (Review).

3   Schäfer, J. P.; Bloomfield, J. J. *Org. React.* **1967,** *15,* 1–203. (Review).

4   Yoshizawa, K.; Toyota, S.; Toda, F. *Tetrahedron Lett.* **2001,** *42,* 7983–7985.

5   Heath, R. J.; Rock, C. O. *Nat. Prod. Rep.* **2002,** *19,* 581–596. (Review).

6   Honda, Y.; Katayama, S.; Kojima, M.; Suzuki, T. *Org. Lett.* **2002,** *4,* 447–449.

7   Mogilaiah, K.; Reddy, N. V. *Synth. Commun.* **2003,** *33,* 73–78.

8   Linderberg, M. T.; Moge, M.; Sivadasan, S. *Org. Process Res. Dev.* **2004,** *8,* 838–845.

9   Kawata, A.; Takata, K.; Kuninobu, Y.; Takai, K. *Angew. Chem. Int. Ed.* **2007,** *46,*
    7793–7795.

10 Iida, K.; Ohtaka, K.; Komatsu, T.; Makino, T.; Kajiwara, M. *J. Labelled Compd. Ra-
    diopharm.* **2008,** *51,* 167–169.

11 Song, Y. Y.; He, H. G.; Li, Y.; Deng, Y. *Tetrahedron Lett.* **2013,** *54,* 2658–2660.

12 Reber, K. P.; Burdge, H. E. *J. Nat. Prod.* **2018,** *81,* 292–297.

Claisen Rearrangement

The Claisen, *para*-Claisen rearrangements, Belluš–Claisen rearrangement; Corey–Claisen, Eschenmoser–Claisen rearrangement, Ireland–Claisen, Kazmaier–Claisen, Saucy–Claisen; orthoester Johnson–Claisen, along with the Carroll rearrangement, belong to the category of *[3,3]-sigmatropic rearrangements*. The Claisen rearrangement is a concerted process and the arrow pushing here is merely illustrative.

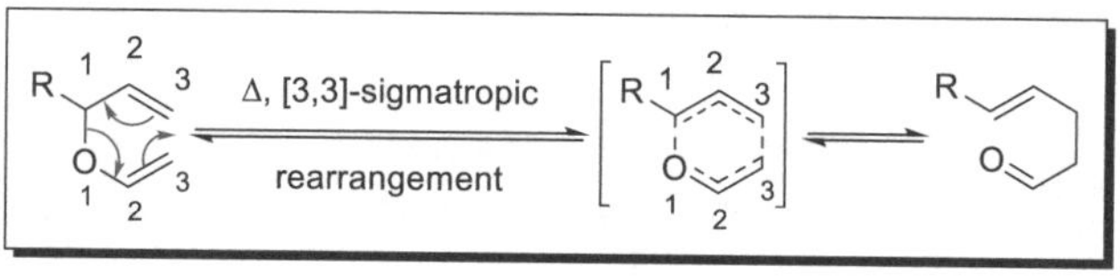

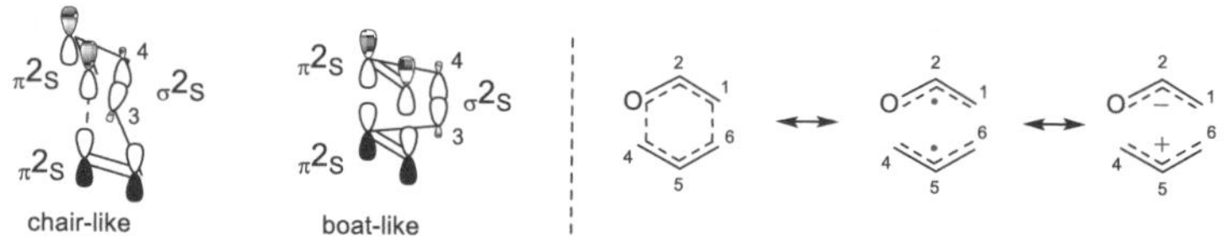

Example 1[7]

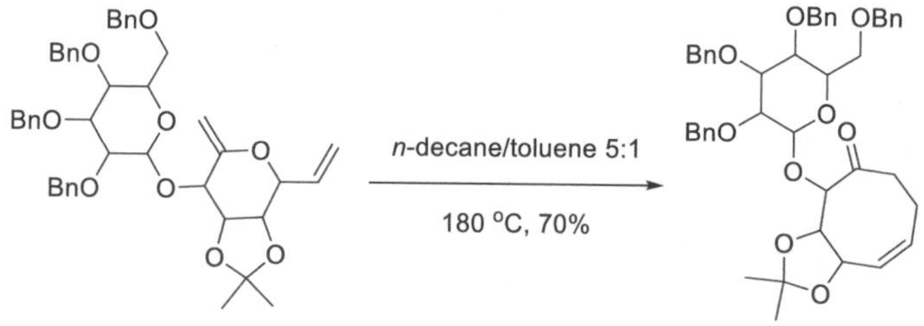

Example 2[8]

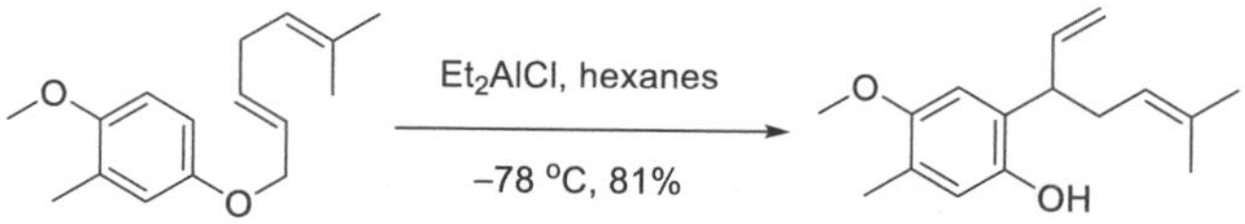

Example 3[9]

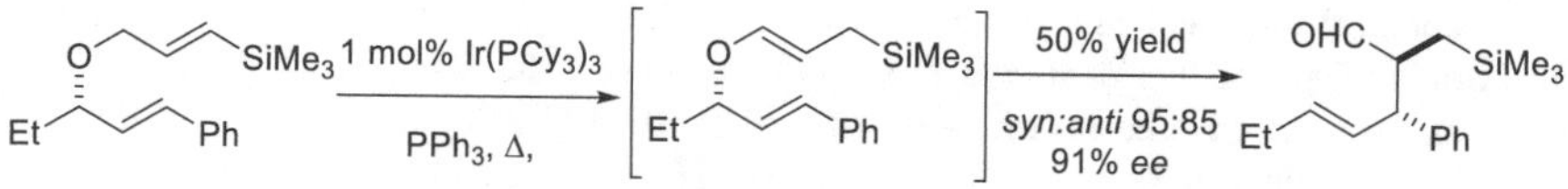

J. J. Li, *Name Reactions*, https://doi.org/10.1007/978-3-030-50865-4_30

Example 4, Asymmetric Claisen rearrangement[10]

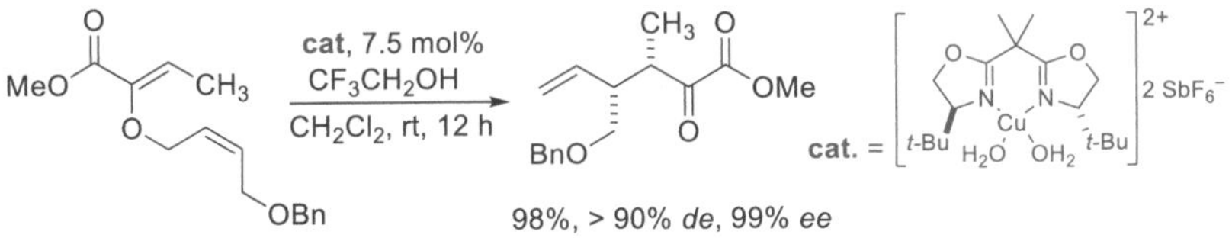

98%, > 90% *de*, 99% *ee*

Example 5, Asymmetric Claisen rearrangement[11]

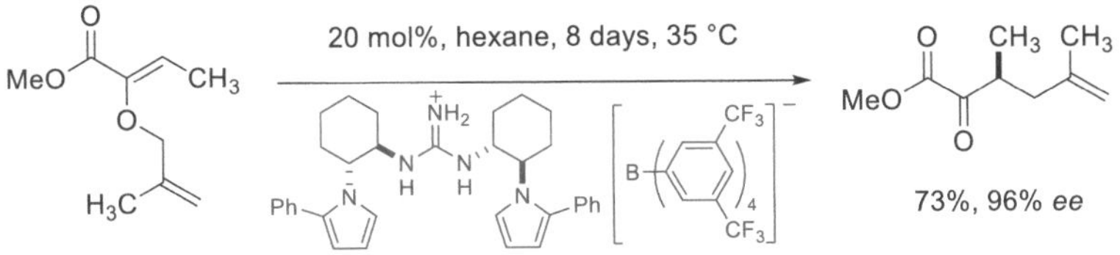

73%, 96% *ee*

Example 6[13]

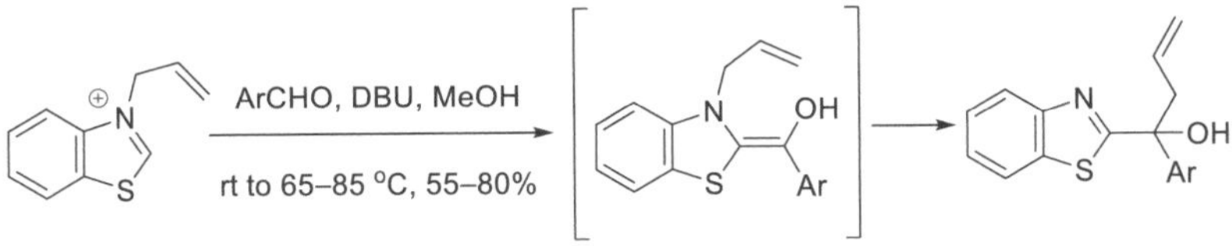

Example 7, Introduction of the prenyl group[14]

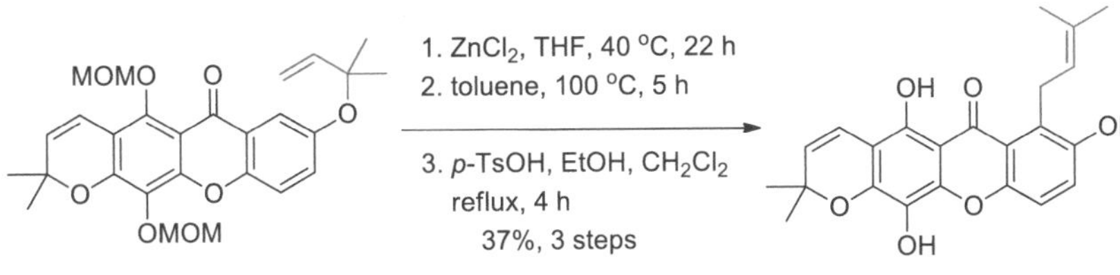

37%, 3 steps

References

1. Claisen, L. *Ber.* **1912**, *45*, 3157–3166.
2. Rhoads, S. J.; Raulins, N. R. *Org. React.* **1975**, *22*, 1–252. (Review).
3. Wipf, P. In *Comprehensive Organic Synthesis;* Trost, B. M.; Fleming, I., Eds.; Pergamon, **1991**, *Vol. 5*, 827–873. (Review).
4. Ganem, B. *Angew. Chem. Int. Ed.* **1996**, *35*, 937–945. (Review).
5. Ito, H.; Taguchi, T. *Chem. Soc. Rev.* **1999**, *28*, 43–50. (Review).
6. Castro, A. M. M. *Chem. Rev.* **2004**, *104*, 2939–3002. (Review).
7. Jürs, S.; Thiem, J. *Tetrahedron: Asymmetry* **2005**, *16*, 1631–1638.
8. Vyvyan, J. R.; Oaksmith, J. M.; Parks, B. W.; Peterson, E. M. *Tetrahedron Lett.* **2005**, *46*, 2457–2460.

9. Nelson, S. G.; Wang, K. *J. Am. Chem. Soc.* **2006,** *128*, 4232–4233.

10. Körner, M.; Hiersemann, M. *Org. Lett.* **2007,** *9*, 4979–4982.

11. Uyeda, C.; Jacobsen, E. N. *J. Am. Chem. Soc.* **2008,** *130*, 9228–9229.

12. Williams, D. R.; Nag, P. P. *Claisen and Related Rearrangements.* In *Name Reactions for Homologations-Part II*; Li, J. J., Ed.; Wiley: Hoboken, NJ, **2009,** pp 33–43. (Review).

13. Alwarsh, S.; Ayinuola, K.; Dormi, S. S.; McIntosh, M. C. *Org. Lett.* **2013,** *15*, 3–5.

14. Ito, S.; Kitamura, T.; Arulmozhiraja, S.; Manabe, K.; Tokiwa, H.; Suzuki, Y. *Org. Lett.* **2019,** *21*, 2777–2781.

15. Miro, J.; Ellwart, M.; Han, S.-J.; Lin, H.-H.; Toste, F. D.; Gensch, T.; Sigman, M. S.; Han, S.-J. *J. Am. Chem. Soc.* **2020,** *142*, 6390–6399.

para-*Claisen Rearrangement*

Further rearrangement of the normal *ortho*-Claisen rearrangement product gives the *para*-Claisen rearrangement product.

Mechanism 1:

Mechanism 2:

Mechanism 3:

Example 1[6]

Example 2[7]

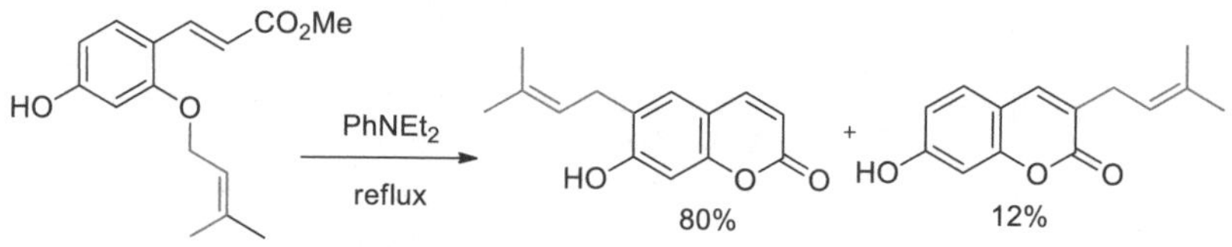

Example 3[8]

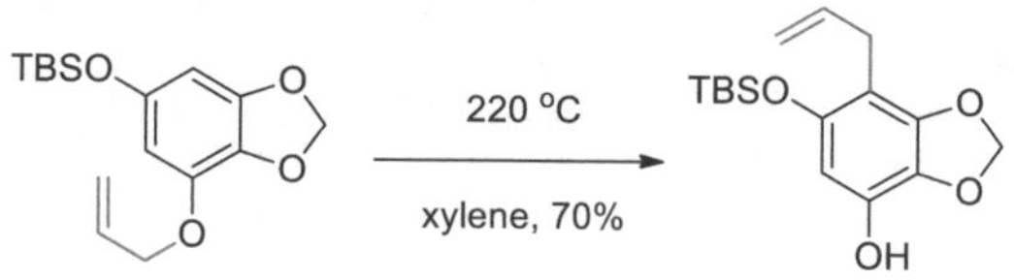

Example 4[10]

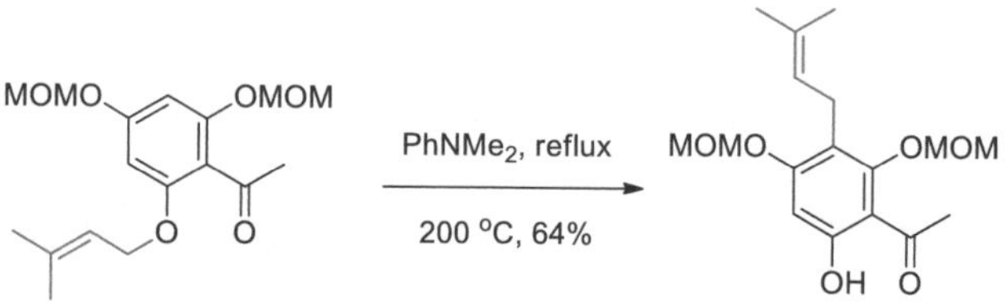

Example 5[11]

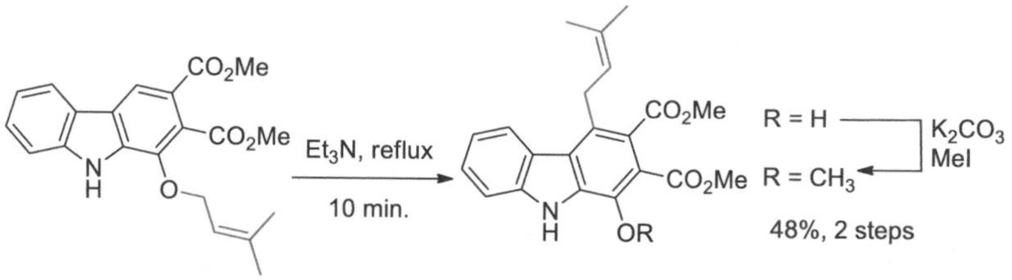

Example 6, fod = 1,1,1,2,2,3,3-heptafluoro-7,7-dimethyl-4,6-octanedionate = Siever's reagent[12]

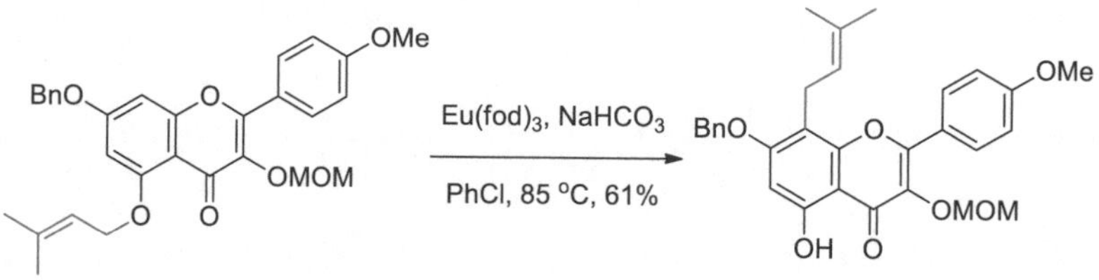

References

1. Alexander, E. R.; Kluiber, R. W. *J. Am. Chem. Soc.* **1951,** *73,* 4304–4306.
2. Rhoads, S. J.; Raulins, R.; Reynolds, R. D. *J. Am. Chem. Soc.* **1953,** *75,* 2531–2532.
3. Dyer, A.; Jefferson, A.; Scheinmann, F. *J. Org. Chem.* **1968,** *33,* 1259–1261.
4. Murray, R. D. H.; Lawrie, K. W. M. *Tetrahedron* **1979,** *35,* 697–699.
5. Cairns, N.; Harwood, L. M.; Astles, D. P. *J. Chem. Soc., Chem. Commun.* **1986,** 1264–1266.
6. Kilényi, S. N.; Mahaux, J.-M.; van Durme, E. *J. Org. Chem.* **1991,** *56,* 2591–2594.
7. Cairns, N.; Harwood, L. M.; Astles, D. P. *J. Chem. Soc., Perkin Trans. 1* **1994,** 3101–3107.
8. Pettus, T. R. R.; Inoue, M.; Chen, X.-T.; Danishefsky, S. J. *J. Am. Chem. Soc.* **2000,** *122,* 6160–6168.
9. Al-Maharik, N.; Botting, N. P. *Tetrahedron* **2003,** *59,* 4177–4181.
10. Khupse, R. S.; Erhardt, P. W. *J. Nat. Prod.* **2007,** *70,* 1507–1509.
11. Jana, A. K.; Mal, D. *Chem. Commun.* **2010,** *46,* 4411–4413.
12. Mei, Q.; Wang, C.; Zhao, Z.; Yuan, W.; Zhang, G. *Beilst. J. Org. Chem.* **2015,** *11,* 1220–1225.
13. Wang, Z.; Wang, H.; Ren, P.; Wang, M. *J. Macromol. Sci. Part A* **2019,** *56,* 794–802.

Abnormal Claisen Rearrangement

Further rearrangement of the normal Claisen rearrangement product with the β-carbon becoming attached to the ring.

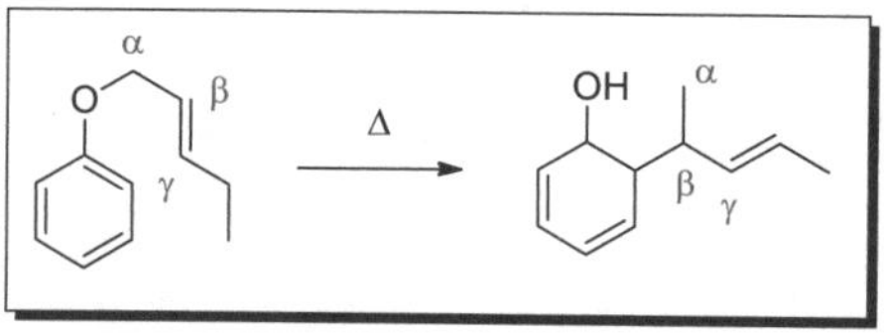

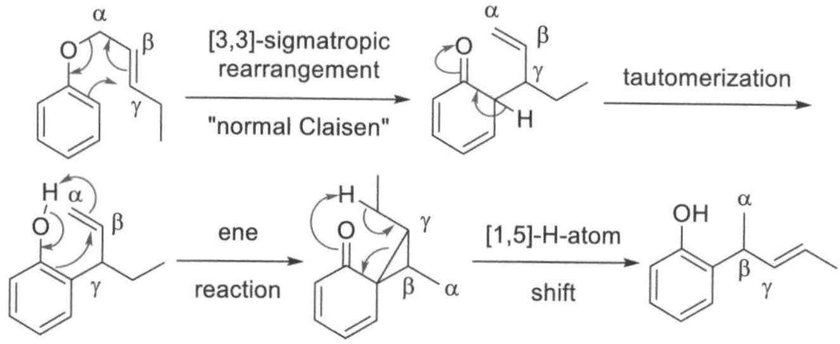

Example 1[3]

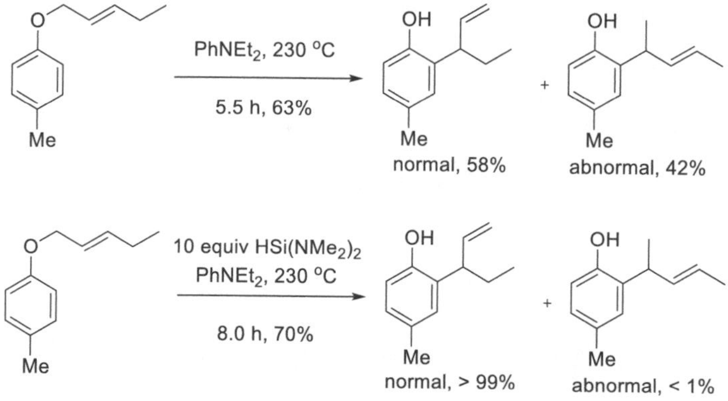

PhNEt$_2$, 230 °C

5.5 h, 63%

normal, 58%   abnormal, 42%

10 equiv HSi(NMe$_2$)$_2$
PhNEt$_2$, 230 °C

8.0 h, 70%

normal, > 99%   abnormal, < 1%

Example 2, Enantioselective aromatic Claisen rearrangement[4]

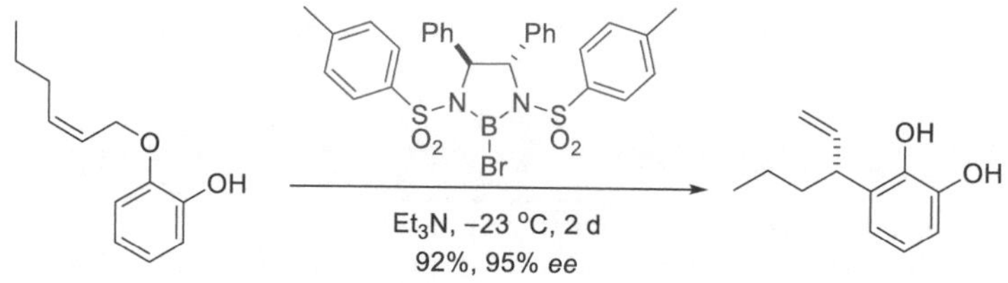

Et$_3$N, –23 °C, 2 d
92%, 95% ee

Example 3[5]

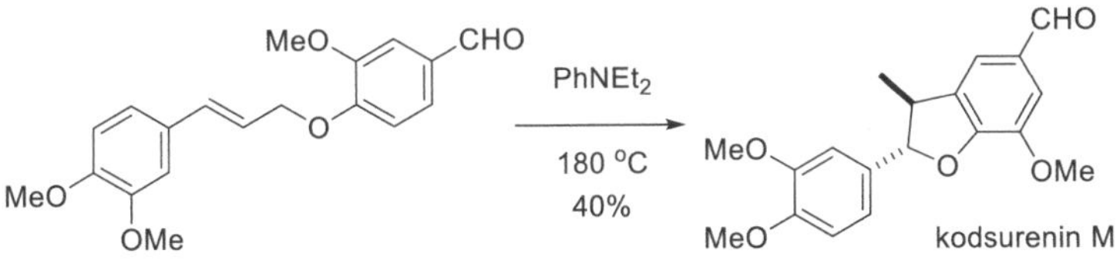

kodsurenin M

Example 4[6]

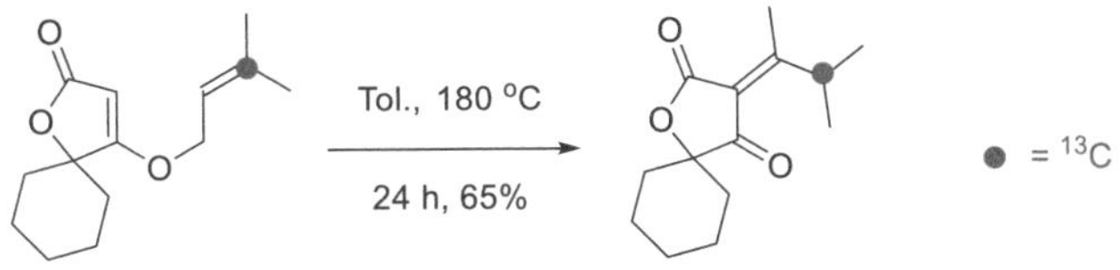

● = ^{13}C

Example 5[7]

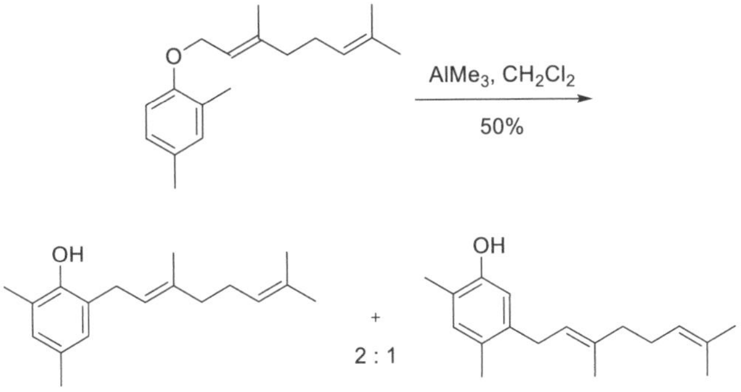

+

2 : 1

Example 6[10]

Microwave irradiation

180 °C, 20 h, 73%

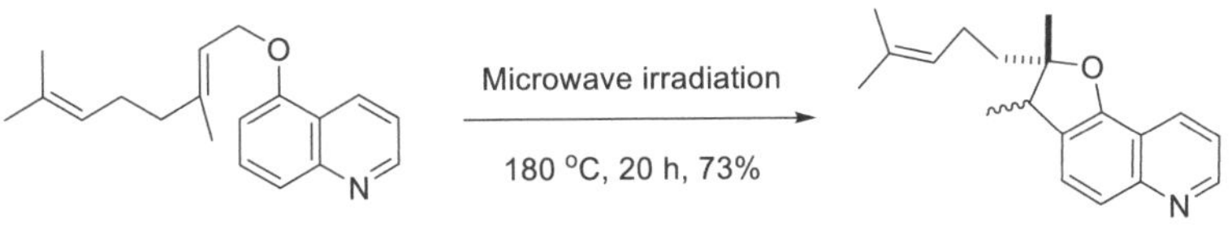

Example 7[11]

185 °C, 26 h, sealed tube

32%

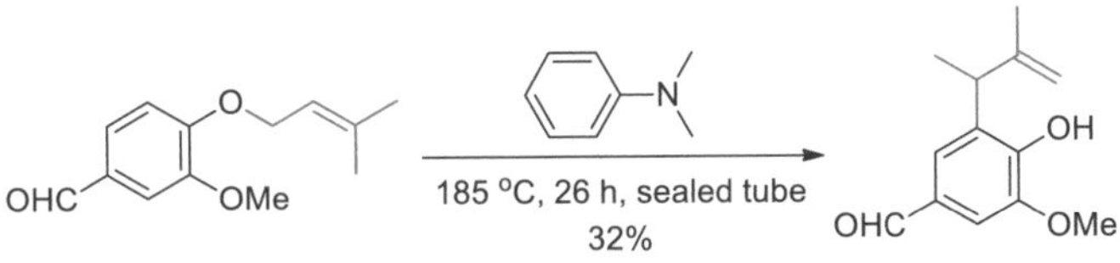

References

1.   Hansen, H.-J. In *Mechanisms of Molecular Migrations;* vol. 3, Thyagarajan, B. S., ed.; Wiley-Interscience: New York, **1971,** pp 177–236. (Review).
2.   Kilényi, S. N.; Mahaux, J.-M.; van Durme, E. *J. Org. Chem.* **1991,** *56,* 2591–2594.
3.   Fukuyama, T.; Li, T.; Peng, G. *Tetrahedron Lett.* **1994,** *35,* 2145–2148.
4.   Ito, H.; Sato, A.; Taguchi, T. *Tetrahedron Lett.* **1997,** *38,* 4815–4818.
5.   Yi, W. M.; Xin, W. A.; Fu, P. X. *J. Chem. Soc., (S),* **1998,** 168.
6.   Schobert, R.; Siegfried, S.; Gordon, G.; Mulholland, D.; Nieuwenhuyzen, M. *Tetrahedron Lett.* **2001,** *42,* 4561–4564.
7.   Wipf, P.; Rodriguez, S. *Ad. Synth. Catal.* **2002,** *344,* 434–440.
8.   Puranik, R.; Rao, Y. J.; Krupadanam, G. L. D. *Indian J. Chem., Sect. B* **2002,** *41B,* 868–870.
9.   Williams, D. R.; Nag, P. P. *Claisen and Related Rearrangements*. In *Name Reactions for Homologations-Part II*; Li, J. J., Ed.; Wiley: Hoboken, NJ, **2009,** pp 33–87. (Review).
10. Torincsi, M.; Kolonits, P.; Fekete, J.; Novak, L. *Synth.Commun.* **2012,** *42,* 3187–3199.
11. He, J.; Li, J.; Liu, Z.-Q. *Med. Chem. Res.* **2013,** *22,* 2847–2854.

Eschenmoser–Claisen Amide Acetal Rearrangement

[3,3]-Sigmatropic rearrangement of *N,O*-ketene acetals to yield γ,δ-unsaturated amides. Since Eschenmoser was inspired by Meerwein's observations on the interchange of amide, the Eschenmoser–Claisen rearrangement is sometimes known as the Meerwein–Eschenmoser–Claisen (MEC) rearrangement.

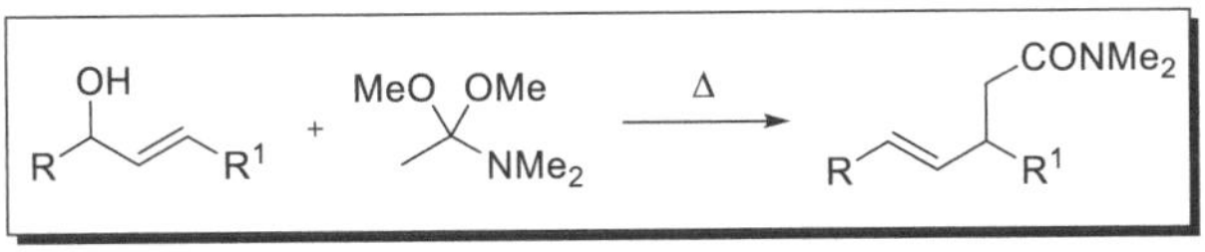

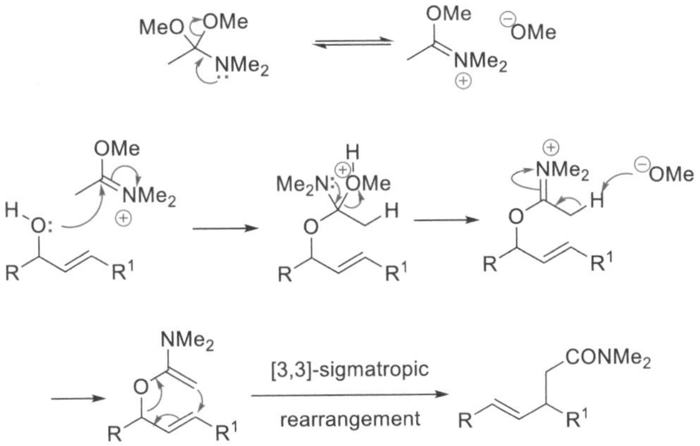

Example 1[4]

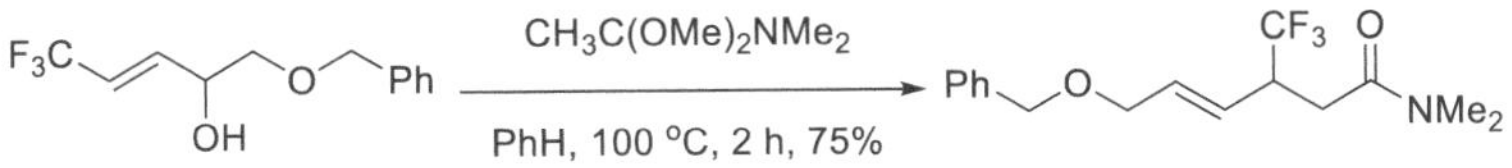

Example 2[5]

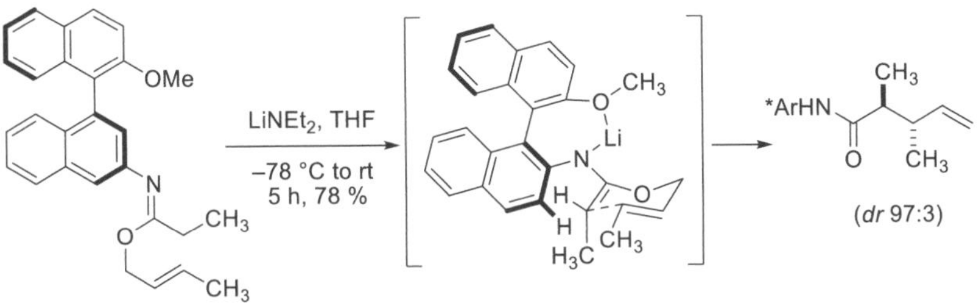

Example 3[6]

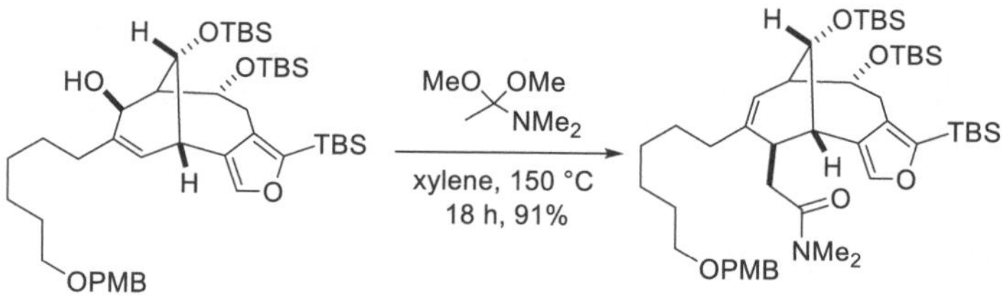

Example 4[8]

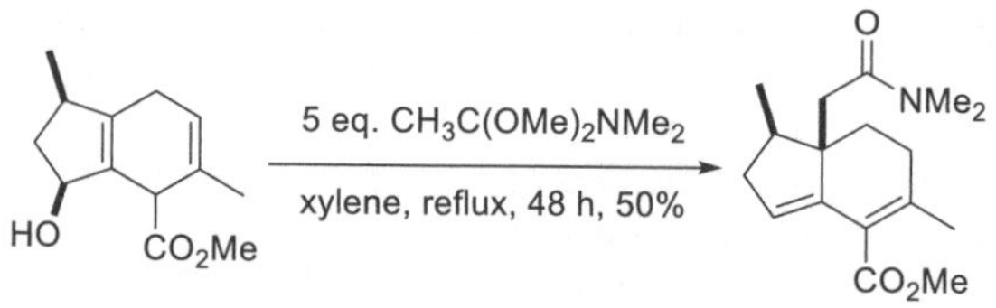

Example 5[9]

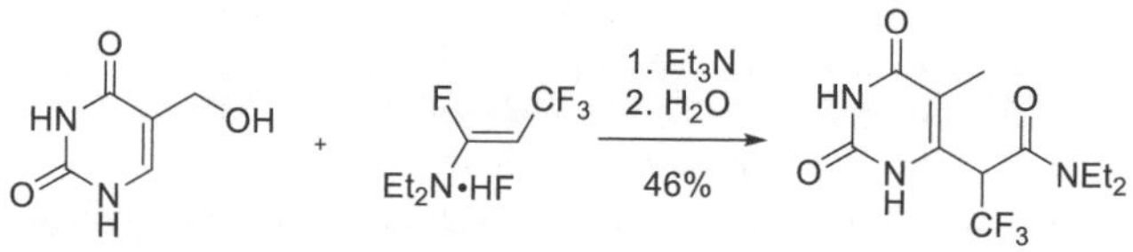

Example 6, Application in total synthesis[11]

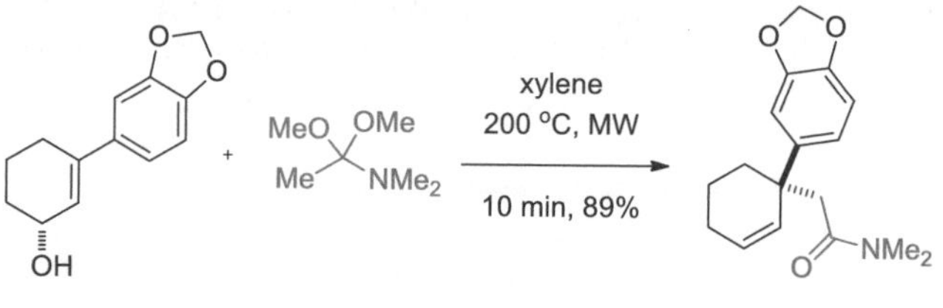

Example 7, A one-pot diastereoselective Meerwein–Eschenmoser–Claisen (MEC) rearrangement[15]

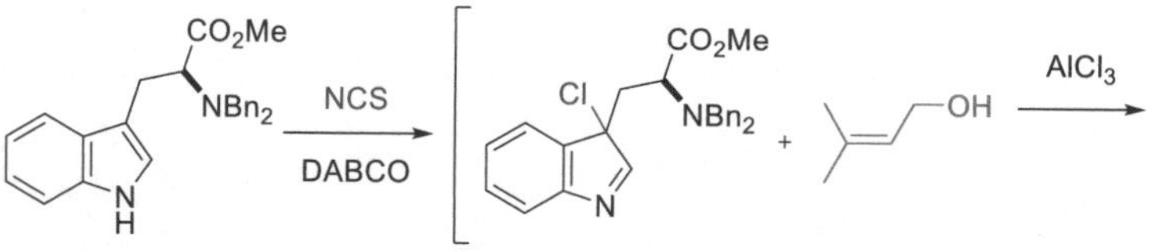

References

1. Meerwein, H.; Florian, W.; Schön, N.; Stopp, G. *Ann.* **1961,** *641,* 1−39.
2. Wick, A. E.; Felix, D.; Steen, K.; Eschenmoser, A. *Helv. Chim. Acta* **1964,** *47,* 2425−2429. Albert Eschenmoser (Switzerland, 1925−) is known for his work on, among many others, the monumental total synthesis of Vitamin B_{12} with R. B. Woodward in 1973. He now holds dual appointments at both ETH Zürich and the Scripps Research Institute in La Jolla, CA.
3. Wipf, P. In *Comprehensive Organic Synthesis;* Trost, B. M.; Fleming, I., Eds.; Pergamon, **1991,** *Vol. 5,* 827−873. (Review).
4. Konno, T.; Nakano, H.; Kitazume, T. *J. Fluorine Chem.* **1997,** *86,* 81−87.
5. Metz, P.; Hungerhoff, B. *J. Org. Chem.* **1997,** *62,* 4442−4448.
6. Kwon, O. Y.; Su, D. S.; Meng, D. F.; Deng, W.; D'Amico, D. C.; Danishefsky, S. J. *Angew. Chem. Int. Ed.* **1998,** *37,* 1877−1880.
7. Ito, H.; Taguchi, T. *Chem. Soc. Rev.* **1999,** *28,* 43−50. (Review).
8. Loh, T.-P.; Hu, Q.-Y. *Org. Lett.* **2001,** *3,* 279−281.
9. Castro, A. M. M. *Chem. Rev.* **2004,** *104,* 2939−3002. (Review).
10. Williams, D. R.; Nag, P. P. *Claisen and Related Rearrangements.* In *Name Reactions for Homologations-Part II*; Li, J. J., Ed.; Wiley: Hoboken, NJ, **2009,** pp 60−68. (Review).
11. Walkowiak, J.; Tomas-Szwaczyk, M.; Haufe, G.; Koroniak, H. *J. Fluorine Chem.* **2012,** *143,* 189−197.
12. Yoshida, M.; Kasai, T.; Mizuguchi, T.; Namba, K. *Synlett* **2014,** *25,* 1160−1162.
13. Das, M. K.; De, S.; Shubhashish; B., A. *Org. Biomol. Chem.* **2015,** *13,* 3585−3588.
14. Zhang, X.; Cai, X.; Huang, B.; Guo, L.; Gao, Z.; Jia, Y. *Angew. Chem. Int. Ed.* **2019,** *58,* 13380−13384.
15. Yu, H.; Zong, Y.; Xu, T. *Chem. Sci.* **2020,** *511,* 656−660.

Ireland–Claisen (Silyl Ketene Acetal) Rearrangement

Rearrangement of allyl trimethylsilyl ketene acetal, prepared by reaction of allylic ester enolates with trimethylsilyl chloride, to yield γ,δ-unsaturated carboxylic acids. The Ireland–Claisen rearrangement seems to be advantageous to the other variants of the Claisen rearrangement in terms of *E/Z* geometry control and mild conditions.

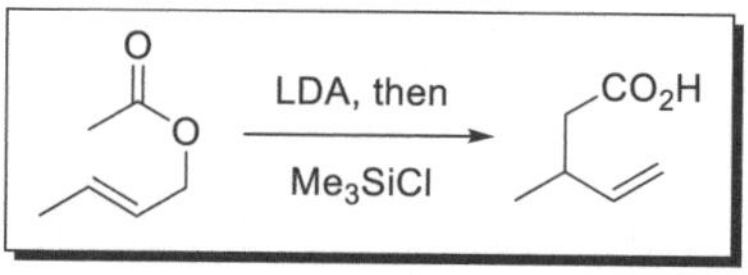

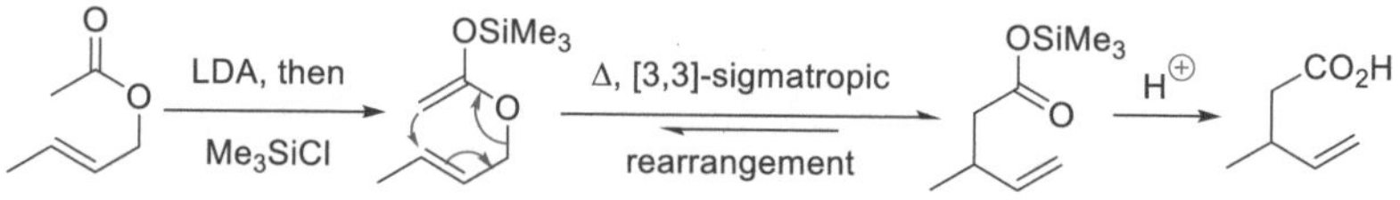

Example 1[2]

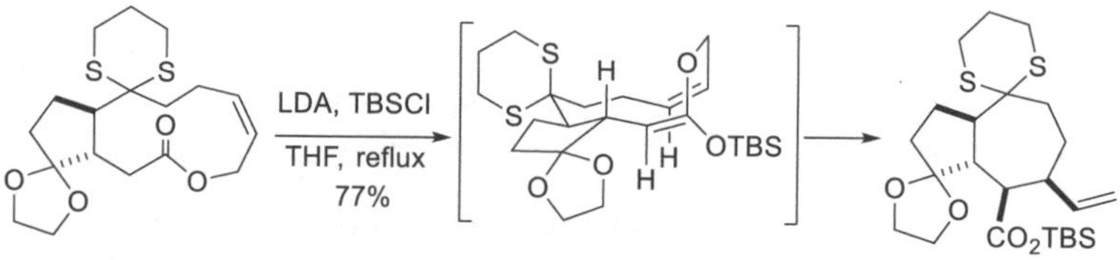

Example 2[3]

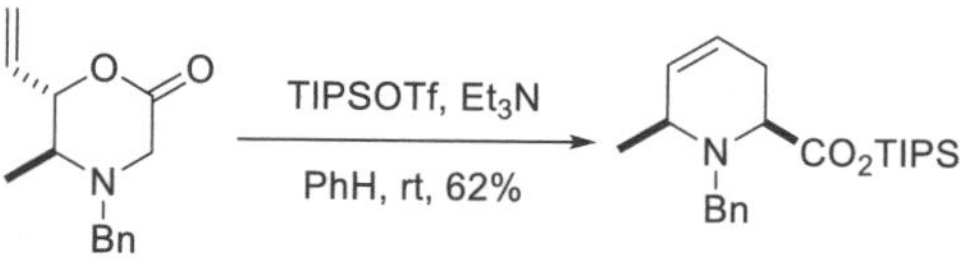

Example 3, Enantioselective ester enolate–Claisen Rearrangement[6]

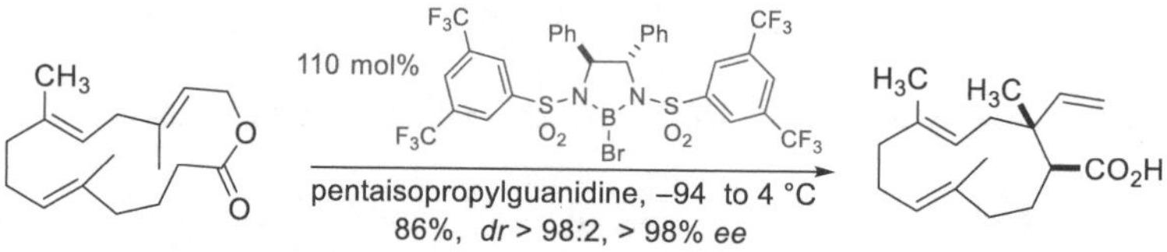

Example 4, A modified Ireland–Claisen rearrangement[8]

Example 5[9]

Example 6, Chirality-transferring Ireland–Claisen rearrangement[11]

Example 7, Stereodivergence in the Ireland–Claisen rearrangement of α-alkoxy esters[12]

Example 8, A highly diastereoselective Ireland–Claisen rearrangement[15]

References

1. Ireland, R. E.; Mueller, R. H. *J. Am. Chem. Soc.* **1972,** *94*, 5897–5898. Also *J. Am. Chem. Soc.* **1976,** *98*, 2868–2877. Robert E. Ireland obtained his Ph.D. from William S. Johnson before becoming a professor at the University of Virginia and later at the California Institute of Technology. He is now retired.

2. Begley, M. J.; Cameron, A. G.; Knight, D. W. *J. Chem. Soc., Perkin Trans. 1* **1986,** 1933–1938.

3. Angle, S. R.; Breitenbucher, J. G. *Tetrahedron Lett.* **1993,** *34*, 3985–3988.

4. Pereira, S.; Srebnik, M. *Aldrichimica Acta* **1993,** *26*, 17–29. (Review).

5. Ganem, B. *Angew. Chem. Int. Ed.* **1996,** *35*, 936–945. (Review).

6. Corey, E.; Kania, R. S. *J. Am. Chem. Soc.* **1996,** *118,* 1229–1230.

7. Chai, Y.; Hong, S.-p.; Lindsay, H. A.; McFarland, C.; McIntosh, M. C. *Tetrahedron* **2002,** *58*, 2905–2928. (Review).

8. Churcher, I.; Williams, S.; Kerrad, S.; Harrison, T.; Castro, J. L.; Shearman, M. S.; Lewis, H. D.; Clarke, E. E.; Wrigley, J. D. J.; Beher, D.; Tang, Y. S.; Liu, W. *J. Med. Chem.* **2003,** *46*, 2275–2278.

9. Fujiwara, K.; Goto, A.; Sato, D.; Kawai, H.; Suzuki, T. *Tetrahedron Lett.* **2005,** *46*, 3465–3468.

10. Williams, D. R.; Nag, P. P. *Claisen and Related Rearrangements.* In *Name Reactions for Homologations-Part II*; Li, J. J., Ed.; Wiley: Hoboken, NJ, **2009,** pp 45–51. (Review).

11. Nogoshi, K.; Domon, D.; Fujiwara, K.; Kawamura, N.; Katoono, R.; Kawai, H.; Suzuki, T. *Tetrahedron Lett.* **2013,** *54*, 676–680.

12. Crimmins, M. T.; Knight, J. D.; Williams, P. S.; Zhang, Y. *Org. Lett.* **2014,** *16*, 2458–2461.

13. Anugu, R. R.; Mainkar, P. S.; Sridhar, B.; Chandrasekhar, S. *Org. Biomol. Chem.* **2016,** *14*, 1332–1337.

14. Podunavac, M.; Lacharity, J. J.; Jones, K. E.; Zakarian, A. *Org. Lett.* **2018,** *20*, 4867–4870.

15. Zavesky, B. P.; De Jesus Cruz, P.; Johnson, J. S. *Org. Lett.* **2020,** *22*, 3537–3541.

Johnson–Claisen Orthoester Rearrangement

Heating of an allylic alcohol with an excess of trialkyl orthoacetate in the presence of trace amounts of a weak acid gives a mixed orthoester. Mechanistically, the orthoester loses alcohol to generate the ketene acetal, which undergoes [3,3]-sigmatropic rearrangement to give a γ,δ-unsaturated ester.

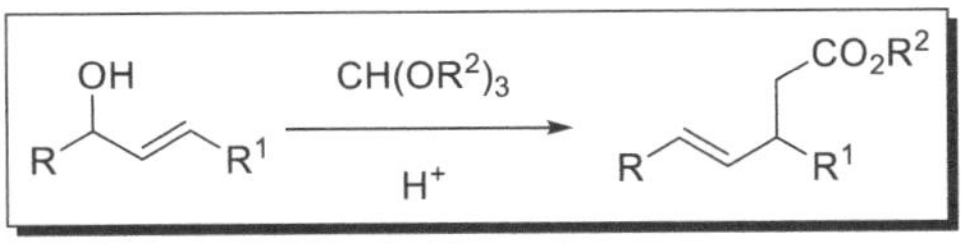

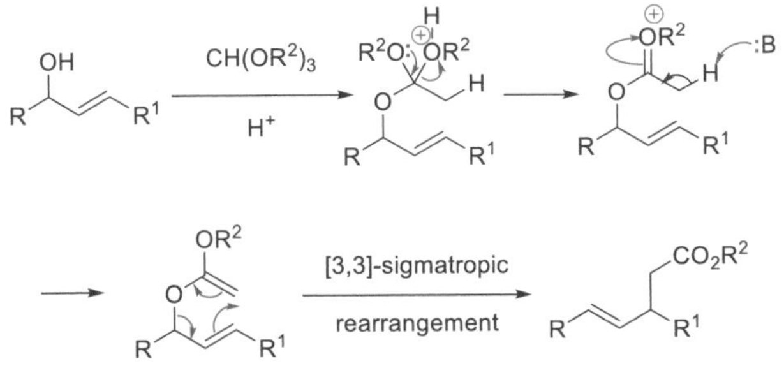

Example 1[2]

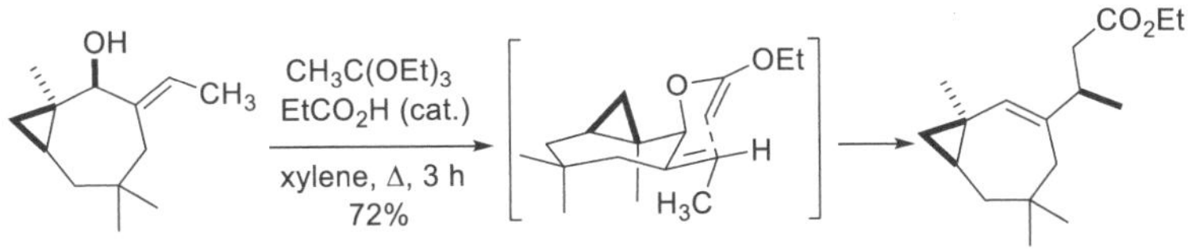

Example 2[3]

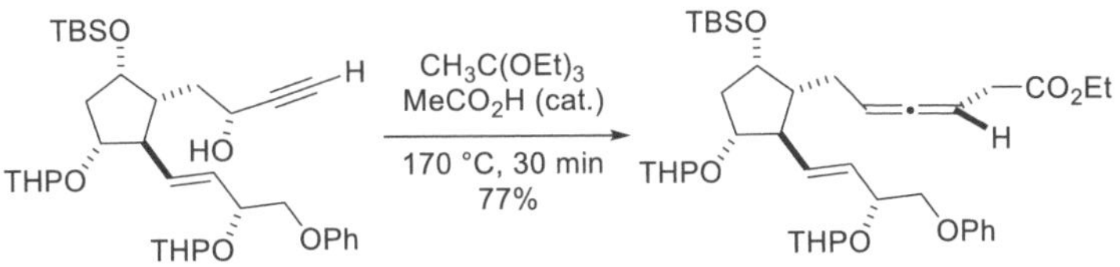

Example 3[4]

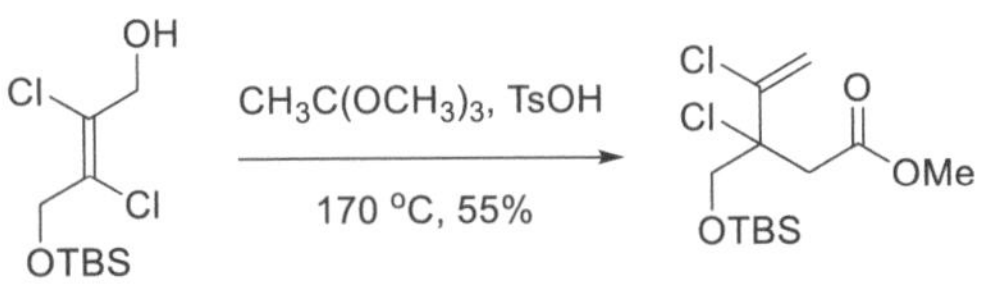

Example 4[9]

$CH_3C(OCH_3)_3$
cat. $CH_3CH_2CO_2H$

reflux, 77%

Example 5[10]

$(EtO)_3CCH_3$

pivalic acid
xylene, 140 °C
54%, E:Z > 95:5

Example 6. Scalable microwave-assisted Johnson–Claisen rearrangement[11]

AcOH, microwave
200 W, 2.5 MPa

flow rate: 4.5 mL/min
224 °C, 92%

References

1. Johnson, W. S.; Werthemann, L.; Bartlett, W. R.; Brocksom, T. J.; Li, T.-T.; Faulkner, D. J.; Peterson, M. R. *J. Am. Chem. Soc.* **1970**, *92*, 741–743. William S. Johnson (1913–1995) was born in New Rochelle, New York. He earned his Ph.D. in only two years at Harvard under Louis Fieser. He was a professor at the University of Wisconsin for 20 years before moving to Stanford University, where he was credited with building the modern-day Stanford Chemistry Department.
2. Paquette, L.; Ham, W. H. *J. Am. Chem. Soc.* **1987**, *109*, 3025–3036.
3. Cooper, G. F.; Wren, D. L.; Jackson, D. Y.; Beard, C. C.; Galeazzi, E.; Van Horn, A. R.; Li, T. T. *J. Org. Chem.* **1993**, *58*, 4280–4286.
4. Schlama, T.; Baati, R.; Gouverneur, V.; Valleix, A.; Falck, J. R.; Mioskowski, C. *Angew. Chem. Int. Ed.* **1998**, *37*, 2085–2087.
5. Giardiná, A.; Marcantoni, E.; Mecozzi, T.; Petrini, M. *Eur. J. Org. Chem.* **2001**, 713–718.
6. Funabiki, K.; Hara, N.; Nagamori, M.; Shibata, K.; Matsui, M. *J. Fluorine Chem.* **2003**, *122*, 237–242.
7. Montero, A.; Mann, E.; Herradón, B. *Eur. J. Org. Chem.* **2004**, 3063–3073.
8. Scaglione, J. B.; Rath, N. P.; Covey, D. F. *J. Org. Chem.* **2005**, *70*, 1089–1092.

9. Zartman, A. E.; Duong, L. T.; Fernandez-Metzler, C.; Hartman, G. D.; Leu, C.-T.; Prueksaritanont, T.; Rodan, G. A.; Rodan, S. B.; Duggan, M. E.; Meissner, R. S. *Bioorg. Med. Chem. Lett.* **2005,** *15*, 1647–1650.

10. Hicks, J. D.; Roush, W. R. *Org. Lett.* **2008,** *10*, 681–684.

11. Williams, D. R.; Nag, P. P. *Claisen and Related Rearrangements.* In *Name Reactions for Homologations-Part II*; Li, J. J., Ed.; Wiley: Hoboken, NJ, **2009,** pp 68–72. (Review).

12. Sydlik, S. A.; Swager, T. M. *Adv. Funct. Mater.* **2013,** *23,* 1873–1882.

13. Egami, H.; Tamaoki, S.; Abe, M.; Ohneda, N.; Yoshimura, T.; Okamoto, T.; Odajima, H.; Mase, N.; Takeda, K.; Hamashima, Y. *Org. Process Res. Dev.* **2018,** *22,* 1029–1033.

14. Zhou, Y.-G.; Wong, H. N. C.; Peng, X.-S. *J. Org. Chem.* **2020,** *85*, 967–976.

Clemmensen Reduction

Reduction of aldehydes or ketones to the corresponding methylene compounds using amalgamated zinc in hydrochloric acid.

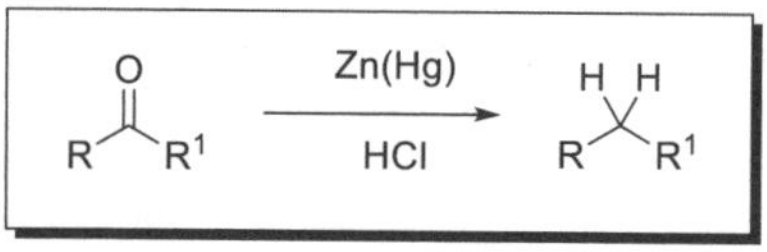

The zinc-carbenoid mechanism:[3]

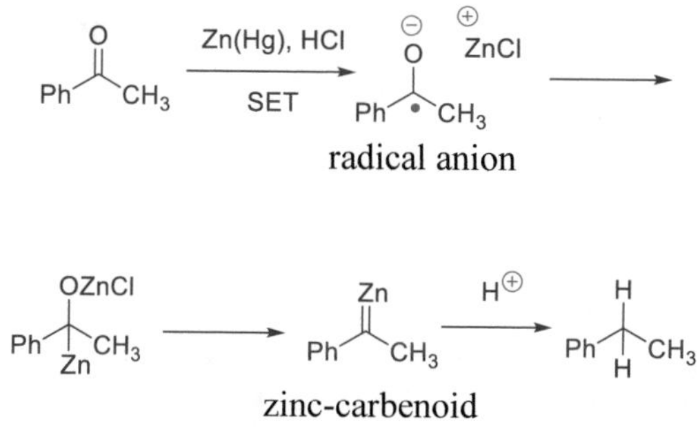

radical anion

zinc-carbenoid

The radical anion mechanism:

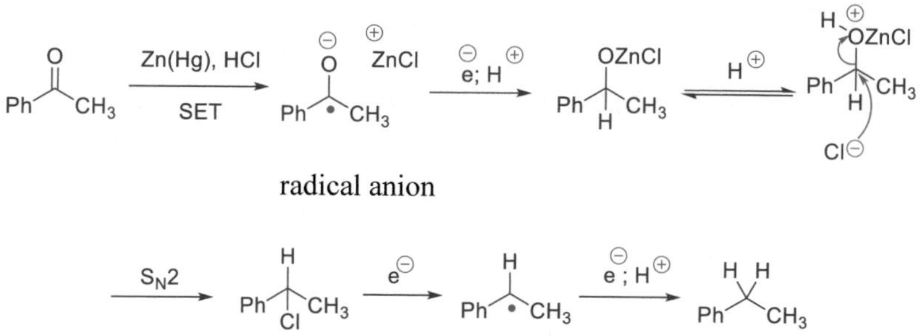

radical anion

Example 1[5]

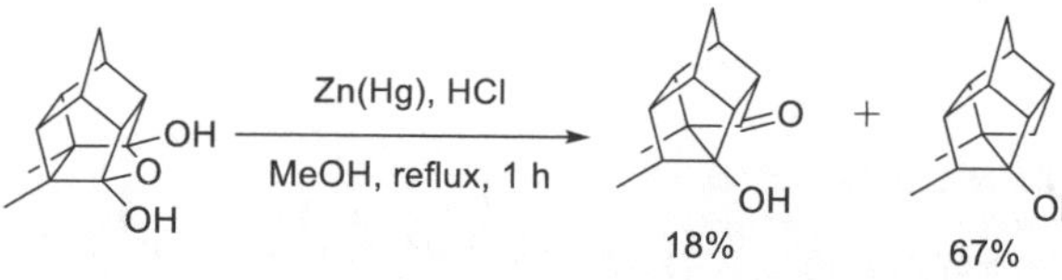

18%   67%

J. J. Li, *Name Reactions*, https://doi.org/10.1007/978-3-030-50865-4_31

Example 2[6]

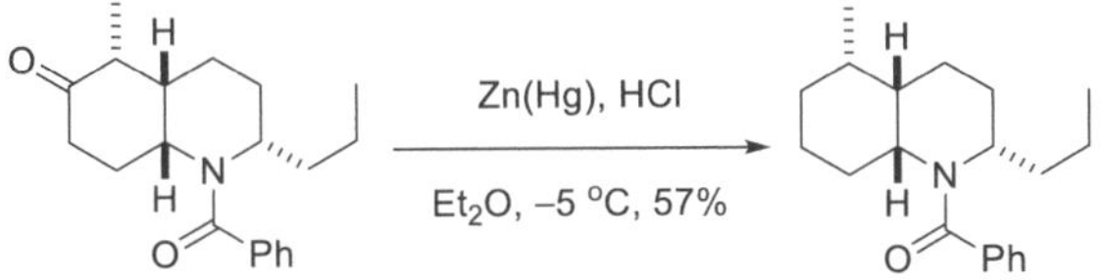

Example 3[7]

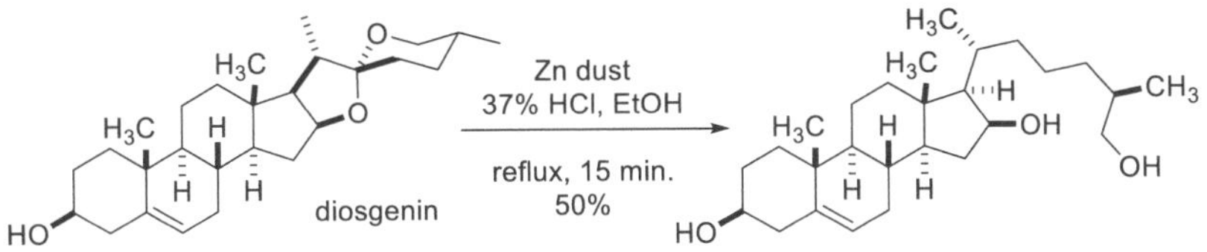

diosgenin

Example 4[9]

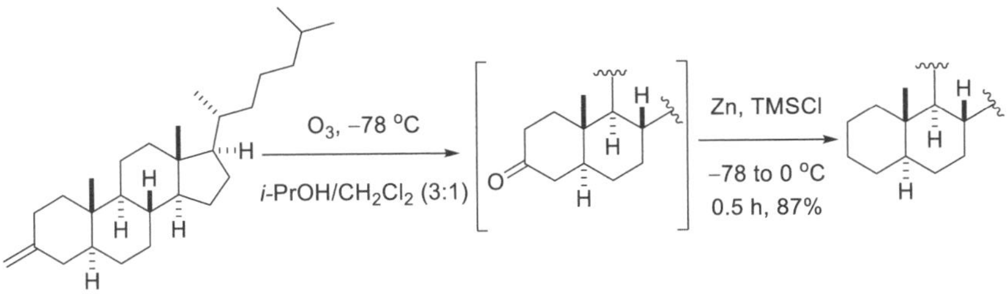

Example 5, Clemensen reductive rearrangement[10]

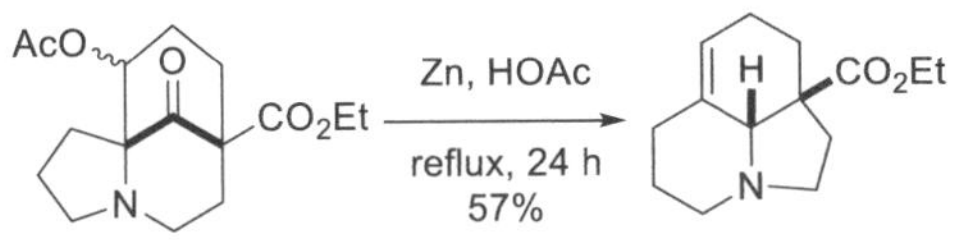

Example 6, Reuctive lactone alkylation[11]

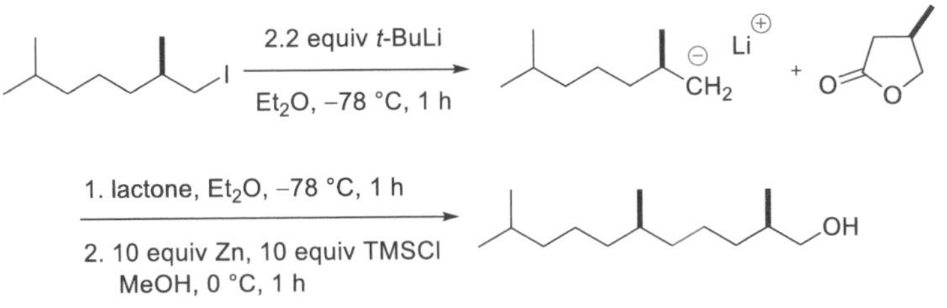

Example 7, Toward the synthesis of dibarrelane[12]

References

1. Clemmensen, E. *Ber.* **1913,** *46,* 1837–1843. Erik C. Clemmensen (1876–1941) was born in Odense, Denmark. He received the M.S. degree from the Royal Polytechnic Institute in Copenhagen. In 1900, Clemmensen immigrated to the United States, and worked at Parke, Davis and Company in Detroit (coincidently, this author's first employer!) as a research chemist for 14 years, where he discovered the reduction of carbonyl compounds with amalgamated zinc. Clemmensen later founded a few chemical companies and was the president of one of them, the Clemmensen Chemical Corporation in Newark, New Jersey.
2. Martin, E. L. *Org. React.* **1942,** *1,* 155–209. (Review).
3. Vedejs, E. *Org. React.* **1975,** *22,* 401–422. (Review).
4. Talpatra, S. K.; Chakrabarti, S.; Mallik, A. K.; Talapatra, B. *Tetrahedron* **1990,** *46,* 6047–6052.
5. Martins, F. J. C.; Viljoen, A. M.; Coetzee, M.; Fourie, L.; Wessels, P. L. *Tetrahedron* **1991,** *47,* 9215–9224.
6. Naruse, M.; Aoyagi, S.; Kibayashi, C. *J. Chem. Soc., Perkin Trans. 1* **1996,** 1113–1124.
7. Alessandrini, L.; et al. *Steroids* **2004,** *69,* 789–794.
8. Dey, S. P.; et al. *J. Indian Chem. Soc.* **2008,** *85,* 717–720.
9. Xu, S.; Toyama, T.; Nakamura, J.; Arimoto, H. *Tetrahedron Let.* **2010,** *51,* 4534–4537.
10. Zhang, J.; Wang, Y.-Q.; Wang, X.-W.; Li, W.-D. Z. *J. Org. Chem.* **2013,** *78,* 6154–6162.
11. Cao, J.; Perlmutter, P. *Org. Lett.* **2015,** *15,* 4327–4329.
12. Suzuki, T.; Okuyama, H.; Takano, A.; Suzuki, S.; Shimizu, I.; Kobayashi, S. *J. Org. Chem.* **2014,** *79,* 2803–2808.
13. Sanchez-Viesca, F.; Berros, M.; Gomez, R. *Am. J. Chem.* **2018,** *8,* 8–12.
14. Oyama, K.-i.; Kimura, Y.; Iuchi, S.; Koga, N.; Yoshida, K.; Kondo, T. *RSC Adv.* **2019,** *9,* 31435–31439.

Cope Elimination

Thermal elimination of *N*-oxides to olefins and *N*-hydroxyl amines.

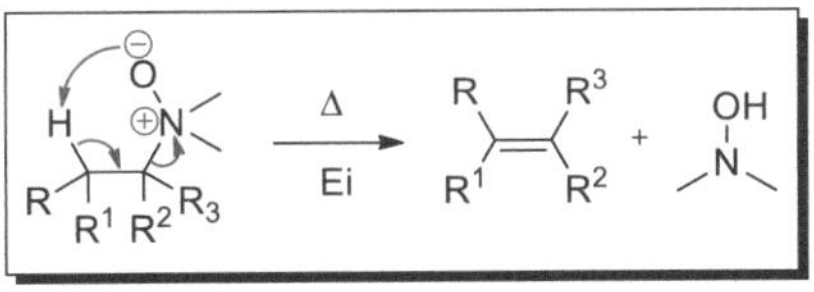

Example 1, Solid-phase Cope elimination[5]

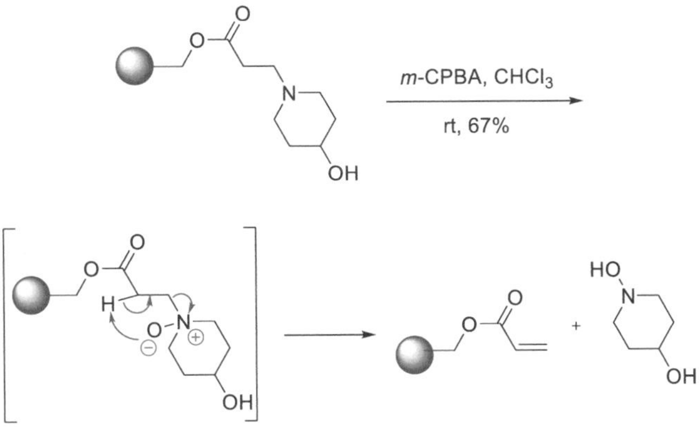

Example 2[6]

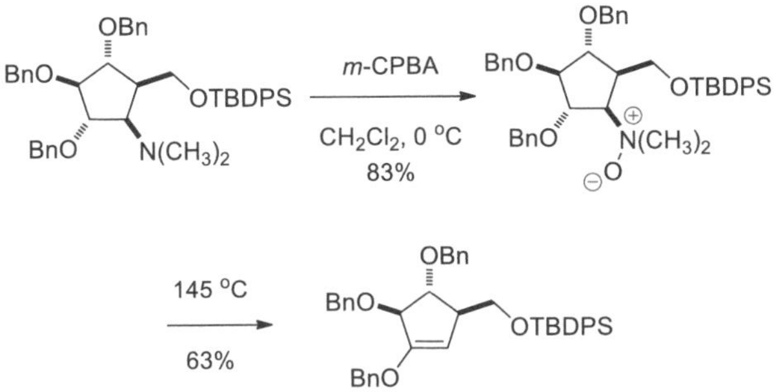

Example 3[8]

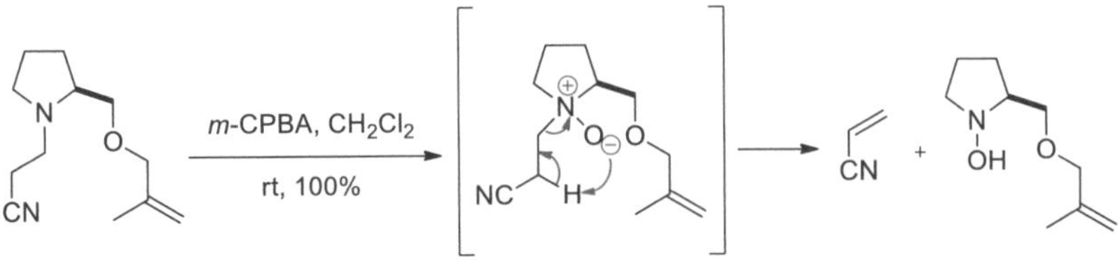

© Springer Nature Switzerland AG 2021
J. J. Li, *Name Reactions*, https://doi.org/10.1007/978-3-030-50865-4_32

Example 4, Retro-Cope elimination[9]

Example 5[12]

Example 6, Application in medicinal chemistry[13]

Example 7[13]

Example 8, Similar to Example 3[14]

1. 1.1 equiv m-CPBA
   CH$_2$Cl$_2$, −40 °C → rt, 12 h

2. BzCl, Et$_3$N, DMAP
   CH$_2$Cl$_2$, 0 °C, 30 min
   52%

References

1. Cope, A. C.; Foster, T. T.; Towle, P. H. *J. Am. Chem. Soc.* **1949,** *71*, 3929−3934. Arthur Clay Cope (1909−1966) was born in Dunreith, Indiana. He was a professor and head at MIT where he discovered the Cope elimination reaction after he taught at Bryn Mawr and Columbia where he discovered the Cope rearrangement. The Arthur Cope Award is a prestigious award in organic chemistry administered by the American Chemical Society.

2. Cope, A. C.; Trumbull, E. R. *Org. React.* **1960,** *11*, 317−493. (Review).

3. DePuy, C. H.; King, R. W. *Chem. Rev.* **1960,** *60*, 431−457. (Review).

4. Gallagher, B. M.; Pearson, W. H. *Chemtracts: Org. Chem.* **1996,** *9*, 126−130. (Review).

5. Sammelson, R. E.; Kurth, M. J. *Tetrahedron Lett.* **2001,** *42*, 3419−3422.

6. Vasella, A.; Remen, L. *Helv. Chim. Acta.* **2002,** *85*, 1118−1127.

7. Garcia Martinez, A.; Teso Vilar, E.; Garcia Fraile, A.; de la Moya Cerero, S.; Lora Maroto, B. *Tetrahedron: Asymmetry* **2002,** *13*, 17−19.

8. O'Neil, I. A.; Ramos, V. E.; Ellis, G. L.; Cleator, E.; Chorlton, A. P.; Tapolczay, D. J.; Kalindjian, S. B. *Tetrahedron Lett.* **2004,** *45*, 3659−3661.

9. Henry, N.; O'Meil, I. A. *Tetrahedron Lett.* **2007,** *48*, 1691−1694.

10. Fuchter, M. J. *Cope Elimination Reaction.* In *Name Reactions for Functional Group Transformations*; Li, J. J., Ed.; Wiley: Hoboken, NJ, **2007,** pp 342−353. (Review).

11. Bourgeois, J.; Dion, I.; Cebrowski, P. H.; Loiseau, F.; Bedard, A.-C.; Beauchemin, A. M. *J. Am. Chem. Soc.* **2009,** *131*, 874−875.

12. Miyatake-Ondozabal, H.; Bannwart, L. M.; Gademann, K. *Chem. Commun.* **2013,** *49*, 1921−1923.

13. Chrovian, C. C.; Soyode-Johnson, A.; Peterson, A. A.; Gelin, C. F.; Deng, X.; Dvorak, C. A.; Carruthers, N. I.; Lord, B.; Fraser, I.; Aluisio, L.; et al. *J. Med. Chem.* **2018,** *61*, 207−223.

14. Hegmann, N.; Prusko, L.; Diesendorf, N.; Heinrich, M. R. *Org. Lett.* **2018,** *20*, 7825−7829.

15. Grassl, S.; Chen, Y.-H.; Hamze, C.; Tuellmann, C. P.; Knochel, P. *Org. Lett.* **2019,** *21*, 494−497.

16. Grassl, S.; Knochel, P. *Org. Lett.* **2020,** *22*, 1947−1950.

Cope Rearrangement

The Cope, aza-Cope, anionic oxy-Cope, and oxy-Cope rearrangements belong to the category of *[3,3]-sigmatropic rearrangements*. Since it is a concerted process, the arrow pushing here is only illustrative. This reaction is an equilibrium process. *Cf.* Claisen rearrangement.

Example 1[4]

Example 2[6]

Example 3[9]

Example 4[10]

Example 5[11]

© Springer Nature Switzerland AG 2021
J. J. Li, *Name Reactions*, https://doi.org/10.1007/978-3-030-50865-4_33

Example 6[12]

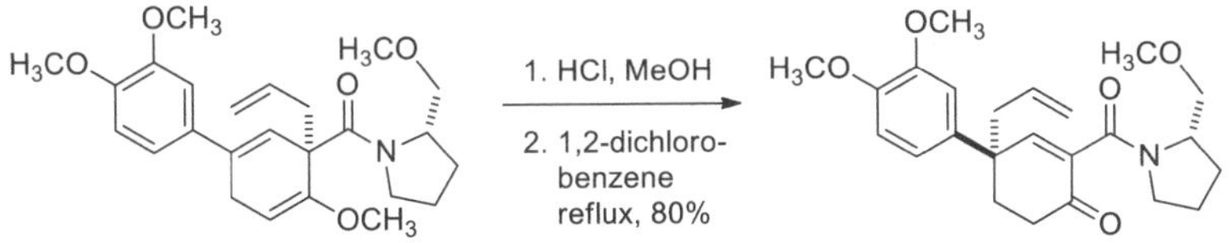

Example 7[14]

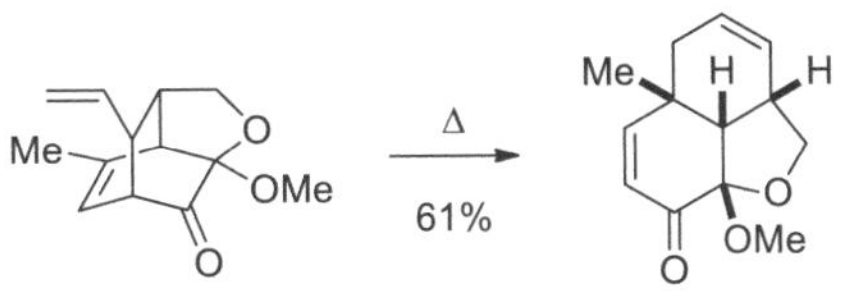

Example 8[15]

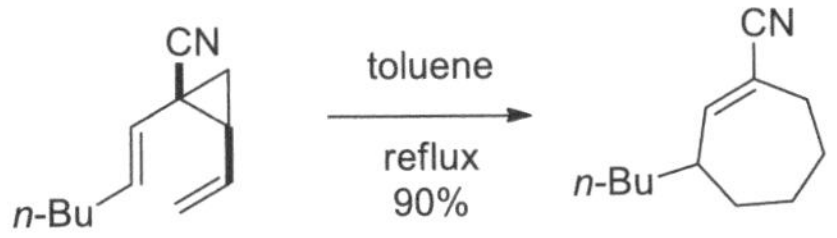

Example 9, Allylation, followed by 2-aza-Cope rearrangement[16]

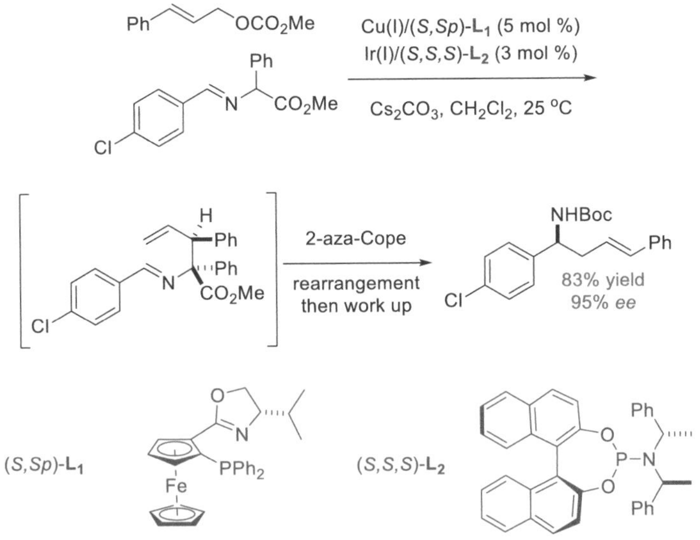

References

1. Cope, A. C.; Hardy, E. M. *J. Am. Chem. Soc.* **1940,** *62*, 441–444.
2. Frey, H. M.; Walsh, R. *Chem. Rev.* **1969,** *69*, 103–124. (Review).
3. Rhoads, S. J.; Raulins, N. R. *Org. React.* **1975,** *22*, 1–252. (Review).
4. Wender, P. A.; Schaus, J. M. White, A. W. *J. Am. Chem. Soc.* **1980,** *102*, 6159–6161.
5. Hill, R. K. In *Comprehensive Organic Synthesis* Trost, B. M.; Fleming, I., Eds.; Pergamon, **1991,** *Vol. 5*, 785–826. (Review).
6. Chou, W.-N.; White, J. B.; Smith, W. B. *J. Am. Chem. Soc.* **1992,** *114*, 4658–4667.
7. Davies, H. M. L. *Tetrahedron* **1993,** *49*, 5203–5223. (Review).
8. Miyashi, T.; Ikeda, H.; Takahashi, Y. *Acc. Chem. Res.* **1999,** *32*, 815–824. (Review).
9. Von Zezschwitz, P.; Voigt, K.; Lansky, A.; Noltemeyer, M.; De Meijere, A. *J. Org. Chem.* **1999,** *64*, 3806–3812.
10. Lo, P. C.-K.; Snapper, M. L. *Org. Lett.* **2001,** *3*, 2819–2821.
11. Clive, D. L. J.; Ou, L. *Tetrahedron Lett.* **2002,** *43*, 4559–4563.
12. Malachowski, W. P.; Paul, T.; Phounsavath, S. *J. Org. Chem.* **2007,** *72*, 6792–6796.
13. Mullins, R. J.; McCracken, K. W. *Cope and Related Rearrangements*. In *Name Reactions for Homologations-Part II*; Li, J. J., Ed.; Wiley: Hoboken, NJ, **2009,** pp 88–135. (Review).
14. Ren, H.; Wulff, W. D. *Org. Lett.* **2013,** *15*, 242–245.
15. Yamada, T.; Yoshimura, F.; Tanino, K. *Tetrahedron Lett.* **2013,** *54*, 522–525.
16. Wei, L.; Zhu, Q.; Xiao, L.; Tao, H.-Y.; Wang, C.-J. *Nat. Commun.* **2019,** *10*, 1–12.
17. Wang, Y.; Cai, P.-J.; Yu, Z.-X. *J. Am. Chem. Soc.* **2020,** *142*, 2777–2786.

Anionic Oxy-Cope Rearrangement

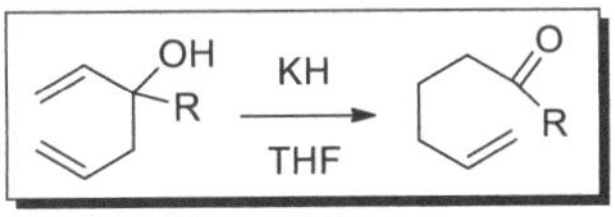
OH
R
KH
THF
O
R

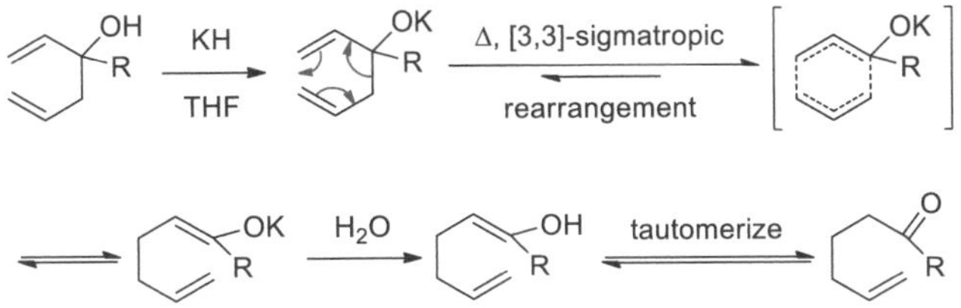
KH
THF
Δ, [3,3]-sigmatropic
rearrangement
OK
H2O
tautomerize

Example 1[1]

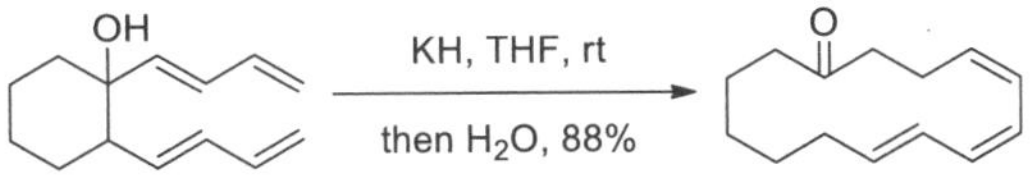
KH, THF, rt
then H2O, 88%

Example 2[4]

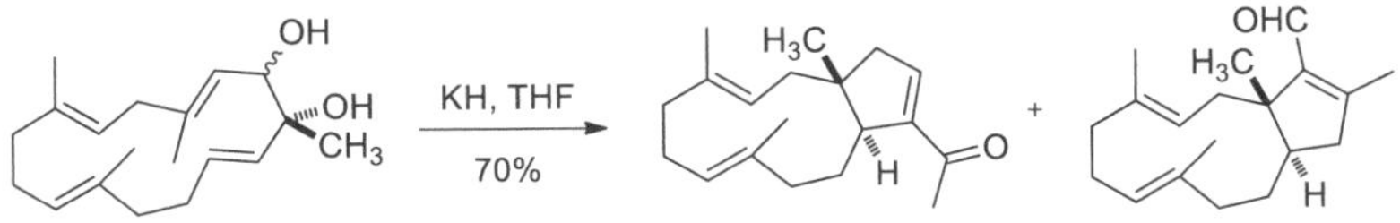
KH, THF
70%
OHC

Example 3[5]

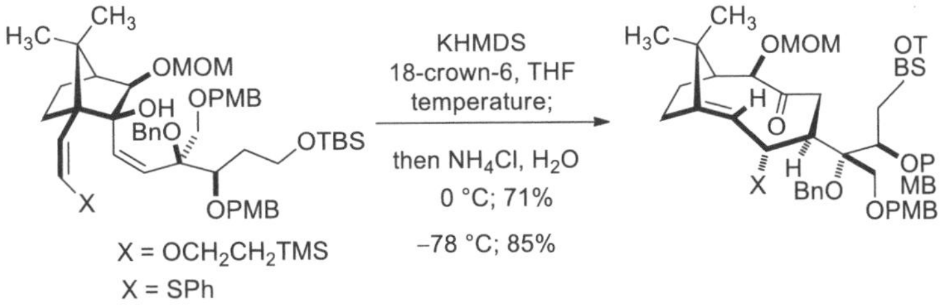
KHMDS
18-crown-6, THF
temperature;
then NH4Cl, H2O
0 °C; 71%
–78 °C; 85%
X = OCH2CH2TMS
X = SPh

Example 4[8]

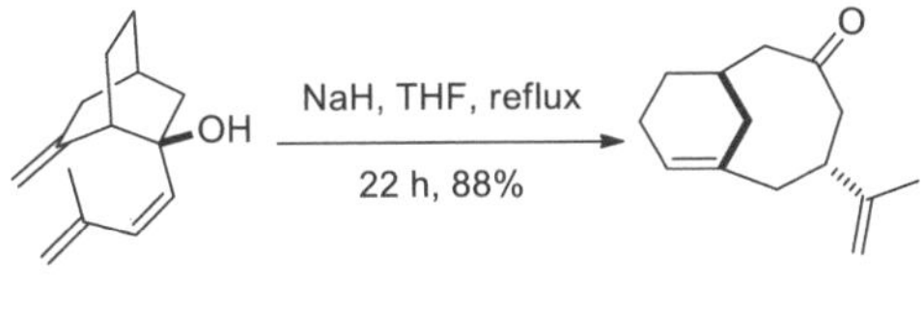
NaH, THF, reflux
22 h, 88%

Example 5[9]

Example 6[11]

References

1. Wender, P. A.; Sieburth, S. M.; Petraitis, J. J.; Singh, S. K. *Tetrahedron* **1981,** *37,* 3967–3975.
2. Wender, P. A.; Ternansky, R. J.; Sieburth, S. M. *Tetrahedron Lett.* **1985,** *26,* 4319–4322.
3. Paquette, L. A. *Tetrahedron* **1997,** *53,* 13971–14020. (Review).
4. Corey, E. J.; Kania, R. S. *Tetrahedron Lett.* **1998,** *39,* 741–744.
5. Paquette, L. A.; Reddy, Y. R.; Haeffner, F.; Houk, K. N. *J. Am. Chem. Soc.* **2000,** *122,* 740–741.
6. Voigt, B.; Wartchow, R.; Butenschon, H. *Eur. J. Org. Chem.* **2001,** 2519–2527.
7. Hashimoto, H.; Jin, T.; Karikomi, M.; Seki, K.; Haga, K.; Uyehara, T. *Tetrahedron Lett.* **2002,** *43,* 3633–3636.
8. Gentric, L.; Hanna, I.; Huboux, A.; Zaghdoudi, R. *Org. Lett.* **2003,** *5,* 3631–3634.
9. Jones, S. B.; He, L.; Castle, S. L. *Org. Lett.* **2006,** *8,* 3757–3760.
10. Mullins, R. J.; McCracken, K. W. *Cope and Related Rearrangements.* In *Name Reactions for Homologations-Part II*; Li, J. J., Ed.; Wiley: Hoboken, NJ, **2009,** pp 88–135. (Review).
11. Taber, D. F.; Gerstenhaber, D. A.; Berry, J. F. *J. Org. Chem.* **2013,** *76,* 7614–7617.
12. Roosen, P. C.; Vanderwal, C. D. *Org. Lett.* **2014,** *16,* 368–4371.
13. Anagnostaki, E. E.; Demertzidou, V. P.; Zografos, A. L. *Chem. Commun.* **2015,** *51,* 2364–2367.
14. Fujimoto, Y.; Yanai, H.; Matsumoto, T. *Synlett* **2016,** *27,* 2229–2232.
15. Simek, M.; Bartova, K.; Pohl, R.; Cisarova, I.; Jahn, U. *Angew. Chem. Int. Ed.* **2020,** *59,* 6160–6165.

Oxy-Cope Rearrangement

While the anionic oxy-Cope rearrangements work at low temperature, the oxy-Cope rearrangements require high temperature but provide a thermodynamic sink.

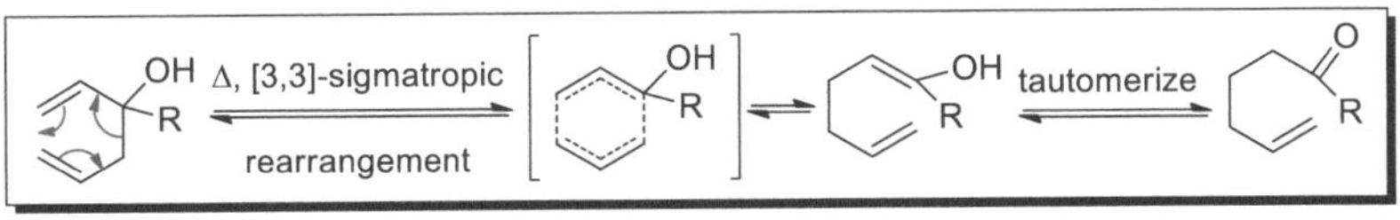
OH
R
Δ, [3,3]-sigmatropic
rearrangement
OH
R
OH
R
tautomerize
O
R

Example 1[2]

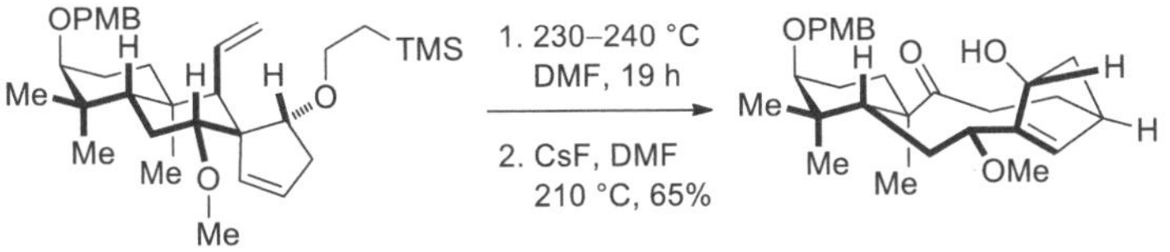
OPMB
H
Me
Me
H
H
Me
O
Me
O
TMS
1. 230–240 °C
DMF, 19 h
2. CsF, DMF
210 °C, 65%
OPMB
H
O
HO
H
Me
Me
Me
OMe
H

Example 2[3]

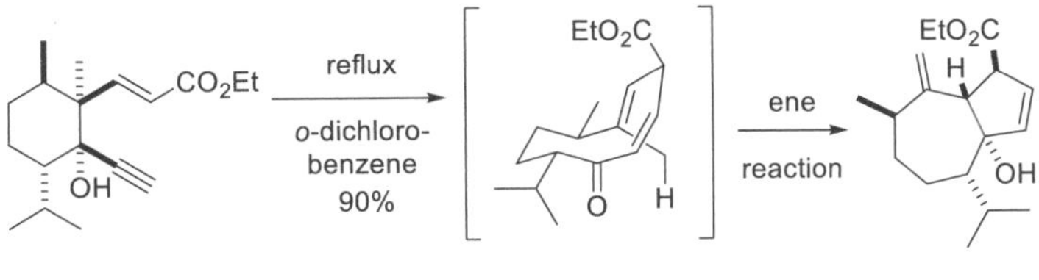
CO2Et
OH
reflux
o-dichloro-
benzene
90%
EtO2C
O
H
ene
reaction
EtO2C
H
OH

Example 3[4]

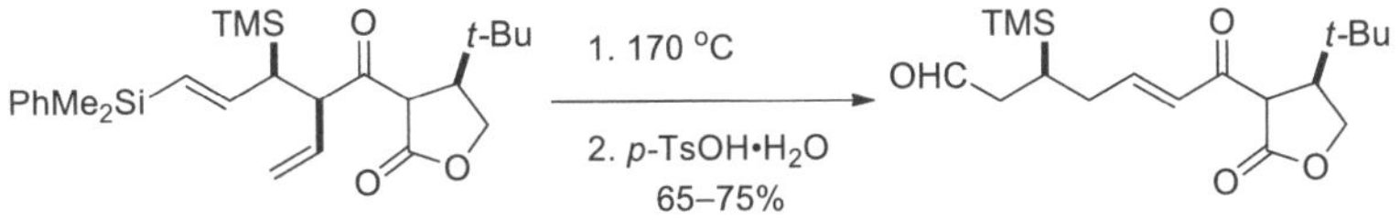
TMS
O
t-Bu
PhMe2Si
O
O
1. 170 °C
2. p-TsOH•H2O
65–75%
TMS
OHC
O
t-Bu
O
O

Example 4[6]

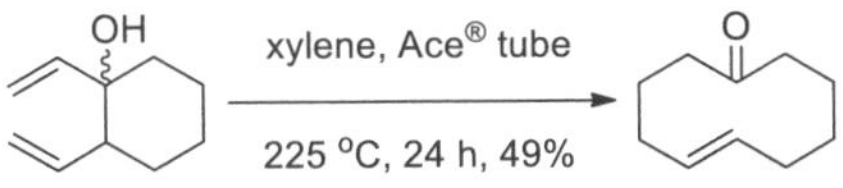
OH
xylene, Ace® tube
225 °C, 24 h, 49%
O

Example 5[8]

furanogermenone

Example 6, Thermally induced oxy-Cope ring-expansion[8]

References

1. Paquette, L. A. *Angew. Chem. Int. Ed.* **1990**, *29*, 609–626. (Review).
2. Paquette, L. A.; Backhaus, D.; Braun, R. *J. Am. Chem. Soc.* **1996**, *118*, 11990–11991.
3. Srinivasan, R.; Rajagopalan, K. *Tetrahedron Lett.* **1998**, *39*, 4133–4136.
4. Schneider, C.; Rehfeuter, M. *Chem. Eur. J.* **1999**, *5*, 2850–2858.
5. Schneider, C. *Synlett* **2001**, 1079–1091. (Review on siloxy-Cope rearrangement).
6. DiMartino, G.; Hursthouse, M. B.; Light, M. E.; Percy, J. M.; Spencer, N. S.; Tolley, M. *Org. Biomol. Chem.* **2003**, *1*, 4423–4434.
7. Mullins, R. J.; McCracken, K. W. *Cope and Related Rearrangements*. In *Name Reactions for Homologations-Part II*; Li, J. J., Ed.; Wiley: Hoboken, NJ, **2009**, pp 88–135. (Review).
8. Anagnostaki, E. E.; Zografos, A. L. *Org. Lett.* **2013**, *15*, 152–155.
9. Massaro, N. P.; Stevens, J. C.; Chatterji, A.; Sharma, I. *Org. Lett.* **2018**, *20*, 7585–7589.
10. Tang, Q.; Fu, K.; Ruan, P.; Dong, S.; Su, Z.; Liu, X.; Feng, X. *Angew. Chem. Int. Ed.* **2019**, *58*, 11846–11851.
11. Emmetiere, F.; Grenning, A. J. *Org. Lett.* **2020**, *22*, 842–847.

Siloxy-Cope Rearrangement

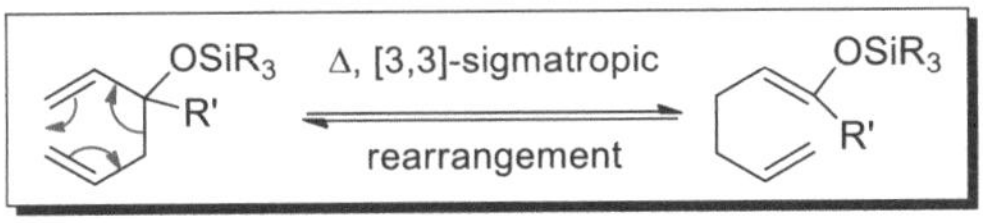

Example 1[1]

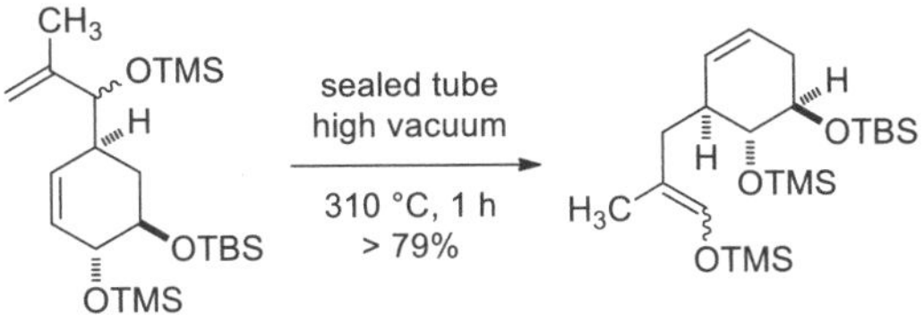

Example 2[2]

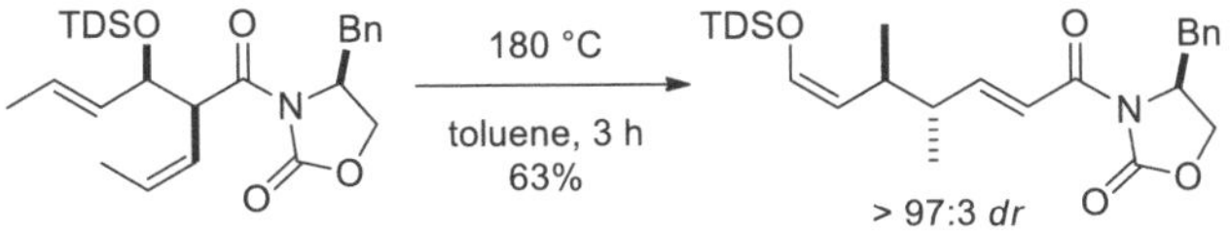

TDS = thexyldimethylsilyl

Example 3[3]

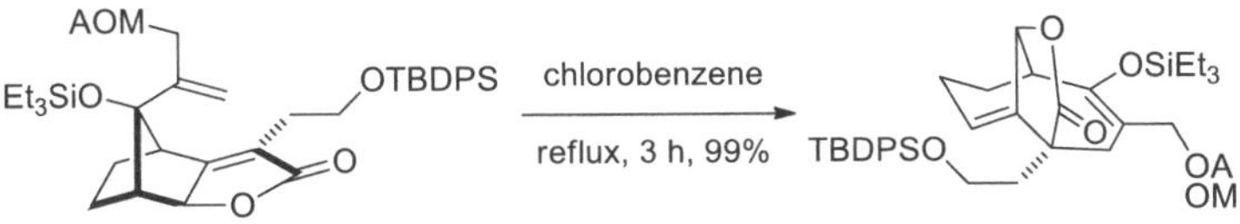

AOM = *p*-Anisyloxymethyl = *p*-MeOC₆H₄OCH₂-

Example 4[4]

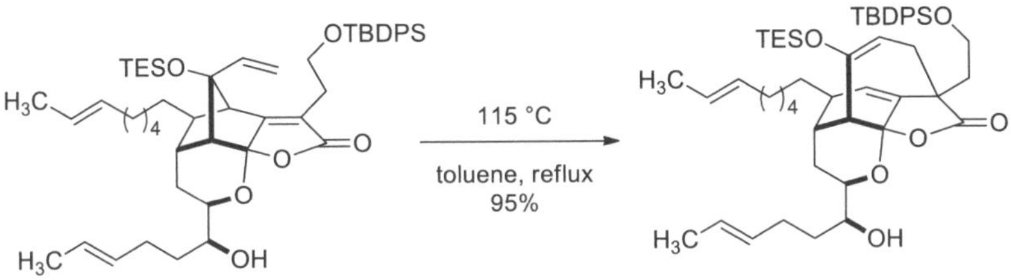

Example 5, Tandem aldol reaction/siloxy-Cope rearrangement[6]

References

1. Askin, D.; Angst, C.; Danishefsky, D. J. *J. Org. Chem.* **1987,** *52*, 622–635.
2. Schneider, C. *Eur. J. Org. Chem.* **1998,** 1661–1663.
3. Clive, D. L. J.; Sun, S.; Gagliardini, V.; Sano, M. K. *Tetrahedron Lett.* **2000,** *41*, 6259–6263.
4. Bio, M. M.; Leighton, J. L. *J. Org. Chem.* **2003,** *68*, 1693–1700.
5. Mullins, R. J.; McCracken, K. W. *Cope and Related Rearrangements*. In *Name Reactions for Homologations-Part II*; Li, J. J., Ed.; Wiley: Hoboken, NJ, **2009,** pp 88–135. (Review).
6. Davies, H. M. L.; Lian, Y. *Acc. Chem. Res.* **2012,** *45*, 923–935. (Review).

Corey–Bakshi–Shibata (CBS) Reduction

The CBS (Corey–Bakshi–Shibata) reagent is a chiral catalyst derived from proline. Also known as Corey's oxazaborolidine, it is used in enantioselective borane reduction of ketones, asymmetric Diels–Alder reactions and [3 + 2] cycloadditions.

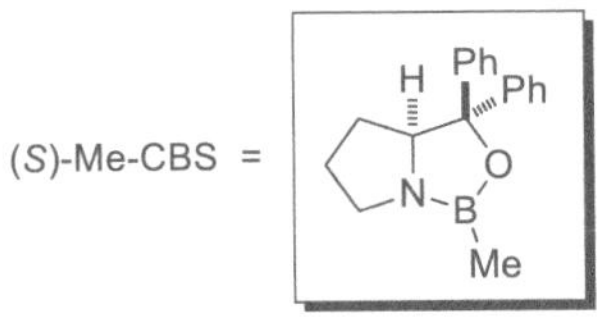
(S)-Me-CBS =

Preparation[1,3]

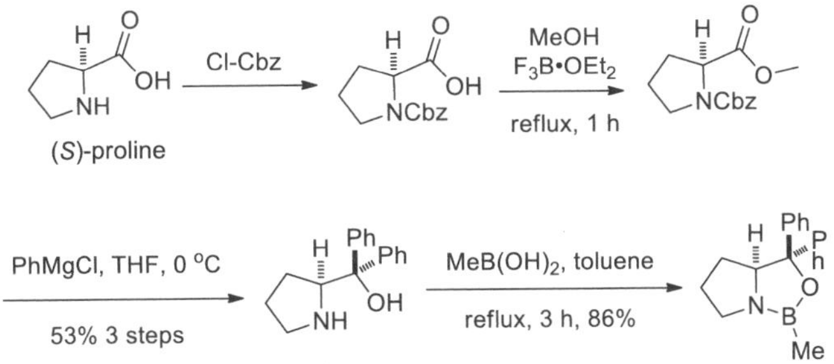
(S)-proline
Cl-Cbz
MeOH
$F_3B{\bullet}OEt_2$
reflux, 1 h
PhMgCl, THF, 0 °C
53% 3 steps
$MeB(OH)_2$, toluene
reflux, 3 h, 86%

Example 1[6]

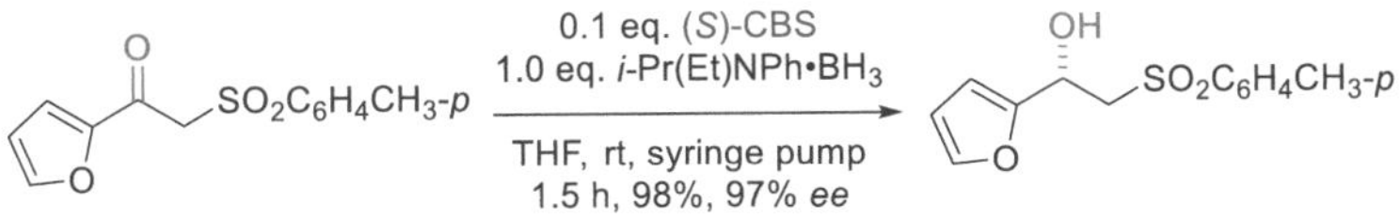
0.1 eq. (S)-CBS
1.0 eq. i-Pr(Et)NPh•BH_3
THF, rt, syringe pump
1.5 h, 98%, 97% ee

Example 2[9]

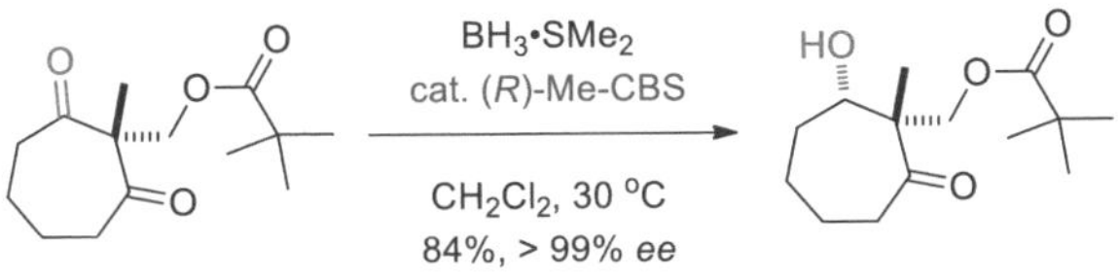
$BH_3{\bullet}SMe_2$
cat. (R)-Me-CBS
CH_2Cl_2, 30 °C
84%, > 99% ee

J. J. Li, *Name Reactions*, https://doi.org/10.1007/978-3-030-50865-4_34

The mechanism and catalytic cycle:[1,3]

Example 3[11]

Example 4, Asymmetric [3 + 2]-cycloaddition[10]

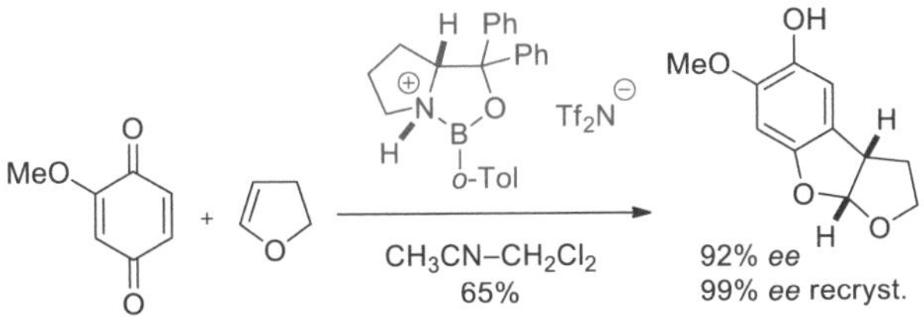

Example 5[13]

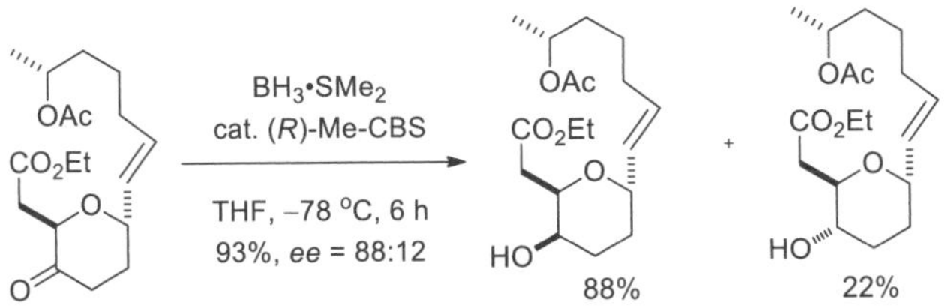

Example 6, Application in total synthesis of alkaloid[14]

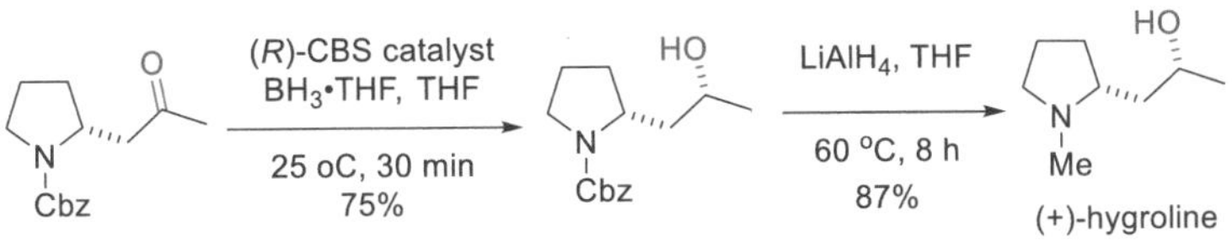

Example 7, CBS reduction in flow microreactor system[15]

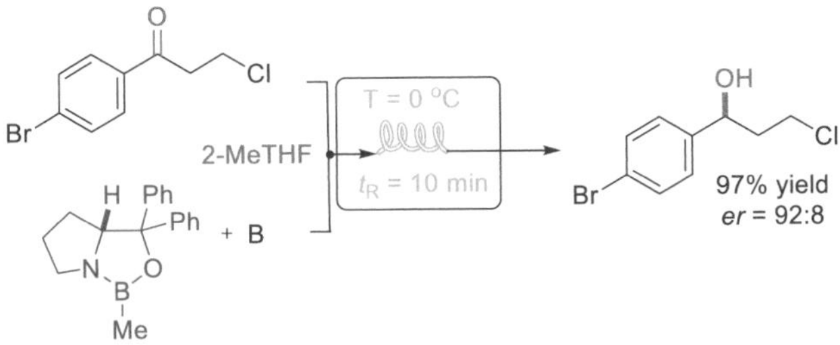

References

1.   (a) Corey, E. J.; Bakshi, R. K.; Shibata, S. *J. Am. Chem. Soc.* **1987**, *109*, 5551–5553.
     (b) Corey, E. J.; Bakshi, R. K.; Shibata, S.; Chen, C.-P.; Singh, V. K. *J. Am. Chem.*
     Soc. **1987**, *109*, 7925–7926. (c) Corey, E. J.; Shibata, S.; Bakshi, R. K. *J. Org. Chem.*
     1988, *53*, 2861–2863.

2. Reviews: (a) Corey, E. J. *Pure Appl. Chem.* **1990,** *62*, 1209–1216. (b) Wallbaum, S.; Martens, J. *Tetrahedron: Asymm.* **1992,** *3*, 1475–1504. (c) Singh, V. K. *Synthesis* **1992,** 605–617. (d) Deloux, L.; Srebnik, M. *Chem. Rev.* **1993,** *93*, 763–784. (e) Taraba, M.; Palecek, J. *Chem. Listy* **1997,** *91*, 9–22. (f) Corey, E. J.; Helal, C. J. *Angew. Chem. Int. Ed.* **1998,** *37*, 1986–2012. (g) Corey, E. J. *Angew. Chem. Int. Ed.* **2002,** *41*, 1650–1667. (h) Itsuno, S. *Org. React.* **1998,** *52*, 395–576. (i) Cho, B. T. *Aldrichimica Acta* **2002,** *35*, 3–16. (j) Glushkov, V. A.; Tolstikov, A. G. *Russ. Chem. Rev.* **2004,** *73*, 581–608. (k) Cho, B .T. *Tetrahedron* **2006,** *62*, 7621–7643.

3. (a) Mathre, D. J.; Thompson, A. S.; Douglas, A. W.; Hoogsteen, K.; Carroll, J. D.; Corley, E. G.; Grabowski, E. J. J. *J. Org. Chem.* **1993,** *58*, 2880–2888. (b) Xavier, L. C.; Mohan, J. J.; Mathre, D. J.; Thompson, A. S.; Carroll, J. D.; Corley, E. G.; Desmond, R. *Org. Synth.* **1997,** *74*, 50–71.

4. Corey, E. J.; Helal, C. J. *Tetrahedron Lett.* **1996,** *37*, 4837–4840.

5. Clark, W. M.; Tickner-Eldridge, A. M.; Huang, G. K.; Pridgen, L. N.; Olsen, M. A.; Mills, R. J.; Lantos, I.; Baine, N. H. *J. Am. Chem. Soc.* **1998,** *120*, 4550–4551.

6. Cho, B. T.; Kim, D. J. *Tetrahedron: Asymmetry* **2001,** *12*, 2043–2047.

7. Price, M. D.; Sui, J. K.; Kurth, M. J.; Schore, N. E. *J. Org. Chem.* **2002,** *67*, 8086–8089.

8. Degni, S.; Wilen, C.-E.; Rosling, A. *Tetrahedron: Asymmetry* **2004,** *15*, 1495–1499.

9. Watanabe, H.; Iwamoto, M.; Nakada, M. *J. Org. Chem.* **2005,** *70*, 4652–4658.

10. Zhou, G.; Corey, E. J. *J. Am. Chem. Soc.* **2005,** *127*, 11958–11959.

11. Yeung, Y.-Y.; Hong, S.; Corey, E. J. *J. Am. Chem. Soc.* **2006,** *128*, 6310–6311.

12. Patti, A.; Pedotti, S. *Tetrahedron: Asymmetry* **2008,** *19*, 1891–1897.

13. Sridhar, Y.; Srihari, P. *Eur. J. Org. Chem.* **2013,** 578–587.

14. Bhoite, S. P.; Kamble, R. B.; Suryavanshi, G. M. *Tetrahedron Lett.* **2015,** *56,* 4704–4705.

15. De Angelis, S.; De Renzo, M.; Carlucci, C.; Degennaro, L.; Luisi, R. *Org. Biomol. Chem.* **2016,** *14*, 4304–4311.

16. Hughes, D. L. *Org. Process Res. Dev.* **2018,** *22,* 574–584. (Review).

17. Cannon, J. S. *Org. Lett.* **2018,** *20,* 3883–3887.

18. Zhou, Y.-G.; Wong, H. N. C.; Peng, X.-S. *J. Org. Chem.* **2020,** *85,* 967–976.

Corey–Chaykovsky Reaction

The Corey–Chaykovsky reaction entails the reaction of a sulfur ylide, either dimethylsulfoxonium methylide **1** (Corey's ylide) or dimethylsulfonium methylide **2**, with electrophile **3** such as carbonyl, olefin, imine, or thiocarbonyl, to offer **4** as the corresponding epoxide, cyclopropane, aziridine, or thiirane.

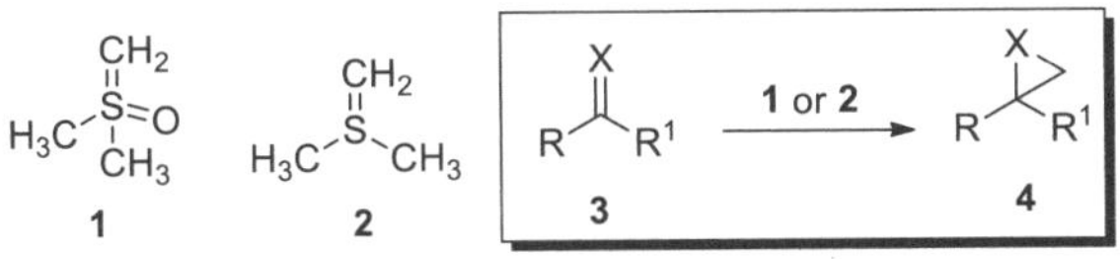

$X = O, CH_2, NR^2, S, CHCOR^3,$
$CHCO_2R^3, CHCONR_2, CHCN$

Preparation[1]

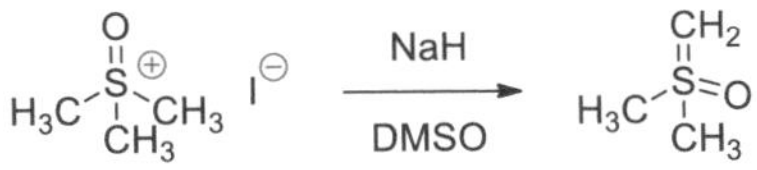

Mechanism[1]

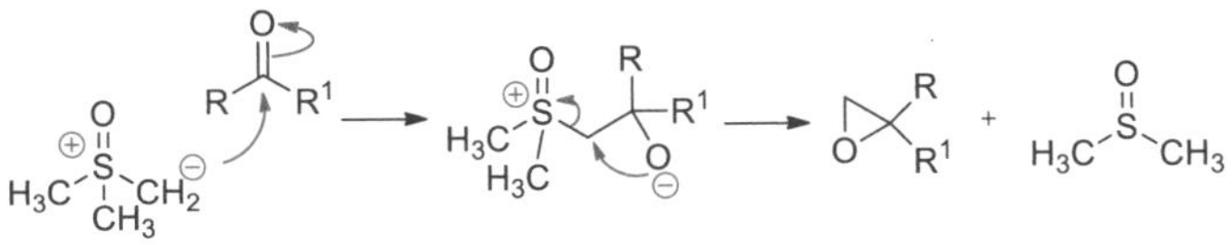

Example 1, Epoxide from ketone[11]

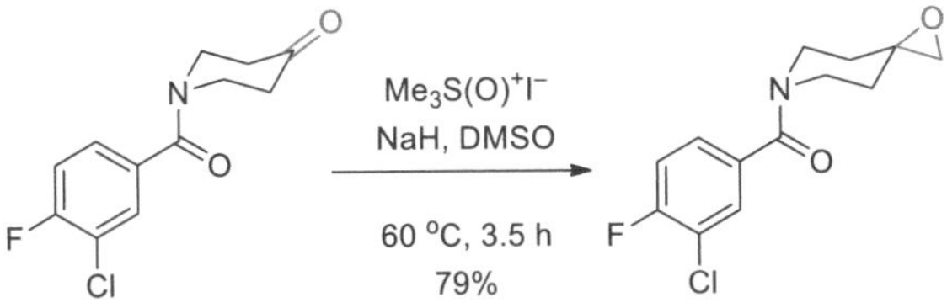

Example 2, Cyclopropane from olefin[9]

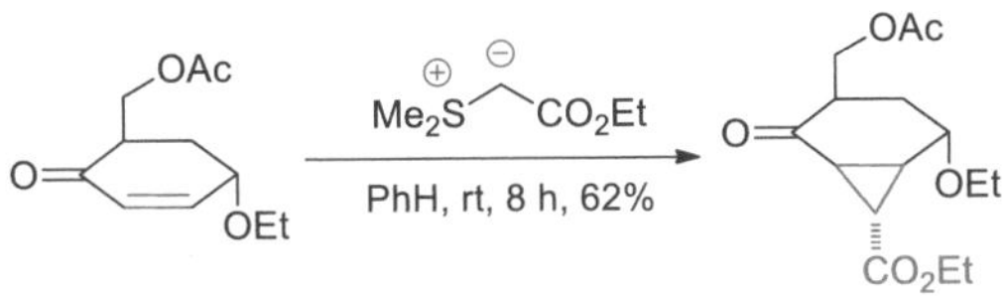

J. J. Li, *Name Reactions*, https://doi.org/10.1007/978-3-030-50865-4_35

Example 3, Aziridine from aza-Corey–Chaykovsky reaction[9]

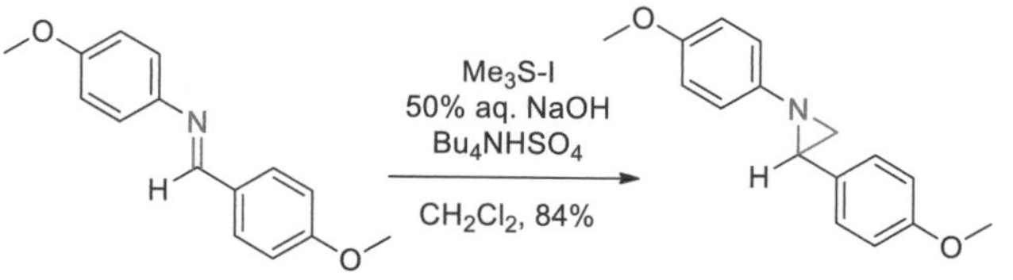

Me₃S-I
50% aq. NaOH
Bu₄NHSO₄

CH₂Cl₂, 84%

Example 4[12]

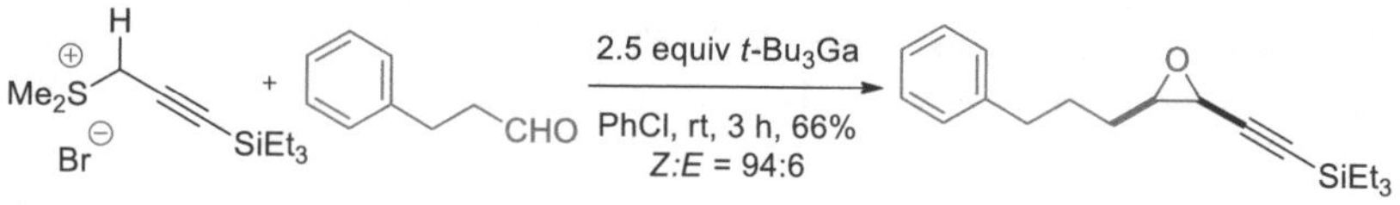

2.5 equiv t-Bu₃Ga

PhCl, rt, 3 h, 66%
$Z:E$ = 94:6

Example 5[13]

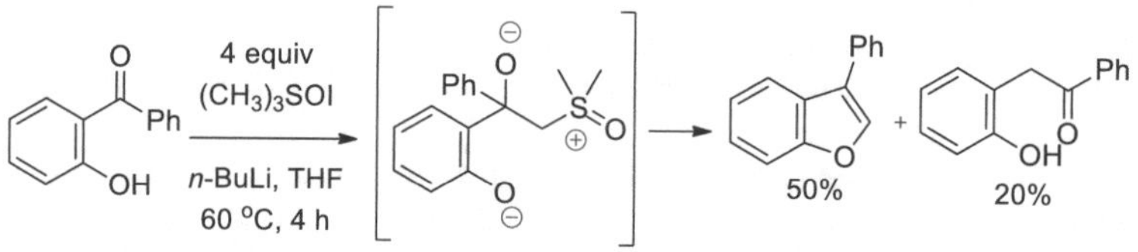

4 equiv
(CH₃)₃SOI

n-BuLi, THF
60 °C, 4 h

50%   20%

Example 6, Cyclopropane from olefin[14]

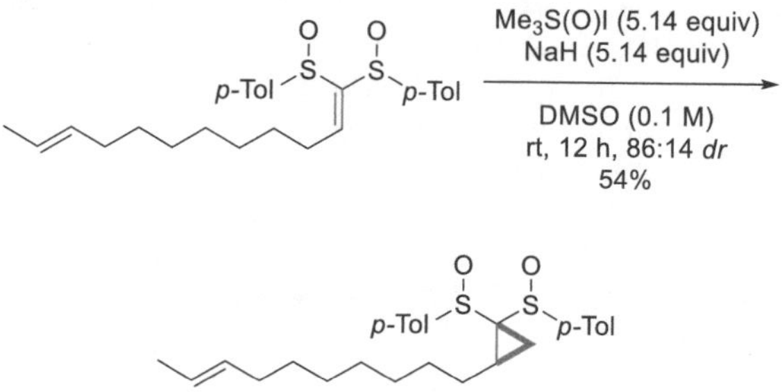

Me₃S(O)I (5.14 equiv)
NaH (5.14 equiv)

DMSO (0.1 M)
rt, 12 h, 86:14 dr
54%

Example 7, Diastereoselective aziridination of ketimino ester[15]

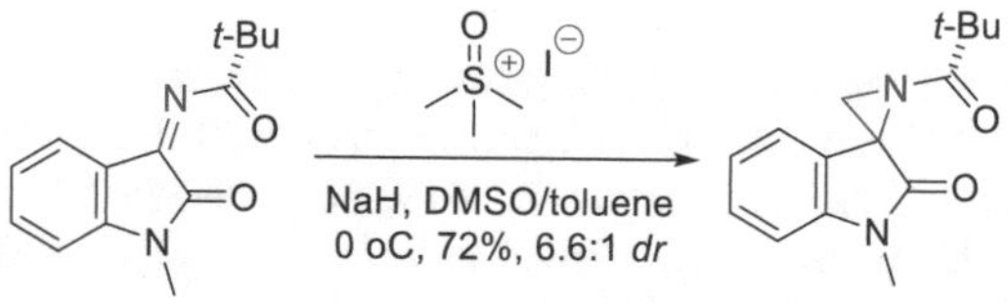

NaH, DMSO/toluene
0 oC, 72%, 6.6:1 dr

Example 8, Corey–Chaykovsky reaction for one-pot access to spirocyclopropyl oxindoles[17]

References

1   (a) Corey, E. J.; Chaykovsky, M. *J. Am. Chem. Soc.* **1962,** *84*, 867–868. (b) Corey, E. J.; Chaykovsky, M. *J. Am. Chem. Soc.* **1962,** *84*, 3782. (c) Corey, E. J.; Chaykovsky, M. *Tetrahedron Lett.* **1963,** 169–171. (d) Corey, E. J.; Chaykovsky, M. *J. Am. Chem. Soc.* **1964,** *86*, 1639–1640. (e) Corey, E. J.; Chaykovsky, M. *J. Am. Chem. Soc.* **1965,** *87*, 1353–1364.

2   Okazaki, R.; Tokitoh, N. In *Encyclopedia of Reagents in Organic Synthesis;* Paquette, L. A., Ed.; Wiley: New York, **1995,** pp 2139–2141. (Review).

3   Ng, J. S.; Liu, C. In *Encyclopedia of Reagents in Organic Synthesis;* Paquette, L. A., Ed.; Wiley: New York, **1995,** pp 2159–2165. (Review).

4   Trost, B. M.; Melvin, L. S., Jr. *Sulfur Ylides;* Academic Press: New York, **1975**. (Review).

5   Block, E. *Reactions of Organosulfur Compounds* Academic Press: New York, **1978**. (Review).

6   Gololobov, Y. G.; Nesmeyanov, A. N. *Tetrahedron* **1987,** *43*, 2609–2651. (Review).

7   Aubé, J. In *Comprehensive Organic Synthesis;* Trost, B. M.; Fleming, I., Ed.; Pergamon: Oxford, **1991,** *Vol. 1*, pp 820–825. (Review).

8   Li, A.-H.; Dai, L.-X.; Aggarwal, V. K. *Chem. Rev.* **1997,** *97*, 2341–2372. (Review).

9   Tewari, R. S.; Awatsthi, A. K.; Awasthi, A. *Synthesis* **1983,** 330–331.

10   Vacher, B.; Bonnaud, B. Funes, P.; Jubault, N.; Koek, W.; Assie, M.-B.; Cosi, C.; Kleven, M. *J. Med. Chem.* **1999,** *42*, 1648–1660.

11   Li, J. J. *Corey–Chaykovsky Reaction.* In *Name Reactions in Heterocyclic Chemistry*; Li, J. J., Ed.; Wiley: Hoboken, NJ, **2005,** pp 1–14. (Review).

12   Nishimura, Y.; Shiraishi, T.; Yamaguchi, M. *Tetrahedron Lett.* **2008,** *49*, 3492–3495.

13   Chittimalla, S. K.; Chang, T.-C.; Liu, T.-C.; Hsieh, H.-P.; Liao, C.-C. *Tetrahedron* **2008,** *64*, 2586–2595.

14   Palko, J. W.; Buist, P. H.; Manthorpe, J. M. *Tetrahedron: Asymmetry* **2013,** *24*, 165–168.

15   Marsini, M. A.; Reeves, J. T.; Desrosiers, J.-N.; Herbage, M. A.; Savoie, J.; Li, Z.; Fandrick, K. R.; Sader, C. A.; McKibben, B.; Gao, D. A.; et al. *Org. Lett.* **2015,** *17*, 5614–5617.

16   Yarmoliuk, D. V.; Serhiichuk, D.; Smyrnov, V.; Tymtsunik, A. V.; Hryshchuk, O. V.; Kuchkovska, Y.; Grygorenko, O. O. *Tetrahedron Lett.* **2018,** *59*, 4611–4615.

17   Hajra, S.; Roy, S.; Saleh, S. A. *Org. Lett.* **2018,** *20*, 4540–54544.

18   Zhang, Z.-W.; Li, H.-B.; Li, J.; Wang, C.-C.; Feng, J.; Yang, Y.-H.; Liu, S. *J. Org. Chem.* **2020,** *85*, 537–547.

Corey–Fuchs Reaction

One-carbon homologation of an aldehyde to dibromoolefin, which is then treated with *n*-BuLi to produce a terminal alkyne.

Wittig reaction

$$Br_2 + Zn \longrightarrow ZnBr_2$$

Example 1[3]

© Springer Nature Switzerland AG 2021
J. J. Li, *Name Reactions*, https://doi.org/10.1007/978-3-030-50865-4_36

Example 2[7]

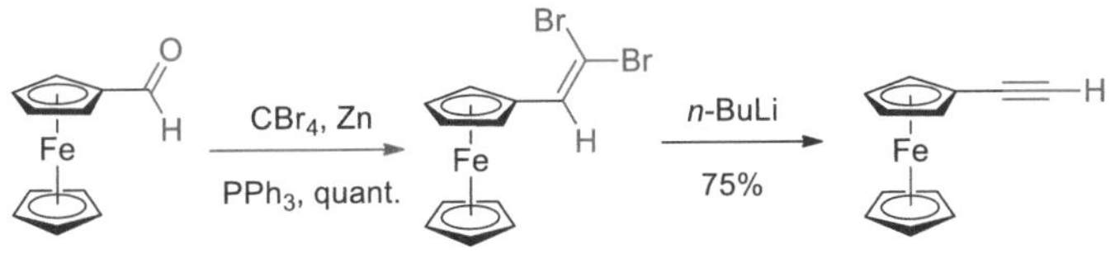

Example 3[8]

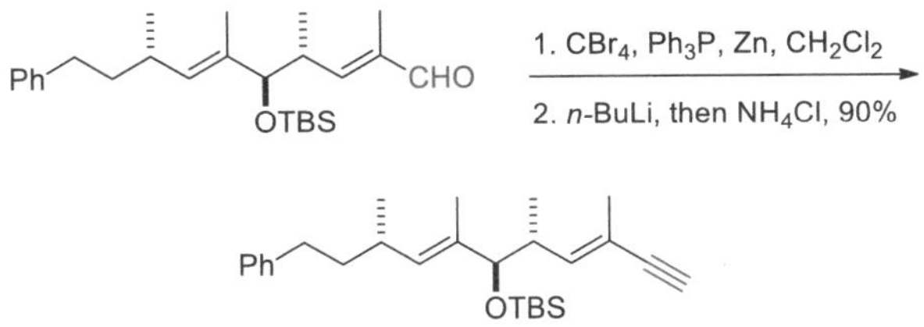

Example 4[10]

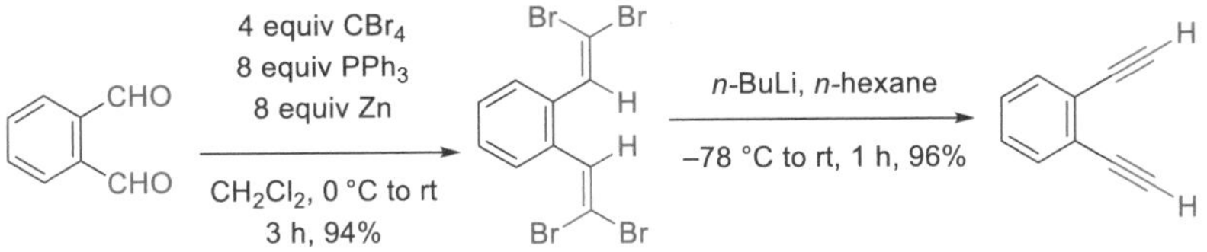

Example 5[12]

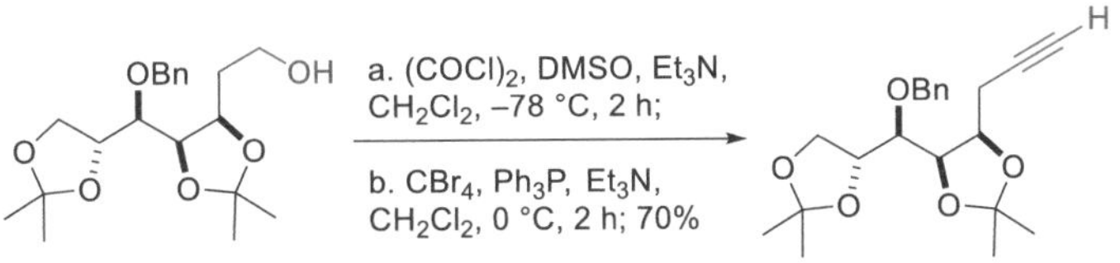

Example 6[12]

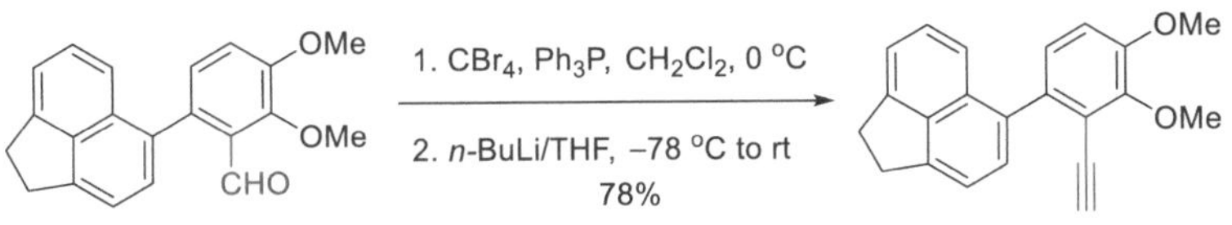

Example 7, Prepation of terminal alkyne as a Sonogashira precursor[14]

Example 8, Large-scale process Corey–Fuchs reaction[16]

References

1   Corey, E. J.; Fuchs, P. L. *Tetrahedron Lett.* **1972**, *13*, 3769–3772. Phil Fuchs is a professor at Purdue University.

2   For the synthesis of 1-bromalkynes see Grandjean, D.; Pale, P.; Chuche, J. *Tetrahedron Lett.* **1994**, *35*, 3529–3530.

3   Gilbert, A. M.; Miller, R.; Wulff, W. D. *Tetrahedron* **1999**, *55*, 1607–1630.

4   Muller, T. J. J. *Tetrahedron Lett.* **1999**, *40*, 6563–6566.

5   Serrat, X.; Cabarrocas, G.; Rafel, S.; Ventura, M.; Linden, A.; Villalgordo, J. M. *Tetrahedron: Asymmetry* **1999**, *10*, 3417–3430.

6   Okamura, W. H.; Zhu, G.-D.; Hill, D. K.; Thomas, R. J.; Ringe, K.; Borchardt, D. B.; Norman, A. W.; Mueller, L. J. *J. Org. Chem.* **2002**, *67*, 1637–1650.

7   Tsuboya, N.; Hamasaki, R.; Ito, M.; Mitsuishi, M. *J. Mater. Chem.* **2003**, *13*, 511–513

8   Zeng, X.; Zeng, F.; Negishi, E.-i. *Org. Lett.* **2004**, *6*, 3245–3248.

9   Quéron, E.; Lett, R. *Tetrahedron Lett.* **2004**, *45*, 4527–4531.

10 Sahu, B.; Muruganantham, R.; Namboothiri, I. N. N. *Eur. J. Org. Chem.* **2007**, 2477–2489.

11 Han, X. *Corey–Fuchs Reaction.* In *Name Reactions for Homologations-Part I*; Li, J. J., Ed.; Wiley: Hoboken, NJ, **2009**, pp 393–403. (Review).

12 Pradhan, T. K.; Lin, C. C.; Mong, K. K. T. *Synlett* **2013**, *24*, 219–222.

13 Thomson, P. F.; Parrish, D.; Pradhan, P.; Lakshman, M. K. *J. Org. Chem.* **2015**, *80*, 7435–7446

14 Dumpala, M.; Theegala, S.; Palakodety, R. K. *Tetrahedron Lett.* **2017**, *58*, 1273–1275.

15 Martynow, J.; Hanselmann, R.; Duffy, E.; Bhattacharjee, A. *Org. Process Res. Dev.* **2019**, *23*, 1026–1033.

Curtius Rearrangement

Alkyl-, vinyl-, and aryl-substituted acyl azides undergo thermal 1,2-carbon-to-nitrogen migration with extrusion of dinitrogen — the Curtius rearrangement — producing isocyantes. Reaction of the isocyanate products with nucleophiles, often *in situ*, provides carbamates, ureas, and other *N*-acyl derivatives. Alternatively, hydrolysis of the isocyanates leads to primary amines.

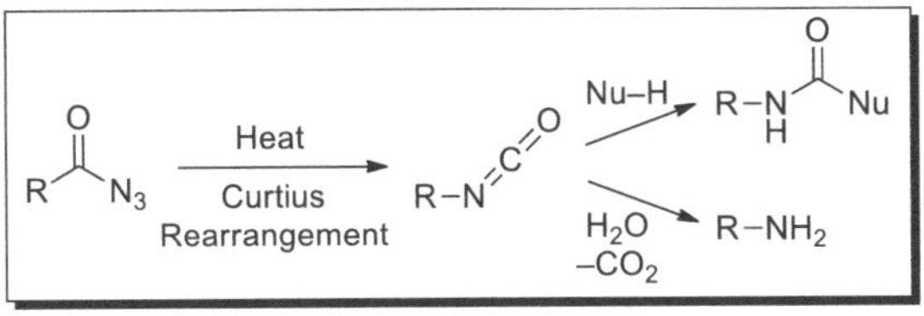

The thermal rearrangement:

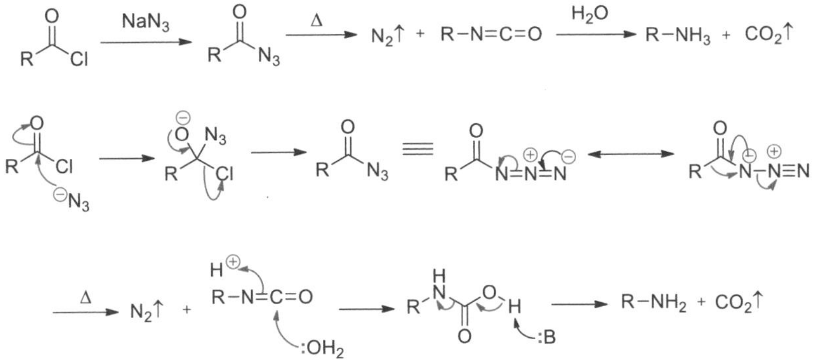

isocyanate intermediate

The photochemical rearrangement:

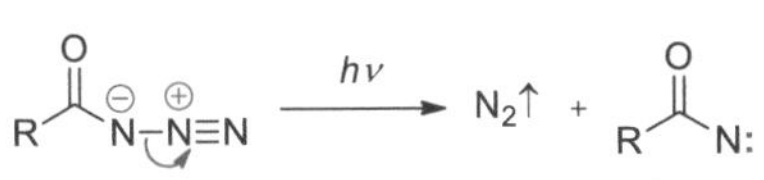

nitrene

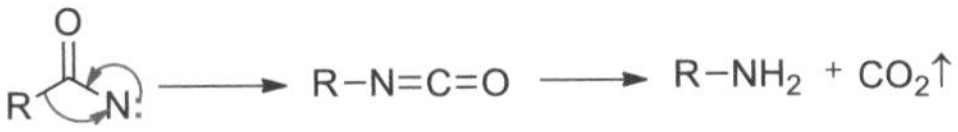

J. J. Li, *Name Reactions*, https://doi.org/10.1007/978-3-030-50865-4_37

Example 1, The Shioiri–Ninomiya–Yamada modification[2]

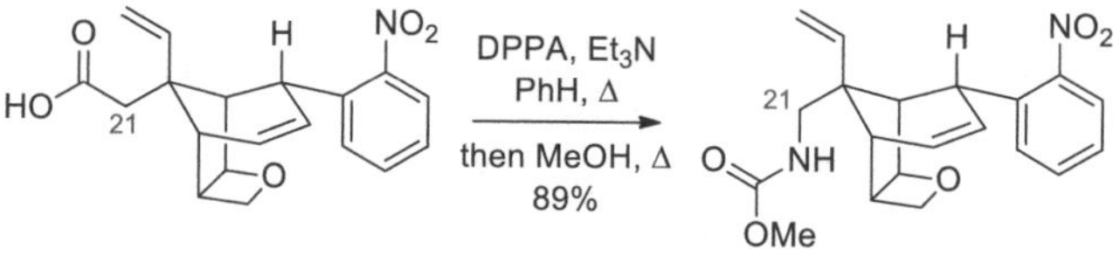

DPPA = diphenylphosphoryl azide

Example 2[3]

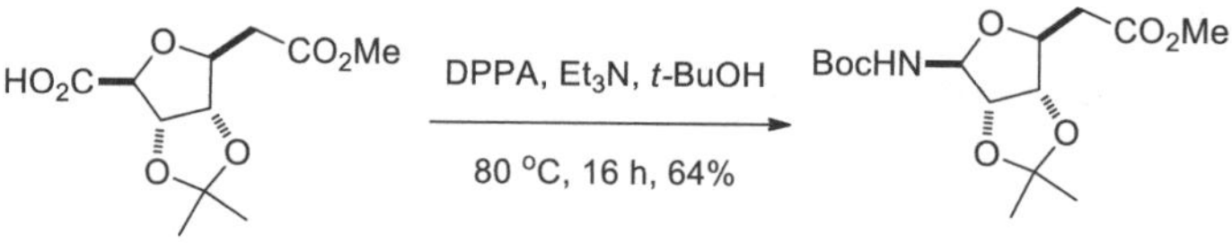

Example 3[4]

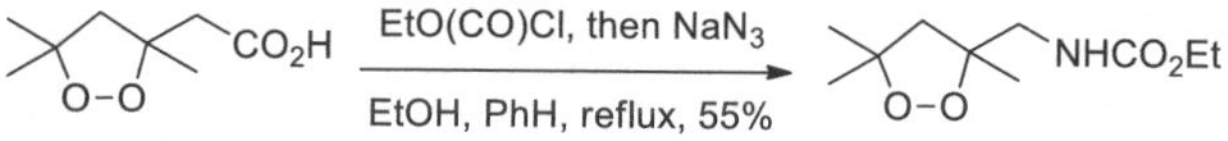

Example 4, The Weinstock variant of the Curtius rearrangement[6]

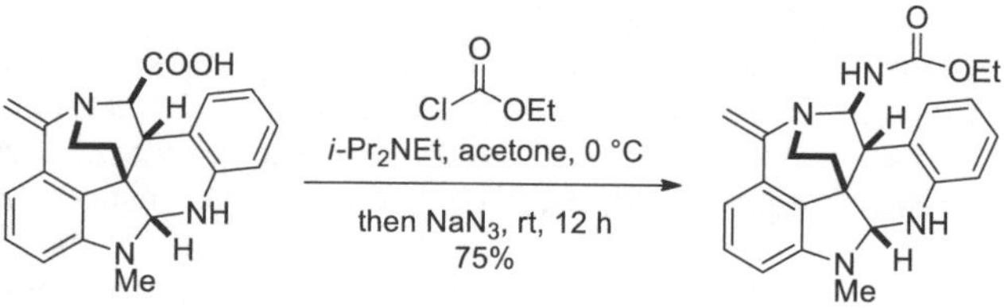

Example 5[7]

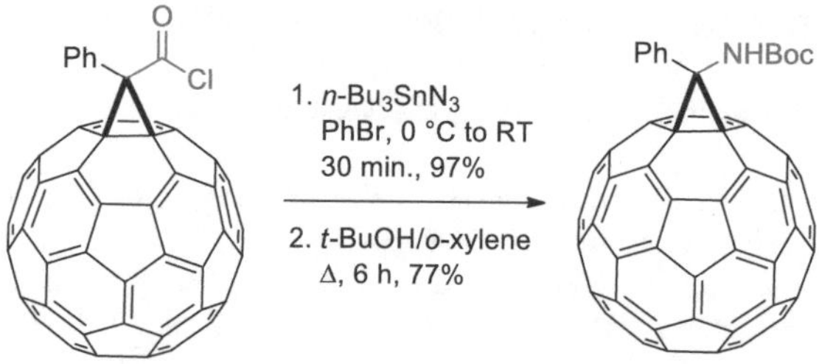

Example 6, The Lebel modification[8]

Example 7, Utility in total synthesis[9]

References

1. Curtius, T. *Ber.* **1890**, *23*, 3033–3041. Theodor Curtius (1857–1928) was born in Duisburg, Germany. He studied music before switching to chemistry under Bunsen, Kolbe, and von Baeyer before succeeding Victor Meyer as a Professor of Chemistry at Heidelberg. He discovered diazoacetic ester, hydrazine, pyrazoline derivatives, and many nitrogen-heterocycles. Curtius also sang in concerts and composed music.

2. Ng, F. W.; Lin, H.; Danishefsky, S. J. *J. Am. Chem. Soc.* **2002**, *124*, 9812–9824.

3. van Well, R. M.; Overkleeft, H. S.; van Boom, J. H.; Coop, A.; Wang, J. B.; Wang, H.; van der Marel, G. A.; Overhand, M. *Eur. J. Org. Chem.* **2003**, 1704–1710.

4. Dussault, P. H.; Xu, C. *Tetrahedron Lett.* **2004**, *45*, 7455–7457.

5. Holt, J.; Andreassen, T.; Bakke, J. M. *J. Heterocycl. Chem.* **2005**, *42*, 259–264.

6. Crawley, S. L.; Funk, R. L. *Org. Lett.* **2006**, *8*, 3995–3998.

7. Tada, T.; Ishida, Y.; Saigo, K. *Synlett* **2007**, 235–238.

8. Sawada, D.; Sasayama, S.; Takahashi, H.; Ikegami, S. *Eur. J. Org. Chem.* **2007**, 1064–1068.

9. Rojas, C. M. *Curtius Rearrangements*. In *Name Reactions for Homologations-Part II*; Li, J. J., Ed.; Wiley: Hoboken, NJ, **2009**, pp 136–163. (Review).

10. Koza, G.; Keskin, S.; Özer, M. S.; Cengiz, B. *Tetrahedron* **2013**, *69*, 395–409.

11. Ghosh, A. K.; Sarkar, A.; Brindisi, M. *Org. Biomol. Chem.* **2018**, *16*, 2006–2027. (Review).

12. Ghosh, A. K.; Brindisi, M.; Sarkar, A. *ChemMedChem* **2018**, *13*, 2351–2373. (Review).

13. Hartrampf, N.; Winter, N.; Pupo, G.; Stoltz, B. M.; Trauner, D. *J. Am. Chem. Soc.* **2018**, *140*, 8675–8680.

Dakin Oxidation

Oxidation of aryl aldehydes or aryl ketones to phenols using basic hydrogen peroxide conditions. *Cf.* A variant of the Baeyer–Villiger oxidation.

Example 1[6]

Example 2[7]

Example 3, Improved solvent-free Dakin oxidation protocol[9]

© Springer Nature Switzerland AG 2021

J. J. Li, *Name Reactions*, https://doi.org/10.1007/978-3-030-50865-4_38

Example 4[10]

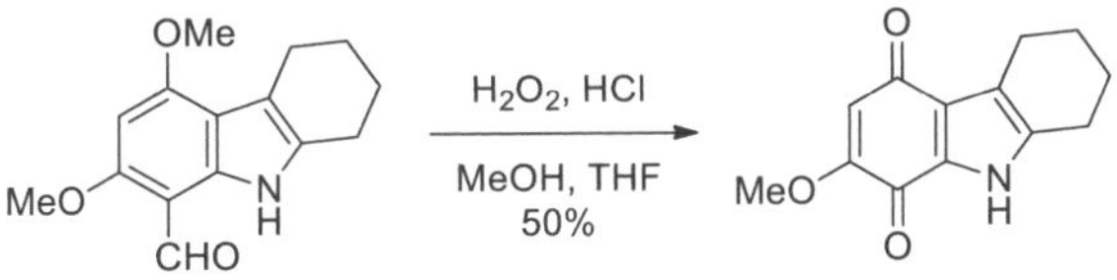

Example 5, Aerobic organocatalytic oxidation: flavin catalyst turnover by Hantsch's ester[11]

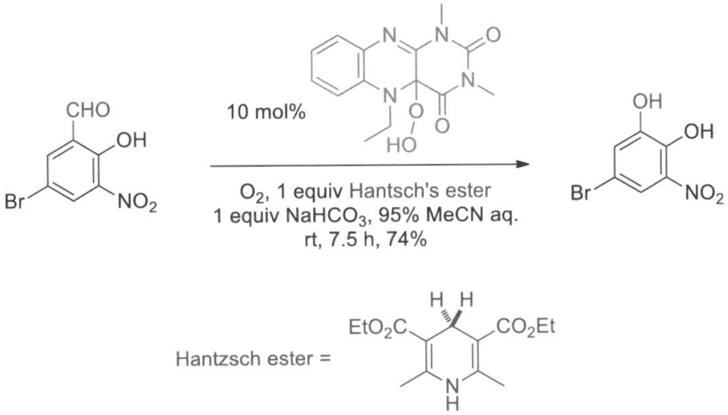

Hantzsch ester =

Example 6, One-pot synthesis of tryptanthrin[12]

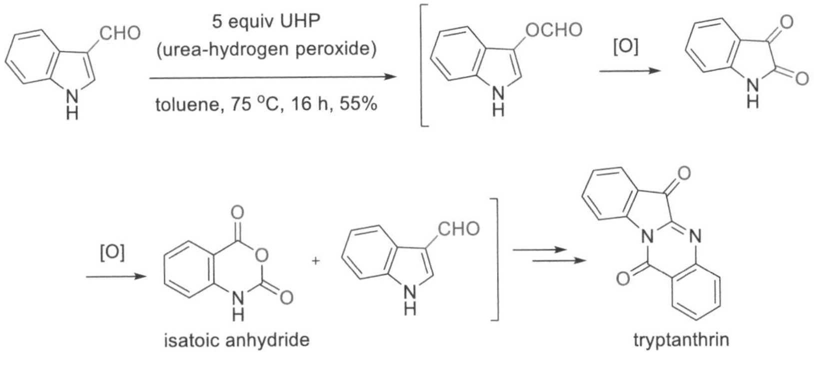

isatoic anhydride

tryptanthrin

Example 7, Employing "Water Extract of Rice Straw Ash" (WERSA) as catalysts (for real!)[13]

References

1. Dakin, H. D. *Am. Chem. J.* **1909**, *42*, 477–498. Henry D. Dakin (1880–1952) was born in London, England. During WWI, he invented his hypochlorite solution (Dakin's solution), which became a popular antiseptic for the treatment of wounds. After the Great War, he emmigrated to New York, where he investigated the B vitamins.
2. Hocking, M. B.; Bhandari, K.; Shell, B.; Smyth, T. A. *J. Org. Chem.* **1982**, *47*, 4208–4215.
3. Matsumoto, M.; Kobayashi, H.; Hotta, Y. *J. Org. Chem.* **1984**, *49*, 4740–4741.
4. Zhu, J.; Beugelmans, R.; Bigot, A.; Singh, G. P.; Bois-Choussy, M. *Tetrahedron Lett.* **1993**, *34*, 7401–7404.
5. Guzmán, J. A.; Mendoza, V.; García, E.; Garibay, C. F.; Olivares, L. Z.; Maldonado, L. A. *Synth. Commun.* **1995**, *25*, 2121–2133.
6. Jung, M. E.; Lazarova, T. I. *J. Org. Chem.* **1997**, *62*, 1553–1555.
7. Varma, R. S.; Naicker, K. P. *Org. Lett.* **1999**, *1*, 189–191.
8. Lawrence, N. J.; Rennison, D.; Woo, M.; McGown, A. T.; Hadfield, J. A. *Bioorg. Med. Chem. Lett.* **2001**, *11*, 51–54.
9. Teixeira da Silva, E.; Camara, C. A.; Antunes, O. A. C.; Barreiro, E. J.; Fraga, C. A. M. *Synth. Commun.* **2008**, *38*, 784–788.
10. Alamgir, M.; Mitchell, P. S. R.; Bowyer, P. K.; Kumar, N.; Black, D. St. C. *Tetrahedron* **2008**, *64*, 7136–7142.
11. Chen, S.; Foss, F. W. *Org. Lett.* **2012**, *14*, 5150–5153.
12. Abe, T.; Itoh, T.; Choshi, T.; Hibino, S.; Ishikura, M. *Tetrahedron Lett.* **2014**, *55*, 5268–5270.
13. Saikia, B.; Borah, P. *RSC Adv.* **2015**, *5*, 105583–105586.
14. Pak, Y. L.; Park, S. J.; Song, G.; Yim, Y.; Kang, Hyuk; K., Hwan M.; Bouffard, J.; Yoon, J. *Anal. Chem.* **2018**, *90*, 12937–12943.
15. Gao, D.; Jin, F.; Lee, J. K.; Zare, R. N. *Chem. Sci.* **2019**, *10*, 10974–10978.

Dakin–West Reaction

The direct conversion of an α-amino acid into the corresponding α-acetylamino-alkyl methyl ketone, *via* oxazolone (azalactone) intermediates. The reaction proceeds in the presence of acetic anhydride and a base, such as pyridine, with the evolution of CO_2. The reaction racemizes that chiral center as shown.

a.   Mechanism proposed by Dakin and West[1]

oxazolone (azalactone) intermediate

b.   Mechanism proposed by Levene and Steiger[2]

© Springer Nature Switzerland AG 2021
J. J. Li, *Name Reactions*, https://doi.org/10.1007/978-3-030-50865-4_39

c.   Mechanism proposed by Wiley for the reaction of *N*-acetyl sarcosine[3]

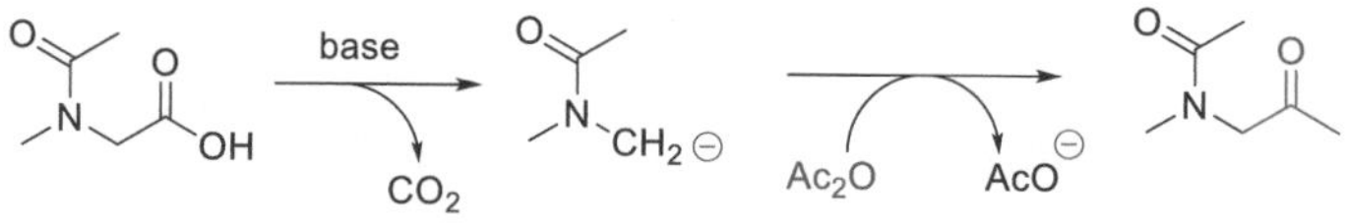

Example 1[5]

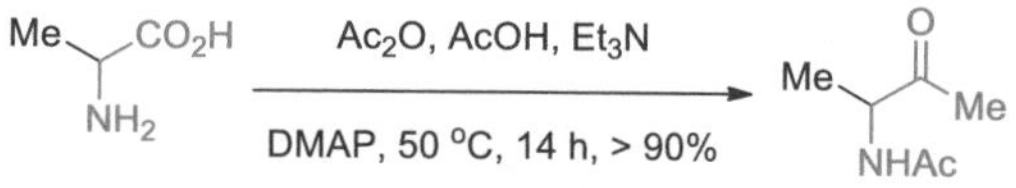

Example 2[7]

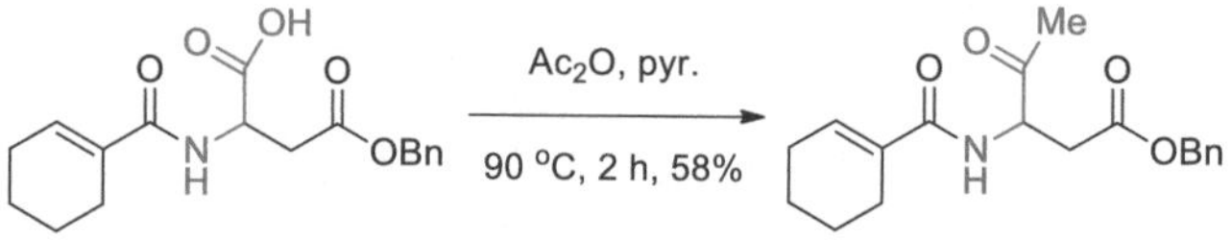

Example 3, A green Dakin–West reaction using the heteropoly acid catalyst, acetonitrile is a reactant[9]

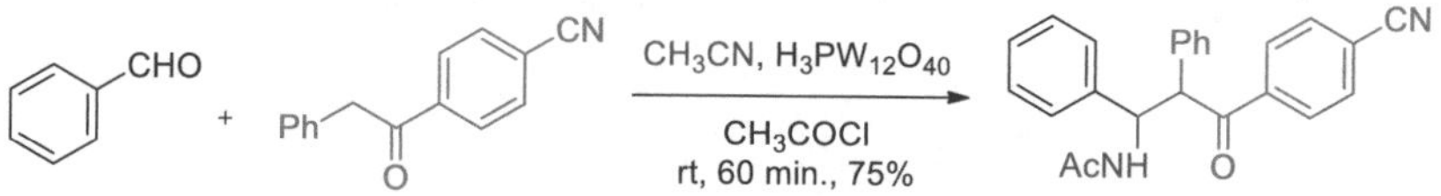

Example 4, A diasteroselective trifluoroacetylation of hghly substituted pyrrolidines[12]

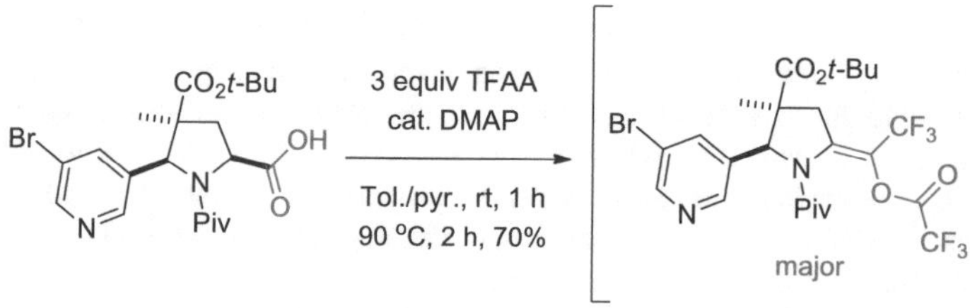

Example 5, Enantioselective Dakin–West reaction[13]

Example 6, Dakin–West reaction to make trifluoromethyl acyloin[14]

References

1. Dakin, H. D.; West, R. *J. Biol. Chem.* **1928**, *78,* 91, 745, and 757. In 1928, Henry Dakin and Rudolf West, a clinician, reported on the reaction of α-amino acids with acetic anhydride to give α-acetamido ketones *via* azalactone intermediates. Interestingly, one year before this paper by Dakin and West, Levene and Steiger had observed both tyrosine and α-phenylananine gave "abnormal" products when acetylated under these conditions.[2,3] Unfortunately, they were slow to identify the products and lost an opportunity to be immortalized by a name reaction.

2. Levene, P. A.; Steiger, R.E. *J. Biol. Chem.* **1928**, *79,* 95–103.

3. Wiley, R. H. *Sci.* **1950**, *79,* 95–103.

4. Buchanan, G. L. *Chem. Soc. Rev.* **1988**, *17,* 91–109. (Review).

5. Kawase, M.; Hirabayashi, M.; Koiwai, H.; Yamamoto, K.; Miyamae, H. *Chem. Commun.* **1998**, 641–642.

6. Fischer, R. W.; Misun, M. *Org. Process Res. Dev.* **2001**, *5,* 581–588.

7. Godfrey, A. G.; Brooks, D. A.; Hay, L. A.; Peters, M.; McCarthy, J. R.; Mitchell, D. *J. Org. Chem.* **2003**, *68,* 2623–2632.

8. Khodaei, M. M.; Khosropour, A. R.; Fattahpour, P. *Tetrahedron Lett.* **2005,** *46*, 2105–2108.
9. Rafiee, E.; Tork, F.; Joshaghani, M. *Bioorg. Med. Chem. Lett.* **2006,** *16*, 1221–1226.
10. Tiwari, A. K.; Kumbhare, R. M.; Agawane, S. B.; Ali, A. Z.; Kumar, K. V. *Bioorg. Med. Chem. Lett.* **2008,** *18*, 4130–4132.
11. Dalla-Vechia, L.; Santos, V. G.; Godoi, M. N.; Cantillo, D.; Kappe, C. O.; Eberlin, M. N.; de Souza, R. O. M. A.; Miranda, L. S. M. *Org. Biomol. Chem.* **2012,** *10*, 9013–9020. (Mechanism).
12. Baumann, M.; Baxendale, I. R. *J. Org. Chem.* **2016,** *81*, 11898–11908.
13. Wende, R. C.; Seitz, A.; Niedek, D.; Schuler, S. M. M.; Hofmann, C.; Becker, J.; Schreiner, P. R. *Angew. Chem. Int. Ed.* **2016,** *55,* 2719–2723.
14. Allison, Brett D.; Mani, Neelakandha S. *ACS Omega* **2017,** *2,* 397–408.
15. Dalla Vechia, L.; de Souza, R. O. M. A.; de Mariz e Miranda, L. S. *Tetrahedron* **2018,** *74*, 4359–4371. (Review).

Darzens Condensation

α,β-Epoxy esters (glycidic esters) from base-catalyzed condensation of α-haloesters with carbonyl compounds.

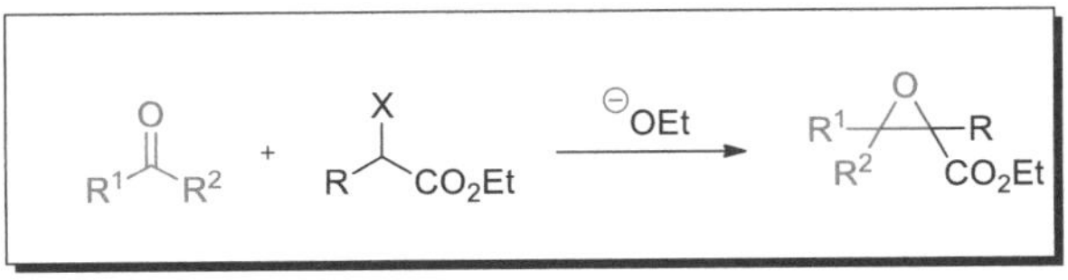

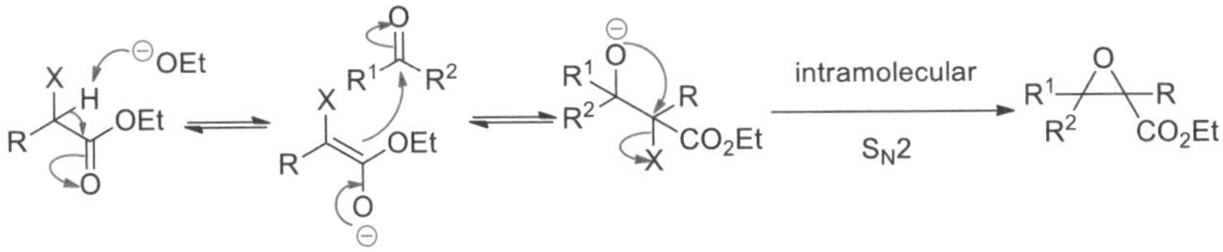

Example 1[4]

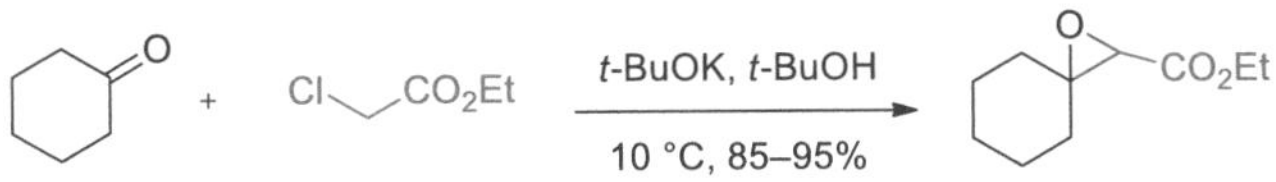

Example 2[6]

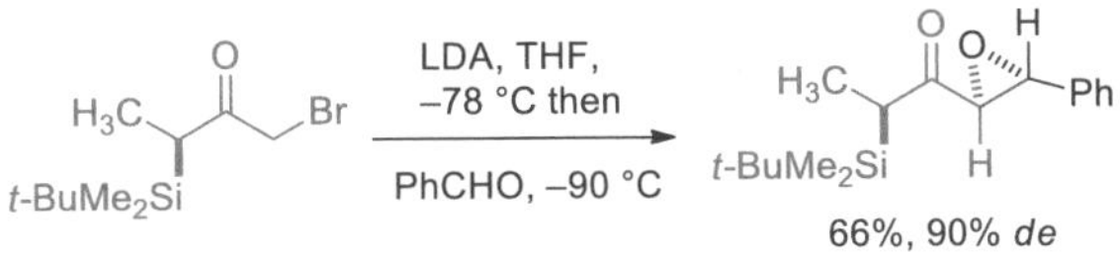

66%, 90% de

Example 3, the phenyl ring substituting for the carbonyl to acidify the protons[10]

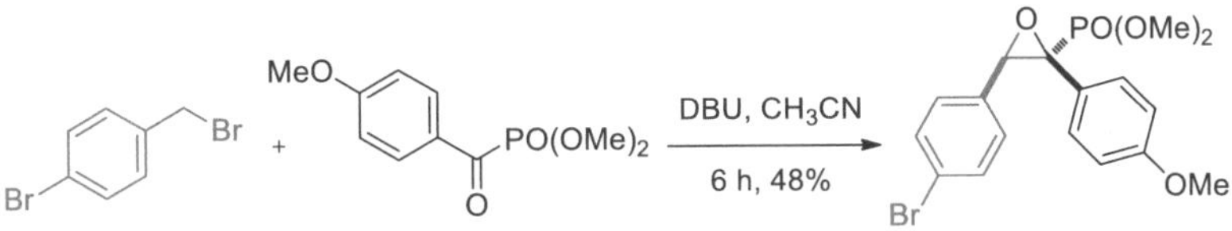

Example 4, A variation[11]

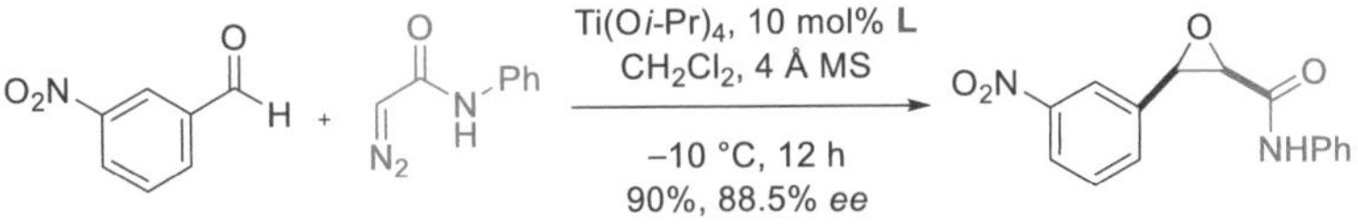

90%, 88.5% ee

© Springer Nature Switzerland AG 2021
J. J. Li, *Name Reactions*, https://doi.org/10.1007/978-3-030-50865-4_40

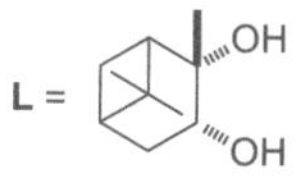

L =

Example 5, Substrate-controlled stereoselective Darzens reaction[12]

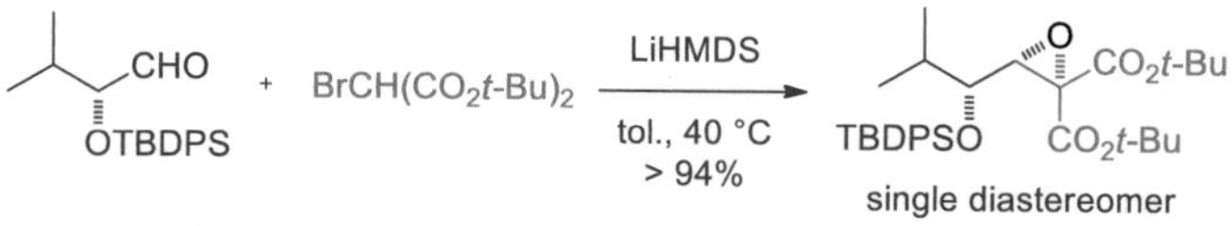

CHO + BrCH(CO$_2$t-Bu)$_2$ → (LiHMDS, tol., 40 °C, > 94%) → single diastereomer

OTBDPS

CO$_2$t-Bu, TBDPSO, CO$_2$t-Bu

Example 6, Asymmetric Darzens reaction between isatin and diazoacetamide[13]

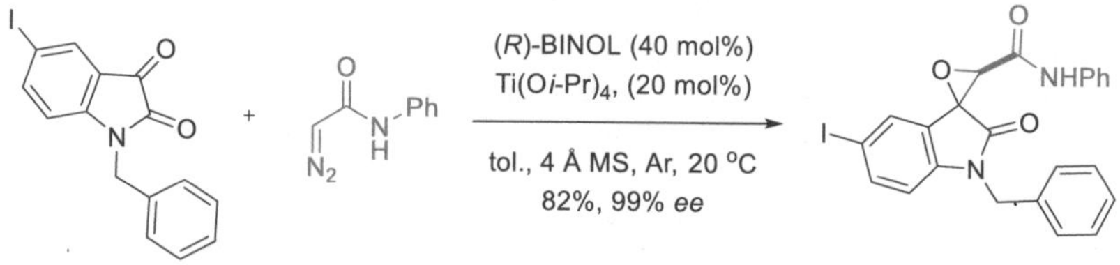

(R)-BINOL (40 mol%)
Ti(Oi-Pr)$_4$, (20 mol%)

tol., 4 Å MS, Ar, 20 °C
82%, 99% ee

NHPh

Example 7, Darzens condensation/Friedel–Crafts alkylation cascade[14]

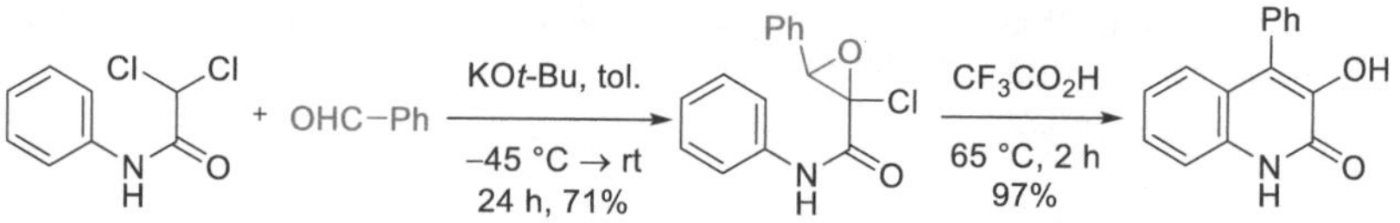

+ OHC–Ph → (KOt-Bu, tol., −45 °C → rt, 24 h, 71%) → (CF$_3$CO$_2$H, 65 °C, 2 h, 97%)

Example 8, Asymmetric vinylogous aza-Darzens approach to vinyl aziridines[15]

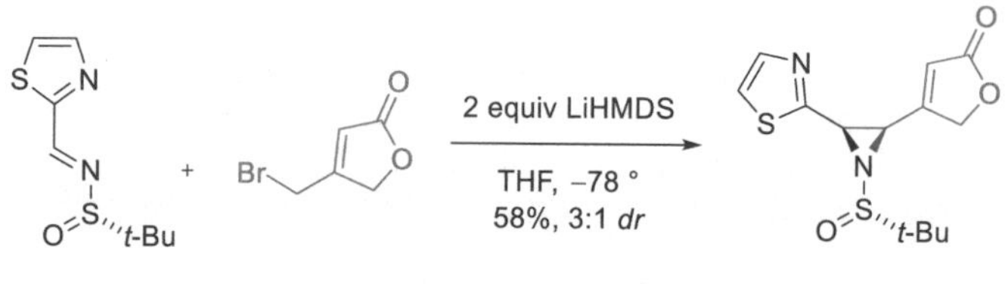

+ Br → (2 equiv LiHMDS, THF, −78 °, 58%, 3:1 dr)

Example 9, Enantioselective aza-Darzens reaction[16]

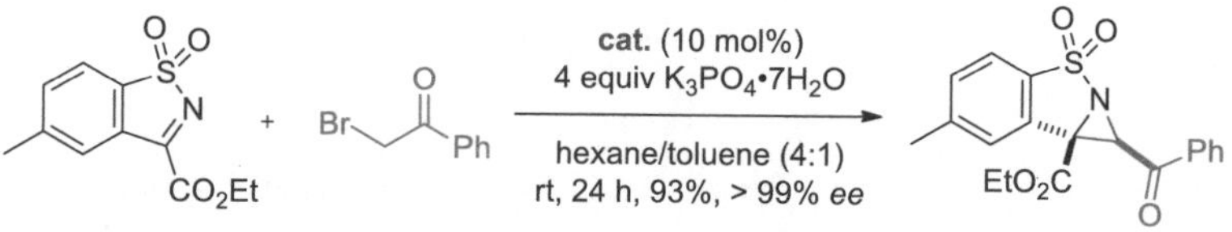

+ Br—Ph → (cat. (10 mol%), 4 equiv K$_3$PO$_4$·7H$_2$O, hexane/toluene (4:1), rt, 24 h, 93%, > 99% ee)

CO$_2$Et, EtO$_2$C

cat. =

References

1   Darzens, G. A. *Compt. Rend. Acad. Sci.* **1904,** *139,* 1214–1217. George Auguste Dar-
    zens (1867–1954), born in Moscow, Russia, studied at École Polytechnique in Paris
    and stayed there as a professor.

2   Newman, M. S.; Magerlein, B. J. *Org. React.* **1949,** *5,* 413–441. (Review).

3   Ballester, M. *Chem. Rev.* **1955,** *55,* 283–300. (Review).

4   Hunt, R. H.; Chinn, L. J.; Johnson, W. S. *Org. Syn. Coll. IV,* **1963,** 459.

5   Rosen, T. *Darzens Glycidic Ester Condensation* In *Comprehensive Organic Synthesis;*
    Trost, B. M.; Fleming, I., Eds.; Pergamon: Oxford, **1991,** *Vol. 2,* pp 409–439. (Re-
    view).

6   Enders, D.; Hett, R. *Synlett* **1998,** 961–962.

7   Davis, F. A.; Wu, Y.; Yan, H.; McCoull, W.; Prasad, K. R. *J. Org. Chem.* **2003,** *68,*
    2410–2419.

8   Myers, B. J. *Darzens Glycidic Ester Condensation.* In *Name Reactions in Heterocyclic
    Chemistry*; Li, J. J., Ed.; Wiley: Hoboken, NJ, **2005,** pp 15–21. (Review).

9   Achard, T. J. R.; Belokon, Y. N.; Ilyin, M.; Moskalenko, M.; North, M.; Pizzato, F.
    Tetrahedron Lett. **2007,** *48,* 2965–2969.

10 Demir, A. S.; Emrullahoglu, M.; Pirkin, E.; Akca, N. *J. Org. Chem.* **2008,** *73,* 8992–
    8997.

11 Liu, G.; Zhang, D.; Li, J.; Xu, G.; Sun, J. *Org. Biomol. Chem.* **2013,** *11,* 900–904.

12 Tanaka, K.; Kobayashi, K.; Kogen, H. *Org. Lett.* **2016,** *18,* 1920–1923.

13 Chai, G.-L.; Han, J.-W.; Wong, H. N. C. *J. Org. Chem.* **2017,** *82,* 12647–12654.

14 Chogii, I.; Das, P.; Delost, M. D.; Crawford, M. N.; Njardarson, J. T. *Org. Lett.* **2018,**
    20, 4942–4945.

15 Mamedov, V. A.; Mamedova, V. L.; Kadyrova, S. F.; Galimullina, V. R.; Khikmatova,
    Gul′naz Z.; Korshin, D. E.; Gubaidullin, A. T.; Krivolapov, D. B.; Rizvanov, I. Kh.;
    Bazanova, O. B.; et al. *J. Org. Chem.* **2018,** *83,* 13132–13145.

16 Pan, J.; Wu, J.-H.; Zhang, H.; Ren, X.; Tan, J.-P.; Zhu, L.; Zhang, H.-S.; Jiang, C.;
    Wang, T. *Angew. Chem. Int. Ed.* **2019,** *58,* 7425–7430.

17 Bierschenk, S. M.; Bergman, R. G.; Raymond, K. N.; Toste, F. D. *J. Am. Chem. Soc.*
    2020, *142,* 733–737.

de Mayo Reaction

[2 + 2]-Photochemical cyclization of enones with olefins is followed by a retro-aldol reaction to give 1,5-diketones.

Head-to-tail alignment gives the major product:[1b]

Head-to-head alignment gives the minor regioisomer:

© Springer Nature Switzerland AG 2021
J. J. Li, *Name Reactions*, https://doi.org/10.1007/978-3-030-50865-4_41

Example 1[3]

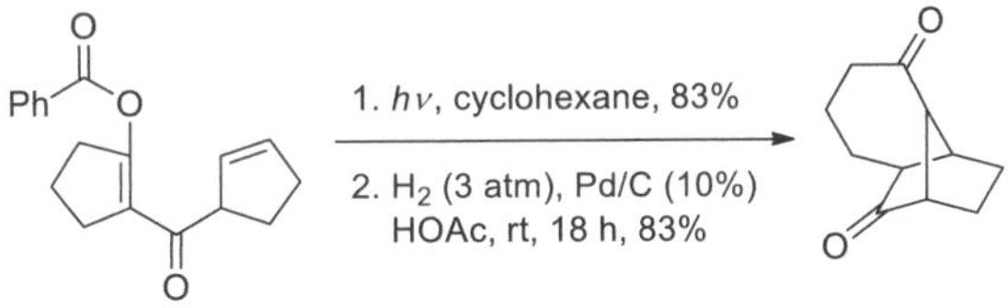
Ph
O
1. $h\nu$, cyclohexane, 83%
2. H_2 (3 atm), Pd/C (10%)
HOAc, rt, 18 h, 83%

Example 2[6]

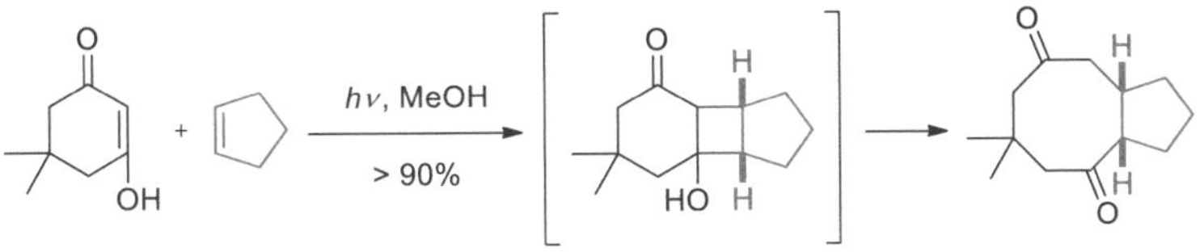
OH
$h\nu$, MeOH
> 90%
HO
H

Example 3[9]

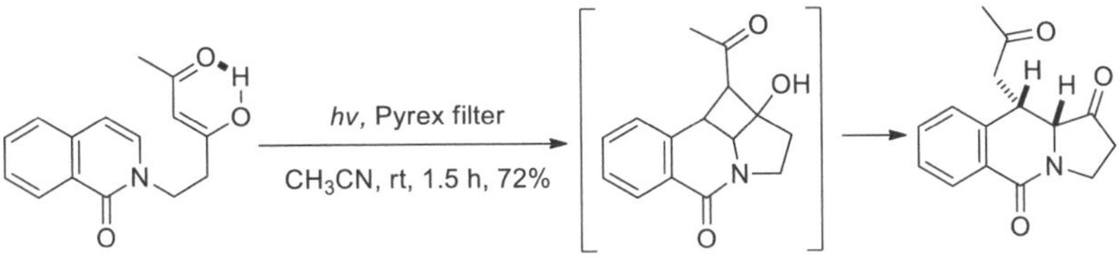
$h\nu$, Pyrex filter
CH_3CN, rt, 1.5 h, 72%
OH
N
H

Example 4[10]

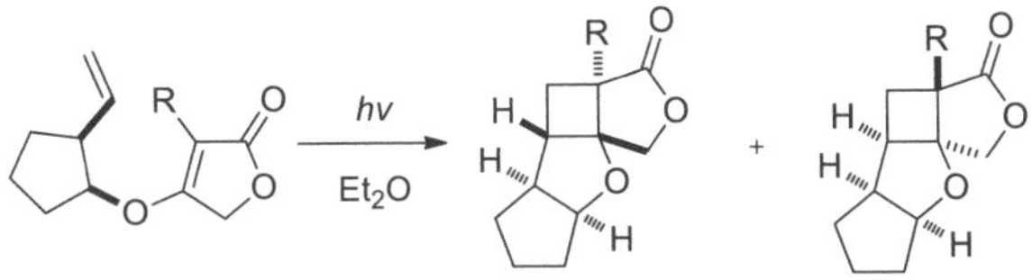
R
$h\nu$
Et_2O
H

R = H	70%	100	:	0
R = Me	58%	50	:	50
R = *t*-Bu	72%	0	:	100

Example 5, Alkyne de Mayo reaction followed by ring expansion[15]

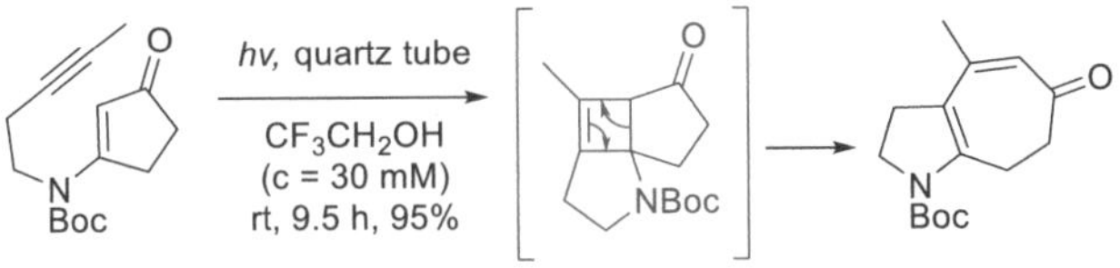
N
Boc
$h\nu$, quartz tube
CF_3CH_2OH
(c = 30 mM)
rt, 9.5 h, 95%
NBoc

Example 6, via [2 + 2]-cycloaddition[16]

Example 7, Intermolecular [2 + 2]-photocycloaddition[17]

References

1. (a) de Mayo, P.; Takeshita, H.; Sattar, A. B. M. A. *Proc. Chem. Soc., London* **1962**, 119. Paul de Mayo received his doctorate from Sir Derek Barton at Birkbeck College, University of London. He later became a professor at the University of Western Ontario in London, Ontario, Canada, where he discovered the de Mayo reaction. (b) Challand, B. D.; Hikino, H.; Kornis, G.; Lange, G.; de Mayo, P. *J. Org. Chem.* **1969**, *34*, 794–806.

2. de Mayo, P. *Acc. Chem. Res.* **1971**, *4*, 41–48. (Review).

3. Oppolzer, W.; Godel, T. *J. Am. Chem. Soc.* **1978**, *100*, 2583–2584.

4. Oppolzer, W. *Pure Appl. Chem.* **1981**, *53*, 1181–1201. (Review).

5. Kaczmarek, R.; Blechert, S. *Tetrahedron Lett.* **1986**, *27*, 2845–2848.

6. Disanayaka, B. W.; Weedon, A. C. *J. Org. Chem.* **1987**, *52*, 2905–2910.

7. Crimmins, M. T.; Reinhold, T. L. *Org. React.* **1993**, *44*, 297–588. (Review).

8. Quevillon, T. M.; Weedon, A. C. *Tetrahedron Lett.* **1996**, *37*, 3939–3942.

9. Minter, D. E.; Winslow, C. D. *J. Org. Chem.* **2004**, *69*, 1603–1606.

10. Kemmler, M.; Herdtweck, E.; Bach, T. *Eur. J. Org. Chem.* **2004**, 4582–4595.

11. Wu, Y.-J. *de Mayo Reaction* in *Name Reactions in Carbocyclic Ring Formations*, Li, J. J., Ed., Wiley: Hoboken, NJ, 2010; pp 451–488. (Review).

12. Kärkäs, M. D.; Porco, J. A.; Stephenson, C. R. *J. Chem. Rev.* **2016**, *116*, 9683–9747. (Review).

13. Poplata, S.; Tröster, A.; Zou, Y.-Q.; Bach, T. *Chem. Rev.* **2016**, *116*, 9748–9815. (Review).

14. Tymann, D.; Tymann, D. C.; Bednarzick, U.; Iovkova-Berends, L.; Rehbein, J.; Hiersemann, M. *Angew. Chem. Int. Ed.* **2018,** *57,* 15553–15557.
15. Martinez-Haya, R.; Marzo, L.; König, B. *Chem. Comm.* **2018,** *54,* 11602–11605.
16. Petz, S.; Allmendinger, L.; Mayer, P.; Wanner, K. T. *Tetrahedron* **2019,** *75,* 2755–2762.
17. Gu, J.-H.; Wang, W.-J.; Chen, J.-Z.; Liu, J.-S.; Li, N.-P.; Cheng, M.-Ji.; Hu, L.-J.; Li, C.-C.; Ye, W.-C.; Wang, L. *Org. Lett.* **2020,** *22,* 1796–1800.

Demjanov Rearrangement

Carbocation rearrangement of primary amines *via* diazotization to give alcohols through C–C bond migration. The Demjanov rearrangement has been largely replaced by the Tiffeneau–Demjanov rearrangement due to the latter's convenience.

Example 1[3]

© Springer Nature Switzerland AG 2021
J. J. Li, *Name Reactions*, https://doi.org/10.1007/978-3-030-50865-4_42

Example 2[6]

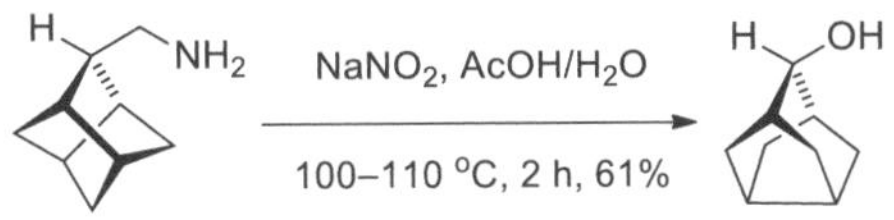

H
NH2
NaNO2, AcOH/H2O
100–110 °C, 2 h, 61%
H
OH

Example 3[7]

O
NH2
NaNO2, 0–4 °C
0.25 M H2SO4/H2O
O
OH
+
O
HO

Example 4[8]

H2N
H
H
MeO
O
O
5 equiv NaNO2
5 equiv AcOH
H2O/THF (1:1)
0 °C, 2 h, 61%
HO
H
H
MeO
O
O

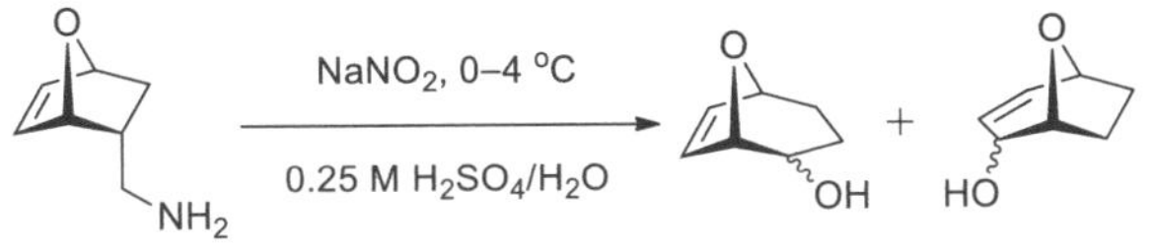

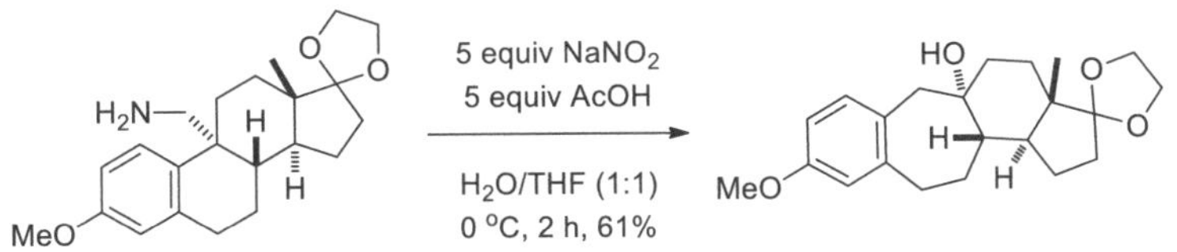

References

1. Demjanov, N. J.; Lushnikov, M. *J. Russ. Phys. Chem. Soc.* **1903,** *35*, 26–42. Nikolai J. Demjanov (1861–1938) was a Russian chemist.
2. Smith, P. A. S.; Baer, D. R. *Org. React.* **1960,** *11*, 157–188. (Review).
3. Diamond, J.; Bruce, W. F.; Tyson, F. T. *J. Org. Chem.* **1965,** *30*, 1840–184.
4. Kotani, R. *J. Org. Chem.* **1965,** *30*, 350–354.
5. Diamond, J.; Bruce, W. F.; Tyson, F. T. *J. Org. Chem.* **1965,** *30*, 1840–1844.
6. Nakazaki, M.; Naemura, K.; Hashimoto, M. *J. Org. Chem.* **1983,** *48*, 2289–2291.
7. Fattori, D.; Henry, S.; Vogel, P. *Tetrahedron* **1993,** *49*, 1649–1664.
8. Kürti, L.; Czakó, B.; Corey, E. J. *Org. Lett.* **2008,** *10*, 5247–5250.
9. Curran, T. T. *Demjanov and Tiffeneau–Demjanov Rearrangement*. In *Name Reactions for Homologations-Part II*; Li, J. J., Ed.; Wiley: Hoboken, NJ, **2009,** pp 2–32. (Review).

Tiffeneau–Demjanov Rearrangement

Carbocation rearrangement of β-aminoalcohols *via* diazotization to afford carbonyl compounds through C–C bond migration.

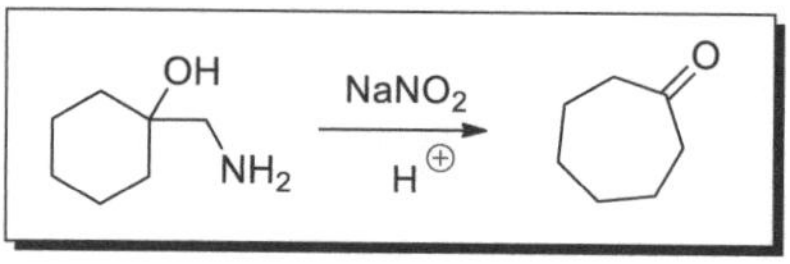
OH
NaNO2
NH2
H⊕
O

Step 1, Generation of N_2O_3

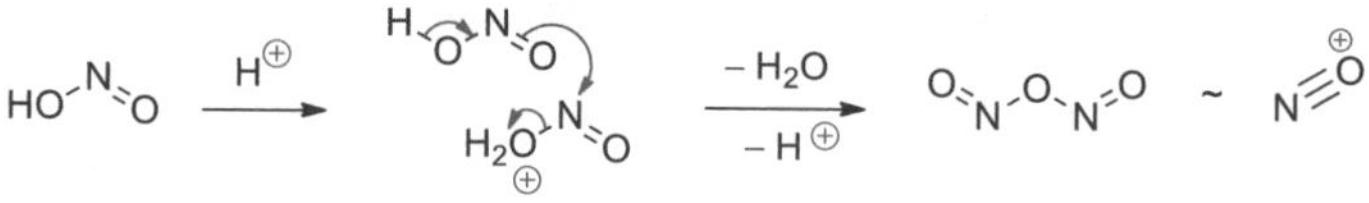
HO
H⊕
– H2O
– H⊕

N-nitrosonium ion

Step 2, Transformation of amine to diazonium salt

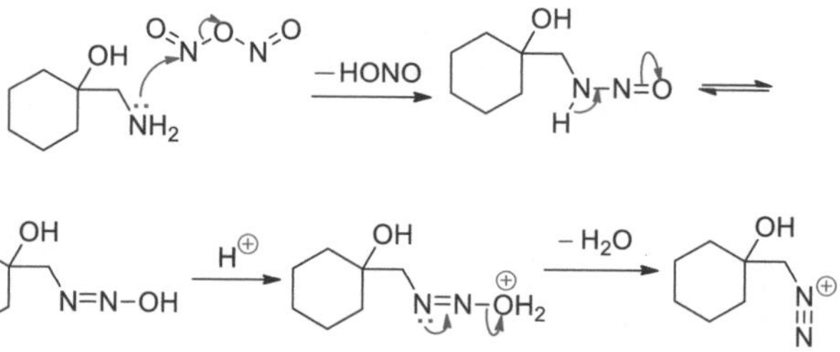
OH
NH2
–HONO
N=N–OH
H⊕
– H2O

Step 3, Ring-expansion *via* rearrangement

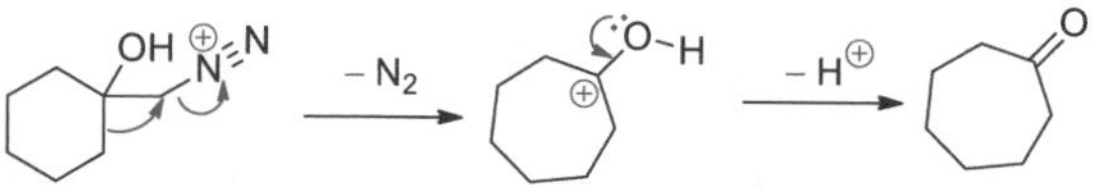
– N2
– H⊕

Example 1[5]

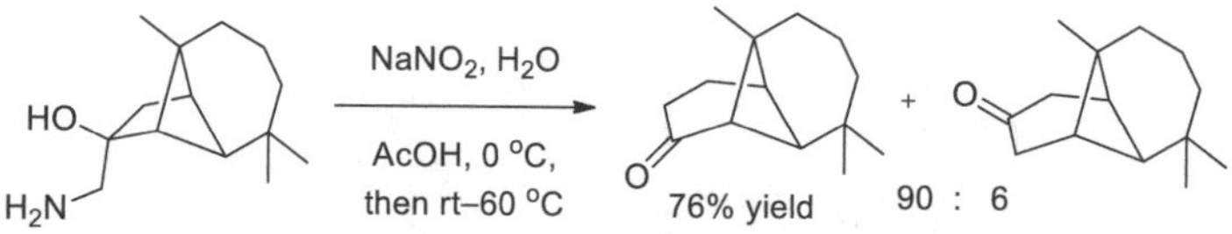
HO
H2N
NaNO2, H2O
AcOH, 0 °C,
then rt–60 °C
76% yield
90 : 6

Example 2[6]

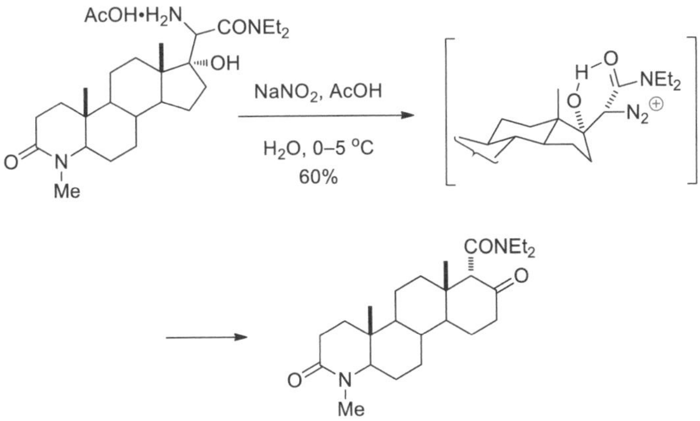

Example 3[7]

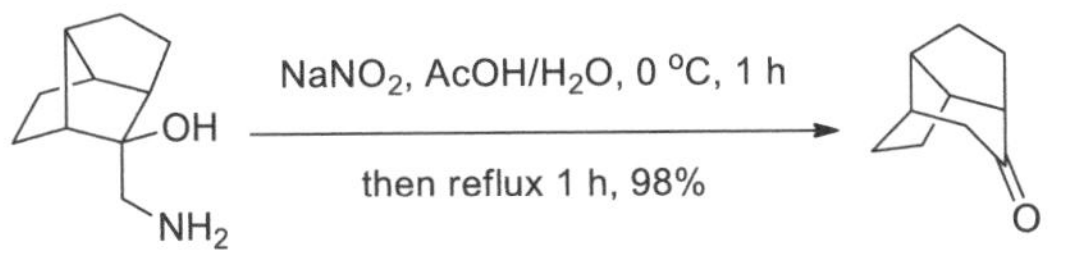

NaNO$_2$, AcOH/H$_2$O, 0 °C, 1 h

then reflux 1 h, 98%

Example 4[9]

NaNO$_2$

AcOH

72%   SiMe$_3$ 28%

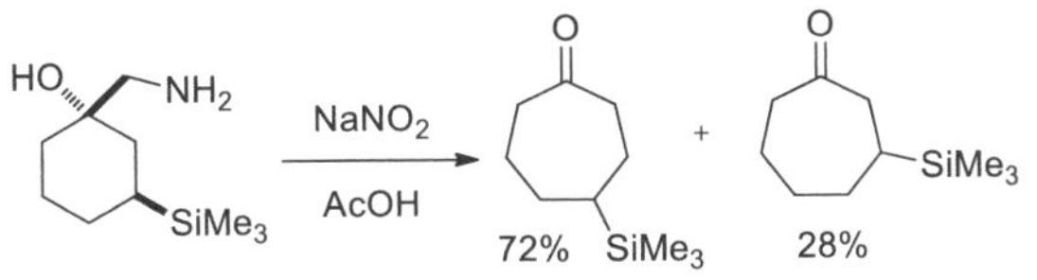

Example 5, In the total synthesis of echinopines[12]

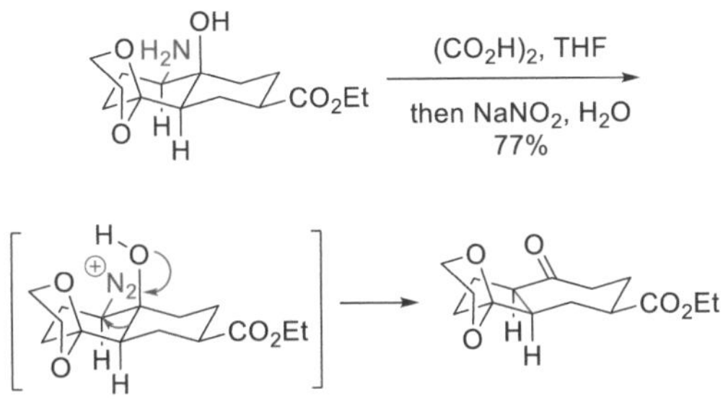

(CO$_2$H)$_2$, THF

then NaNO$_2$, H$_2$O

77%

Example 6, Ring expansion of the *N*-Boc-piperidone[13]

Example 7, Chemo- and region-selective Tiffeneau–Demjanov rearrangement[14]

Example 8, One-pot reaction of cyclic *N*-sulfonylimine with diazo intermediate generated from *N*-tosylhydrazone[15]

Example 9, From seven- to eight-membered ring[16]

References

1. Tiffeneau, M.; Weill, P.; Tehoubar, B. *Compt. Rend.* **1937**, *205*, 54–56.
2. Smith, P. A. S.; Baer, D. R. *Org. React.* **1960**, *11*, 157–188. (Review).
3. Parham, W. E.; Roosevelt, C. S. *J. Org. Chem.* **1972**, *37*, 1975–1979.
4. Jones, J. B.; Price, P. *Tetrahedron* **1973**, *29*, 1941–1947.
5. Miyashita, M.; Yoshikoshi, A. *J. Am. Chem Soc.* **1974**, *96*, 1917–1925.
6. Steinberg, N. G.; Rasmusson, G. H. *J. Org. Chem.* **1984**, *49*, 4731–4733.
7. Stern, A. G.; Nickon, A. *J. Org. Chem.* **1992**, *57*, 5342–5352.
8. Fattori, D.; Henry, S.; Vogel, P. *Tetrahedron* **1993**, *49*, 1649–1664. (Review).
9. Chow, L.; McClure, M.; White, J. *Org. Biomol. Chem.* **2004**, *2*, 648–650.

10. Curran, T. T. *Demjanov and Tiffeneau–Demjanov Rearrangement*. In *Name Reactions for Homologations-Part II*; Li, J. J., Ed.; Wiley: Hoboken, NJ, **2009,** pp 293–304. (Review).
11. Shi, L.; Meyer, K.; Greaney, M. F. *Angew. Chem. Int. Ed.* **2010,** *49,* 9250–9253.
12. Xu, W.; Wu, S.; Zhou, L.; Liang, G. *Org. Lett.* **2013,** *15,* 1978–1981.
13. Nortcliffe, A.; Moody, C. J. *Bioorg. Med. Lett.* **2015,** *23,* 2730–2735.
14. Alves, L. C.; Ley, S. V.; Brocksom, T. J. *Org. Biomol. Chem.* **2015,** *13,* 7633–7642.
15. Xia, A.-J.; Kang, T.-R.; He, L. *Angew. Chem. Int. Ed.* **2016,** *55,* 1441–1444.
16. Liu, J.; Zhou, X.; Wang, C.; Fu, W.; Chu, W. *Chem. Comm.* **2016,** *52,* 5152–5155.
17. Kohlbacher, S. M.; Ionasz, V.-S.; Ielo, L.; Pace, V. *Monat. Chem.* **2019,** *150,* 2011–2019.

Dess–Martin Periodinane Oxidation

Oxidation of alcohols to the corresponding carbonyl compounds using triacetoxy-periodinane. The Dess–Martin periodinane (DMP), 1,1,1-triacetoxy-1,1-dihydro-1,2-benziodoxol-3(1H)-one, is one of the most useful oxidant for the conversion of primary and secondary alcohols to their corresponding aldehyde or ketone products, respectively.

Preparation,[1,2] the oxone preparation is much safer and easier than $KBrO_3$. The IBX intermediate that comes out of it has proven to be far less explosive[12]

However, The Dess–Martin periodinane is hydrolyzed by moisture to o-iodoxybenzoic acid (IBX), which is a more powerful oxidizing agent[3]

Mechanism[1]

J. J. Li, *Name Reactions*, https://doi.org/10.1007/978-3-030-50865-4_43

Example 1[6]

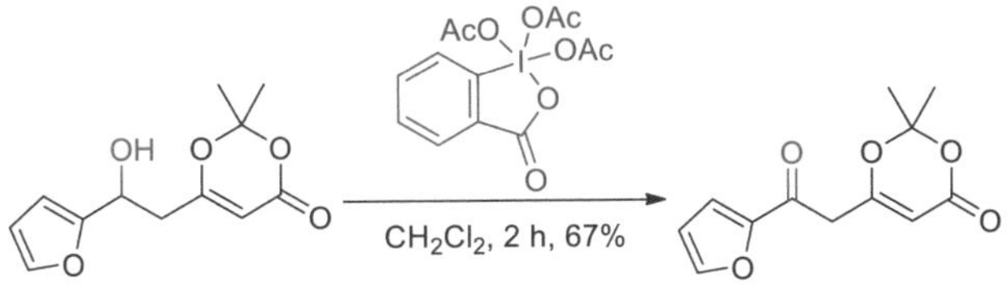

Example 2, An atypical Dess–Martin periodinane (DMP) reactivity[7]

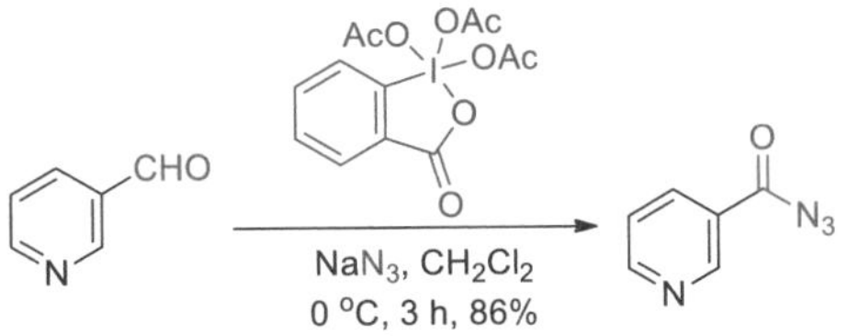

Example 3[10]

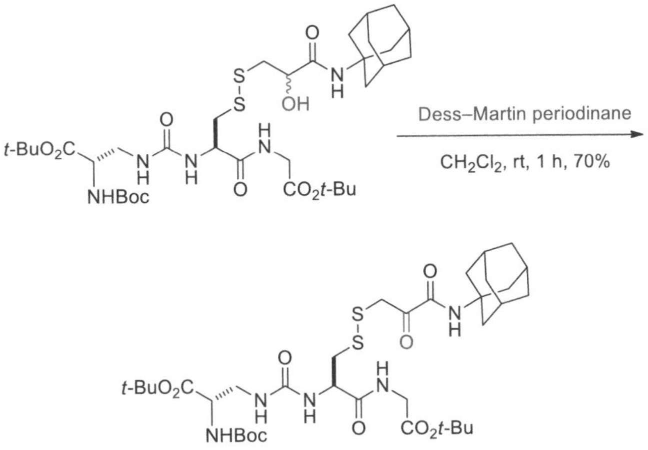

Example 4[11]

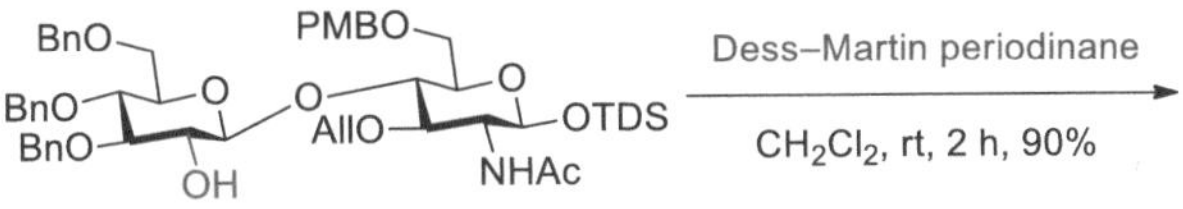

Example 5[12]

Example 6, In the total synthesis of (−)-maoecrystal[13]

Example 7, As a terminal oxidant in Wacker-type oxidation[14]

Example 8, Iodoxybenzoic acid tosylate (IBX-OTs) as an alternative to DMP[15]

IBX   DMP-OTs   IBX-OTs

Example 9, In the total synthesis of hestisine-type C_{20}-diterpenoid alkaloids: *The Dess–Martin oxidation was run at 80 °C!*[16]

Example 10, In the total synthesis of cephalostatin 1: dilution was key to achieve selectivity against the two secondary alcohols (0.08 M, 86:3)[17]

References

1. (a) Dess, D. B.; Martin, J. C. *J. Org. Chem.* **1983**, *48*, 4155–4156. James Cullen (J. C.) Martin (1928–1999) had a distinguished career spanning 36 years both at the University of Illinois at Urbana-Champaign and Vanderbilt University. J. C.'s formal training in physical organic chemistry with Don Pearson at Vanderbilt and P. D. Bartlett at Harvard prepared him well for his early studies on carbocations and radicals. However, it was his interest in understanding the limits of chemical bonding that led to his landmark investigations into hypervalent compounds of the main group elements. Over a 20-year period the Martin laboratories successfully prepared unprecedented chemical structures from sulfur, phosphorus, silicon and bromine while the ultimate "Holy Grail" of stable pentacoordinate carbon remained elusive. Although most of these studies were driven by J. C.'s fascination with unusual bonding schemes, they were not without practical value. Two hypervalent compounds, Martin's sulfurane (for dehydration) and the Dess–Martin periodinane (DMP) have found widespread application in synthetic organic chemistry. J. C. Martin and his student Daniel Dess developed this methodology at the University of Illinois at Urbana. (Martin's biography was kindly supplied by Prof. Scott E. Denmark). (b) Dess, D. B.; Martin, J. C. *J. Am. Chem. Soc.* **1991**, *113*, 7277–7287.
2. Ireland, R. E.; Liu, L. *J. Org. Chem.* **1993**, *58*, 2899.
3. Meyer, S. D.; Schreiber, S. L. *J. Org. Chem.* **1994**, *59*, 7549–7552.

4. Frigerio, M.; Santagostino, M.; Sputore, S. *J. Org. Chem.* **1999,** *64*, 4537–4538.
5. Nicolaou, K. C.; Zhong, Y.-L.; Baran, P. S. *Angew. Chem. Int. Ed.* **2000,** *39,* 622–625.
6. Bach, T.; Kirsch, S. *Synlett* **2001,** 1974–1976.
7. Bose, D. S.; Reddy, A. V. N. *Tetrahedron* **2003,** *44*, 3543–3545.
8. Tohma, H.; Kita, Y. *Adv. Synth. Cat.* **2004,** *346*, 111–124. (Review).
9. Holsworth, D. D. *Dess–Martin oxidation.* In *Name Reactions for Functional Group Transformations*; Li, J. J., Ed.; Wiley: Hoboken, NJ; **2007,** pp 218–236. (Review).
10. More, S. S.; Vince, R. *J. Med. Chem.* **2008,** *51*, 4581–4588.
11. Crich, D.; Li, M.; Jayalath, P. *Carbohydrate Res.* **2009,** *344*, 140–144.
12. Howard, J. K.; Hyland, C. J. T.; Just, J.; J. A. *Org. Lett.* **2013,** *15*, 1714–1717.
13. Cernijenko, A.; Risgaard, R.; Baran, P. S. *J. Am. Chem Soc.* **2016,** *138*, 9425–9428.
14. Chaudhari, D. A.; Fernandes, R. A. *J. Org. Chem.* **2016,** *81*, 2113–2121.
15. Yusubov, M. S.; Postnikov, P. S.; Yusubova, R. Ya.; Yoshimura, A.; Juerjens, G.; Kirschning, A.; Zhdankin, V. V. *Adv. Synth. Cat.* **2017,** *359*, 3207–3216.
16. Pflueger, J. J.; Morrill, L. C.; de Gruyter, J. N.; Perea, M. A.; Sarpong, R. *Org. Lett.* **2017,** *19*, 4632–4635.
17. Shi, Y.; Xiao, Q.; Lan, Q.; Wang, D.-H.; Jia, L.-Q.; Tang, X.-H.; Zhou, T.; Li, M.; Tian, W.-S. *Tetrahedron* **2019,** *75*, 1722–1738.
18. Zheng, Q.; Maksimovic, I.; Upad, A.; Guber, D.; David, Y. *J. Org. Chem.* **2020,** *85*, 1691–1697.

Dieckmann Condensation

The Dieckmann condensation is the intramolecular version of the Claisen condensation.

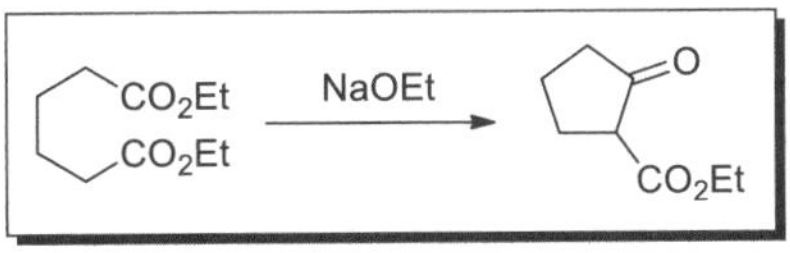

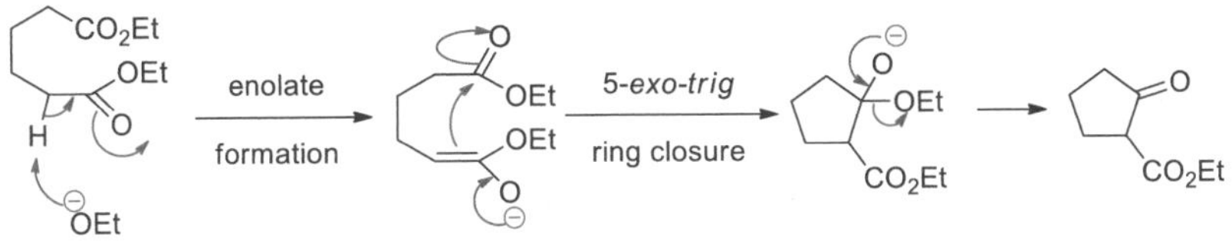

Example 1[4]

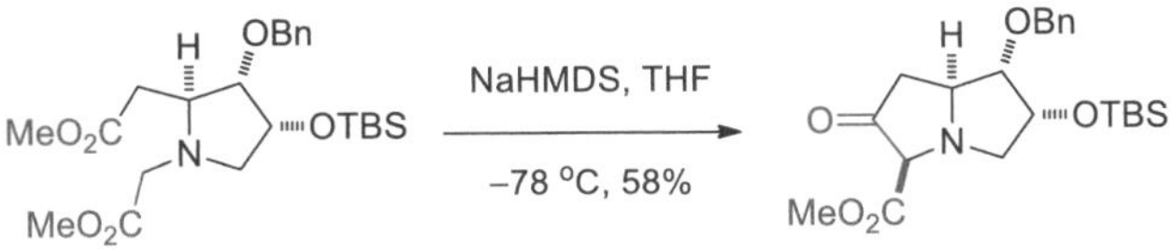

Example 2, Followed by decarboxylation[6]

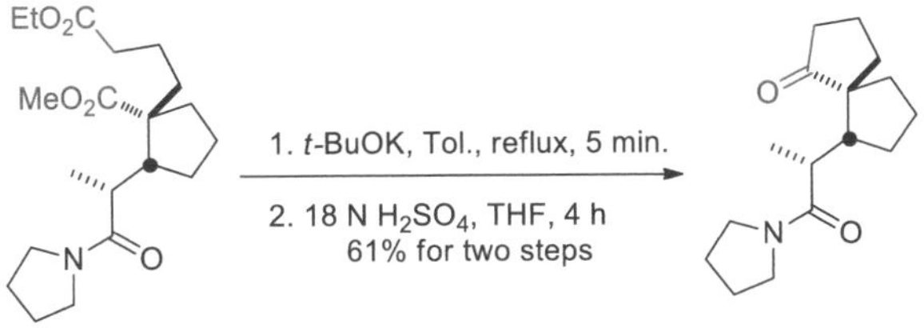

Example 3[7]

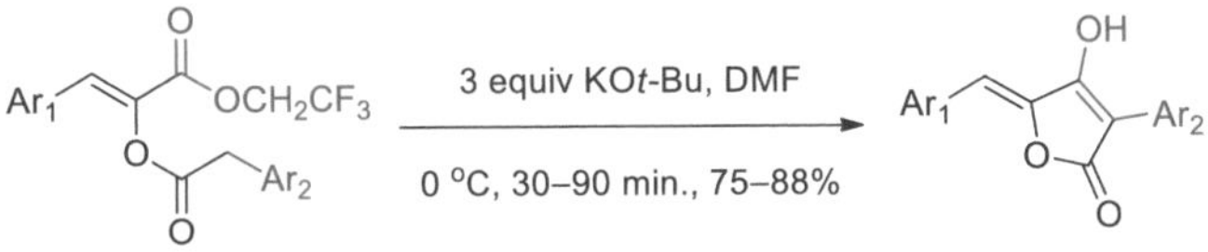

© Springer Nature Switzerland AG 2021
J. J. Li, *Name Reactions*, https://doi.org/10.1007/978-3-030-50865-4_44

Example 4[8]

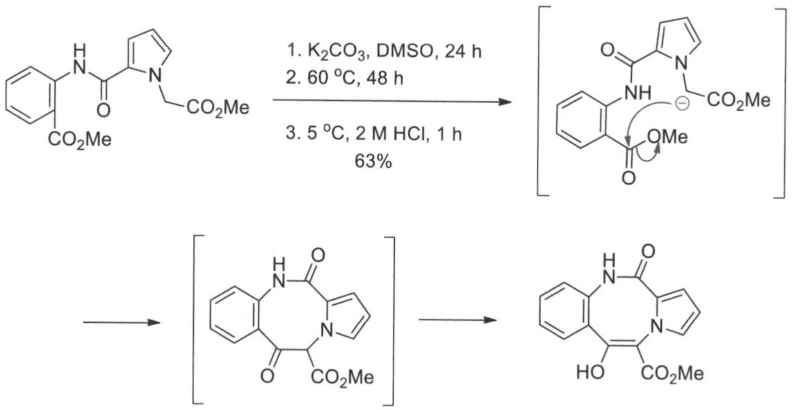

Example 5, Michael–Dieckmann condensation[10]

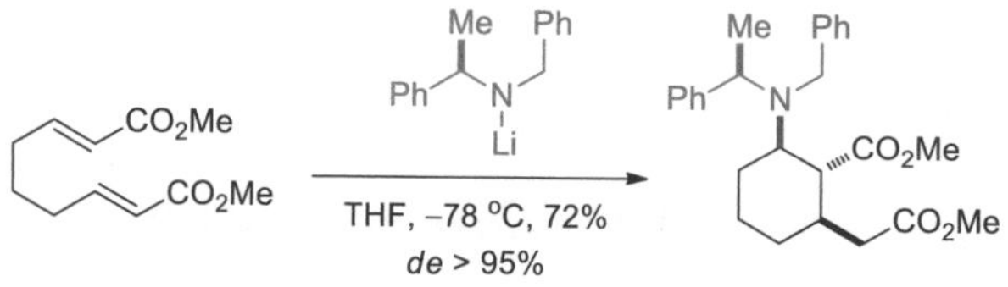

THF, −78 °C, 72%
de > 95%

Example 6, Toward preparation of oseltamivir[10]

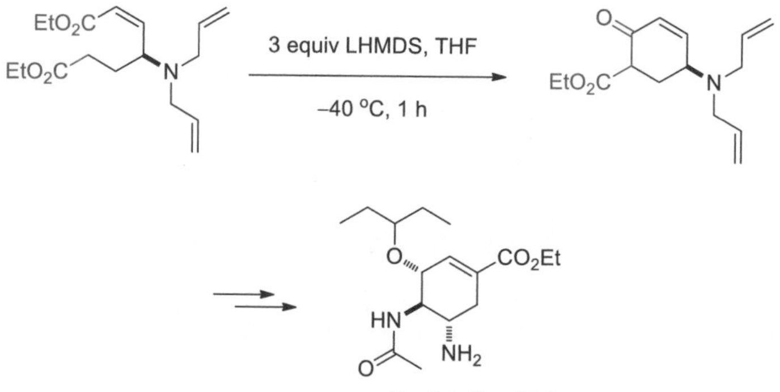

3 equiv LHMDS, THF

−40 °C, 1 h

oseltamivir (Tamiflu)

Example 7, On surface, the double bond rotated. But in reality, it rotated as a single bond in one of the resonance structures[11]

Example 8, From an enol and an ester[12]

Example 8, Large (5.44 Kg)-scale Dieckmann condensation[13]

Example 9, Gram-scale synthesis of a tricycle[14]

References

1. Dieckmann, W. *Ber.* **1894**, *27*, 102. Walter Dieckman (1869–1925), born in Hamburg, Germany, studied with E. Bamberger at Munich. After serving as an assistant to von Baeyer in his private laboratory, he became a professor at Munich. At age 56, he died while working in his chemical laboratory at the Barvarian Academy of Science.
2. Davis, B. R.; Garratt, P. J. *Comp. Org. Synth.* **1991**, *2*, 795–863. (Review).

3. Shindo, M.; Sato, Y.; Shishido, K. *J. Am. Chem. Soc.* **1999**, *121*, 6507–6508.

4. Rabiczko, J.; Urbańczyk-Lipkowska, Z.; Chmielewski, M. *Tetrahedron* **2002**, *58*, 1433–1441.

5. Ho, J. Z.; Mohareb, R. M.; Ahn, J. H.; Sim, T. B.; Rapoport, H. *J. Org. Chem.* **2003**, *68*, 109–114.

6. de Sousa, A. L.; Pilli, R. A. *Org. Lett.* **2005**, *7*, 1617–1617.

7. Bernier, D.; Brueckner, R. *Synthesis* **2007**, 2249–2272.

8. Koriatopoulou, K.; Karousis, N.; Varvounis, G. *Tetrahedron* **2008**, *64*, 10009–10013.

9. Takao, K.-i.; Kojima, Y.; Miyashita, T.; Yashiro, K.; Yamada, T.; Tadano, K.-i. *Heterocycles* **2009**, *77*, 167–172.

10. Garrido, N. M.; Nieto, C. T.; Diez, D. *Synlett* **2013**, *24*, 169–172.

11. Ohashi, T.; Hosokawa, S. *Org. Lett.* **2018**, *20*, 3021–3024.

12. Bruckner, S.; Weise, M.; Schobert, R. *J. Org. Chem.* **2018**, *83*, 10805–10812.

13. Xu, H.; Yin, W.; Liang, H.; Nan, Y.; Qiu, F.; Jin, Y. *Org. Process Res. Dev.* **2019**, *23*, 990–997.

14. Hugelshofer, C. L.; Palani, V.; Sarpong, R. *J. Am. Chem. Soc.* **2019**, *141*, 8431–8435.

15. Gao, J.; Rao, P.; Xu, K.; Wang, S.; Wu, Y.; He, C.; Ding, H. *J. Am. Chem. Soc.* **2020**, *142*, 4592–4597.

Diels–Alder Reaction

The Diels–Alder reaction, inverse electronic demand Diels–Alder reaction, as well as the hetero-Diels–Alder reaction, belong to the category of *[4+2]-cycloaddition reactions*, which are concerted processes. The arrow pushing here is merely mnemonic.

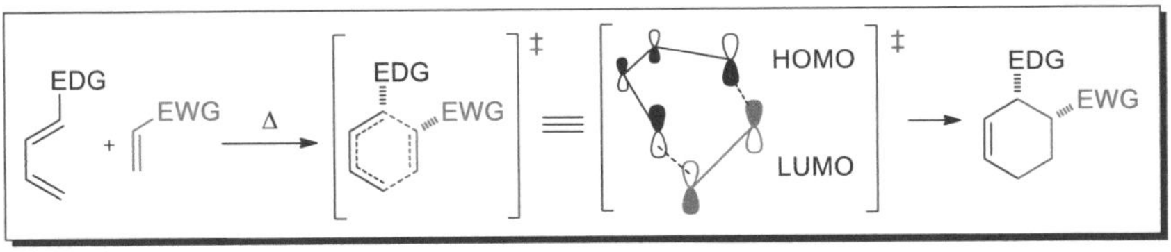

diene dienophile   adduct

EDG = electron-donating group; EWG = electron-withdrawing group

Example 1, Intramolecular Diels–Alder reaction to prepare an intermediate for Scherring–Plough's thrombin receptor (also known as protease activated receptor-1, PAR-1) antagonist, vorapaxar (Zontivity)[6]

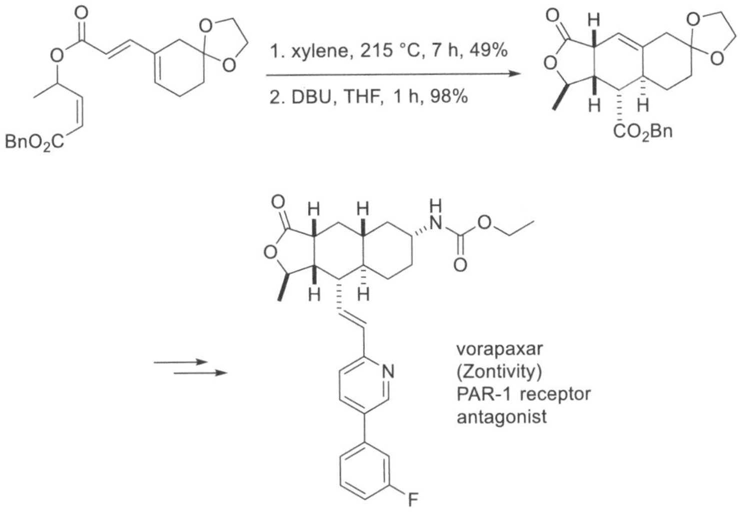

1. xylene, 215 °C, 7 h, 49%

2. DBU, THF, 1 h, 98%

vorapaxar
(Zontivity)
PAR-1 receptor
antagonist

Example 2[7]

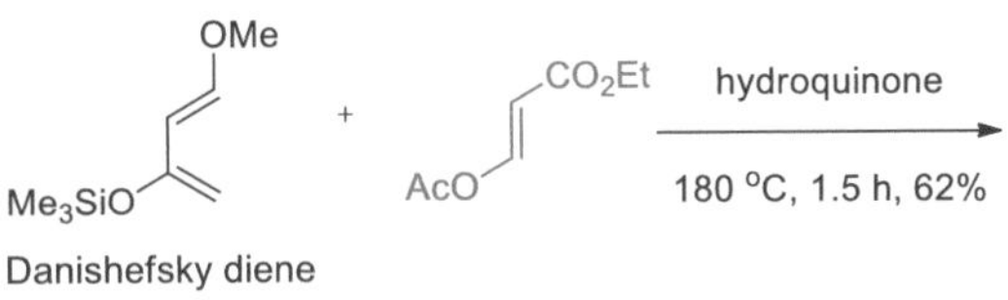

Danishefsky diene

hydroquinone

180 °C, 1.5 h, 62%

© Springer Nature Switzerland AG 2021
J. J. Li, *Name Reactions*, https://doi.org/10.1007/978-3-030-50865-4_45

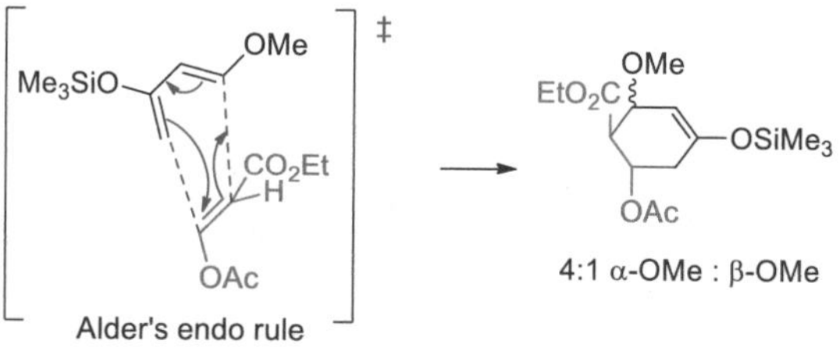

Alder's endo rule

4:1 α-OMe : β-OMe

Example 3, Intramolecular Diels–Alder reaction[8]

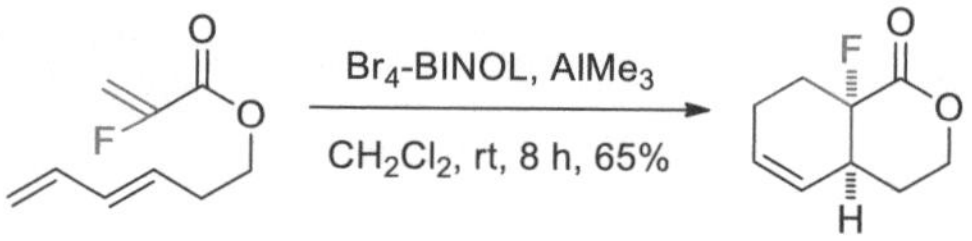

Br$_4$-BINOL, AlMe$_3$

CH$_2$Cl$_2$, rt, 8 h, 65%

Example 4, Asymmetric Diels–Alder reaction employing a CBS-like catalyst[9]

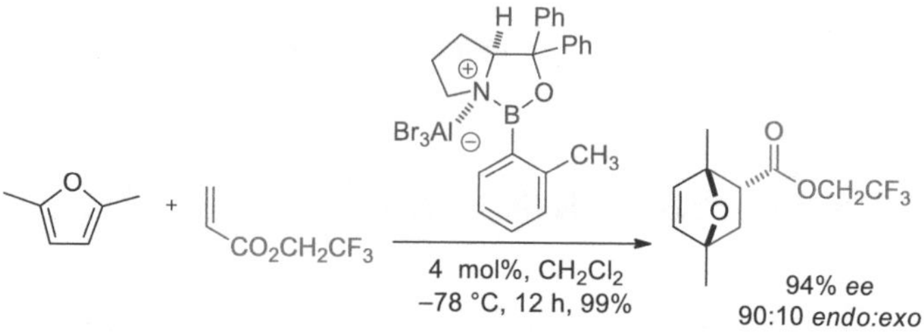

+

CO$_2$CH$_2$CF$_3$

4 mol%, CH$_2$Cl$_2$
−78 °C, 12 h, 99%

OCH$_2$CF$_3$

94% ee
90:10 endo:exo

Example 5, Retro-Diels–Alder reaction[10]

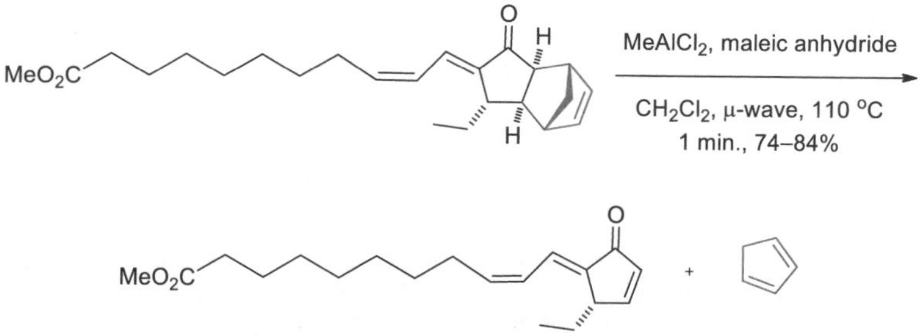

MeO$_2$C

MeAlCl$_2$, maleic anhydride

CH$_2$Cl$_2$, μ-wave, 110 °C
1 min., 74–84%

MeO$_2$C

+

Example 6, Intramolecular Diels–Alder reaction[11]

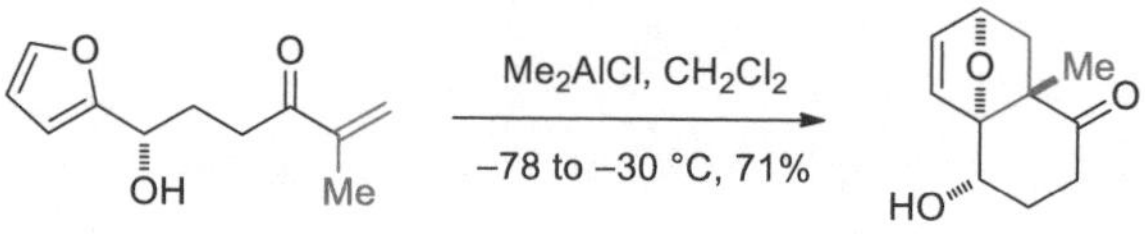

Me$_2$AlCl, CH$_2$Cl$_2$

−78 to −30 °C, 71%

Example 7[12]

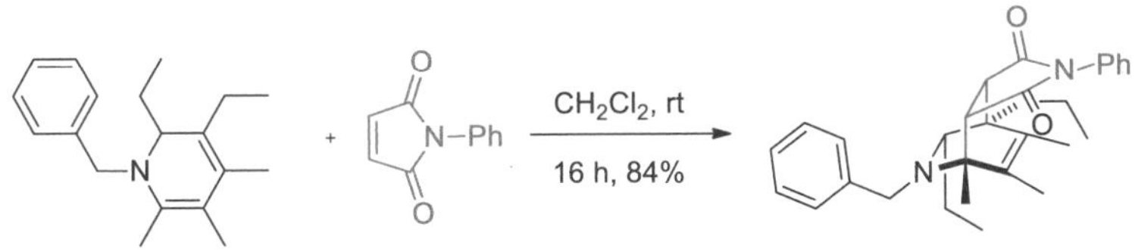

Example 8, Intramolecular Diels–Alder cyclization[13]

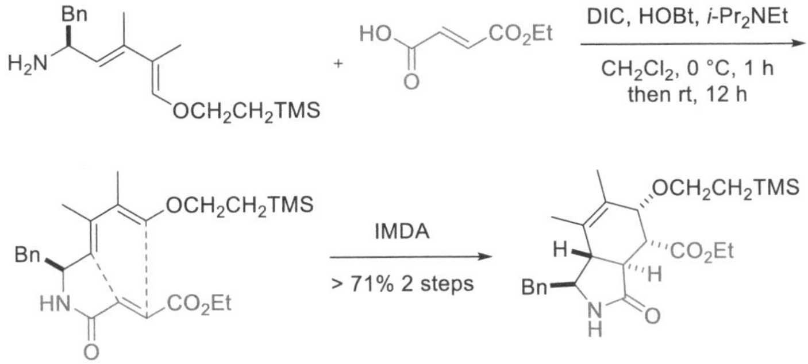

Example 9, Toward the total synthesis of catharidin[14]

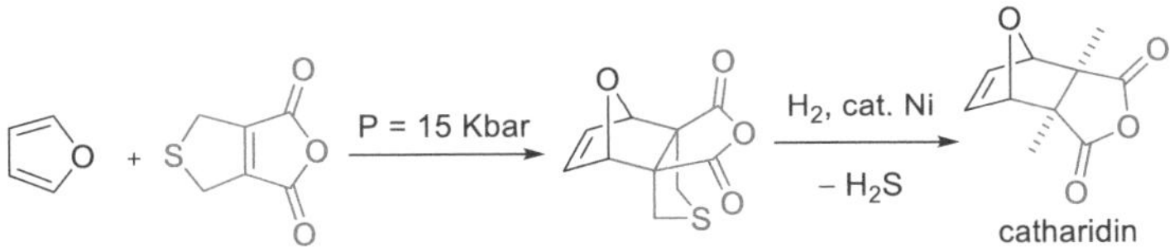

catharidin

Example 10, *exo*-Selective (bulky silatrane helped the selectivity) intermolecular Diels–Alder reaction[15]

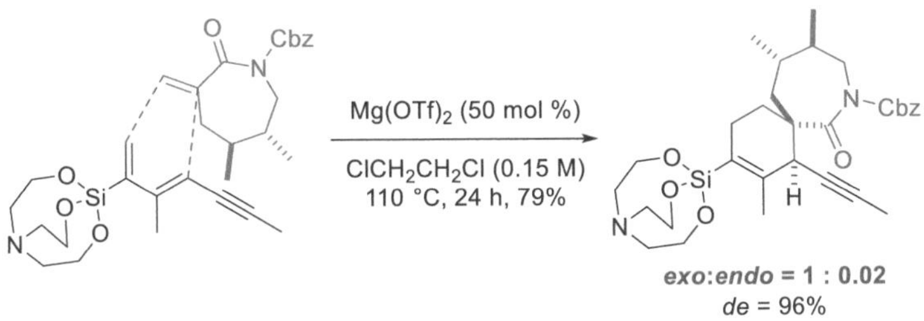

exo:endo = 1 : 0.02
de = 96%

References

1. Diels, O.; Alder, K. *Ann.* **1928,** *460,* 98–122. Otto Diels (Germany, 1876–1954) and his student, Kurt Alder (Germany, 1902–1958), shared the Nobel Prize in Chemistry in 1950 for development of the diene synthesis. In this article they claimed their territory in applying the Diels–Alder reaction in total synthesis: "We explicitly reserve for ourselves the application of the reaction developed by us to the solution of such problems."

2. Oppolzer, W. In *Comprehensive Organic Synthesis;* Trost, B. M.; Fleming, I., Eds.; Pergamon, **1991,** *Vol. 5,* 315–399. (Review).

3. Weinreb, S. M. In *Comprehensive Organic Synthesis;* Trost, B. M.; Fleming, I., Eds.; Pergamon, **1991,** *Vol. 5,* 401–449. (Review).

4. (a) Rickborn, B. The *retro-Diels–Alder reaction. Part I. C–C dienophiles* in *Org. React.* Wiley: Hoboken, NJ, **1998,** *52.* (b) Rickborn, B. *The retro-Diels–Alder reaction. Part II. Dienophiles with one or more heteroatom* in *Org. React.* Wiley: Hoboken, NJ, **1998,** *53.*

5. Corey, E. J. *Angew. Chem. Int. Ed.* **2002,** *41,* 1650–1667. (Review).

6. (a) Chackalamannil, S.; Asberon, T.; Xia, Y.; Doller, D.; Clasby, M. C.; Czarniecki, M. F. US Patent 6,063,847 (2000). (b) Chelliah, M. V.; Chackalamannil, S.; Xia, Y.; Eagen, K.; Clasby, M. C.; Gao, X.; Greenlee, W.; Ahn, H.-S.; Agans-Fantuzzi, J.; Boykow, G.; et al. *J. Med. Chem.* **2007,** *50,* 5147–5160.

7. Wang, J.; Morral, J.; Hendrix, C.; Herdewijn, P. *J. Org. Chem.* **2001,** *66,* 8478–8482.

8. Saito, A.; Yanai, H.; Sakamoto, W.; Takahashi, K.; Taguchi, T. *J. Fluorine Chem.* **2005,** *126,* 709–714.

9. Liu, D.; Canales, E.; Corey, E. J. *J. Am. Chem. Soc.* **2007,** *129,* 1498–1499.

10. Iqbal, M.; Duffy, P.; Evans, P.; Cloughley, G.; Allan, B.; Lledo, A.; Verdaguer, X.; Riera, A. *Org. Biomol. Chem.* **2008,** *6,* 4649–4661.

11. Gao, S.; Wang, Q.; Chen, C. *J. Am. Chem. Soc.* **2009,** *131,* 1410–1412.

12. Martin, R. M.; Bergman, R. G.; Ellman, J. A. *Org. Lett.* **2013,** *15,* 444–447.

13. Xu, J.; Lin, B.; Jiang, X.; Jia, Z.; Wu, J.; Dai, W.-M. *Org. Lett.* **2019,** *21,* 830–834.

14. Davidson, M. G.; Eklov, B. M.; Wuts, P.; Loertscher, B. M.; Schow, S. R. WO2019070980 (2019).

15. Minamino, K.; Murata, M.; Tsuchikawa, H. *Org. Lett.* **2019,** *21,* 8970–8975.

16. Dyan, O. T.; Borodkin, G. I.; Zaikin, P. A. *Eur. J. Org. Chem.* **2019,** 7271–7306. (Review).

17. Farley, C. M.; Sasakura, K.; Zhou, Y.-Y.; Kanale, V. V.; Uyeda, C. *J. Am. Chem. Soc.* **2020,** *142,* 4598–4603.

Hetero-Diels–Alder Reaction

Heterodiene addition to dienophile or heterodienophile addition to diene. Typical hetero-Diels–Alder reactions are aza-Diels–Alder reaction and oxo-Diels–Alder reaction.

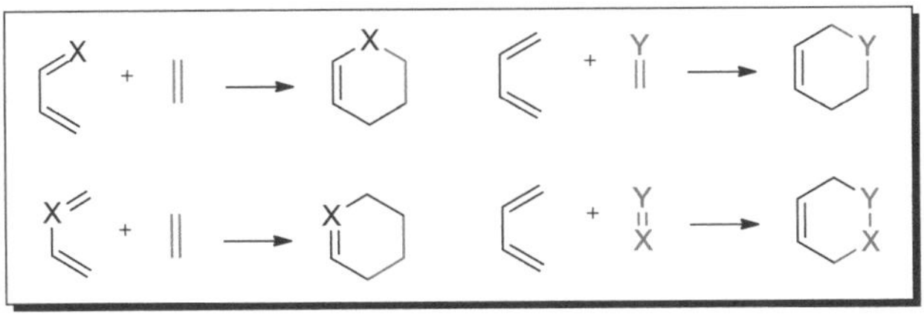

Example 1,

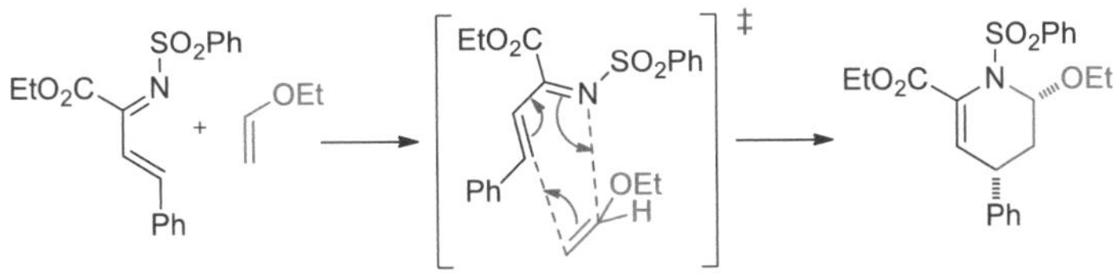

Example 2, Heterodienophile addition to diene[1]

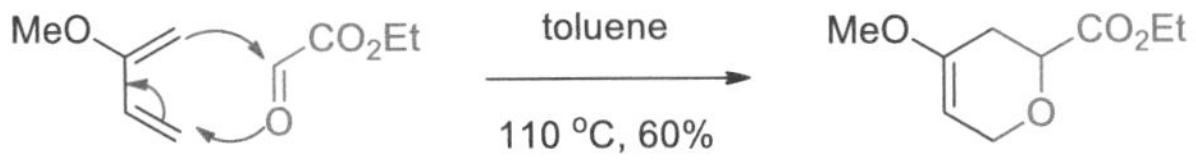

Example 3, Similar to the Boger pyridine synthesis[2]

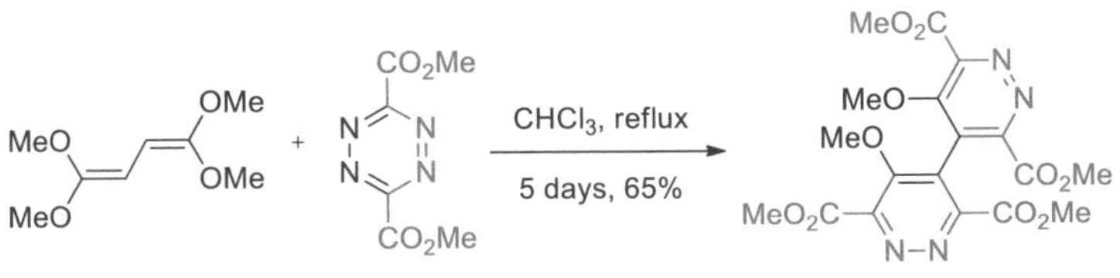

Example 4, Using the Rawal diene[4]

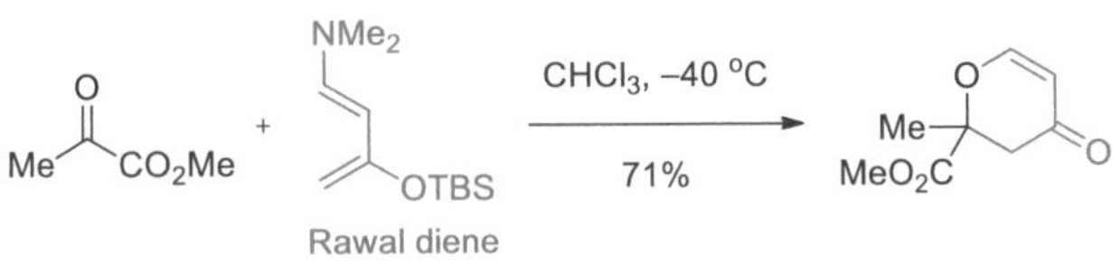

Rawal diene

Example 5, Also similar to the Boger pyridine synthesis[6]

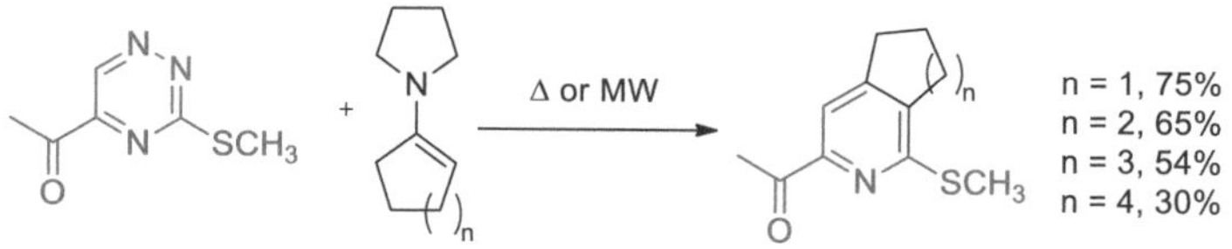

Δ or MW
SCH3
n = 1, 75%
n = 2, 65%
n = 3, 54%
n = 4, 30%

Example 6, Asymmetric hetero-Diels–Alder reaction[7]

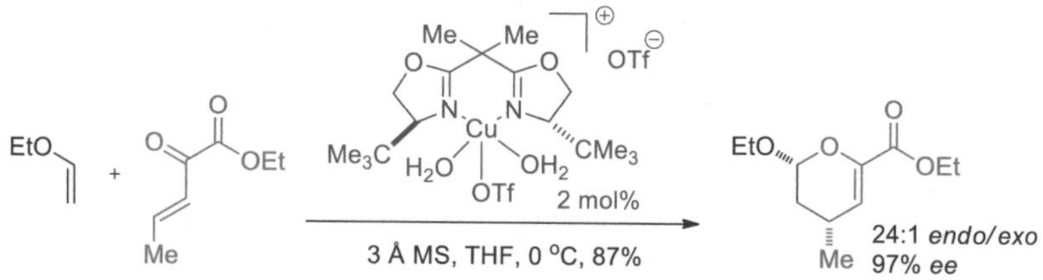

EtO
OEt
Me
Me3C
H2O
OH2
CMe3
OTf
2 mol%
3 Å MS, THF, 0 °C, 87%
24:1 endo/exo
97% ee

Example 7, Asymmetric hetero-Diels–Alder reaction[8]

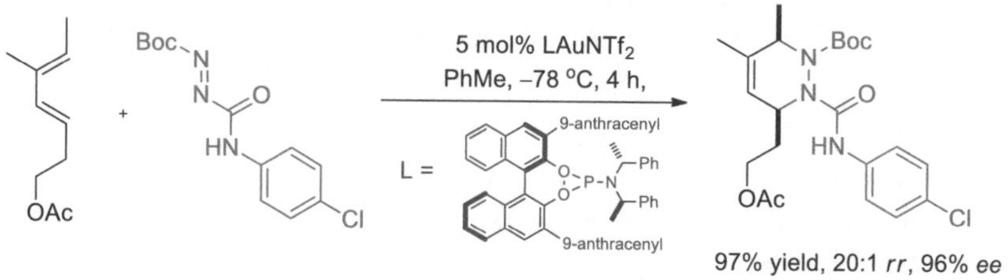

Boc
5 mol% LAuNTf2
PhMe, –78 °C, 4 h,
L =
9-anthracenyl
Ph
OAc
HN
Cl
97% yield, 20:1 rr, 96% ee

rr = *r*egioisomeric *r*atio

Example 8, Asymmetric hetero-Diels–Alder reaction[10]

TBSO
OPMB
OTMS
cat. (3 mol %)
4 Å MS, rt, 20.5 h
60%
cat. =
Cr
Cl

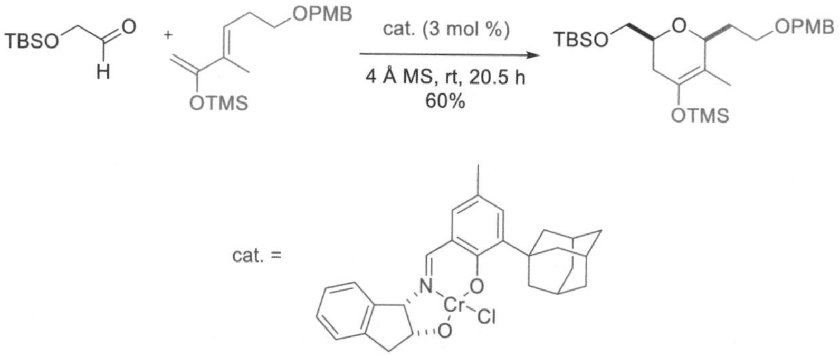

Example 9, Enantioselective intramolecular oxa-Diels–Alder reaction[11]

References

1. Wender, P. A.; Keenan, R. M.; Lee, H. Y. *J. Am. Chem. Soc.* **1987,** *109*, 4390–4392.
2. Boger, D. L. In *Comprehensive Organic Synthesis;* Trost, B. M.; Fleming, I., Eds.; Pergamon, **1991,** *Vol. 5*, 451–512. (Review).
3. Boger, D. L.; Baldino, C. M. *J. Am. Chem. Soc.* **1993,** *115*, 11418–11425.
4. Huang, Y.; Rawal, V. H. *Org. Lett.* **2000,** *2*, 3321–3323.
5. Jørgensen, K. A. *Eur. J. Org. Chem.* **2004,** 2093–2102. (Review).
6. Lipińska, T. M. *Tetrahedron* **2006,** *62*, 5736–5747.
7. Evans, D. A.; Kvaerno, L.; Dunn, T. B.; Beauchemin, A.; Raymer, B.; Mulder, J. A.; Olhava, E. J.; Juhl, M.; Kagechika, K.; Favor, D. A. *J. Am. Chem. Soc.* **2008,** *130*, 16295–16309.
8. Liu, B.; Li, K.-N.; Luo, S.-W.; Huang, J.-Z.; Pang, H.; Gong, L.-Z. *J. Am. Chem. Soc.* **2013,** *135*, 3323–3326.
9. Heravi, M. M.; Ahmadi, T.; Ghavidel, M.; Heidari, B.; Hamidi, H. *RSC Adv.* **2015,** *123*, 101999–102075. (Review).
10. Iwasaki, K.; Sasaki, S.; Kasai, Y.; Kawashima, Y.; Sasaki, S.; Ito, T.; Yotsu-Yamashita, M.; Sasaki, M. *J. Org. Chem.* **2017,** *82*, 13204–13219.
11. Ukis, R.; Schneider, C. *J. Org. Chem.* **2019,** *84*, 7175–7188.

Inverse-electron-demand Diels–Alder Reaction

Inverse-electron-demand Diels–Alder reaction (IEDDA) employs dienes with electron-withdrawing groups (EWG) and dienophile with electron-donating groups (EDG).

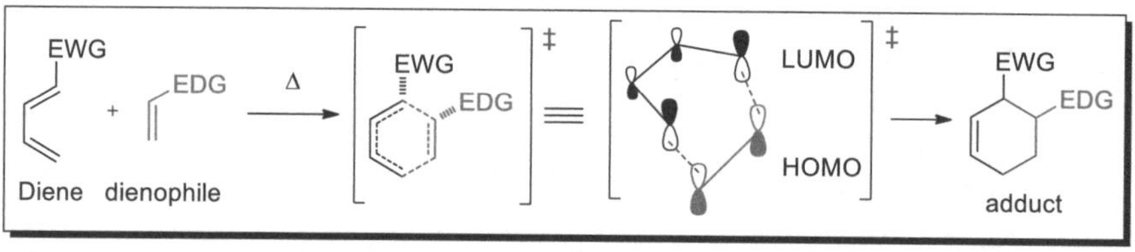
EWG
EDG
Δ
Diene
dienophile
EWG
EDG
‡
LUMO
HOMO
‡
EWG
EDG
adduct

Example 1, Catalytic asymmetric IEDDA[2]

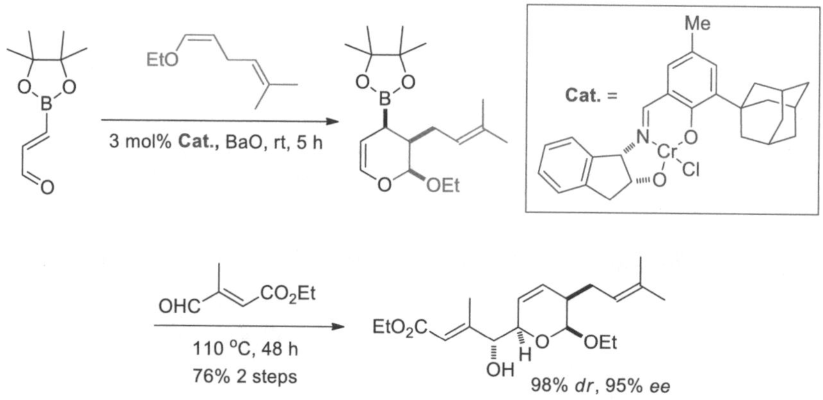
EtO
3 mol% Cat., BaO, rt, 5 h
OEt
Cat. =
Me
OHC
CO2Et
110 °C, 48 h
76% 2 steps
EtO2C
OH
H
OEt
98% dr, 95% ee

Example 2, Catalytic asymmetric IEDDA[3]

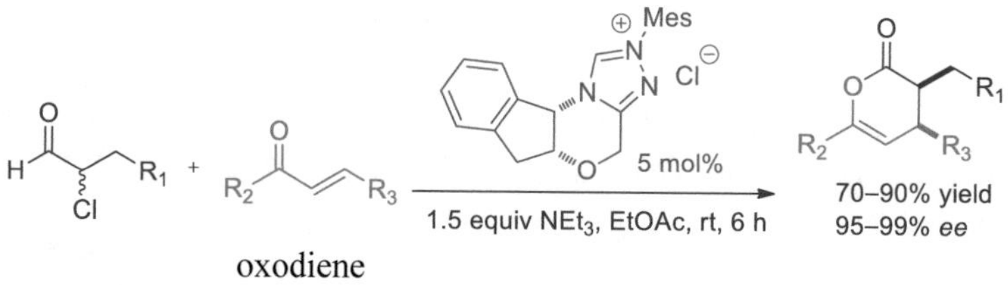
Mes
Cl
5 mol%
H
Cl
R1
R2
R3
oxodiene
1.5 equiv NEt3, EtOAc, rt, 6 h
70–90% yield
95–99% ee

Example 3, Catalytic asymmetric IEDDA[4]

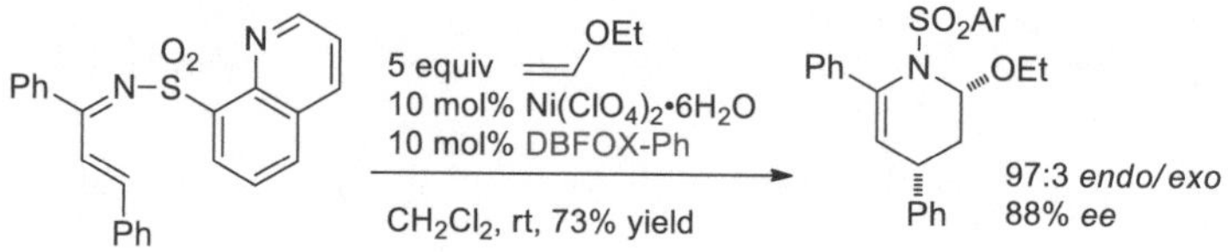
Ph
Ph
5 equiv
OEt
10 mol% Ni(ClO4)2•6H2O
10 mol% DBFOX-Ph
CH2Cl2, rt, 73% yield
SO2Ar
OEt
97:3 endo/exo
88% ee

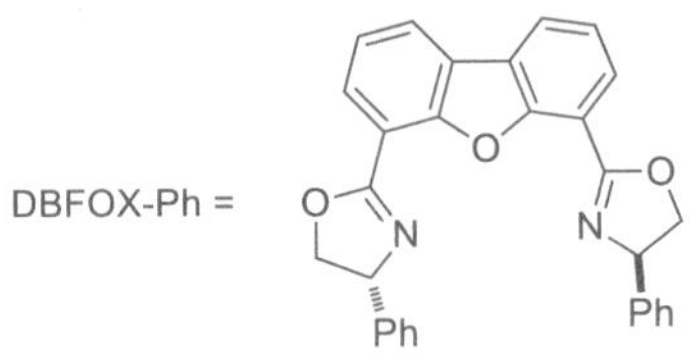
DBFOX-Ph =
O
O
N
O
N
O
Ph
Ph

Example 4, IEDDA[5]

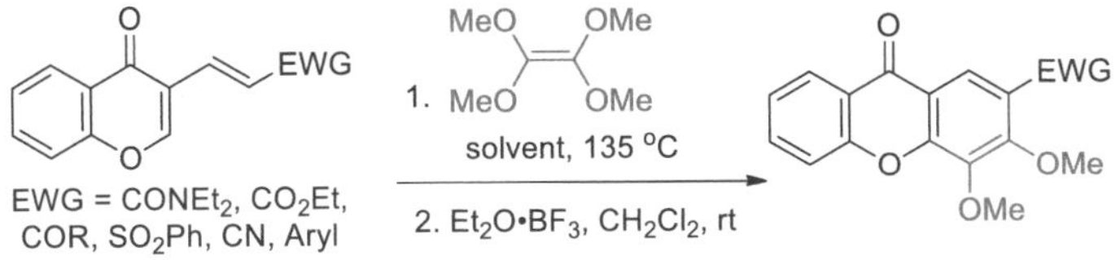
EWG
EWG = CONEt2, CO2Et,
COR, SO2Ph, CN, Aryl
MeO OMe
1. MeO OMe
solvent, 135 °C
2. Et2O•BF3, CH2Cl2, rt
EWG
OMe
OMe

Example 5, IEDDA[6]

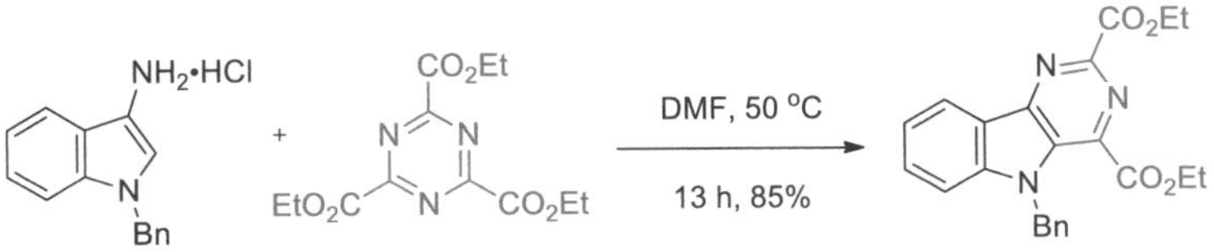
NH2•HCl
Bn
CO2Et
EtO2C
CO2Et
DMF, 50 °C
13 h, 85%
CO2Et
CO2Et
Bn

Example 6, Application of IEDDA in DNA-encoded library (DEL) synthesis[7]

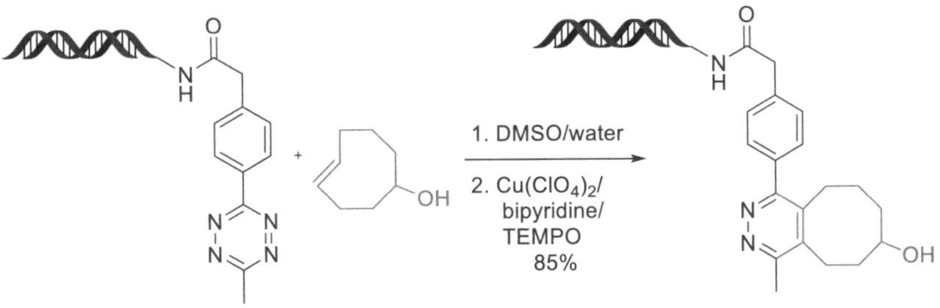
OH
1. DMSO/water
2. Cu(ClO4)2/
bipyridine/
TEMPO
85%
OH

Example 7, Inverse electron-demand oxa-Diels–Alder reaction (oxa-IEDDA)[8]

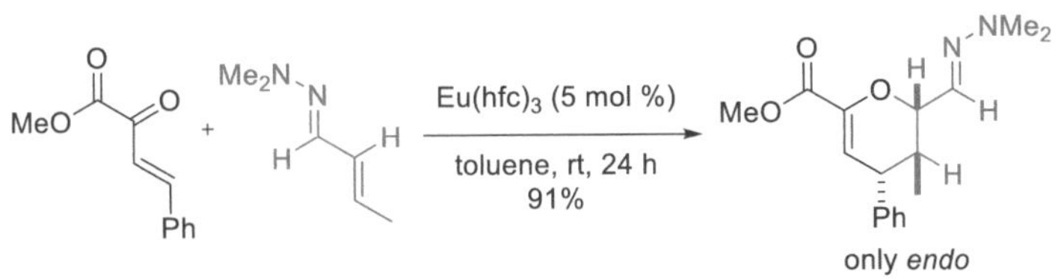
MeO
Ph
Me2N
Eu(hfc)3 (5 mol %)
toluene, rt, 24 h
91%
NMe2
MeO
Ph
only endo

Example 8, IEDDA carried out in liposomes[9]

References

1. Boger, D. L.; Patel, M. *Prog. Heterocycl. Chem.* **1989**, *1*, 30–64. (Review).
2. Gao, X.; Hall, D. G. *J. Am. Chem. Soc.* **2005**, *127*, 1628–1629.
3. He, M.; Uc, G. J.; Bode, J. W. *J. Am. Chem. Soc.* **2006**, *128*, 15088–15089.
4. Esquivias, J.; Gomez Arrayas, R.; Carretero, J. C. *J. Am. Chem. Soc.* **2007**, *129*, 1480–1481.
5. Dang, A.-T.; Miller, D. O.; Dawe, L. N.; Bodwell, G. J. *Org. Lett.* **2008**, *10*, 233–236.
6. Xu, G.; Zheng, L.; Dang, Q.; Bai, X. *Synthesis* **2013**, *45*, 743–752.
7. Li, H.; Sun, Z.; Wu, W.; Wang, X.; Zhang, M.; Lu, X.; Zhong, W.; Dai, D. *Org. Lett.* **2018**, *20*, 7186–7191.
8. Hashimoto, Y.; Ikeda, T.; Ida, A.; Morita, N.; Tamura, O. *Org. Lett.* **2019**, *21*, 4245–4249.
9. Kannaka, K.; Sano, K.; Hagimori, M.; Yamasaki, T. *Bioorg. Med. Chem.* **2019**, *27*, 3613–3618.
10. Zhang, J.; Shukla, V.; Boger, D. L. *J. Org. Chem.* **2019**, *84*, 9397–9445. (Review).
11. Saktura, M.; Grzelak, P.; Dybowska, J.; Albrecht, L. *Org. Lett.* **2020**, *22*, 1813–1817.

Dienone–Phenol Rearrangement

Acid-promoted rearrangement of 4,4-disubstituted cyclohexadienones to 3,4-disubstituted phenols.

Example 1, Intramolecular dienone–phenol rearrangement[4]

Example 2, Classic dienone–phenol rearrangement[5]

Example 3, Intramolecular dienone–phenol rearrangement[9]

J. J. Li, *Name Reactions*, https://doi.org/10.1007/978-3-030-50865-4_46

Example 4, Via a common intermediate[10]

Example 5, Dienone–phenol rearrangement using an optically active substrate[11]

66% ee   65% ee

Example 6, Intramolecular dienone–phenol rearrangement[12]

Nf = nonafluorobutanesulfonyl, $-SO_2CF_2CF_2CF_2CF_3$, a protecting group for $-OH$

Example 7, An unprecedented *decarboxylative* dienone–phenol rearrangement[13]

References

1. Shine, H. J. In *Aromatic Rearrangements;* Elsevier: New York, **1967**, pp 55–68. (Review).
2. Schultz, A. G.; Hardinger, S. A. *J. Org. Chem.* **1991**, *56*, 1105–1111.
3. Schultz, A. G.; Green, N. J. *J. Am. Chem. Soc.* **1992**, *114*, 1824–1829.

4. Hart, D. J.; Kim, A.; Krishnamurthy, R.; Merriman, G. H.; Waltos, A.-M. *Tetrahedron* **1992,** *48*, 8179–8188.
5. Frimer, A. A.; Marks, V.; Sprecher, M.; Gilinsky-Sharon, P. *J. Org. Chem.* **1994,** *59*, 1831–1834.
6. Oshima, T.; Nakajima, Y.-i.; Nagai, T. *Heterocycles* **1996,** *43*, 619–624.
7. Draper, R. W.; Puar, M. S.; Vater, E. J.; Mcphail, A. T. *Steroids* **1998,** *63*, 135–140.
8. Kodama, S.; Takita, H.; Kajimoto, T.; Nishide, K.; Node, M. *Tetrahedron* **2004,** *60*, 4901–4907.
9. Bru, C.; Guillou, C. *Tetrahedron* **2006,** *62*, 9043–9048.
10. Sauer, A. M.; Crowe, W. E.; Henderson, G.; Laine, R. A. *Tetrahedron Lett.* **2007,** *48*, 6590–6593.
11. Yoshida, M.; Nozaki, T.; Nemoto, T.; Hamada, Y. *Tetrahedron* **2013,** *69*, 9609–9615.
12. Takubo, K.; Mohamed, A. A. B.; Ide, T.; Saito, K.; Ikawa, T.; Yoshimitsu, T.; Akai, S. *J. Org. Chem.* **2017,** *82*, 13141–13151.
13. Zentar, H.; Arias, F.; Haidour, A.; Alvarez-Manzaneda, R.; Chahboun, R.; Alvarez-Manzaneda, E. *Org. Lett.* **2018,** *20*, 7007–7010.

Dötz Reaction

Also known as the Dötz benzannulation, the Dötz reaction is the Cr(CO)$_3$-coordinated hydroquinone from vinylic alkoxy pentacarbonyl chromium carbene (Fischer carbene) complex and alkynes.

Example 1[5]

Example 3[8]

J. J. Li, *Name Reactions*, https://doi.org/10.1007/978-3-030-50865-4_47

Example 3[8]

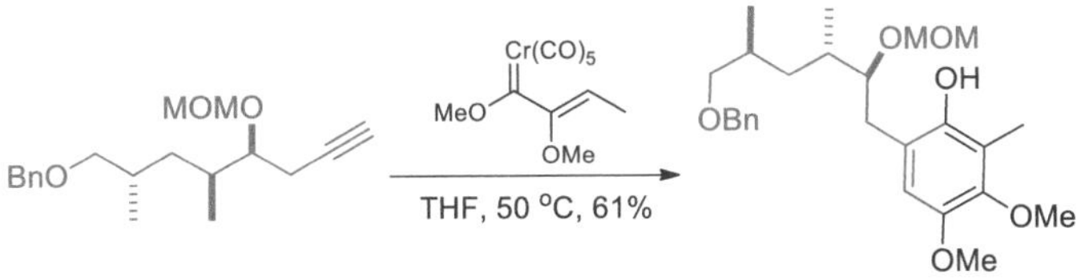
MOMO
BnO
Cr(CO)5
MeO
OMe
THF, 50 °C, 61%
OMOM
OH
OBn
OMe
OMe

Example 3[9]

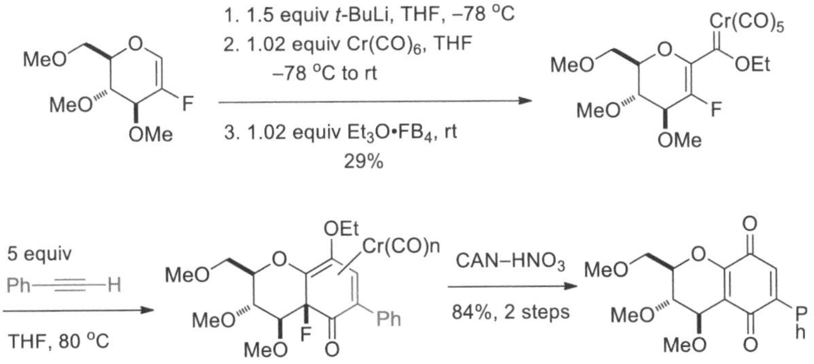
MeO
MeO
OMe
F
1. 1.5 equiv t-BuLi, THF, –78 °C
2. 1.02 equiv Cr(CO)6, THF
–78 °C to rt
3. 1.02 equiv Et3O•FB4, rt
29%
Cr(CO)5
OEt
5 equiv
Ph—≡—H
THF, 80 °C
Cr(CO)n
Ph
CAN–HNO3
84%, 2 steps

Example 4[10]

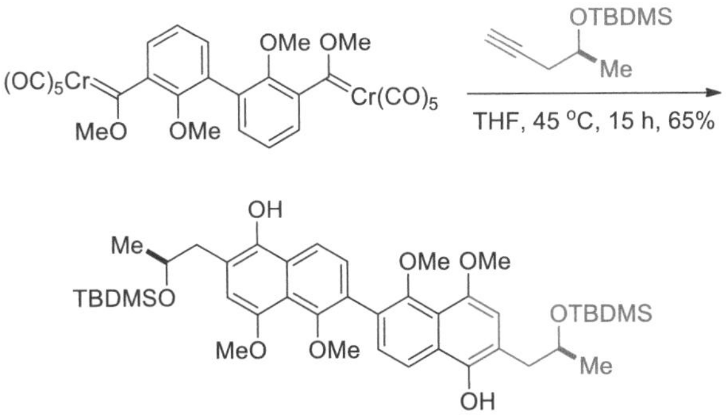
(OC)5Cr
MeO
OMe
OMe OMe
Cr(CO)5
OTBDMS
Me
THF, 45 °C, 15 h, 65%
OH
TBDMSO
MeO
OMe
OH

Example 5, Assembly of polysubstituted and highly oxygenated phenols, a tungsten (W) variant[11]

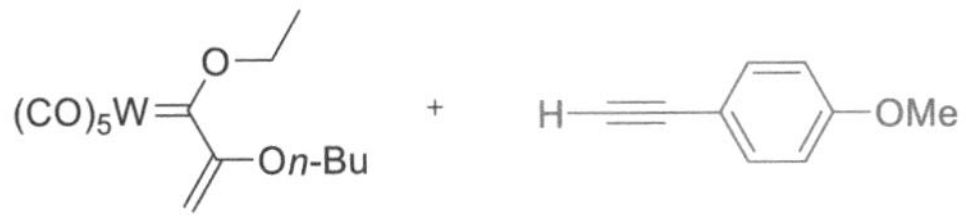
(CO)5W
On-Bu
+
H
OMe

α-alkoxyvinyl(ethoxy) carbene complex

1. PhH, 60 °C, 4 h

2. air, H$_2$O/THF
40%

n-BuO

OH

OMe

H

n-BuO

O

+

O

O

On-Bu

Example 6, Dötz benzannulation[12]

MeO

OMe

Br

OMe

1. n-BuLi, THF, –78 °C, 10–12 min
2. Cr(CO)$_6$, 0 °C, 1.5 h; rt, 1.5 h

3. Me$_3$OBF$_4$, CH$_2$Cl$_2$, 0 °C; rt, 1.5 h
54%

MeO

OMe

MeO

OMe

Cr(CO)$_5$

+

TBSO

OTBS

Ac$_2$O, THF

45 °C, 14 h
32%

MeO

OMe

MeO

OTBS

OTBS

MeO

OH

References

1. Dötz, K. H. *Angew. Chem. Int. Ed.* **1975**, *14*, 644–645. Karl H. Dötz (1943–) was a professor at the University of Munich in Germany.
2. Wulff, W. D. In *Advances in Metal-Organic Chemistry*; Liebeskind, L. S., Ed.; JAI Press, Greenwich, CT; **1989**; *Vol. 1.* (Review).
3. Wulff, W. D. In *Comprehensive Organometallic Chemistry II*; Abel, E. W., Stone, F. G. A., Wilkinson, G., Eds.; Pergamon Press: Oxford, **1995**; *Vol. 12.* (Review).
4. Torrent, M.; Solá, M.; Frenking, G. *Chem. Rev.* **2000**, *100*, 439–494. (Review).
5. Caldwell, J. J.; Colman, R.; Kerr, W. J.; Magennis, E. J. *Synlett* **2001**, 1428–1430.
6. Solá, M.; Duran, M.; Torrent, M. *The Dötz reaction: A chromium Fischer carbene-mediated benzannulation reaction.* In *Computational Modeling of Homogeneous Catalysis* Maseras, F.; Lledós, eds.; Kluwer Academic: Boston; **2002**, 269–287. (Review).
7. Pulley, S. R.; Czakó, B. *Tetrahedron Lett.* **2004**, *45*, 5511–5514.
8. White, J. D.; Smits, H. *Org. Lett.* **2005**, *7*, 235–238.
9. Boyd, E.; Jones, R. V. H.; Quayle, P.; Waring, A. J. *Tetrahedron Lett.* **2005**, *47*, 7983–7986.
10. Fernandes, R. A.; Mulay, S. V. *J. Org. Chem.* **2010**, *75*, 7029–7032.
11. Montenegro, M. M.; Vega-Baez, J. L.; Vazquez, M. A.; Flores-Conde, M. I.; Sanchez, A.; Gonzalez-Tototzin, M.A.; Gutierrez, R. U.; Lazcano-Seres, J. M.; Ayala, F.; Zepeda, L. G.; et al. *J. Organomet. Chem.* **2016**, *825–826*, 41–54.
12. Kotha, S.; Aswar, V. R.; Manchoju, A. *Tetrahedron* **2016**, *72*, 2306–2315.
13. Hirose, T.; Kojima, Y.; Matsui, H.; Hanaki, H.; Iwatsuki, M.; Shiomi, K.; Omura, S.; Sunazuka, T. *J. Antibiot.* **2017**, *70*, 574–581.
14. Fernandes, R. A.; Kumari, A.; Pathare, R. S. *Synlett* **2020**, *31*, 403–420. (Review).

Eschweiler–Clarke Reductive Amination

Reductive methylation of primary or secondary amines using formaldehyde and formic acid. *Cf.* Leuckart–Wallach reaction.

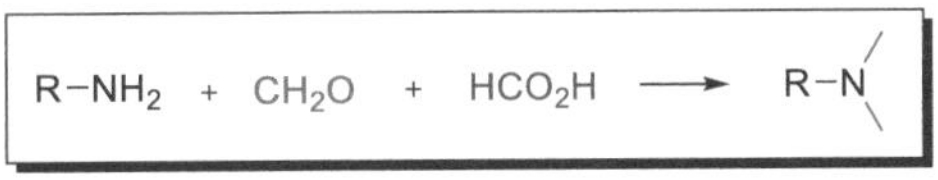

$$R-NH_2 \ + \ CH_2O \ + \ HCO_2H \ \longrightarrow \ R-N\diagup$$

formic acid is the hydride source, serving as a reducing agent

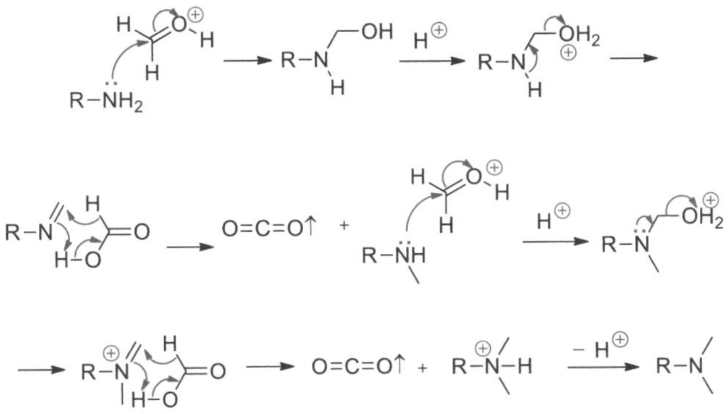

Example 1[7]

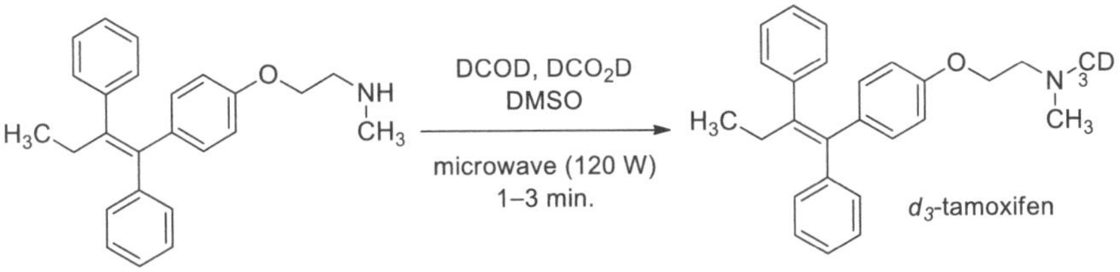

DCOD, DCO$_2$D
DMSO

microwave (120 W)
1–3 min.

d_3-tamoxifen

Example 2[9]

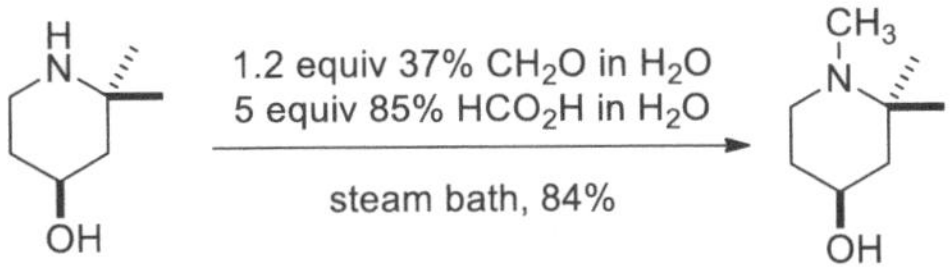

1.2 equiv 37% CH$_2$O in H$_2$O
5 equiv 85% HCO$_2$H in H$_2$O

steam bath, 84%

Example 3[10]

varenicline (Chantix)

CHO

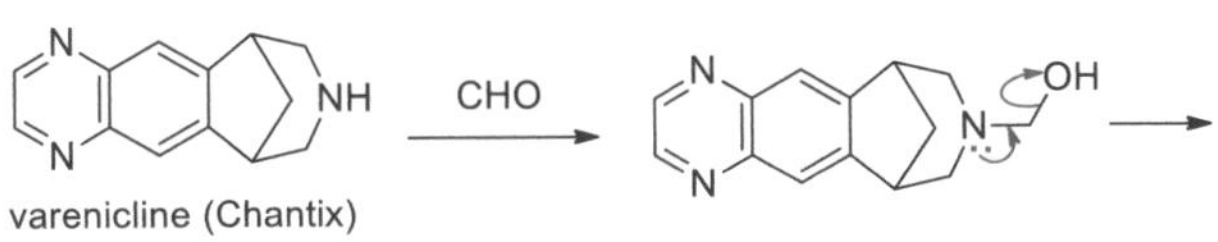

© Springer Nature Switzerland AG 2021
J. J. Li, *Name Reactions*, https://doi.org/10.1007/978-3-030-50865-4_48

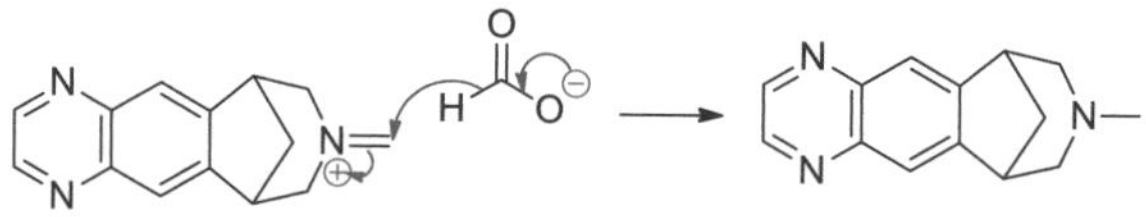

Example 4, Preparation of a selective serotonin reuptake inhibitor (SSRI) for treatment of premature ejaculation (PE)[11]

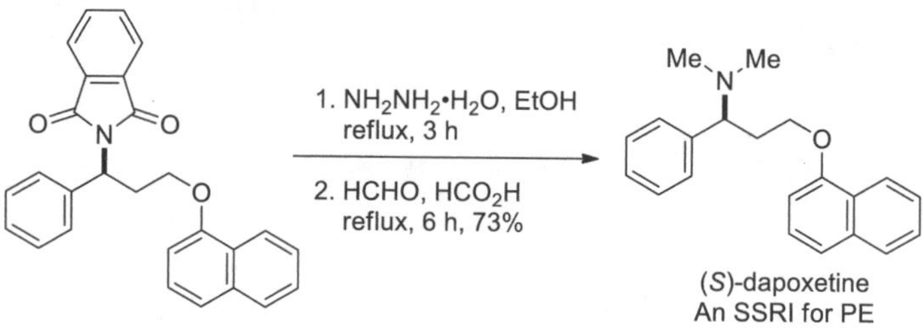

(S)-dapoxetine
An SSRI for PE

Example 5, Toward synthesis of evogliptin (Suganon), a dipeptidyl peptidase IV (DPP-4) inhibitor[12]

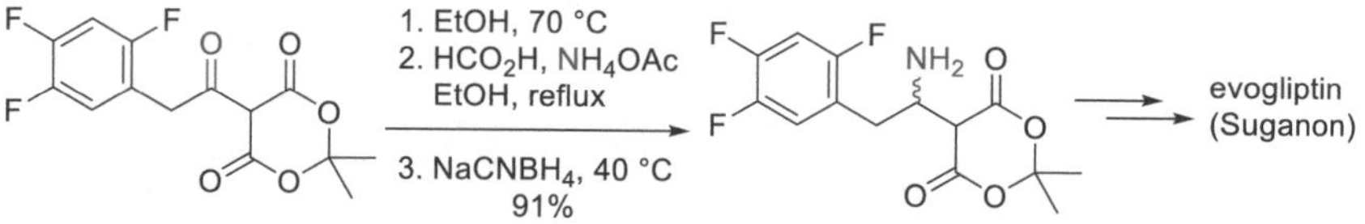

Example 6, Toward synthesis of rucaparib (Rubraca), a poly(ADP-ribosyl) polymerase (PARP) inhibitor[13]

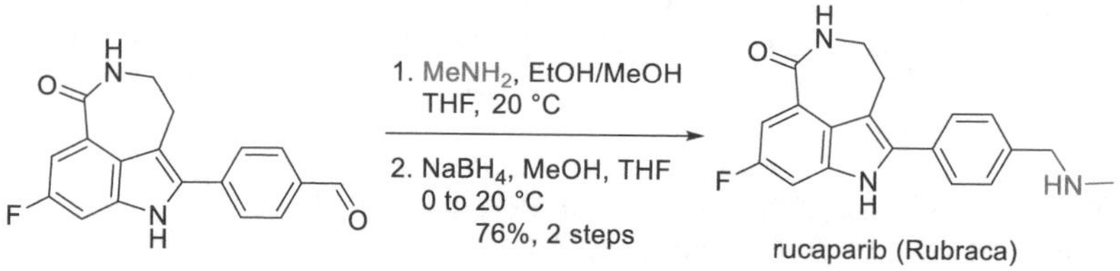

rucaparib (Rubraca)

Example 7, Toward the synthesis of abemaciclib (Verzanio), a cyclin-dependent kinase 4/6 (CDK4/6) inhibitor[14]

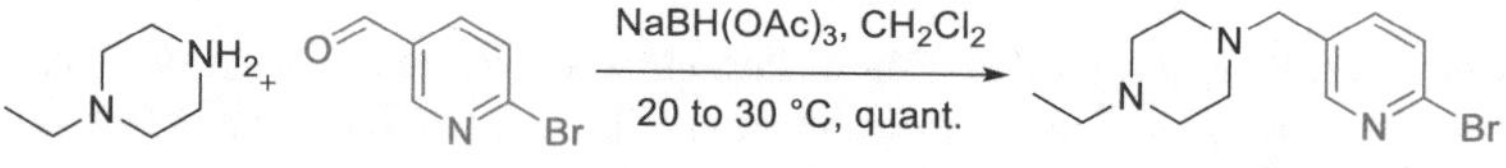

abemaciclib (Verzanio)
Lilly, 2017
CDK4/6 inhibitor

Example 8, Dynamic kinetic resolution (DKR)–asymmetric reductive amination (ARA) protocol[15]

1.25 equiv MeNH$_2$: AcOH
Ir/BiPheP (S : C = 1000)

70 °C, 50 bar H$_2$, 18 h
IPA, 86%

(3R,4R)

tofacitinib (Xeljanz)
Pfizer, 2018 (for RA)
JAK1/2 inhibitor

BiPheP =

Ar = 3,5-di-t-Bu, 4-OMe

References

1   (a) Eschweiler, W. *Chem. Ber.* **1905,** *38*, 880–892. Wilhelm Eschweiler (1860–1936) was born in Euskirchen, Germany. (b) Clarke, H. T.; Gillespie, H. B.; Weisshaus, S. Z. *J. Am. Chem. Soc.* **1933,** *55*, 4571–4587. Hans T. Clarke (1887–1927) was born in Harrow, England.

2   Moore, M. L. *Org. React.* **1949,** *5*, 301–330. (Review).

3   Pine, S. H.; Sanchez, B. L. *J. Org. Chem.* **1971,** *36*, 829–832.

4   Bobowski, G. *J. Org. Chem.* **1985,** *50*, 929–931.

5   Alder, R. W.; Colclough, D.; Mowlam, R. W. *Tetrahedron Lett.* **1991,** *32*, 7755–7758.

6   Bulman Page, P. C.; Heaney, H.; Rassias, G. A.; Reignier, S.; Sampler, E. P.; Talib, S. *Synlett* **2000,** 104–106.

7   Harding, J. R.; Jones, J. R.; Lu, S.-Y.; Wood, R. *Tetrahedron Lett.* **2002,** *43*, 9487–9488.

8   Brewer, A. R. E. *Eschweiler–Clarke Reductive Alkylation of Amine.* In *Name Reactions for Functional Group Transformations*; Li, J. J., Ed.; Wiley: Hoboken, NJ, **2007,** pp 86–111. (Review).

9   Weis, R.; Faist, J.; di Vora, U.; Schweiger, K.; Brandner, B.; Kungl, A. J.; Seebacher, W. *Eur. J. Med. Chem.* **2008,** *43*, 872–879.

10   Waterman, K. C.; Arikpo, W. B.; Fergione, M. B.; Graul, T. W.; Johnson, B. A.; Mac-
     donald, B. C.; Roy, M. C.; Timpano, R. J. *J. Pharm. Sci.* **2008,** *97,* 1499–1507.

11   Sasikumar, M.; Nikalje, Milind D. *Synth. Commun.* **2012,** *42,* 3061–3067.

12   Kwak, W. Y.; Kim, H. J.; Mi, J. P.; Yoon, T. H.; Shim, H. J.; Yoo, M. EP 2,415,754
     (2012).

13   Gillmore, A. T.; Badland, M.; Crook, C. L.; Castro, N. M.; Critcher, D. J.; Fussell, S.
     J.; Jones, K. J.; Jones, M. C.; Kougoulos, E.; Mathew, J. S.; et al. *Org. Process Res.
     Dev.* **2012,** *16,* 1897–1904.

14   Verzijl, G. K. M.; Schuster, C.; Dax, T.; de Vries, A. H. M.; Lefort, L. *Org. Process
     Res. Dev.* **2018,** *22,* 1817–1822.

15   Afanasyev, O. I.; Kuchuk, E.; Usanov, D. L.; Chusov, D. *Chem. Rev.* **2019,** *119,*
     11857–11911. (Review).

16   Hu, L.; Zhang, Y.; Zhang, Q.-W.; Yin, Q.; Zhang, X. *Angew. Chem. Int. Ed.* **2020,** *59,*
     5321–5325.

Favorskii Rearrangement

Transformation of enolizable α-haloketones to esters, carboxylic acids, or amides *via* alkoxide-, hydroxide-, or amine-catalyzed rearrangements, respectively.

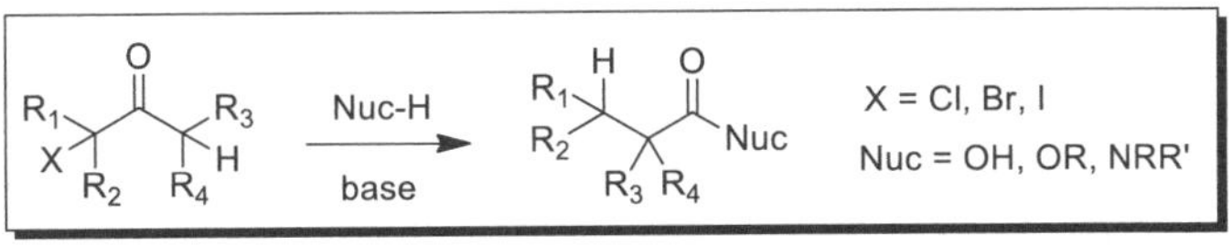

The intramolecular Favorskii Rearrangement:

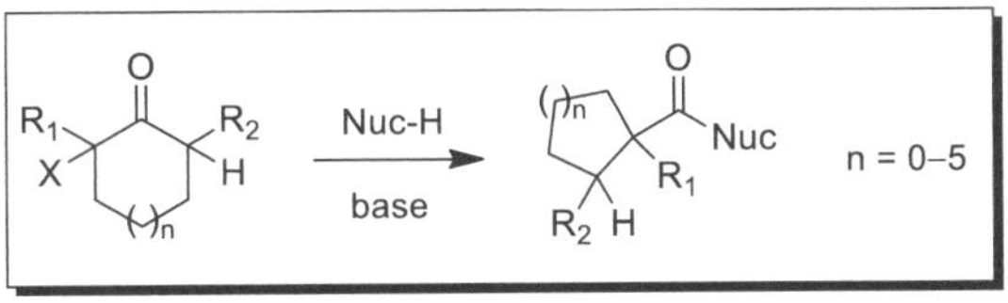

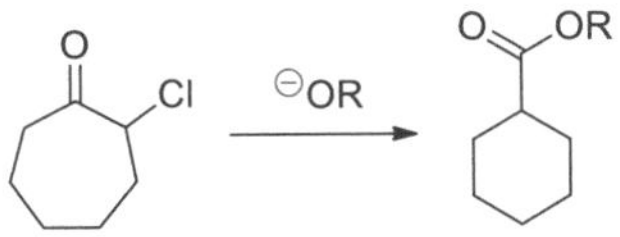

enolizable α-haloketone

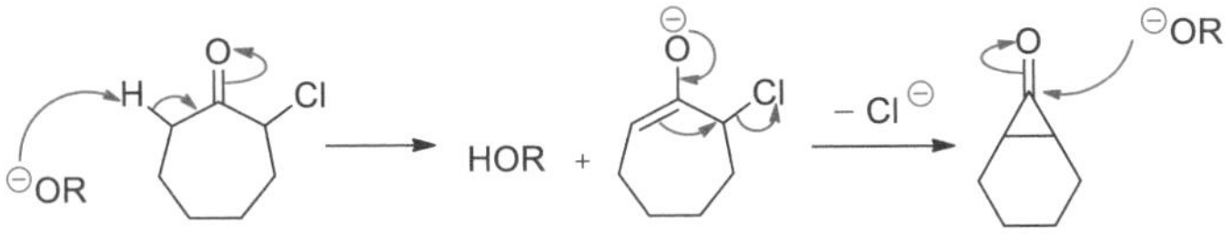

cyclopropanone intermediate

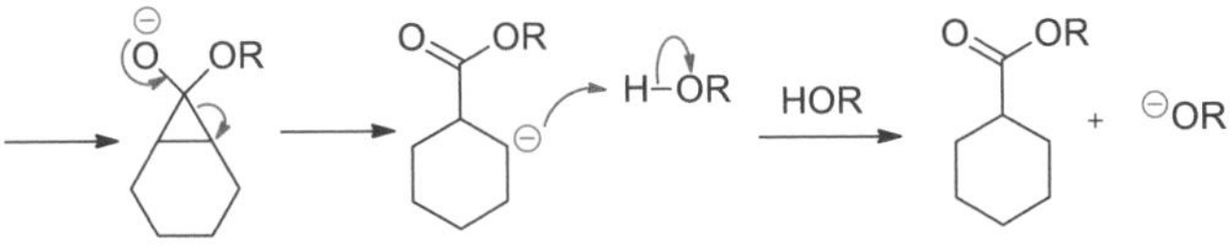

Example 1[2]

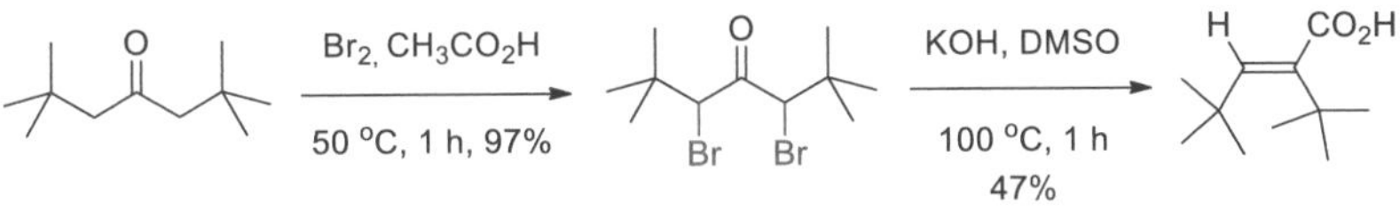

© Springer Nature Switzerland AG 2021
J. J. Li, *Name Reactions*, https://doi.org/10.1007/978-3-030-50865-4_49

Example 2, Homo-Favorskii rearrangement[3]

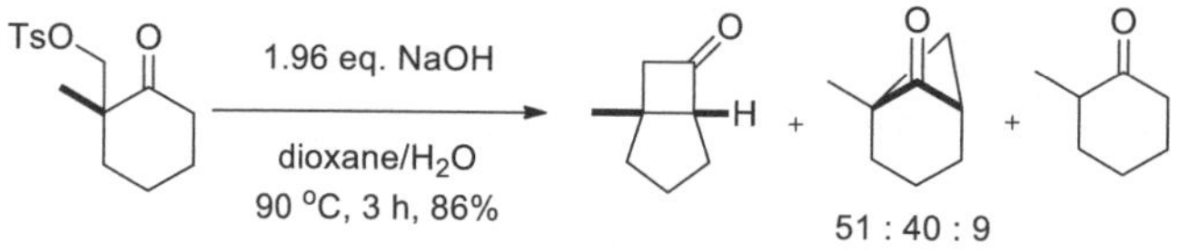

51 : 40 : 9

Example 3[6]

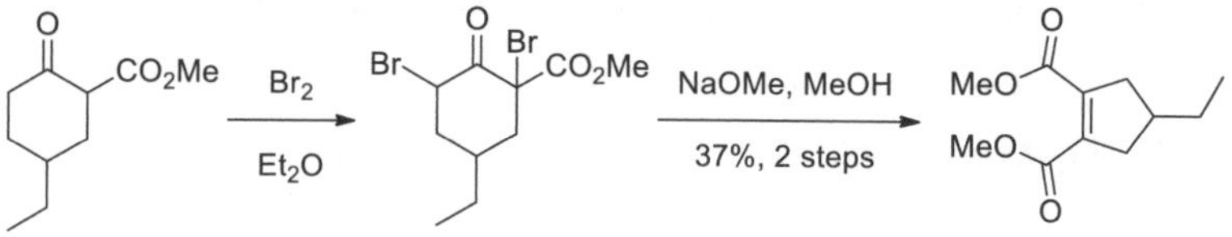

Example 4, Photo-Favorskii Rearrangement[7]

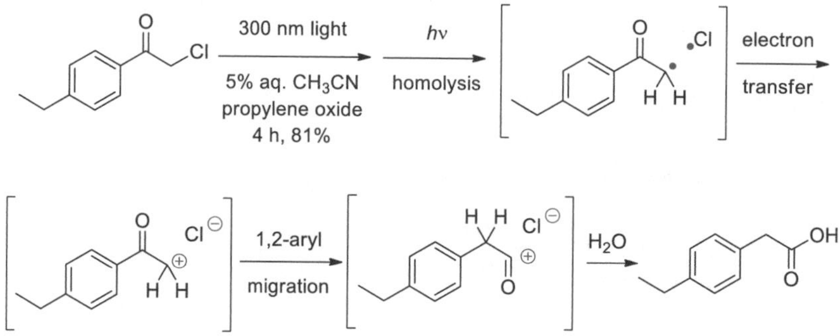

Example 5[8]

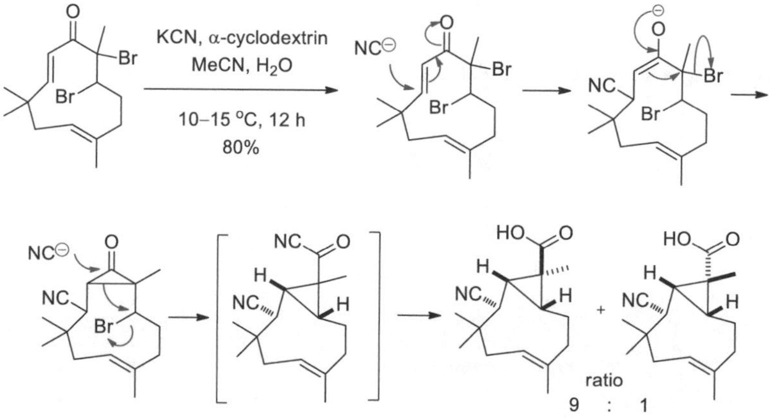

ratio
9 : 1

Example 6[10]

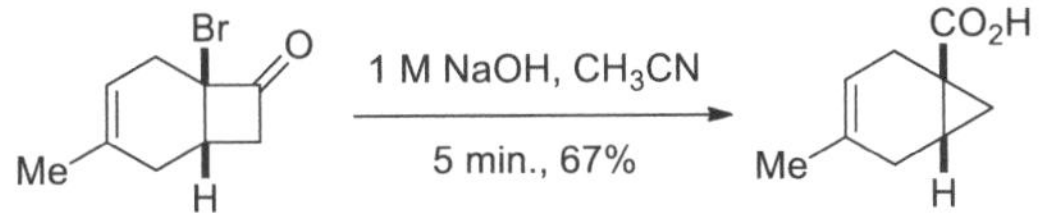

Example 7[11]

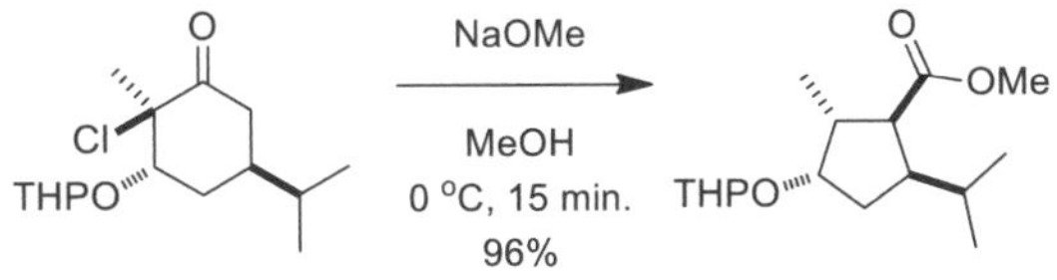

Example 8, Process scale (5 kg)[14]

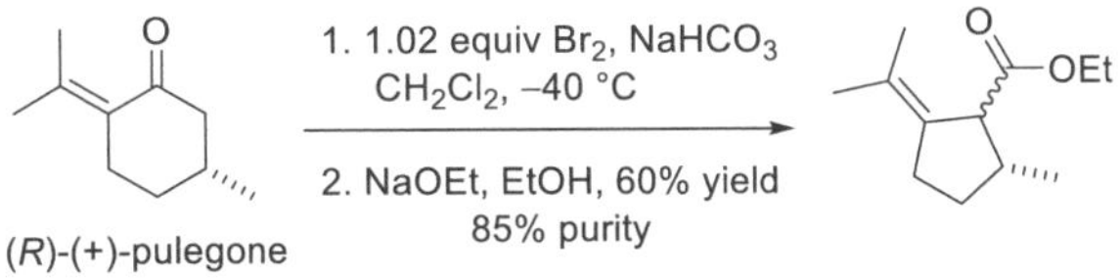

(R)-(+)-pulegone

Example 9, Process scale (3.5 kg)[15]

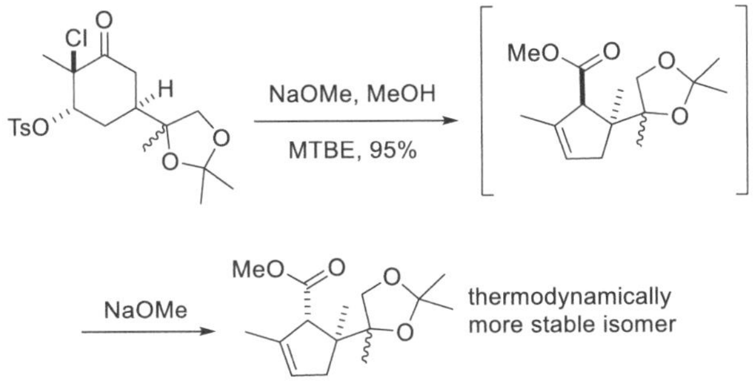

thermodynamically
more stable isomer

Example 10, A semi-Favorskii Rearrangement[16]

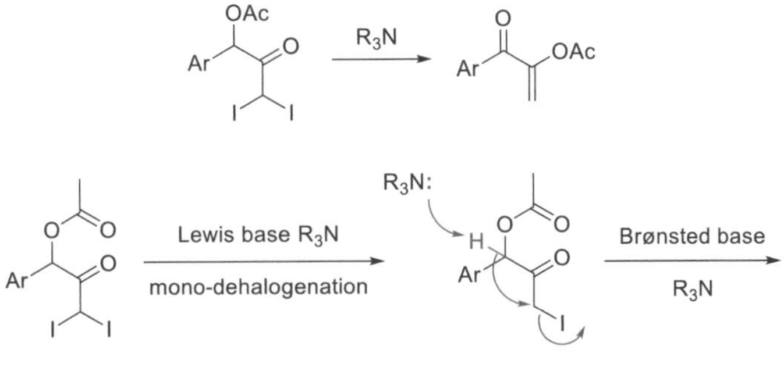

oxy-allyl cation

References

1. (a) Favorskii, A. E. *J. Prakt. Chem.* **1895**, *51*, 533–563. Aleksei E. Favorskii (1860–1945), born in Selo Pavlova, Russia, studied at St. Petersburg State University, where he became a professor since 1900. (b) Favorskii, A. E. *J. Prakt. Chem.* **1913,** *88*, 658.
2. Wagner, R. B.; Moore, J. A. *J. Am. Chem. Soc.* **1950,** *72*, 3655–3658.
3. Wenkert, E.; Bakuzis, P.; Baumgarten, R. J.; Leicht, C. L.; Schenk, H. P. *J. Am. Chem. Soc.* **1971,** *93*, 3208–3216.
4. Chenier, P. J. *J. Chem. Ed.* **1978,** *55*, 286–291. (Review).
5. Barreta, A.; Waegell, B. In *Reactive Intermediates*; Abramovitch, R. A., ed.; Plenum Press: New York, **1982,** *2*, pp 527–585. (Review).
6. White, J. D.; Dillon, M. P.; Butlin, R. J. *J. Am. Chem. Soc.* **1992,** *114*, 9673–9674.
7. Dhavale, D. D.; Mali, V. P.; Sudrik, S. G.; Sonawane, H. R. *Tetrahedron* **1997,** *53*, 16789–16794.
8. Kitayama, T.; Okamoto, T. *J. Org. Chem.* **1999,** *64*, 2667–2672.
9. Mamedov, V. A.; Tsuboi, S.; Mustakimova, L. V.; Hamamoto, H.; Gubaidullin, A. T.; Litvinov, I. A.; Levin, Y. A. *Chem. Heterocyclic Compd.* **2001,** *36*, 911. (Review).
10. Harmata, M.; Wacharasindhu, S. *Org. Lett.* **2005,** *7*, 2563–2565.
11. Pogrebnoi, S.; Saraber, F. C. E.; Jansen, B. J. M.; de Groot, A. *Tetrahedron* **2006,** *62*, 1743–1748.
12. Filipski, K. J.; Pfefferkorn, J. A. *Favorskii Rearrangement.* In *Name Reactions for Homologations-Part II*; Li, J. J., Ed.; Wiley: Hoboken, NJ, **2009,** pp 238–252. (Review).
13. Kammath, V. B.; Šolomek, T.; Ngoy, B. P.; Heger, D.; Klán, P.; Rubina, M.; Givens, R. S. *J. Org. Chem.* **2013,** *78*, 1718–1729.
14. Lane, J. W.; Spencer, K. L.; Shakya, S. R.; Kallan, N. C.; Stengel, P. J.; Remarchuk, T. *Org. Process Res. Dev.* **2014,** *18*, 31641–1651.
15. Xu, H.; Wang, F.; Xue, W.; Zheng, Y.; Wang, Q.; Qiu, F. G.; Jin, Y. *Org. Process Res. Dev.* **2018,** *22*, 377–384.
16. Sadhukhan, S.; Baire, B. *Org. Lett.* **2018,** *20*, 1748–1751.
17. Shuai, B.; Fang, P.; Mei, T.-S. *Synlet* **2020,** in press.

Quasi-Favorskii Rearrangement

If there are no enolizable hydrogens present, the classical Favorskii rearrangement is not possible. Instead, a semi-benzylic mechanism can lead to a rearrangement referred to as quasi-Favorskii.

Example 1, Arthur C. Cope's initial discovery[1]

non-enolizable ketone

Example 2[5]

Example 3[6]

References
1. Cope, A. C.; Graham, E. S. *J. Am. Chem. Soc.* **1951,** *73*, 4702–4706.
2. Smissman, E. E.; Diebold, J. L. *J. Org. Chem.* **1965,** *30*, 4005–4007.
3. Sasaki, T.; Eguchi, S.; Toru, T. *J. Am. Chem. Soc.* **1969,** *91*, 3390–3391.
4. Baudry, D.; Begue, J. P.; Charpentier-Morize, M. *Tetrahedron Lett.* **1970,** 2147–2150.
5. Stevens, C. L.; Pillai, P. M.; Taylor, K. G. *J. Org. Chem.* **1974,** *39*, 3158–3161.
6. Harmata, M.; Wacharasindhu, S. *Org. Lett.* **2005,** *7,* 2563–2565.
7. Filipski, K.J.; Pfefferkorn, J. A. *Favorskii Rearrangement*. In *Name Reactions for Homologations-Part II*; Li, J. J., Ed.; Wiley: Hoboken, NJ, **2009,** pp 438–452. (Review).
8. Harmata, M.; Wacharasindhu, S. *Synthesis* **2007,** 2365–2369.
9. Ross, A. G.; Townsend, S. D.; Danishefsky, S. J. *J. Org. Chem.* **2013,** *78*, 204–210.
10. Behnke, N. E.; Siitonen, J. H.; Chamness, S. A.; Kürti, L. *Org. Lett.* **2020,** *78*, 204–210.

Ferrier Carbocyclization

This process has proved to be of considerable value for the efficient, one-step conversion of 5,6-unsaturated hexopyranose derivatives into functionalized cyclohexanones useful for the preparation of such enantiomerically pure compounds as inositols and their amino, deoxy, unsaturated and selectively O-substituted derivatives, notably phosphate esters. In addition, the products of the carbocyclization have been incorporated into many complex compounds of interest in biological and medicinal chemistry.[1,2]

General examples:[3]

More complex products:

© Springer Nature Switzerland AG 2021
J. J. Li, *Name Reactions*, https://doi.org/10.1007/978-3-030-50865-4_50

Complex bioactive compounds made following the application of the reaction:

paniculide A[9]   pancratistatin[10]   calystegine B$_2$[11]

Modified hex-5-enopyranosides and reactions

a, Hg(OCOCF$_3$)$_2$, Me$_2$CO, H$_2$O, 0 °C; b, NaBH(OAc)$_3$, AcOH, MeCN, rt; c, i-
Bu$_3$Al, PhMe, 40 °C; d, Ti(Oi-Pr)Cl$_3$, CH$_2$Cl$_2$, −78 °C, 15 min. (Note: The aglycon
is retained in the Al- and Ti-induced reactions).

A recent example of a novel Ferrier-type carbocyclization[19]

References

1. Ferrier, R. J.; Middleton, S. *Chem. Rev.* **1993,** *93,* 2779–2831. (Review).
2. Ferrier, R. J. *Top. Curr. Chem.* **2001,** *215,* 277–291 (Review).
3. Ferrier, R. J. *J. Chem. Soc., Perkin Trans. 1* **1979,** 1455–1458. The discovery (1977) was made in the Pharmacology Department, University of Edinburgh, while R. J. Ferrier was on leave from Victoria University of Wellington, New Zealand where he was Professor of Organic Chemistry. He is now a consultant with Industrial Research Ltd., Lower Hutt, New Zealand.
4. Blattner, R.; Ferrier, R. J.; Haines, S. R. *J. Chem. Soc., Perkin Trans. 1,* **1985,** 2413–2416.
5. Chida, N.; Ohtsuka, M.; Ogura, K.; Ogawa, S. *Bull. Chem. Soc. Jpn.* **1991,** *64,* 2118–2121.
6. Machado, A. S.; Olesker, A.; Lukacs, G. *Carbohydr. Res.* **1985,** *135,* 231–239.
7. Sato, K.-i.; Sakuma, S.; Nakamura, Y.; Yoshimura, J.; Hashimoto, H. *Chem. Lett.* **1991,** 17–20.
8. Ermolenko, M. S.; Olesker, A.; Lukacs, G. *Tetrahedron Lett.* **1994,** *35,* 711–714.
9. Amano, S.; Takemura, N.; Ohtsuka, M.; Ogawa, S.; Chida, N. *Tetrahedron* **1999,** *55,* 3855–3870.
10. Park, T. K.; Danishefsky, S. J. *Tetrahedron Lett.* **1995,** *36,* 195–196.
11. Boyer, F.-D.; Lallemand, J.-Y. *Tetrahedron* **1994,** *50,* 10443–10458.
12. Das, S. K.; Mallet, J.-M.; Sinaÿ, P. *Angew. Chem. Int. Ed.* **1997,** *36,* 493–496.
13. Sollogoub, M.; Mallet, J.-M.; Sinaÿ, P. *Tetrahedron Lett.* **1998,** *39,* 3471–3472.
14. Bender, S. L.; Budhu, R. J. *J. Am. Chem. Soc.* **1991,** *113,* 9883–9884.
15. Estevez, V. A.; Prestwich, E. D. *J. Am. Chem. Soc.* **1991,** *113,* 9885–9887.
16. Yadav, J. S.; Reddy, B. V. S.; Narasimha Chary, D.; Madavi, C.; Kunwar, A. C. *Tetrahedron Lett.* **2009,** *50,* 81–84.
17. Chen, P.; Wang, S. *Tetrahedron* **2013,** *69,* 583–588.
18. Chen, P.; Lin, L. *Tetrahedron* **2013,** *69,* 4524–4531.
19. Hedberg, C.; Estrup, M.; Eikeland, E. Z.; Jensen, H. *J. Org. Chem.* **2018,** *83,* 2154–2165.
20. Ausmus, A. P.; Hogue, M.; Snyder, J. L.; Rundell, S. R.; Bednarz, K. M.; Banahene, N.; Swarts, B. M. *J. Org. Chem.* **2020,** *85,* 3182–3191.

Ferrier Glycal Allylic Rearrangement

In the presence of Lewis acid catalysts *O*-substituted glycal derivatives can react with *O*-, *S*-, *C*- and, less frequently, *N*-, *P*- and halide nucleophiles to give 2,3-unsaturated glycosyl products.[1,2] This allylic transformation has been "Ferrier Rearrangement". However, the reaction was first noted by Emil Fischer when he heated tri-*O*-acetyl-*D*-glucal in water.[3] When carbon nucleophiles are involved, the term "Carbon Ferrier Reaction" has been used,[4] although the only contribution the Ferrier group made in this area was to find that tri-*O*-acetyl-*D*-glucal dimerizes under acid catalysis to give a *C*-glycosidic product.[5] The general reaction is illustrated by the separate conversions of tri-*O*-acetyl-*D*-glucal with *O*-, *S*- and *C*-nucleophiles to the corresponding 2,3-unsaturated glycosyl derivatives. Normally, Lewis acids are used as catalysts, boron trifluoride etherate being the most common. Allyloxycarbenium ions are involved as intermediates, high yields of products are obtained, and glycosidic compounds with quasi-axial bonds (as illustrated) predominate (commonly in the α,β-ratio of about 7:1). The examples illustrated[4,6,7] are typical of a very large number of literature reports.[1]

General examples[4]

More complex products made directly from the corresponding glycols:

benzene, BF$_3$•OEt$_2$,
5 °C, 10 min, (67%,
α-anomer).[8]

PhCOCH$_2$CO$_2$Et,
BF$_3$•OEt$_2$,
rt, 15 min,
(81% α-anomer).[9]

By spontaneous sigmatropic
rearrangement of the glycal
3-trichloroacetimidate made
with NaH, Cl$_3$CCN,
(78% α-anomer).[10]

J. J. Li, *Name Reactions*, https://doi.org/10.1007/978-3-030-50865-4_51

Products formed without acid catalysts:

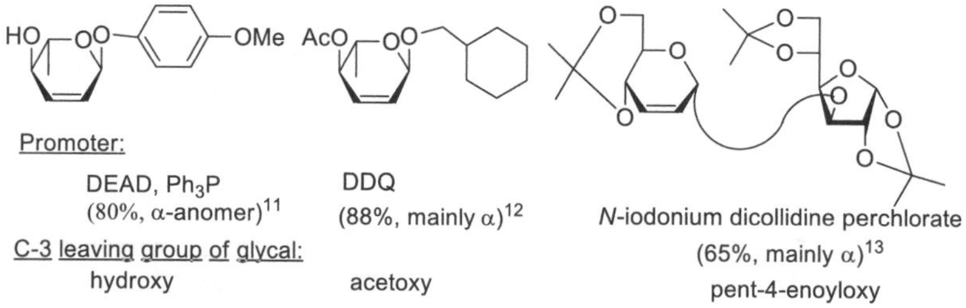

HO
OMe
AcO
Promoter:
DEAD, Ph3P
(80%, α-anomer)11
DDQ
(88%, mainly α)12
N-iodonium dicollidine perchlorate
(65%, mainly α)13
C-3 leaving group of glycal:
hydroxy
acetoxy
pent-4-enoyloxy

Modified glycols and their reactions:

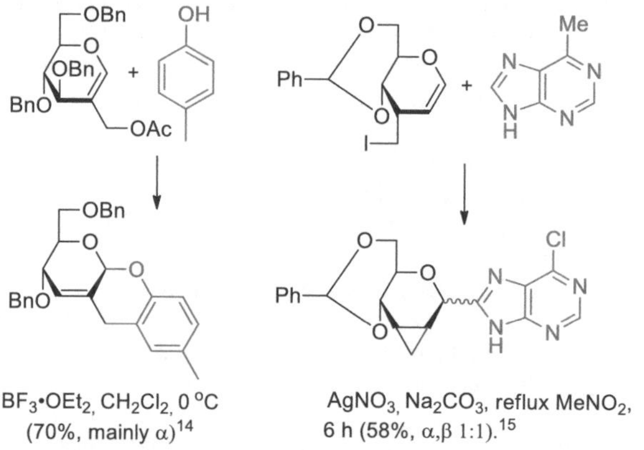

OBn
OH
BnO
OAc
Ph
Me
I
Cl
BF3•OEt2, CH2Cl2, 0 °C
(70%, mainly α)14
AgNO3, Na2CO3, reflux MeNO2,
6 h (58%, α,β 1:1).15

A variant using inexpensive Montmorillonite K-10 clay as the catalyst:

OBn
BnO
OBn OH
OH
Me
Me
cat. Mont. K10
ClCH2CH2Cl
rt, 6 h, 51%
OBn
BnO
Me
Me
OH

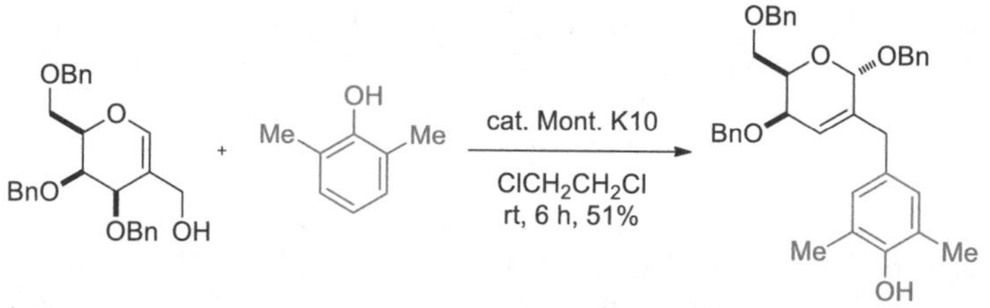

A recent example, A gold(I)-catalyzed tandem 1,3-acyloxy migration/Ferrier rearrangement[21]

OAc
AcO
AcO
Ph3PAuCl, AgOTf
Ferrier rearranegment
R1
R2
Ph3PAuCl, AgOTf
1,3-migration

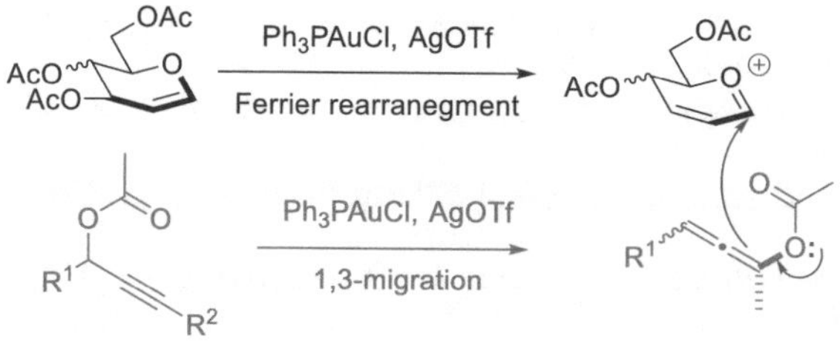

References

1. Ferrier, R. J.; Zubkov, O. A. Transformation of glycals into 2,3-unsaturated glycosyl derivatives, In *Org. React.* **2003**, *62*, 569–736. (Review). It was almost 50 years after Fischer's seminal finding that water took part in the reaction[3] that Ann Ryan, working in George Overend's Department in Birkbeck College, University of London, found, by chance, that *p*-nitrophenol likewise participates.[16] Robin Ferrier, her immediate supervisor, who suggested her experiment, then found that simple alcohols at high temperatures also take part,[17] and with other students, notably Nagendra Prasad and George Sankey, he explored the reaction extensively. They did not apply it to make the very important *C*-glycosides.

2. Ferrier, R. J. *Top. Curr. Chem.* **2001**, *215*, 153–175. (Review).

3. Fischer, E. *Chem. Ber.* **1914**, *47*, 196–210.

4. Herscovici, J.; Muleka, K.; Boumaïza, L.; Antonakis, K. *J. Chem. Soc., Perkin Trans. 1* **1990**, 1995–2009.

5. Ferrier, R. J.; Prasad, N. *J. Chem. Soc. (C)* **1969**, 581–586.

6. Moufid, N.; Chapleur, Y.; Mayon, P. *J. Chem. Soc., Perkin Trans. 1* **1992**, 999–1007.

7. Whittman, M. D.; Halcomb, R. L.; Danishefsky, S. J.; Golik, J.; Vyas, D. *J. Org. Chem.* **1990**, *55*, 1979–1981.

8. Klaffke, W.; Pudlo, P.; Springer, D.; Thiem, J. *Ann.* **1991**, 509–512.

9. Yougai, S.; Miwa, T. *J. Chem. Soc., Chem. Commun.* **1983**, 68–69.

10. Armstrong, P. L.; Coull, I. C.; Hewson, A. T.; Slater, M. J. *Tetrahedron Lett.* **1995**, *36*, 4311–4314.

11. Sobti, A.; Sulikowski, G. A. *Tetrahedron Lett.* **1994**, *35*, 3661–3664.

12. Toshima, K.; Ishizuka, T.; Matsuo, G.; Nakata, M.; Kinoshita, M. *J. Chem. Soc., Chem. Commun.* **1993**, 704–705.

13. López, J. C.; Gómez, A. M.; Valverde, S.; Fraser-Reid, B. *J. Org. Chem.* **1995**, *60*, 3851–3858.

14. Booma, C.; Balasubramanian, K. K. *Tetrahedron Lett.* **1993**, *34*, 6757–6760.

15. Tam, S. Y.-K.; Fraser-Reid, B. *Can. J. Chem.* **1977**, *55*, 3996–4001.

16. Ferrier, R. J.; Overend, W. G.; Ryan, A. E. *J. Chem. Soc. (C)* **1962**, 3667–3670.

17. Ferrier, R. J. *J. Chem. Soc.* **1964**, 5443–5449.

18. De, K.; Legros, J.; Crousse, B.; Bonnet-Delpon, D. *Tetrahedron* **2008**, *64*, 10497–10500.

19. Kumaran, E.; Santhi, M., Balasubramanian, K. K.; Bhagavathy, S. *Carbohydr. Res.* **2011**, *346*, 1654–1661.

20. Okazaki, H.; Hanaya, K.; Shoji, M.; Hada, N.; Sugai, T. *Tetrahedron* **2013**, *69*, 7931–7935.

21. Huang, N.; Liao, H.; Yao, H.; Xie, T.; Zhang, S.; Zou, K.; Liu, X.-W. *Org. Lett.* **2018**, *20*, 16–19.

22. Bhardwaj, M.; Rasool, F.; Tatina, M. B.; Mukherjee, D. *Org. Lett.* **2019**, *21*, 3038–3042.

Fischer Indole Synthesis

Cyclization of arylhydrazones to indoles.

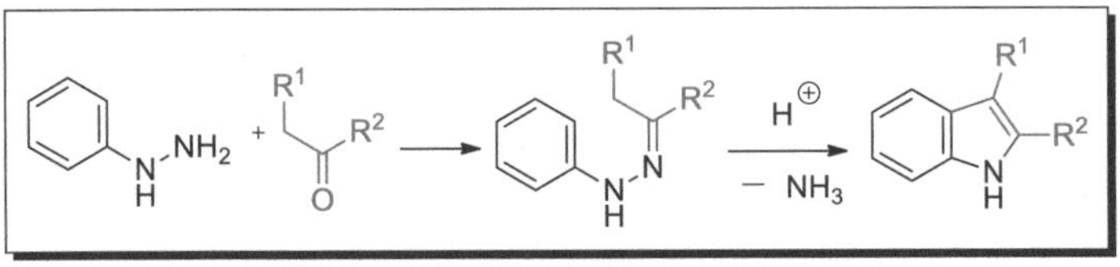

phenylhydrazine   phenylhydrazone

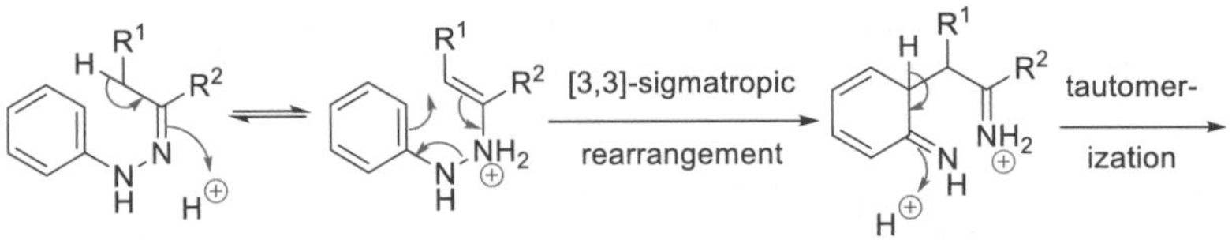

protonation   ene-hydrazine   double imine

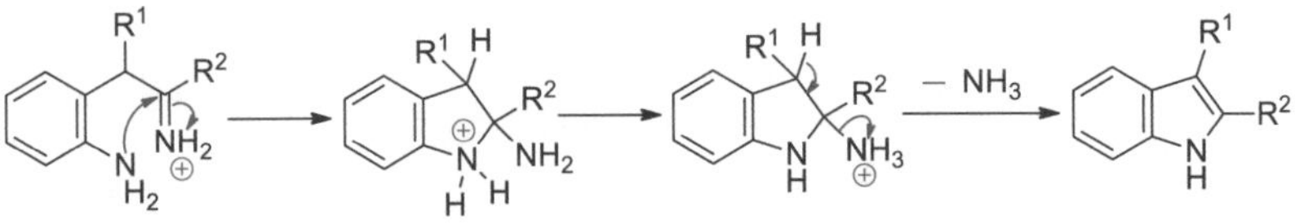

Example 1[3]

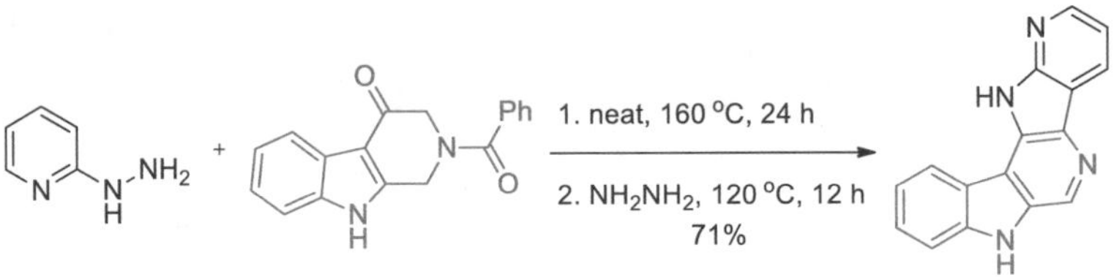

1. neat, 160 °C, 24 h

2. NH2NH2, 120 °C, 12 h
71%

Example 2[3]

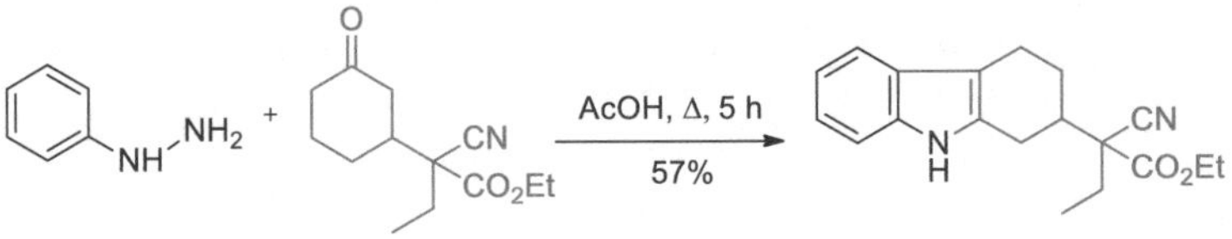

AcOH, Δ, 5 h

57%

© Springer Nature Switzerland AG 2021
J. J. Li, *Name Reactions*, https://doi.org/10.1007/978-3-030-50865-4_52

Example 3[10]

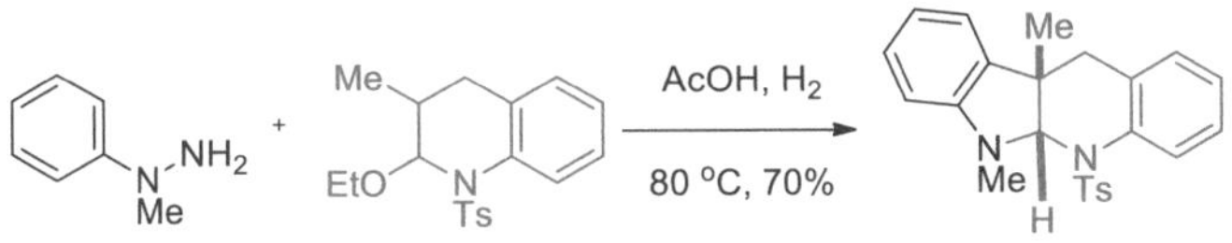

Example 4[12]

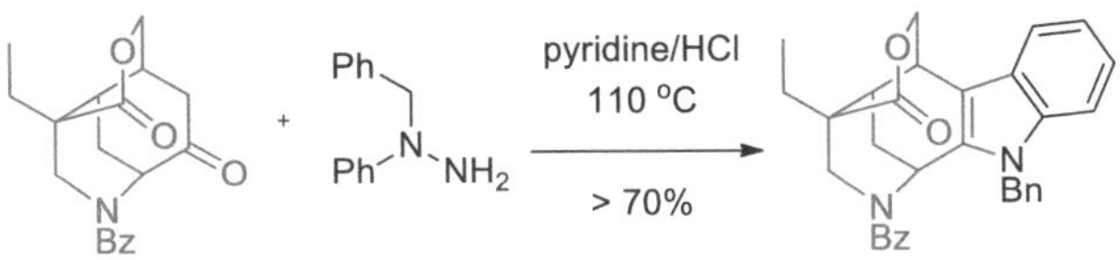

Example 5, An eco-friendly industrial scale Fischer indole cyclization (3 kg scale, PPA = phosphoric acid)[13]

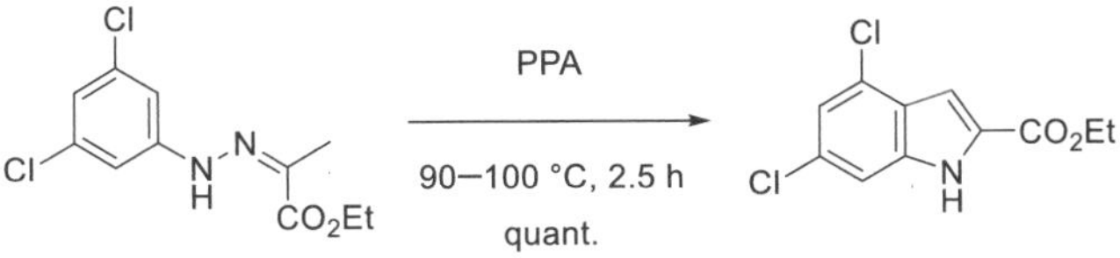

Example 6, Reductive interrupted Fischer indolization[14]

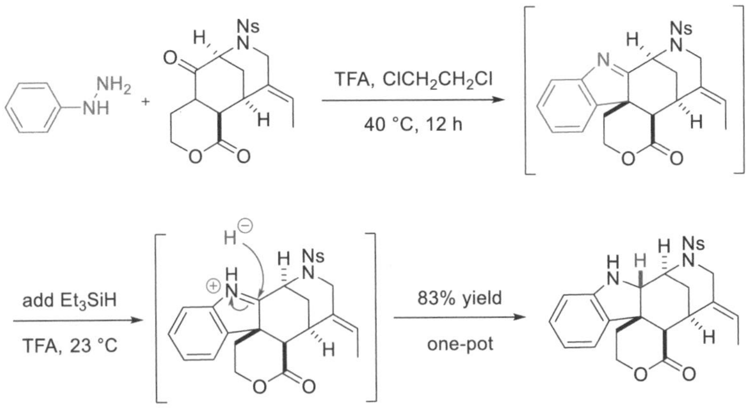

Example 7, Fused-indoline via reductive interrupted indolization in a microfluidic reactor[15]

References

1. (a) Fischer, E.; Jourdan, F. *Ber.* **1883**, *16*, 2241–2245. H. Emil Fischer (1852–1919) is arguably the greatest organic chemist ever. He was born in Euskirchen, near Bonn, Germany. When he was a boy, his father, Lorenz, said about him: "The boy is too stupid to go in to business; so in God's name, let him study." Fischer studied at Bonn and then Strassburg under Adolf von Baeyer. Fischer won the Nobel Prize in Chemistry in 1902 (three years ahead of his master, von Baeyer) for his synthetic studies in the area of sugar and purine groups. Sadly, Fischer committed suicide after WWI after his son died during the war and his fortunes completely gone. (b) Fischer, E.; Hess, O. *Ber.* **1884**, *17*, 559.

2. Robinson, B. *The Fisher Indole Synthesis,* Wiley: New York, NY, **1982**. (Book).

3. Martin, M. J.; Trudell, M. L.; Arauzo, H. D.; Allen, M. S.; LaLoggia, A. J.; Deng, L.; Schultz, C. A.; Tan, Y.; Bi, Y.; Narayanan, K.; Dorn, L. J.; Koehler, K. F.; Skolnick, P.; Cook, J. M. *J. Med. Chem.* **1992**, *35*, 4105–4117.

4. Hughes, D. L. *Org. Prep. Proc. Int.* **1993**, *25*, 607–632. (Review).

5. Bosch, J.; Roca, T.; Armengol, M.; Fernández-Forner, D. *Tetrahedron* **2001**, *57*, 1041–1048.

6. Ergün, Y.; Patir, S.; Okay, G. *J. Heterocycl. Chem.* **2002**, *39*, 315–317.

7. Pete, B.; Parlagh, G. *Tetrahedron Lett.* **2003**, *44*, 2537–2539.

8. Li, J.; Cook, J. M. *Fischer Indole Synthesis.* In *Name Reactions in Heterocyclic Chemistry*; Li, J. J., Ed.; Wiley: Hoboken, NJ, **2005**, pp 116–127. (Review).

9. Borregán, M.; Bradshaw, B.; Valls, N.; Bonjoch, J. *Tetrahedron: Asymmetry* **2008**, *19*, 2130–2134.

10. Boal, B. W.; Schammel A. W.; Garg, N. K. *Org. Lett.* **2013**, *11*, 3458–3461.

11. Donald, J. R.; Taylor, R. J. K. *Synlett* **2009**, 59–62.

12. Adams, G. L.; Carroll, P. J.; Smith, A. B. III *J. Am. Chem. Soc.* **2013**, *135*, 519–523.

13. Yang, X.; Zhang, X.; Yin, D. *Org. Process Res. Dev.* **2018**, *22,* 1115–1118.

14. Picazo, E.; Morrill, L. A.; Susick, R. B.; Moreno, J.; Smith, J. M.; Garg, N. K. *J. Am. Chem. Soc.* **2018**, *149*, 6483–56492.

15. Duong, A. T.-H.; Simmons, B. J.; Alam, M. P.; Campagna, J.; Garg, N. K.; John, V. *Tetrahedron Lett.* **2019**, *60*, 322–326.

16. Ghiyasabadi, Z.; Bahadorikhalili, S.; Saeedi, M.; Karimi-Niyazagheh, M.; Mirfazli, S. S. *J. Heterocycl. Chem.* **2020**, *57*, 606–610.

Friedel–Crafts Reaction

Friedel–Crafts Acylation Reaction:

Introduction of an acyl group onto an aromatic substrate by treating the substrate with an acyl halide or anhydride in the presence of a Lewis acid.

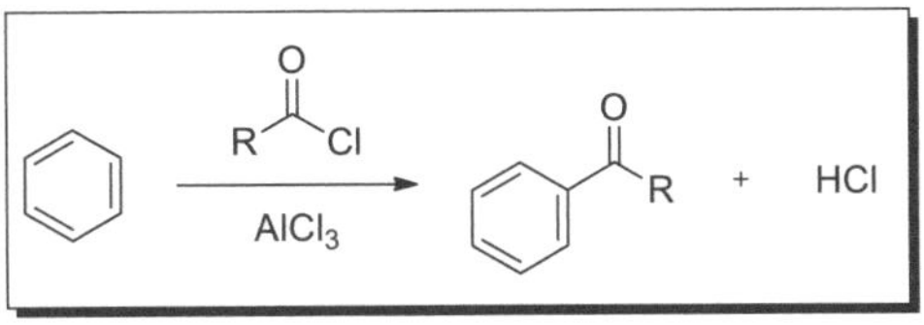

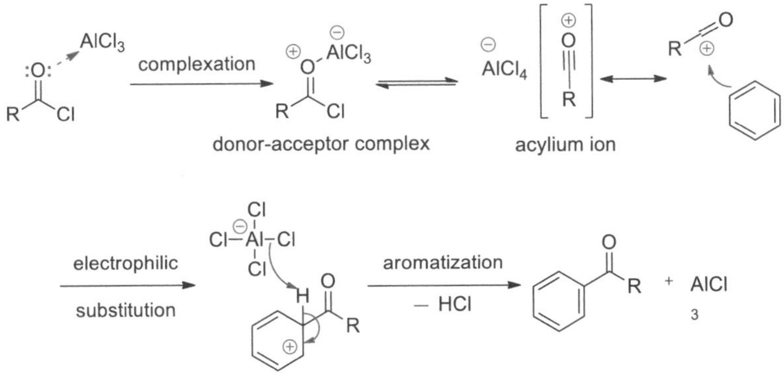

Example 1, Intermolecular Friedel–Crafts acylation[6]

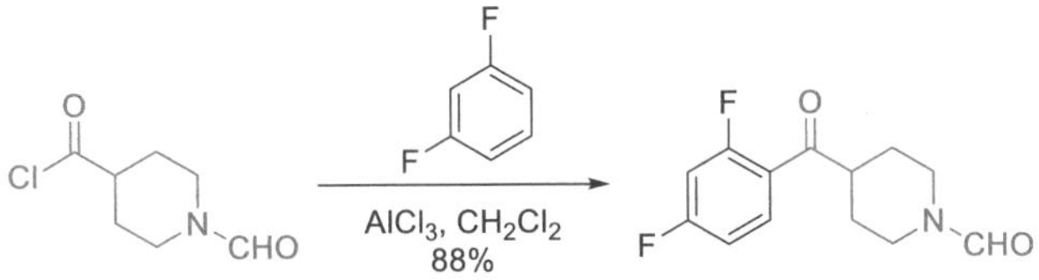

Example 2, Intramolecular Friedel–Crafts acylation[7]

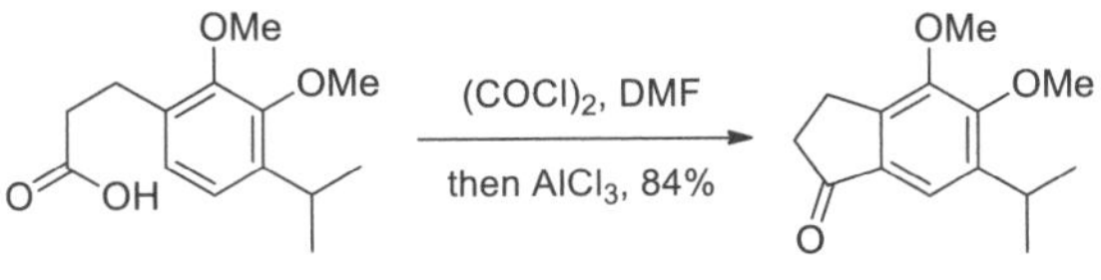

© Springer Nature Switzerland AG 2021
J. J. Li, *Name Reactions*, https://doi.org/10.1007/978-3-030-50865-4_53

Example 3, Intramolecular Friedel–Crafts acylation[8]

PPSE = Trimethylsilyl polyphosphate

Example 4, Intramolecular Friedel–Crafts acylation[9]

Example 5, "Kinetic capture" of acylium ion[11]

donor–acceptor complex   acylium ion

Example 6, Introduction of an acrolyl group by Friedel–Crafts acylation followed by elimination under mild conditions[12]

Example 7, Intramolecular Friedel–Crafts acylation "*in situ*"[13]

phthalonitrile   tetrafluorophthalonitrile

41%

References

1. Friedel, C.; Crafts, J. M. *Compt. Rend.* **1877,** *84*, 1392–1395. Charles Friedel (1832–1899) was born in Strasbourg, France. He earned his Ph.D. In 1869 under Wurtz at Sorbonne and became a professor and later chair (1884) of organic chemistry at Sorbonne. Friedel was one of the founders of the French Chemical Society and served as its president for four terms. James Mason Crafts (1839–1917) was born in Boston, Massachusetts. He studied under Bunsen and Wurtz in his youth and became a professor at Cornell and MIT. From 1874 to 1891, Crafts collaborated with Friedel at École de Mines in Paris, where they discovered the Friedel–Crafts reaction. He returned to MIT in 1892 and later served as its president. The discovery of the Friedel–Crafts reaction was the fruit of serendipity and keen observation. In 1877, both Friedel and Crafts were working in Charles A. Wurtz's laboratory. In order to prepare amyl iodide, they treated amyl chloride with aluminum and iodide using benzene as the solvent. Instead of amyl iodide, they ended up with amylbenzene! Unlike others before them who may have simply discarded the reaction, they thoroughly investigated the Lewis acid-catalyzed alkylations and acylations and published more than 50 papers and patents on the Friedel–Crafts reaction, which has become one of the most useful organic reactions.

2. Pearson, D. E.; Buehler, C. A. *Synthesis* **1972,** 533–542. (Review).

3. Hermecz, I.; Mészáros, Z. *Adv. Heterocycl. Chem.* **1983,** *33*, 241–330. (Review).

4. Metivier, P. *Friedel-Crafts Acylation.* In *Friedel-Crafts Reaction* Sheldon, R. A.; Bekkum, H., eds.; Wiley-VCH: New York. **2001,** pp 161–172. (Review).

5. Basappa; Mantelingu, K.; Sadashira, M. P.; Rangappa, K. S. *Indian J. Chem. B.* **2004,** *43B*, 1954–1957.

6. Olah, G. A.; Reddy, V. P.; Prakash, G. K. S. *Chem. Rev.* **2006,** *106*, 1077–1104. (Review).
7. Simmons, E.M.; Sarpong, R. *Org. Lett.* **2006,** *8*, 2883–2886.
8. Bourderioux, A.; Routier, S.; Beneteau, V.; Merour, J.-Y. *Tetrahedron* **2007,** *63*, 9465–9475.
9. Fillion, E.; Dumas, A. M. *J. Org. Chem.* **2008,** *73*, 2920–2923.
10. de Noronha, R. G.; Fernandes, A. C.; Romao, C. C. *Tetrahedron Lett.* **2009,** *50*, 1407–1410.
11. Huang, Z.; Jin, L.; Han, H.; Lei, A. *Org. Biomol. Chem.* **2013,** *11*, 1810–1814.
12. Allu, S. R.; Banne, S.; Jiang, J.; Qi, N.; He, Y. *J. Org. Chem.* **2019,** *84*, 7227–7237.
13. Tejerina, L.; Martínez-Díaz, M. V.; Torres, T. *Org. Lett.* **2019,** *21*, 2908–2912.
14. Patil, D. V.; Kim, H. Y.; Oh, K. *Org. Lett.* **2020,** *22*, 3018–3022.

Friedel–Crafts Alkylation Reaction:

Introduction of an alkyl group onto an aromatic substrate by treating the substrate with an alkylating agent such as alkyl halide, alkene, alkyne and alcohol in the presence of a Lewis acid.

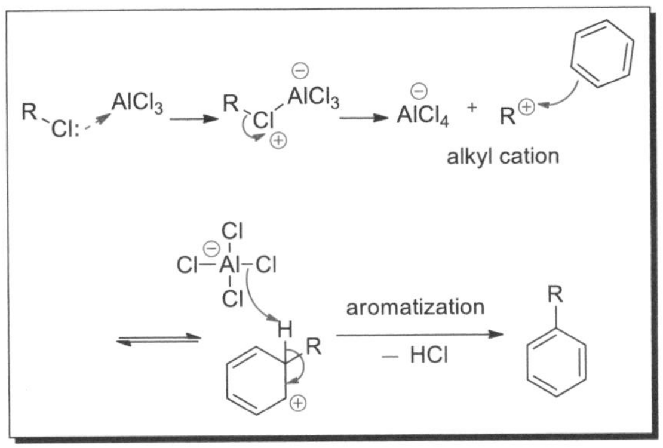

AlCl3
AlCl3
AlCl4
alkyl cation
aromatization
— HCl

Example 1[1]

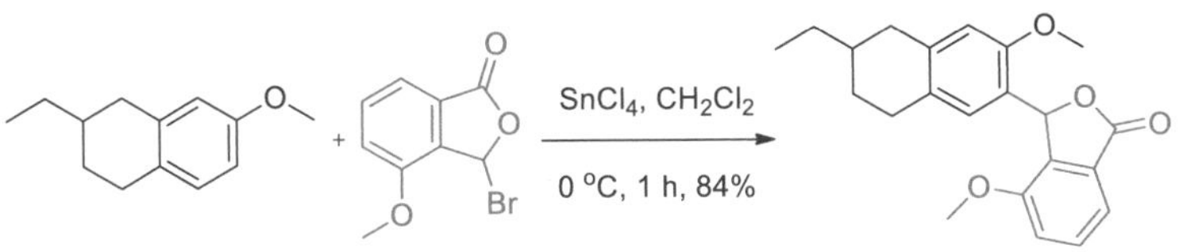

SnCl4, CH2Cl2
0 °C, 1 h, 84%
Br

Example 2, An intramolecular Friedel–Crafts alkylation[6]

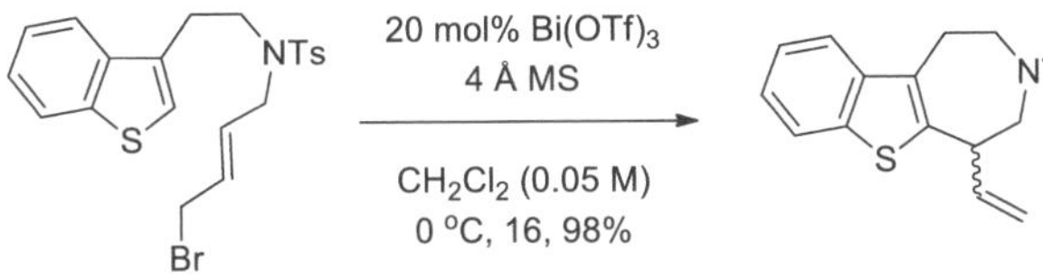

20 mol% Bi(OTf)3
4 Å MS
CH2Cl2 (0.05 M)
0 °C, 16, 98%
NTs
Br

Example 3, Diastereoselctive Friedel–Crafts alkylation[7]

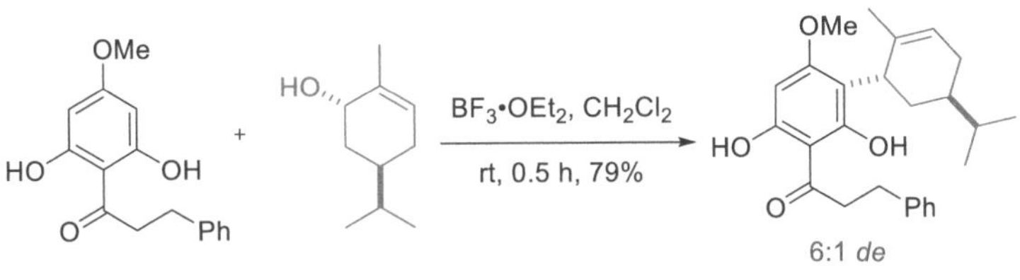

BF3•OEt2, CH2Cl2
rt, 0.5 h, 79%
6:1 de

Example 4, Friedel–Crafts alkylation to set up a quaternary carbon center (DTBP = di-*tert*-butyl peroxide)[8]

Example 5, Friedel–Crafts alkylation followed by ring contraction[9]

References

1. Patil, M. L.; Borate, H. B.; Ponde, D. E. *Tetrahedron Lett.* **1999,** *40*, 4437–4438.
2. Meima, G. R.; Lee, G. S.; Garces, J. M. *Friedel–Crafts Alkylation.* In *Friedel–Crafts Reaction* Sheldon, R. A.; Bekkum, H., eds.; Wiley-VCH: New York. **2001,** pp 550–556. (Review).
3. Bandini, M.; Melloni, A. *Angew. Chem. Int. Ed.* **2004,** *43*, 550–556. (Review).
4. Poulsen, T. B.; Jorgensen, K. A. *Chem. Rev.* **2008,** *108*, 2903–2915. (Review).
5. Silvanus, A. C.; Heffernan, S. J.; Liptrot, D. J.; Kociok-Kohn, G.; Andrews, B. I.; Carbery, D. R. *Org. Lett.* **2009,** *11*, 1175–1178.
6. Kargbo, R. B.; Sajjadi-Hashemi, Z.; Roy, S.; Jin, X.; Herr, R. J. *Tetrahedron Lett.* **2013,** *54*, 2018–2021.
7. Dethe, D. H.; Dherange, B. D. *J. Org. Chem.* **2018,** *83*, 3392–3396.
8. Hodges, T. R.; Benjamin, N. M.; Martin, S. F. *Org. Lett.* **2017,** *19*, 2254–2257.
9. Turnu, F.; Luridiana, A.; Cocco, A. *Org. Lett.* **2019,** *21*, 7329–7332.
10. Gallo, R. D. C.; Momo, P. B.; Day, D. P.; Burtoloso, A. C. B. *Org. Lett.* **2020,** *22*, 2339–2343.

Friedländer Quinoline Synthesis

Also known as the Friedländer condensation, it combines an α-amino aldehyde or ketone with another aldehyde or ketone with at least one methylene α adjacent to the carbonyl to furnish a substituted quinoline. The reaction can be promoted by either acid, base, or heat.

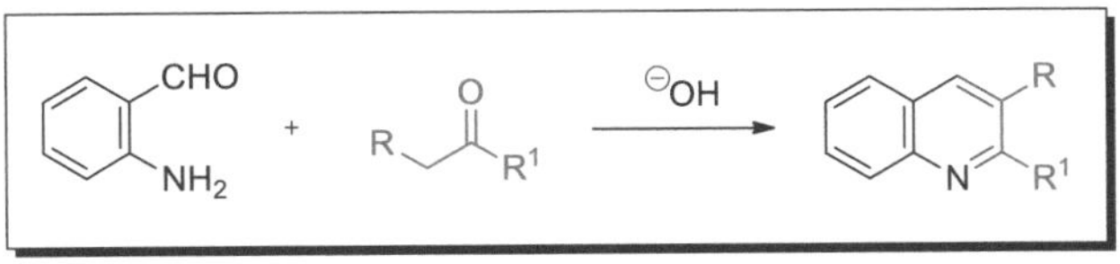

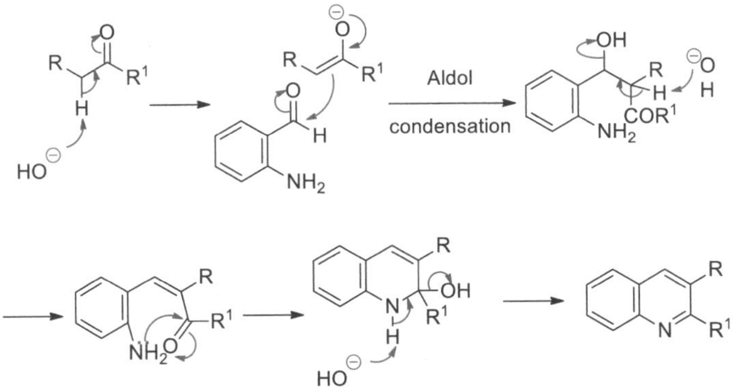

Example 1[5]

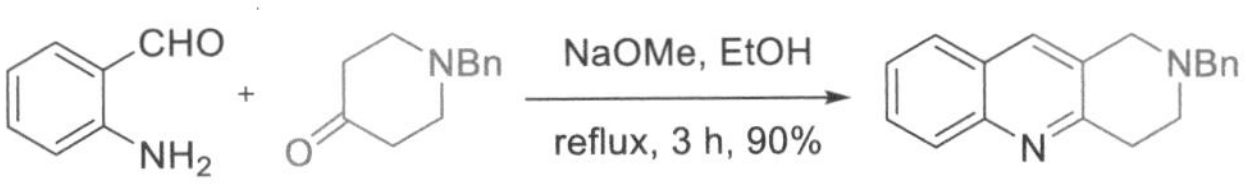

Example 2[7]

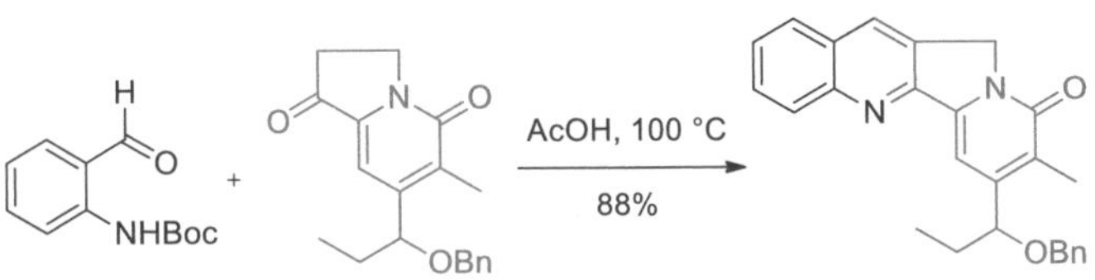

© Springer Nature Switzerland AG 2021
J. J. Li, *Name Reactions*, https://doi.org/10.1007/978-3-030-50865-4_54

Example 3[8]

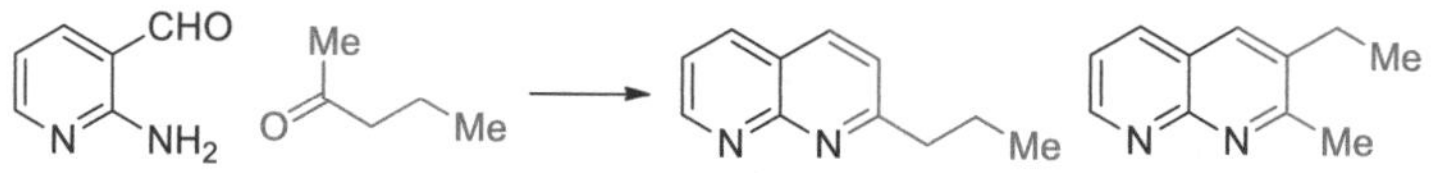

Conditions	Conversion	Ratio
NaOH, rt	> 99%	37:63
pyrrolidine, 5% H_2SO_4, rt	97%	86:14
TBAO, 5% H_2SO_4, rt	> 99%	87:13
TBAO, 5% H_2SO_4, slow addition, 65 °C	> 99%	94:6

TBAO = 1,3,3-trimethyl-6-azabicyclo[3.2.1]octane

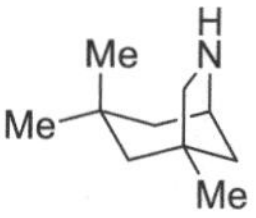

Example 4[10]

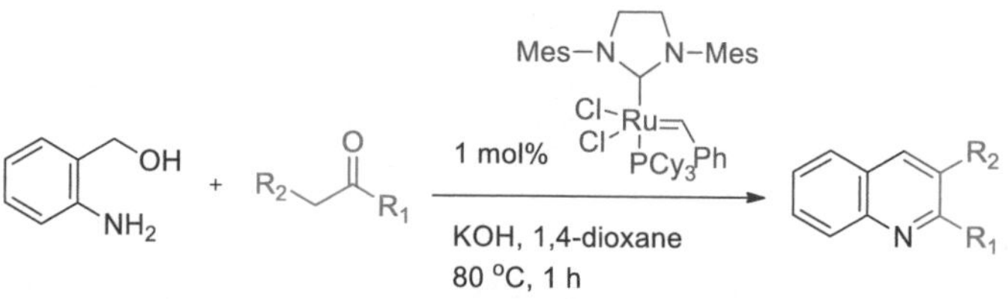

Example 5, Using propylphosphonic anhydride (T3P) as the coupling agent[11]

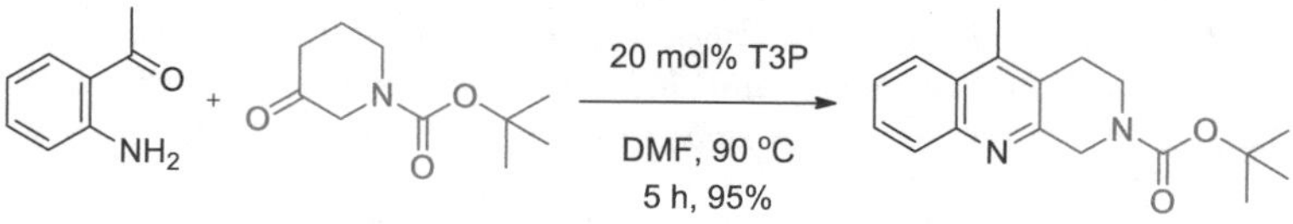

Example 6, NHC-Cu(I)-catalyzed Friedländer-type annulation of fluorinated *o*-aminophenones with alkynes on water[12]

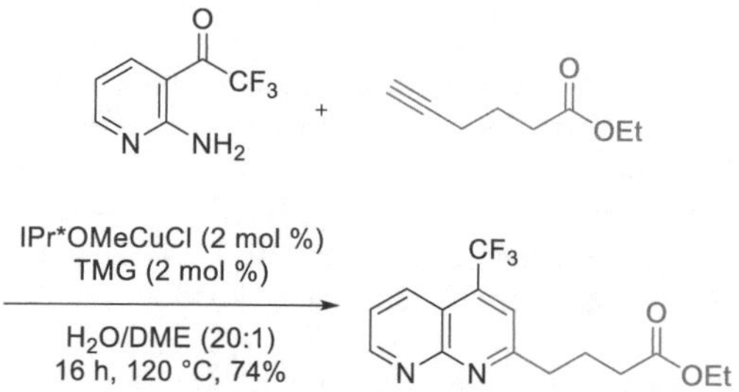

Example 7, Toward total synthesis of (+)-eburnamonine[13]

Example 8, Organocatalytic atroposelective Friedländer quinoline heteroannulation[14]

References

1. Friedländer, P. *Ber.* **1882**, *15*, 2572–2575. Paul Friedländer (1857–1923), born in Königsberg, Prussia, apprenticed under Carl Graebe and Adolf von Baeyer. He was interested in music and was an accomplished pianist.
2. Elderfield, R. C. In *Heterocyclic Compounds*, Elderfield, R. C., ed.; Wiley: New York, **1952**, *4*, *Quinoline, Isoquinoline and Their Benzo Derivatives*, 45–47. (Review).
3. Jones, G. In *Heterocyclic Compounds*, Quinolines, vol. 32, **1977**; Wiley: New York, pp 181–191. (Review).
4. Cheng, C.-C.; Yan, S.-J. *Org. React.* **1982**, *28*, 37–201. (Review).
5. Shiozawa, A.; Ichikawa, Y.-I.; Komuro, C. *Chem. Pharm. Bull.* **1984**, *32*, 2522–2529.
6. Gladiali, S.; Chelucci, G.; Mudadu, M. S. *J. Org. Chem.* **2001**, *66*, 400–405.
7. Henegar, K. E.; Baughman, T. A. *J. Heterocycl. Chem.* **2003**, *40*, 601–605.
8. Dormer, P. G.; Eng, K. K.; Farr, R. N. *J. Org. Chem.* **2003**, *68*, 467–477.
9. Pflum, D. A. *Friedländer Quinoline Synthesis*. In *Name Reactions in Heterocyclic Chemistry*; Li, J. J., Ed.; Wiley: Hoboken, NJ, **2005**, 411–415. (Review).
10. Vander Mierde, H.; Van Der Voot, P. *Eur. J. Org. Chem.* **2008**, 1625–1631.
11. Augustine, J. K.; Bombrun, A. *Tetrahedron Lett.* **2011**, *52*, 6814–6818.
12. Czerwiński, P.; Michalak, M. *J. Org. Chem.* **2017**, *82*, 7980–7997.
13. Pandey, G.; Mishra, A.; Khamrai, J. *Org. Lett.* **2017**, *19*, 3267–3270.
14. Shao, Y-D.; Dong, M. M.; Wang, Y.-A.; Cheng, P.-M.; Wang, T. *Org. Lett.* **2019**, *21*, 4831–4836.
15. Nainwal, L. M.; Tasneem, S.; Akhtar, W.; Verma, G.; Khan, M. F.; Parvez, S. *Eur. J. Med. Chem.* **2019**, *164*, 121–170. (Review).

Fries Rearrangement

Lewis acid-catalyzed rearrangement of phenol esters and lactams to 2- or 4-ketophenols. Also known as the Fries–Finck rearrangement.

aluminum phenolate, acylium ion

Example 1[5]

© Springer Nature Switzerland AG 2021

J. J. Li, *Name Reactions*, https://doi.org/10.1007/978-3-030-50865-4_55

Example 2[6]

Example 3, Photo-Fries rearrangement[7]

Example 4, *ortho*-Fries rearrangement[8]

Example 5, Thia-Fries rearrangement[9]

Example 6, Remote anionic thia-Fries rearrangement[10]

Example 7, A consecutive Snieckus–Fries rearrangement, anionic Si→C alkyl re-arrangement, and Claisen–Schmidt condensation[11]

4 equiv LDA, THF, –78 °C → rt;
1.2 equiv TMSCl, 0 °C → rt, 30 min;
1.2 equiv MeI, 1.2 equiv RCHO
35 °C, 10 h, 51%

Example 8, Orthosodiations employing sodium diisopropylamide (NaDA) and Snieckus–Fries Rearrangement of aryl carbamates[12]

NaDA, THF

79%

References

1. Fries, K.; Finck, G. *Ber.* **1908**, *41*, 4271–4284. Karl Theophil Fries (1875–1962) was born in Kiedrich near Wiesbaden on the Rhine. He earned his doctorate under Theodor Zincke. Although G. Finck co-discovered the rearrangement of phenolic esters, some-how his name has been forgotten by history. In all fairness, the Fries rearrangement should really be the Fries–Finck rearrangement.

2. Martin, R. *Org. Prep. Proced. Int.* **1992**, *24*, 369–435. (Review).

3. Boyer, J. L.; Krum, J. E.; Myers, M. C. *J. Org. Chem.* **2000**, *65*, 4712–4714.

4. Guisnet, M.; Perot, G. *The Fries rearrangement.* In *Fine Chemicals through Hetero-geneous Catalysis* **2001**, 211–216. (Review).

5. Tisserand, S.; Baati, R.; Nicolas, M. *J. Org. Chem.* **2004**, *69*, 8982–8983.

6. Ollevier, T.; Desyroy, V.; Asim, M.; Brochu, M.-C. *Synlett* **2004**, 2794–2796.

7. Ferrini, S.; Ponticelli, F.; Taddei, M. *Org. Lett.* **2007**, *9*, 69–72.

8. Macklin, T. K.; Panteleev, J.; Snieckus, V. *Angew. Chem. Int. Ed.* **2008**, *47*, 2097–2101.

9. Dyke, A. M.; Gill, D. M.; Harvey, J. N.; Hester, A. J.; Lloyd-Jones, G. C.; Munoz, M. P.; Shepperson, I. R. *Angew. Chem. Int. Ed.* **2008**, *47*, 5067–5070.

10. Xu, X.-H.; Taniguchi, M.; Azuma, A.; Liu, G. K.; Tokunaga, E.; Shibata, N. *Org. Lett.* **2013**, *15*, 686–689.

11. Kumar, S. N.; Bavikar, S. R.; Kumar, C. N. S. S. P.; Yu, I, F.; Chein, R.-J. *Org. Lett.* **2018**, *20*, 5362–5366.

12. Ma, Y.; Woltornist, R. A.; Algera, R. F.; Collum, D. B. *J. Org. Chem.* **2019**, *84*, 9051–9051.

13. Alessi, M.; Patel, J. J.; Zumbansen, K.; Snieckus, V. *Org. Lett.* **2020**, *22*, 2147–2151.

Gabriel Synthesis

Synthesis of primary amines using potassium phthalimide and alkyl halides.

Example 1[2]

Example 2[6]

© Springer Nature Switzerland AG 2021
J. J. Li, *Name Reactions*, https://doi.org/10.1007/978-3-030-50865-4_56

Example 3[8]

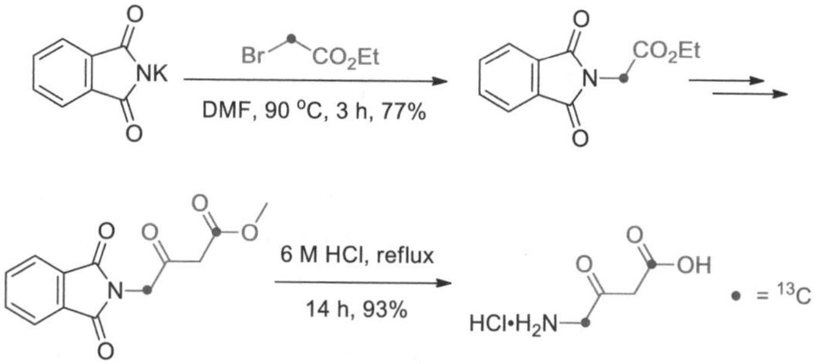

Example 4[9]

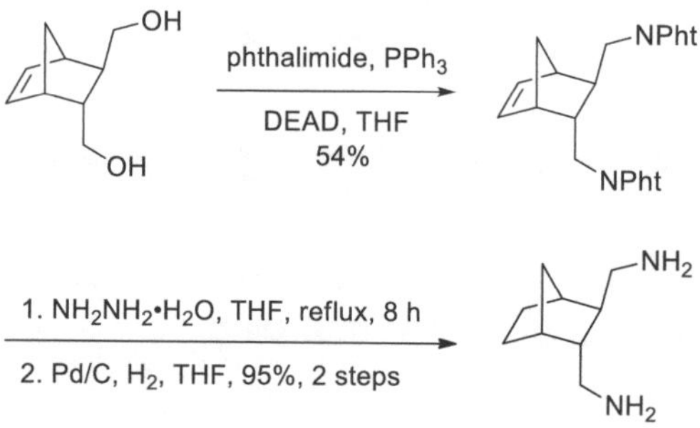

Example 5, Utility in medicinal chemistry[14]

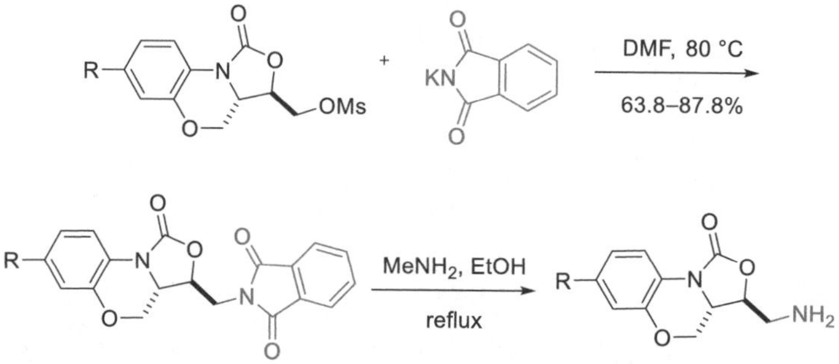

Example 6, Enantioselective Gabriel synthesis[15]

Example 7, Intermolecular cyclization after Gabriel synthesis[16]

References

1. Gabriel, S. *Ber.* **1887**, *20*, 2224–2226. Siegmund Gabriel (1851–1924), born in Berlin, Germany, studied under Hofmann at Berlin and Bunsen in Heidelberg. He taught at Berlin, where he discovered the Gabriel synthesis of amines. Gabriel, a good friend of Emil Fischer, often substituted for Fischer in his lectures.
2. Sheehan, J. C.; Bolhofer, V. A. *J. Am. Chem. Soc.* **1950**, *72*, 2786–2788.
3. Han, Y.; Hu, H. *Synthesis* **1990**, 122–124.
4. Ragnarsson, U.; Grehn, L. *Acc. Chem. Res.* **1991**, *24*, 285–289. (Review).
5. Toda, F.; Soda, S.; Goldberg, I. *J. Chem. Soc., Perkin Trans. 1* **1993**, 2357–2361.
6. Sen, S. E.; Roach, S. L. *Synthesis*, **1995**, 756–758.

7.   Khan, M. N. *J. Org. Chem.* **1996,** *61*, 8063–8068.
8.   Iida, K.; Tokiwa, S.; Ishii, T.; Kajiwara, M. *J. Labelled. Compd. Radiopharm.* **2002,** *45*, 569–570.
9.   Tanyeli, C.; Özçubukçu, S. *Tetrahedron Asymmetry* **2003,** *14,* 1167–1170.
10. Ahmad, N. M. *Gabriel synthesis*. In *Name Reactions for Functional Group Transformations*; Li, J. J., Ed.; Wiley: Hoboken, NJ, **2007,** pp 438–450. (Review).
11. Al-Mousawi, S. M.; El-Apasery, M. A.; Al-Kanderi, N. H. *ARKIVOC* **2008,** *(16),* 268–278.
12. Richter, J. M. *Name Reactions in Heterocyclic Chemistry-II*, Li, J. J., Ed.; Wiley: Hoboken, NJ, 2011, pp 11–20. (Review).
13. Cytlak, T.; Marciniak, B.; Koroniak, H. In *Efficient Preparations of Fluorine Compounds*; Roesky, H. W., ed.; Wiley: Hoboken, NJ, (2013), pp 375–378. (Review).
14. Xue, T.; Ding, S.; Guo, B.; Zhou, Y.; Sun, P.; Wang, H.; Chu, W.; Gong, G.; Wang, Y.; Chen, X.; Yang, Y. *J. Med. Chem.* **2014,** *57*, 7770–7791.
15. Avidan-Shlomovich, S.; Ghosh, H.; Szpilman, A. M. *ACS Catal.* **2015,** *5*, 336–342.
16. Fernandez, S.; Ganiek, M. A.; Karpacheva, M.; Hanusch, F. C.; Reuter, S.; Bein, T.; Auras, F.; Knochel, P. *Org. Lett.* **2016,** *18*, 3158–3161.
17. Chen, J.; Park, J.; Kirk, S. M.; Chen, H.-C.; Li, X.; Lippincott, D. J.; Melillo, B.; Smith, A. B. *Org. Process Res. Dev.* **2019,** *23*, 2464–2469.

Ing–Manske Procedure

A variant of Gabriel amine synthesis where hydrazine is used to release the amine from the corresponding phthalimide:

Example 1[6]

Example 2, Toward preparation of human cluster of differentiation 4 (CD4) receptor modulators[10]

Example 3, Toward preparation of D_3 dopamine receptor agonist[11]

$$\xrightarrow[\substack{\text{reflux, overnight} \\ 89\%}]{N_2H_4 \cdot H_2O, \text{ EtOH}}$$

References

1. Ing, H. R.; Manske, R. H. F. *J. Chem. Soc.* **1926**, 2348–2351. H. R. Ing was a professor of pharmacological chemistry at Oxford. R. H. F. Manske, Ing's collaborator at Oxford, was of German origin but trained in Canada before studying at Oxford. Manske left England to return to Canada, eventually to become Director of Research in the Union Rubber Company, Guelph, Ontario, Canada.
2. Ueda, T.; Ishizaki, K. *Chem. Pharm. Bull.* **1967**, *15*, 228–237.
3. Khan, M. N. *J. Org. Chem.* **1995**, *60*, 4536–4541.
4. Hearn, M. J.; Lucas, L. E. *J. Heterocycl. Chem.* **1984**, *21*, 615–622.
5. Khan, M. N. *J. Org. Chem.* **1996**, *61*, 8063–8063.
6. Tanyeli, C.; Özçubukçu, S. *Tetrahedron: Asymmetry* **2003**, *14*, 1167–1170.
7. Ariffin, A.; Khan, M. N.; Lan, L. C.; May, F. Y.; Yun, C. S. *Synth. Commun.* **2004**, *34*, 4439–4445.
8. Ali, M. M.; Woods, M.; Caravan, P.; Opina, A. C. L.; Spiller, M.; Fettinger, J. C.; Sherry, A. D. *Chem. Eur. J.* **2008**, *14*, 7250–7258.
9. Nagarapu, L.; Apuri, S.; Gaddam, C.; Bantu, R. *Org. Prep. Proc. Int.* **2009**, *41*, 243–247.
10. Chawla, R.; Van Puyenbroeck, V.; Pflug, N. C.; Sama, A.; Ali, R.; Schols, D.; Vermeire, K.; Bell, T. W. *J. Med. Chem.* **2016**, *59*, 2633–2647.
11. Battiti, F. O.; Cemaj, S. L.; Guerrero, A. M.; Shaik, A. B.; Lam, J.; Rais, R.; Slusher, B. S.; Deschamps, J. R.; Imler, G. H.; Newman, A. H.; Bonifazi, A. *J. Med. Chem.* **2019**, *62*, 6287–6314.

Gewald Aminothiophene Synthesis

Base-promoted aminothiophene formation from ketone, α-active methylene nitrile and elemental sulfur.

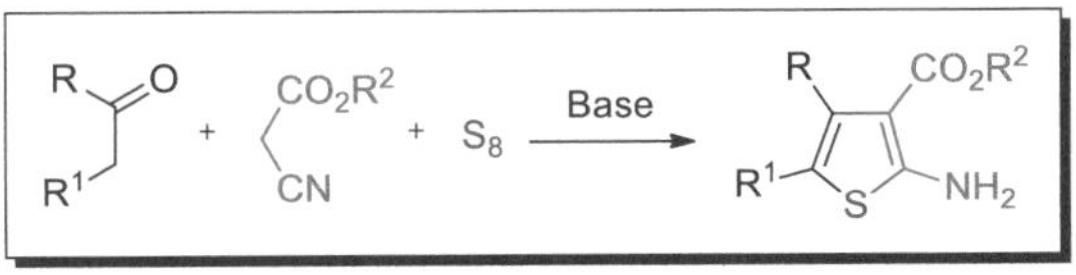

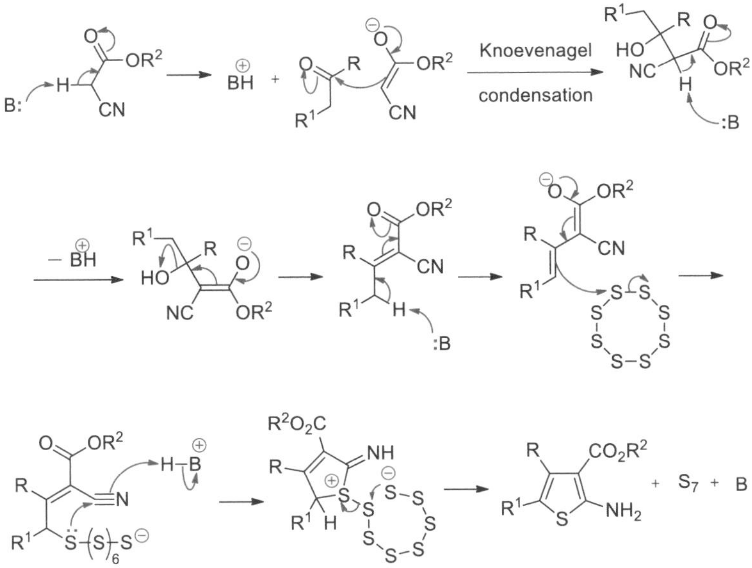

ylidene-sulfur adduct

Example 1[4]

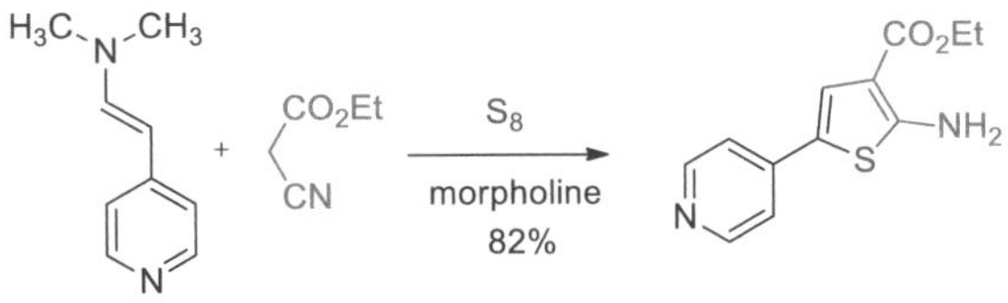

Example 2[7]

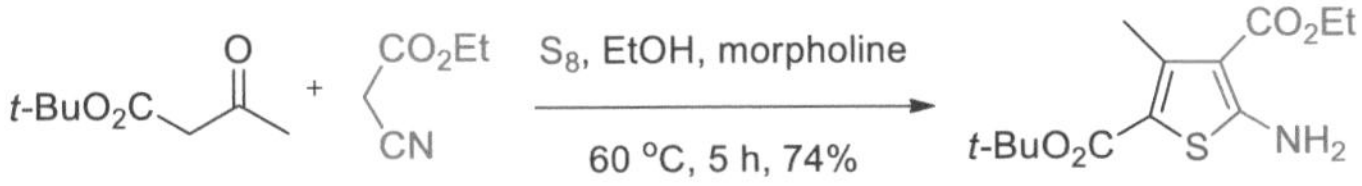

© Springer Nature Switzerland AG 2021
J. J. Li, *Name Reactions*, https://doi.org/10.1007/978-3-030-50865-4_57

Example 3[9]

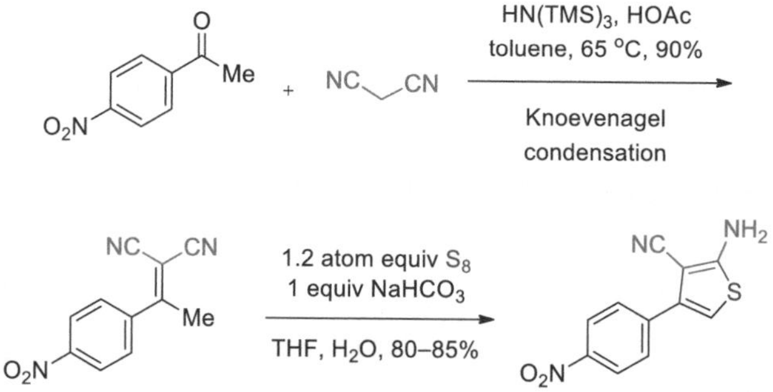

Example 4[10]

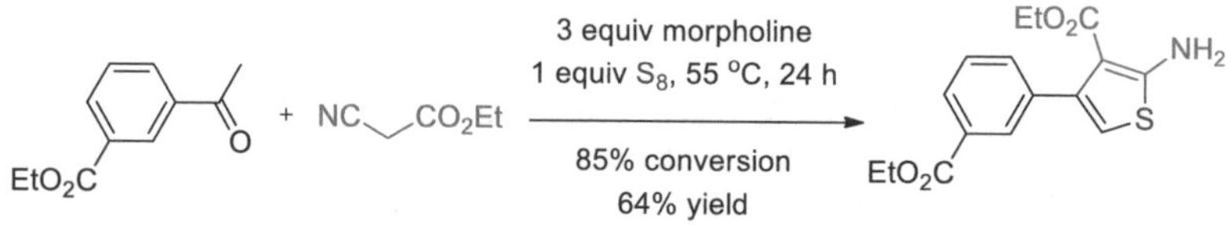

Example 5[11]

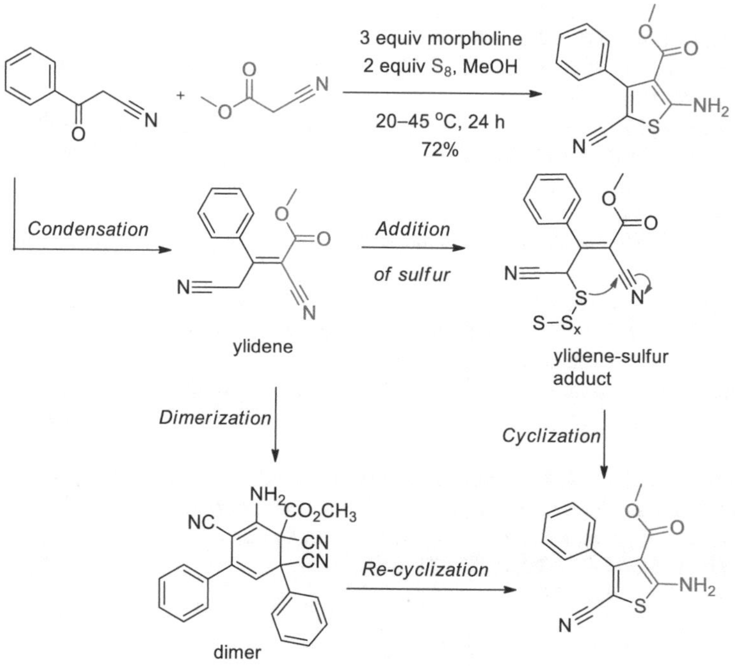

Example 6, *N*-Methylpiperazine-functionalized polyacrylonitrile fiber catalyst[12]

Example 7, NaAlO₂ as an environmentally-friendly and inexpensive catalyst[13]

References

1. (a) Gewald, K. *Z. Chem.* **1962,** *2*, 305–306. (b) Gewald, K.; Schinke, E.; Böttcher, H. *Chem. Ber.* **1966,** *99*, 94–100. (c) Gewald, K.; Neumann, G.; Böttcher, H. *Z. Chem.* **1966,** *6*, 261. (d) Gewald, K.; Schinke, E. *Chem. Ber.* **1966,** *99*, 271–275. Karl Z. Gewald (1930–2017) is a professor at Technical University of Dresden.
2. Mayer, R.; Gewald, K. *Angew. Chem. Int. Ed.* **1967,** *6*, 294–306. (Review).
3. Gewald, K. *Chimia* **1980,** *34*, 101–110. (Review).
4. Bacon, E. R.; Daum, S. J. *J. Heterocycl. Chem.* **1991,** *28*, 1953–1955.
5. Sabnis, R. W. *Sulfur Rep.* **1994,** *16*, 1–17. (Review).
6. Sabnis, R. W.; Rangnekar, D. W.; Sonawane, N. D. *J. Heterocycl. Chem.* **1999,** *36*, 333–345. (Review).
7. Gütschow, M.; Kuerschner, L.; Neumann, U.; Pietsch, M.; Löser, R.; Koglin, N.; Eger, K. *J. Med. Chem.* **1999,** *42*, 5437.
8. Tinsley, J. M. *Gewald Aminothiophene Synthesis.* In *Name Reactions in Heterocyclic Chemistry*; Li, J. J., Ed.; Wiley: Hoboken, NJ, **2005,** pp 193–198. (Review).
9. Barnes, D. M.; Haight, A. R.; Hameury, T.; McLaughlin, M. A.; Mei, J.; Tedrow, J. S.; Dalla Riva Toma, J. *Tetrahedron* **2006,** *62*, 11311–11319.
10. Tormyshev, V. M.; Trukhin, D. V.; Rogozhnikova, O. Yu.; Mikhalina, T. V.; Troitskaya, T. I.; Flinn, A. *Synlett* **2006,** 2559–2564.
11. Puterová, Z.; Andicsová, A.; Végh, D. *Tetrahedron* **2008,** *64*, 11262–11269.
12. Ma, L.; Yuan, L.; Xu, C.; Li, G.; Tao, M.; Zhang, W. *Synthesis* **2013,** *45*, 45–52.
13. Bai, R.; Liu, P.; Yang, J.; Liu, C.; Gu, Y. *ACS Sustainable Chem. Eng.* **2015,** *3*, 1292–1297.
14. Bozorov, K.; Nie, L. F.; Zhao, J.; Aisa, H. A. *Eur. J. Med. Chem.* **2017,** *140*, 465–493.
15. Shipilovskikh, S. A.; Rubtsov, A. E. *J. Org. Chem.* **2019,** *84*, 15788–15796.
16. Madacsi, R.; Traj, P.; Hackler, L. Jr.; Nagy, L. I.; Kari, B.; Puskas, L. G.; Kanizsai, I. *J. Heterocycl. Chem.* **2020,** *57*, 635–652.

Glaser Coupling

Sometimes known as the Glaser–Hay coupling, it is the oxidative homo-coupling of terminal alkynes using copper catalyst in the presence of oxygen.

Alternatively, the radical mechanism is also operative:

Example 1[1]

© Springer Nature Switzerland AG 2021
J. J. Li, *Name Reactions*, https://doi.org/10.1007/978-3-030-50865-4_58

Example 2, Homo-coupling[2]

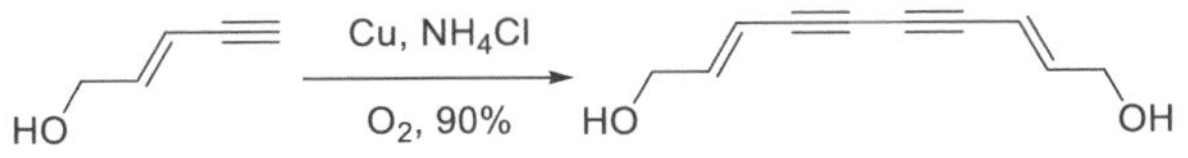
Cu, NH4Cl
O2, 90%
HO
HO
OH

Example 3[7]

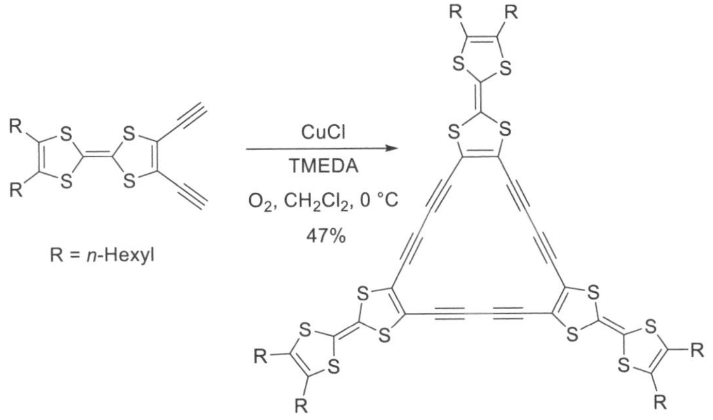
CuCl
TMEDA
O2, CH2Cl2, 0 °C
47%
R = n-Hexyl

Example 4[9]

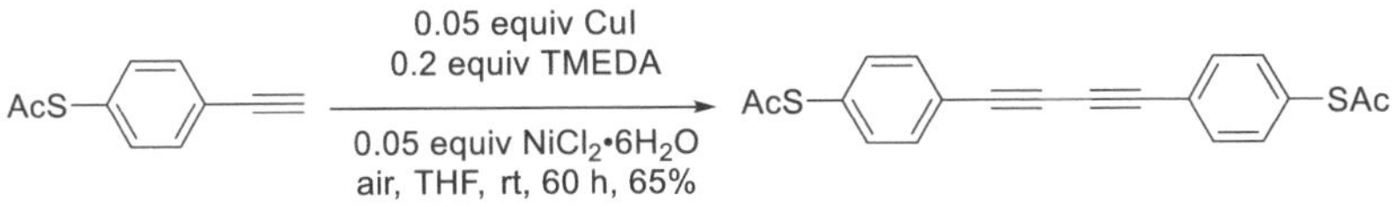
0.05 equiv CuI
0.2 equiv TMEDA
0.05 equiv NiCl2•6H2O
air, THF, rt, 60 h, 65%
AcS
AcS
SAc

Example 5, Macrocyclic Glaser–Hay coupling[10]

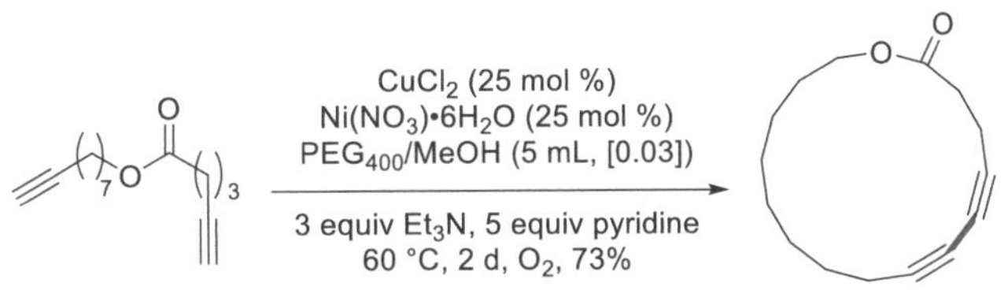
CuCl2 (25 mol %)
Ni(NO3)•6H2O (25 mol %)
PEG400/MeOH (5 mL, [0.03])
3 equiv Et3N, 5 equiv pyridine
60 °C, 2 d, O2, 73%

Example 6, Macrocyclic Glaser–Hay coupling to prepare amino acids[13]

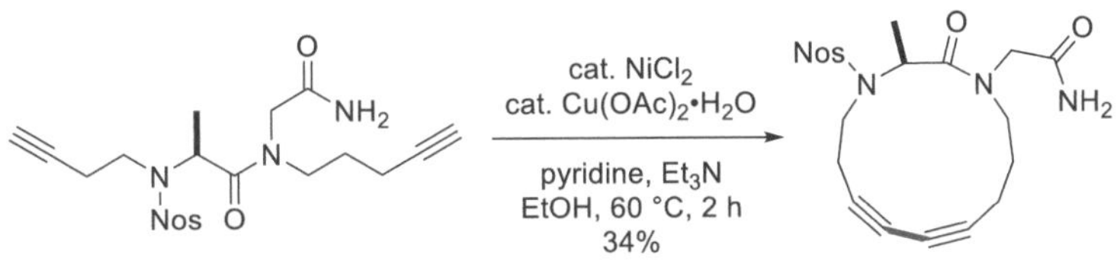
cat. NiCl2
cat. Cu(OAc)2•H2O
pyridine, Et3N
EtOH, 60 °C, 2 h
34%
Nos
NH2

References

1. Glaser, C. *Ber.* **1869,** *2,* 422–424. Carl Andreas Glaser (1841–1935) studied under Justus von Liebig and Adolph Strecker. He became a professor in 1869 when the Glaser coupling was discovered. He became the Chairman of the Board of BASF after WWI.

2. Bowden, K.; Heilbron, I.; Jones, E. R. H.; Sondheimer, F. *J. Chem. Soc.* **1947,** 1583–1590.

3. Hoeger, S.; Meckenstock, A.-D.; Pellen, H. *J. Org. Chem.* **1997,** *62,* 4556–4557.

4. Siemsen, P.; Livingston, R. C.; Diederich, F. *Angew. Chem. Int. Ed.* **2000,** *39,* 2632–2657. (Review).

5. Youngblood, W. J.; Gryko, D. T.; Lammi, R. K.; Bocian, D. F.; Holten, D.; Lindsey, J. S. *J. Org. Chem.* **2002,** *67,* 2111–2117.

6. Moriarty, R. M.; Pavlovic, D. *J. Org. Chem.* **2004,** *69,* 5501–5504.

7. Andersson, A. S.; Kilsa, K.; Hassenkam, T.; Gisselbrecht, J.-P.; Boudon, C.; Gross, M.; Nielsen, M. B.; Diederich, F. *Chem. Eur. J.* **2006,** *12,* 8451–8459.

8. Gribble, G. W. *Glaser Coupling.* In *Name Reactions for Homologations-Part I*; Li, J. J., Ed.; Wiley: Hoboken, NJ, **2009,** pp 236–257. (Review).

9. Muesmann, T. W. T.; Wickleder, M. S.; Christoffers, J. *Synthesis* **2011,** 2775–2780.

10. Bédard, A.-C.; Collins, S. K. *J. Am. Chem. Soc.* **2011,** *133,* 19976–19981.

11. Sindhu, K. S.; Anilkumar, G. *RSC Adv.* **2014,** *4,* 27867–27887. (Review).

12. Godin, É.; Bédard, A.-C.; Raymond, M.; Collins, S. K. *J. Org. Chem.* **2017,** *82,* 7576–7582.

13. Okorochenkov, S.; Krchňák, V. *ACS Comb. Sci.* **2019,** *21,* 316–322.

Eglinton Coupling

Oxidative homo-coupling of terminal alkynes mediated by stoichiometric (or often excess) Cu(OAc)$_2$. A variant of the Glaser coupling reaction.

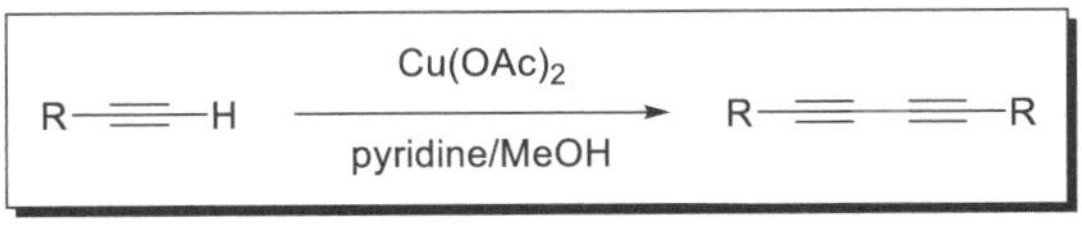

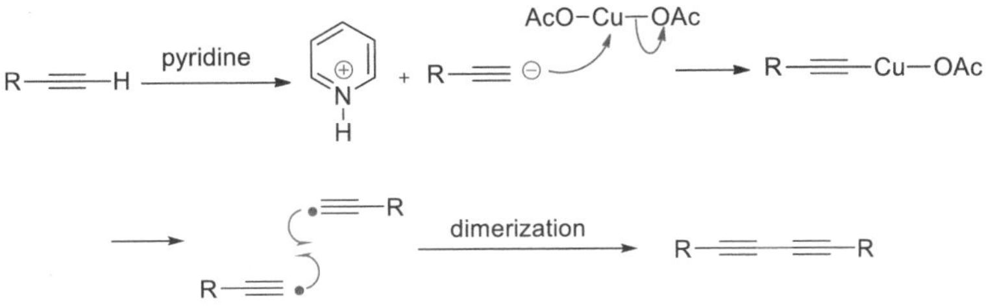

Example 1, Homo-coupling[2]

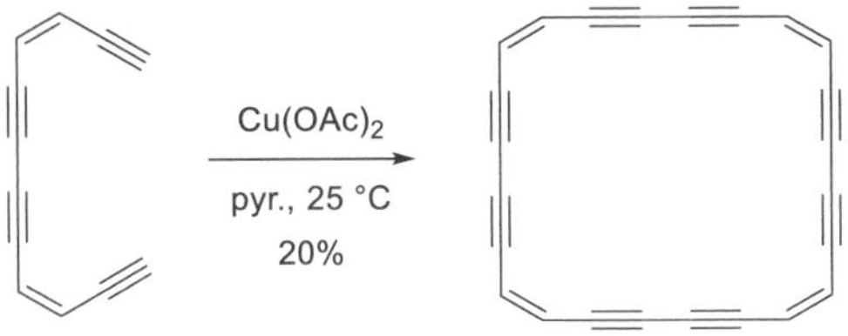

Example 2, Cross-coupling[3]

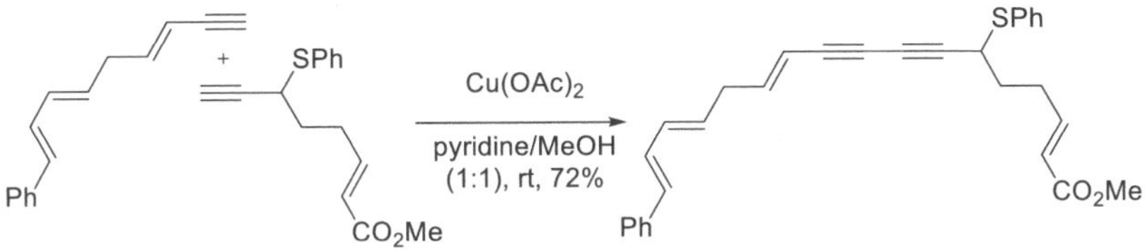

Example 3, Homo-coupling[4]

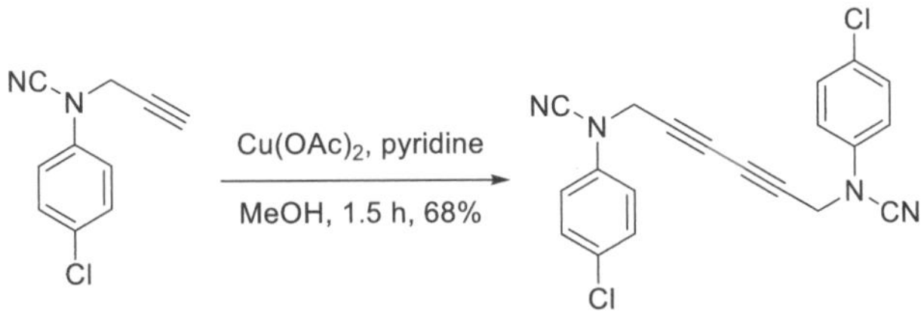

Example 4[5]

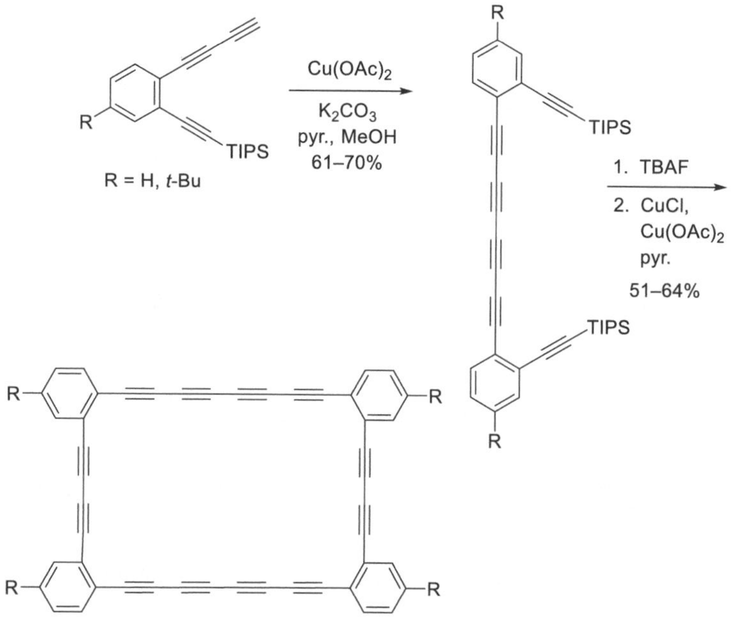

Cu(OAc)$_2$

K$_2$CO$_3$
pyr., MeOH
61–70%

R = H, *t*-Bu

R

TIPS

1. TBAF

2. CuCl,
Cu(OAc)$_2$
pyr.

51–64%

TIPS

R

R

R

R

R

Example 5[11]

CN

Fe

O

2 equiv Cu(OAc)$_2$•H$_2$O

1 : 1 Pyr : MeOH
70 °C, 12 h, 41%

CN

Fe

O

O

Fe

CN

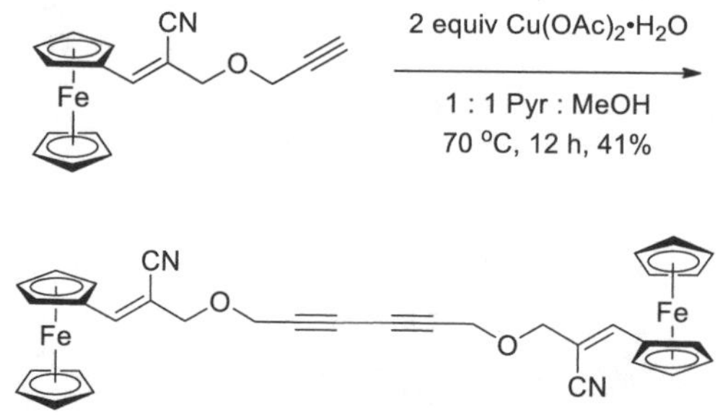

Example 6[12]

Cu(OAc)$_2$

Pyr./MeOH/Et$_2$O
62%

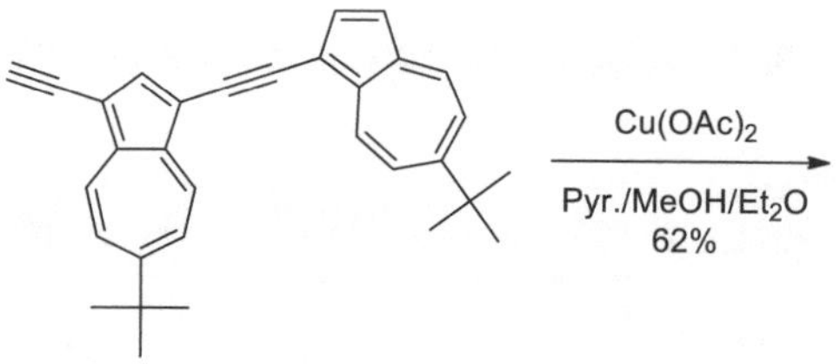

Example 7[13]

Example 8, Cu(OAc)$_2$-catalyzed intramolecular Eglington coupling[14]

Example 9, Employing dicobalt masking group to protect alkynes[15]

References

1. (a) Eglinton, G.; Galbraith, A. R. *Chem. Ind.* **1956,** 737–738. Geoffrey Eglinton (1927–2016), born in Cardiff, Wales, is a Professor Emeritus at Bristol University. (b) Behr, O. M.; Eglinton, G.; Galbraith, A. R.; Raphael, R. A. *J. Chem. Soc.* **1960,** 3614–3625. (c) Eglinton, G.; McRae, W. *Adv. Org. Chem.* **1963,** *4,* 225–328. (Review).

2. McQuilkin, R. M.; Garratt, P. J.; Sondheimer, F. *J. Am. Chem. Soc.* **1970,** *92,* 6682–6683.

3. Nicolaou, K. C.; Petasis, N. A.; Zipkin, R. E.; Uenishi, J. *J. Am. Chem. Soc.* **1982,** *104,* 5558–5560.

4. Srinivasan, R.; Devan, B.; Shanmugam, P.; Rajagopalan, K. *Indian J. Chem., Sect. B* **1997,** *36B,* 123–125.

5. Haley, M. M.; Bell, M. L.; Brand, S. C.; Kimball, D. B.; Pak, J. J.; Wan, W. B. *Tetrahedron Lett.* **1997,** *38,* 7483–7486.

6. Nakanishi, H.; Sumi, N.; Aso, Y.; Otsubo, T. *J. Org. Chem.* **1998,** *63,* 8632–8633.

7. Kaigtti-Fabian, K. H. H.; Lindner, H.-J.; Nimmerfroh, N.; Hafner, K. *Angew. Chem. Int. Ed.* **2001,** *40,* 3402–3405.

8. Siemsen, P.; Livingston, R. C.; Diederich, F. *Angew. Chem. Int. Ed.* **2000,** *39,* 2632–2657. (Review).

9. Inouchi, K.; Kabashi, S.; Takimiya, K.; Aso, Y.; Otsubo, T. *Org. Lett.* **2002,** *4,* 2533–2536.

10. Xu, G.-L.; Zou, G.; Ni, Y.-H.; DeRosa, M. C.; Crutchley, R. J.; Ren, T. *J. Am. Chem. Soc.* **2003,** *125,* 10057–10065.

11. Shanmugam, P.; Vaithiyananthan, V.; Viswambharan, B.; Madhavan, S. *Tetrahedron Lett.* **2007,** *48,* 9190–9194.

12. Miljanic, O. S.; Dichtel, W. R.; Khan, S. I.; Mortezaei, S.; Heath, J. R.; Stoddart, J. F. *J. Am. Chem. Soc.* **2007,** *129,* 8236–8246.

13. White, N. G.; Beer, P. D. *Beilst. J. Org. Chem.* **2012,** *8,* 246–252.

14. Peng, L.; Xu, F.; Suzuma, Y.; Orita, A.; Otera, J. *J. Org. Chem.* **2013,** *78,* 12802–12808.

15. Kohn, D. R.; Gawel, P.; Xiong, Y.; Christensen, K. E.; Anderson, H. L. *J. Org. Chem.* **2018,** *83,* 2077–2086.

16. Zhang, S.; Zhao, L. *Nat. Commun.* **2019,** *10,* 1–10.

17. Gu, M.-D.; Lu, Y.; Wang, M.-X. *J. Org. Chem.* **2020,** *85,* 2312–2320.

Gould–Jacobs Reaction

The Gould–Jacobs reaction is a sequence of the following reactions:

a.   Substitution of an aniline with either alkoxy methylenemalonic ester or acyl malonic ester providing the anilinomethylenemalonic ester;

b.   Cyclization of to the 4-hydroxy-3-carboalkoxyquinoline (4-hydroxy-quinolines exist predominantly in 4-oxoform);

c.   Saponification to form acid;

d.   Decarboxylation to give the 4-hydroxyquinoline. Extension could lead to unsubstituted parent heterocycles with fused pyridine ring of Skraup type.

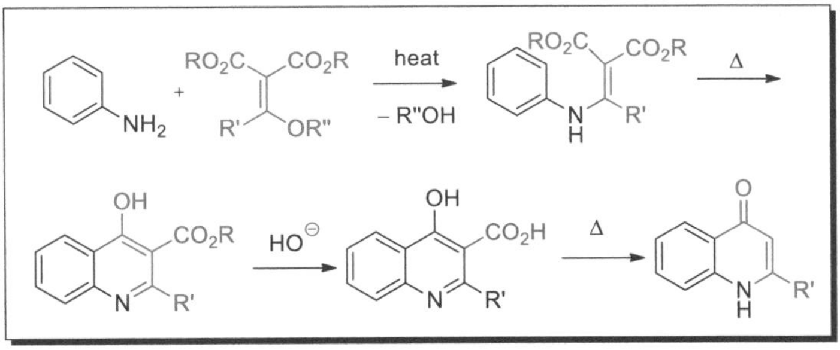

R = alkyl; R' = alkyl, aryl, or H; R'' = alkyl or H

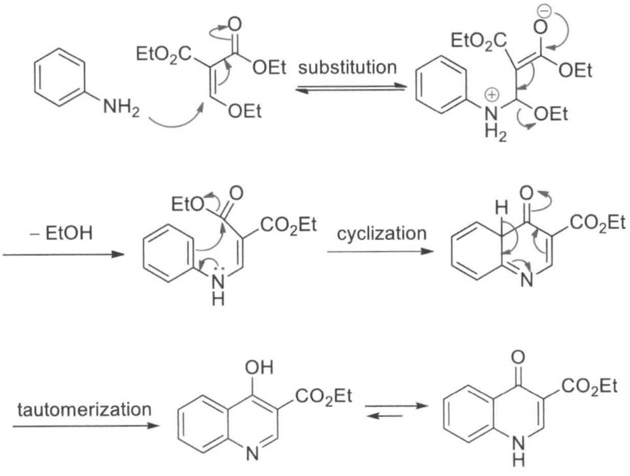

Example 1[3]

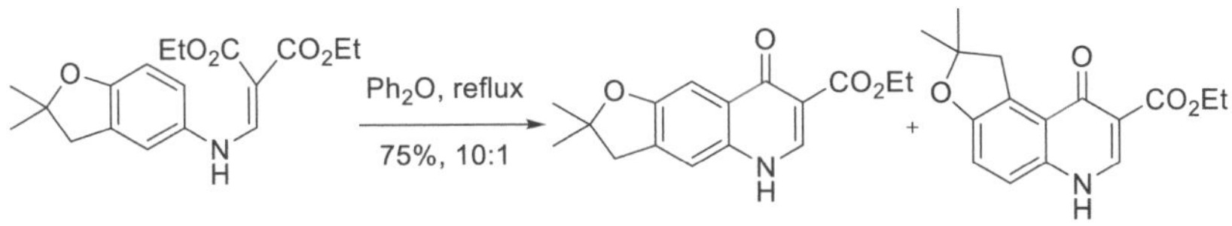

© Springer Nature Switzerland AG 2021
J. J. Li, *Name Reactions*, https://doi.org/10.1007/978-3-030-50865-4_59

Example 2[7]

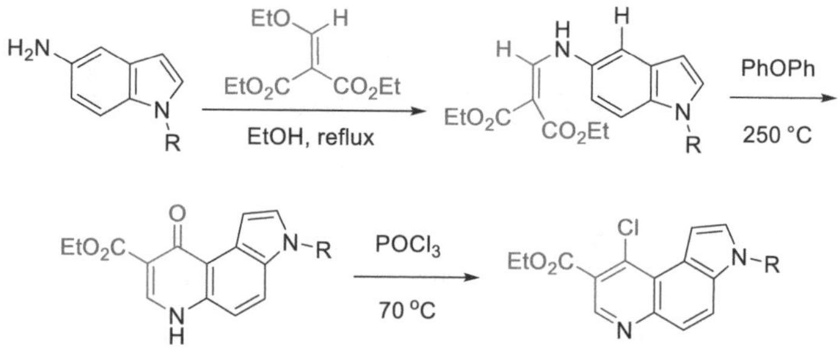

Example 3, Microwave-assisted Gould–Jacobs reaction[8]

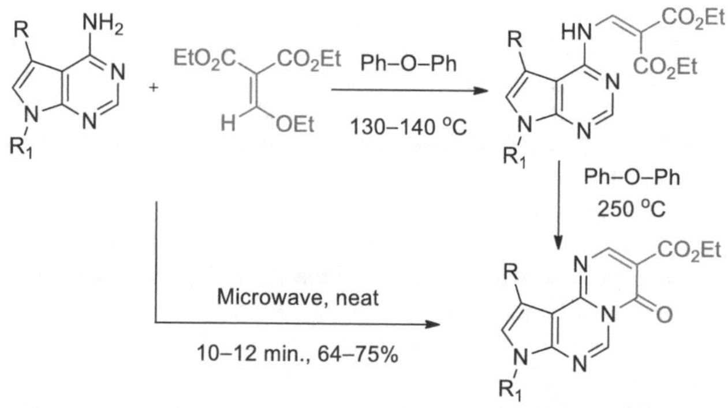

Example 4[9]

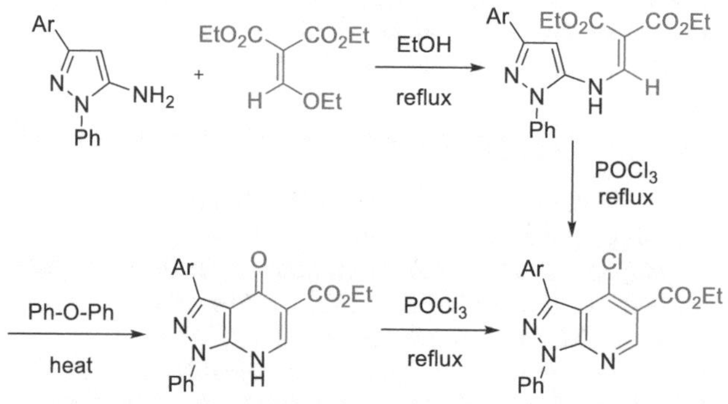

Example 5, Gould–Jacobs reaction in a novel three-mode pyrolysis reactor[11]

References

1. Gould, R. G.; Jacobs, W. A. *J. Am. Chem. Soc.* **1939**, *61*, 2890–2895. R. Gordon Gould was born in Chicago in 1909. He earned his Ph.D. at Harvard University in 1933. After serving as an instructor at Harvard and Iowa, Gould worked at Rockefeller Institute for Medical Research where he discovered the Gould–Jacobs reaction with his colleague Walter A. Jacobs.
2. Reitsema, R. H. *Chem. Rev.* **1948,** *53*, 43–68. (Review).
3. Cruickshank, P. A., Lee, F. T., Lupichuk, A. *J. Med. Chem.* **1970**, *13*, 1110–1114.
4. Elguero J., Marzin C., Katritzky A. R., Linda P., *The Tautomerism of Heterocycles*, Academic Press, New York, **1976,** pp 87–102. (Review).
5. Milata, V.; Claramunt, R. M.; Elguero, J.; Zálupský, P. *Targets in Heterocyclic Systems* **2000**, *4*, 167–203. (Review).
6. Curran, T. T. *Gould–Jacobs Reaction*. In *Name Reactions in Heterocyclic Chemistry*; Li, J. J., Ed.; Wiley: Hoboken, NJ, **2005,** 423–436. (Review).
7. Ferlin, M. G.; Chiarelotto, G.; Dall'Acqua, S.; Maciocco, E.; Mascia, M. P.; Pisu, M. G.; Biggio, G. *Bioorg. Med. Chem.* **2005**, *13*, 3531–3541.
8. Desai, N. D. *J. Heterocycl. Chem.* **2006**, *43*, 1343–1348.
9. Kendre, D. B.; Toche, R. B.; Jachak, M. N. *J. Heterocycl. Chem.* **2008**, *45*, 1281–1286.
10. Lengyel, L.; Nagy, T. Z.; Sipos, G.; Jones, R.; Dormán, G.; Üerge, L.; Darvas, F. *Tetrahedron Lett.* **2012**, *53*, 738–743.
11. Lengyel, L. C.; Sipos, G.; Sipőcz, T.; Vágó, T.; Dormán, G.; Gerencsér, J.; Makara, G.; Darvas, F. *Org. Process Res. Dev.* **2015**, *19*, 399–409.
12. Malvacio, I.; Moyano, E. L.; Vera, D. M. A. *RSC Adv.* **2016**, *6*, 83973–83981.
13. Trah, S.; Lamberth, C. *Tetrahedron Lett.* **2017**, *58*, 794–796.
14. Milata, V.; Vaculka, M. *Monat. Chem.* **2019**, *5150*, 711–719.
15. Orozco, D.; Kouznetsov, V. V.; Bermudez, A.; Vargas Mendez, L. Y.; Mendoza Salgado, A. R.; Melendez Gomez, C. M. *RSC Adv.* **2020**, *10*, 4876–4898.

Grignard Reaction

Addition of organomagnesium compounds (Grignard reagents), generated from organohalides and magnesium metal, to electrophiles.

Formation of the Grignard reagent:

Grignard reaction, ionic mechanism:

Grignard reaction, radical mechanism,

Example 1[4]

This reaction is known as the *Hoch–Campbell aziridine synthesis*, which entails treatment of ketoximes with excess Grignard reagents and subsequent hydrolysis of the organometallic complex to produce aziridines.

© Springer Nature Switzerland AG 2021
J. J. Li, *Name Reactions*, https://doi.org/10.1007/978-3-030-50865-4_60

Example 2[5]

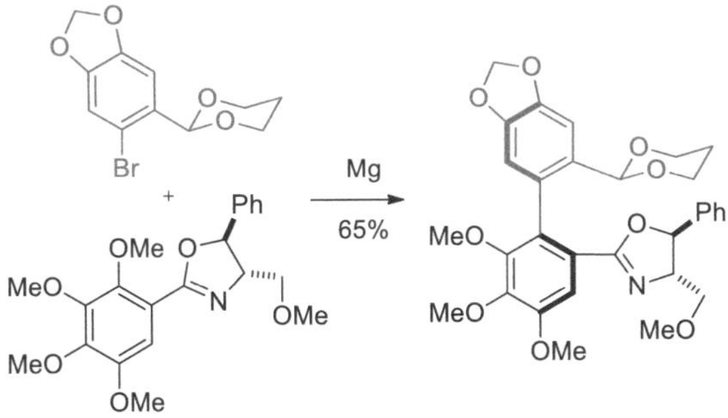

Example 5[10]

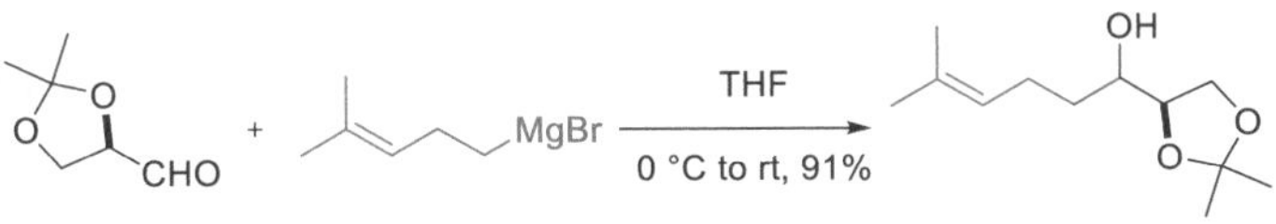

Example 6[11]

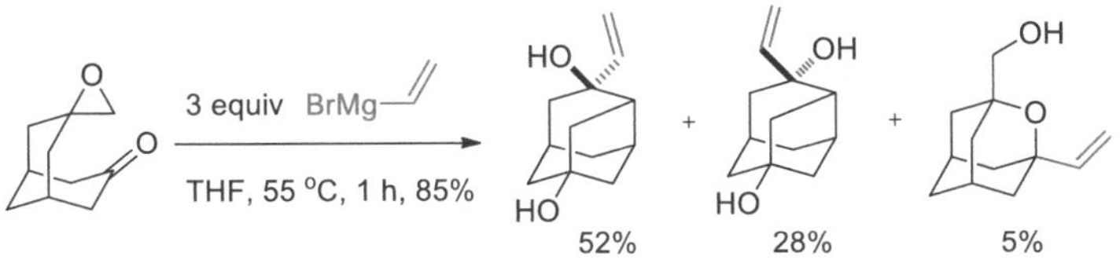

Example 7, Asymmetric conjugate addition[12]

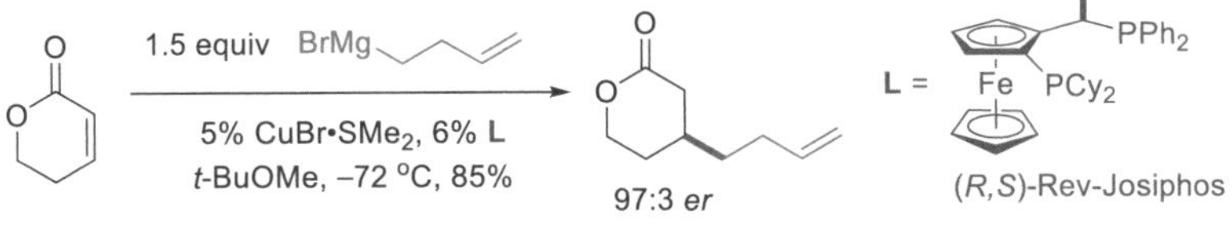

Example 7, Regio- and enantioselective copper-catalyzed allylic alkylation of with Grignard reagents[13]

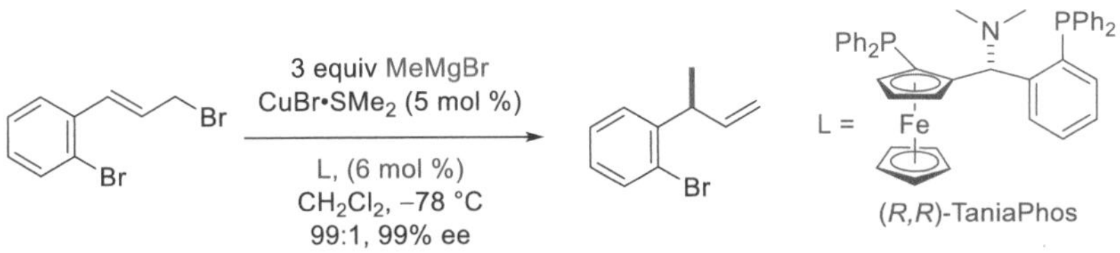

Example 8, Additions of Grignard reagents to aliphatic aldehydes do not involve single-electron-transfer (SET) processes[14]

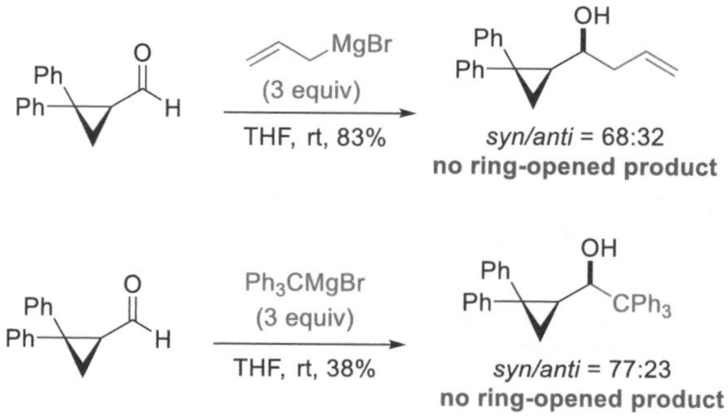

MgBr
(3 equiv)
THF, rt, 83%
syn/anti = 68:32
no ring-opened product
Ph3CMgBr
(3 equiv)
THF, rt, 38%
syn/anti = 77:23
no ring-opened product

Example 9, Sodium methyl carbonate (SMC) as an effective C1 synthon[15]

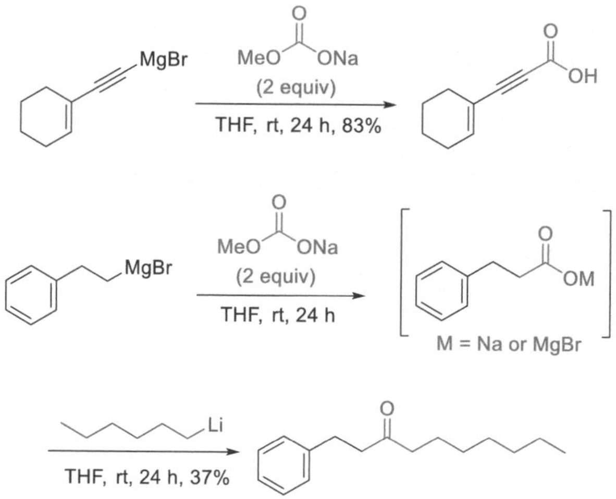

MgBr
MeO ONa
(2 equiv)
THF, rt, 24 h, 83%
OH
THF, rt, 24 h
OM
M = Na or MgBr
Li
THF, rt, 24 h, 37%

Example 10, Grignard carboxylation[16]

1. i-PrMgCl (1.05 equiv), THF, 4 °C
2. CO2, −25 °C, 97%
CO2H

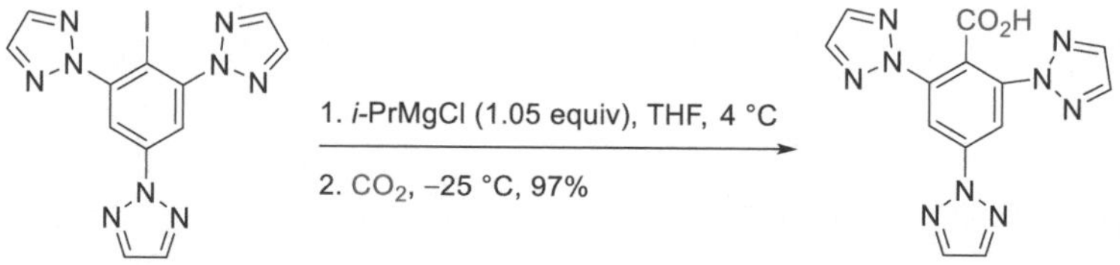

References

1. Grignard, V. *C. R. Acad. Sci.* **1900,** *130*, 1322–1324. Victor Grignard (France, 1871–1935) was a colleague of Philippe Barbier (of the Barbier reaction fame) and won the Nobel Prize in Chemistry in 1912 for his discovery of the Grignard reagent.
2. Ashby, E. C.; Laemmle, J. T.; Neumann, H. M. *Acc. Chem. Res.* **1974,** *7*, 272–280. (Review).
3. Ashby, E. C.; Laemmle, J. T. *Chem. Rev.* **1975,** *75*, 521–546. (Review).
4. Sasaki, T.; Eguchi, S.; Hattori, S. *Heterocycles* **1978,** *11*, 235–242.
5. Meyers, A. I.; Flisak, J. R.; Aitken, R. A. *J. Am. Chem. Soc.* **1987,** *109*, 5446–5452.
6. *Grignard Reagents* Richey, H. G., Jr., Ed.; Wiley: New York, **2000**. (Book).
7. Holm, T.; Crossland, I. In *Grignard Reagents* Richey, H. G., Jr., Ed.; Wiley: New York, **2000,** Chapter 1, pp 1–26. (Review).
8. Shinokubo, H.; Oshima, K. *Eur. J. Org. Chem.* **2004,** 2081–2091. (Review).
9. Graden, H.; Kann, N. *Cur. Org. Chem.* **2005,** *9*, 733–763. (Review).
10. Babu, B. N.; Chauhan, K. R. *Tetrahedron Lett.* **2008,** *50*, 66–67.
11. Mlinaric-Majerski, K.; Kragol, G.; Ramljak, T. S. *Synlett* **2008,** 405–409.
12. Mao, B.; Fanãnás-Mastral, M.; Feringa, B. L. *Org. Lett.* **2013,** *15*, 286–289.
13. van der Molen, N. C.; Tiemersma-Wegman, T. D.; Fanãnás-Mastral, M.; Feringa, B. L. *J. Org. Chem.* **2015,** *80*, 4981–4984.
14. Otte, D. A. L.; Woerpel, K. A. *Org. Lett.* **2015,** *17*, 3906–3909.
15. Hurst, T. E.; Deichert, J. A.; Kapeniak, L.; Lee, R.; Harris, J.; Jessop, P. G.; Snieckus, V. *Org. Lett.* **2019,** *21*, 3882–3885.
16. Roth, R.; Schmidt, G.; Prud'homme, A.; Abele, S. *Org. Process Res. Dev.* **2019,** *23*, 234–243.
17. Hosoya, M.; Nishijima, S.; Kurose, N. *Org. Process Res. Dev.* **2020,** *24*, 405–414.

Grob Fragmentation

The C–C bond cleavage primarily via a concerted process involving a five-atom system.
General scheme:

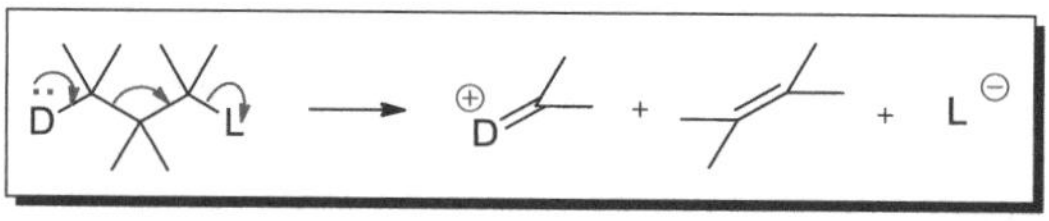

$$D = O^-, NR_2; L = OH_2^+, OTs, I, Br, Cl$$

Example 1[2]

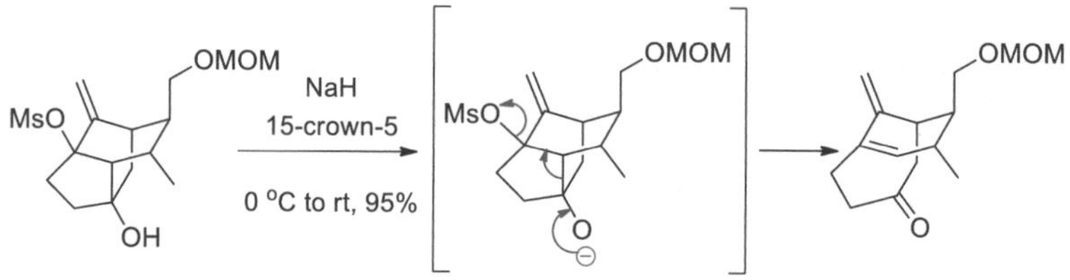

Example 2, Aza-Grob fragmentation[3]

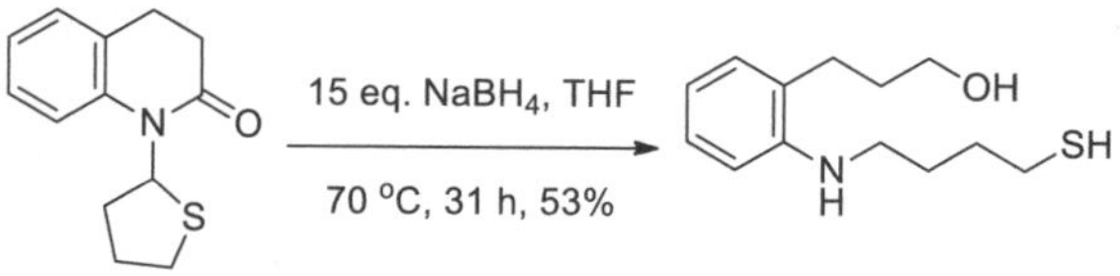

Example 3[7]

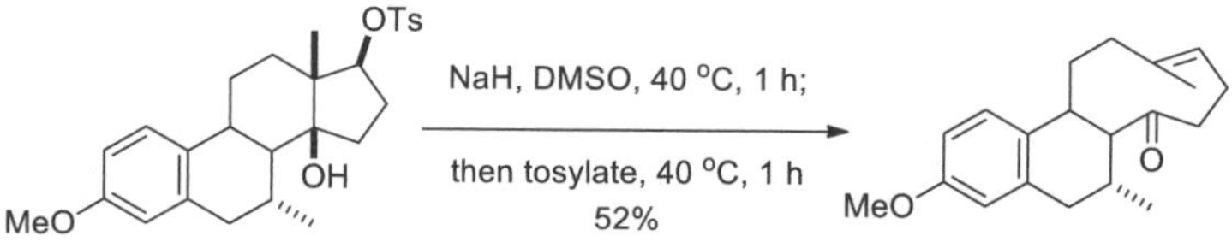

Example 4[8]

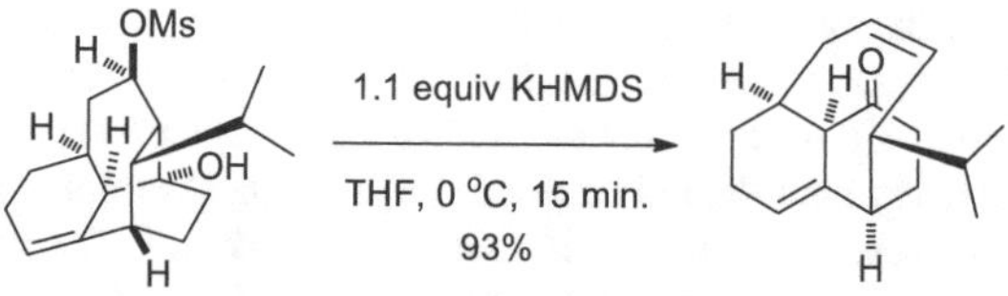

© Springer Nature Switzerland AG 2021
J. J. Li, *Name Reactions*, https://doi.org/10.1007/978-3-030-50865-4_61

Example 5[8]

Example 6, Grob-type fragmentation releases paracyclophane ring strain in a late-stage precursor of haouamine A[12]

Example 7, Grob-fragmentation-enabled approach to clavulactone analogs[13]

References

1.   (a) Grob, C. A.; Baumann, W. *Helv. Chim. Acta* **1955**, *38*, 594–603. (b) Grob, C. A.; Schiess, P. W. *Angew. Chem. Int. Ed.* **1967**, *6*, 1–15. Cyril A. Grob (1917–2003) was born in London (UK) to Swiss parents, studied chemistry at ETH Zürich and completed his PhD in 1943 under the guidance of Leopold Ruzicka (Nobel laureate) on artificial steroidal antigens. He then moved to Basel to work with Taddeus Reichstein (another Nobel laureate) first at the pharmaceutical institute and from 1947 at the organic chemistry institute of the university, where he moved up the academic career ladder to become the director of the institute and holder of the chair there as Reichstein's suc-

cessor in 1960. An investigation of the reductive elimination of bromine from 1,4-dibromides in the presence of zinc led in 1955 to the recognition of heterolytic fragmentation as a general reaction principle. The heterolytic fragmentation has now entered textbooks under his name. Experimental evidence for vinyl cations as discrete reactive intermediates was also first provided by Grob. Cyril Grob never acted impulsively, but always calmly and deliberately. He never sought attention in public, but fulfilled his social duties efficiently, reliably, and without a fuss. He died in his home in Basel (Switzerland) on December 15, 2003 at the age of 86. (Schiess, P. *Angew. Chem. Int. Ed.* **2004,** *43,* 4392.) A review[10] revealed that Grob was not even the first to investigate such reactions.

2. Yoshimitsu, T.; Yanagiya, M.; Nagaoka, H. *Tetrahedron Lett.* **1999,** *40*, 5215–5218.
3. Hu, W.-P.; Wang, J.-J.; Tsai, P.-C. *J. Org. Chem.* **2000,** *65*, 4208–4029.
4. Molander, G. A.; Le Huerou, Y.; Brown, G. A. *J. Org. Chem.* **2001,** *66*, 4511–4516.
5. Paquette, L. A.; Yang, J.; Long, Y. O. *J. Am. Chem. Soc.* **2002,** *124*, 6542–6543.
6. Barluenga, J.; Alvarez-Perez, M.; Wuerth, K.; *et al. Org. Lett.* **2003,** *5*, 905–908.
7. Khripach, V. A.; Zhabinskii, V. N.; Fando, G. P.; *et al. Steroids* **2004,** *69*, 495–499.
8. Maimone, T. J.; Voica, A.-F.; Baran, P. S. *Angew. Chem. Int. Ed.* **2008,** *47*, 3054–3056.
9. Barbe, G.; St-Onge, M.; Charette, A. B. *Org. Lett.* **2008,** *10*, 5497–5499.
10. Prantz, K.; Mulzer, J. *Chem. Rev.* **2010,** *110,* 3741–4766. (Review).
11. Umland, K.-D.; Palisse, A.; Haug, T. T.; Kirsch, S. F. *Angew. Chem. Int. Ed.* **2011,** *50*, 9965–9968.
12. Cao, L.; Wang, C.; Wipf. P. *Org. Lett.* **2019,** *21*, 1538–1541.
13. Gu, Q.; Wang, X.; Sun, B.; Lin, G. *Org. Lett.* **2019,** *21*, 5082–5085.
14. Rivero-Crespo, M. A.; Tejeda-Serrano, M.; Perez-Sanchez, H.; Ceron-Carrasco, J. P.; Leyva-Perez, A. *Angew. Chem. Int. Ed.* **2020,** *59*, 3846–3849.

Hajos–Wiechert Reaction

Asymmetric Robinson annulation catalyzed by (S)-$(-)$-proline.

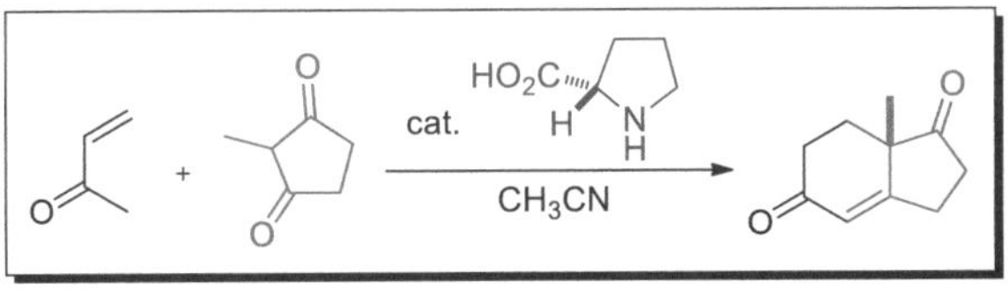

Hajos–Wiechert ketone

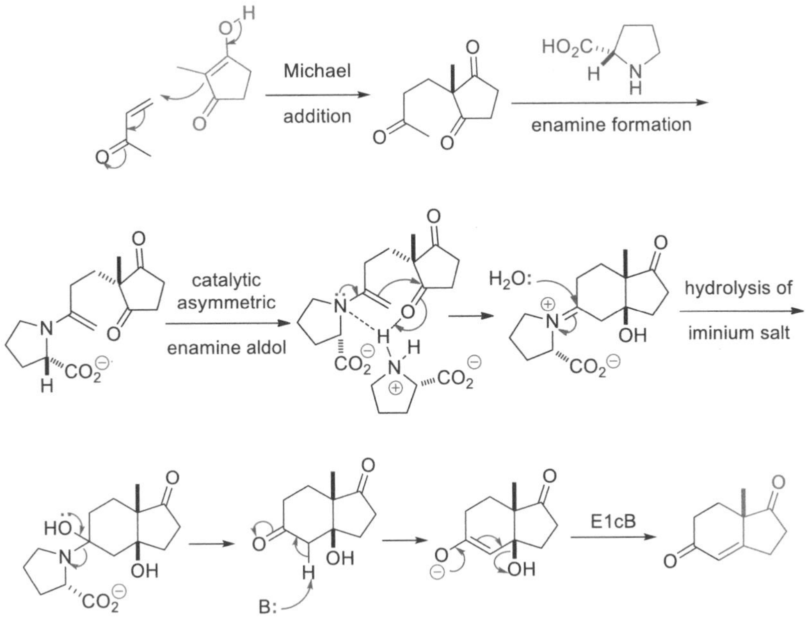

Example 1, Intramolecular Hajos–Wiechert reaction[1a]

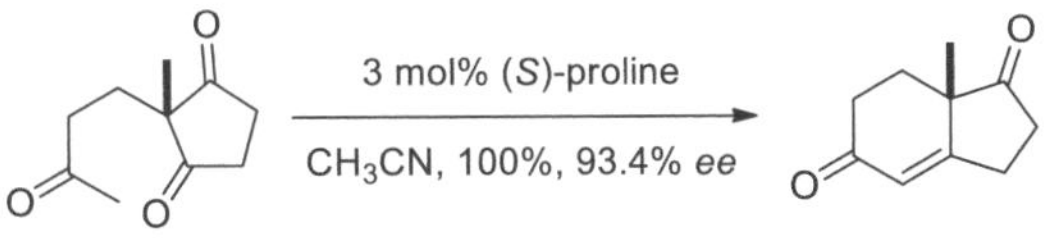

Hajos–Wiechert ketone

© Springer Nature Switzerland AG 2021
J. J. Li, *Name Reactions*, https://doi.org/10.1007/978-3-030-50865-4_62

Example 2[3]

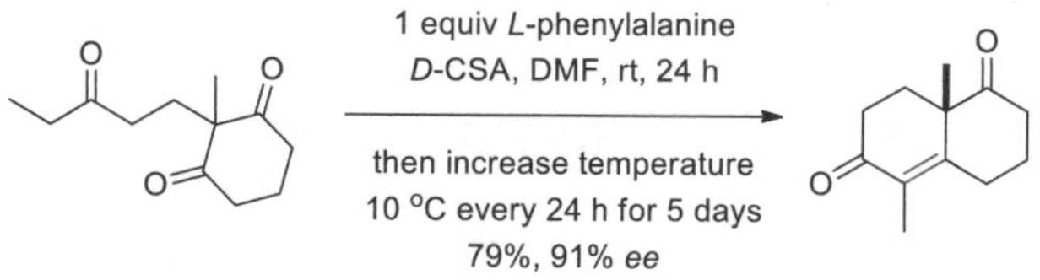

Wieland–Miescher ketone

Example 3[8]

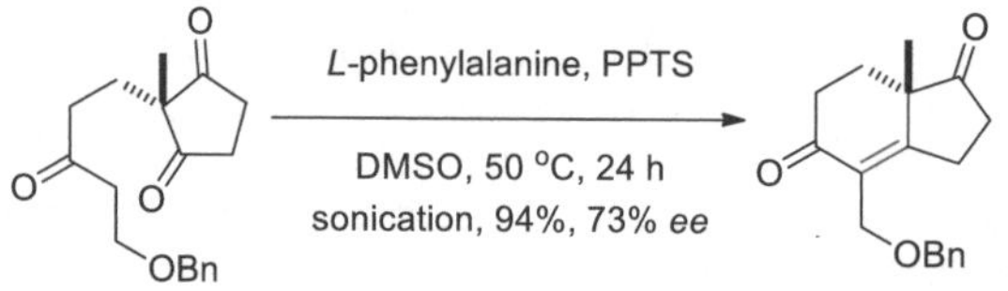

Hajos–Wiechert ketone

Example 4[9]

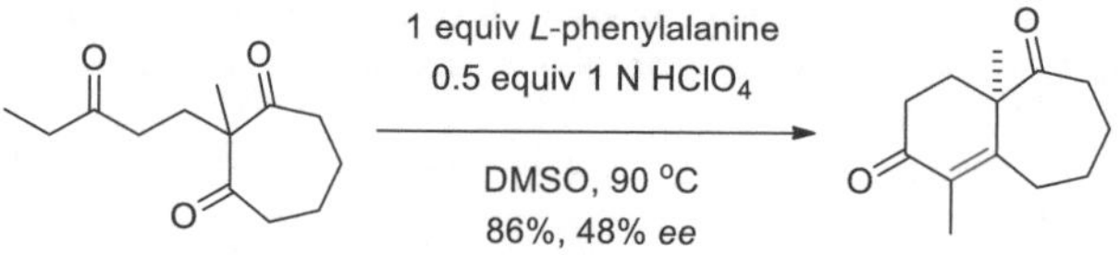

Example 5, Bifunctional organocatalysts based on a carbazole scaffold[14]

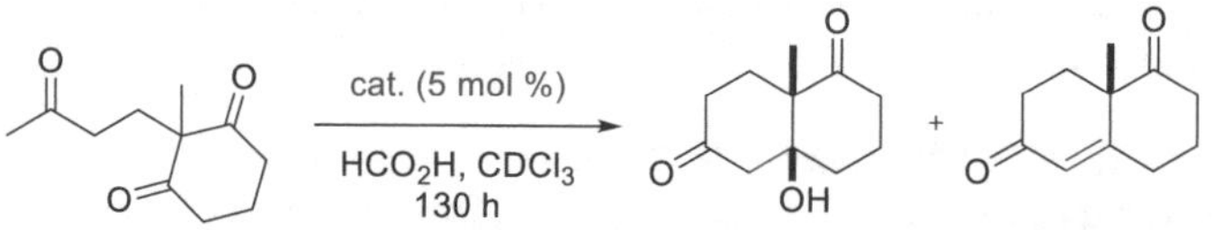

15 (99.9% ee) : 75 (99.9% ee)

Wieland–Miescher ketones

cat. =

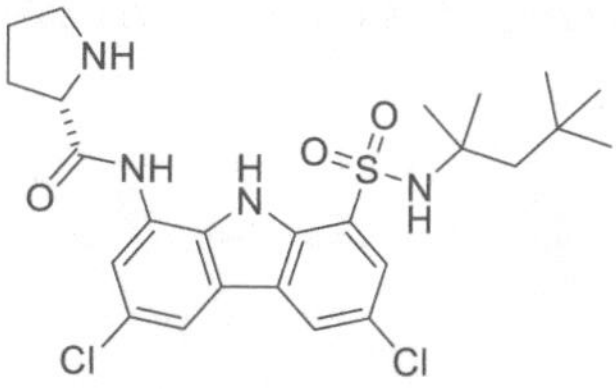

Example 5, Hajos–Parrish–Eder–Sauer–Wiechert-type reaction (a mouthful?!)[15]

Hajos–Wiechert ketone

References

1. (a) Hajos, Z. G.; Parrish, D. R. *J. Org. Chem.* **1974,** *39*, 1615–1621. Hajos and Parrish were chemists at Hoffmann–La Roche. (b) Eder, U.; Sauer, G.; Wiechert, R. *Angew. Chem. Int. Ed.* **1971,** *10*, 496–497.
2. Brown, K. L.; Dann, L.; Duntz, J. D.; Eschenmoser, A.; Hobi, R.; Kratky, C. *Helv. Chim. Acta* **1978,** *61*, 3108–3135.
3. Hagiwara, H.; Uda, H. *J. Org.Chem.* **1998,** *53*, 2308–2311.
4. Nelson, S. G. *Tetrahedron: Asymmetry* **1998,** *9*, 357–389.
5. List, B.; Lerner, R. A.; Barbas, C. F., III. *J. Am. Chem. Soc.* **2000,** *122*, 2395–2396.
6. List, B.; Pojarliev, P.; Castello, C. *Org. Lett.* **2001,** *3*, 573–576.
7. Hoang, L.; Bahmanyar, S.; Houk, K. N.; List, B. *J. Am. Chem. Soc.* **2003,** *125*, 16–17.
8. Shigehisa, H.; Mizutani, T.; Tosaki, S.-y.; Ohshima, T.; Shibasaki, M. *Tetrahedron* **2005,** *61*, 5057–5065.
9. Nagamine, T.; Inomata, K.; Endo, Y.; Paquette, L. A. *J. Org. Chem.* **2007,** *72*, 123–131.
10. Kennedy, J. W. J.; Vietrich, S.; Weinmann, H.; Brittain, D. E. A. *J. Org. Chem.* **2009,** *73*, 5151–5154.
11. Christen, D. P. *Hajos–Wiechert Reaction.* In *Name Reactions for Homologations-Part II*; Li, J. J., Ed.; Wiley: Hoboken, NJ, **2009,** pp 554–582. (Review).
12. Zhu, H.; Clemente, F. R.; Houk, K. N.; Meyer, M. P. *J. Am. Chem. Soc.* **2009,** *131*, 1632–1633.
13. Bradshaw, B.; Bonjoch, J. *Synlett* **2012,** *23*, 337–356. (Review).
14. Rubio, O. H.; de Arriba, Á. L.; Monleón, L. M.; Sanz, F.; Simón, L.; Alcázae, V.; Morán, J. R. *Tetrahedron* **2015,** *71*, 1297–1303.
15. Schneider, L. M.; Schmiedel, V. M.; Pecchioli, T.; Lentz, D.; Merten, C.; Christmann, M. *Org. Lett.* **2017,** *19*, 2310–2313.
16. Yadav, G. D.; Deepa; Singh, S. *ChemistrySelect* **2019,** *14*, 5591–5618.

Hantzsch Dihydropyridine Synthesis

1,4-Dihydropyridine from the condensation of aldehyde, β-ketoester and ammonia. Hantzsch 1,4-dihydropyridines are popular reducing reagents in organocatalysis. Well-known calcium channel blockers (CCBs) nifedipine (Adalat), felodipine (Plendil), and amlodipine (Norvasc) for treating hypertension all have 1,4-dihydropyridine core structures.

Example 1[2]

nifedipine, the first calcium channel blocker

Example 2[10]

Example 3, Covalently anchored sulfonic acid on silica gel as catalyst[10]

Example 4, Intermolecular aryne *ene reaction*[13]

Example 5, Hantzsch ester for radical alkynylation (photochemistry)[14]

References

1. Hantzsch, A. *Ann.* **1882,** *215*, 1–83.
2. Bossert, F.; Vater, W. *Naturwissenschaften* **1971,** *58*, 578–585.
3. Balogh, M.; Hermecz, I.; Naray-Szabo, G.; Simon, K.; Meszaros, Z. *J. Chem. Soc., Perkin Trans. 1* **1986,** 753–757.
4. Katritzky, A. R.; Ostercamp, D. L.; Yousaf, T. I. *Tetrahedron* **1987,** *43*, 5171–5187.
5. Menconi, I.; Angeles, E.; Martinez, L.; Posada, M. E.; Toscano, R. A.; Martinez, R. *J. Heterocycl. Chem.* **1995,** *32*, 831–833.
6. Raboin, J.-C.; Kirsch, G.; Beley, M. *J. Heterocycl. Chem.* **2000,** *37*, 1077–1080.
7. Sambongi, Y.; Nitta, H.; Ichihashi, K.; Futai, M.; Ueda, I. *J. Org. Chem.* **2002,** *67*, 3499–3501.
8. Wang, L.-M.; Sheng, J.; Zhang, L.; Han, J.-W.; Fan, Z.-Y.; Tian, H.; Qian, C.-T. *Tetrahedron* **2005,** *61*, 1539–1543.
9. Galatsis, P. *Hantzsch Dihydro-Pyridine Synthesis.* In *Name Reactions in Heterocyclic Chemistry*; Li, J. J., Ed.; Wiley: Hoboken, NJ, **2005,** pp 304–307. (Review).
10. Gupta, R.; Gupta, R.; Paul, S.; Loupy, A. *Synthesis* **2007,** 2835–2838.
11. Snyder, N. L.; Boisvert, C. J. *Hantzsch Synthesis*, in *Name Reactions in Heterocyclic Chemistry II,* Li, J. J., Ed.; Wiley: Hoboken, NJ, **2011,** pp 591–644. (Review).
12. Ghosh, S.; Saikh, F.; Das, J.; Pramanik, A. K. *Tetrahedron Lett.* **2013,** *54*, 58–62.
13. Trinchera, P.; Sun, W.; Smith, J. E.; Palomas, D.; Crespo-Otero, R.; Jones, C. R. *Org. Lett.* **2017,** *19*, 4644–4647.
14. Liu, X.; Liu, R.; Dai, J.; Cheng, X.; Li, G. *Org. Lett.* **2018,** *20*, 6906–6909.
15. Zeynizadeh, B.; Rahmani, S. *RSC Adv.* **2019,** *9*, 8002–8015.
16. Li, J.; Fang, X.; Ming, X. *J. Org. Chem.* **2020,** *85*, 4602–4610.

Heck Reaction

The palladium-catalyzed alkenylation or arylation of olefins.

$R^1 = $ aryl, alkenyl, alkyl (with no β-hydrogen)
$X = $ Cl, Br, I, OTf, OTs, N_2^+

The catalytic cycle:

A: Oxidative addition
B: Migratory insertion (*syn*)
C: C–C bond rotation

D: *syn*-β-elimination
E: Reductive elimination

Example 1, Asymmetric intermolecular Heck reaction[6]

J. J. Li, *Name Reactions*, https://doi.org/10.1007/978-3-030-50865-4_64

Example 2, Intramolecular Heck[7]

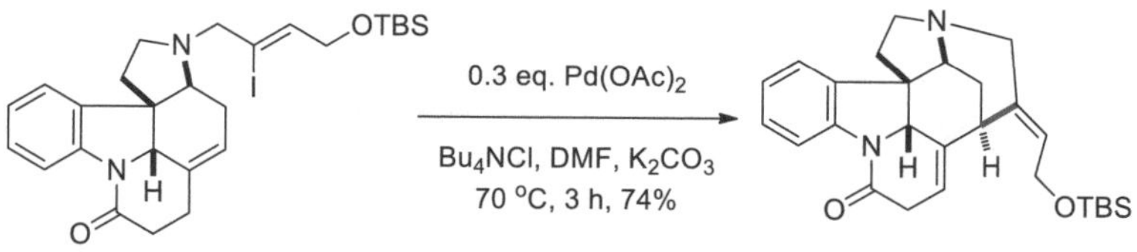

0.3 eq. Pd(OAc)$_2$

Bu$_4$NCl, DMF, K$_2$CO$_3$
70 °C, 3 h, 74%

Example 3[8]

1.5% Pd$_2$(dba)$_3$, 6% P(t-Bu)$_3$
1.1. eq. Cs$_2$CO$_3$, dioxane
120 °C, 24 h, 82%

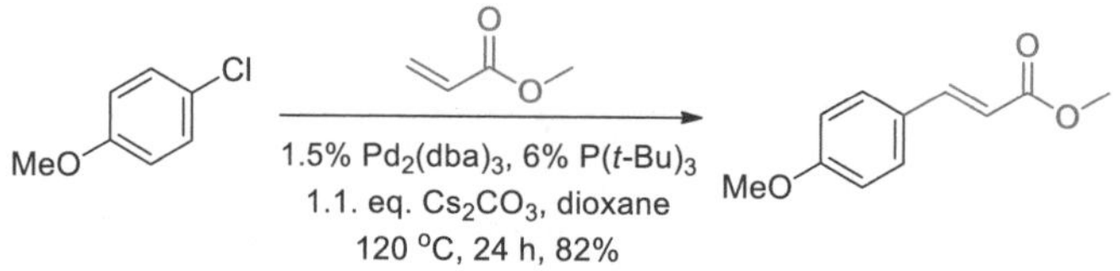

Example 4, Intramolecular Heck[9]

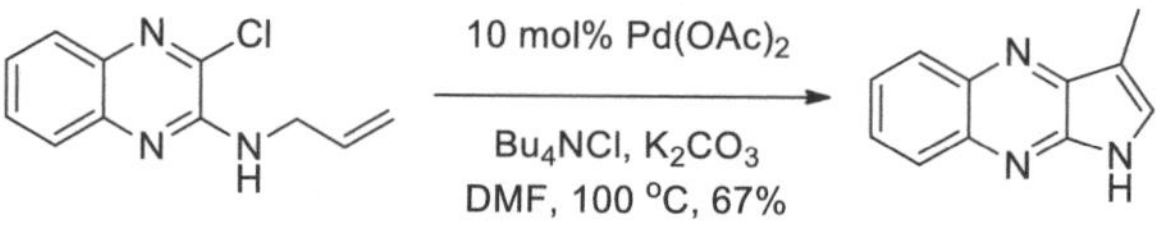

10 mol% Pd(OAc)$_2$

Bu$_4$NCl, K$_2$CO$_3$
DMF, 100 °C, 67%

Example 5, Intramolecular Heck[13]

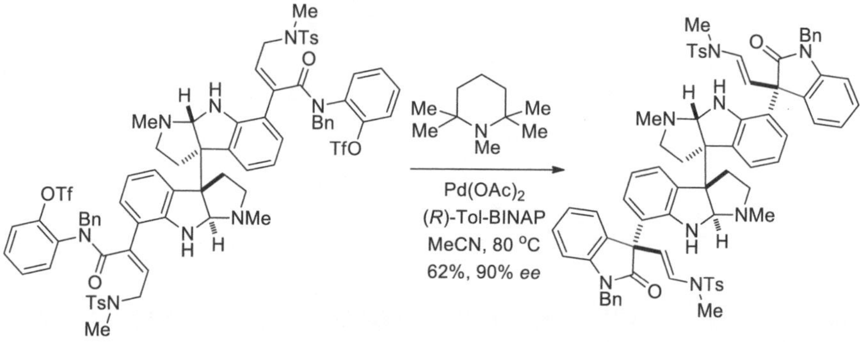

Pd(OAc)$_2$
(R)-Tol-BINAP
MeCN, 80 °C
62%, 90% ee

Example 6, Reductive Heck reaction[17]

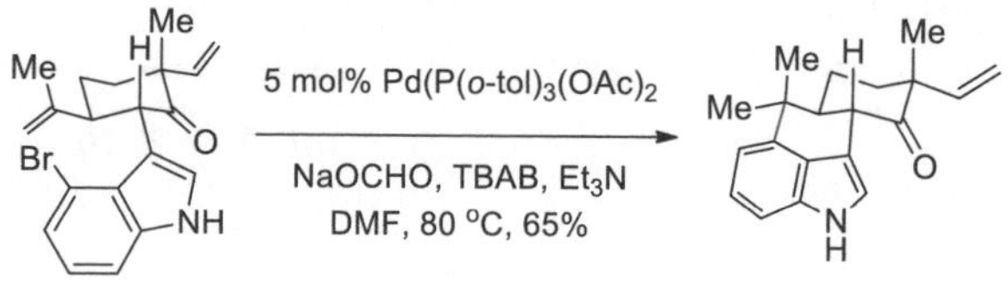

5 mol% Pd(P(o-tol)$_3$(OAc)$_2$

NaOCHO, TBAB, Et$_3$N
DMF, 80 °C, 65%

Example 7, Intramolecular Heck[20]

Example 8, Converting FK506 to its non-immunosuppressive analogs[21]

FK506

Example 9, Asymmetric Heck carbonylation to prepare quaternary carbon enantiaoselectively[22]

62%, 93% *ee*

References

1. Heck, R. F.; Nolley, J. P., Jr. *J. Am. Chem. Soc.* **1968**, *90*, 5518–5526. Richard F. Heck (1931–2015) discovered the Heck reaction when he was at Hercules Corp. Heck won Nobel Prize in 2010 along with Akira Suzuki and Ei-ichi Negishi "for palladium-catalyzed cross-couplings in organic synthesis".
2. Heck, R. F. *Acc. Chem. Res.* **1979**, *12*, 146–151. (Review).
3. Heck, R. F. *Org. React.* **1982**, *27*, 345–390. (Review).
4. Heck, R. F. *Palladium Reagents in Organic Synthesis,* Academic Press, London, **1985**. (Book).
5. Hegedus, L. S. *Transition Metals in the Synthesis of Complex Organic Molecule* **1994**, University Science Books: Mill Valley, CA, pp 103–113. (Book).
6. Ozawa, F.; Kobatake, Y.; Hayashi, T. *Tetrahedron Lett.* **1993**, *34*, 2505–2508.
7. Rawal V. H.; Iwasa, H. *J. Org. Chem.* **1994**, *59*, 2685–2686.
8. Littke, A. F.; Fu, G. C. *J. Org. Chem.* **1999**, *64*, 10–11.
9. Li, J. J. *J. Org. Chem.* **1999**, *64*, 8425–8427.
10. Beletskaya, I. P.; Cheprakov, A. V. *Chem. Rev.* **2000**, *100*, 3009–3066. (Review).
11. Amatore, C.; Jutand, A. *Acc. Chem. Res.* **2000**, *33*, 314–321. (Review).
12. Link, J. T. *Org. React.* **2002**, *60*, 157–534. (Review).
13. Lebsack, A. D.; Link, J. T.; Overman, L. E.; Stearns, B. A. *J. Am. Chem. Soc.* **2002**, *124*, 9008–9009.
14. Dounay, A. B.; Overman, L. E. *Chem. Rev.* **2003**, *103*, 2945–2963. (Review).
15. Beller, M.; Zapf, A.; Riermeier, T. H. *Transition Metals for Organic Synthesis* (2nd edn.) **2004**, *1*, 271–305. (Review).
16. Oestreich, M. *Eur. J. Org. Chem.* **2005**, 783–792. (Review).
17. Baran, P. S.; Maimone, T. J.; Richter, J. M. *Nature* **2007**, *446*, 404–406.
18. Fuchter, M. J. *Heck Reaction*. In *Name Reactions for Homologations-Part I*; Li, J. J., Ed.; Wiley: Hoboken, NJ, **2009**, pp 2–32. (Review).
19. *The Mizoroki–Heck Reaction*; Oestreich, M., Ed.; Wiley: Hoboken, NJ, **2009**.
20. Bennasar, M.-L.; Solé, D.; Zulaica, E.; Alonso, S. *Tetrahedron* **2013**, *69*, 2534–2541.
21. Wang, Y.; Peiffer, B. J.; Su, Q.; Liu, J. O. *ACS Med. Chem. Lett.* **2019**, *69*, 2534–2541.
22. Cheng, C.; Wan, B.; Zhou, B.; Gu, Y.; Zhang, Y. *Chem. Sci.* **2019**, *10*, 9853–9858.
23. Okita, T.; Asahara, K. K.; Muto, K.; Yamaguchi, J. *Org. Lett.* **2020**, *22*, 3205–3208.

Henry Nitroaldol Reaction

The nitroaldol condensation reaction involving aldehydes and nitronates, derived from deprotonation of nitroalkanes by bases.

Example 1[4]

Example 2, Retro-Henry reaction[5]

Example 3, Aza-Henry reaction[8]

© Springer Nature Switzerland AG 2021
J. J. Li, *Name Reactions*, https://doi.org/10.1007/978-3-030-50865-4_65

Example 4, Intramolecular Henry reaction[10]

Example 5, A highly asymmetric Henry reaction catalyzed by chiral copper(II) complexes[12]

Example 6, 2,6-*cis*-Substituted tetrahydropyrans using a one-pot sequential catalysis[13]

Example 7, Vinylogous Henry reaction, also works for trifluoromethylketone substrates[14]

Example 8, An asymmetric Henry reaction catalyzed by a Nd/Na heterobimetallic catalyst[15]

Nd/Na heterobimetallic catalyst (6:1:1) =

References

1. Henry, L. *Compt. Rend.* **1895,** *120,* 1265–1268.
2. Barrett, A. G. M.; Robyr, C.; Spilling, C. D. *J. Org. Chem.* **1989,** *54,* 1233–1234.
3. Rosini, G. In *Comprehensive Organic Synthesis;* Trost, B. M.; Fleming, I., Eds.; Pergamon, **1991,** *2,* 321–340. (Review).
4. Chen, Y.-J.; Lin, W.-Y. *Tetrahedron Lett.* **1992,** *33,* 1749–1750.
5. Saikia, A. K.; Hazarika, M. J.; Barua, N. C.; Bezbarua, M. S.; Sharma, R. P.; Ghosh, A. C. *Synthesis* **1996,** 981–985.
6. Luzzio, F. A. *Tetrahedron* **2001,** *57,* 915–945. (Review).
7. Westermann, B. *Angew. Chem. Int. Ed.* **2003,** *42,* 151–153. (Review on aza-Henry reaction).
8. Bernardi, L.; Bonini, B. F.; Capito, E.; Dessole, G.; Comes-Franchini, M.; Fochi, M.; Ricci, A. *J. Org. Chem.* **2004,** *69,* 8168–8171.
9. Palomo, C.; Oiarbide, M.; Laso, A. *Angew. Chem. Int. Ed.* **2005,** *44,* 3881–3884.
10. Kamimura, A.; Nagata, Y.; Kadowaki, A.; Uchidaa, K.; Uno, H. *Tetrahedron* **2007,** *63,* 11856–11861.
11. Wang, A. X. *Henry Reaction.* In *Name Reactions for Homologations-Part I*; Li, J. J., Ed.; Wiley: Hoboken, NJ, **2009,** pp 404–419. (Review).
12. Ni, B.; He, J. *Tetrahedron Lett.* **2013,** *54,* 462–465.
13. Dai, Q.; Rana, N. K.; Zhao, J. C.-G. *Org. Lett.* **2013,** *15,* 2922–2925.
14. Zhang, Y.; Wei, B.-W.; Zou, L.-N.; Kang, M.-L.; Luo, H.-Q. *Tetrahedron* **2016,** *72,* 2472–2475.
15. Karasawa, T.; Oriez, R.; Kumagai, N.; Shibasaki, M. *J. Am. Chem. Soc.* **2018,** *140,* 12290–12295.
16. Araki, Y.; Miyoshi, N.; Morimoto, K.; Kudoh, T.; Mizoguchi, H.; Sakakura, A. *J. Org. Chem.* **2020,** *85,* 798–805.

Hiyama Reaction

Palladium-catalyzed cross-coupling reaction of organosilicons with organic halides, triflates, *etc.* In the presence of an activating agent such as fluoride or hydroxide (transmetalation is reluctant to occur without the effect of an activating agent). For the catalytic cycle, see the Kumada coupling.

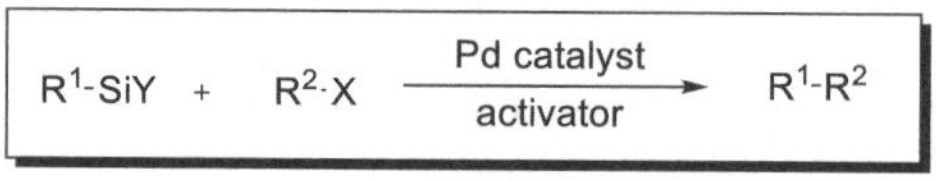

$$R^1\text{-SiY} \;+\; R^2\text{-X} \quad \xrightarrow[\text{activator}]{\text{Pd catalyst}} \quad R^1\text{-}R^2$$

R^1 = alkenyl, aryl, alkynyl, alkyl
R^2 = aryl, alkyl, alkenyl
Y = $(OR)_3$, Me_3, Me_2OH, $Me_{(3-n)}F_{(n+3)}$
X = Cl, Br, I, OTf
activator = TBAF, base

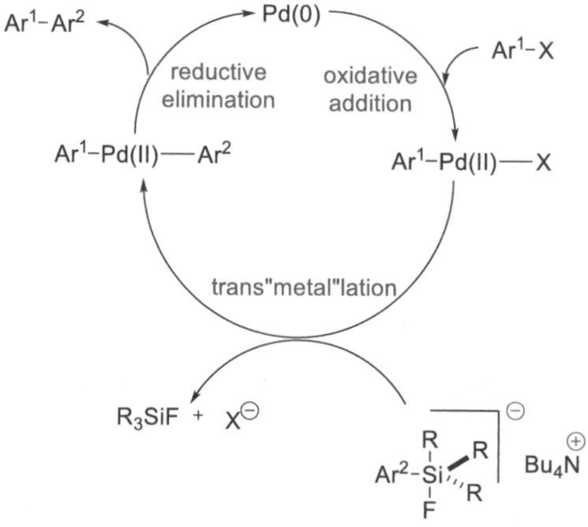

Example 1[1a]

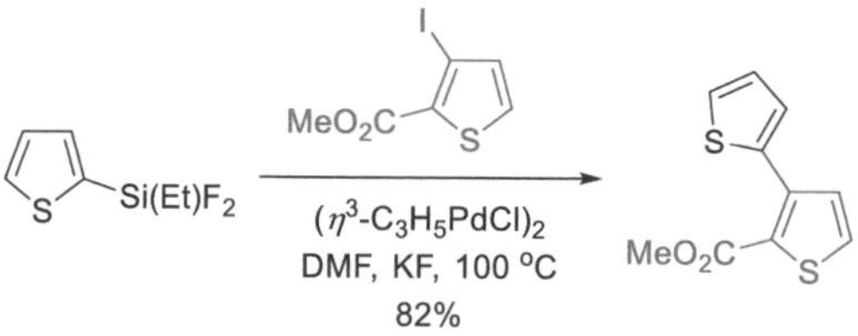

MeO_2C

$(\eta^3\text{-}C_3H_5PdCl)_2$
DMF, KF, 100 °C
82%

Example 2[2]

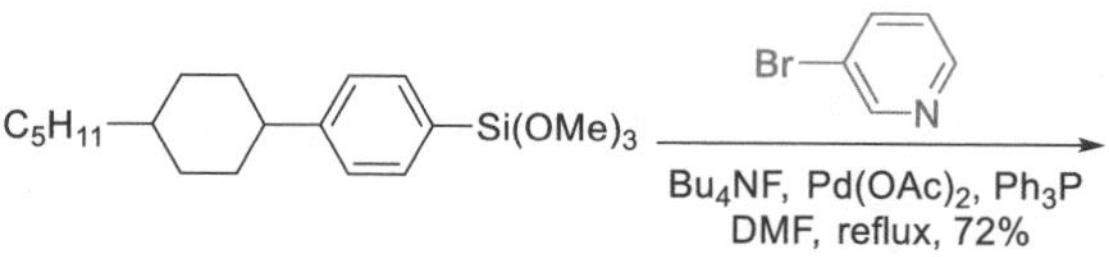

C_5H_{11} — cyclohexyl — phenyl — $Si(OMe)_3$

Br — pyridine-N

Bu_4NF, $Pd(OAc)_2$, Ph_3P
DMF, reflux, 72%

© Springer Nature Switzerland AG 2021
J. J. Li, *Name Reactions*, https://doi.org/10.1007/978-3-030-50865-4_66

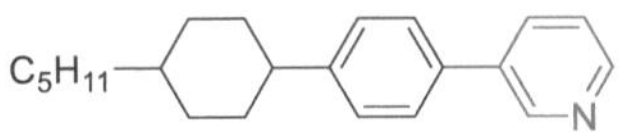

C_5H_{11}

Example 3[7]

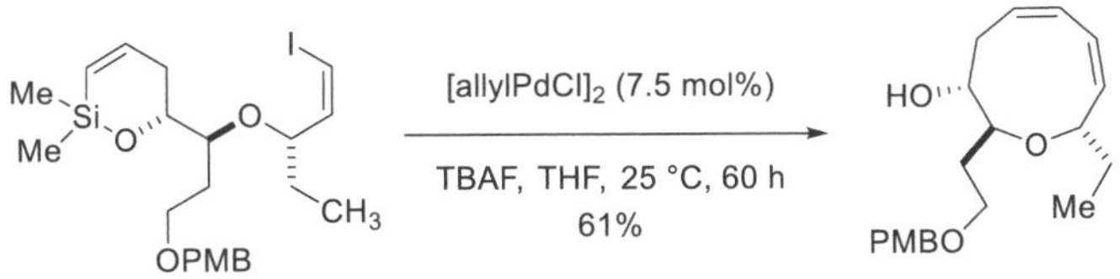

[allylPdCl]$_2$ (7.5 mol%)

TBAF, THF, 25 °C, 60 h
61%

Example 4[9]

[allylPdCl]$_2$
TBAF (12 mol%)

THF, 60 °C, 20 h
20%

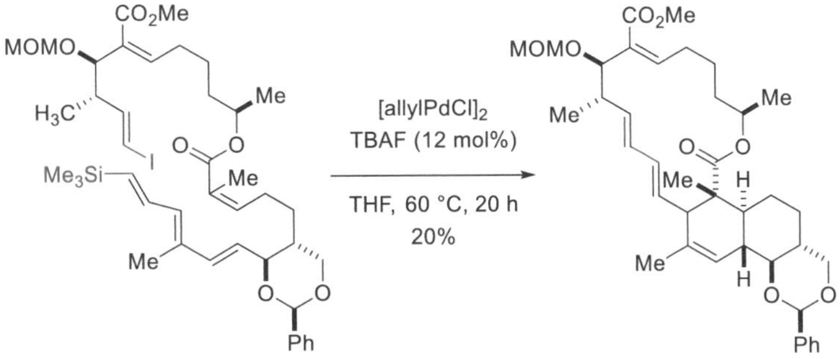

Example 5, Re-usable polystyrene-supported palladium catalyst[11]

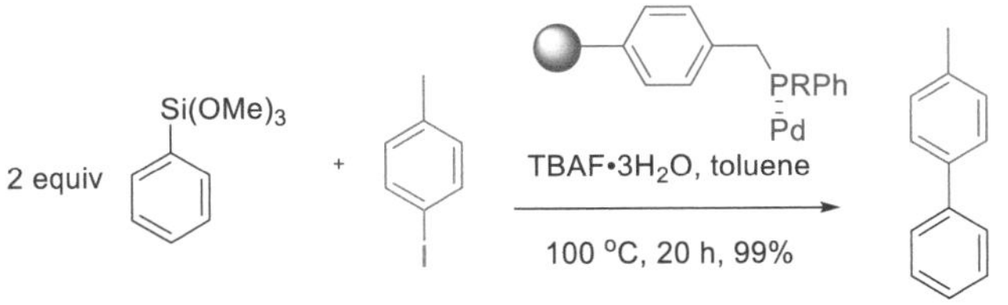

2 equiv

Si(OMe)$_3$

+

PRPh
Pd

TBAF•3H$_2$O, toluene

100 °C, 20 h, 99%

Example 6, Nickel-catalyzed monofluoroalkylation[12]

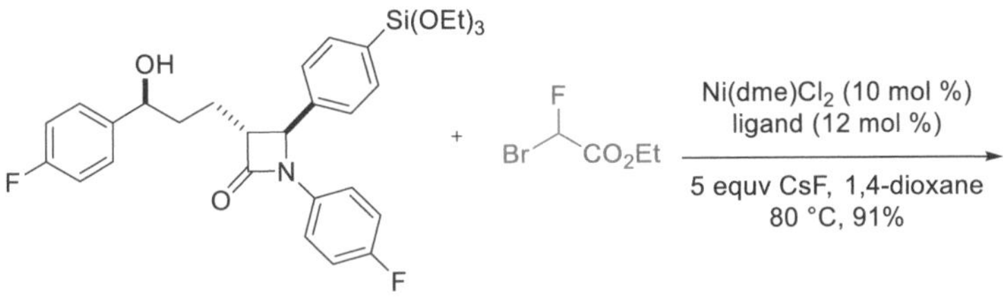

Si(OEt)$_3$

+

Br   CO$_2$Et
F

Ni(dme)Cl$_2$ (10 mol %)
ligand (12 mol %)

5 equv CsF, 1,4-dioxane
80 °C, 91%

ligand =

Example 7, 3-Arylazetidine from 3-iodoazetidine[13]

Pd(OAc)$_2$ (5 mol %)
dppf (10 mol %)
2.5 equiv TBAF

dioxane, 60 °C, 12 h, Ar
80%

References

1. (a) Hatanaka, Y.; Fukushima, S.; Hiyama, T. *Heterocycles* **1990,** *30,* 303–306. (b) Hiyama, T.; Hatanaka, Y. *Pure Appl. Chem.* **1994,** *66,* 1471–1478. (c) Matsuhashi, H.; Kuroboshi, M.; Hatanaka, Y.; Hiyama, T. *Tetrahedron Lett.* **1994,** *35,* 6507–6510.
2. Shibata, K.; Miyazawa, K.; Goto, Y. *Chem. Commun.* **1997,** 1309–1310.
3. Hiyama, T. In *Metal-Catalyzed Cross-Coupling Reactions;* **1998,** Diederich, F.; Stang, P. J., Eds.; Wiley–VCH: Weinheim, Germany, pp 421–53. (Review).
4. Denmark, S. E.; Wang, Z. *J. Organomet. Chem.* **2001,** *624,* 372–375.
5. Hiyama, T. *J. Organomet. Chem.* **2002,** *653,* 58–61.
6. Pierrat, P.; Gros, P.; Fort, Y. *Org. Lett.* **2005,** *7,* 697–700.
7. Denmark, S. E.; Yang, S.-M. *J. Am. Chem. Soc.* **2004,** *126,* 12432–12440.
8. Domin, D.; Benito-Garagorri, D.; Mereiter, K.; Froehlich, J.; Kirchner, K. *Organometallics* **2005,** *24,* 3957–3965.
9. Anzo, T.; Suzuki, A.; Sawamura, K.; Motozaki, T.; Hatta, M.; Takao, K.-i.; Tadano, K.-i. *Tetrahedron Lett.* **2007,** *48,* 8442–8448.
10. Yet L. *Hiyama Cross-Coupling Reaction.* In *Name Reactions for Homologations-Part I*; Li, J. J., Ed.; Wiley: Hoboken, NJ, **2009,** pp 33–416. (Review).
11. Diebold, C.; Derible, A.; Becht, J.-M.; Drian, C. L. *Tetrahedron* **2013,** *69,* 264–267.
12. Wu, Y.; Zhang, H.-R.; Cao, Y.-X.; Lan, Q.; Wang, X.-S. *Org. Lett.* **2016,** *18,* 5564–5567.
13. Liu, Z.; Luan, N.; Shen, L.; Li, J.; Zou, D.; Wu, Y.; Wu, Y. *J. Org. Chem.* **2019,** *84,* 12358–12365.
14. Lu, M.-Z.; Ding, X.; Shao, C.; Hu, Z.; Luo, H.; Zhi, S.; Hu, H.; Kan, Y.; Loh, T.-P. *Org. Lett.* **2020,** *22,* 2663–2668.

Hofmann Elimination

Elimination reaction of alkyl trialkyl amines proceeds with *anti*-stereochemistry, furnishing the least highly substituted olefins.

Example 1, Amine released from the resin by Hofmann elimination[10]

Example 2, Isomerization to a thermodynamically more stable olefin also occurred below[11]

Example 3, Hofmann elimination product, the olefin, is the substrate for C–H activation[12]

J. J. Li, *Name Reactions*, https://doi.org/10.1007/978-3-030-50865-4_67

Example 4, Double "open and shut" transformation of γ-carbolines triggered by ammonium salts[13]

References

1. Hofmann, A. W. *Ber.* **1881**, *14*, 659–669.
2. Eubanks, J. R. I.; Sims, L. B.; Fry, A. *J. Am. Chem. Soc.* **1991**, *113*, 8821–8829.
3. Bach, R. D.; Braden, M. L. *J. Org. Chem.* **1991**, *56*, 7194–7195.
4. Lai, Y. H.; Eu, H. L. *J. Chem. Soc., Perkin Trans. 1* **1993**, 233–237.
5. Sepulveda-Arques, J.; Rosende, E. G.; Marmol, D. P.; Garcia, E. Z.; Yruretagoyena, B.; Ezquerra, J. *Monatsh. Chem.* **1993**, *124*, 323–325.
6. Woolhouse, A. D.; Gainsford, G. J.; Crump, D. R. *J. Heterocycl. Chem.* **1993**, *30*, 873–880.
7. Bhonsle, J. B. *Synth. Commun.* **1995**, *25*, 289–300.
8. Berkes, D.; Netchitailo, P.; Morel, J.; Decroix, B. *Synth. Commun.* **1998**, *28*, 949–956.
9. Morphy, J. R.; Rankovic, Z.; York, M. *Tetrahedron Lett.* **2002**, *43*, 6413–6415.
10. Liu, Z.; Medina-Franco, J. L.; Houghten, R. A.; Giulianotti, M. A. *Tetrahedron Lett.* **2010**, *51*, 5003–5004.
11. Arava, V. R.; Malreddy, S.; Thummala, S. R. *Synth. Commun.* **2012**, *42*, 3545–3552.
12. Spettel, M.; Pollice, R.; Schnürch, M. *Org. Lett.* **2017**, *19*, 4287–4290.
13. Abe, T.; Shimizu, H.; Takada, S.; Tanaka, T.; Yoshikawa, M.; Yamada, K. *Org. Lett.* **2018**, *20*, 1589–1592.
14. Schoenbauer, D.; Spettel, M.; Pollice, R.; Pittenauer, E.; Schnuerch, M. *Org. Biomol. Chem.* **2019**, *17*, 4024–4030.
15. Tayama, E.; Hirano, K.; Baba, S. *Tetrahedron* **2020**, *76*, 131064.

Hofmann Rearrangement

Upon treatment of primary amides with hypohalites, primary amines with one less carbon are obtained *via* the intermediacy of isocyanate. Also known as the Hofmann degradation reaction.

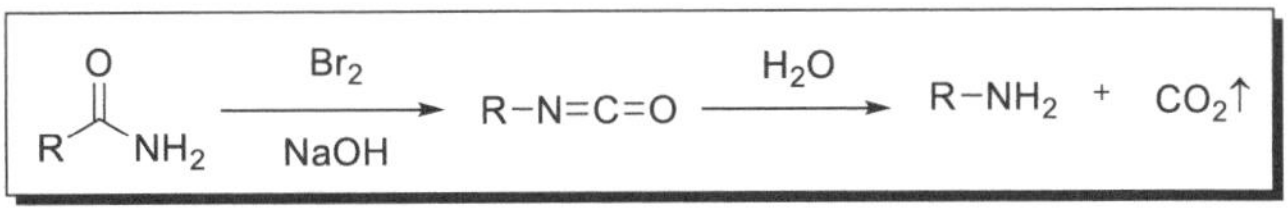

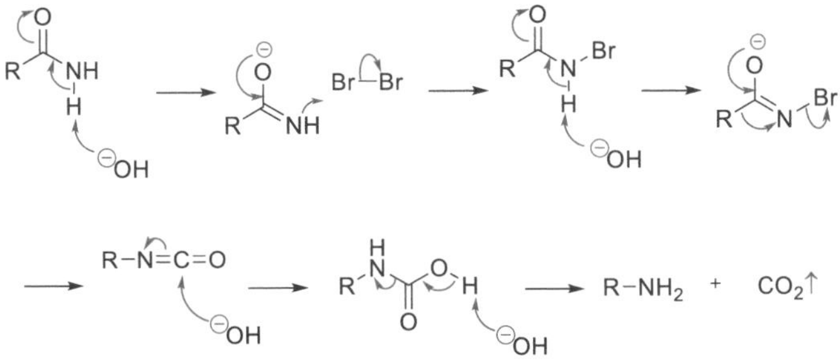

isocyanate intermediate

Example 1, An NBS variant[2]

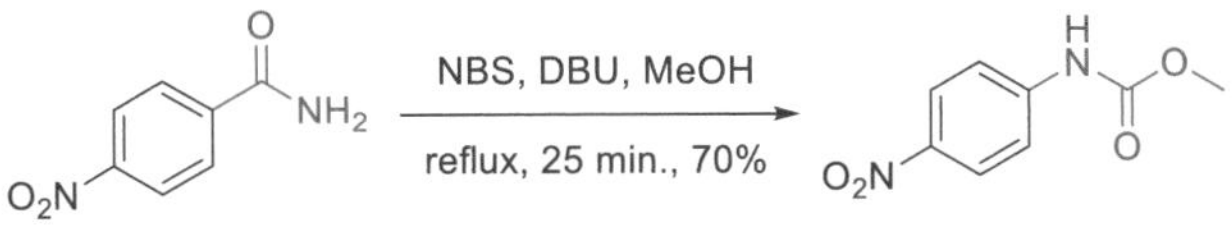

Example 2, Iodosobenzene ditrifluoroacetate[5]

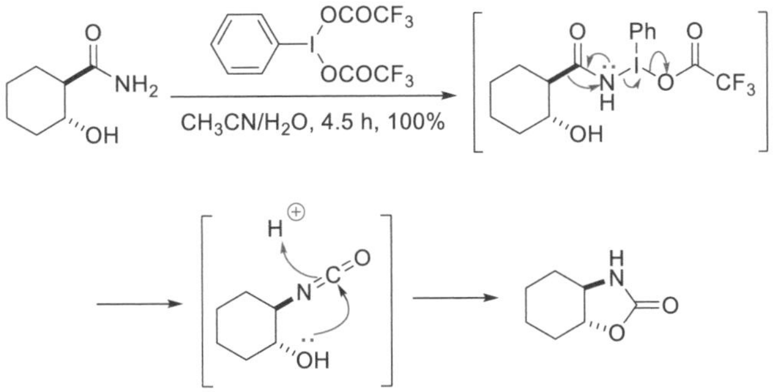

© Springer Nature Switzerland AG 2021
J. J. Li, *Name Reactions*, https://doi.org/10.1007/978-3-030-50865-4_68

Example 3, Bromine and alkoxide[6]

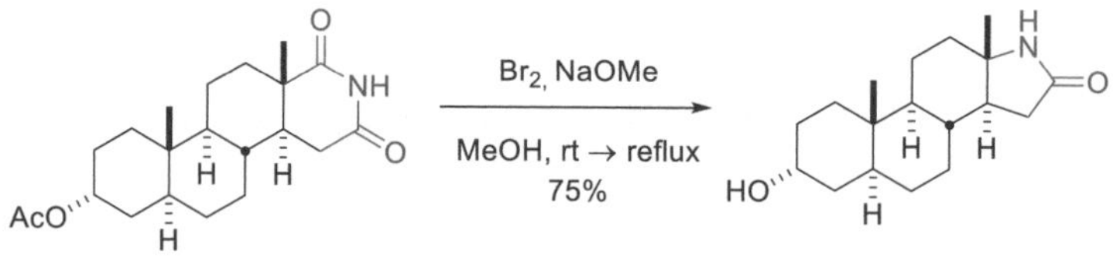

Br$_2$, NaOMe

MeOH, rt → reflux
75%

Example 4, Sodium hypochlorite[7]

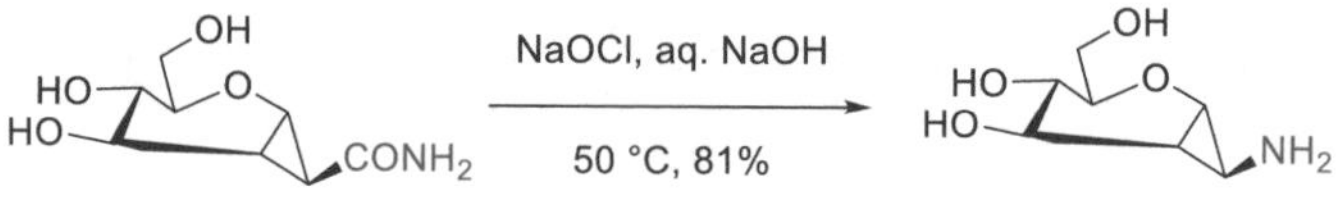

NaOCl, aq. NaOH

50 °C, 81%

Example 5, The original conditions, bromine and hydroxide[9]

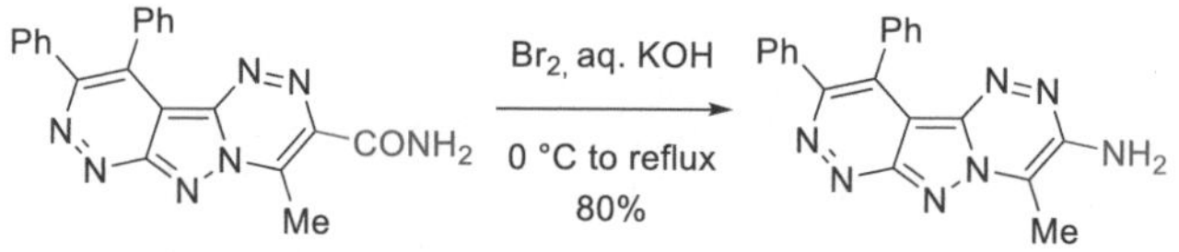

Br$_2$, aq. KOH

0 °C to reflux
80%

Example 6, Lead tetraacetate[10]

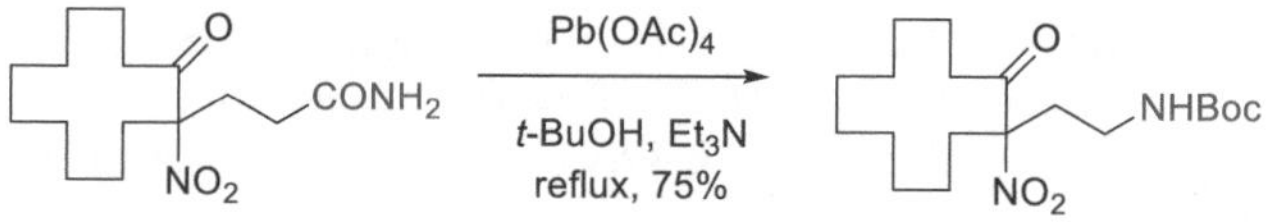

Pb(OAc)$_4$

t-BuOH, Et$_3$N
reflux, 75%

Example 7, Iodosobenzene diacetate (IBDA)[13]

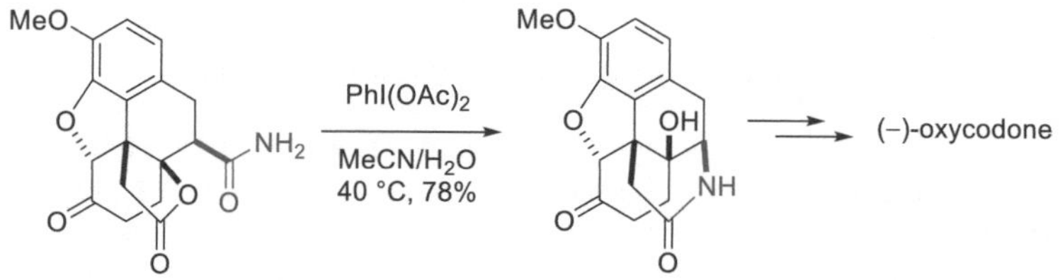

PhI(OAc)$_2$

MeCN/H$_2$O
40 °C, 78%

(−)-oxycodone

Example 8, A 100 g-scale reaction[14]

Example 9, A 100 g-scale reaction[15]

References

1. Hofmann, A. W. *Ber.* **1881,** *14,* 2725–2736.
2. Jew, S.-s.; Kang, M.-h. *Arch. Pharmacol Res.* **1994,** *17,* 490–491.
3. Huang, X.; Seid, M.; Keillor, J. W. *J. Org. Chem.* **1997,** *62,* 7495–7496.
4. Togo, H.; Nabana, T.; Yamaguchi, K. *J. Org. Chem.* **2000,** *65,* 8391–8394.
5. Yu, C.; Jiang, Y.; Liu, B.; Hu, L. *Tetrahedron Lett.* **2001,** *42,* 1449–1452.
6. Jiang, X.; Wang, J.; Hu, J.; Ge, Z.; Hu, Y.; Hu, H.; Covey, D. F. *Steroids* **2001,** *66,* 655–662.
7. Stick, R. V.; Stubbs, K. A. *J. Carbohydr. Chem.* **2005,** *24,* 529–547.
8. Moriarty, R. M. *J. Org. Chem.* **2005,** *70,* 2893–2903. (Review).
9. El-Mariah, F.; Hosney, M.; Deeb, A. *Phosphorus, Sulfur Silicon Relat. Elem.* **2006,** *181,* 2505–2517.
10. Jia, Y.-M.; Liang, X.-M.; Chang, L.; Wang, D.-Q. *Synthesis* **2007,** 744–748.
11. Gribble, G. W. *Hofmann Rearrangement.* In *Name Reactions for Homologations-Part II*; Li, J. J., Ed.; Wiley: Hoboken, NJ, **2009,** pp 164–199. (Review).
12. Yoshimura, A.; Luedtke, M. W.; Zhdankin, V. V. *J. Org. Chem.* **2012,** *77,* 2087–2091.
13. Kimishima, A.; Umihara, H.; Mizoguchi, A.; Yokoshima, S.; Fukuyama, T. *Org. Lett.* **2014,** *16,* 6244–6247.
14. Daver, S.; Rodeville, N.; Pineau, F.; Arlabosse, J.-M.; Moureou, C.; Muller, F.; Pierre, R.; Bouquet, K.; Dumais, L.; Boiteau, J.-G.; et al. *Org. Process Res. Dev.* **2017,** *21,* 231–240.
15. Chang, Z.; Boyaud, F.; Guillot, R.; Boddaet, T.; Aitken, D. *J. Org. Chem.* **2018,** *83,* 527–534.
16. Ohmi, K.; Miura, Y.; Nakao, Y.; Goto, A.; Yoshimura, S.; Ouchi, H.; Inai, M.; Asakawa, T.; Yoshimura, F.; Kondo, M.; et al. *Eur. J. Org. Chem.* **2020,** 488–491.

Hofmann–Löffler–Freytag Reaction

Formation of pyrrolidines or piperidines by thermal or photochemical decomposition of protonated *N*-haloamines.

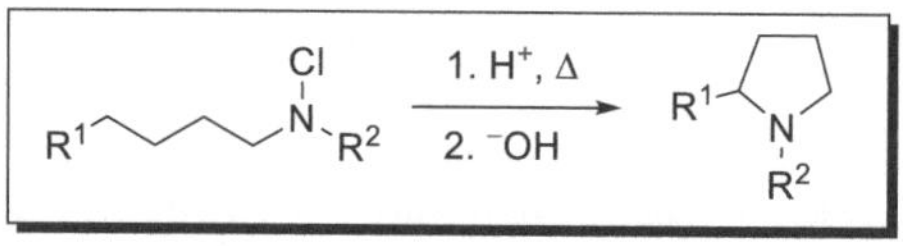

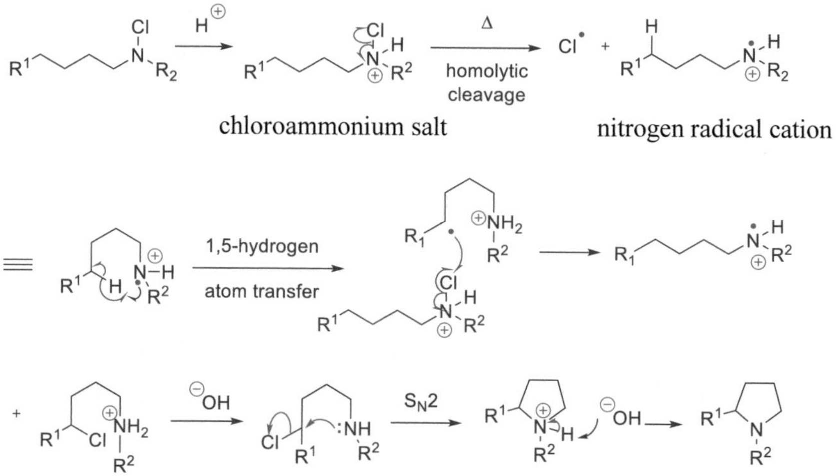

chloroammonium salt   nitrogen radical cation

Example 1[2]

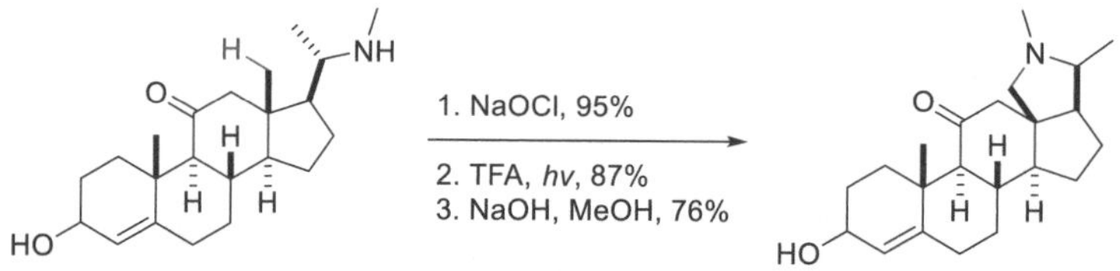

1. NaOCl, 95%

2. TFA, *hv*, 87%
3. NaOH, MeOH, 76%

Example 2[4]

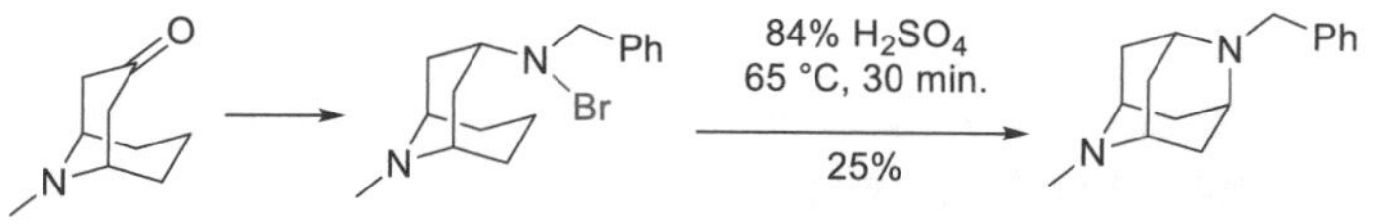

84% H₂SO₄
65 °C, 30 min.

25%

© Springer Nature Switzerland AG 2021
J. J. Li, *Name Reactions*, https://doi.org/10.1007/978-3-030-50865-4_69

Example 3[5]

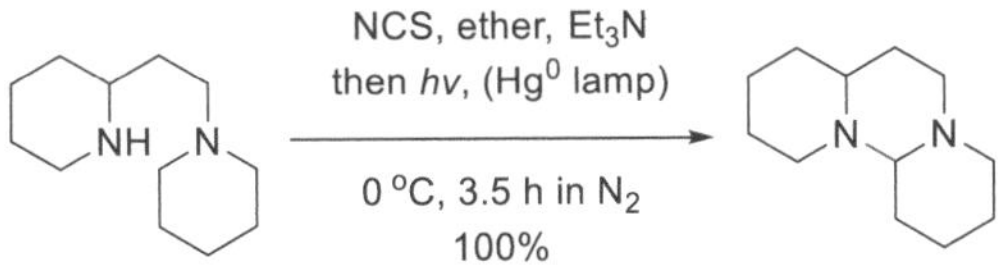

Example 4, Suárez modification of the Hofmann–Löffler–Freytag reaction[7]

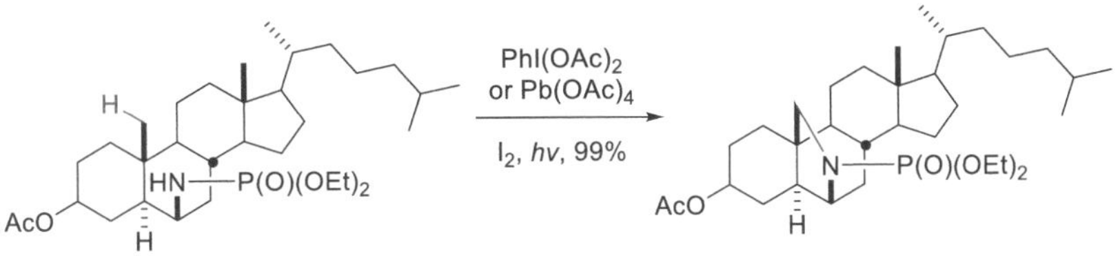

Example 5[12]

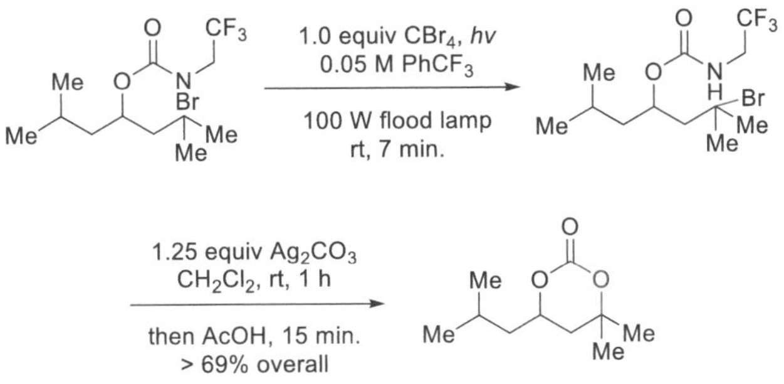

Example 6[13]

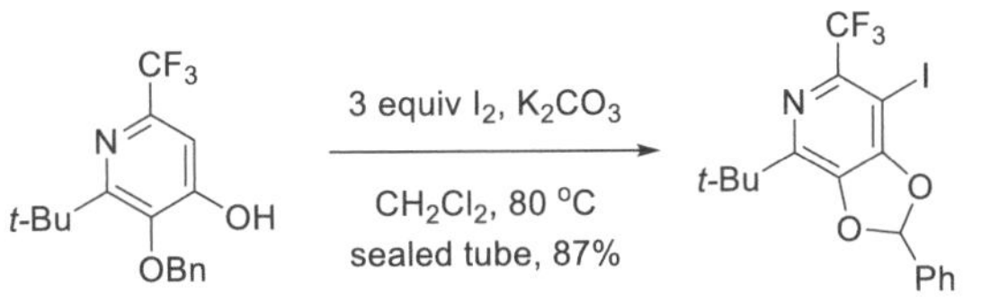

Example 7, NIS-promoted Hofmann–Löffler–Freytag reaction of sulfonimides under visible light (Ns = nosylate)[14]

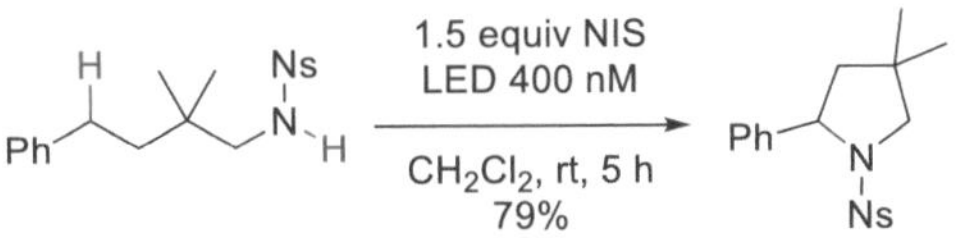

Example 8, Suárez modification of the Hofmann–Löffler–Freytag reaction (PIDA phenyliodine diacetate = IBDA = iodosobenzene diacetate), similar to Example 4[15]

Example 8, Copper-catalyzed remote C(sp^3)-H oxidative trifluoromethylation[16]

References

1. (a) Hofmann, A. W. *Ber.* **1883**, *16*, 558–560. (b) Löffler, K.; Freytag, C. *Ber.* **1909**, *42*, 3727.
2. Wolff, M. E.; Kerwin, J. F.; Owings, F. F.; Lewis, B. B.; Blank, B.; Magnani, A.; Karash, C.; Georgian, V. *J. Am. Chem. Soc.* **1960**, *82*, 4117–4118.
3. Wolff, M. E. *Chem. Rev.* **1963**, *63*, 55–64. (Review).
4. Dupeyre, R.-M.; Rassat, A. *Tetrahedron Lett.* **1973**, 2699–2701.
5. Kimura, M.; Ban, Y. *Synthesis* **1976**, 201–202.
6. Stella, L. *Angew. Chem. Int. Ed.* **1983**, *22*, 337–422. (Review).
7. Betancor, C.; Concepcion, J. I.; Hernandez, R.; Salazar, J. A.; Suárez, E. *J. Org. Chem.* **1983**, *48*, 4430–4432.
8. Majetich, G.; Wheless, K. *Tetrahedron* **1995**, *51*, 7095–7129. (Review).
9. Togo, H.; Katohgi, M. *Synlett* **2001**, 565–581. (Review).
10. Pellissier, H.; Santelli, M. *Org. Prep. Proced. Int.* **2001**, *33*, 455–476. (Review).
11. Li, J. J. *Hofmann–Löffler–Freytag Reaction*. In *Name Reactions in Heterocyclic Chemistry*; Li, J. J., Ed.; Wiley: Hoboken, NJ, **2005**, pp 89–97. (Review).
12. Chen, K.; Richter, J. M.; Baran, P. S. *J. Am. Chem. Soc.* **2008**, *130*, 17247–17249.
13. Lechel, T.; Podolan, G.; Brusilowskij, B.; Schalley, C. A.; Reissig, H.-U. *Eur. J. Org. Chem.* **2012**, 5685–5692.
14. O'Broin, C. Q.; Fernández, P.; Martínez, C.; Muñiz, K. *Org. Lett.* **2016**, *18*, 436–439.
15. (a) Cherney, E. C.; Lopchuk, J. M.; Green, J. C.; Baran, P. S. *J. Am. Chem. Soc.* **2014**, *136*, 12592–12595. (b) Francisco, C. G.; Herrera, A. J.; Suárez, E. *J. Org. Chem.* **2003**, *68*, 1012–1017.
16. Bao, X.; Wang, Q.; Zhu, J. *Nat. Commun.* **2019**, *10*, 1–7.

Horner–Wadsworth–Emmons Reaction

Olefin formation from aldehydes and phosphonates. Workup is more advantageous than the corresponding Wittig reaction because the phosphate by-product can be washed away with water. Typically gives the *trans-* rather than the *cis-* olefins.

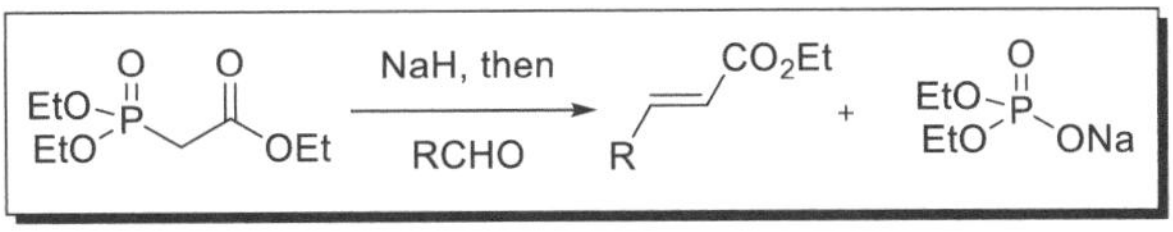

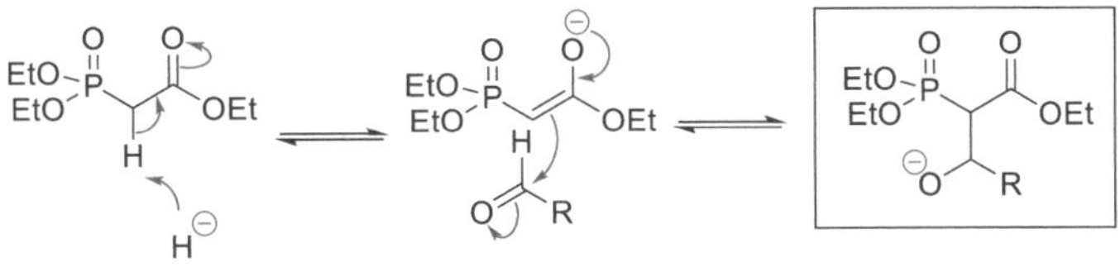

The stereochemical outcome: *erythro* (kinetic) or *threo* (thermodynamic)

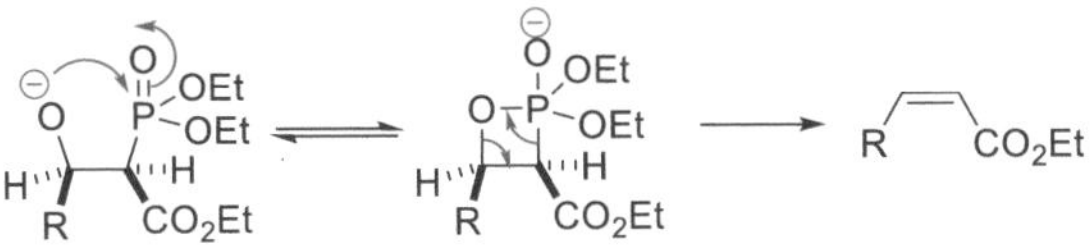

erythro, kinetic adduct

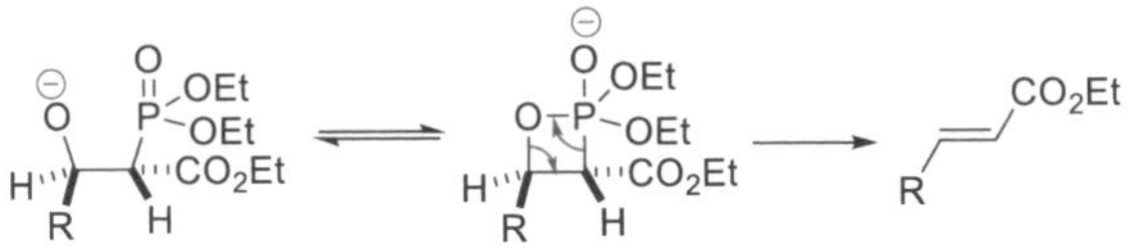

threo, thermodynamic adduct

Example 1[3]

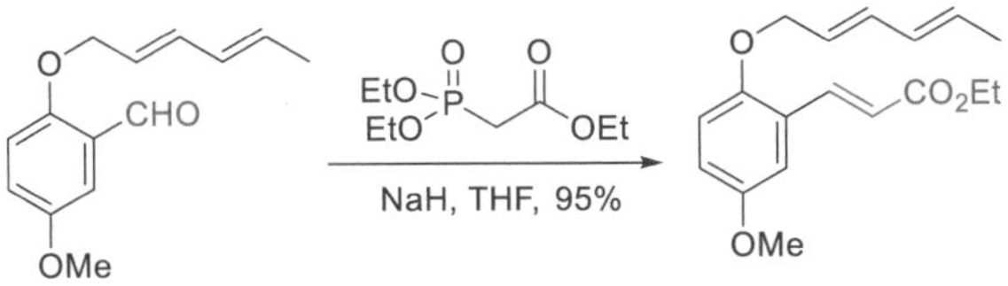

© Springer Nature Switzerland AG 2021
J. J. Li, *Name Reactions*, https://doi.org/10.1007/978-3-030-50865-4_70

Example 2[4]

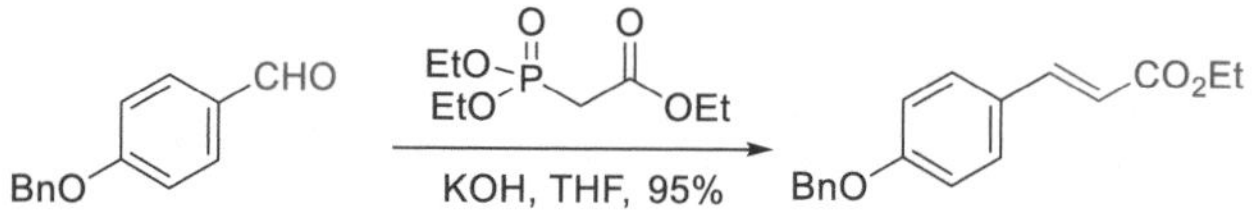

Example 3, Weinreb amide[7]

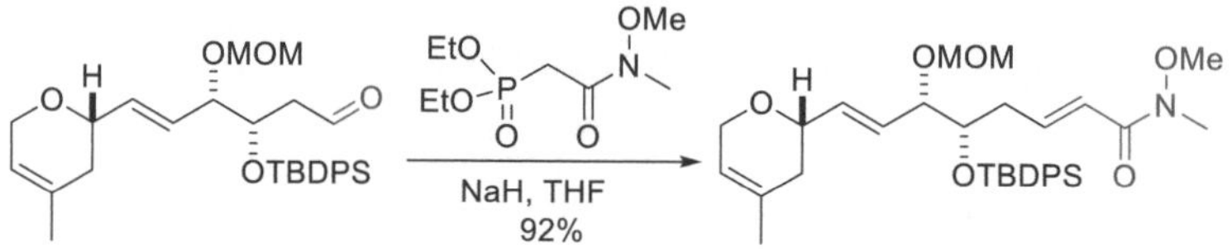

Example 4, Intramolecular Horner–Wadsworth–Emmons[9]

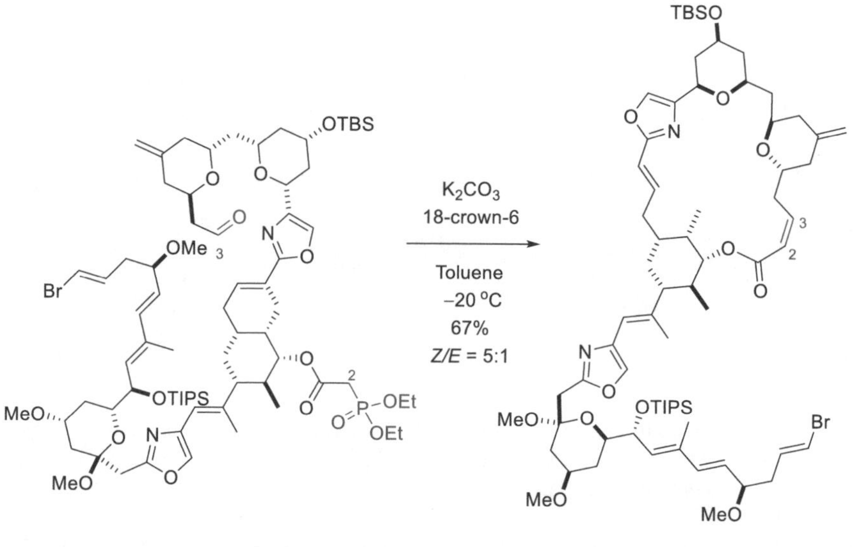

Example 4[11]

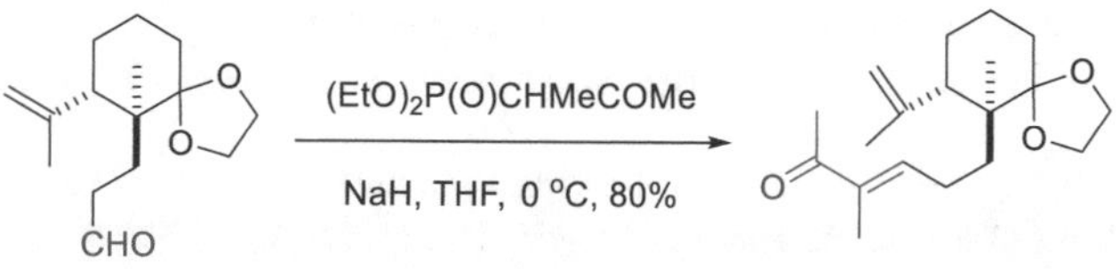

Example 5, MWI = Microwave irradiation[12]

Example 6, Intramolecular Horner–Wadsworth–Emmons to form a 14-membered lactone[13]

References

1. (a) Horner, L.; Hoffmann, H.; Wippel, H. G.; Klahre, G. *Chem. Ber.* **1959**, *92*, 2499–2505. (b) Wadsworth, W. S., Jr.; Emmons, W. D. *J. Am. Chem. Soc.* **1961**, *83*, 1733–1738. (c) Wadsworth, D. H.; Schupp, O. E.; Seus, E. J.; Ford, J. A., Jr. *J. Org. Chem.* **1965**, *30*, 680–685.
2. Maryanoff, B. E.; Reitz, A. B. *Chem. Rev.* **1989**, *89*, 863–927. (Review).
3. Shair, M. D.; Yoon, T. Y.; Mosny, K. K.; Chou, T. C.; Danishefsky, S. J. *J. Am. Chem. Soc.* **1996**, *118*, 9509–9525.
4. Nicolaou, K. C.; Boddy, C. N. C.; Li, H.; Koumbis, A. E.; Hughes, R. J.; Natarajan, S.; Jain, N. F.; Ramanjulu, J. M.; Bräse, S.; Solomon, M. E. *Chem. Eur. J.* **1999**, *5*, 2602–2621.
5. Comins, D. L.; Ollinger, C. G. *Tetrahedron Lett.* **2001**, *42*, 4115–4118.
6. Lattanzi, A.; Orelli, L. R.; Barone, P.; Massa, A.; Iannece, P.; Scettri, A. *Tetrahedron Lett.* **2003**, *44*, 1333–1337.
7. Ahmed, A.; Hoegenauer. E. K.; Enev, V. S.; Hanbauer, M.; Kaehlig, H.; Öhler, E.; Mulzer, J. *J. Org. Chem.* **2003**, *68*, 3026–3042.
8. Blasdel, L. K.; Myers, A. G. *Org. Lett.* **2005**, *7*, 4281–4283.

9. Li, D.-R.; Zhang, D.-H.; Sun, C.-Y.; Zhang, J.-W.; Yang, L.; Chen, J.; Liu, B.; Su, C.; Zhou, W.-S.; Lin, G.-Q. *Chem. Eur. J.* **2006,** *12,* 1185–1204.

10. Rong, F. *Horner–Wadsworth–Emmons reaction* In *Name Reactions for Homologations-Part I*; Li, J. J., Ed.; Wiley: Hoboken, NJ, **2009,** pp 420–466. (Review).

11. Okamoto, R.; Takeda, K.; Tokuyama, H.; Ihara, M.; Toyota, M. *J. Org. Chem.* **2013,** *78,* 93–103.

12. Krzyzanowski, A.; Saleeb, M.; Elofsson, M. *Org. Lett.* **2018,** *20,* 6650–6654.

13. Paul, D.; Saha, S.; Goswami, R. K. **2018,** *20,* 4606–4609.

14. Everson, J.; Kiefel, M. J. *J. Org. Chem.* **2019,** *84,* 15226–15235.

15. Iwanejko, J.; Sowinski, M.; Wojaczynska, E.; Olszewski, T. K.; Gorecki, M. *RSC Adv.* **2020,** *10,* 14618–14629.

Still–Gennari Phosphonates

A variant of the Horner–Wadsworth–Emmons reaction using bis(trifluoroethyl)-phosphonates (Still–Gennari phosphonates) to give predominantly Z-olefins.

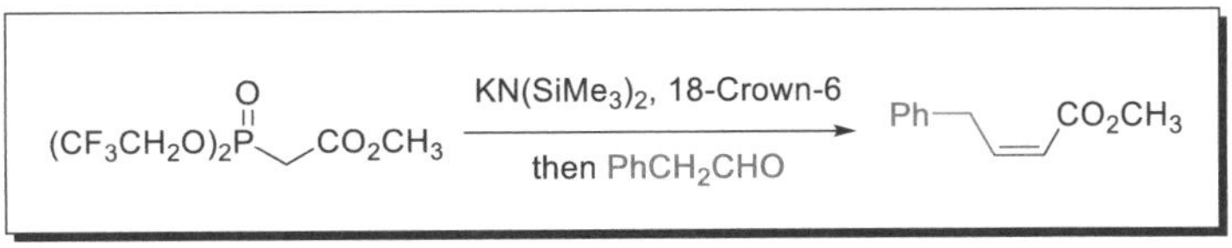

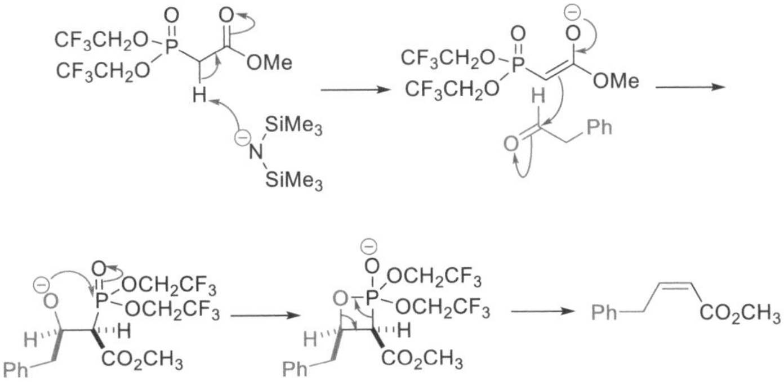

erythro isomer, kinetic adduct

Example 1[2]

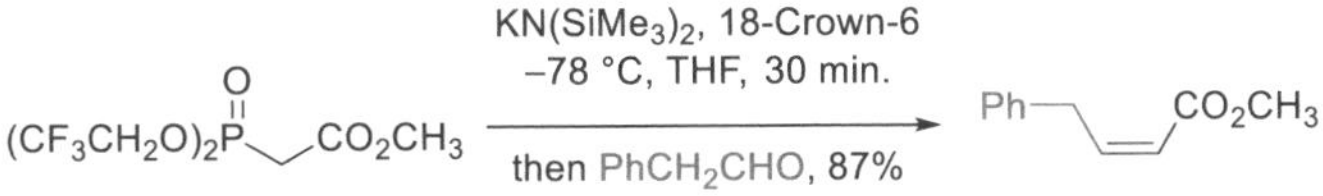

Example 2[3]

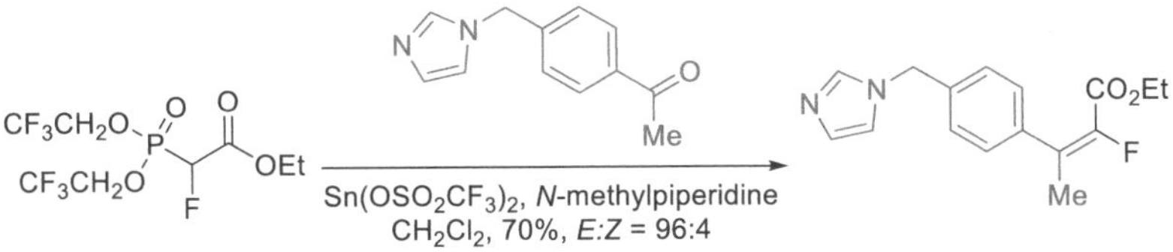

Example 3[4]

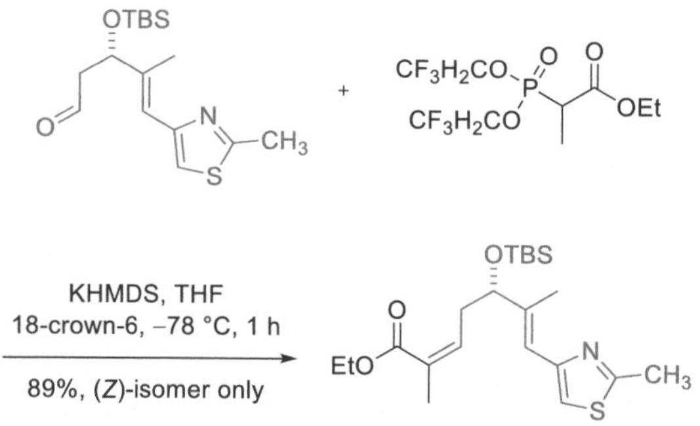

KHMDS, THF
18-crown-6, −78 °C, 1 h
→
89%, (Z)-isomer only

Example 4[9]

NaH, THF
→
73%, Z:E 5:1

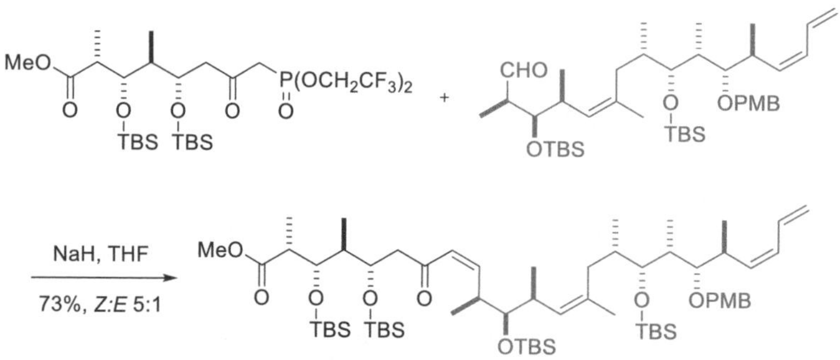

Example 5, An expedient access to Still–Gennari phosphonates[11]

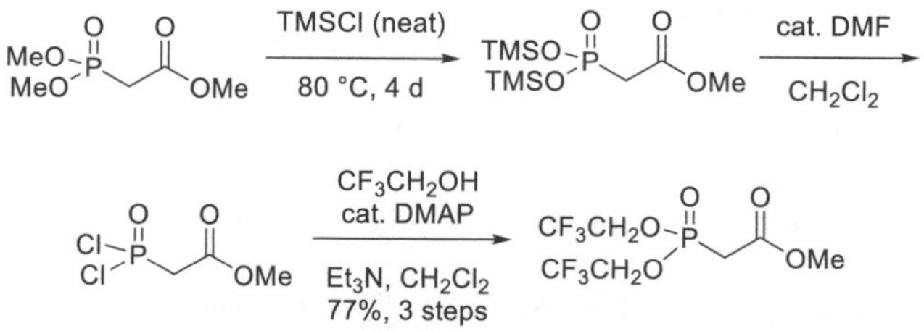

TMSCl (neat)
80 °C, 4 d
→

cat. DMF
CH₂Cl₂
→

CF₃CH₂OH
cat. DMAP
Et₃N, CH₂Cl₂
77%, 3 steps

Example 6, Toward synthesis of aglycone of lycoperdinosides[12]

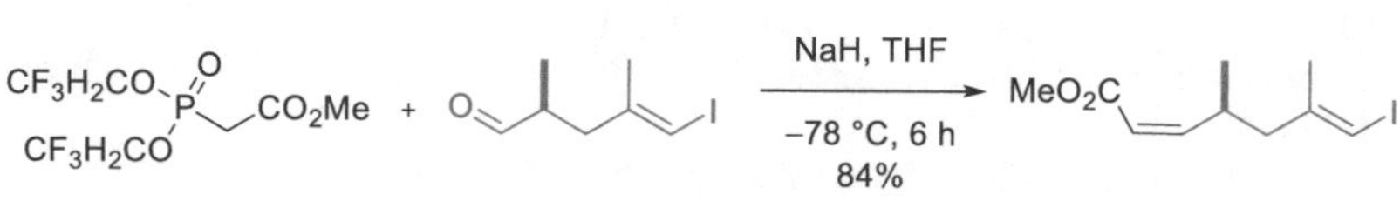

NaH, THF
−78 °C, 6 h
84%

Example 7, For (Z)-α,β-unsaturated phosphonates[13]

Z, 64%   E, 30%

Example 8, Methylated *Ando-type* Horner–Wadsworth–Emmons reagent is significantly cheaper than the Still–Gennari phosphonates[15]

Z:E = 97:3

Ar =

References

1. Still, W. C.; Gennari, C. *Tetrahedron Lett.* **1983**, *24,* 4405–4408. W. Clark Still (1946–) was born in Augusta, Georgia. He was a professor at Columbia University.
2. Nicolaou, K. C.; Nadin, A.; Leresche, J. E.; LaGreca, S.; Tsuri, T.; Yue, E. W.; Yang, Z. *Chem. Eur. J.* **1995**, *1*, 467–494.
3. Sano, S. Yokoyama, K.; Shiro, M.; Nagao, Y. *Chem. Pharm. Bull.* **2002**, *50*, 706–709.
4. Mulzer, J.; Mantoulidis, A.; Öhler, E. *Tetrahedron Lett.* **1998**, *39*, 8633–8636.
5. Paterson, I.; Florence, G. J.; Gerlach, K.; Scott, J. P.; Sereinig, N. *J. Am. Chem. Soc.* **2001**, *123*, 9535–9544.
6. Mulzer, J.; Ohler, E. *Angew. Chem. Int. Ed.* **2001**, *40*, 3842–3846.
7. Beaudry, C. M.; Trauner, D. *Org. Lett.* **2002**, *4*, 2221–2224.
8. Dakin, L. A.; Langille, N. F.; Panek, J. S. *J. Org. Chem.* **2002**, *67*, 6812–6815.
9. Paterson, I.; Lyothier, I. *J. Org. Chem.* **2005**, *70*, 5494–5507.
10. Rong, F. *Horner–Wadsworth–Emmons reaction*. In *Name Reactions for Homologations-Part I*; Li, J. J., Ed.; Wiley: Hoboken, NJ, **2009**, pp 420–466. (Review).
11. Messik, F.; Oberthür, M. *Synthesis* **2013**, *45*, 167–170.
12. Chandrasekhar, B.; Athe, S.; Reddy, P. P.; Ghosh, S. *Org. Biomol. Chem.* **2015**, *13*, 115–124.
13. Janicki, I.; Kielbasinski, P. *Synthesis* **2018**, *50*, 4140–4144.
14. (a) Bressin, R. K.; Driscoll, J. L.; Wang, Y.; Koide, K. *Org. Process Res. Dev.* **2019**, *23*, 274–277. (b) Ando, K. *Tetrahedron Lett.* **1995**, *36*, 4105–4108.

Houben–Hoesch Reaction

Acid-catalyzed acylation of phenols as well as phenolic ethers using nitriles to offer imines, which are hydrolyzed to the corresponding ketones.

Example 1, Intramolecular Houben–Hoesch reaction[3]

Example 2[6]

J. J. Li, *Name Reactions*, https://doi.org/10.1007/978-3-030-50865-4_71

Example 3[8]

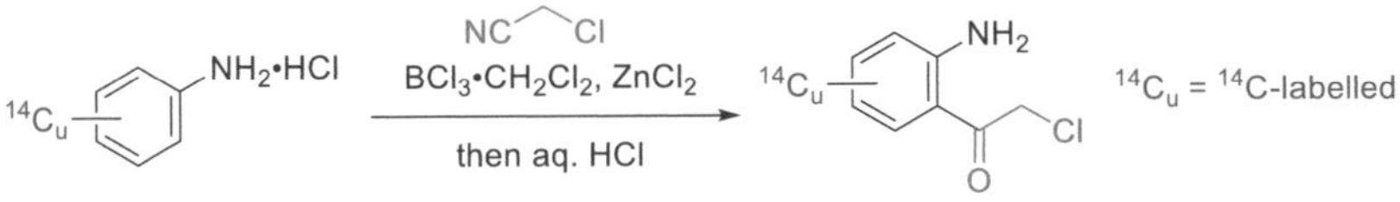

$^{14}C_u = {}^{14}C$-labelled

Example 4[9]

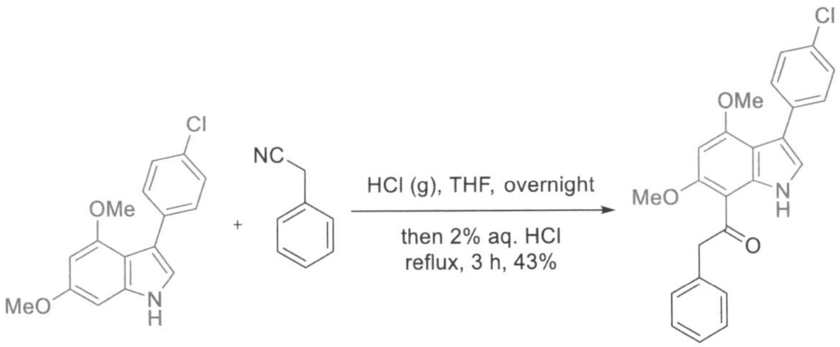

Example 5, Intramolecular Houben–Hoesch reaction[10]

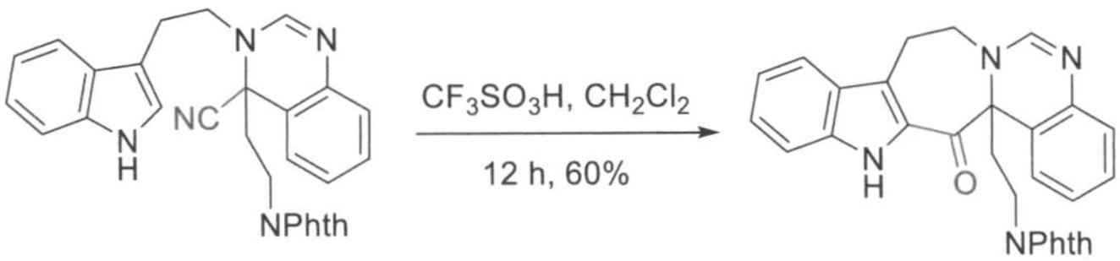

Example 6, Intramolecular Houben–Hoesch reaction, the product was "stuck" as the aniline[11]

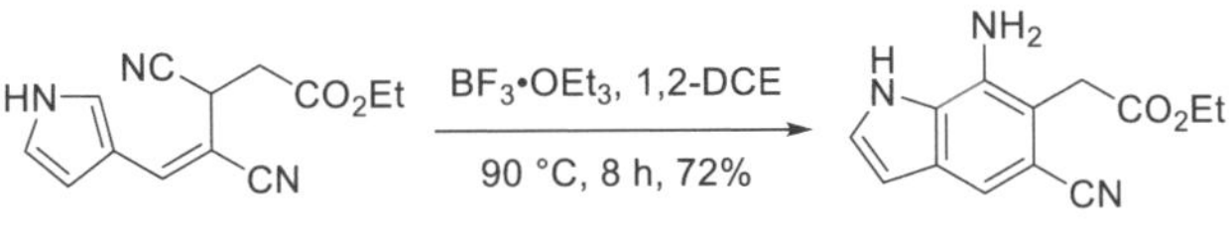

Example 7[12]

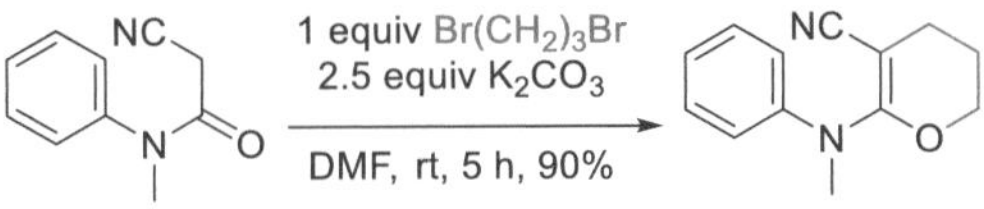

Example 8, Toward synthesis of genistein[13]

References

1. (a) Hoesch, K. *Ber.* **1915**, *48*, 1122–1133. Kurt Hoesch (1882–1932) was born in Kre-zau, Germany. He studied at Berlin under Emil Fischer. During WWI, Hoesch was Professor of Chemistry at the University of Istanbul, Turkey. After the war he gave up his scientific activities to devote himself to the management of a family business. (b) Houben, J. *Ber.* **1926**, *59*, 2878–2891.
2. Yato, M.; Ohwada, T.; Shudo, K. *J. Am. Chem. Soc.* **1991**, *113*, 691–692.
3. Rao, A. V. R.; Gaitonde, A. S.; Prakash, K. R. C.; Rao, S. P. *Tetrahedron Lett.* **1994**, *35*, 6347–6350.
4. Sato, Y.; Yato, M.; Ohwada, T.; Saito, S.; Shudo, K. *J. Am. Chem. Soc.* **1995**, *117*, 3037–3043.
5. Kawecki, R.; Mazurek, A. P.; Kozerski, L.; Maurin, J. K. *Synthesis* **1999**, 751–753.
6. Udwary, D. W.; Casillas, L. K.; Townsend, C. A. *J. Am. Chem. Soc.* **2002**, *124*, 5294–5303.
7. Sanchez-Viesca, F.; Gomez, M. R.; Berros, M. *Org. Prep. Process Int.* **2004**, *36*, 135–140.
8. Wager, C. A. B.; Miller, S. A. *J. Labelled Compd. Radiopharm.* **2006**, *49*, 615–622.
9. Black, D. St. C.; Kumar, N.; Wahyuningsih, T. D. *ARKIVOC* **2008**, *(6)*, 42–51.
10. Zhao, B.; Hao, X.-Y.; Zhang, J.-X.; Liu, S.; Hao, X.-J. *J. Org. Chem.* **2013**, *15*, 528–530.
11. Outlaw, V. K.; Townsend, C. A. *Org. Lett.* **2014**, *16*, 6334–6337.
12. Wu, C.; Huang, P.; Sun, Z.; Lin, M.; Jiang, Y.; Tong, J.; Ge, C. *Tetrahedron* **2016**, *72*, 1461–1466.
13. Filip, K.; Kleczkowska-Plichta, E.; Araźny, Z.; Grynkiewicz, G.; Polowczyk, M.; Ga-barski, K.; Trzcińska, K. *Org. Process Res. Dev.* **2016**, *20*, 1354–1362.

Hunsdiecker–Borodin Reaction

Conversion of silver carboxylate to halide by treatment with halogen.

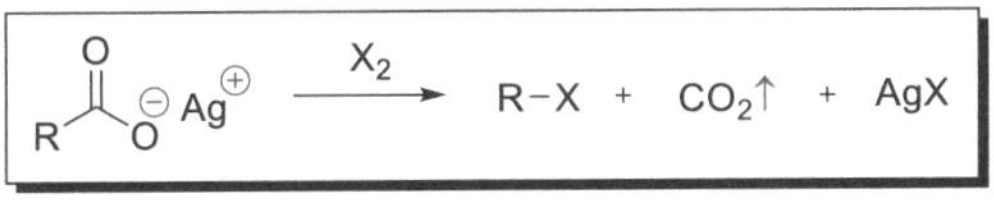

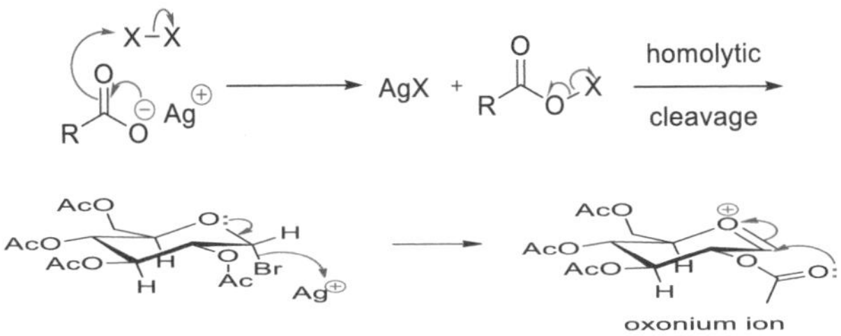

oxonium ion

Example 1[5]

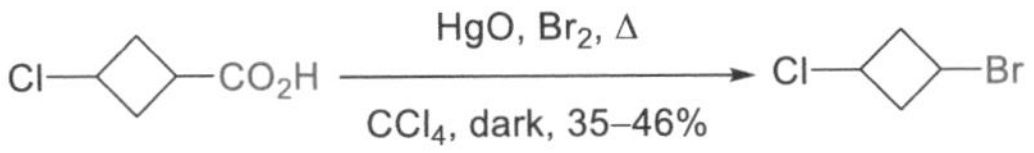

$$\text{Cl}-\square-\text{CO}_2\text{H} \xrightarrow[\text{CCl}_4,\ \text{dark, 35–46\%}]{\text{HgO, Br}_2,\ \Delta} \text{Cl}-\square-\text{Br}$$

Example 2[6]

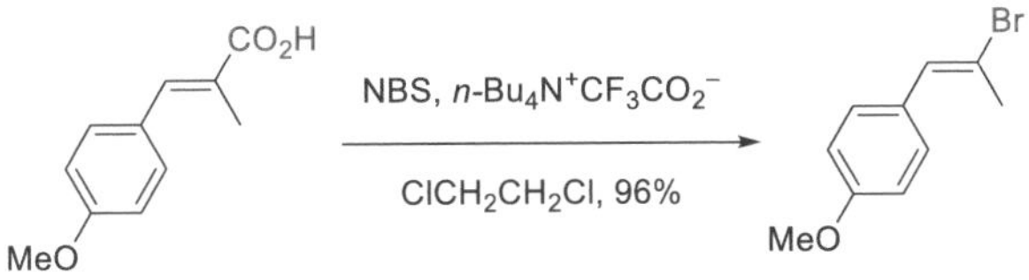

NBS, n-Bu$_4$N$^+$CF$_3$CO$_2^-$

ClCH$_2$CH$_2$Cl, 96%

Example 3[8]

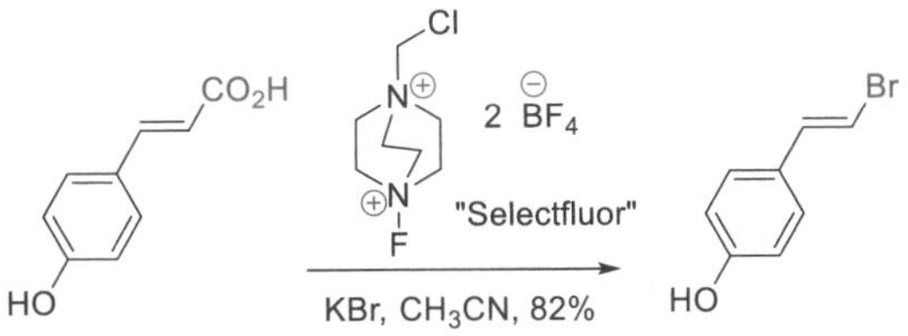

2 BF$_4$

"Selectfluor"

KBr, CH$_3$CN, 82%

J. J. Li, *Name Reactions*, https://doi.org/10.1007/978-3-030-50865-4_72

Example 4, One-pot microwave-Hunsdiecker–Borodin followed by Suzuki[10]

Example 5[11]

Example 6, For cinnamic acid substrates[12]

X = Br (83%); X = I (80%)

Example 7, Metal-free decarboxylative iodination, an aromatic Hunsdiecker reaction[13]

References

1. (a) Borodin, A. *Ann.* **1861**, *119*, 121–123. Aleksandr Porfirevič Borodin (1833–1887) was born in St. Petersburg, the illegitimate son of a prince. He prepared methyl bromide from silver acetate in 1861, but another eighty years elapsed before Heinz and Cläre Hunsdiecker converted Borodin's synthesis into a general method, the Hunsdiecker or Hunsdiecker–Borodin reaction. Borodin was also an accomplished composer and is now best known for his musical masterpiece, opera Prince Igor. He kept a piano outside his laboratory. (b) Hunsdiecker, H.; Hunsdiecker, C. *Ber.* **1942**,

    75, 291–297. Cläre Hunsdiecker was born in 1903 and educated in Cologne. She developed the bromination of silver carboxylate alongside her husband, Heinz.

2. Sheldon, R. A.; Kochi, J. K. *Org. React.* **1972,** *19*, 326–421. (Review).

3. Barton, D. H. R.; Crich, D.; Motherwell, W. B. *Tetrahedron Lett.* **1983,** *24,* 4979–4982.

4. Crich, D. In *Comprehensive Organic Synthesis;* Trost, B. M.; Steven, V. L., Eds.; Pergamon, **1991,** *Vol. 7*, pp 723–734. (Review).

5. Lampman, G. M.; Aumiller, J. C. *Org. Synth.* **1988,** *Coll. Vol. 6*, 179.

6. Naskar, D.; Chowdhury, S.; Roy, S. *Tetrahedron Lett.* **1998,** *39*, 699–702.

7. Das, J. P.; Roy, S. *J. Org. Chem.* **2002,** *67*, 7861–7864.

8. Ye, C.; Shreeve, J. M. *J. Org. Chem.* **2004,** *69*, 8561–8563.

9. Li, J. J. *Hunsdiecker Reaction.* In *Name Reactions for Functional Group Transformations*; Li, J. J., Corey, E. J., Eds., Wiley: Hoboken, NJ, **2007,** pp 623–629. (Review).

10. Bazin, M.-A.; El Kihel, L.; Lancelot, J.-C.; Rault, S. *Tetrahedron Lett.* **2007,** *48*, 4347–4351.

11. Wang, Z.; Zhu, L.; Yin, F.; Su, Z.; Li, Z.; Li, C. *J. Am. Chem. Soc.* **2012,** *134*, 4258–4263.

12. Lorentzen, M.; Bayer, A.; Sydnes, M. O.; Jøgensen, K. B. *Tetrahedron* **2015,** *71*, 8278–8284.

13. Perry, G. J. P.; Quibell, J. M.; Panigrahi, A.; Larrosa, I. *J. Am. Chem. Soc.* **2017,** *139*, 11527–11536.

14. Zarei, M.; Noroozizadeh, E.; Moosavi-Zare, A. R.; Zolfigol, M. A. *J. Org. Chem.* **2018,** *83*, 3645–3650.

Jacobsen–Katsuki Epoxidation

Mn(III)salen-catalyzed asymmetric epoxidation of (Z)-olefins.

1. Concerted oxygen transfer (*cis*-epoxide):

2. Oxygen transfer *via* radical intermediate (*trans*-epoxide):

3. Oxygen transfer *via* manganaoxetane intermediate (*cis*-epoxide):

Example 1[2]

J. J. Li, *Name Reactions*, https://doi.org/10.1007/978-3-030-50865-4_73

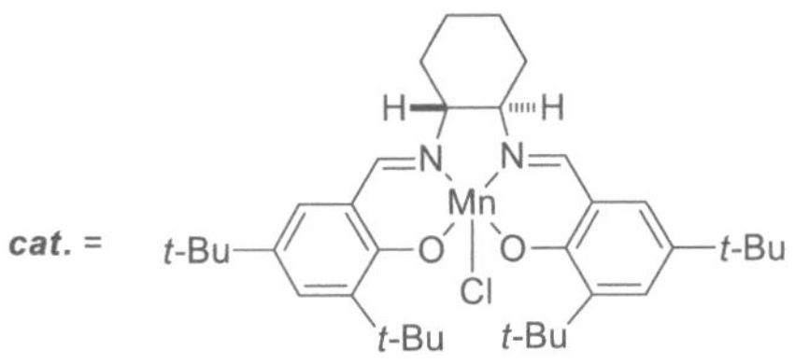

cat. =
t-Bu
t-Bu
t-Bu
t-Bu
Mn
Cl

Example 2[5]

OMe
Ph
N
cat., NaOCl
58% yield, 89% ee

Example 3[6]

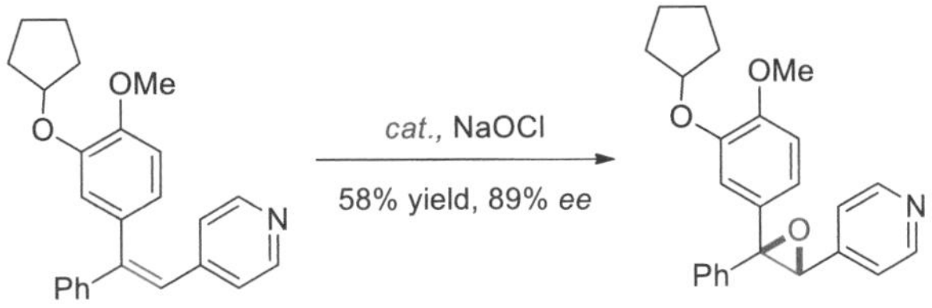

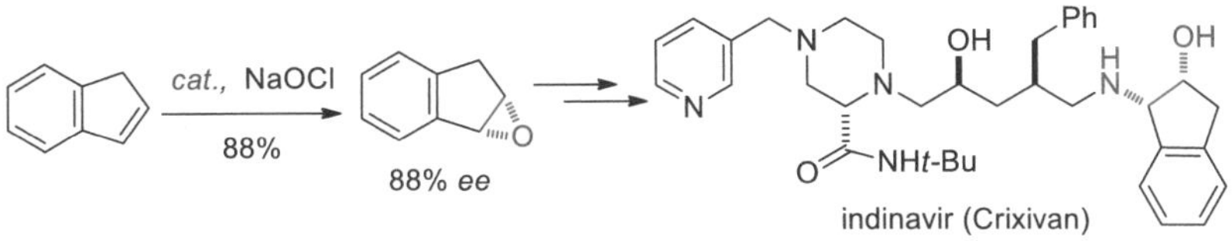

cat., NaOCl
88%
88% ee
NHt-Bu
indinavir (Crixivan)

Example 4, Jacobsen hydrolytic kinetic resolution (HKR)[9]

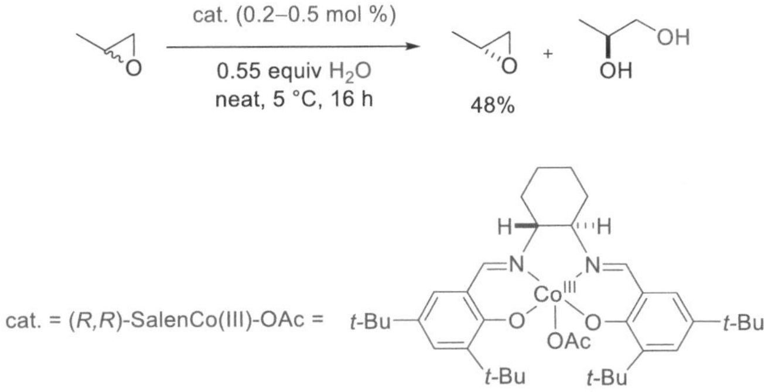

cat. (0.2–0.5 mol %)
0.55 equiv H2O
neat, 5 °C, 16 h
48%
cat. = (R,R)-SalenCo(III)-OAc =
t-Bu
OAc

Example 5, Jacobsen hydrolytic kinetic resolution (HKR)[13]

Example 6, Hydrolytic kinetic resolution (HKR) using Jacobsen's dimeric (second generation) salen-Co(III) catalyst[14]

References

1. (a) Zhang, W.; Loebach, J. L.; Wilson, S. R.; Jacobsen, E. N. *J. Am. Chem. Soc.* **1990,** *112,* 2801–2903. (b) Irie, R.; Noda, K.; Ito, Y.; Matsumoto, N.; Katsuki, T. *Tetrahedron Lett.* **1990,** *31,* 7345–7348. (c) Irie, R.; Noda, K.; Ito, Y.; Katsuki, T. *Tetrahedron Lett.* **1991,** *32,* 1055–1058. (d) Deng, L.; Jacobsen, E. N. *J. Org. Chem.* **1992,** 57, 4320–4323. (e) Palucki, M.; McCormick, G. J.; Jacobsen, E. N. *Tetrahedron Lett.* **1995,** *36,* 5457–5460.
2. Zhang, W.; Jacobsen, E. N. *J. Org. Chem.* **1991,** *56,* 2296–2298.
3. Jacobsen, E. N. In *Catalytic Asymmetric Synthesis;* Ojima, I., Ed.; VCH: Weinheim, New York, **1993,** Ch. 4.2. (Review).
4. Jacobsen, E. N. In *Comprehensive Organometallic Chemistry II*, Eds. G. W. Wilkinson, G. W.; Stone, F. G. A.; Abel, E. W.; Hegedus, L. S., Pergamon, New York, **1995,** vol 12, Chapter 11.1. (Review).

5.   Lynch, J. E.; Choi, W.-B.; Churchill, H. R. O.; Volante, R. P.; Reamer, R. A.; Ball, R. G. *J. Org. Chem.* **1997**, *62*, 9223–9228.

6.   Senananyake, C. H. *Aldrichimica Acta* **1998**, *31*, 3–15. (Review).

7.   Jacobsen, E. N.; Wu, M. H. In *Comprehensive Asymmetric Catalysis*, Jacobsen, E. N.; Pfaltz, A.; Yamamoto, H. Eds.; Springer: New York; 1999, Chapter 18.2. (Review).

8.   Katsuki, T. In *Catalytic Asymmetric Synthesis;* 2nd edn.; Ojima, I., Ed.; Wiley-VCH: New York, **2000,** 287. (Review).

9.   Schaus, S. E.; Brandes, B. D.; Larrow, J. F.; Tokunaga, M.; Hansen, K. B.; Gould, A. E.; Furrow, M. E.; Jacobsen, E. N. *J. Am. Chem. Soc.* **2002,** *128*, 6790–6791.

10. Katsuki, T. *Synlett* **2003,** 281–297. (Review).

11. Palucki, M. *Jacobsen–Katsuki epoxidation.* In *Name Reactions in Heterocyclic Chemistry*; Li, J. J., Ed.; Wiley: Hoboken, NJ, **2005,** pp 29–43. (Review).

12. Olson, J. A.; Shea, K. M. *Acc. Chem. Res.* **2011,** *44*, 311–321. (Review).

13. Njiojob, C. N.; Rhinehart, J. L.; Bozell, J. J.; Long, B. K. *J. Org. Chem.* **2015,** *80*, 1771–1780.

14. Mower, M. P.; Blackmond, D. G. *ACS Catal.* **2018,** *8*, 5977–5982.

15. Day, A. J.; Lee, J. H. Z.; Phan, Q. D.; Lam, H. C.; Ametovski, A.; Sumby, C. J.; Bell, S. G.; George, J. H. *Angew. Chem. Int. Ed.* **2019,** *58*, 1427–1431.

Jones Oxidation

The **Collins/Sarett oxidation** (chromium trioxide-pyridine complex), and **Corey´s PCC** (pyridinium chlorochromate) and **PDC** (pyridinium dichromate) **oxidations** follow a similar pathway as the **Jones oxidation** (chromium trioxide and sulfuric acid in acetone). All these oxidants have a chromium (VI), normally orange or yellow, which is reduced to Cr(III), often green.

Jones Oxidation

By the Jones oxidation, the primary alcohols are oxidized to the corresponding aldehyde or carboxylic acids, whereas the secondary alcohols are oxidized to the corresponding ketones.

$$CrO_3 + H_2O \longrightarrow H_2CrO_4$$

chromate ester

The intramolecular mechanism is also operative:

© Springer Nature Switzerland AG 2021

J. J. Li, *Name Reactions*, https://doi.org/10.1007/978-3-030-50865-4_74

Example 1[6]

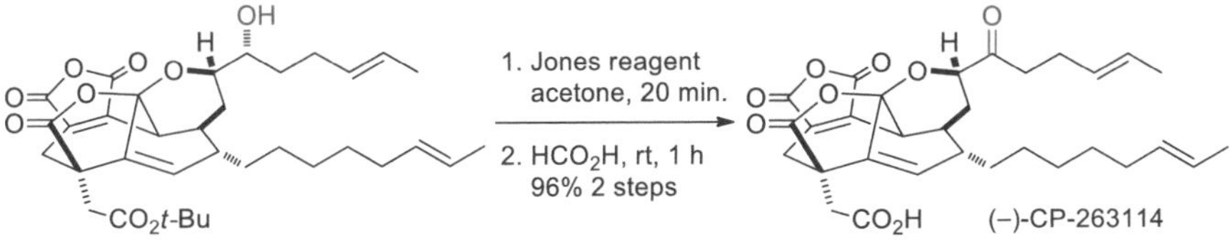

1. Jones reagent
   acetone, 20 min.

2. HCO_2H, rt, 1 h
   96% 2 steps

CO_2t-Bu

CO_2H (–)-CP-263114

Example 2[7]

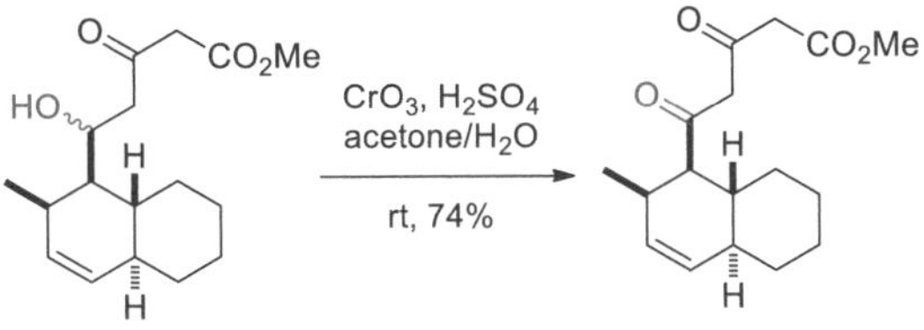

CO_2Me

HO

CrO_3, H_2SO_4
acetone/H_2O

rt, 74%

CO_2Me

Example 3[9]

CrO_3, H_2SO_4

acetone, 0 °C
1–2 h, 86%

HO

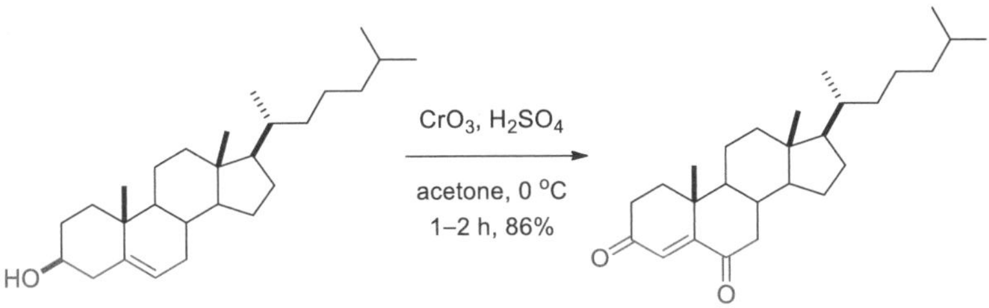

Example 4, Working up the reaction mixture with cold ice–water to precipitate out the pure carboxylic acid[12]

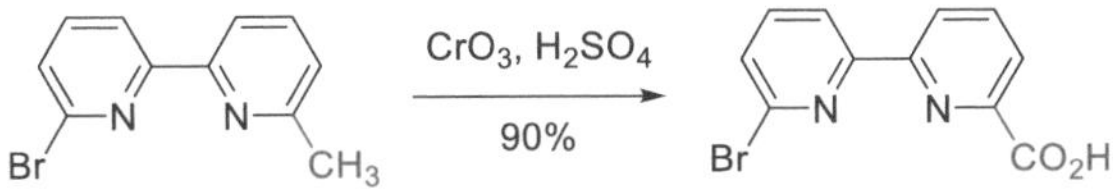

CrO_3, H_2SO_4

90%

Br CH_3

Br CO_2H

Example 5, Boc protection survived the Jones oxidation conditions (at least 40% of it)[13]

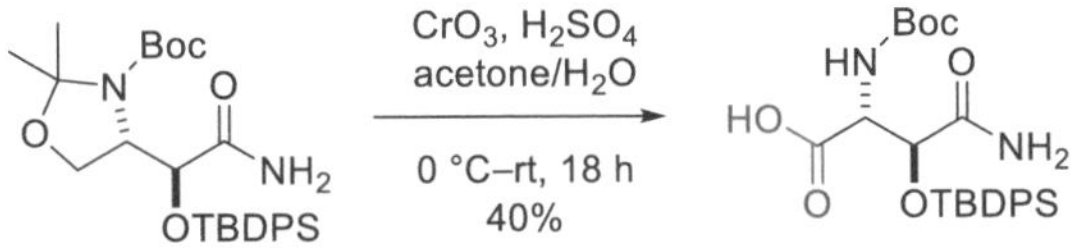

Boc

NH_2
OTBDPS

CrO_3, H_2SO_4
acetone/H_2O

0 °C–rt, 18 h
40%

Boc
HN
HO

NH_2
O   OTBDPS

References

1. Bowden, K.; Heilbron, I. M., Jones, E. R. H.; Weedon, B. C. L. *J. Chem. Soc.* **1946,** 39–45. Ewart R. H. (Tim) Jones worked with Ian M. Heilbron at Imperial College. Jones later succeeded Robert Robinson to become the prestigious Chair of Organic Chemistry at Manchester. *The recipe for the Jones reagent: 25 g CrO₃, 25 mL conc. H₂SO₄, and 70 mL H₂O.*

2. Ratcliffe, R. W. *Org. Synth.* **1973,** *53*, 1852.

3. Vanmaele, L.; De Clerq, P.; Vandewalle, M. *Tetrahedron Lett.* **1982,** *23*, 995–998.

4. Luzzio, F. A. *Org. React.* **1998,** *53*, 1–222. (Review).

5. Zhao, M.; Li, J.; Song, Z.; Desmond, R. J.; Tschaen, D. M.; Grabowski, E. J. J.; Reider, P. J. *Tetrahedron Lett.* **1998,** *39*, 5323–5326. (Catalytic CrO₃ oxidation).

6. Waizumi, N.; Itoh, T.; Fukuyama, T. *J. Am. Chem. Soc.* **2000,** *122*, 7825–7826.

7. Hagiwara, H.; Kobayashi, K.; Miya, S.; Hoshi, T.; Suzuki, T.; Ando, M. *Org. Lett.* **2001,** *3*, 251–254.

8. Fernandes, R. A.; Kumar, P. *Tetrahedron Lett.* **2003,** *44*, 1275–1278.

9. Hunter, A. C.; Priest, S.-M. *Steroids* **2006,** *71*, 30–33.

10. Kim, D.-S.; Bolla, K.; Lee, S.; Ham, J. *Tetrahedron* **2013,** *67*, 1062–1070.

11. Marshall, A. J.; Lin, J.-M.; Grey, A.; Reid, I. R; Cornish, J.; Denny, W. A *Bioorg. Med. Chem.* **2013,** *21*, 4112–4119.

12. Almaliti, J.; Al-Hamashi, A. A.; Negmeldin, A. T.; Hanigan, C. L.; Perera, L.; Pflum, M. K. H.; Casero, R. A.; Tillekeratne, L. M. V. *J. Med. Chem.* **2016,** *59*, 10642–10660.

13. Esgulian, M.; Buchotte, M.; Guillot, R.; Deloisy, S.; Aitken, D. J. *Org. Lett.* **2019,** *21*, 2378–2382.

14. Liu, S.; Gellman, S. H. *Org. Lett.* **2020,** *85*, 1718–1724.

Collins Oxidation

Different from the Jones oxidation, the Collins oxidation, also known as the Collins–Sarett oxidation, converts primary alcohols to the corresponding aldehydes. $CrO_3 \cdot 2Pyr$ is known as the *Collins reagent*.

Example 1[5]

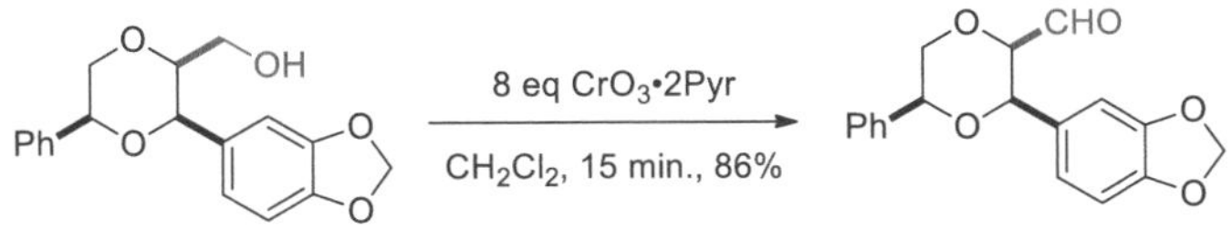

8 eq $CrO_3 \cdot 2Pyr$

CH_2Cl_2, 15 min., 86%

Example 2[7]

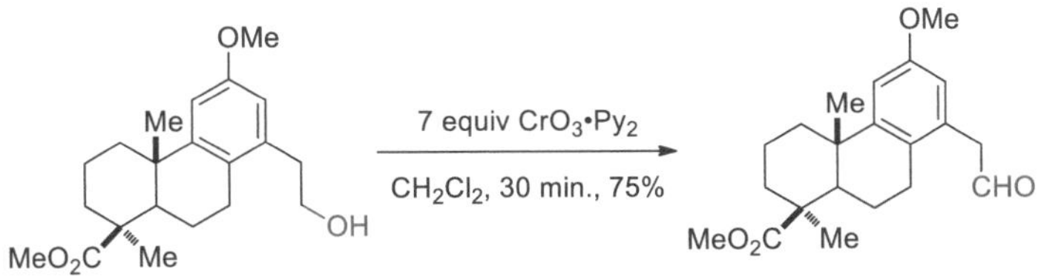

7 equiv $CrO_3 \cdot Py_2$

CH_2Cl_2, 30 min., 75%

Example 3[9]

CrO_3, Pyr.

50–60 °C, 48 h

50–65%

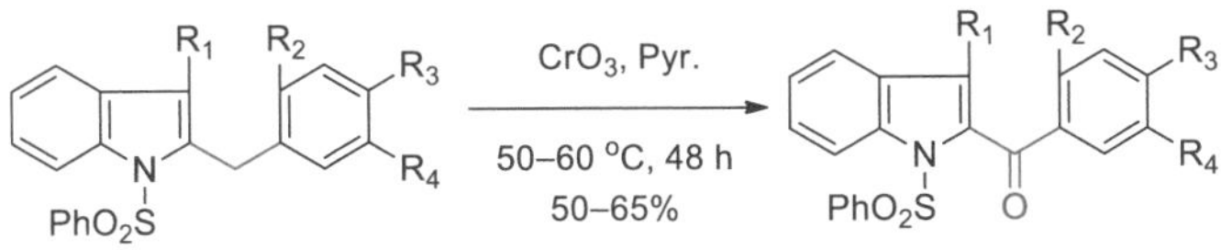

Example 4, TBS and MOM-ether protective groups survived Collins reagent[10]

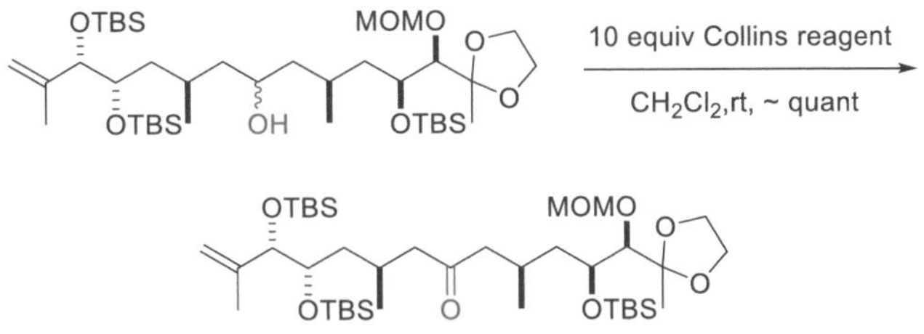

10 equiv Collins reagent

CH_2Cl_2, rt, ~ quant

Example 5, Gemcitabine analog[11]

Example 6[12]

References

1. Poos, G. I.; Arth, G. E.; Beyler, R. E.; Sarett, L. H. *J. Am. Chem. Soc.* **1953,** *75*, 422–429.
2. Collins, J. C; Hess, W. W.; Frank, F. J. *Tetrahedron Lett.* **1968,** 3363–3366. J. C. Collins was a chemist at Sterling–Winthrop in Rensselaer, New York.
3. Collins, J. C; Hess, W. W. *Org. Synth.* **1972,** *Coll. Vol. V*, 310.
4. Hill, R. K.; Fracheboud, M. G.; Sawada, S.; Carlson, R. M.; Yan, S.-J. *Tetrahedron Lett.* **1978,** 945–948.
5. Krow, G. R.; Shaw, D. A.; Szczepanski, S.; Ramjit, H. *Synth. Commun.* **1984,** *14*, 429–433.
6. Li, M.; Johnson, M. E. *Synth. Commun.* **1995,** *25*, 533–537.
7. Harris, P. W. R.; Woodgate, P. D. *Tetrahedron* **2000,** *56*, 4001–4015.
8. Nguyen-Trung, N. Q.; Botta, O.; Terenzi, S.; Strazewski, P. *J. Org. Chem.* **2003,** *68*, 2038–2041.
9. Arumugam, N.; Srinivasan, P. C. *Synth. Commun.* **2003,** *33*, 2313–2320.
10. Zhang, F.-M.; Peng, L.; Li, H.; Ma, A.-J.; Peng, J.-B.; Guo, J.-Ji.; Yang, D.; Hou, S.-H.; Tu, Y.-Q.; Kitching, W. *Angew. Chem. Int. Ed.* **2012,** *51*, 10846–10850.
11. Gonzalez, C.; de Cabrera, M.; Wnuk, S. F. *Nucleosides, Nucleotides Nucleic Acids* **2018,** *37*, 248–260.
12. Tagirov, A. R.; Fayzullina, L. K.; Enikeeva, D. R.; Galimova, Yu. S.; Salikhov, S. M.; Valeev, F. A. *Russ. J. Org. Chem.* **2018,** *54*, 726–733.

PCC Oxidation

Alcohols are oxidized by pyridinium chlorochromate (PCC) to the corresponding aldehydes or ketones. They are not further oxidized to the corresponding carboxylic acids because the reaction was done in organic solvents, not in water. If water existed, the carbonyls would form *aldehyde hydrates* or *ketone hydrates*, which are then oxidized to acids.

Example 1, One-pot PCC–Wittig reactions[2]

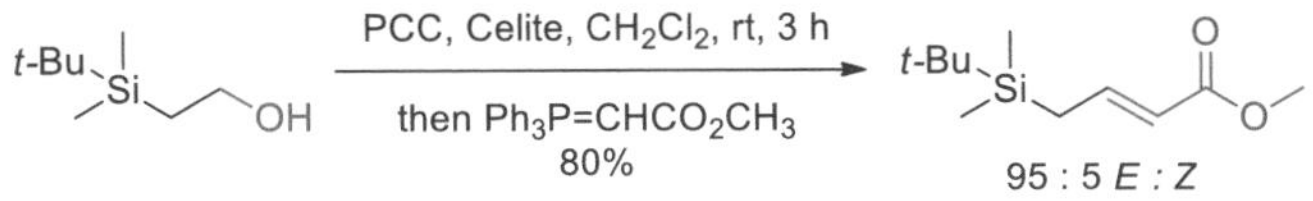

Example 2[3]

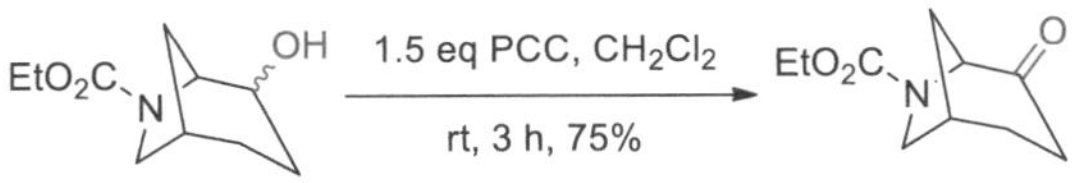

Example 3, Allylic oxidation[4]

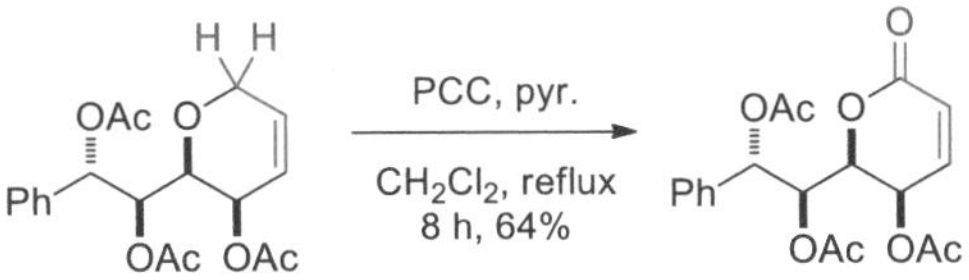

Example 4, Hemiacetal oxidation[5]

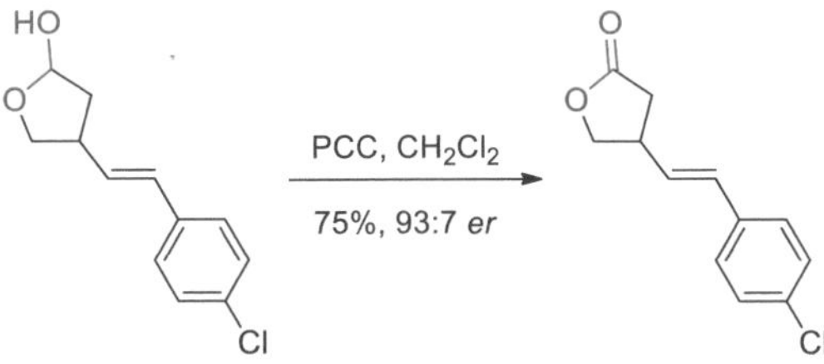

Example 5[8]

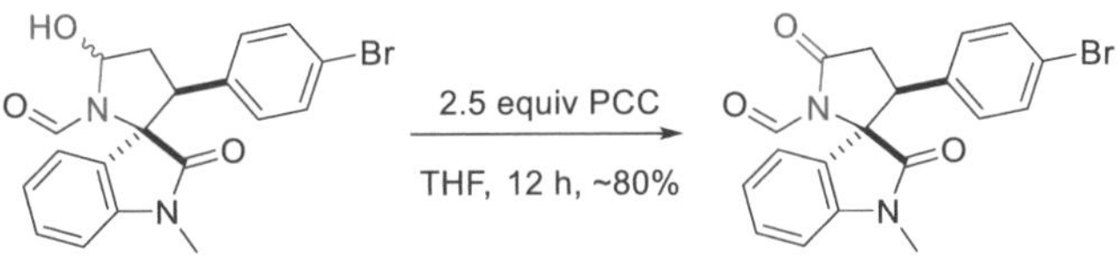

Example 6[9]

Example 7[10]

References

1. Corey, E. J.; Suggs, W. *Tetrahedron Lett.* **1975,** *16,* 2647–2650.
2. Bressette, A. R.; Glover, L. C., IV *Synlett* **2004,** 738–740.
3. Breining, S. R.; Bhatti, B. S.; Hawkins, G. D.; Miao, L. WO2005037832 (**2005**).
4. Srikanth, G. S. C.; Krishna, U. M. *Tetrahedron* **2006,** *62,* 11165–11171.
5. Kim, S.-G. *Tetrahedron Lett.* **2008,** *49,* 6148–6151.
6. Mehta, G.; Bera, M. K. *Tetrahedron* **2013,** *69,* 1815–1821.
7. Fowler, K. J.; Ellis, J. L.; Morrow, G. W. *Synth. Commun.* **2013,** *43,* 1676–1682.
13. Yang, P.; Wang, X.; Chen, F.; Zhang, Z.-B.; Chen, C.; Peng, L.; Wang, L.-X. *J. Org. Chem.* **2017,** *82,* 3908–3916.
14. Hasimujiang, B.; Zeng, J.; Zhang, Y.; Abudu Rexit, A. *Synth. Commun.* **2018,** *48,* 887–891.
15. Dhotare, B. B.; Kumar, M.; Nayak, S. K. *J. Org. Chem.* **2018,** *83,* 10089–10096.

PDC Oxidation

Pyridinium dichromate (PDC) may oxidize alcohols all the way to the corresponding carboxylic acids instead of aldehydes and ketones as PCC does.

Example 1[2]

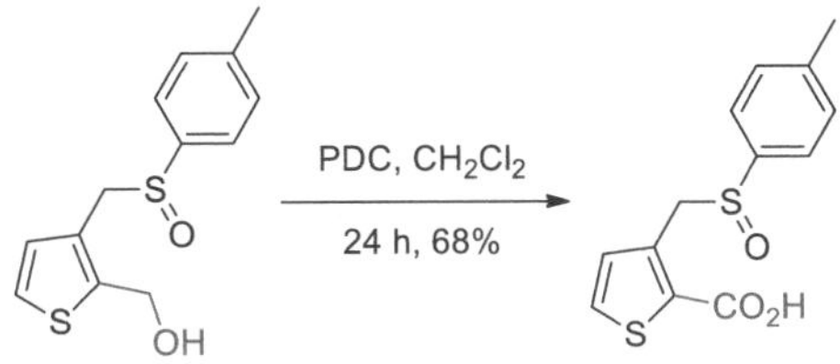

PDC, CH$_2$Cl$_2$

24 h, 68%

Example 2, Cleavage of primary carbon–boron bond[3]

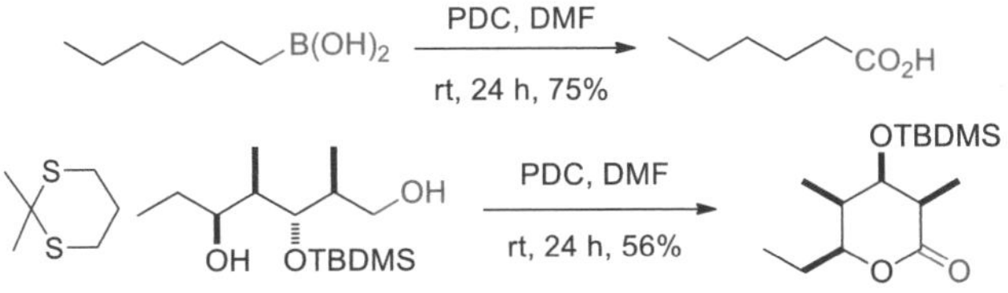

PDC, DMF

rt, 24 h, 75%

PDC, DMF

rt, 24 h, 56%

Example 4, PDC oxidation provided the β-keto-amide, which was hydrolyzed off after protection of the ketone as dithiane[9]

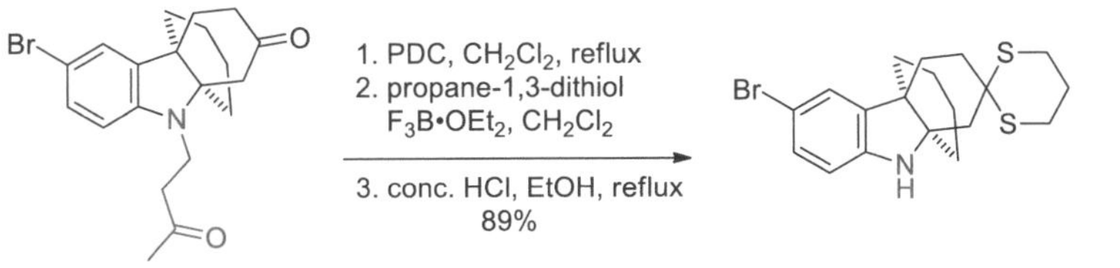

1. PDC, CH$_2$Cl$_2$, reflux
2. propane-1,3-dithiol
   F$_3$B•OEt$_2$, CH$_2$Cl$_2$

3. conc. HCl, EtOH, reflux
   89%

Example 5, Oxidation of lactol to lactone[10]

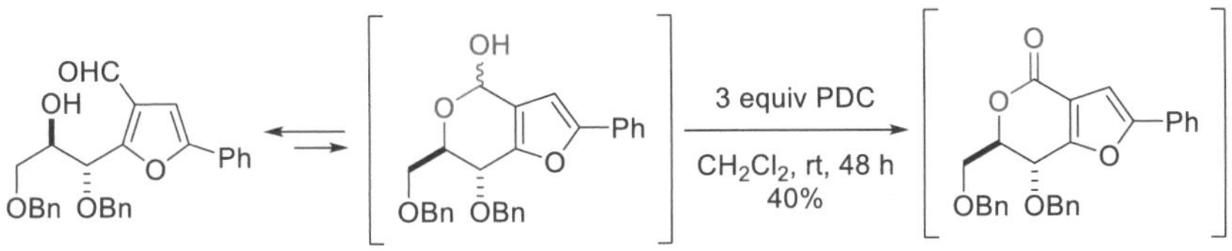

3 equiv PDC

CH$_2$Cl$_2$, rt, 48 h
40%

Example 6, The 10% impurity was the result of the Piancatelli rearrangement with the aid of water after exposure to silica gel[11]

60%   10%

Example 7, Chemoselective oxidation of 2*H*-chromenes to coumarins[12]

References

1   Corey, E. J.; Schmidt, G. *Tetrahedron Lett.* **1979**, 399–402.
2   Terpstra, J. W.; Van Leusen, A. M. *J. Org. Chem.* **1986**, *51*, 230–208.
3   Brown, H. C.; Kulkarni, S. V.; Khanna, V. V.; Patil, V. D.; Racherla, U. S. *J. Org. Chem.* **1992**, *57*, 6173–6177.
4   Nakamura, M.; Inoue, J.; Yamada, T. *Bioorg. Med. Chem. Lett.* **2000**, *10*, 2807–2810.
5   Chênevert, R. Courchene, G.; Caron, D. *Tetrahedron: Asymmetry* **2003**, 2567–2571.
6   Jordão, A. K *Synlett* **2006**, 3364–3365. (Review).
7   Xu, G.; Hou, A.-J.; Wang, R.-R.; Liang, G.-Y.; Zheng, Y.-T.; Liu, Z.-Y.; Li, X.-L.; Zhao, Y.; Huang, S.-X.; Peng, L.-Y.; et al. *Org. Lett.* **2006**, *8*, 4453–4456.
8   Morzycki, J. W; Perez-Diaz, J. O. H; Santillan, R.; Wojtkielewicz, A. *Steroids* **2010**, *75*, 70–76.
9   Cai, Q.; You, S.-L. *Org. Lett.* **2012**, *14*, 3040–3043.
10 Mal, K.; Sharma, A.; Das, I. *Chem. Eur. J.* **2014**, *20*, 11932–11945.
11 Veits, G. K.; Wenz, D. R.; Palmer, L. I.; St. Amant, A. H.; Hein, J. E.; Read de Alaniz, J. *Org. Biomol. Chem.* **2015**, *13*, 8465–8469.
12 Sharif, S. A. I.; Calder, E. D. D.; Harkiss, A. H.; Maduro, M.; Sutherland, A. *J. Org. Chem.* **2016**, *81*, 9810–9819.
13 Li, C.; Ji, Y.; Cao, Q.; Li, J.; Li, B. *Synth. Commun.* **2017**, *47*, 1301–1306.

Julia–Kocienski Olefination

Modified one-pot Julia olefination to give predominantly (*E*)-olefins from heteroarylsulfones and aldehydes. A sulfone reduction step is *not* required.

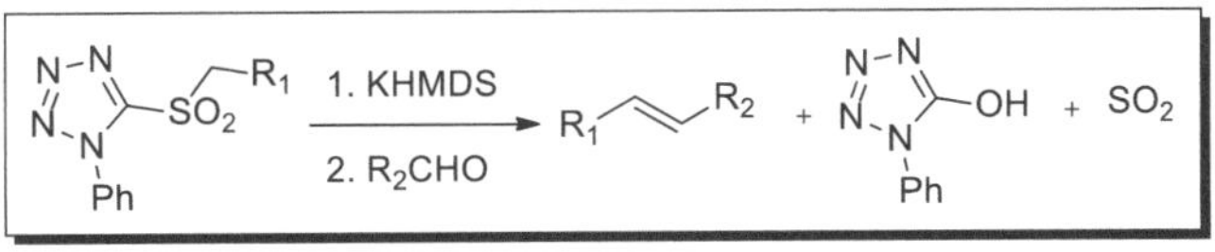

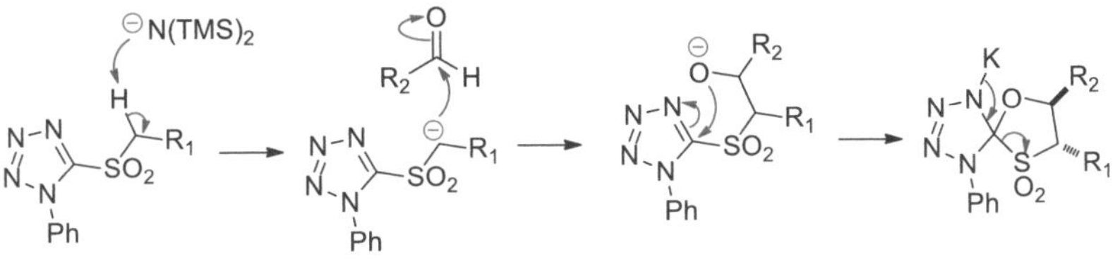

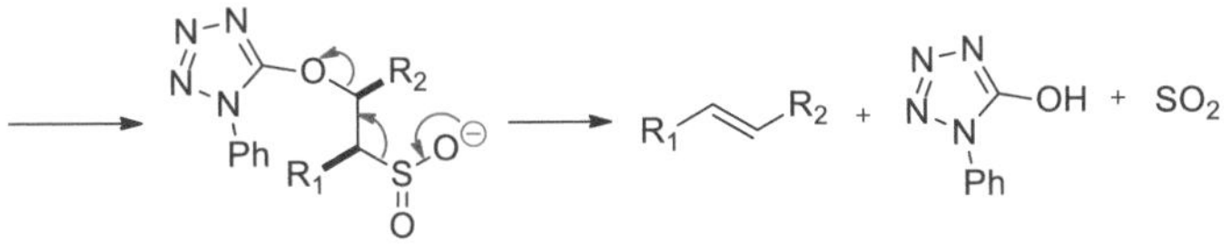

Alternatives to tetrazole:

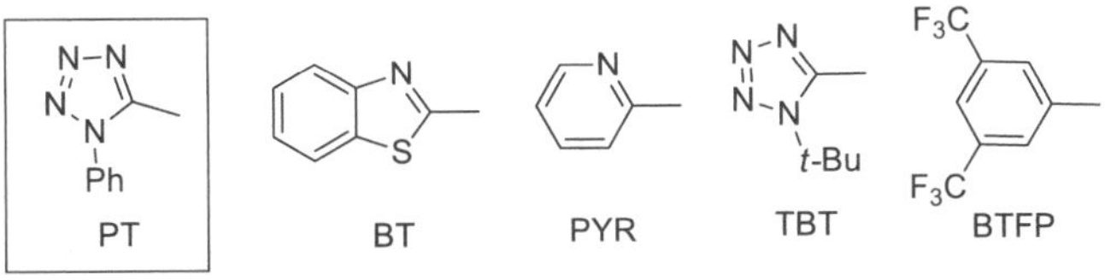

| PT | BT | PYR | TBT | BTFP |

The use of larger counter-ion (such as K^+) and polar solvents (such as DME) favors an open transition state (PT = *N*-phenyltetrazolyl):

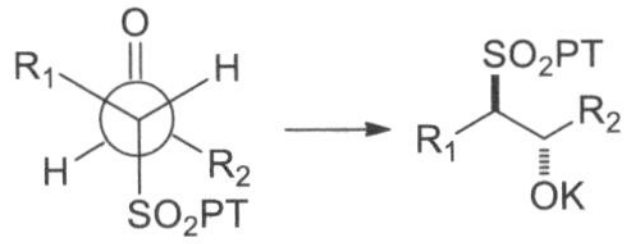

Example 1, (BT = benzothiazole)[2]

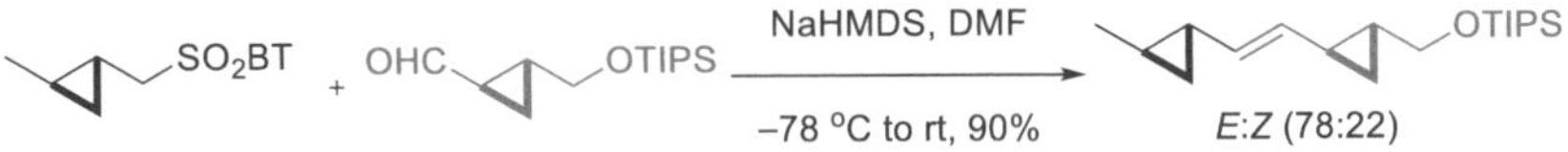

NaHMDS, DMF

−78 °C to rt, 90%

E:Z (78:22)

© Springer Nature Switzerland AG 2021
J. J. Li, *Name Reactions*, https://doi.org/10.1007/978-3-030-50865-4_75

Example 2[3]

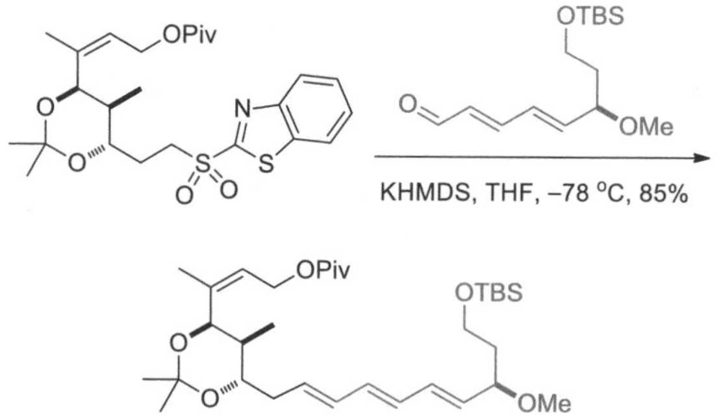

OPiv
OTBS
OMe
KHMDS, THF, –78 °C, 85%

Example 3[7]

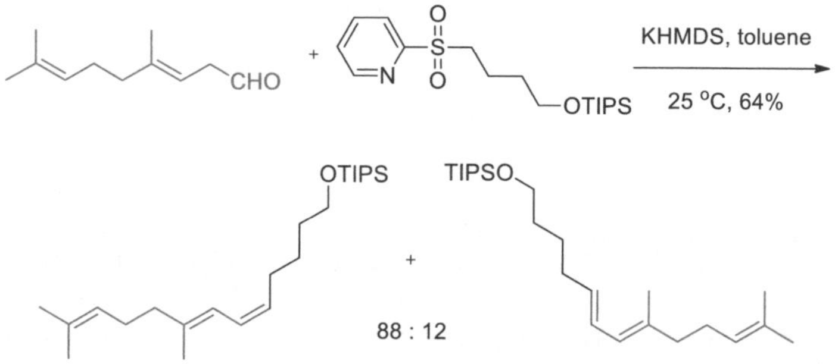

CHO
OTIPS
KHMDS, toluene
25 °C, 64%
TIPSO
88 : 12

Example 4, P4-*t*-Bu is a strong base, but a weak nucleophile[8]

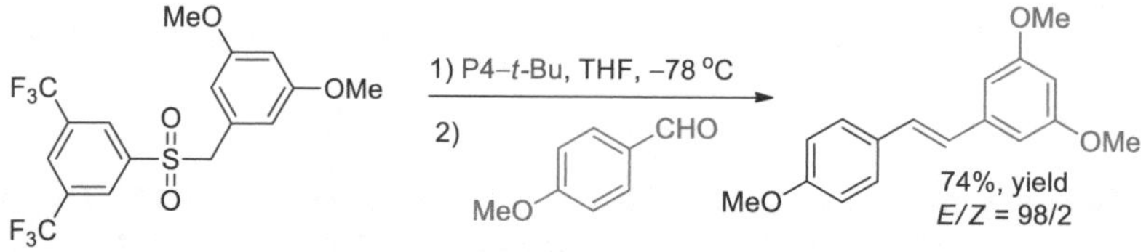

MeO
OMe
F_3C
1) P4–*t*-Bu, THF, –78 °C
2)
CHO
74%, yield
E/*Z* = 98/2

Example 5, (*E*)-Isomer only[12]

TBSO
Boc
Ph
KHMDS, 18-crown-6
DME, –78 °C–rt, 12 h
> 80%

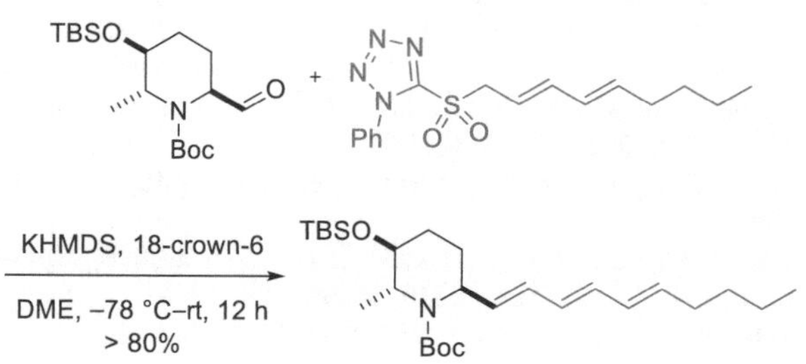

Example 6, (*E*)-Isomer only[13]

Example 7, *En route* to preparation of HCV NS3/4A inhibitors[15]

References

1. (a) Baudin, J. B.; Hareau, G.; Julia, S. A.; Ruel, O. *Tetrahedron Lett.* **1991**, *32*, 1175–1178. (b) Baudin, J. B.; Hareau, G.; Julia, S. A.; Ruel, O. *Bull. Soc. Chim. Fr.* **1993**, *130*, 336–357. (c) Baudin, J. B.; Hareau, G.; Julia, S. A.; Loene, R.; Ruel, O. *Bull. Soc. Chim. Fr.* **1993**, *130*, 856–878. (d) Blakemore, P. R.; Cole, W. J.; Kocienski, P. J.; Morely, A. *Synlett* **1998**, 26–28.
2. Charette, A. B.; Lebel, H. *J. Am. Chem. Soc.* **1996**, *118*, 10327–10328.
3. Blakemore, P. R.; Kocienski, P. J.; Morley, A.; Muir, K. *J. Chem. Soc., Perkin Trans. 1* **1999**, 955–968.
4. Williams, D. R.; Brooks, D. A.; Berliner, M. A. *J. Am. Chem. Soc.* **1999**, *121*, 4924–4925.
5. Kocienski, P. J.; Bell, A.; Blakemore, P. R. *Synlett* **2000**, 365–366.
6. Liu, P.; Jacobsen, E. N. *J. Am. Chem. Soc.* **2001**, *123*, 10772–10773.
7. Charette, A. B.; Berthelette, C.; St-Martin, D. *Tetrahedron Lett.* **2001**, *42*, 5149–5153.
8. Alonso, D. A.; Najera, C.; Varea, M. *Tetrahedron Lett.* **2004**, *45*, 573–577.
9. Alonso, D. A.; Fuensanta, M.; Najera, C.; Varea, M. *J. Org. Chem.* **2005**, *70*, 6404.
10. Rong, F. *Julia–Lythgoe olefination*. In *Name Reactions for Homologations-Part I*; Li, J. J., Ed.; Wiley: Hoboken, NJ, **2009**, pp 447–473. (Review).
11. Davies, S. G.; Fletcher, A. M.; Foster, E. M. *J. Org. Chem.* **2013**, *78*, 2500–2510.
12. Velázuez, F.; Chelliah, M.; Clasby, M.; Guo, Z.; Howe, J.; Miller, R.; Neelamkavil, S.; Shah, U.; Soriano, A.; Xia, Y.; Venkatramann, S.; Chackalamannil, S.; Davies, I. W. *ACS Med. Chem. Lett.* **2016**, *7*, 1173–1178.
13. Friedrich, R.; Sreenilayam, G.; Hackbarth, J.; Friestad, G. K. *J. Org. Chem.* **2018**, *83*, 13636–13649.
14. Blakemore, P. R.; Sephton, S. M.; Ciganek, E. *Org. React.* **2018**, *95*, 1–422. (Review).
15. Macha, L.; Ha, H.-J. *J. Org. Chem.* **2019**, *84*, 94–103.
16. Lood, K.; Schmidt, B. *J. Org. Chem.* **2020**, *85*, 5122–5130.

Julia–Lythgoe Olefination

(*E*)-Olefins from sulfones and aldehydes.

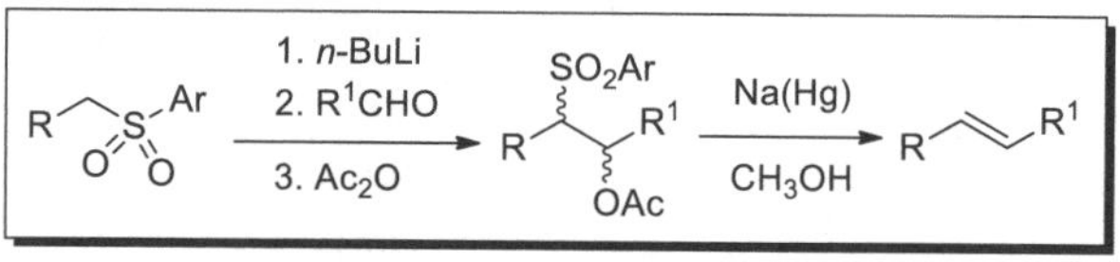

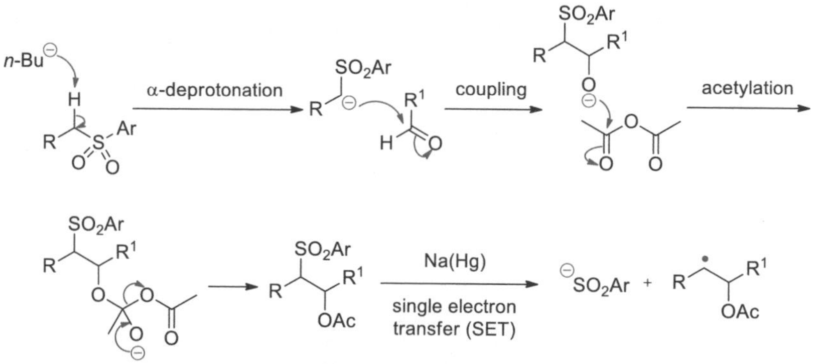

4 possible diastereomers

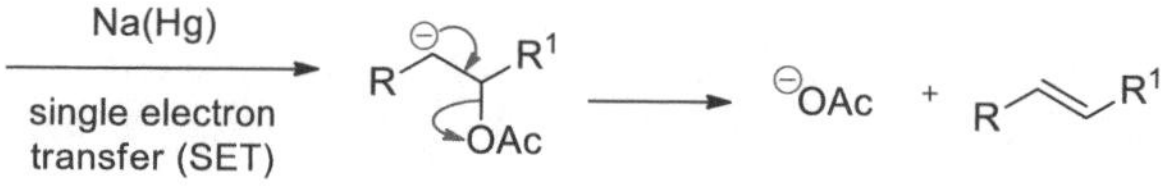

Example 1[2]

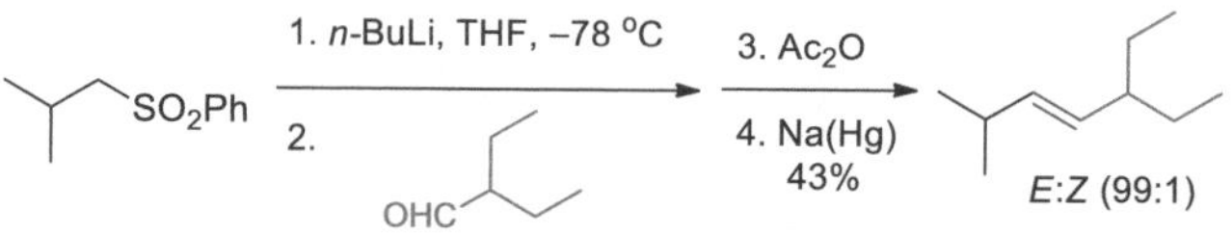

Example 2[3]

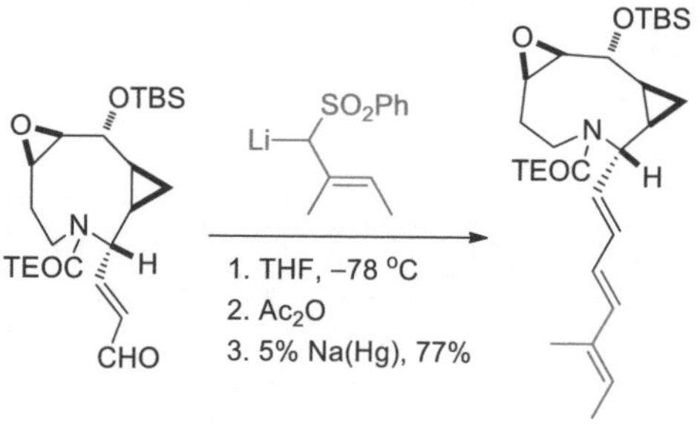

© Springer Nature Switzerland AG 2021
J. J. Li, *Name Reactions*, https://doi.org/10.1007/978-3-030-50865-4_76

Example 3[7]

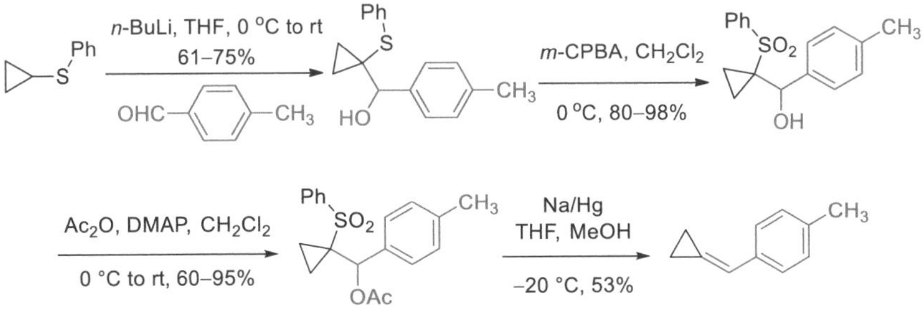

Example 4[8]

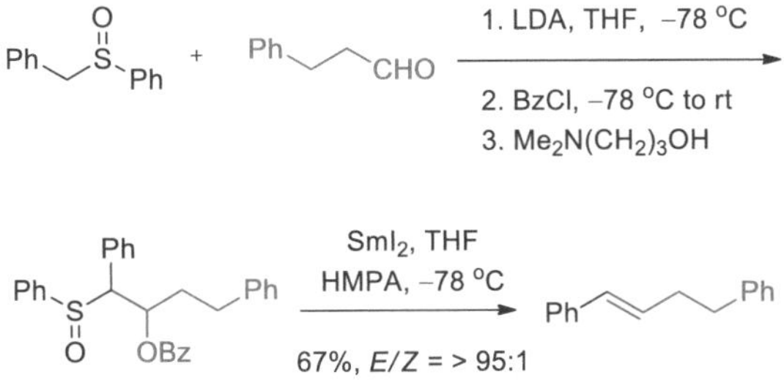

Example 5, Sometimes, the difference between Julia–Kocienski olefination and Julia–Lythgoe olefination is blurred, rightfully so as Kocienski and Lythgoe were coauthors for reference 2[12]

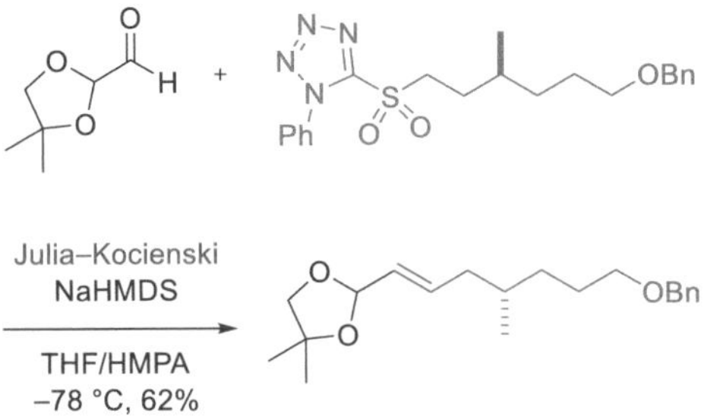

Example 6, As above, this is more appropriate to call it Julia–Kocienski olefination[13]

References

1.   (a) Julia, M.; Paris, J. M. *Tetrahedron. Lett.* **1973**, 4833–4836. (b) Lythgoe, B. *J. Chem. Soc., Perkin Trans. 1* **1978,** 834–837.
2.   Kocienski, P. J.; Lythgoe, B.; Waterhause, I. *J. Chem. Soc., Perkin Trans. 1* **1980,** 1045–1050.
3.   Kim, G.; Chu-Moyer, M. Y.; Danishefsky, S. J. *J. Am. Chem. Soc.* **1990**, *112*, 2003–2005.
4.   Keck, G. E.; Savin, K. A.; Weglarz, M. A. *J. Org. Chem.* **1995,** *60*, 3194–3204.
5.   Breit, B. *Angew. Chem. Int. Ed.* **1998,** 37, 453–456.
6.   Marino, J. P.; McClure, M. S.; Holub, D. P.; Comasseto, J. V.; Tucci, F. C. *J. Am. Chem. Soc.* **2002,** *124*, 1664–1668.
7.   Bernard, A. M.; Frongia, A.; Piras, P. P.; Secci, F. *Synlett* **2004,** *6*, 1064–1068.
8.   Pospíšil, J.; Pospíšil, T, Markó, I. E. *Org. Lett.* **2005,** *7*, 2373–2376.
9.   Gollner, A.; Mulzer, J. *Org. Lett.* **2008,** *10*, 4701–4704.
10. Rong, F. *Julia–Lythgoe olefination*. In *Name Reactions for Homologations-Part I*; Li, J. J., Ed.; Wiley: Hoboken, NJ, **2009,** pp 447–473. (Review).
11. Dams, I.; Chodynski, M.; Krupa, M.; Pietraszek, A.; Zezula, M.; Cmoch, P.; Kosińska, M.; Kutner, A. *Tetrahedron* **2013,** *69*, 1634–1648.
12. Ren, R.-G.; Li, M.; Si, C.-M.; Mao, Z.-Y.; Wei, B.-G. *Tetrahedron Lett.* **2014,** *55*, 6903–6906.
13. Samala, R.; Sharma, S.; Basu, M. K.; Kukkanti, K.; Porstmann, F. *Tetrahedron Lett.* **2016,** *57*, 1309–1312.

Knoevenagel Condensation

Condensation between carbonyl compounds and activated methylene compounds catalyzed by amines.

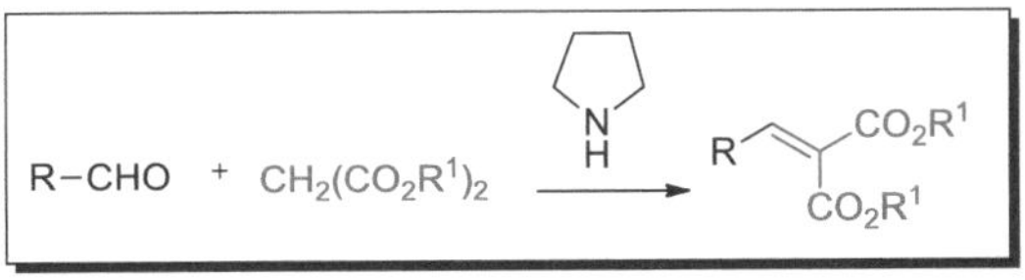

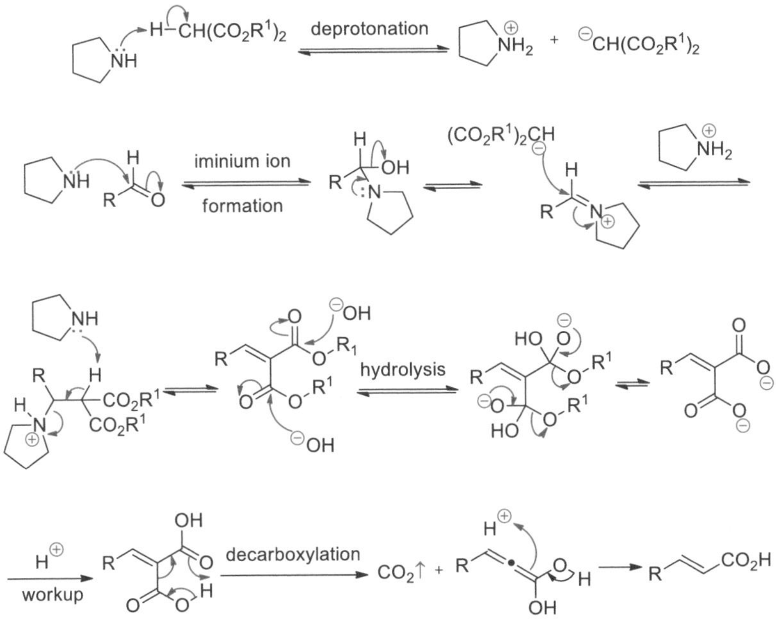

Example 1[3]

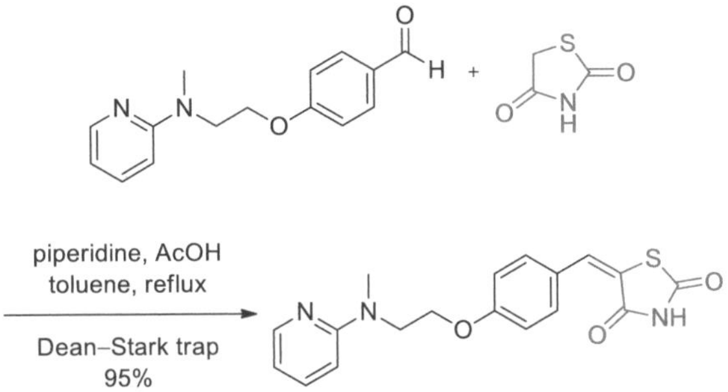

piperidine, AcOH
toluene, reflux
———————→
Dean–Stark trap
95%

© Springer Nature Switzerland AG 2021
J. J. Li, *Name Reactions*, https://doi.org/10.1007/978-3-030-50865-4_77

Example 2, EDDA = Ethylenediamine diacetate[5]

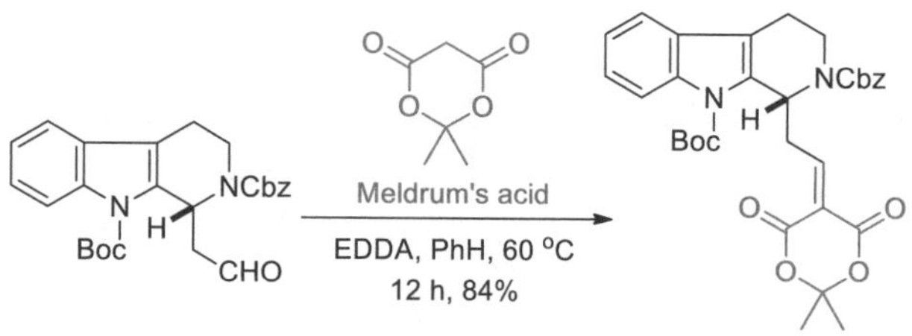

Example 3, Using ionic liquid ethylammonium nitrate (EAN) as solvent[8]

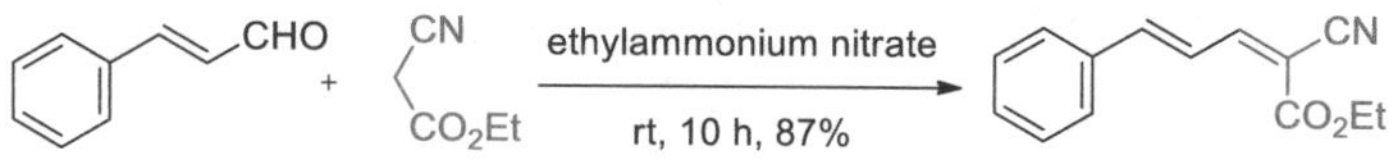

Example 4[9]

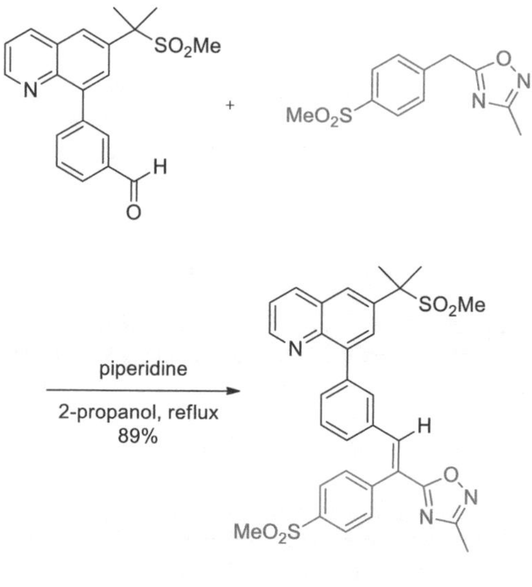

Example 5[11]

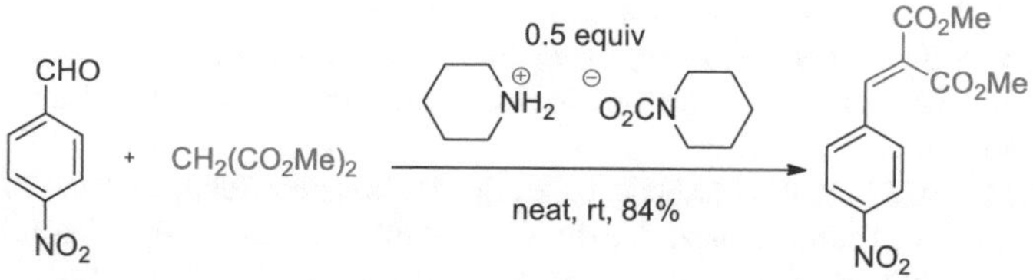

Example 6, Fluoride as the base for an intramolecular Knoevenagel condensation and unexpected decarboxylation[12]

Example 7, EDDA = Ethylenediamine diacetate as the base here[13]

References

1. Knoevenagel, E. *Ber.* **1898**, *31*, 2596–2619. Emil Knoevenagel (1865–1921) was born in Hannover, Germany. He studied at Göttingen under Victor Meyer and Gattermann, receiving a Ph.D. in 1889. He became a full professor at Heidelberg in 1900. When WWI broke out in 1914, Knoevenagel was one of the first to enlist and rose to the rank of staff officer. After the war, he returned to his academic work until his sudden death during an appendectomy.
2. Jones, G. *Org. React.* **1967**, *15*, 204–599. (Review).
3. Cantello, B. C. C.; Cawthornre, M. A.; Cottam, G. P.; Duff, P. T.; Haigh, D.; Hindley, R. M.; Lister, C. A.; Smith, S. A.; Thurlby, P. L. *J. Med. Chem.* **1994**, *37*, 3977–3985.
4. Paquette, L. A.; Kern, B. E.; Mendez-Andino, J. *Tetrahedron Lett.* **1999**, *40*, 4129–4132.
5. Tietze, L. F.; Zhou, Y. *Angew. Chem. Int. Ed.* **1999**, *38*, 2045–2047.
6. Pearson, A. J.; Mesaros, E. F. *Org. Lett.* **2002**, *4*, 2001–2004.
7. Kourouli, T.; Kefalas, P.; Ragoussis, N.; Ragoussis, V. *J. Org. Chem.* **2002**, *67*, 4615–4618.
8. Hu, Y.; Chen, J.; Le, Z.-G.; Zheng, Q.-G. *Synth. Commun.* **2005**, *35*, 739–744.
9. Conlon, D. A.; Drahus-Paone, A.; Ho, G.-J.; Pipik, B.; Helmy, R.; McNamara, J. M.; Shi, Y.-J.; Williams, J. M.; MacDonald, D. *Org. Process Res. Dev.* **2006**, *10*, 36–45.
10. Rong, F. *Knoevenagel Condensation.* In *Name Reactions for Homologations-Part I*; Li, J. J., Ed.; Wiley: Hoboken, NJ, **2009**, pp 474–501. (Review).

11. Mase, N.; Horibe, T. *Org. Lett.* **2013,** *15*, 1854–1857.
12. Lopez, A. M.; Ibrahim, A. A.; Rosenhauer, G. J.; Sirinimal, H. S.; Stockdill, J. L. *Org. Lett.* **2018,** *20*, 2216–2219.
13. Schuppe, A. W.; Zhao, Y.; Liu, Y.; Newhouse, T. R. *J. Am. Chem. Soc.* **2019,** *141*, 9191–9196.
14. Yan Z.; Zhao C.; Gong J.; Yang Z.; Yang Z. *Org. Lett.* **2020,** *22*, 1644–1647.

Knorr Pyrazole Synthesis

Also known as Knorr reaction. Reaction of hydrazine or substituted hydrazine with 1,3-dicarbonyl compounds to provide the pyrazole or pyrazolone ring system. *Cf.* Paal–Knorr pyrrole synthesis.

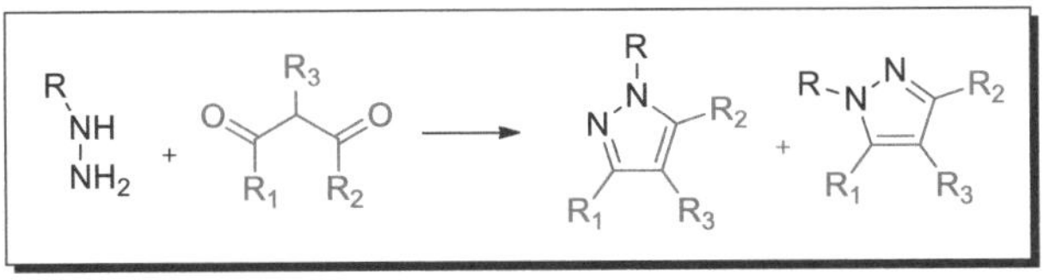

R = H, Alkyl, Aryl, Het-aryl, Acyl, *etc*.

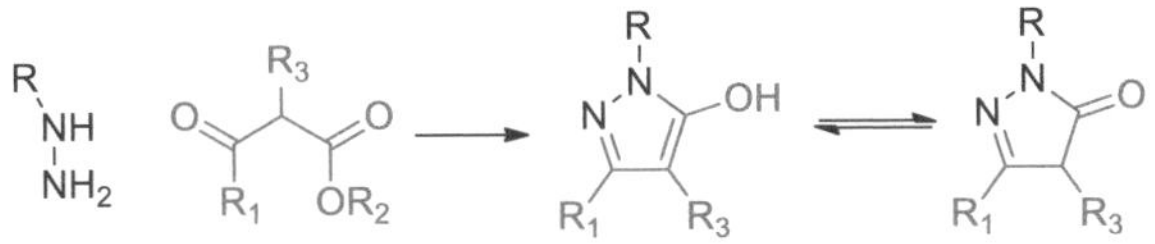

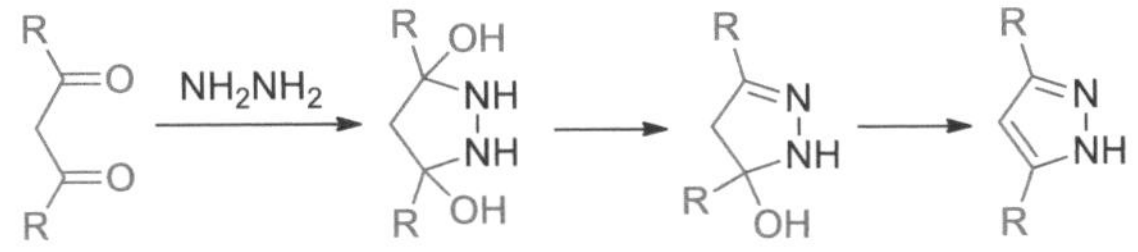
NH₂NH₂

Alternatively,

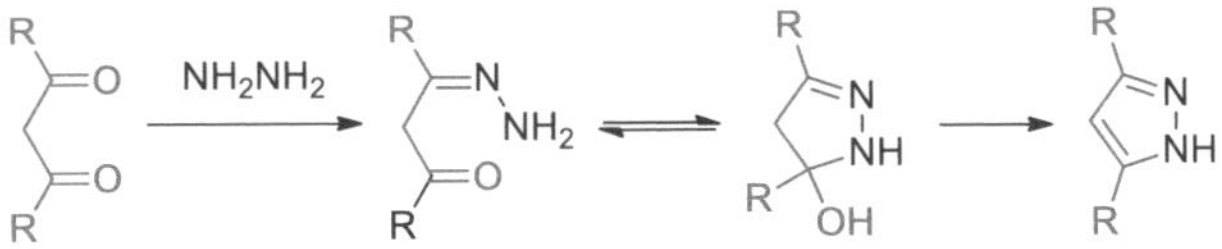
NH₂NH₂

Example 1[2]

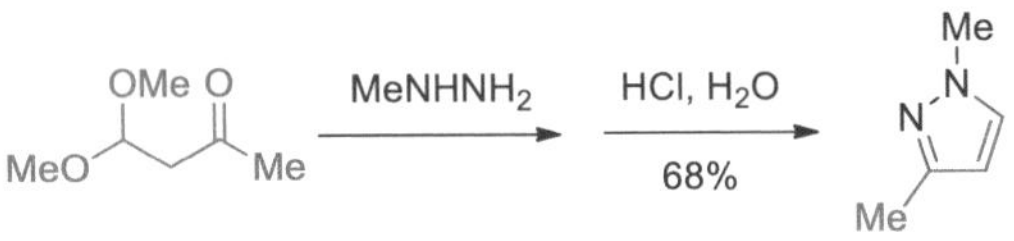
MeNHNH₂
HCl, H₂O
68%

Example 2[8]

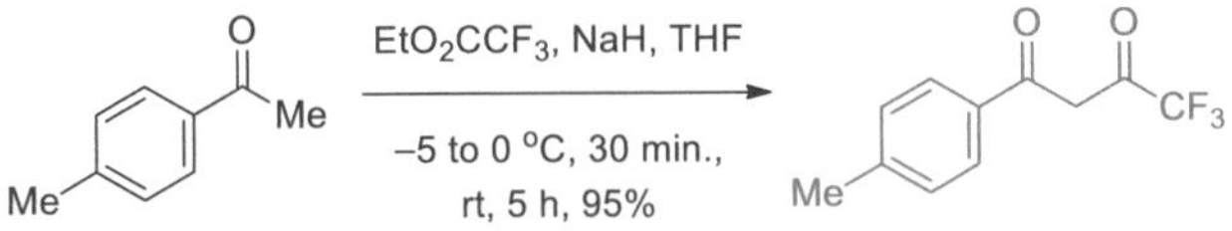
EtO₂CCF₃, NaH, THF
–5 to 0 °C, 30 min.,
rt, 5 h, 95%

J. J. Li, *Name Reactions*, https://doi.org/10.1007/978-3-030-50865-4_78

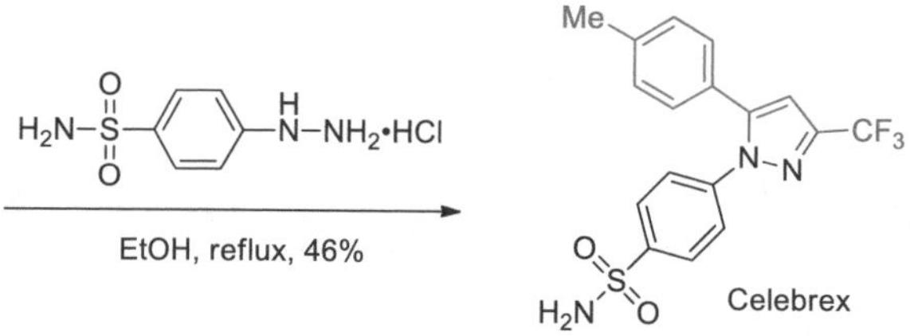

Example 3, Preparation of an intermediate for tenegliptin, a dipeptidyl peptidase-4 (DPP-4) inhibitor[9]

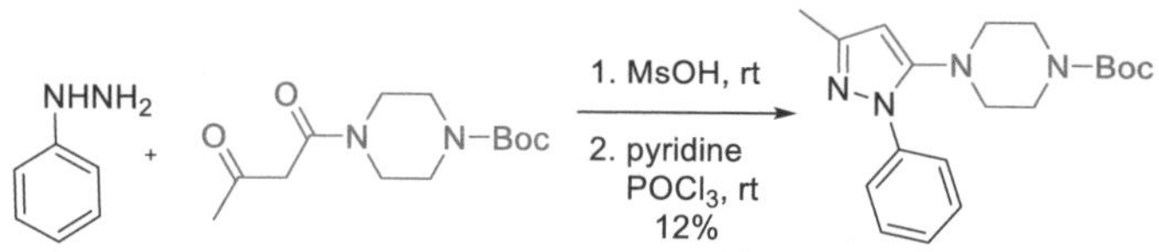

Example 4[10]

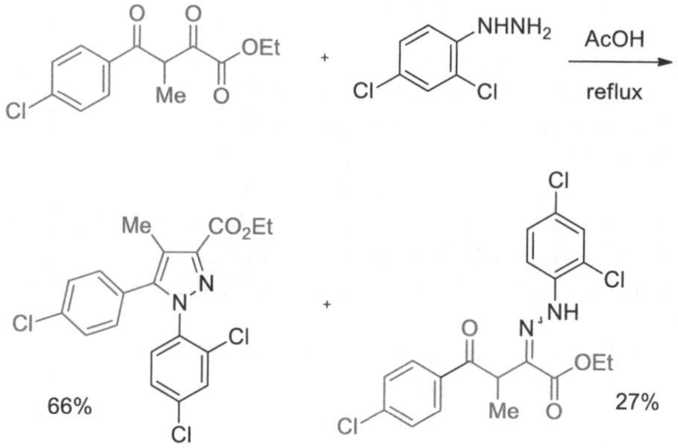

Example 5[11]

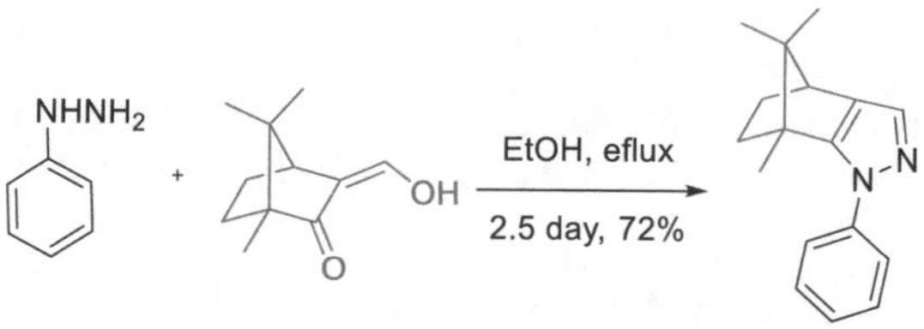

Example 6, Knorr pyrazole-thioester generation employed in the native chemical ligation (NCL) strategies for chemical protein synthesis[12]

References

1   (a) Knorr, L. *Ber* **1883**, *16*, 2597. Ludwig Knorr (1859–1921) was born near Munich, Germany. After studying under Volhard, Emil Fischer, and Bunsen, he was appointed professor of chemistry at Jena. Knorr made tremendous contributions in the synthesis of heterocycles in addition to discovering the important pyrazolone drug, pyrine. (b) Knorr, L. *Ber* **1884**, *17*, 546, 2032. (c) Knorr, L. *Ber.* **1885**, *18*, 311. (d) Knorr, L. *Ann.* **1887**, *238*, 137.

2   Burness, D. M. *J. Org. Chem.* **1956**, *21*, 97–101.

3   Jacobs, T. L. in *Heterocyclic Compounds*, Elderfield, R. C., Ed.; Wiley: New York, **1957**, *5*, 45. (Review).

4   *Houben–Weyl*, **1967**, *10/2*, 539, 587, 589, 590. (Review).

5   Elguero, J., In *Comprehensive Heterocyclic Chemistry II*, Katrizky, A. R.; Rees, C. W.: Scriven. E. F. V., Eds; Elsevier: Oxford, **1996**, *3*, 1. (Review).

6   Stanovnik, E.; Svete, J. In *Science of Synthesis*, **2002**, *12*, 15; Neier, R., Ed.; Thieme. (Review).

7   Sakya, S. M. *Knorr Pyrazole Synthesis*. In *Name Reactions in Heterocyclic Chemistry*; Li, J. J., Corey, E. J., Eds, Wiley: Hoboken, NJ, **2005**, pp 292–300. (Review).

8   Ahlstroem, M. M.; Ridderstroem, M.; Zamora, I.; Luthman, K. *J. Med. Chem.* **2007**, *50*, 4444–4452.

9   Yoshida, T.; Akahoshi, F.; Sakashita, H.; Kitajima, H.; Nakamura, M.; Sonda, S; Takeuchi, M.; Tanaka, Y.; Ueda, N.; Sekiguchi, S.; et al. *Bioorg. Med. Chem.* **2012**, *20*, 5705–5719.

10 Jiang, J. A.; Huang, W. B.; Zhai, J. J.; Liu, H. W.; Cai, Q.; Xu, L. X.; Wang, W.; Ji, Y. F. *Tetrahedron* **2013**, *69*, 627–635.

11 Nozari, M.; Addison, A. W.; Reeves, G. T.; Zeller, M.; Jasinski, J. P.; Kaur, M.; Gilbert, J. G.; Hamilton, C. R.; Popovitch, J. M.; Wolf, L. M.; et al. *J. Heterocycl. Chem.* **2018**, *55*, 1291–1307.

12 Flood, D. T.; Hintzen, J. C. J.; Bird, M. J.; Cistrone, P. A.; Chen, J. S.; Dawson, P. E. *Angew. Chem. Int. Ed.* **2018**, *57*, 11634–11639.

13 Du, Y.; Xu, Y.; Qi, C.; Wang, C. **2019**, *60*, 1999–2004.

Koenig–Knorr Glycosidation

Formation of the β-glycoside from α-halocarbohydrate under the influence of silver salt.

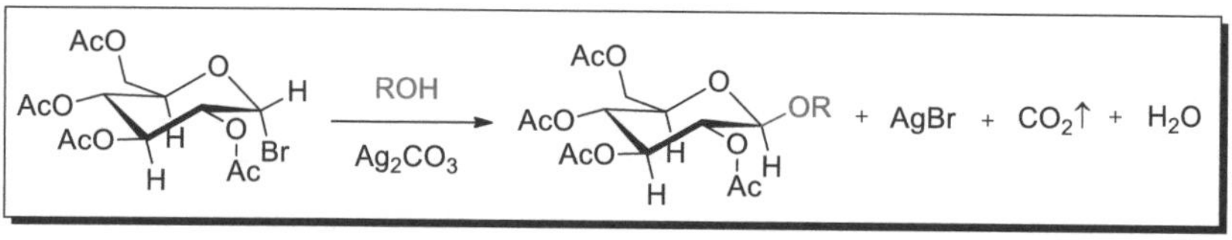

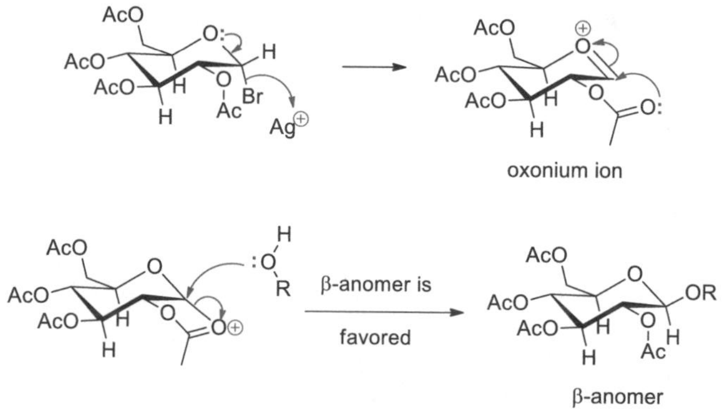

oxonium ion

β-anomer is

favored

β-anomer

Example 1[7]

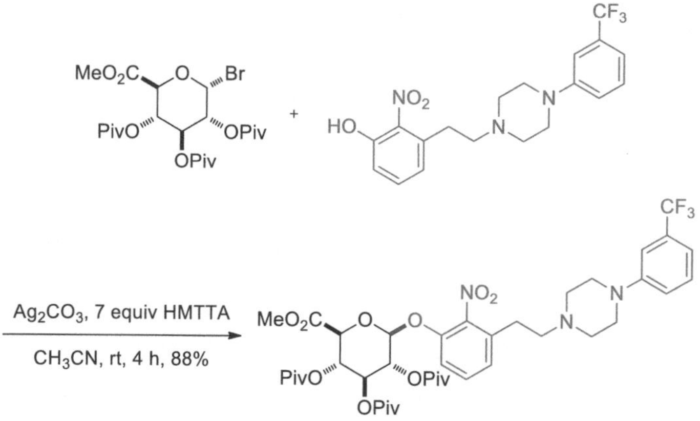

Ag₂CO₃, 7 equiv HMTTA

CH₃CN, rt, 4 h, 88%

Example 2[8]

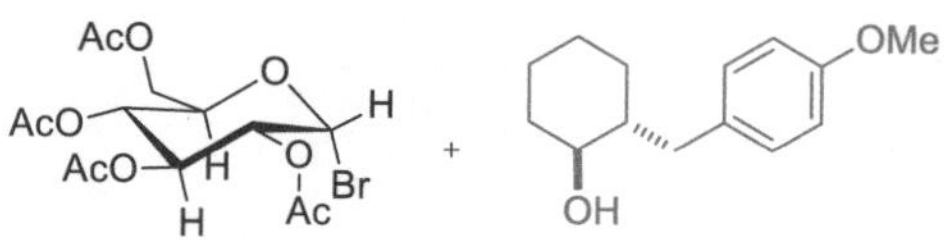

© Springer Nature Switzerland AG 2021
J. J. Li, *Name Reactions*, https://doi.org/10.1007/978-3-030-50865-4_79

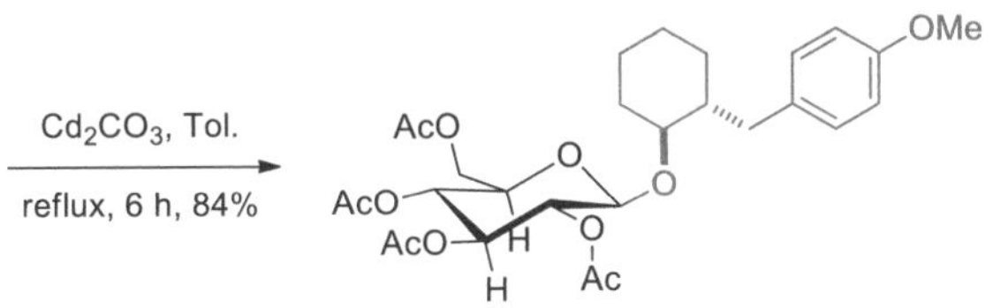

Example 3, TMU = tetramethylurea[9]

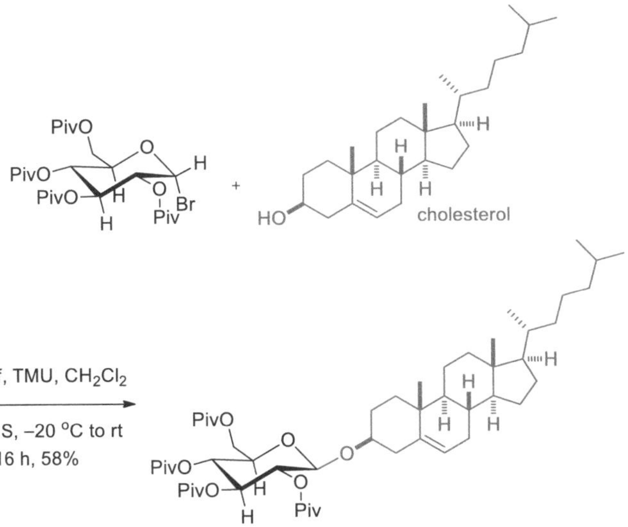

AgOTf, TMU, CH$_2$Cl$_2$

4 Å MS, –20 °C to rt
16 h, 58%

Example 4[11]

Ag$_2$CO$_3$
4 Å MS

quinoline,
0 °C to rt
overnight
18%

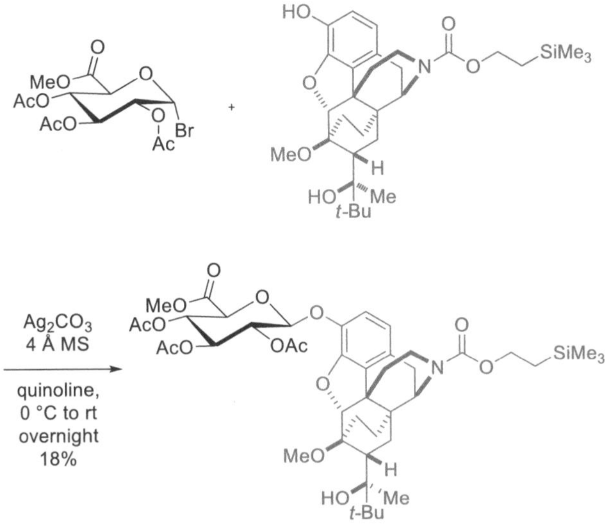

Example 5, Antitumor agents[12]

Example 6, C$_3$-Neoglycosides of digoxigenin with anticancer activities[13]

digoxigenin

Example 7, Macrophage-inducible C-type lectin (Mincle) receptor agonists[14]

References

1. Koenig, W.; Knorr, E. *Ber.* **1901,** *34,* 957–981.
2. Igarashi, K. *Adv. Carbohydr. Chem. Biochem.* **1977,** *34,* 243–83. (Review).
3. Schmidt, R. R. *Angew. Chem.* **1986,** *98,* 213–236.
4. Smith, A. B., III; Rivero, R. A.; Hale, K. J.; Vaccaro, H. A. *J. Am. Chem. Soc.* **1991,** *113,* 2092–2112.
5. Fürstner, A.; Radkowski, K.; Grabowski, J.; Wirtz, C.; Mynott, R. *J. Org. Chem.* **2000,** *65,* 8758–8762.
6. Yashunsky, D. V.; Tsvetkov, Y. E.; Ferguson, M. A. J.; Nikolaev, A. V. *J. Chem. Soc., Perkin Trans. 1* **2002,** 242–256.
7. Stazi, F.; Palmisano, G.; Turconi, M.; Clini, S.; Santagostino, M. *J. Org. Chem.* **2004,** *69,* 1097–1103.
8. Wimmer, Z.; Pechova, L.; Saman, D. *Molecules* **2004,** *9,* 902–912.
9. Presser, A.; Kunert, O.; Pötschger, I. *Monat. Chem.* **2006,** *137,* 365–374.
10. Schoettner, E.; Simon, K.; Friedel, M.; Jones, P. G.; Lindel, T. *Tetrahedron Lett.* **2008,** *49,* 5580–5582.
11. Fan, J.; Brown, S. M.; Tu, Z.; Kharasch, E. D. *Bioconjugate Chem.* **2011,** *22,* 752–758.
12. Cui, Y.; Xu, M.; Mao, J.; Ouyang, J.; Xu, R.; Xu, Y. *Eur. J. Med. Chem.* **2012,** *54,* 867–872.
13. Li, X.-s.; Ren, Y.-c.; Bao, Y.-z.; Liu, J.; Zhang, X.-k.; Zhang, Y.-w.; Sun, X.-L.; Yao, X.-s.; Tang, J.-S. *Eur. J. Med. Chem.* **2018,** *145,* 252–262.
14. Van Huy, L.; Tanaka, C.; Imai, T.; Yamasaki, S.; Miyamoto, T. *ACS Med. Chem. Lett.* **2019,** *10,* 44–49.
15. Singh, Y.; Demchenko, A. V. *Chem. Eur. J.* **2020,** *26,* 1042–1051.

Krapcho Reaction

Nucleophilic decarboxylation of β-ketoesters, malonate esters, α-cyanoesters, or α-sulfonylesters.

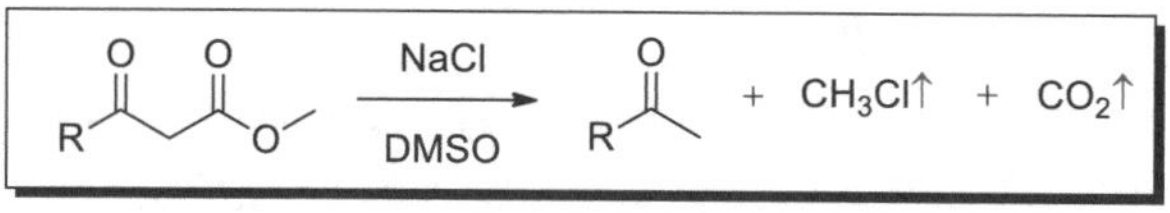

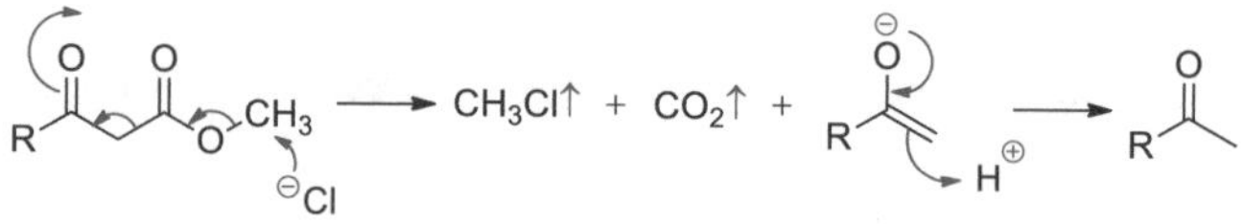

Example 1[5]

Example 2[10]

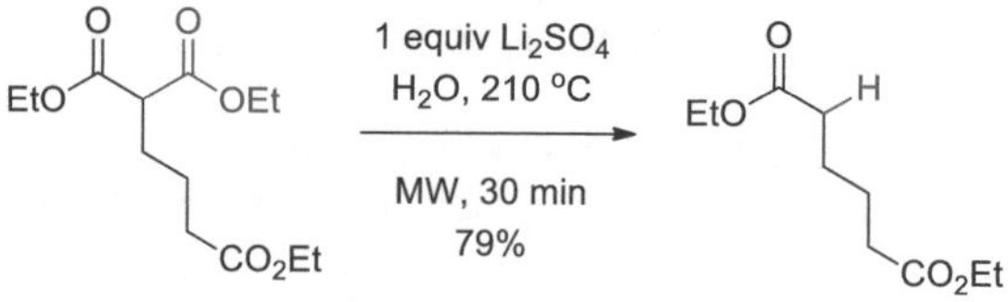

Example 3, Synthesis of homoallylic chiral nitriles[11]

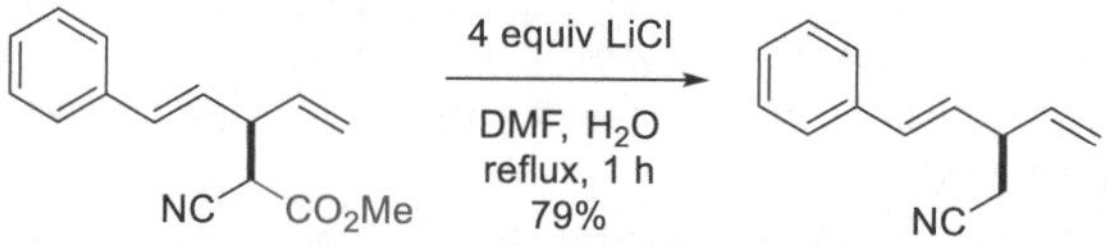

© Springer Nature Switzerland AG 2021
J. J. Li, *Name Reactions*, https://doi.org/10.1007/978-3-030-50865-4_80

Example 4, Toward production of rebamipide, an antiulcer drug[12]

Example 5, A reaction cascade with a Krapcho decarboxylation step[13]

References

1. Krapcho, A. P.; Glynn, G. A.; Grenon, B. J. *Tetrahedron Lett.* **1967**, 215–217. A. Paul Krapcho is a professor at the University of Vermont.
2. Duval, O.; Gomes, L. M. *Tetrahedron* **1989**, *45*, 4471–4476.
3. Flynn, D. L.; Becker, D. P.; Nosal, R.; Zabrowski, D. L. *Tetrahedron Lett.* **1992**, *33*, 7283–7286.
4. Martin, C. J.; Rawson, D. J.; Williams, J. M. J. *Tetrahedron: Asymmetry* **1998**, *9*, 3723–3730.
5. Gonzalez-Gomez, J. C.; Uriarte, E. *Synlett* **2002**, 2095–2097.
6. Bridges, N. J.; Hines, C. C.; Smiglak, M.; Rogers, R. D. *Chem. Eur. J.* **2007**, *13*, 207–5212.
7. Poon, P. S.; Banerjee, A. K.; Laya, M. S. *J. Chem. Res.* **2011**, *35*, 67–73. (Review).
8. Farran, D.; Bertrand, P. *Synth. Commun.* **2012**, *42*, 989–1001.
9. Adepu, R.; Rambabu, D.; Prasad, B.; Meda, C. L. T.; Kandale, A.; Rama Krishna, G.; Malla Reddy, C.; Chennuru, L. N. *Org. Biomol. Chem.* **2012**, *10*, 5554–5569.
10. Mason, J. D.; Murphree, S. S. *Synlett* **2013**, *24*, 1391–1394.
11. Matsunami, A.; Takizawa, K.; Sugano, S.; Yano, Y.; Sato, H.; Takeuchi, R. *J. Org. Chem.* **2018**, *83*, 12239–12246.
12. Babu, P. K.; Bodireddy, M. R.; Puttaraju, Re. C.; Vagare, D.; Nimmakayala, R.; Surineni, N.; Gajula, M. R.; Kumar, P. *Org. Process Res. Dev.* **2018**, *22*, 773–779.
13. Sundaravelu, N.; Sekar, G. *Org. Lett.* **2019**, *21*, 6648–6652.
14. Alvarenga, N.; Payer, S. E.; Petermeier, P.; Kohlfuerst, C.; Meleiro Porto, A. L.; Schrittwieser, J. H.; Kroutil, W. *ACS Catal.* **2020**, *10*, 1607–1620.

Kröhnke Pyridine Synthesis

Pyridines from α-pyridinium methyl ketone salts and α,β-unsaturated ketones.

The ketone is more reactive than the enone

Example 1[1b]

© Springer Nature Switzerland AG 2021

J. J. Li, *Name Reactions*, https://doi.org/10.1007/978-3-030-50865-4_81

Example 2[4]

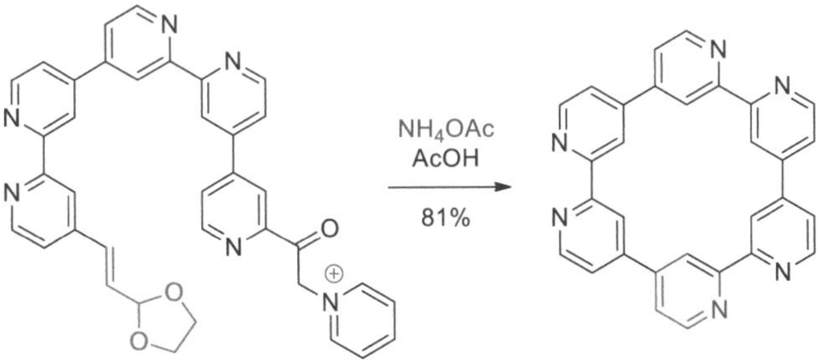

NH₄OAc
AcOH
────────→
81%

Example 3[6]

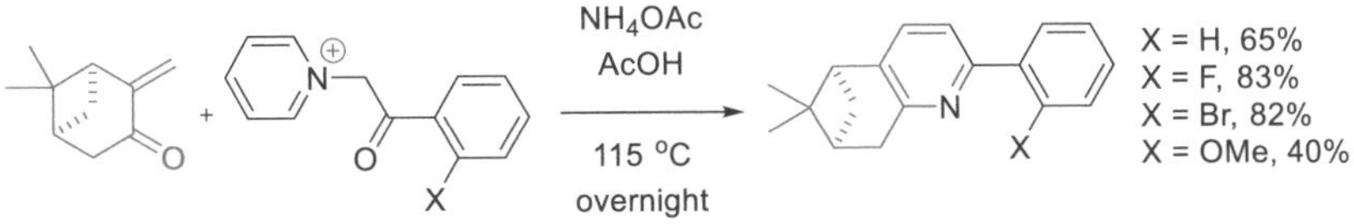

NH₄OAc
AcOH
────────→
115 °C
overnight

X = H, 65%
X = F, 83%
X = Br, 82%
X = OMe, 40%

Example 4, For total synthesis of lycopodium alkaloid (+)-lycopladin A[10]

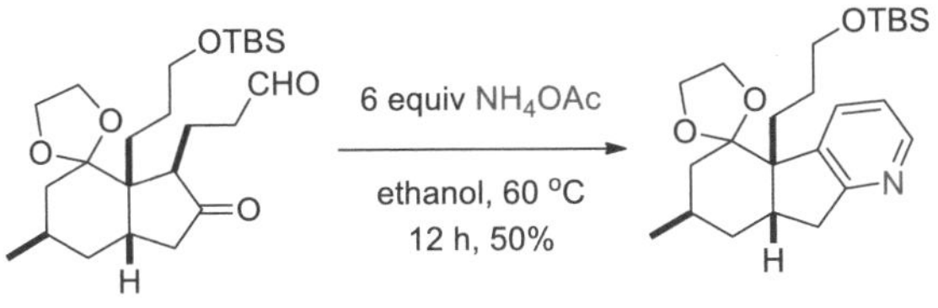

6 equiv NH₄OAc
────────→
ethanol, 60 °C
12 h, 50%

Example 5, Mining a Kröhnke pyridine library for anti-arenavirus activities[12]

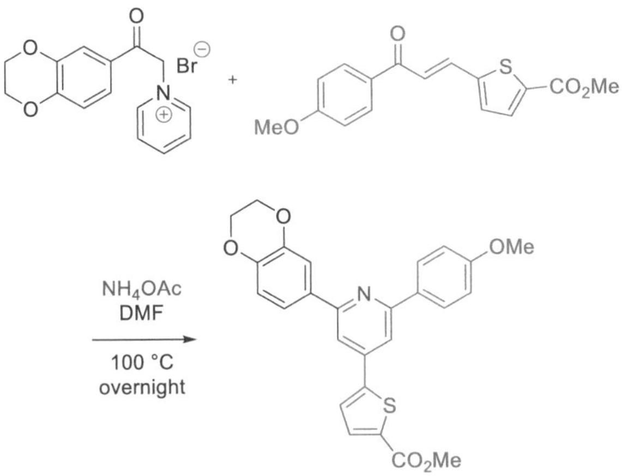

NH₄OAc
DMF
────────→
100 °C
overnight

Example 6, Bipyrido-fused coumarins with antimicrobial activities[14]

References

1. (a) Zecher, W.; Kröhnke, F. *Ber.* **1961**, *94*, 690–697. (b) Kröhnke, F.; Zecher, W. *Angew. Chem.* **1962**, *74*, 811–817. (c) Kröhnke, F. *Synthesis* **1976**, 1–24. (Review).
2. Potts, K. T.; Cipullo, M. J.; Ralli, P.; Theodoridis, G. *J. Am. Chem. Soc.* **1981**, *103*, 3584–3585, 3585–3586.
3. Newkome, G. R.; Hager, D. C.; Kiefer, G. E. *J. Org. Chem.* **1986**, *51*, 850–853.
4. Kelly, T. R.; Lee, Y.-J.; Mears, R. J. *J. Org. Chem.* **1997**, *62*, 2774–2781.
5. Bark, T.; Von Zelewsky, A. *Chimia* **2000**, *54*, 589–592.
6. Malkov, A. V.; Bella, M.; Stara, I. G.; Kocovsky, P. *Tetrahedron Lett.* **2001**, *42*, 3045–3048.
7. Cave, G. W. V.; Raston, C. L. *J. Chem. Soc., Perkin Trans. 1* **2001**, 3258–3264.
8. Malkov, A. V.; Bell, M.; Vassieu, M.; Bugatti, V.; Kocovsky, P. *J. Mol. Cat. A: Chem.* **2003**, *196*, 179–186.
9. Galatsis, P. *Kröhnke Pyridine Synthesis*. In *Name Reactions in Heterocyclic Chemistry*; Li, J. J., Ed.; Wiley: Hoboken, NJ, **2005**, 311–313. (Review).
10. Xu, T.; Luo, X.-L.; Yang, Y.-R. *Tetrahedron Lett.* **2013**, *54*, 2858–2860.
11. Allais, C.; Grassot, J.-M.; Rodriguez, J.; Constantieux, T. *Chem. Rev.* **2014**, *114*, 10829–10868. (Review).
12. Miranda, P. O.; Cubitt, B.; Jacob, N. T.; Janda, K. D.; de la Torre, J. C. *ACS Infect. Dis.* **2018**, *4*, 815–824.
13. Conlon, I. L.; Van Eker, D.; Abdelmalak, S.; Murphy, W. A.; Bashir, H.; Sun, M.; Chauhan, J.; Varney, K. M.; Dodoy-Ruiz, R.; Wilder, P. T.; Fletcher, S. *Bioorg. Med. Chem. Lett.* **2018**, *28*, 1949–1953.
14. Giri, R. R.; Brahmbhatt, D. I. *J. Heterocycl. Chem.* **2019**, *56*, 2630–2636.
15. Bentzinger, G.; Pair, E.; Guillon, J.; Marchivie, M.; Mullie, C.; Agnamey, P.; Dassonville-Klimpt, A.; Sonnet, P. *Tetrahedron* **2020**, *76*, 131088.

Kumada Reaction

The Kumada cross-coupling reaction (also known as Kumada–Tamao–Corriu coupling, also occasionally known as the Kharasch cross-coupling reaction) was originally reported as the nickel-catalyzed cross-coupling of Grignard reagents with aryl- or alkenyl halides. It has subsequently been developed to encompass the coupling of organolithium or organomagnesium compounds with aryl-, alkenyl- or alkyl halides, catalyzed by either nickel or palladium. These reactions follow a general mechanistic catalytic cycle as shown below. There are slight variations for the Hiyama and Suzuki reactions, for which an additional activation step is required for the transmetalation to occur.

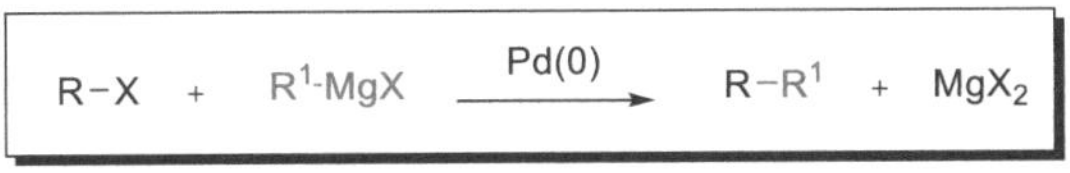

$$R-X \ + \ R^1\text{-MgX} \ \xrightarrow{\ Pd(0)\ } \ R-R^1 \ + \ MgX_2$$

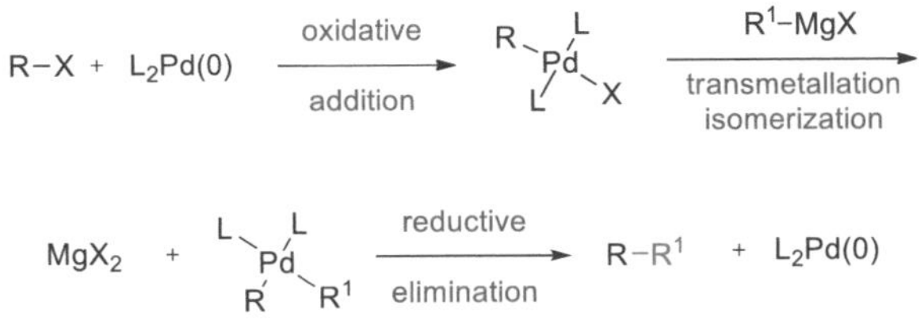

$R-X \ + \ L_2Pd(0) \ \xrightarrow[\text{addition}]{\text{oxidative}} \ R\underset{L}{\overset{L}{\diagdown}}\!\!Pd\!\!\diagdown X \ \xrightarrow[\substack{\text{transmetallation} \\ \text{isomerization}}]{R^1\text{-MgX}}$

$MgX_2 \ + \ \underset{R\ \ R^1}{\overset{L\ \ L}{Pd}} \ \xrightarrow[\text{elimination}]{\text{reductive}} \ R-R^1 \ + \ L_2Pd(0)$

The catalytic cycle:

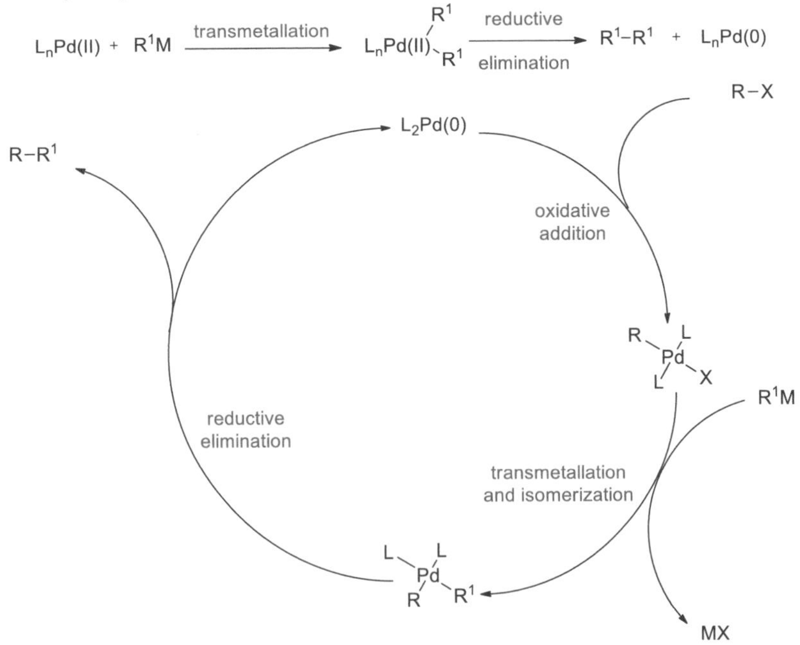

$L_nPd(II) \ + \ R^1M \ \xrightarrow{\text{transmetallation}} \ L_nPd(II)\!\!\underset{R^1}{\overset{R^1}{\diagup}} \ \xrightarrow[\text{elimination}]{\text{reductive}} \ R^1\!-\!R^1 \ + \ L_nPd(0)$

Example 1[2]

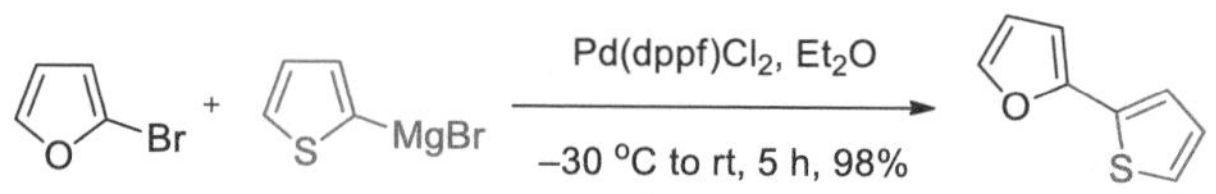

Example 2[3]

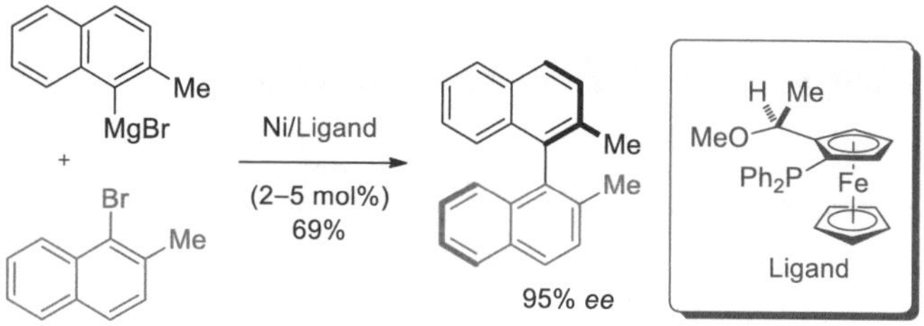

95% *ee*

Example 3[5]

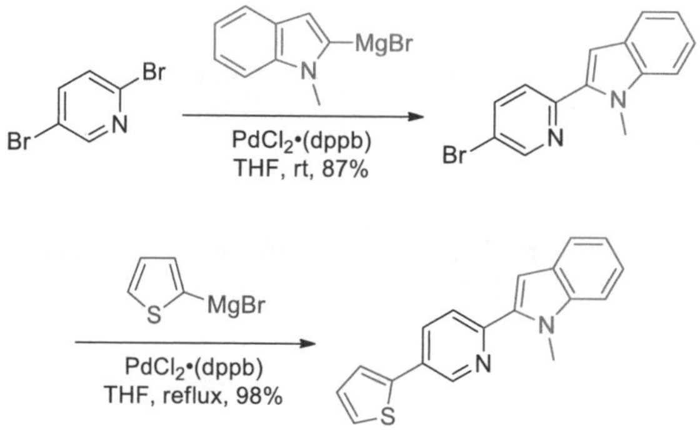

Example 4[8]

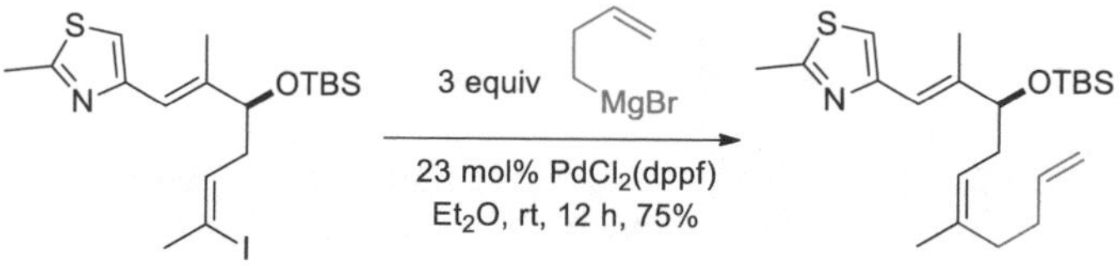

Example 5[9]

Example 6, Nickel-catalyzed Kumada reaction of tosylalkanes[11]

Example 7, Redox-active esters in Fe-catalyzed C–C coupling (DMPU = N,N'-dimethylpropyleneurea)[13]

Example 8, Toward the synthesis of GDC-0994, an extracellular signal-regulated kinase (ERK) inhibitor[14]

PEPPSI = pyridine-enhanced pre-catalyst preparation, stabilization, and initiation;
IPr = diidopropyl-phenylimidazolium derivative

Example 9, Benzyl ethers as the substrates of the Kumada reaction[15]

inversion of configuration
at the benzylic position

References

1. Tamao, K.; Sumitani, K.; Kiso, Y.; Zembayashi, M.; Fujioka, A.; Kodma, S.-i.; Nakajima, I.; Minato, A.; Kumada, M. *Bull. Chem. Soc. Jpn.* **1976**, *49*, 1958–1969.
2. Carpita, A.; Rossi, R.; Veracini, C. A. *Tetrahedron* **1985**, *41*, 1919–1929.
3. Hayashi, T.; Hayashizaki, K.; Kiyoi, T.; Ito, Y. *J. Am. Chem. Soc.* **1988**, *110*, 8153–8156.
4. Kalinin, V. N. *Synthesis* **1992**, 413–432. (Review).
5. Meth-Cohn, O.; Jiang, H. *J. Chem. Soc., Perkin Trans. 1* **1998**, 3737–3746.
6. Stanforth, S. P. *Tetrahedron* **1998**, *54*, 263–303. (Review).
7. Huang, J.; Nolan, S. P. *J. Am. Chem. Soc.* **1999**, *121*, 9889–9890.
8. Rivkin, A.; Njardarson, J. T.; Biswas, K.; Chou, T.-C.; Danishefsky, S. J. *J. Org. Chem.* **2002**, *67*, 7737–7740.
9. William, A. D.; Kobayashi, Y. *J. Org. Chem.* **2002**, *67*, 8771–8782.
10. Fuchter, M. J. *Kumada Cross-Coupling Reaction*. In *Name Reactions for Homologations-Part I*; Li, J. J., Ed.; Wiley: Hoboken, NJ, **2009**, pp 47–69. (Review).
11. Wu, J.-C.; Gong, L.-B.; Xia, Y.; Song, R.-J.; Xie, Y.-X.; Li, J.-H. *Angew. Chem. Int. Ed.* **2012**, *51*, 9909–9913.
12. Handa, S.; Arachchige, Y. L. N. M.; Slaughter, L. M. *J. Org. Chem.* **2013**, *78*, 5694–5699.
13. Toriyama, F.; Cornella, J.; Wimmer, L.; Chen, T.-G.; Dixon, D. D.; Creech, G.; Baran, P. S. *J. Am. Chem. Soc.* **2016**, *138*, 11132–11135.
14. Xin, L.; Wong, N.; Jost, V.; Fantasia, S.; Sowell, C. G.; Gosselin, F. *Org. Process Res. Dev.* **2017**, *21*, 1320–1325.
15. Chen, P.-P.; Lucas, E. L.; Greene, M. A.; Zhang, S.-Q.; Tollefson, E. J.; Erickson, L. W.; Taylor, B. L. H.; Jarvo, E. R.; Hong, X. *J. Am. Chem. Soc.* **2019**, *141*, 5835–5855.
16. Dawson, D. D.; Oswald, V. F.; Borovik, A. S.; Jarvo, E. R. *Chem. Eur. J.* **2020**, *26*, 3044–3048.

Lawesson's Reagent

2,4-Bis(4-methoxyphenyl)-1,3-dithiadiphosphetane-2,4-disulfide transforms the carbonyl groups of aldehydes, ketones, amides, lactams, esters and lactones into the corresponding thiocarbonyl compounds. *Cf.* Knorr thiophene synthesis.

Example 1, Double Lawesson reaction[4]

Example 2[5]

© Springer Nature Switzerland AG 2021
J. J. Li, *Name Reactions*, https://doi.org/10.1007/978-3-030-50865-4_83

Example 3, Thiophene from dione[8]

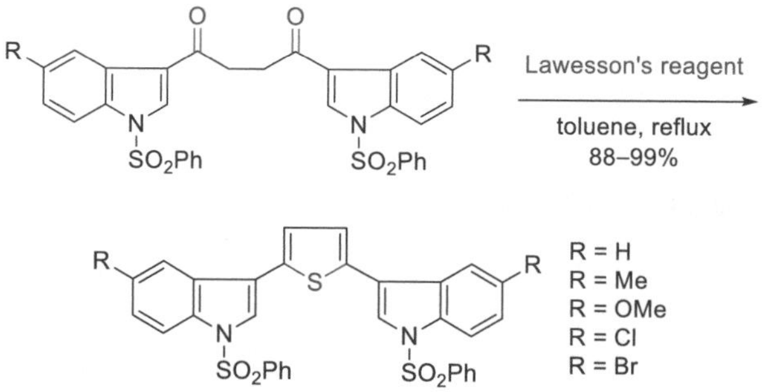

R = H
R = Me
R = OMe
R = Cl
R = Br

Example 4, Latam is more reactive than lactone and enone[10]

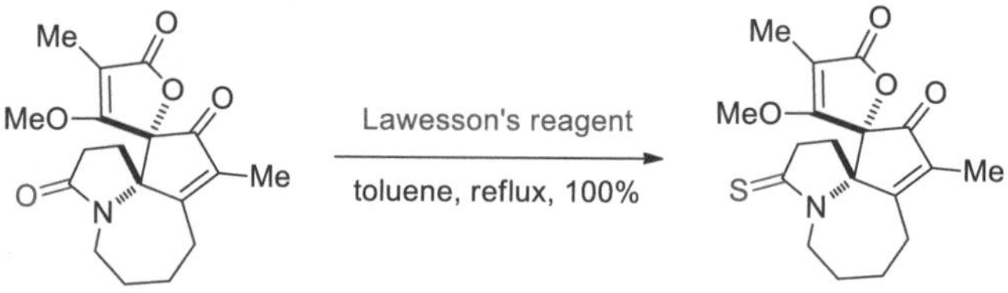

Example 5[11]

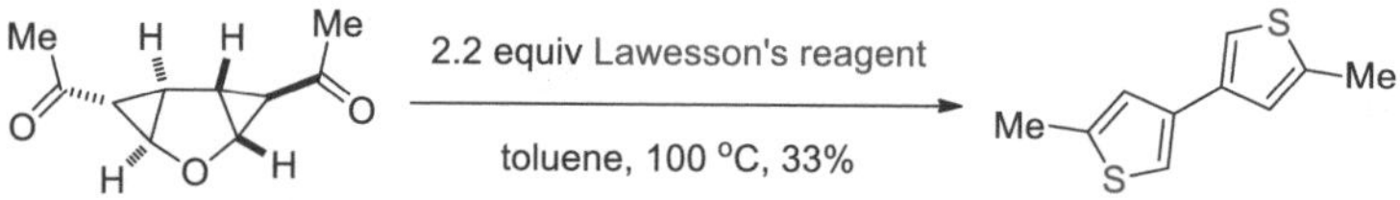

Example 6, Thiophosphinyl dipeptide isosteres (TDIs) to replace phosphinyl dipeptide isosteres (PDIs)[12]

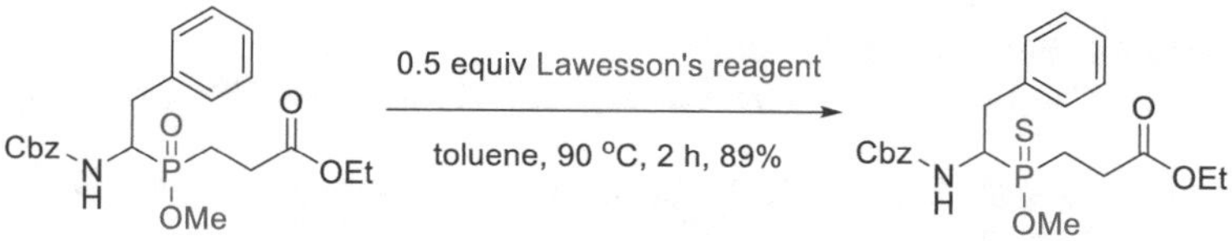

Example 7, A reagent prepared from P_4S_{10} and pyridine as a thionation reagent[13]

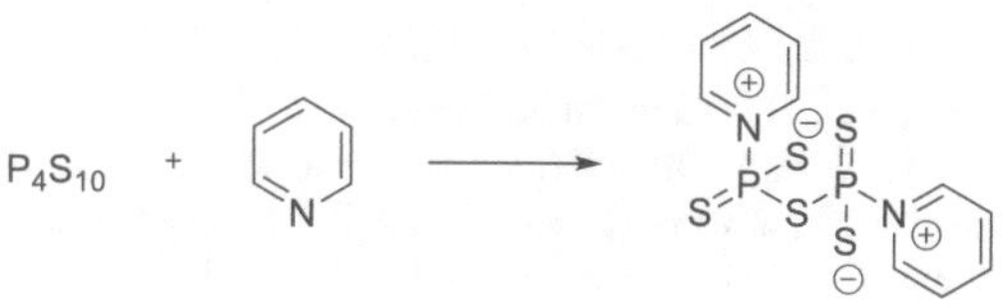

Example 8, Synthesis of 1,4-thiazepines[14]

References

1. Scheibye, S.; Shabana, R.; Lawesson, S. O.; Rømming, C. *Tetrahedron* **1982,** *38,* 993–1001. Sven-Olov Lawesson (1926–1985) was a Swidish chemist.
2. Navech, J.; Majoral, J. P.; Kraemer, R. *Tetrahedron Lett.* **1983,** *24,* 5885–5886.
3. Cava, M. P.; Levinson, M. I. *Tetrahedron* **1985,** *41,* 5061–5087. (Review).
4. Nicolaou, K. C.; Hwang, C.-K.; Duggan, M. E.; Nugiel, D. A.; Abe, Y.; et al. *J. Am. Chem. Soc.* **1995,** *117,* 10227–10238.
5. Kim, G.; Chu-Moyer, M. Y.; Danishefsky, S. J. *J. Am. Chem. Soc.* **1990,** *112,* 2003–2005.
6. Luheshi, A.-B. N.; Smalley, R. K.; Kennewell, P. D.; Westwood, R. *Tetrahedron Lett.* **1990,** *31,* 123–127.
7. Ishii, A.; Yamashita, R.; Saito, M.; Nakayama, J. *J. Org. Chem.* **2003,** *68,* 1555–1558.
8. Diana, P.; Carbone, A.; Barraja, P.; Montalbano, A.; Martorana, A.; Dattolo, G.; Gia, O.; Dalla Via, L.; Cirrincione, G. *Bioorg. Med. Chem. Lett.* **2007,** *17,* 2342–2346.
9. Ozturk, T.; Ertas, E.; Mert, O. *Chem. Rev.* **2007,** *107,* 5210–5278. (Review).
10. Taniguchi, T.; Ishibashi, H. *Tetrahedron* **2008,** *64,* 8773–8779.
11. de Moreira, D. R. M. *Synlett* **2008,** 463–464. (Review).
12. Vassiliou, S.; Tzouma, E. *J. Org. Chem.* **2013,** *78,* 10069–10076.
13. Kingi, N.; Bergman, J. *J. Org. Chem.* **2016,** *81,* 7711–7716.
14. Kelgokmen, Y.; Zora, M. *J. Org. Chem.* **2018,** *83,* 8376–8389.

Leuckart–Wallach Reaction

Amine synthesis from reductive amination of a ketone and an amine in the presence of excess formic acid, which serves as the reducing reagent by delivering a hydride. When the ketone is replaced by formaldehyde, it becomes the Eschweiler–Clarke reductive amination.

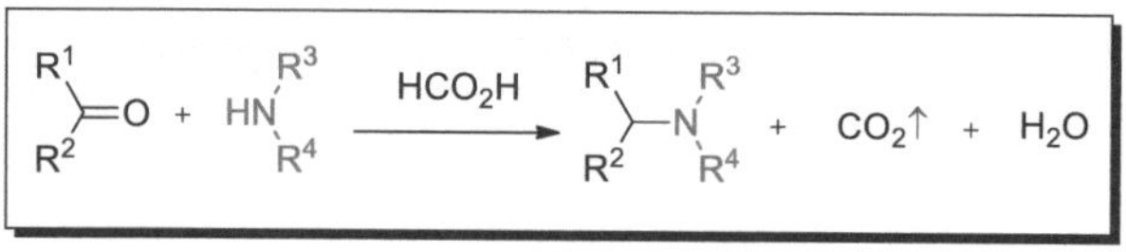

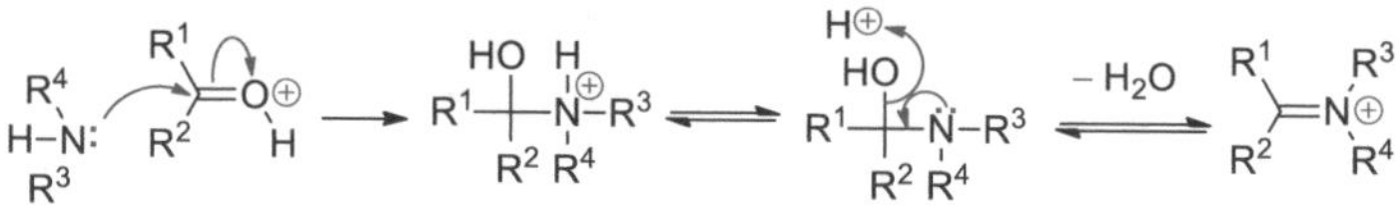

gem-aminoalcohol; iminium ion intermediate

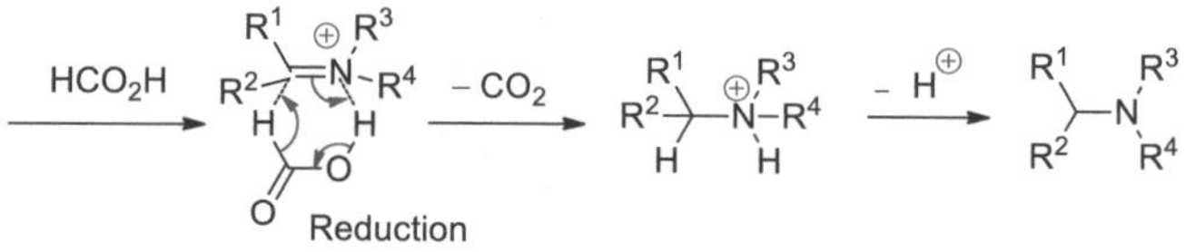

Example 1[4]

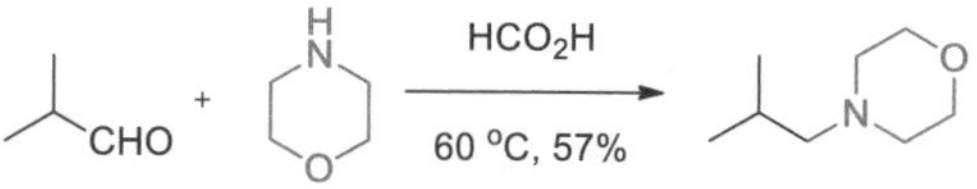

Example 2[6]

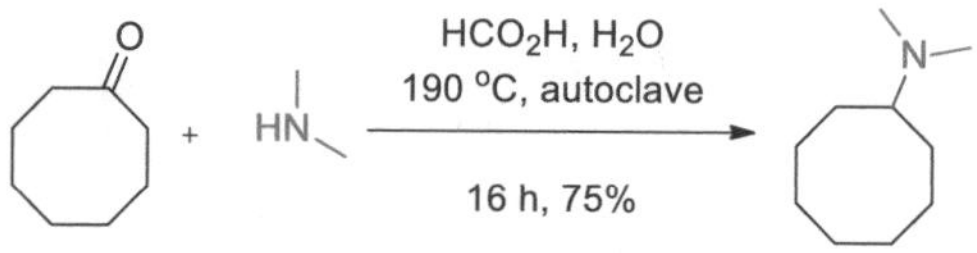

Example 3[7]

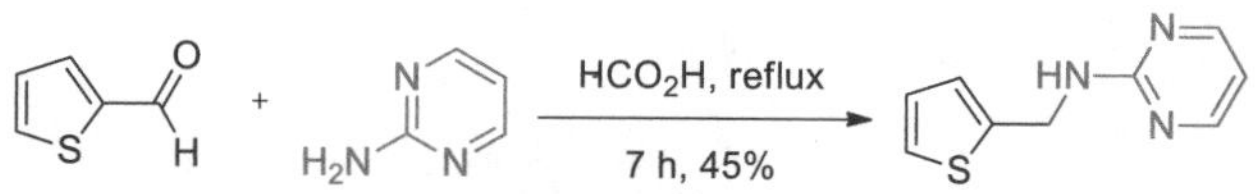

© Springer Nature Switzerland AG 2021
J. J. Li, *Name Reactions*, https://doi.org/10.1007/978-3-030-50865-4_84

Example 4[8]

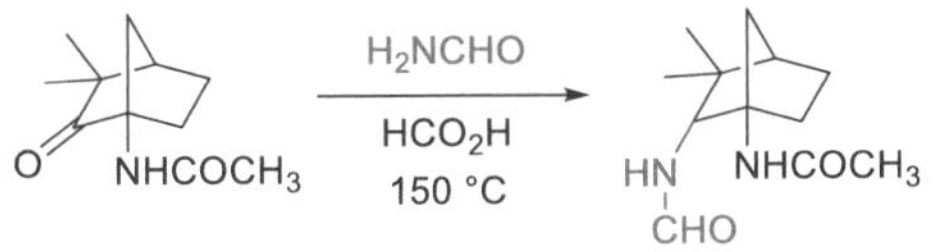

An unexpected intramolecular transamidation *via* a Wagner–Meerwein shift after the Leuckart–Wallach reaction

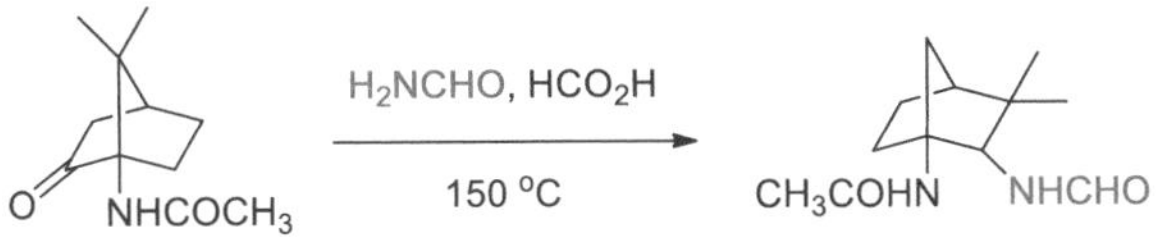

Example 5[12]

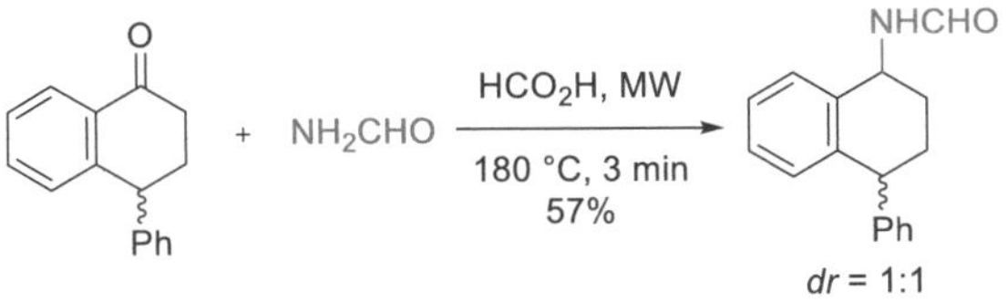

Example 6[13]

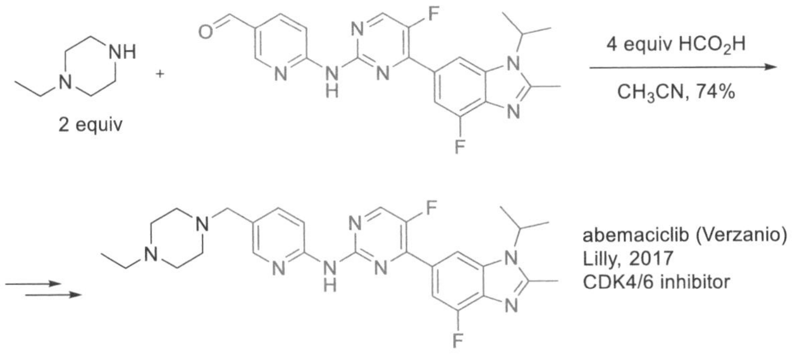

abemaciclib (Verzanio)
Lilly, 2017
CDK4/6 inhibitor

Example 7, Flow chemistry[14]

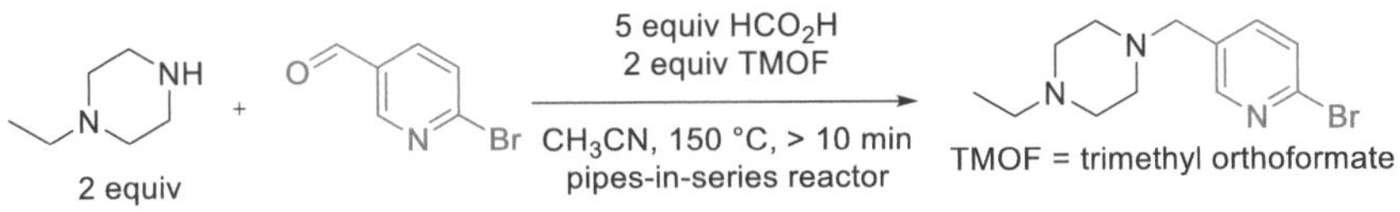

TMOF = trimethyl orthoformate

abemaciclib (Verzanio)
Lilly, 2017
CDK4/6 inhibitor

References

1. Leuckart, R. *Ber.* **1885,** *18,* 2341–2344. Carl L. R. A. Leuckart (1854–1889) was born in Giessen, Germany. After studying under Bunsen, Kolbe, and von Baeyer, he became an assistant professor at Göttingen. Unfortunately, chemistry lost a brilliant contributor by his sudden death at age 35 as a result of a fall in his parent's house.
2. Wallach, O. *Ann.* **1892,** *272,* 99. Otto Wallach (1847–1931), born in Königsberg, Prussia, studied under Wöhler and Hofmann. He was the director of the Chemical Institute at Göttingen from 1889 to 1915. His book "Terpene und Kampfer" served as the foundation for future work in terpene chemistry. Wallach was awarded the Nobel Prize in Chemistry in 1910 for his work on alicyclic compounds.
3. Moore, M. L. *Org. React.* **1949,** *5,* 301–330. (Review).
4. DeBenneville, P. L.; Macartney, J. H. *J. Am. Chem. Soc.* **1950,** *72,* 3073–3075.
5. Lukasiewicz, A. *Tetrahedron* **1963,** *19,* 1789–1799. (Mechanism).
6. Bach, R. D. *J. Org. Chem.* **1968,** *33,* 1647–1649.
7. Musumarra, G.; Sergi, C. *Heterocycles* **1994,** *37,* 1033–1039.
8. Martínez, A. G.; Vilar, E. T.; Fraile, A. G.; Ruiz, P. M.; San Antonio, R. M.; Alcazar, M. P. M. *Tetrahedron: Asymmetry* **1999,** *10,* 1499–1505.
9. Kitamura, M.; Lee, D.; Hayashi, S.; Tanaka, S.; Yoshimura, M. *J. Org. Chem.* **2002,** *67,* 8685–8687.
10. Brewer, A. R. E. *Leuckart–Wallach reaction.* In *Name Reactions for Functional Group Transformations;* Li, J. J., Ed.; Wiley: Hoboken, NJ, **2007,** pp 451–455. (Review).
11. Muzalevskiy, V. M.; Nenajdenko, V. G.; Shastin, A. V.; Balenkova, E. S.; Haufe, G. *J. Fluorine Chem.* **2008,** *129,* 1052–1055.
12. Neochoritis, C.; Stotani, S.; Mishra, B.; Dömling, A. *Org. Lett.* **2015,** *17,* 2002–2005.
13. Frederick, M. O.; Kjell, D. P. *Tetrahedron Lett.* **2015,** *56,* 949–951.
14. Frederick, M. O.; Pietz, M. A.; Kjell, D. P.; Richey, R. N.; Tharp, G. A.; Touge, T.; Yokoyama, N.; Kida, M.; Matsuo, T. *Org. Process Res. Dev.* **2017,** *21,* 1447–1451.

Lossen Rearrangement

The Lossen rearrangement involves the generation of an isocyanate via thermal or base-mediated rearrangement of an activated hydroxamate which can be generated from the corresponding hydroxamic acid. Activation of the hydroxamic acid can be achieved through O-acylation, O-arylation, chlorination, or O-sulfonylation. Such hydroxamic acids can also be activated using polyphosphoric acid, carbodiimide, Mitsunobu conditions, or silyation. The product of the Lossen rearrangement, an isocyanate can be subsequently converted to a urea or an amine resulting in the net loss of one carbon atom relative to the starting hydroxamic acid.

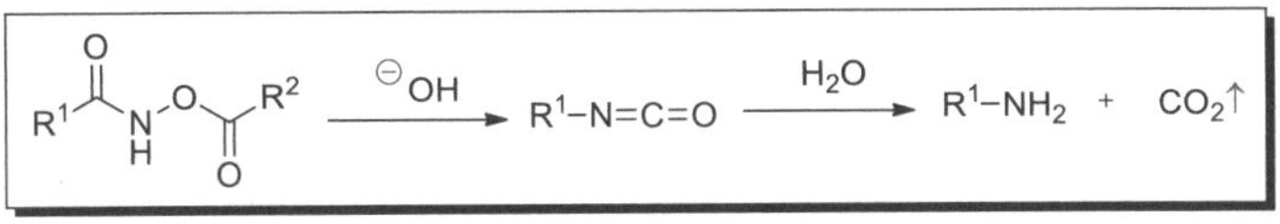

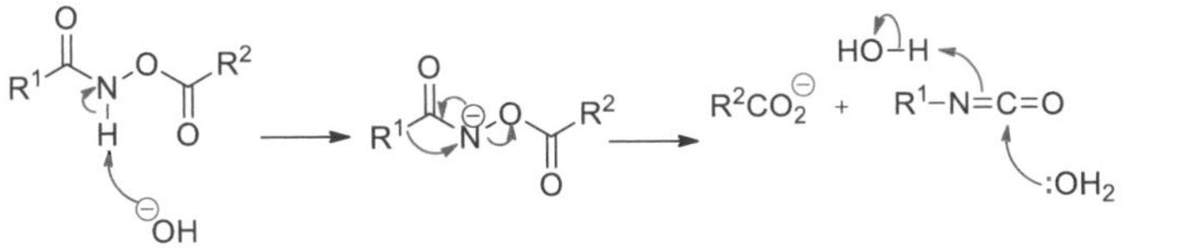

isocyanate intermediate

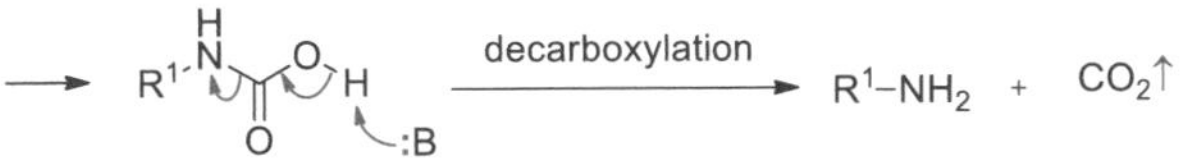

Example 1[6]

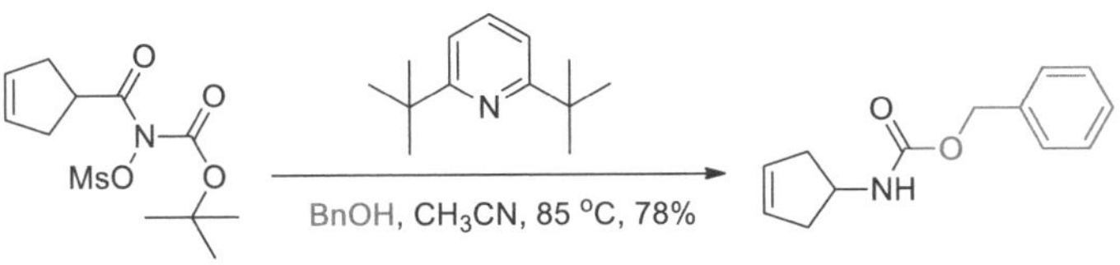

BnOH, CH₃CN, 85 °C, 78%

Example 2[7]

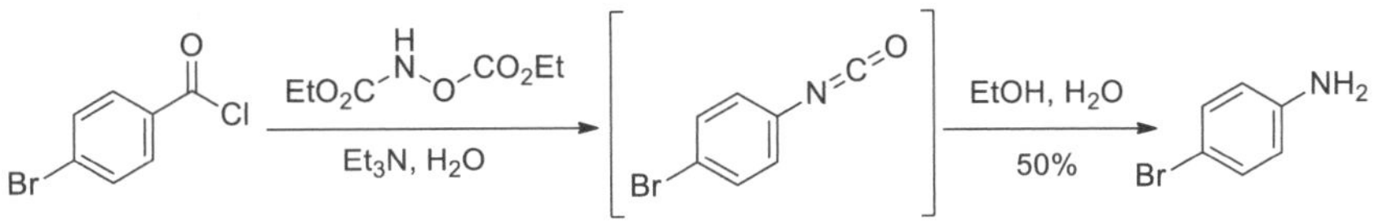

Example 3[8]

Example 4[9]

Example 5[11]

Example 6, Tandem S_NAr/Lossen rearrangement sequence[12]

Example 7[13]

betulin derivative

46.5 kg scale!

Example 8, Aza-Lossen rearrangement[14]

References

1. Lossen, W. *Ann.* **1872,** *161,* 347. Wilhelm C. Lossen (1838–1906) was born in Kreuznach, Germany. After his Ph.D. studies at Göttingen in 1862, he embarked on his independent academic career, and his interests centered on hydroxyamines.
2. Bauer, L.; Exner, O. *Angew. Chem. Int. Ed.* **1974,** *13,* 376.
3. Lipczynska-Kochany, E. *Wiad. Chem.* **1982,** *36,* 735–756.
4. Casteel, D. A.; Gephart, R. S.; Morgan, T. *Heterocycles* **1993,** *36,* 485–495.
5. Zalipsky, S. *Chem. Commun.* **1998,** 69–70.
6. Stafford, J. A.; Gonzales, S. S.; Barrett, D. G.; Suh, E. M.; Feldman, P. L. *J. Org. Chem.* **1998,** 63, 10040–10044.
7. Anilkumar, R.; Chandrasekhar, S.; Sridhar, M. *Tetrahedron Lett.* **2000,** *41,* 5291–5293.
8. Abbady, M. S.; Kandeel, M. M.; Youssef, M. S. K. *Phosphorous, Sulfur and Silicon* **2000,** *163,* 55–64.
9. Ohmoto, K.; Yamamoto, T.; Horiuchi, T.; Kojima, T.; Hachiya, K.; Hashimoto, S.; Kawamura, M.; Nakai, H.; Toda, M. *Synlett* **2001,** 299–301.
10. Choi, C.; Pfefferkorn, J. A. *Lossen rearrangement.* In *Name Reactions for Homologations-Part II*; Li, J. J., Ed.; Wiley: Hoboken, NJ, **2009,** pp 200–209. (Review).
11. Yoganathan, S.; Miller, S. J. *Org. Lett.* **2013,** *15,* 602–605.
12. Morrison, A. E.; Hoang, T. T.; Birepinte, M.; Dudley, G. B. *Org. Lett.* **2017,** *19,* 858–861.
13. Strotman, N. A.; Ortiz, A.; Savage, S. A.; Wilbert, C. R.; Ayers, S.; Kiau, S. *J. Org. Chem.* **2017,** *82,* 4044–4049.
14. Polat, D. E.; Brzezinski, D. D.; Beauchemin, A. M. *Org. Lett.* **2019,** *21,* 4849–4852.
15. Tan, J.-F.; Bormann, C. T.; Severin, K.; Cramer, N. *ACS Catal.* **2020,** *10,* 3790–3796.

McMurry Coupling

Olefination of carbonyls with low-valent titanium such as Ti(0) derived from TiCl$_3$/LiAlH$_4$. A single-electron process.

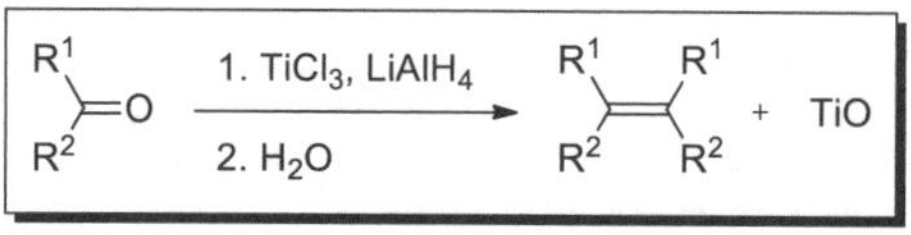

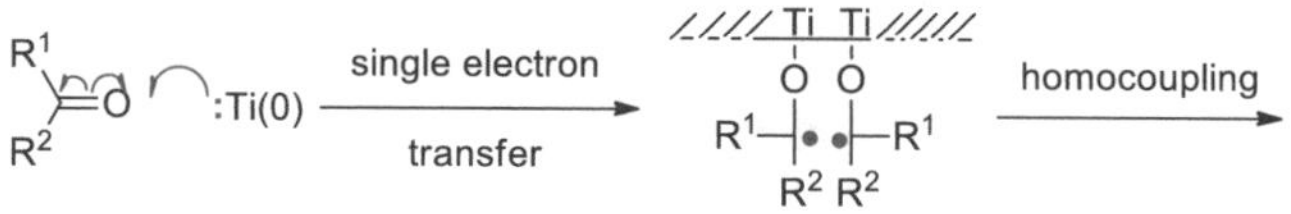

radical anion intermediate

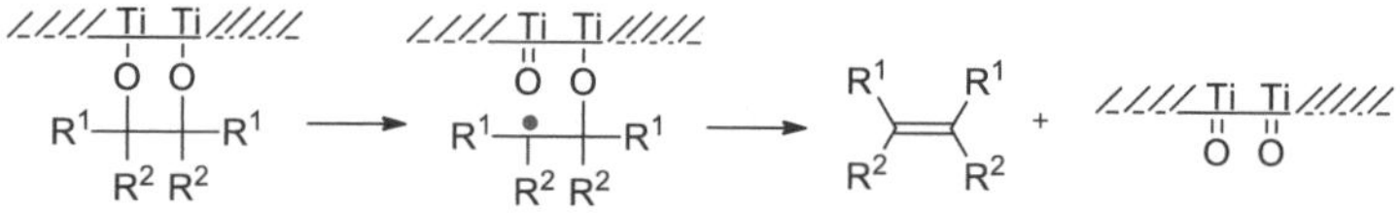

oxide-coated titanium surface

Example 1, Cross-McMurry coupling[7]

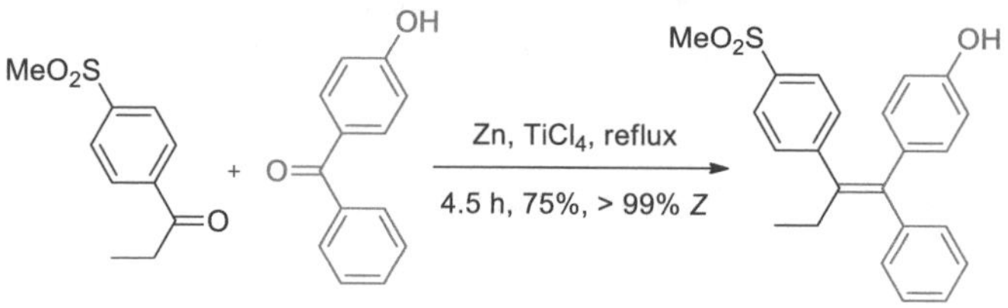

Example 2, Homo-McMurry coupling[8]

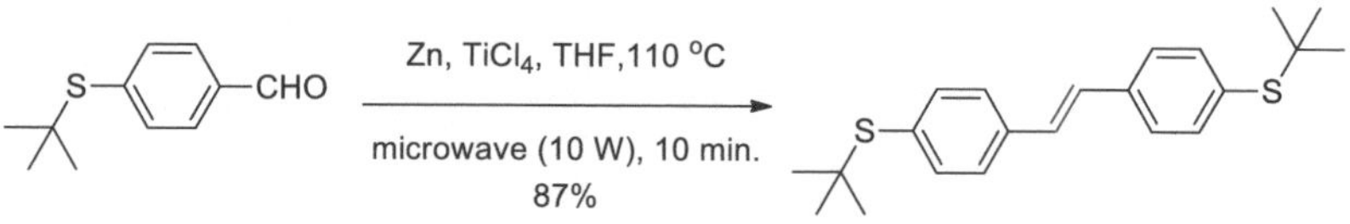

J. J. Li, *Name Reactions*, https://doi.org/10.1007/978-3-030-50865-4_86

Example 3, Cross-McMurry coupling[9]

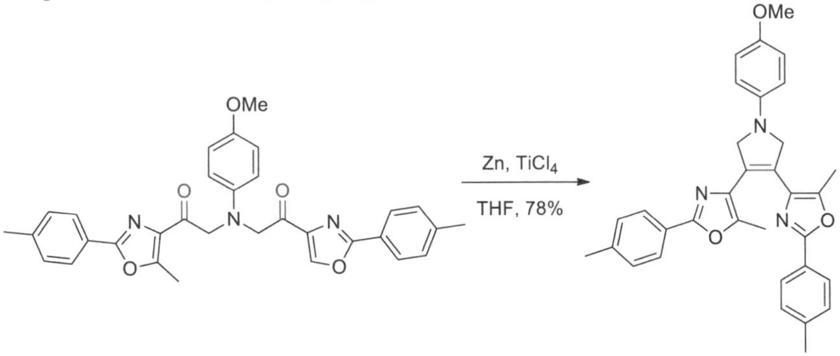

Example 4, Cross-McMurry coupling[10]

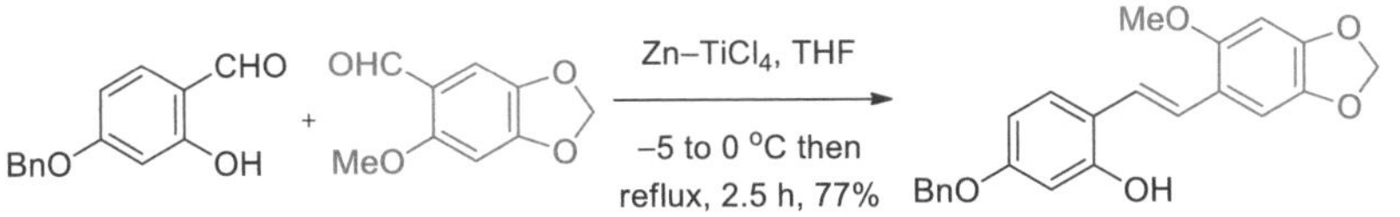

Example 5[12]

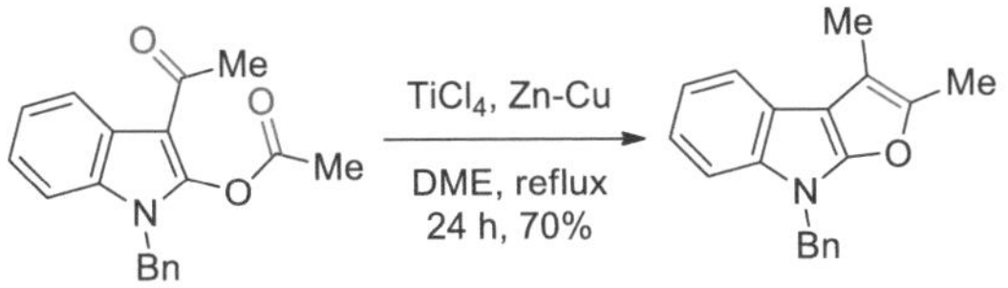

Example 6, Intramolecular McMurry coupling[13]

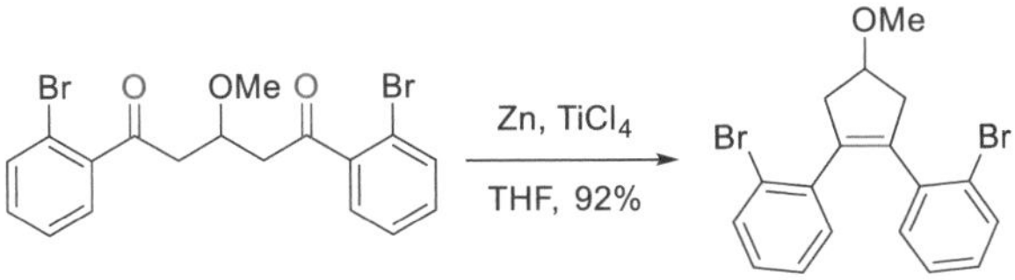

Example 7[14]

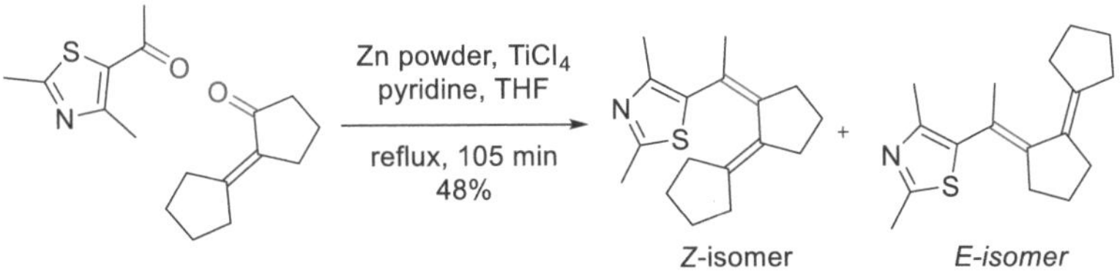

Z-isomer   E-isomer

Example 8, Tetra-indole synthesis[15]

6 equiv Zn powder
3 equiv TiCl$_4$, THF
reflux, 56%

References

1. (a) McMurry, J. E.; Fleming, M. P. *J. Am. Chem. Soc.* **1974**, *96*, 4708–4712. (b) McMurry, J. E. *Chem. Rev.* **1989**, *89*, 1513–1524. (Review).
2. Hirao, T. *Synlett* **1999**, 175–181.
3. Sabelle, S.; Hydrio, J.; Leclerc, E.; Mioskowski, C.; Renard, P.-Y. *Tetrahedron Lett.* **2002**, *43*, 3645–3648.
4. Williams, D. R.; Heidebrecht, R. W., Jr. *J. Am. Chem. Soc.* **2003**, *125*, 1843–1850.
5. Honda, T.; Namiki, H.; Nagase, H.; Mizutani, H. *Tetrahedron Lett.* **2003**, *44*, 3035–3038.
6. Ephritikhine, M.; Villiers, C. In *Modern Carbonyl Olefination* Takeda, T., Ed.; Wiley-VCH: Weinheim, Germany, **2004**, 223–285. (Review).
7. Uddin, M. J.; Rao, P. N. P.; Knaus, E. E. *Synlett* **2004**, 1513–1516.
8. Stuhr-Hansen, N. *Tetrahedron Lett.* **2005**, *46*, 5491–5494.
9. Zeng, D. X.; Chen, Y. *Synlett* **2006**, 490–492.
10. Duan, X.-F.; Zeng, J.; Zhang, Z.-B.; Zi, G.-F. *J. Org. Chem.* **2007**, *72*, 10283–10286.
11. Debroy, P.; Lindeman, S. V.; Rathore, R. *J. Org. Chem.* **2009**, *74*, 2080–2087.
12. Kumar, A. S.; Nagarajan, R. *Synthesis* **2013**, *45*, 1235–1246.
13. Connors, D. M.; Goroff, N. S. *Org. Lett.* **2016**, *18*, 4262–4265.
14. Kochi, J.-i.; Ubukata, T.; Yokoyama, Y. *J. Org. Chem.* **2018**, *83*, 10695–10700.
15. Zheng, X.; Su, R.; Wang, T.; Bin, Z.; She, Z.; Gao, G.; Yong, J. *Org. Lett.* **2019**, *21*, 797–802.
16. Tong, J.; Xia, T.; Wang, B. *Org. Lett.* **2020**, *22*, 2730–2734.

Mannich Reaction

Three-component aminomethylation from amine, aldehyde and a compound with an acidic methylene moiety.

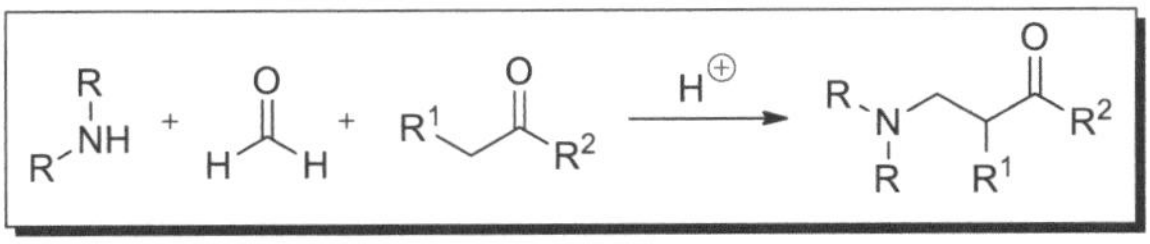

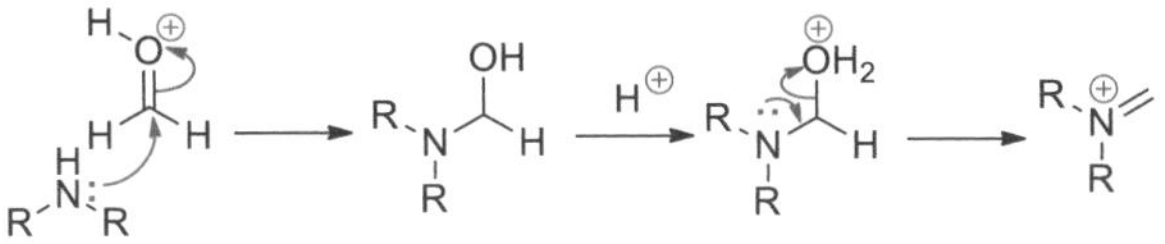

When R = Me, the $^+Me_2N=CH_2$ salt is known as *Eschenmoser's salt*

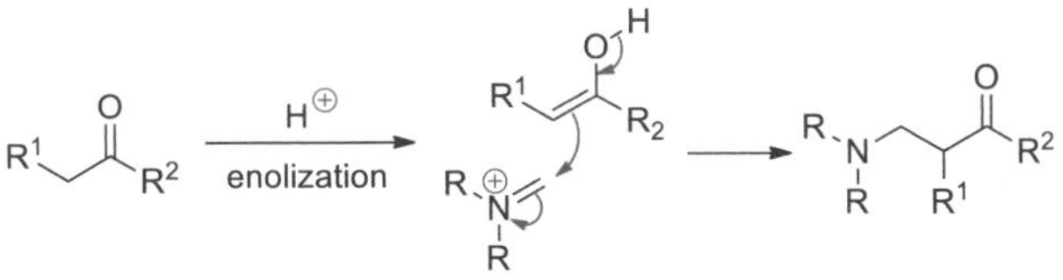

The Mannich reaction can also operate under basic conditions:

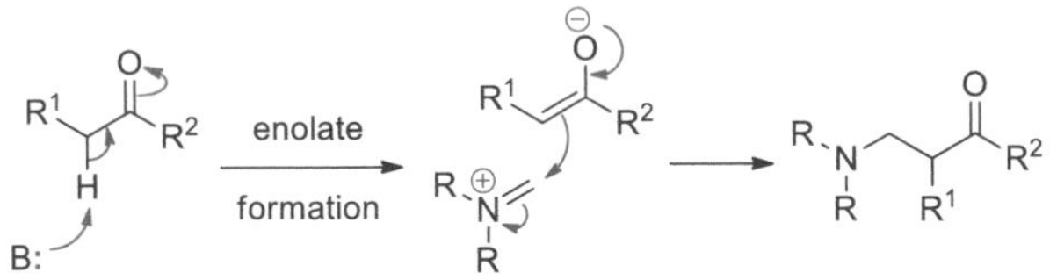

Mannich Base

Example 1, Asymmetric Mannich reaction[2]

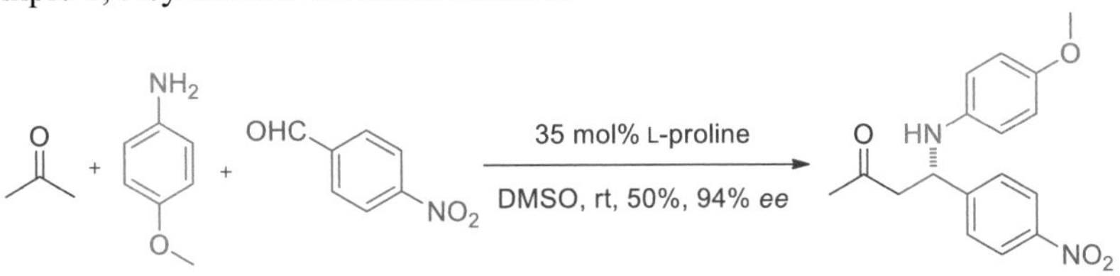

35 mol% L-proline

DMSO, rt, 50%, 94% ee

Example 2, Asymmetric Mannich-type reaction[9]

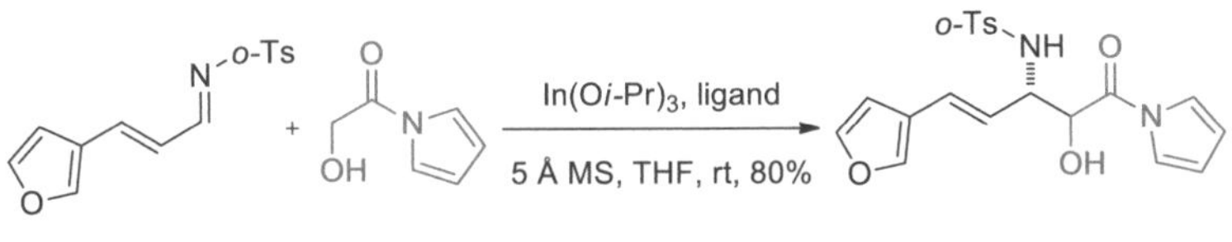

In(O*i*-Pr)$_3$, ligand

5 Å MS, THF, rt, 80%

© Springer Nature Switzerland AG 2021
J. J. Li, *Name Reactions*, https://doi.org/10.1007/978-3-030-50865-4_87

Example 3, Asymmetric Mannich-type reaction[10]

Example 4[11]

Example 5, Vinylogous Mannich reaction (VMR)[13]

Example 6, Asymmetric Mannich reaction[15]

Example 7, A zinc-mediated Mannich-type transformation of 2,2,2-trifluorodiazo-ethane[16]

Example 8, Using preformed imine[17]

dr, 12:1; 97% ee

(S,S)-L (5 mol %) =

References

1. Mannich, C.; Krösche, W. *Arch. Pharm.* **1912,** *250,* 647–667. Carl U. F. Mannich (1877–1947) was born in Breslau, Germany. After receiving a Ph.D. at Basel in 1903, he served on the faculties of Göttingen, Frankfurt and Berlin. Mannich synthesized many esters of *p*-aminobenzoic acid as local anesthetics.
2. List, B. *J. Am. Chem. Soc.* **2000,** *122,* 9336–9337.
3. Schlienger, N.; Bryce, M. R.; Hansen, T. K. *Tetrahedron* **2000,** *56,* 10023–10030.
4. Bur, S. K.; Martin, S. F. *Tetrahedron* **2001,** *57,* 3221–3242. (Review).
5. Martin, S. F. *Acc. Chem. Res.* **2002,** *35,* 895–904. (Review).
6. Padwa, A.; Bur, S. K.; Danca, D. M.; Ginn, J. D.; Lynch, S. M. *Synlett* **2002,** 851–862. (Review).
7. Notz, W.; Tanaka, F.; Barbas, C. F., III. *Acc. Chem. Res.* **2004,** *37,* 580–591. (Review).
8. Córdova, A. *Acc. Chem. Res.* **2004,** *37,* 102–112. (Review).
9. Harada, S.; Handa, S.; Matsunaga, S.; Shibasaki, M. *Angew. Chem. Int. Ed.* **2005,** *44,* 4365–4368.
10. Lou, S.; Dai, P.; Schaus, S. E. *J. Org. Chem.* **2007,** *72,* 9998–10008.
11. Hahn, B. T.; Fröhlich, R.; Harms, K.; Glorius, F. *Angew. Chem. Int. Ed.* **2008,** *47,* 9985–9988.
12. Galatsis, P. *Mannich reaction.* In *Name Reactions for Homologations-Part II*; Li, J. J., Ed.; Wiley: Hoboken, NJ, **2009,** pp 653–670. (Review).
13. Liu, X.-K.; Ye, J.-L.; Ruan, Y.-P.; Li, Y.-X.; Huang, P.-Q. *J. Org. Chem.* **2013,** *78,* 35–41.
14. Karimi, B.; Enders, D.; Jafari, E. *Synthesis* **2013,** *45,* 2769–2812. (Review).
15. Hayashi, Y.; Yamazaki, T.; Kawauchi, G.; Sato, I. *Org. Lett.* **2018,** *20,* 2391–2394.
16. Guo, R.; Lv, N.; Zhang, F.-G.; Ma, J.-A. *Org. Lett.* **2018,** *20,* 6994–6997.
17. Trost, B. M.; Hung, C.-I. J.; Kiao, Z. *J. Am. Chem. Soc.* **2019,** *141,* 16085–16092.
18. Cheng, D.-J.; Shao, Y.-D. *ChemCatChem* **2019,** *11,* 2575–2589. (Review).

Markovnikov's Rule

For addition of HX to olefins, Markovnikov's rule predicts the regiochemistry of HX addition to unsymmetrically substituted alkenes: The halide component of HX bonds preferentially at the more highly substituted carbon, whereas the hydrogen prefers the carbon which already contains more hydrogen atoms.

The intermediate is the more stable secondary cation, and the formal charge is on one carbon.

Regiochemistry favors the benzylic position:

Example 1[3]

© Springer Nature Switzerland AG 2021
J. J. Li, *Name Reactions*, https://doi.org/10.1007/978-3-030-50865-4_88

Example 2, Markovnikov-selective hydrothiolation of styrenes[4]

Example 3, Markovnikov-regioselective hydroboration of alkenes[5]

81   :   19

Example 4, Markovnikov-regioselective despite a radical mechanism[6]

References

1   Markownikoff, W. *Ann. Pharm.* **1870**, *153*, 228–259. Vladimir Vasilyevich Markov-
     nikov (1838–1904) fomulated the rule for addition to alkenes at Moscow University.
     He was one of the most eminant Russian organic chemists in the 19th century. He was
     a prickly individual with strong opinions, and he had no fear in expressing them. His
     tactless outspokenness led to his ouster from professorship at Kazan' and Moscow.
     (Lewis, D. E. *Early Russian Organic Chemists and Their Legacy*, Springer: Heldel-
     berg, Germany, 2012, p 71.).

2   Oparina, L. A.; Artem'ev, A. V.; Vysotskaya, O. V.; Kolyvanov, N. A.; Bagryanskaya,
     Y. I.; Doronina, E. P.; Gusarova, N. K. *Tetrahedron* **2013**, *69*, 6185–6195.
3   Ziyaei Halimehjani, A.; Pasha Zanussi, H. *Synthesis* **2013**, *45*, 1483–1488
4   Savolainen, M. A.; Wu, J. *Org. Lett.* **2013**, *15,* 3802–3804.
5   Zhang, G.; Wu, J.; Li, S.; Cass, S.; Zheng, S. *Org. Lett.* **2018,** *20,* 7893–7897.
6   Neff, R. K.; Su, Y.-L.; Liu, S.; Rosado, M.; Zhang, X.; Doyle, M. P *J. Am. Chem. Soc.*
     2019, *141*, 16643–16650.

Anti-**Markovnikov's Rule**

Some reactions do not follow Markovnikov's rule, and *anti*-Markovnikov products are isolated. The outcome of the regioselectivity may be explained by the relative stability of the radical intermediates.

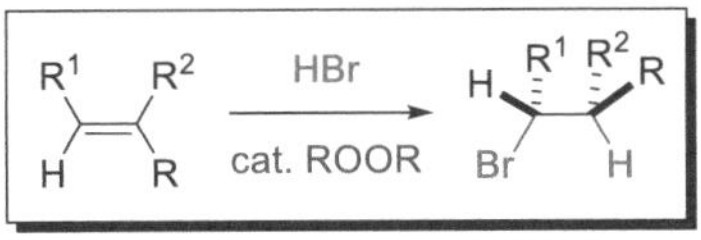

Radical mechanism:

Initiation:

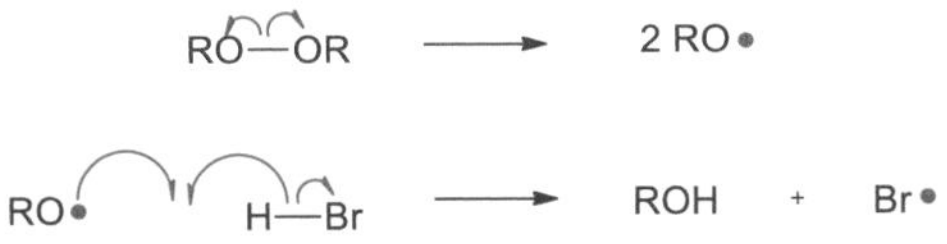

Propagation:

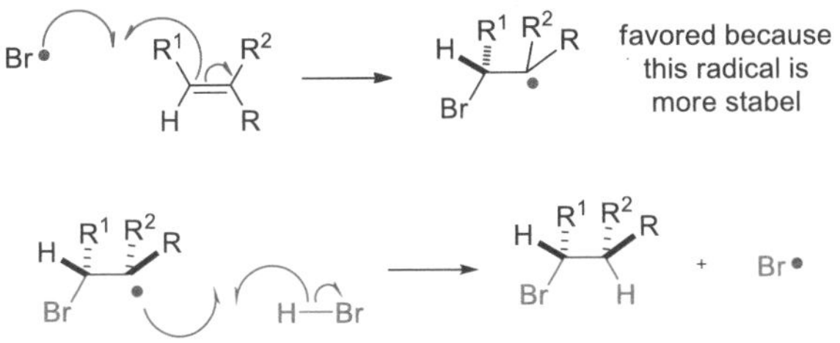

favored because
this radical is
more stabel

Termination:

Example 1, Anti-Markovnikov oxidation of allylic esters[1]

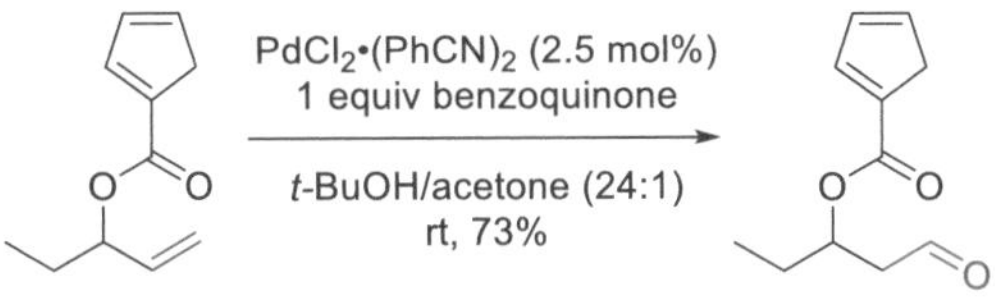

PdCl$_2$•(PhCN)$_2$ (2.5 mol%)
1 equiv benzoquinone

t-BuOH/acetone (24:1)
rt, 73%

Example 2, Anti-Markovnikov hydroamination[3]

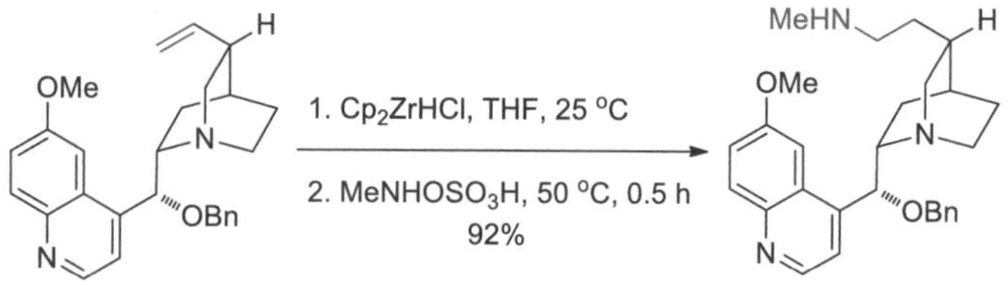
OMe
H
N
OBn
1. Cp_2ZrHCl, THF, 25 °C
2. $MeNHOSO_3H$, 50 °C, 0.5 h
92%
MeHN
H
OMe
N
OBn

Example 3, Anti-Markovnikov hydroheteroarylation of unactivated alkenes[4]

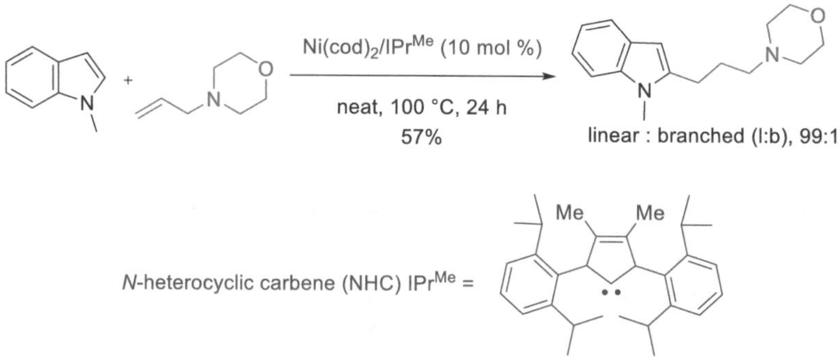
Ni(cod)2/IPrMe (10 mol %)
neat, 100 °C, 24 h
57%
linear : branched (l:b), 99:1
N-heterocyclic carbene (NHC) IPrMe =
Me
Me

Example 4, Anti-Markovnikov addition to alkynes[5]

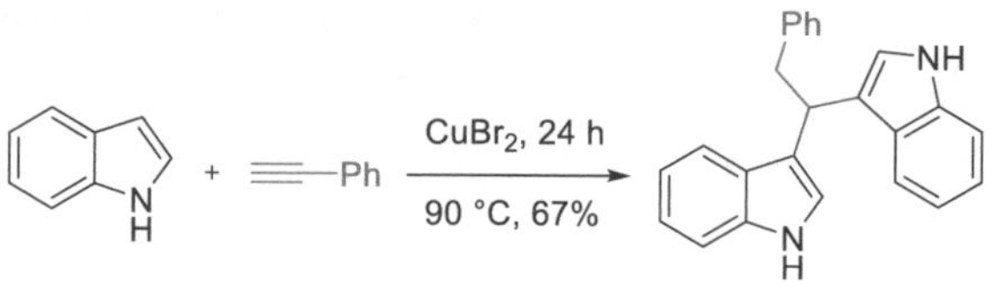
N
H
Ph
CuBr2, 24 h
90 °C, 67%
Ph
NH

Example 5, Anti-Markovnikov hydroamination using *N*-hydroxyphthalimide[6]

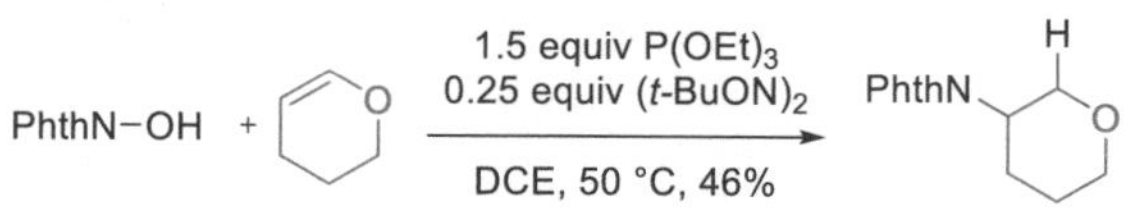
PhthN−OH
1.5 equiv P(OEt)3
0.25 equiv (t-BuON)2
DCE, 50 °C, 46%
PhthN
H
O

Example 6, Anti-Markovnikov hydroamination of unactivated alkenes with primary amines[7]

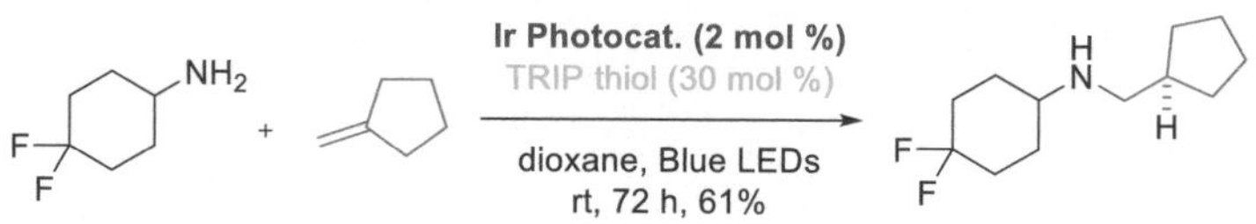
NH2
F
F
Ir Photocat. (2 mol %)
TRIP thiol (30 mol %)
dioxane, Blue LEDs
rt, 72 h, 61%
H
N
H

Ir Photocat. = [Ir(dF(CF$_3$)ppy)$_2$(4,4'-d)(CF$_3$)-bpy)]PF$_6$ =

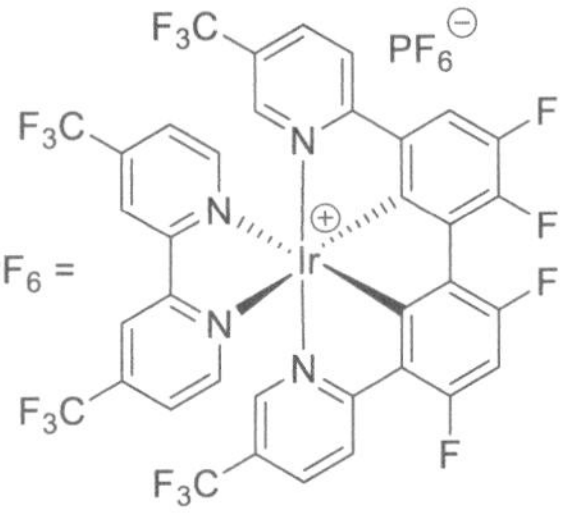

PF$_6^{\ominus}$

TRIP thiol =

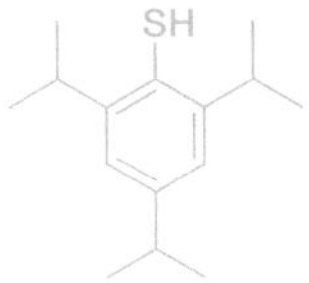

References

1. Nishizawa, M.; Asai, Y.; Imagawa, H. *Org. Lett.* **2006,** *8,* 5793–5796.
2. Dong, J. J.; Fañanás-Mastral, M.; Alsters, P. L.; Browne, W. R.; Feringa, B. L. *Angew. Chem. Int. Ed.* **2008,** *47*, 5561–5565.
3. Strom, A. E.; Hartwig, J. F. *J. Org. Chem.* **2013,** *78*, 8909–8914.
4. Schramm, Y.; Takeuchi, M.; Semba, K.; Nakao, Y.; Hartwig, J. F. *J. Am. Chem. Soc.* **2015,** *137*, 12215–12218.
5. Srivastava, A.; Patel, S. S.; Chandna, N.; Jain, N. *J. Org. Chem.* **2016,** *81*, 11664–11670.
6. Lardy, S. W.; Schmidt, V. A. *J. Am. Chem. Soc.* **2018,** *140*, 12318–12322.
7. Miller, D. C.; Ganley, J. M.; Musacchio, A. J.; Sherwood, T. C.; Ewing, W. R.; Knowles, R. R. *J. Am. Chem. Soc.* **2019,** *141*, 16590–16594.

Martin's Sulfurane Dehydrating Reagent

Dehydrates secondary and tertiary alcohols to give olefins, but forms ethers with primary alcohols. *Cf.* Burgess dehydrating reagent.

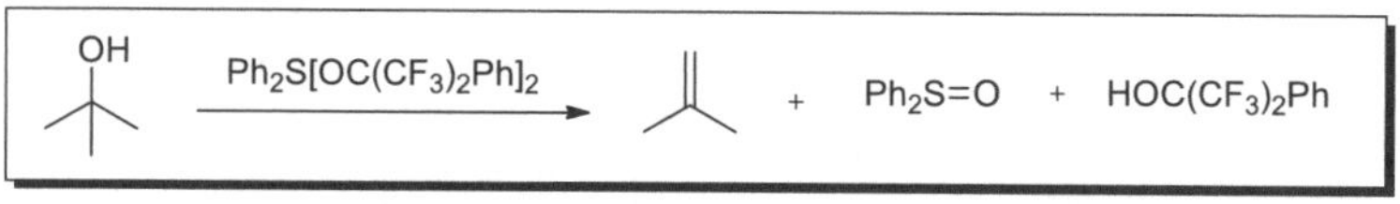

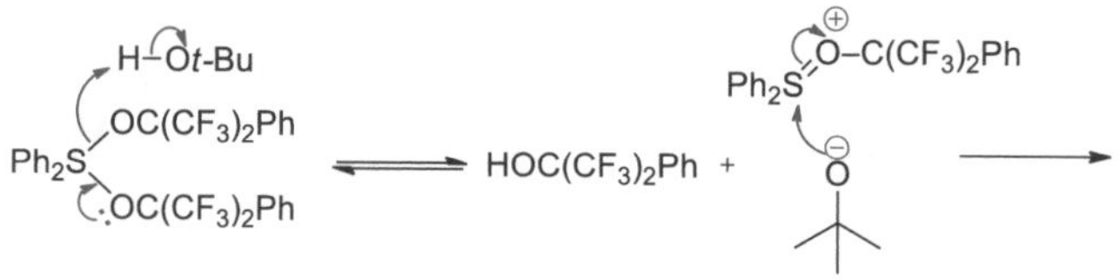

The alcohol is acidic

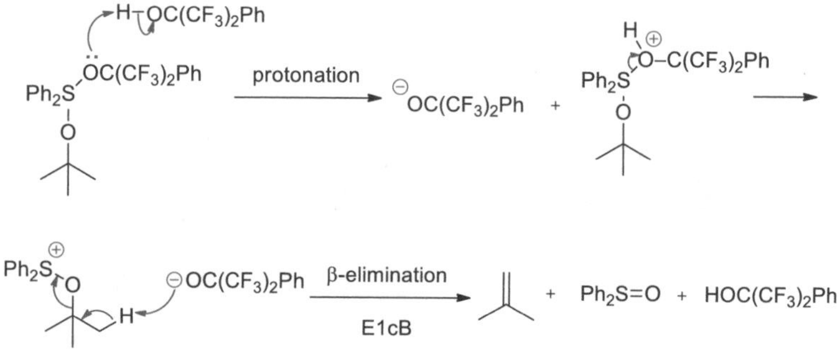

Example 1[5]

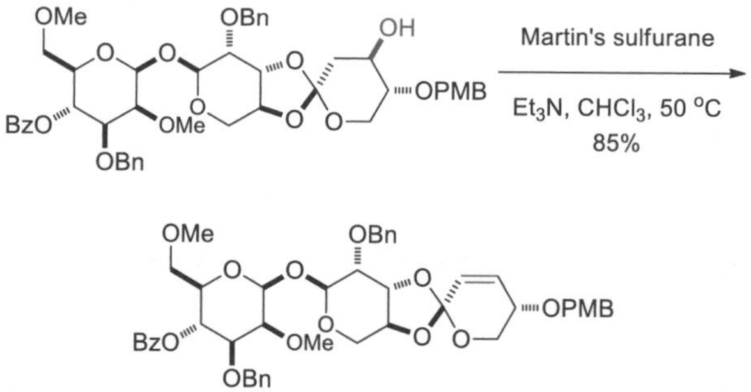

© Springer Nature Switzerland AG 2021
J. J. Li, *Name Reactions*, https://doi.org/10.1007/978-3-030-50865-4_89

Example 2[6]

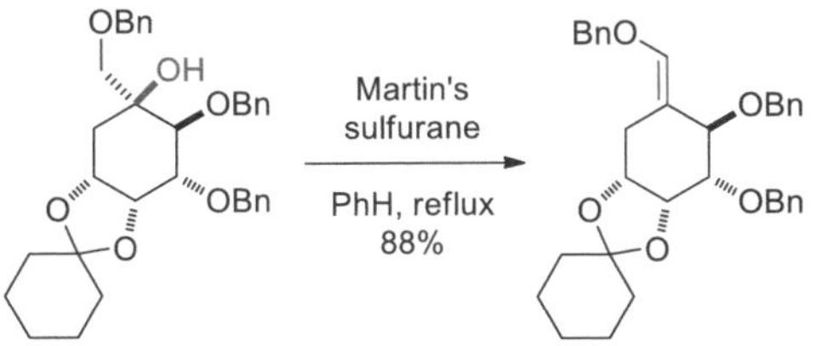

Example 3[7]

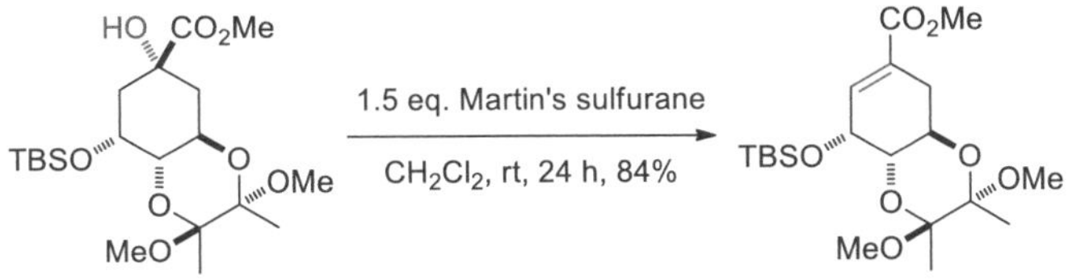

Example 4[9]

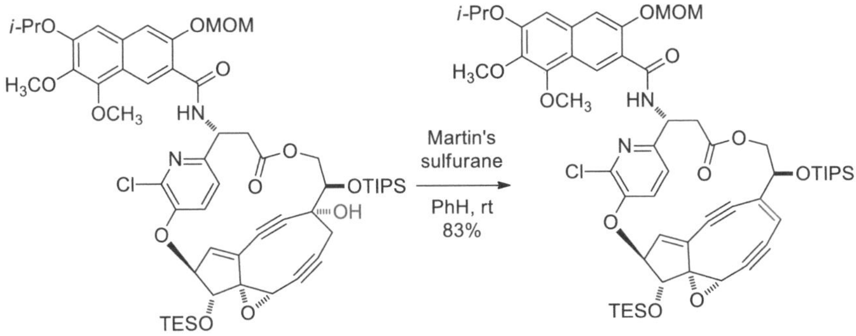

Example 5[12]

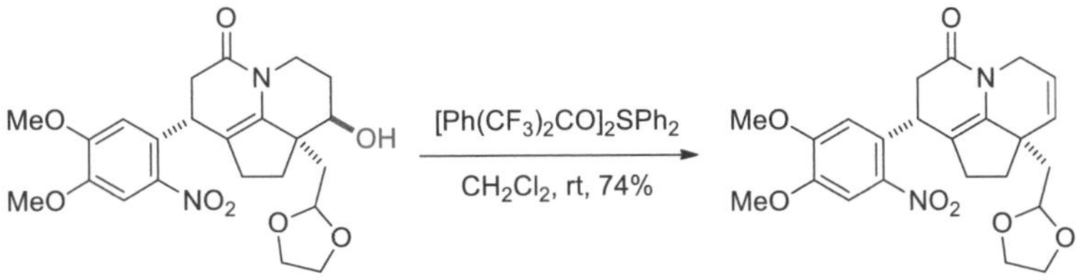

Example 6, Direct ylide transfer in indoles[13]

Example 7[14]

Example 8[15]

Example 9[16]

Example 10[17]

References

1. (a) Martin, J. C.; Arhart, R. J. *J. Am. Chem. Soc.* **1971,** *93,* 2339–2341; (b) Martin, J. C.; Arhart, R. J. *J. Am. Chem. Soc.* **1971,** *93,* 2341–2342; (c) Martin, J. C.; Arhart, R. J. *J. Am. Chem. Soc.* **1971,** *93,* 4327–4329. (d) Martin, J. C.; Arhart, R. J.; Franz, J. A.; Perozzi, E. F.; Kaplan, L. *J. Org. Synth.* **1977,** *57,* 22–26.
2. Gallagher, T. F.; Adams, J. L. *J. Org. Chem.* **1992,** *57,* 3347–3353.
3. Tse, B.; Kishi, Y. *J. Org. Chem.* **1994,** *59,* 7807–7814.
4. Winkler, J. D.; Stelmach, J. E.; Axten, J. *Tetrahedron Lett.* **1996,** *37,* 4317–4320.

5. Nicolaou, K. C.; Rodríguez, R. M.; Fylaktakidou, K. C.; Suzuki, H.; Mitchell, H. J. *Angew. Chem. Int. Ed.* **1999,** *38*, 3340–3345.

6. Kok, S. H. L.; Lee, C. C.; Shing, T. K. M. *J. Org. Chem.* **2001,** *66*, 7184–7190.

7. Box, J. M.; Harwood, L. M.; Humphreys, J. L.; Morris, G. A.; Redon, P. M.; Whitehead, R. C. *Synlett* **2002,** 358–360.

8. Myers, A. G.; Glatthar, R.; Hammond, M.; Harrington, P. M.; Kuo, E. Y.; Liang, J.; Schaus, S. E.; Wu, Y.; Xiang, J.-N. *J. Am. Chem. Soc.* **2002,** *124*, 5380–5401.

9. Myers, A. G.; Hogan, P. C.; Hurd, A. R.; Goldberg, S. D. *Angew. Chem. Int. Ed.* **2002,** *41*, 1062–1067.

10. Shea, K. M. *Martin's Sulfurane Dehydrating Reagent.* In *Name Reactions for Functional Group Transformations*; Li, J. J., Ed.; Wiley: Hoboken, NJ, **2007,** pp 248–264. (Review).

11. Sparling, B. A.; Moslin, R. M.; Jamison, T. F. *Org. Lett.* **2008,** *10*, 1291–1294.

12. Miura, Y.; Hayashi, N.; Yokoshima, S.; Fukuyama, T. *J. Am. Chem. Soc.* **2012,** *134*, 11995–11997.

13. Huang, X.; Patil, M.; Farès, C.; Thiel, W.; Maulide, N. *J. Am. Chem. Soc.* **2013,** *135*, 7313–7323.

14. Ma, Z.; Jiang, J.; Luo, S.; Cai, Y.; cardon, J. M.; Kay, B. M.; Ess, D. H.; Castle, S. L. *Org. Lett.* **2014,** *16*, 4044–4047.

15. Takao, K.-i.; Tsunoda, K.; Kurisu, T.; Sakama, A.; Nishimura, Y.; Yoshida, K.; Tadano, K.-i. *Org. Lett.* **2015,** *17*, 756–759.

16. Klimczyk, S.; Huang, X.; Kählig, H.; Veiros, L. F.; Maulide, N. *J. Org. Chem.* **2015,** *80*, 5719–5729.

17. Zanghi, J. M.; Liu, S.; Meek, S. J. *Org. Lett.* **2019,** *21*, 5172–5177.

Meerwein–Ponndorf–Verley Reduction

Reduction of ketones to the corresponding alcohols using Al(Oi-Pr)$_3$ in isopropanol. Reverse of the Oppernauer oxidation.

cyclic transition state

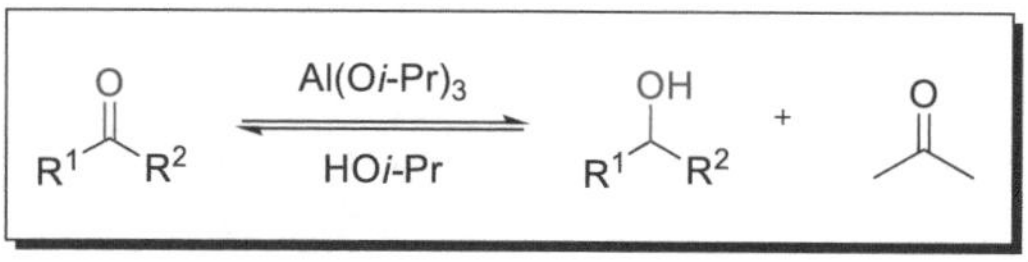

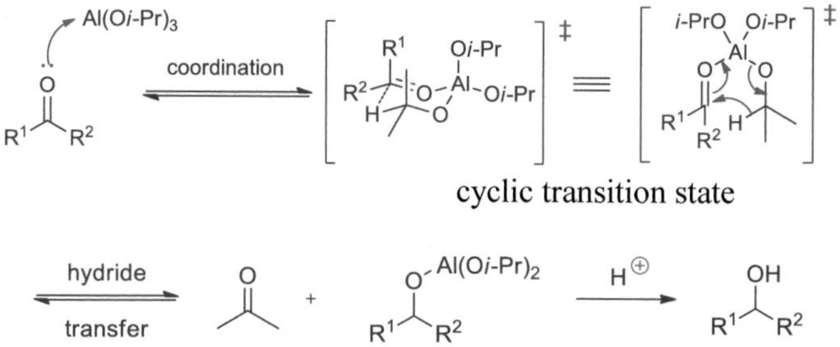

Example 1[2]

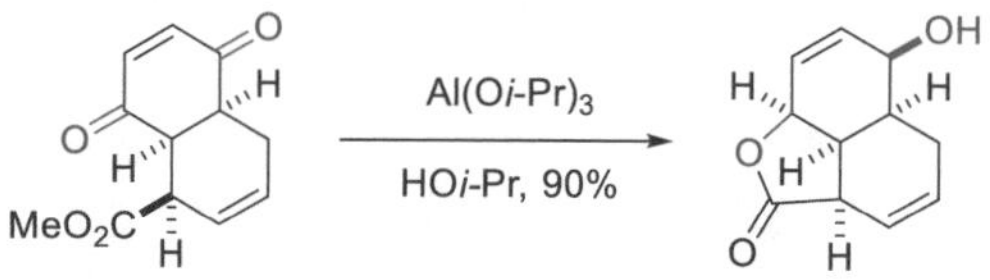

Al(Oi-Pr)$_3$
HOi-Pr, 90%

Example 2[4]

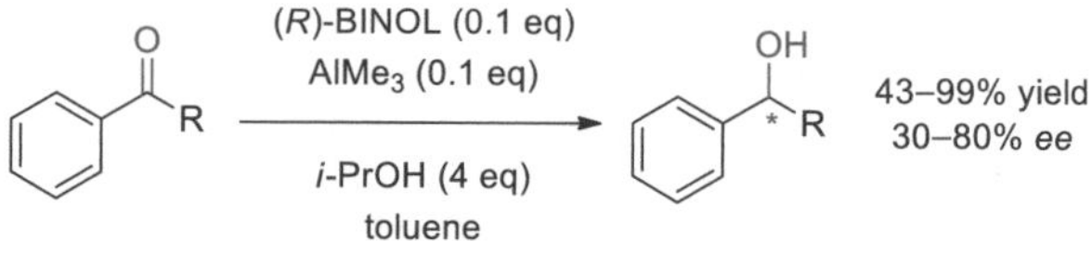

(R)-BINOL (0.1 eq)
AlMe$_3$ (0.1 eq)

i-PrOH (4 eq)
toluene

43–99% yield
30–80% ee

Example 3[7]

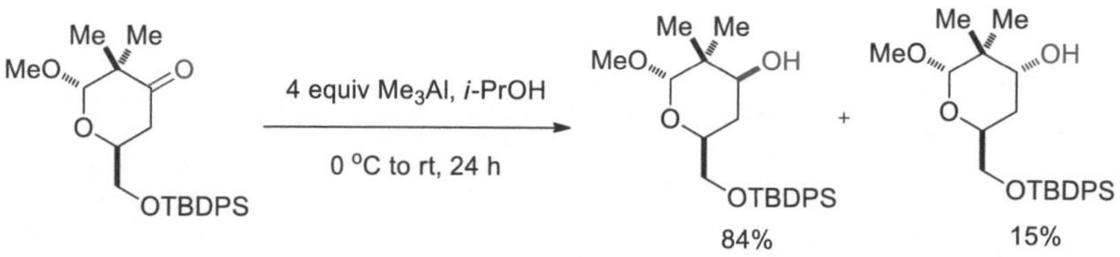

4 equiv Me$_3$Al, i-PrOH

0 °C to rt, 24 h

84%   15%

J. J. Li, *Name Reactions*, https://doi.org/10.1007/978-3-030-50865-4_90

Example 4[9]

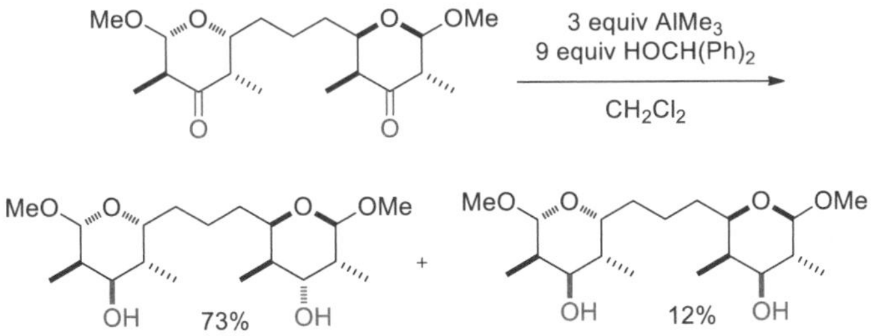
MeO
OMe
3 equiv AlMe3
9 equiv HOCH(Ph)2
CH2Cl2
OH
73%
12%

Example 5[10]

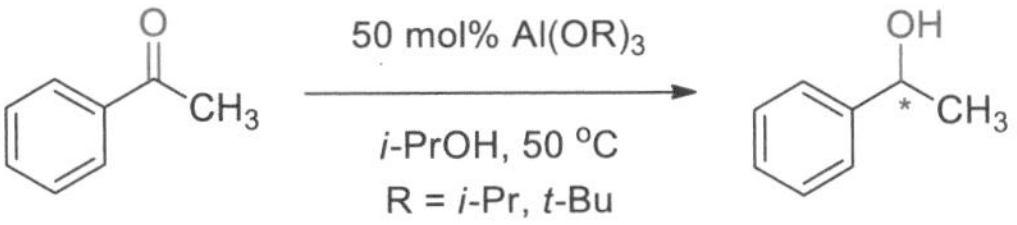
CH3
50 mol% Al(OR)3
i-PrOH, 50 °C
R = i-Pr, t-Bu
OH

Example 6, Reuction of α-silylimines by a chiral lithium amide (a Meerwein–Ponndorf–Verley-*type* reduction)[11]

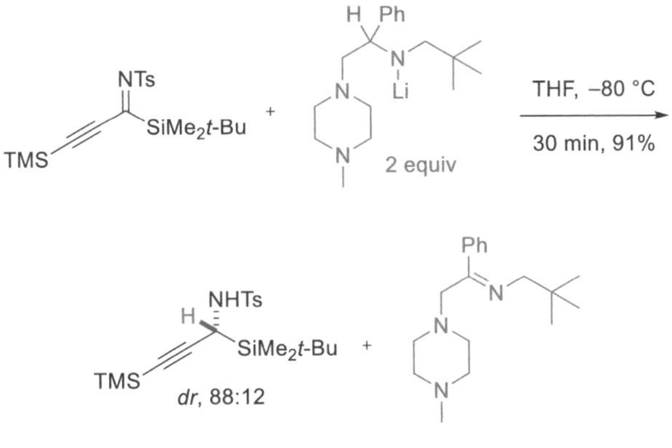
NTs
SiMe2t-Bu
TMS
H Ph
Li
2 equiv
THF, –80 °C
30 min, 91%
NHTs
dr, 88:12

Example 7, A process synthesis of a key intermediate of omarigliptin, a DPP-4 inhibitor[12]

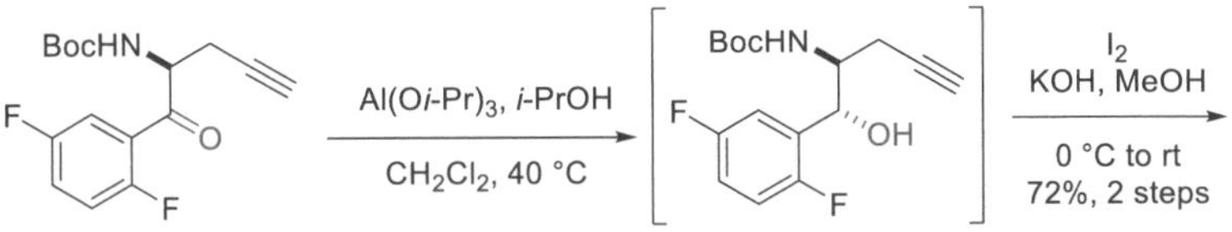
BocHN
Al(Oi-Pr)3, i-PrOH
CH2Cl2, 40 °C
OH
I2
KOH, MeOH
0 °C to rt
72%, 2 steps

omarigliptin (Marizev)
Merck, 2015 (Japan)
DPP-4 inhibitor, once weekly

> 99% ee

Example 8, A base-mediated Meerwein–Ponndorf–Verley reduction[13]

4 equiv K$_3$PO$_4$

dioxane, 130 °C
24 h, 66%

2.5 equiv

References

1. Meerwein, H.; Schmidt, R. *Ann.* **1925**, *444*, 221–238. Hans L. Meerwein, born in Hamburg Germany in 1879, received his Ph.D. at Bonn in 1903. In his long and productive academic career, Meerwein made many notable contributions in organic chemistry.

2. Woodward, R. B.; Bader, F. E.; Bickel, H.; Frey, A. J.; Kierstead, R. W. *Tetrahedron* **1958**, *2*, 1–57.

3. de Graauw, C. F.; Peters, J. A.; van Bekkum, H.; Huskens, J. *Synthesis* **1994**, 1007–1017. (Review).

4. Campbell, E. J.; Zhou, H.; Nguyen, S. T. *Angew. Chem. Int. Ed.* **2002**, *41*, 1020–1022.

5. Sominsky, L.; Rozental, E.; Gottlieb, H.; Gedanken, A.; Hoz, S. *J. Org. Chem.* **2004**, *69*, 1492–1496.

6. Cha, J. S. *Org. Process Res. Dev.* **2006**, *10*, 1032–1053.

7. Manaviazar, S.; Frigerio, M.; Bhatia, G. S.; Hummersone, M. G.; Aliev, A. E.; Hale, K. J. *Org. Lett.* **2006**, *8*, 4477–4480.

8. Clay, J. M. *Meerwein–Ponndorf–Verley reduction.* In *Name Reactions for Functional Group Transformations*; Li, J. J., Ed.; Wiley: Hoboken, NJ, **2007**, pp 123–128. (Review).

9. Dilger, A. K.; Gopalsamuthiram, V.; Burke, S. D. *J. Am. Chem. Soc.* **2007**, *129*, 16273–16277.

10. Flack, K.; Kitagawa, K.; Pollet, P.; Eckert, C. A.; Richman, K.; Stringer, J.; Dubay, W.; Liotta, C. L. *Org. Process Res. Dev.* **2012**, *16*, 1301–1306.

11. Kondo, Y.; Sasaki, M.; Kawahata, M.; Yamaguchi, K.; Takeda, K. *J. Org. Chem.* **2014**, *79*, 3601–3609.

12. Sun, G.; Wei, M.; Luo, Z.; Liu, Y.; Chen, Z.; Wang, Z. *Org. Process Res. Dev.* **2016**, *20*, 2074–2079.

13. Boit, T. B.; Mehta, M. M.; Garg, N. K. *Org. Lett.* **2019**, *21*, 6447–6451.

14. Li, X.; Du, Z.; Wu, Y.; Zhen, Y.; Shao, R.; Li, B.; Chen, C.; Liu, Q.; Zhou, H. *RSC Adv.* **2020**, *10*, 9985–9995.

Meisenheimer Complex

Also known as **the Meisenheimer–Jackson salt**, the stable intermediate for certain $S_N Ar$ reactions.

Sanger's reagent, *ipso* attack   *ipso* substitution

Example 1[7]

Example 2, A fluorescent zwitterionic spirocyclic Meisenheimer complex[9]

The reaction using Sanger's reagent is faster than using the corresponding chloro-, bromo-, and iododinitrobenzene—the fluoro-Meisenheimer complex is the most stabilized because F is the most electron-withdrawing. The reaction rate does not depend upon the capacity of the leaving group.

© Springer Nature Switzerland AG 2021
J. J. Li, *Name Reactions*, https://doi.org/10.1007/978-3-030-50865-4_91

Example 3[10]

Example 4[14]

References

1. Meisenheimer, J. *Ann.* **1902**, *323*, 205–214. In 1902, Jacob Meisenheimer (1876–1934) at the University of Munich reported evidence for the structure of a very intense violet-colored compound that is produced on mixing trinitrobenzene with an alcohol in the presence of alkali.[12]
2. Strauss, M. J. *Acc. Chem. Res.* **1974**, *7*, 181–188. (Review).
3. Bernasconi, C. F. *Acc. Chem. Res.* **1978**, *11*, 147–152. (Review).
4. Terrier, F. *Chem. Rev.* **1982**, *82*, 77–152. (Review).
5. Manderville, R. A.; Buncel, E. *J. Org. Chem.* **1997**, *62*, 7614–7620.
6. Hoshino, K.; Ozawa, N.; Kokado, H.; Seki, H.; Tokunaga, T.; Ishikawa, T. *J. Org. Chem.* **1999**, *64*, 4572–4573.

7. Adam, W.; Makosza, M.; Zhao, C.-G.; Surowiec, M. *J. Org. Chem.* **2000,** *65,* 1099–1101.
8. Gallardo, I.; Guirado, G.; Marquet, J. *J. Org. Chem.* **2002,** *67,* 2548–2555.
9. Al-Kaysi, R. O.; Guirado, G.; Valente, E. J. *Eur. J. Org. Chem.* **2004,** 3408–3411.
10. Um, I.-H.; Min, S.-W.; Dust, J. M. *J. Org. Chem.* **2007,** *72,* 8797–8803.
11. Campodónico, P. R.; Tapia, R. A.; Contreras, R.; Ormazábal-Toledo, R. *Org. Biomol. Chem.* **2013,** *11,* 2302–2309.
12. Lennox, A. J. J. *Angew. Chem. Int. Ed.* **2018,** *57,* 14686–14688. (Review).
13. Liu, R.; Krchnak, V.; Brown, S. N.; Miller, M. J. *ACS Med. Chem. Lett.* **2019,** *10,* 1462–1466.
14. Saaidin, A. S.; Murai, Y.; Ishikawa, T.; Monde, K. *Eur. J. Org. Chem.* **2019,** 7563–7567.
15. Ota, N.; Harada, Y.; Kamitori, Y.; Okada, E. *Heterocycles* **2020,** *101,* 692–700.

Meyer–Schuster Rearrangement

The isomerization of secondary and tertiary α-acetylenic alcohols to α,β-unsaturated carbonyl compounds *via* 1,3-shift. When the acetylenic group is terminal, the products are aldehydes, whereas the internal acetylenes give ketones. *Cf.* Rupe rearrangement.

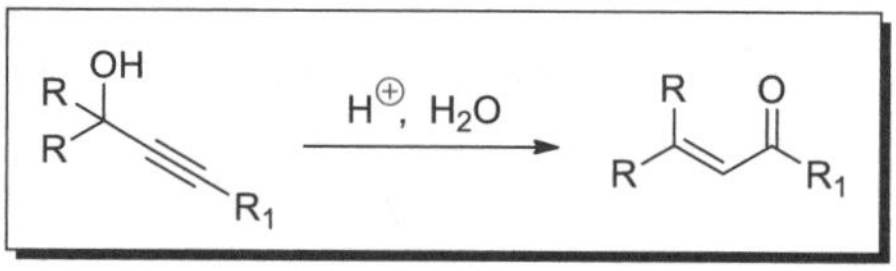

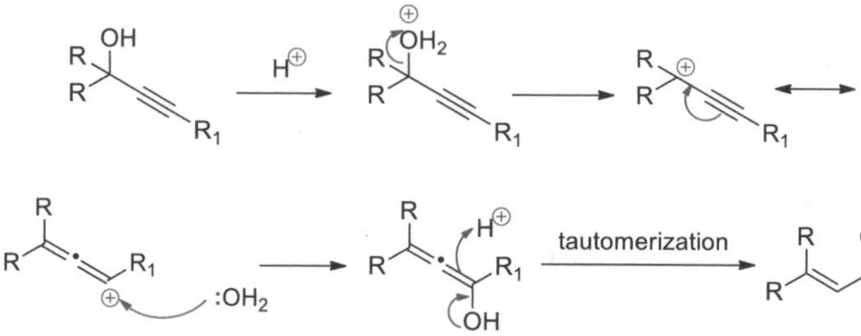

Example 1[6]

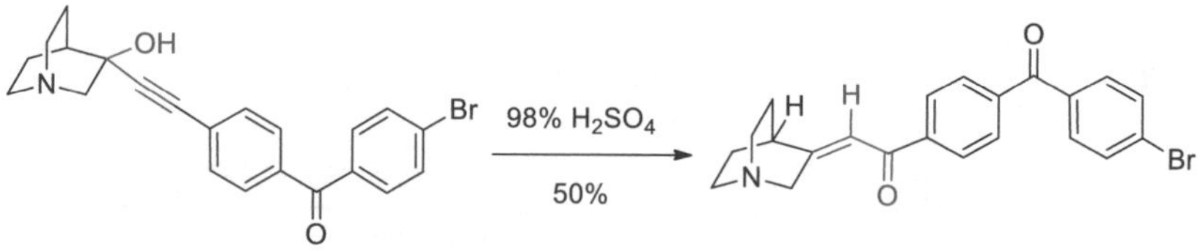

Example 2[7]

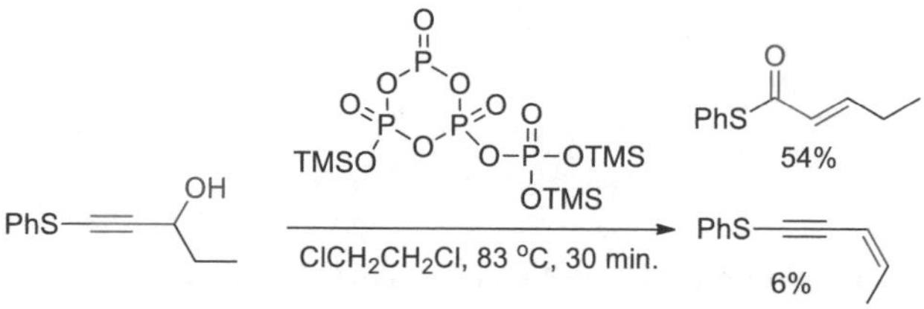

J. J. Li, *Name Reactions*, https://doi.org/10.1007/978-3-030-50865-4_92

Example 3[8]

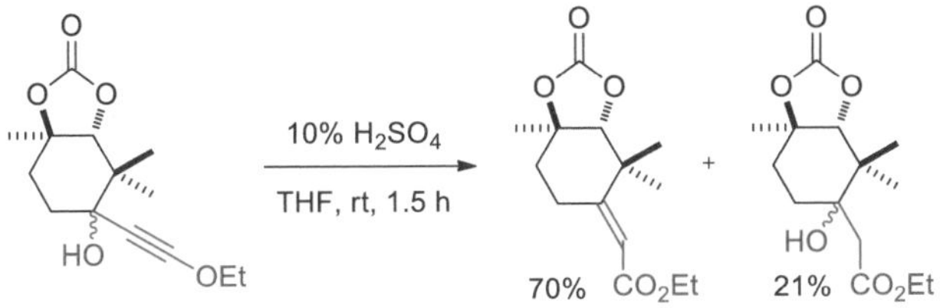

10% H₂SO₄

THF, rt, 1.5 h

70%   CO₂Et   21%   CO₂Et

Example 4[9]

BF₃•Et₂O

TFA, 89%

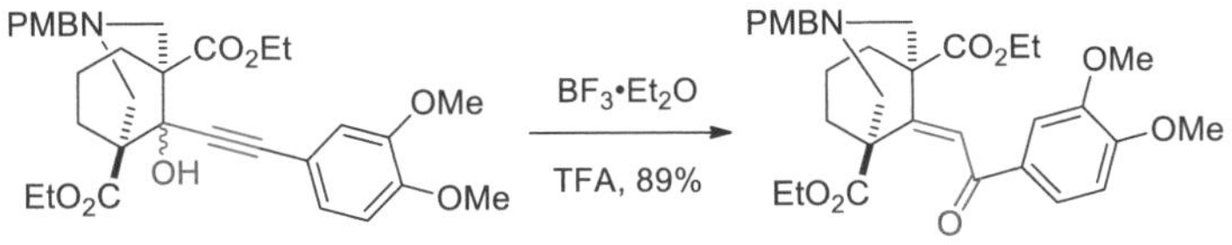

Example 5, DTBP = di-*tert*-butyl peroxide[11]

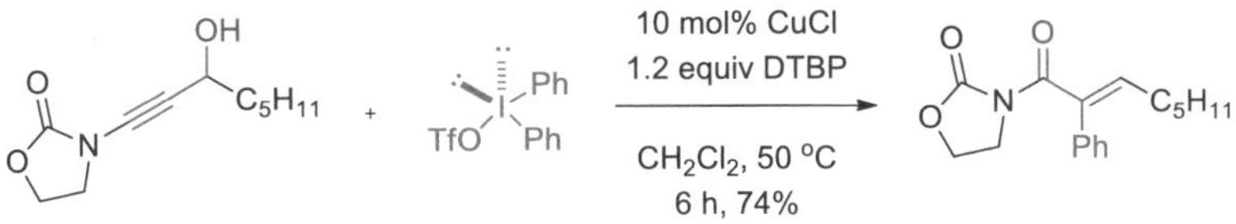

10 mol% CuCl
1.2 equiv DTBP

CH₂Cl₂, 50 °C
6 h, 74%

Example 6[12]

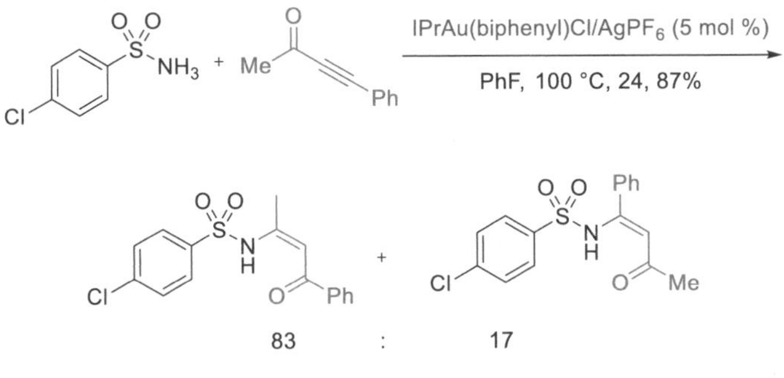

IPrAu(biphenyl)Cl/AgPF₆ (5 mol %)

PhF, 100 °C, 24, 87%

83   :   17

Example 7, Gold-catalyzed Meyer–Schuster rearrangement[13]

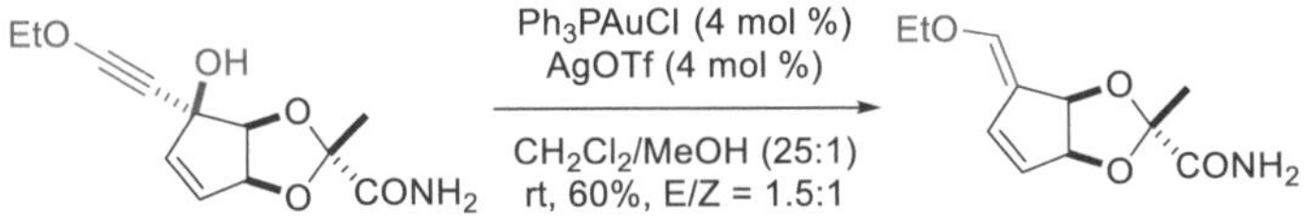

Ph₃PAuCl (4 mol %)
AgOTf (4 mol %)

CH₂Cl₂/MeOH (25:1)
rt, 60%, E/Z = 1.5:1

Example 8[14]

References

1. Meyer, K. H.; Schuster, K. *Ber.* **1922**, *55*, 819–823.
2. Swaminathan, S.; Narayanan, K. V. *Chem. Rev.* **1971**, *71*, 429–438. (Review).
3. Edens, M.; Boerner, D.; Chase, C. R.; Nass, D.; Schiavelli, M. D. *J. Org. Chem.* **1977**, *42*, 3403–3408.
4. Andres, J.; Cardenas, R.; Silla, E.; Tapia, O. *J. Am. Chem. Soc.* **1988**, *110*, 666–674.
5. Tapia, O.; Lluch, J. M.; Cardenas, R.; Andres, J. *J. Am. Chem. Soc.* **1989**, *111*, 829–835.
6. Brown, G. R.; Hollinshead, D. M.; Stokes, E. S.; Clarke, D. S.; Eakin, M. A.; Foubister, A. J.; Glossop, S. C.; Griffiths, D.; Johnson, M. C.; McTaggart, F.; Mirrlees, D. J.; Smith, G. J.; Wood, R. *J. Med. Chem.* **1999**, *42*, 1306–1311.
7. Yoshimatsu, M.; Naito, M.; Kawahigashi, M.; Shimizu, H.; Kataoka, T. *J. Org. Chem.* **1995**, *60*, 4798–4802.
8. Crich, D.; Natarajan, S.; Crich, J. Z. *Tetrahedron* **1997**, *53*, 7139–7158.
9. Williams, C. M.; Heim, R.; Bernhardt, P. V. *Tetrahedron* **2005**, *61*, 3771–3779.
10. Mullins, R. J.; Collins, N. R. *Meyer–Schuster Rearrangement.* In *Name Reactions for Homologations-Part II*; Li, J. J., Ed.; Wiley: Hoboken, NJ, **2009**, pp 305–318. (Review).
11. Collins, B. S. L.; Suero, M. G.; Gaunt, M. J. *Angew. Chem. Int. Ed.* **2013**, *52*, 5799–5802.
12. Lee, D.; Kim, S. M.; Hirao, H.; Hong, S. H. *Org. Lett.* **2017**, *19*, 4734–4737.
13. Chan, W. C.; Koide, K. *Org. Lett.* **2018**, *20*, 7798–7802.
14. Kadiyala, V.; Kumar, P. B.; Balasubramanian, S. *J. Org. Chem.* **2019**, *84*, 12228–12236.
15. Qiu, Y.-F.; Niu, Y.-J.; Song, X.-R.; Wei, X.; Chen, H.; Li, S.-X.; Wang, X.-C.; Huo, C.; Quan, Z.-J.; Liang, Y.-M. *Chem. Commun.* **2020**, *56*, 1421–1424.

Michael Addition

Also known as conjugate addition, Michael addition is the 1,4-addition of a nucleophile to an α,β-unsaturated system.

Example 1, Asymmetric Michael addition[2]

Example 2, Thia-Michael addition[3]

Example 3, Phospha-Michael addition[7]

Example 4, Asymmetric aza-Michael addition[9]

© Springer Nature Switzerland AG 2021
J. J. Li, *Name Reactions*, https://doi.org/10.1007/978-3-030-50865-4_93

Example 5, Intramolecular Michael addition[10]

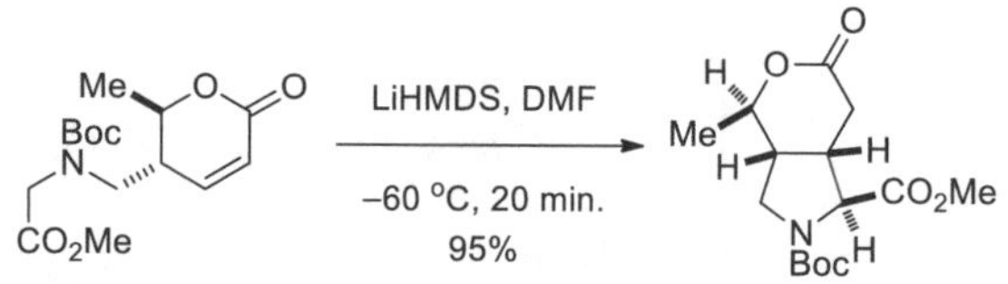

Example 6, Intramolecular Michael addition[11]

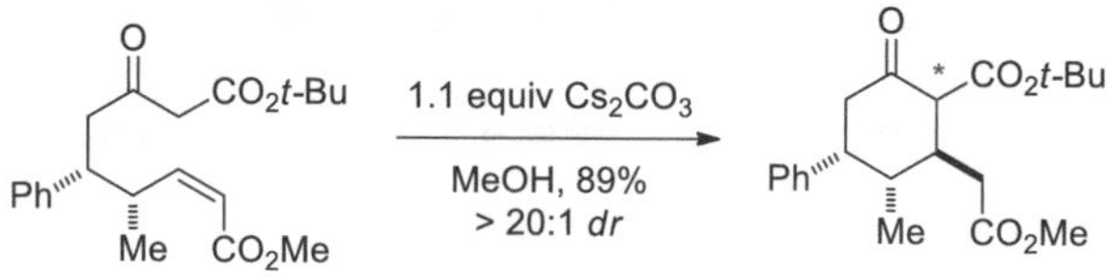

Example 7, Hetero-Michael addition[12]

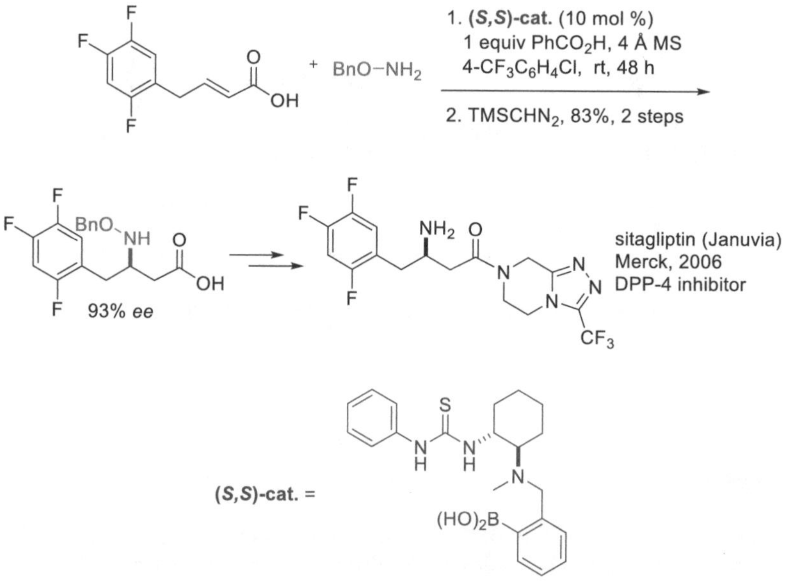

sitagliptin (Januvia)
Merck, 2006
DPP-4 inhibitor

(S,S)-cat. =

Example 8, Cu(II)-catalyzed asymmetric Michael addition[13]

Example 9, Michael addition of phosphonates to ene-nitrosoacetals[14]

Example 10, Organocatalytic asymmetric double Michael addition[15]

References

1. Michael, A. *J. Prakt. Chem.* **1887**, *35*, 349. Arthur Michael (1853–1942) was born in Buffalo, New York. He studied under Robert Bunsen, August Hofmann, Adolphe Wurtz, and Dimitri Mendeleev, but never bothered to take a degree. Back to the United States, Michael became a Professor of Chemistry at Tufts University, where he married one of his students, Helen Abbott, one of the few female organic chemists in this period. Since he failed miserably as an administrator, Michael and his wife set up their own private laboratory at Newton Center, Massachusetts, where the 1,4-addition was discovered.
2. Hunt, D. A. *Org. Prep. Proced. Int.* **1989**, *21*, 705–749.
3. D'Angelo, J.; Desmaële, D.; Dumas, F.; Guingant, A. *Tetrahedron: Asymmetry* **1992**, *3*, 459–505.

4. Lipshutz, B. H.; Sengupta, S. *Org. React.* **1992,** *41*, 135–631. (Review).
5. Hoz, S. *Acc. Chem. Res.* **1993,** *26*, 69–73. (Review).
6. Ihara, M.; Fukumoto, K. *Angew. Chem. Int. Ed.* **1993,** *32*, 1010–1022. (Review).
7. Simoni, D.; Invidiata, F. P.; Manferdini, M.; Lampronti, I.; Rondanin, R.; Roberti, M.; Pollini, G. P. *Tetrahedron Lett.* **1998,** *39*, 7615–7618.
8. Enders, D.; Saint-Dizier, A.; Lannou, M.-I.; Lenzen, A. *Eur. J. Org. Chem.* **2006,** 29–49. (Review on the phospha-Michael addition).
9. Chen, L.-J.; Hou, D.-R. *Tetrahedron: Asymmetry* **2008,** *19*, 715–720.
10. Sakaguchi, H.; Tokuyama, H.; Fukuyama, T. *Org. Lett.* **2008,** *10*, 1711–1714.
11. Kwan, E. E.; Scheerer, J. R.; Evans, D. A. *J. Org. Chem.* **2013,** *78*, 175–203.
12. Hayama, N.; Kuramoto, R.; Földes, T.; Nishibayashi, R.; Kobayashi, Y.; Pápai, I.; Takemoto, Y. *J. Am. Chem. Soc.* **2018,** *140*, 12216–12225.
13. Bhattarai, B.; Nagorny, O. *Org. Lett.* **2018,** *20*, 154–157.
14. Naumovich, Y. A.; Ioffe, S. L.; Sukhorukov, A. Y. *J. Org. Chem.* **2019,** *84*, 7244–7254.
15. Chen, X.-M.; Lei, C.-W.; Yue, D.-F.; Zhao, J.-Q.; Wang, Z.-H.; Zhang, X.-M.; Xu, X.-Y.; Yuan, W.-C. *Org. Lett.* **2019,** *21*, 5452–5456.
16. Ramella, V.; Roosen, P. C.; Vanderwal, C. D. *Org. Lett.* **2020,** *22*, 2883–2886.

Michaelis–Arbuzov Phosphonate Synthesis

Aliphatic phosphonate synthesis from the reaction of alkyl halides with phosphites.
General scheme:

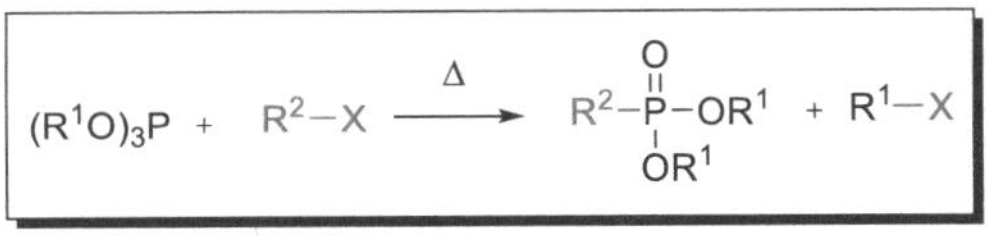

R^1 = alkyl, *etc.*; R^2 = alkyl, acyl, *etc.*; X = Cl, Br, I

For instance:

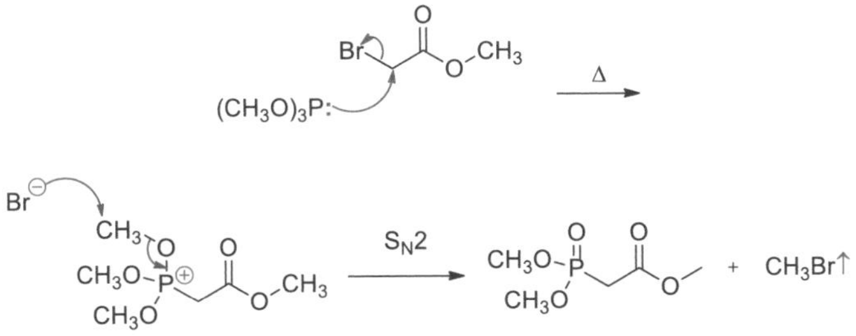

Example 1[2]

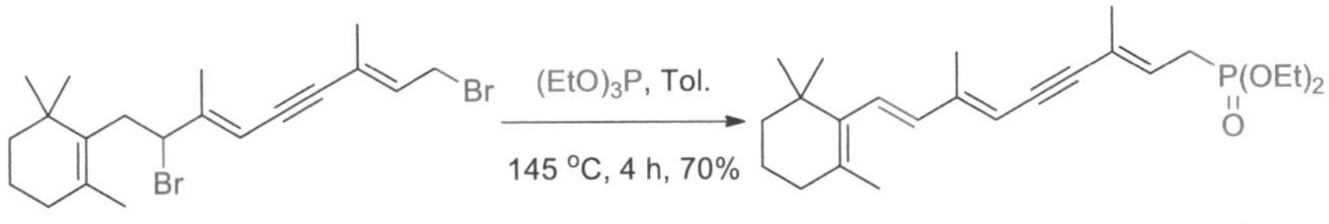

(EtO)$_3$P, Tol.

145 °C, 4 h, 70%

Example 2[6]

140 °C

8 h, 92%

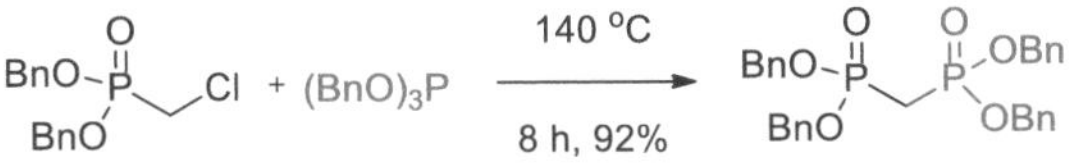

Example 3, Transition metal-catalyzed coupling, not via S_N2[7]

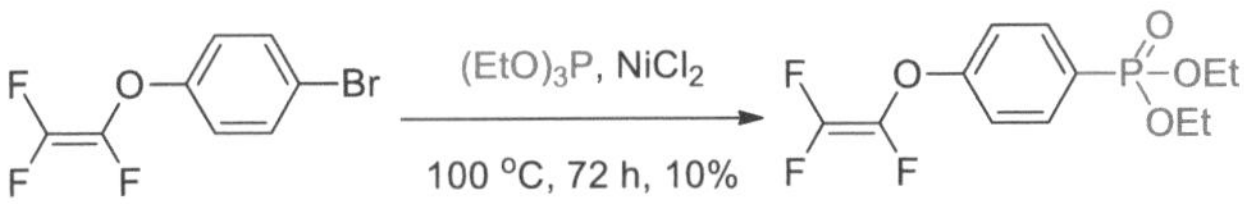

(EtO)$_3$P, NiCl$_2$

100 °C, 72 h, 10%

J. J. Li, *Name Reactions*, https://doi.org/10.1007/978-3-030-50865-4_94

Example 4[9]

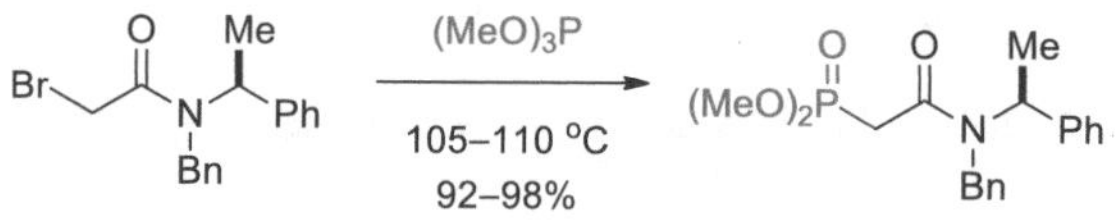

Example 5[10]

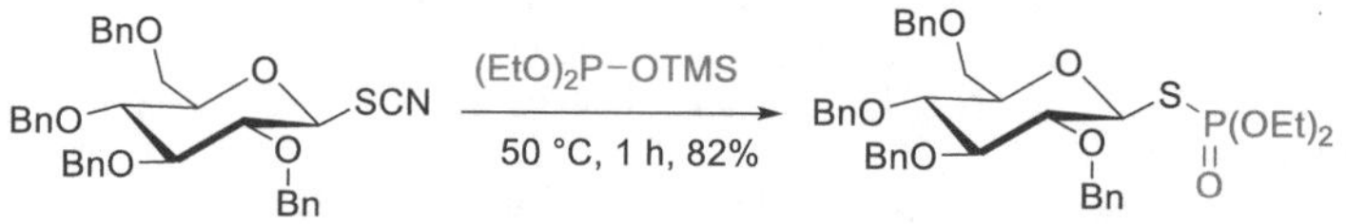

Example 6, An approach to prepare aromatic phosphonates via the intermediacy of benzyne[11]

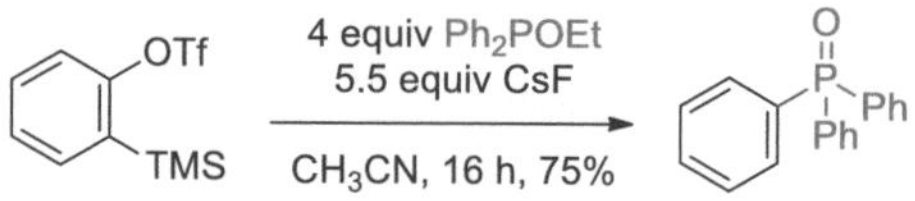

Example 7, Alcohol-based Michaelis–Arbuzov reaction[12]

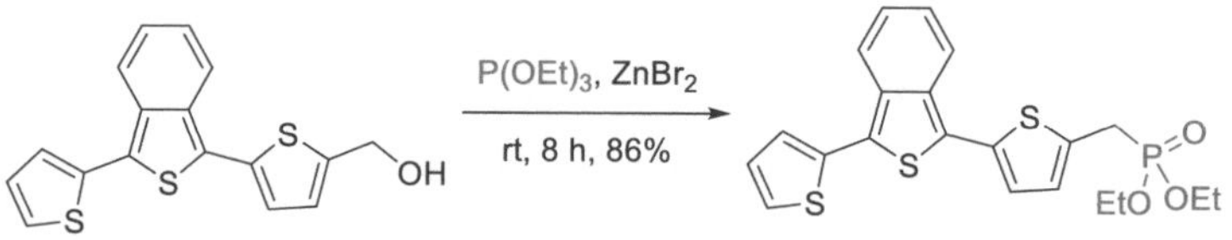

Example 8, Michaelis–Arbuzov-type reaction of 1-imidoalkyltriarylphosphonium salts[13]

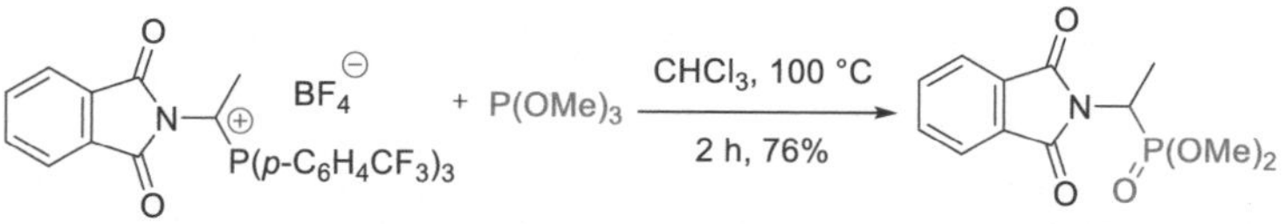

References

1. (a) Michaelis, A.; Kaehne, R. *Ber.* **1898**, *31,* 1048–1055. (b) Arbuzov, A. E. *J. Russ. Phys. Chem. Soc.* **1906**, *38*, 687.
2. Surmatis, J. D.; Thommen, R. *J. Org. Chem.* **1969**, *34*, 559–560.
3. Gillespie, P.; Ramirez, F.; Ugi, I.; Marquarding, D. *Angew. Chem. Int. Ed.* **1973**, *12*, 91–119. (Review).
4. Waschbüsch, R.; Carran, J.; Marinetti, A.; Savignac, P. *Synthesis* **1997**, 727–743.

5.   Bhattacharya, A. K.; Stolz, F.; Schmidt, R. R. *Tetrahedron Lett.* **2001**, *42*, 5393–5395.

6.   Erker, T.; Handler, N. *Synthesis* **2004**, 668–670.

7.   Souzy, R.; Ameduri, B.; Boutevin, B.; Virieux, D. *J. Fluorine Chem.* **2004**, *125*, 1317–1324.

8.   Kadyrov, A. A.; Silaev, D. V.; Makarov, K. N.; Gervits, L. L.; Röschenthaler, G.-V. *J. Fluorine Chem.* **2004**, *125*, 1407–1410.

9.   Ordonez, M.; Hernandez-Fernandez, E.; Montiel-Perez, M.; Bautista, R.; Bustos, P.; Rojas-Cabrera, H.; Fernandez-Zertuche, M.; Garcia-Barradas, O. *Tetrahedron: Asymmetry* **2007**, *18*, 2427–2436.

10.   Piekutowska, M.; Pakulski, Z. *Carbohydrate Res.* **2008**, *343*, 785–792.

11.   Dhokale, R. A.; Mhaske, S. B. *Org. Lett.* **2013**, *15*, 2218–2221.

12.   Nandakumar, M.; Sankar, E.; Mohanakrishnan, A. K. *Synth. Commun.* **2016**, *46*, 1810–1819.

13.   Adamek, J.; Wegrzyk-Schlieter, A.; Stec, K.; Walczak, K.; Erfurt, K. *Molecules* **2019**, *24*, 3405.

14.   Hernandez-Guerra, D.; Kennedy, A. R.; Leon, E. I.; Martin, A.; Perez-Martin, I.; Rodriguez, M. S.; Suarez, E. *J. Org. Chem.* **2020**, *85*, 4861–4880.

Minisci Reaction

Radical-based carbon–carbon bond formation with electron-deficient heteroaromatics. The reaction entails an intermolecular addition of a nucleophilic *radical* to protonated heteroaromatic nucleus.

Example 1[4]

$$S_2O_8^= + CH_3OH \longrightarrow \cdot CH_2OH + H^+ + SO_4^= + SO_4^{\cdot}$$

Example 2[5]

Meerwein's methylating reagent

© Springer Nature Switzerland AG 2021
J. J. Li, *Name Reactions*, https://doi.org/10.1007/978-3-030-50865-4_95

Example 3, Intramolecular Minisci reaction[6]

Example 4[7]

Example 5[10]

Example 6[12]

Example 7[13]

Example 8[14]

DLP = dilauroyl peroxide   product:starting material = 1:1

Example 9[15]

lepidine   dimethylactetamide (DMA)

References

1. Minisci, F, Bernardi. R, Bertini, F, Galli, R, Perchinummo, M. *Tetrahedron* **1971**, *27*, 3575–3579.
2. Minisci, F. *Synthesis* **1973**, 1–24. (Review).
3. Minisci, F. *Acc. Chem. Res.* **1983**, *16*, 27–32. (Review).
4. Katz, R. B.; Mistry, J.; Mitchell, M. B. *Synth. Commun.* **1989**, *19*, 317–325.
5. Biyouki, M. A. A.; Smith, R. A. J. *Synth. Commun.* **1998**, *28*, 3817–3825.
6. Doll, M. K. H. *J. Org. Chem.* **1999**, *64*, 1372–1374.
7. Cowden, C. J. *Org. Lett.* **2003**, *5*, 4497–4499.
8. Kast, O.; Bracher, F. *Synth. Commun.* **2003**, *33*, 3843–3850.
9. Benaglia, M.; Puglisi, A.; Holczknecht, O.; Quici, S.; Pozzi, G. *Tetrahedron* **2005**, *61*, 12058–12064.
10. Palde, P. B.; McNaughton, B. R.; Ross, N. T. *Synthesis* **2007**, 2287–2290.
11. Brebion, F.; Nàjera, F.; Delouvrié, B. *J. Heterocycl. Chem.* **2008**, *45*, 527–532.
12. Presset, M.; Fleury-Brégeot, N.; Oehlrich, D.; Rombouts, F.; Molander, G. A. *J. Org. Chem.* **2013**, *78*, 4615–4619.
13. Lo, J. C.; Kim, D.; Pan, C.-M.; Edwards, J. T.; Yabe, Y.; Gui, J.; Qin, T.; Gutierrez, S.; Giacoboni, J.; Baran, P. S.; et al. *J. Am. Chem. Soc.* **2017**, *139*, 2484–2503.
14. Revil-Baudard, V.; Vors, J.-P.; Zard, S. Z. *Org. Lett.* **2018**, *20*, 3531–3535.
15. Truscello, A. M.; Gambarotti, C. *Org. Process Res. Dev.* **2019**, *23*, 1450–1457.
16. Proctor, R. S. J.; Phipps, R. J. *Angew. Chem. Int. Ed.* **2019**, *58*, 13666–13699.
17. Li, T.; Liang, K.; Zhang, Y.; Hu, D.; Ma, Z.; Xia, C. *Org. Lett.* **2020**, *22*, 2386–2390.

Mitsunobu Reaction

S_N2 inversion of an alcohol by a nucleophile using disubstituted azodicarboxylates (originally, diethyl diazodicarboxylate, or DEAD) and trisubstituted phosphines (originally, triphenylphosphine).

Diethyl azodicarboxylate (DEAD)

Example 1[2]

Example 2[3]

J. J. Li, *Name Reactions*, https://doi.org/10.1007/978-3-030-50865-4_96

Example 3, Glucoronidation of the phenol: preparing a secondary drug metabolite[6]

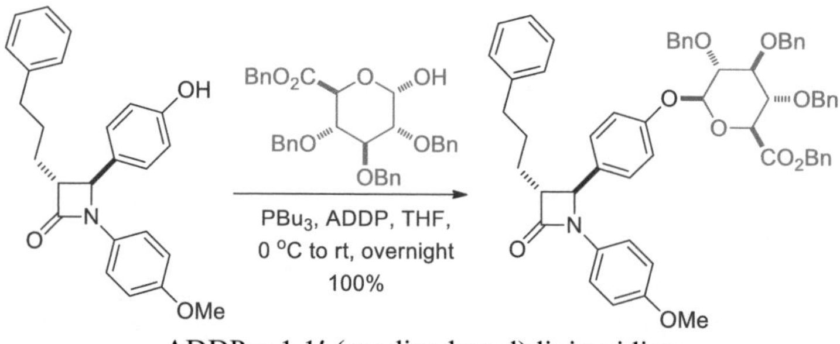

PBu₃, ADDP, THF,
0 °C to rt, overnight
100%

ADDP = 1,1′-(azodicarbonyl)dipiperidine

Example 4, Intramolecular Mitsunobu reaction[7]

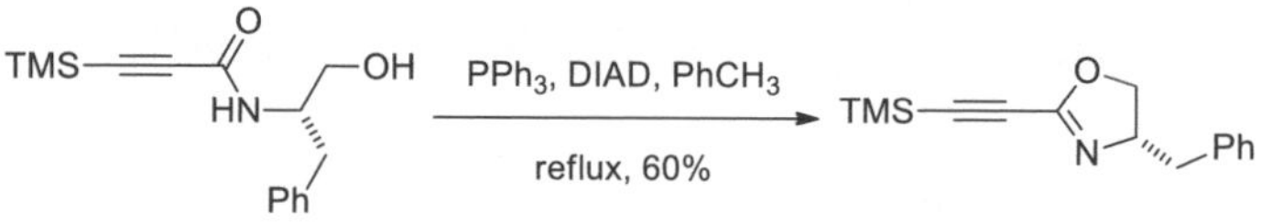

PPh₃, DIAD, PhCH₃
reflux, 60%

Example 5[8]

n-BuLi
then Ph₂PCl

80%

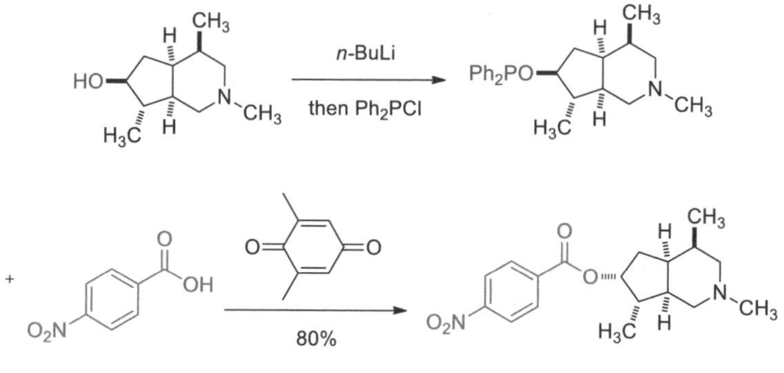

Example 6, Intramolecular Mitsunobu reaction[9]

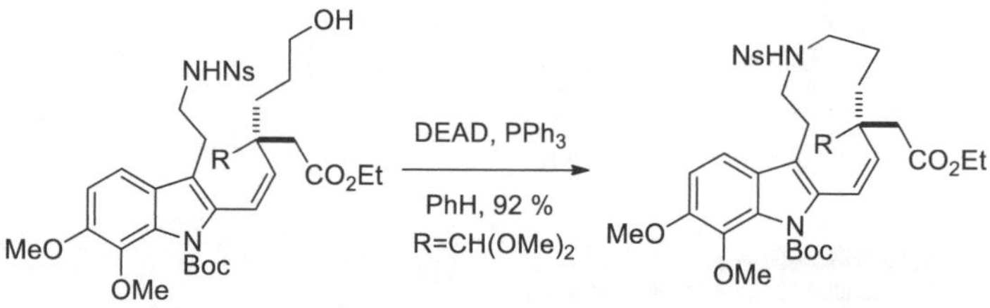

DEAD, PPh₃
PhH, 92 %
R=CH(OMe)₂

Example 7[13]

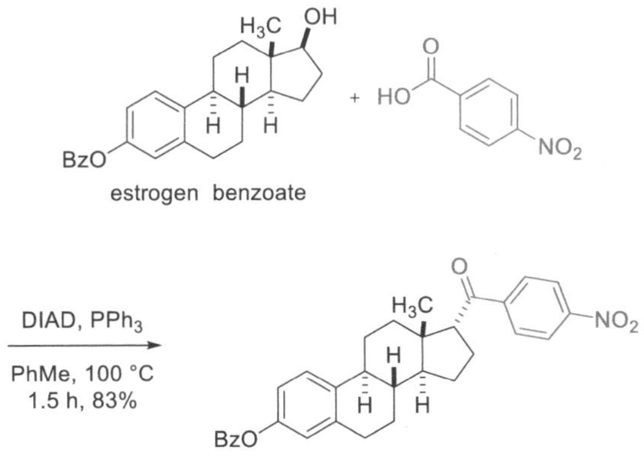

estrogen benzoate

DIAD, PPh₃
―――――――――→
PhMe, 100 °C
1.5 h, 83%

Example 8, Synthesis of an impurity during pramipexole process[14]

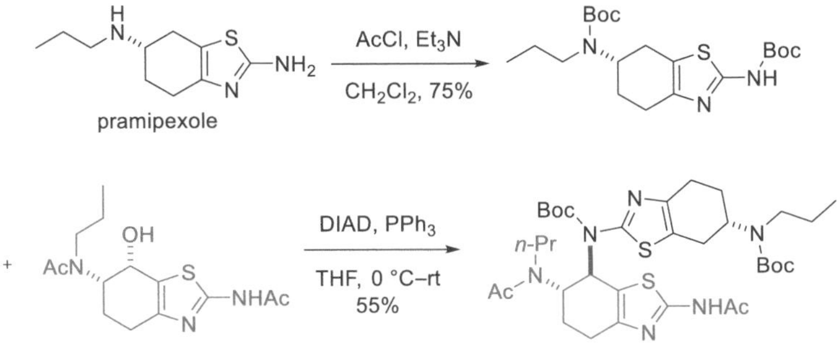

pramipexole

AcCl, Et₃N
―――――――→
CH₂Cl₂, 75%

DIAD, PPh₃
――――――――→
THF, 0 °C–rt
55%

Example 8, Removal of triphenylphosphine oxide (TPPO) by precipitation with zinc chloride in polar solvents (compatible with RNHBoc, but not compatible with compounds with basic amines)[15]

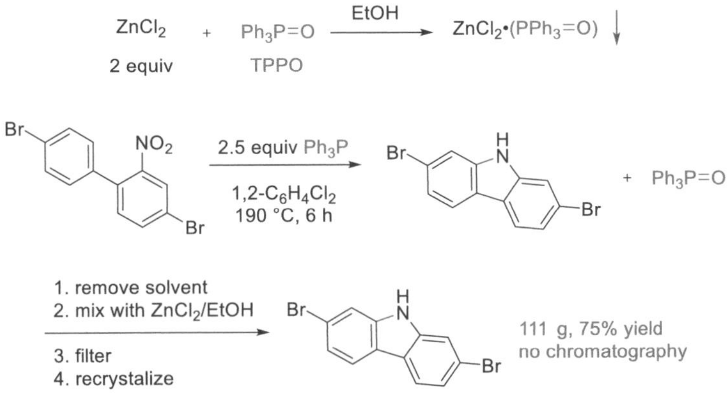

ZnCl₂   +   Ph₃P=O $\xrightarrow{\text{EtOH}}$ ZnCl₂•(PPh₃=O) ↓
2 equiv   TPPO

2.5 equiv Ph₃P
――――――――→
1,2-C₆H₄Cl₂
190 °C, 6 h

+   Ph₃P=O

1. remove solvent
2. mix with ZnCl₂/EtOH
――――――――――――→
3. filter
4. recrystalize

111 g, 75% yield
no chromatography

Example 10, Redox-neutral organocatalytic Mitsunobu reactions[16]

References

1. (a) Mitsunobu, O.; Yamada, M. *Bull. Chem. Soc. Jpn.* **1967**, *40*, 2380–2382. (b) Mitsunobu, O. *Synthesis* **1981**, 1–28. (Review).
2. Smith, A. B., III; Hale, K. J.; Rivero, R. A. *Tetrahedron Lett.* **1986**, *27*, 5813–5816.
3. Kocieński, P. J.; Yeates, C.; Street, D. A.; Campbell, S. F. *J. Chem. Soc., Perkin Trans. 1*, **1987**, 2183–2187.
4. Hughes, D. L. *Org. React.* **1992**, *42*, 335–656. (Review).
5. Hughes, D. L. *Org. Prep. Proc. Int.* **1996**, *28*, 127–164. (Review).
6. Vaccaro, W. D.; Sher, R.; Davis, H. R., Jr. *Bioorg. Med. Chem. Lett.* **1998**, *8*, 35–40.
7. Cevallos, A.; Rios, R.; Moyano, A.; Pericàs, M. A.; Riera, A. *Tetrahedron: Asymmetry* **2000**, *11*, 4407–4416.
8. Mukaiyama, T.; Shintou, T.; Fukumoto, K. *J. Am. Chem. Soc.* **2003**, *125*, 10538–10539.
9. Sumi, S.; Matsumoto, K.; Tokuyama, H.; Fukuyama, T. *Tetrahedron* **2003**, *59*, 8571–8587.
10. Lipshutz, B. H.; Chung, D. W.; Rich, B.; Corral, R. *Org. Lett.* **2006**, *8*, 5069–5072. [Di-*p*-chlorobenzyl azodicarboxylate (DCAD), a stable, solid alternative to DEAD and DIAD].
11. Christen, D. P. *Mitsunobu reaction.* In *Name Reactions for Homologations-Part II*; Li, J. J., Ed.; Wiley: Hoboken, NJ, **2009**, pp 671–748. (Review).
12. Ganesan, M.; Salunke, R. V.; Singh, N.; Ramesh, N. G. *Org. Biomol. Chem.* **2013**, *11*, 559–611.
13. Cardoso, F. S. P.; Mickle, G. E.; da Silva, M. A.; Baraldi, P. T.; Ferreira, F. B. *Org. Process Res. Dev.* **2016**, *20*, 306–311.
14. Hu, T.; Yang, F.; Jiang, T.; Chen, W.; Zhang, J.; Li, J.; Jiang, X.; Shen, J. *Org. Process Res. Dev.* **2016**, *20*, 1899–1905.
15. Batesky, D. C.; Goldfogel, M. J.; Weix, D. J. *J. Org. Chem.* **2017**, *82*, 9931–9936. (Removal of TPPO).
16. Beddoe, R. H.; Andrews, K. G.; Magne, V.; Cuthbertson, J. D.; Saska, J.; Shannon-Little, A. L.; Shanahan, S. E.; Sneddon, H. F.; Denton, R. M. *Science* **2019**, *365*, 910–914. (Redox-neutral organocatalytic Mitsunobu).
17. Howard, E. H.; Cain, C. F.; Kang, C.; Del Valle, J. R. *J. Org. Chem.* **2020**, *85*, 1680–1686.

Miyaura Borylation

Palladium-catalyzed reaction of aryl halides with diboron reagents to produce arylboronates. Also known as Hosomi–Miyaura borylation.

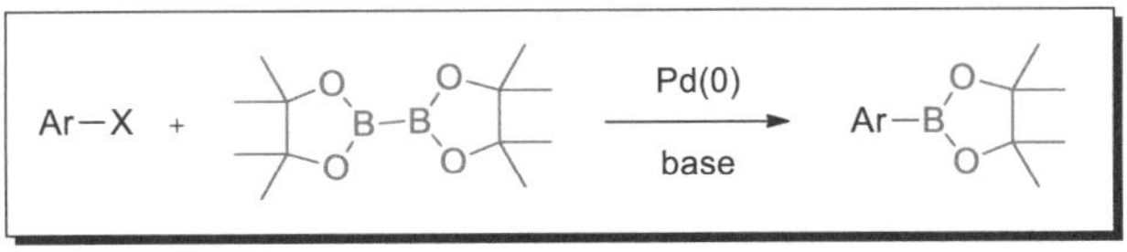

X = I, Br, Cl, OTf.

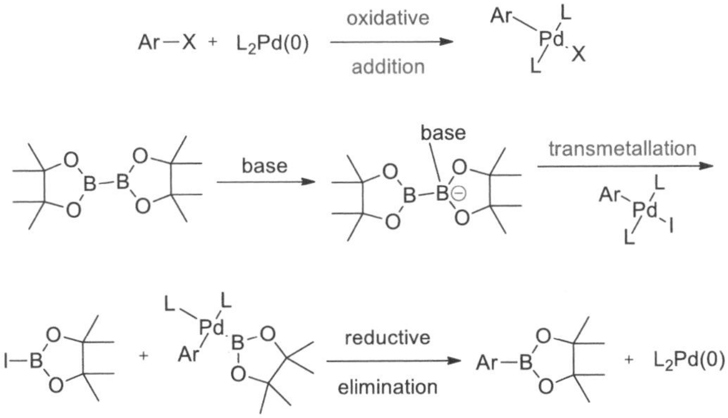

Example 1[7]

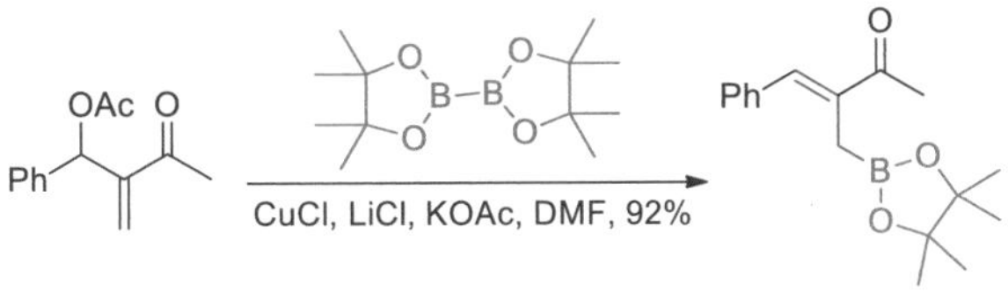

CuCl, LiCl, KOAc, DMF, 92%

Example 2[8]

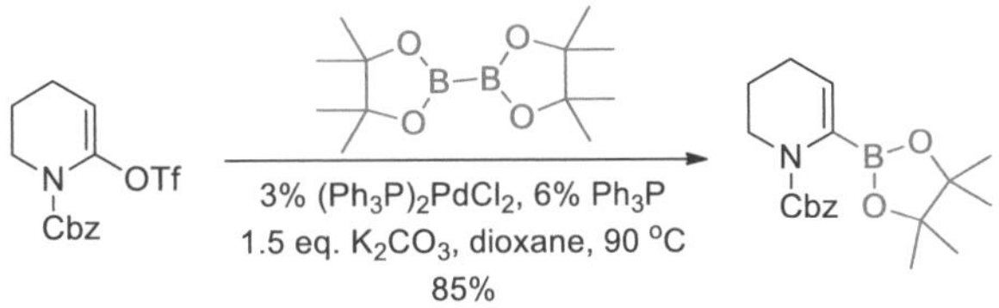

3% (Ph$_3$P)$_2$PdCl$_2$, 6% Ph$_3$P
1.5 eq. K$_2$CO$_3$, dioxane, 90 °C
85%

© Springer Nature Switzerland AG 2021
J. J. Li, *Name Reactions*, https://doi.org/10.1007/978-3-030-50865-4_97

Example 3[9]

Example 4, One-pot synthesis of biindolyl[10]

Example 4, An efficient Miyaura borylation process using tetrahydroxydiboron[13]

Example 5, Synthesis of PI3Kδ inhibitors[4]

Example 6[15]

Example 7, Nickel-catalyzed Miyaura borylation using tetrahydroxydiboron[16]

References

1. Ishiyama, T.; Murata, M.; Miyaura, N. *J. Org. Chem.* **1995,** *60,* 7508−7510.
2. Miyaura, N.; Suzuki, A. *Chem. Rev.* **1995,** *95,* 2457−2483. (Review).
3. Suzuki, A. *J. Organomet. Chem.* **1995,** *576,* 147−168. (Review).
4. Carbonnelle, A.-C.; Zhu, J. *Org. Lett.* **2000,** *2,* 3477−3480.
5. Giroux, A. *Tetrahedron Lett.* **2003,** *44,* 233−235.
6. Kabalka, G. W.; Yao, M.-L. *Tetrahedron Lett.* **2003,** *44,* 7885−7887.
7. Ramachandran, P. V.; Pratihar, D.; Biswas, D.; Srivastava, A.; Reddy, M. V. R. *Org. Lett.* **2004,** *6,* 481−484.
8. Occhiato, E. G.; Lo Galbo, F.; Guarna, A. *J. Org. Chem.* **2005,** *70,* 7324−7330.
9. Skaff, O.; Jolliffe, K. A.; Hutton, C. A. *J. Org. Chem.* **2005,** *70,* 7353−7363.

10. Duong, H. A.; Chua, S.; Huleatt, P. B.; Chai, C. L. L. *J. Org. Chem.* **2008**, *73*, 9177–9180.

11. Jo, T. S.; Kim, S. H.; Shin, J.; Bae, C. *J. Am. Chem. Soc.* **2009**, *131*, 1656–1657.

12. Marciasini, L. D.; Richy, N.; Vaultier, M.; Pucheault, M. *Adv. Synth. Cat.* **2013**, *355*, 1083–1088.

13. Gurung, S. R.; Mitchell, C.; Huang, J.; Jonas, M.; Strawser, J. D.; Daia, E.; Hardy, A.; O'Brien, E.; Hicks, F.; Papageorgiou, C. D. *Org. Process Res. Dev.* **2017**, *21*, 65–74.

14. Edney, D.; Hulcoop, D. G.; Leahy, J. H.; Vernon, L. E.; Wipperman, M. D.; Bream, R. N.; Webb, M. R. *Org. Process Res. Dev.* **2018**, *22*, 368–376.

15. St-Jean, F.; Remarchuk, T.; Angelaud, R.; Carrera, D. E.; Beaudry, D.; Malhotra, S.; McClory, A.; Kumar, A.; Ohlenbusch, G.; Schuster, A. M.; et al. *Org. Process Res. Dev.* **2019**, *23*, 783–793.

16. Fan, C.; Wu, Q.; Zhu, C.; Wu, X.; Li, Y.; Luo, Y.; He, J.-B. *Org. Lett.* **2019**, *21*, 8888–8892.

17. Ring, O. T.; Campbell, A. D.; Hayter, B. R.; Powell, L. *Tetrahedron Lett.* **2020**, *61*, 151589.

Morita–Baylis–Hillman Reaction

Also known as the Baylis–Hillman reaction. It is a carbon–carbon bond-forming transformation of an electron-poor alkene with a carbon electrophile. Electron-poor alkenes include acrylic esters, acrylonitriles, vinyl ketones, vinyl sulfones, and acroleins. On the other hand, carbon electrophiles may be aldehydes, α-alkoxycarbonyl ketones, aldimines, and Michael acceptors.

General scheme:

$X = O, NR_2$, EWG = CO_2R, COR, CHO, CN, SO_2R, SO_3R, PO(OEt)$_2$, CONR$_2$, CH$_2$=CHCO$_2$Me

Catalytic tertiary amines:

DABCO   quinuclidine   indolizine

© Springer Nature Switzerland AG 2021
J. J. Li, *Name Reactions*, https://doi.org/10.1007/978-3-030-50865-4_98

E2 (bimolecular elimination) mechanism is also operative here:

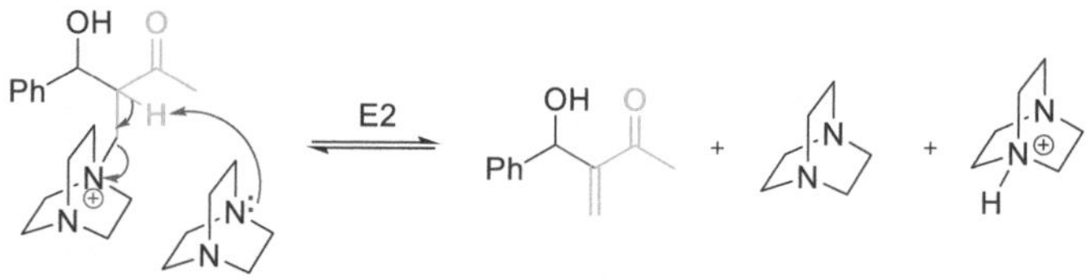

Example 1, Intramolecular Baylis–Hillman reaction[6]

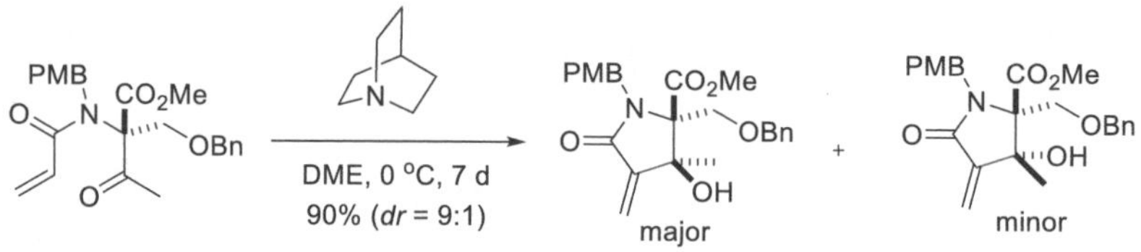

Example 2[7]

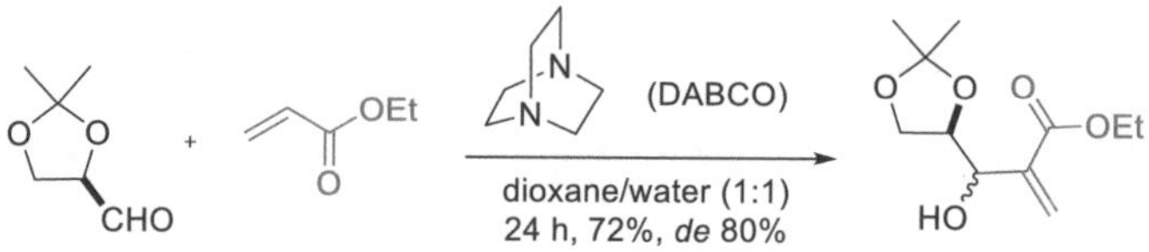

Example 3, TBAI = tetra-*n*-butylammonium iodide[8]

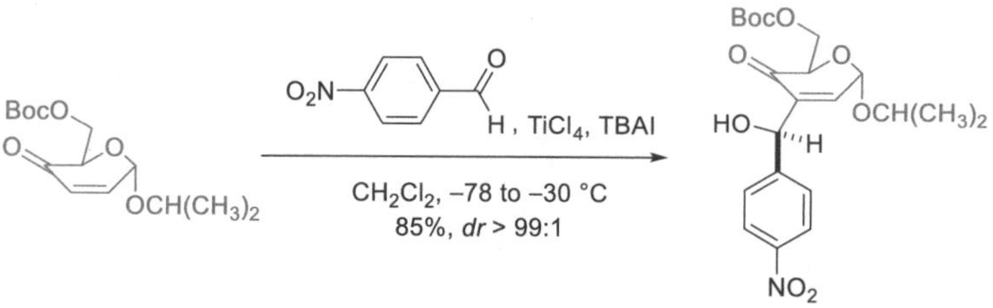

Example 4[9]

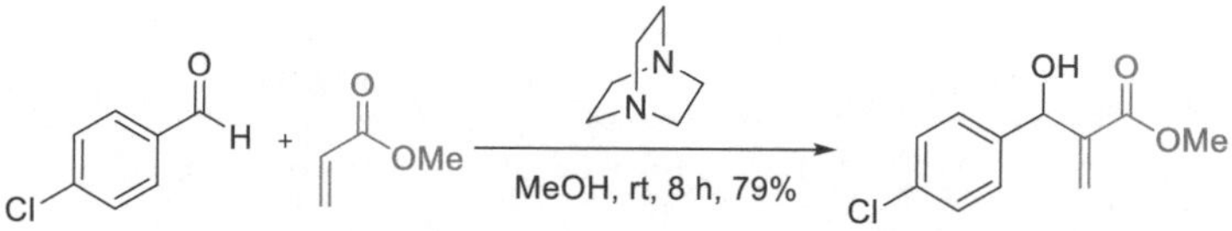

Example 5[10]

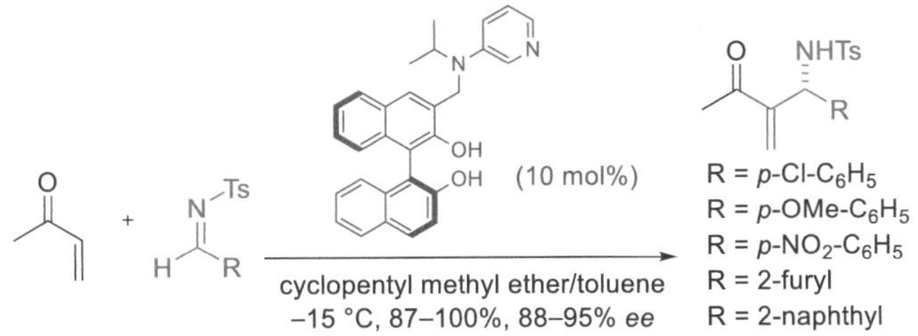

R = p-Cl-C_6H_5
R = p-OMe-C_6H_5
R = p-NO_2-C_6H_5
R = 2-furyl
R = 2-naphthyl

cyclopentyl methyl ether/toluene
−15 °C, 87–100%, 88–95% ee

Example 6[13]

PhSeLi
THF

NH_4Cl
87%

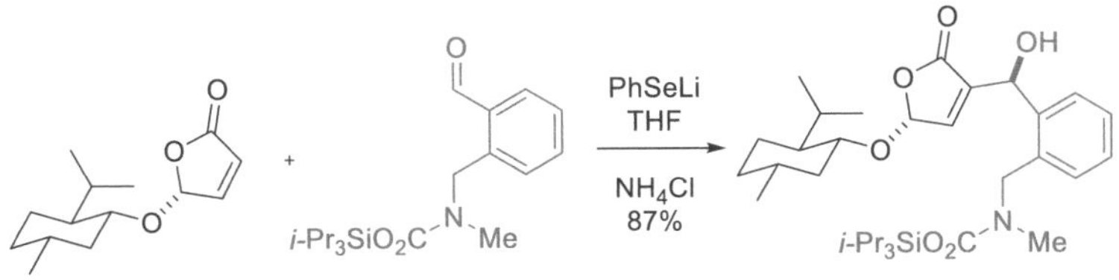

Example 7, Toward the synthesis of a 1,4-oxazepane noradrenaline reuptake inhibitor (NRI)[16]

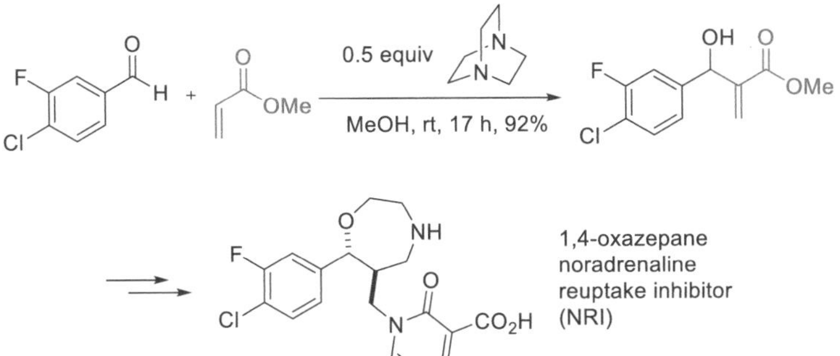

0.5 equiv

MeOH, rt, 17 h, 92%

1,4-oxazepane
noradrenaline
reuptake inhibitor
(NRI)

Example 8, Intramolecular aza-Morita–Baylis–Hillman reaction[17]

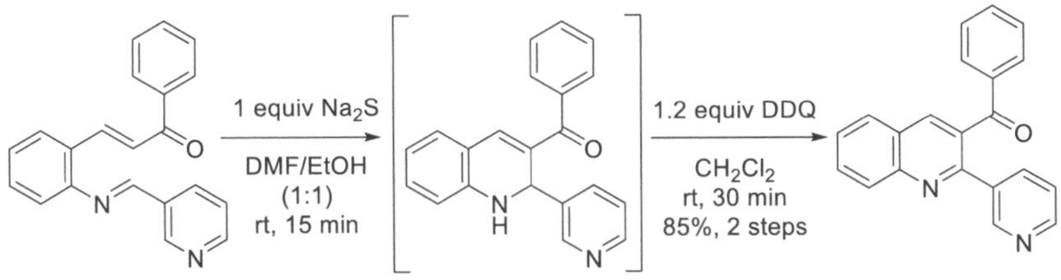

1 equiv Na_2S

DMF/EtOH
(1:1)
rt, 15 min

1.2 equiv DDQ

CH_2Cl_2
rt, 30 min
85%, 2 steps

Example 9, Catalytic enantioselective transannular Morita–Baylis–Hillman reaction[18]

References

1. Baylis, A. B.; Hillman, M. E. D. Ger. Pat. 2,155,113, (**1972**). Both Anthony B. Baylis and Melville E. D. Hillman were chemists at Celanese Corp. USA.
2. Basavaiah, D.; Rao, P. D.; Hyma, R. S. *Tetrahedron* **1996,** *52,* 8001–8062. (Review).
3. Ciganek, E. *Org. React.* **1997,** *51,* 201–350. (Review).
4. Wang, L.-C.; Luis, A. L.; Agapiou, K.; Jang, H.-Y.; Krische, M. J. *J. Am. Chem. Soc.* **2002,** *124,* 2402–2403.
5. Frank, S. A.; Mergott, D. J.; Roush, W. R. *J. Am. Chem. Soc.* **2002,** *124,* 2404–2405.
6. Reddy, L. R.; Saravanan, P.; Corey, E. J. *J. Am. Chem. Soc.* **2004,** *126,* 6230–6231.
7. Krishna, P. R.; Narsingam, M.; Kannan, V. *Tetrahedron Lett.* **2004,** *45,* 4773–4775.
8. Sagar, R,; Pant, C. S.; Pathak, R.; Shaw, A. K. *Tetrahedron* **2004,** *60,* 11399–11406.
9. Mi, X.; Luo, S.; Cheng, J.-P. *J. Org. Chem.* **2005,** *70,* 2338–2341.
10. Matsui, K.; Takizawa, S.; Sasai, H. *J. Am. Chem. Soc.* **2005,** *127,* 3680–3681.
11. Price, K. E.; Broadwater, S. J.; Jung, H. M.; McQuade, D. T. *Org. Lett.* **2005,** *7,* 147–150. A novel mechanism involving a hemiacetal intermediate is proposed.
12. Limberakis, C. *Morita–Baylis–Hillman Reaction.* In *Name Reactions for Homologations-Part I*; Li, J. J., Ed.; Wiley: Hoboken, NJ, **2009,** pp 350–380. (Review).
13. Cheng, P.; Clive, D. L. J. *J. Org. Chem.* **2012,** *77,* 3348–3364.
14. Wei, Y.; Shi, M. *Chem. Rev.* **2013,** *113,* 6659–6690. (Review).
15. Pellissier, H. *Tetrahedron* **2017,** *73,* 2831–2861. (Review).
16. Ishimoto, K.; Yamaguchi, K.; Nishimoto, A.; Murabayashi, M.; Ikemoto, T. *Org. Process Res. Dev.* **2017,** *21,* 2001–2011.
17. Bharadwaj, K. C. *J. Org. Chem.* **2018,** *83,* 14498–14506.
18. Mato, R.; Manzano, R.; Reyes, E.; Carrillo, L.; Uria, U.; Vicario, J. L. *J. Am. Chem. Soc.* **2019,** *141,* 9495–9499.
19. Helberg, J.; Ampssler, T.; Zipse, H. *J. Org. Chem.* **2020,** *85,* 5390–5402.

Mukaiyama Aldol Reaction

Lewis acid-catalyzed aldol condensation of aldehyde and silyl enol ether.

Example 1, Intramolecular Mukaiyama aldol reaction[3]

Example 2, Intermolecular Mukaiyama aldol reaction (DTBMP = 2,6-di-*tert*-butyl-4-methylpyridine)[7]

© Springer Nature Switzerland AG 2021
J. J. Li, *Name Reactions*, https://doi.org/10.1007/978-3-030-50865-4_99

Example 3, Vinylogous Mukaiyama aldol reaction[8]

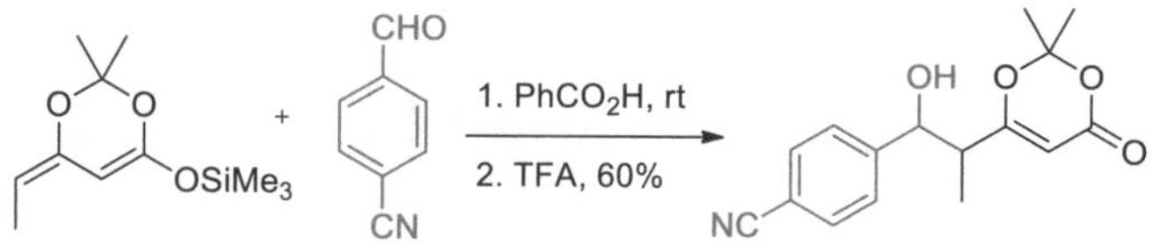

Example 4, Asymmetric Mukaiyama aldol reaction[10]

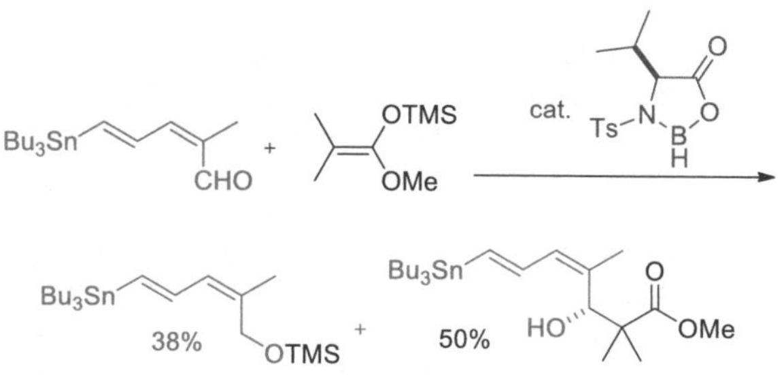

Example 5[12]

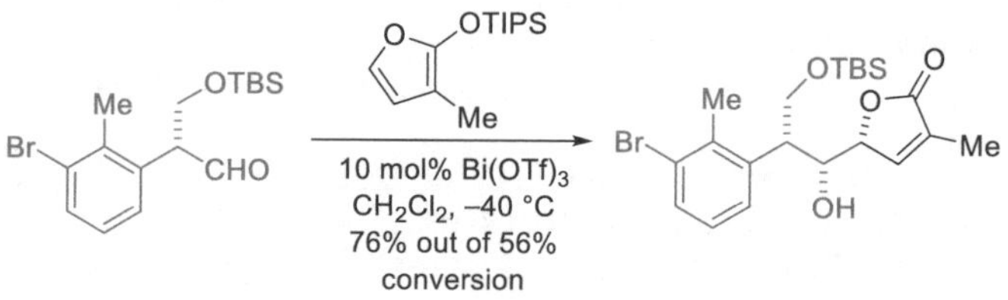

10 mol% Bi(OTf)$_3$
CH$_2$Cl$_2$, −40 °C
76% out of 56%
conversion

Example 6[13]

TMS-OTf, Et$_3$N

CH$_3$CN, 0–22 °C
90%

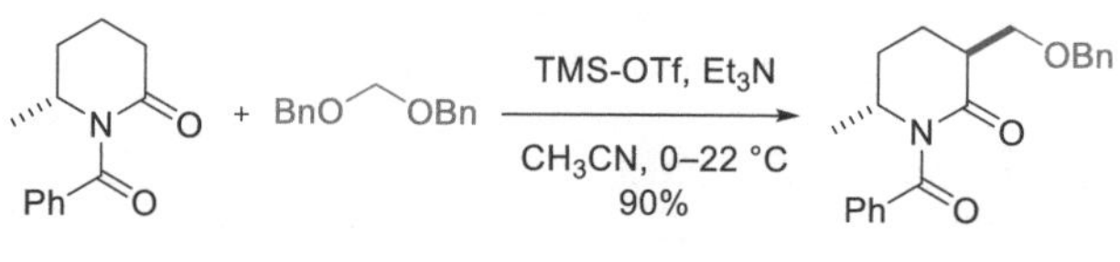

Example 7[14]

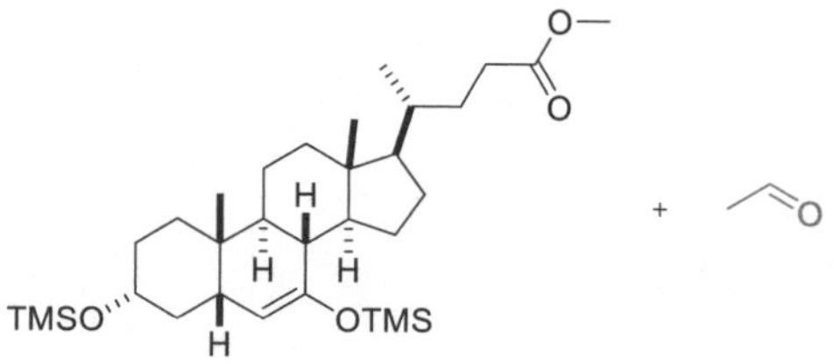

BF₃·CH₃CN, CH₂Cl₂
–60 to –65 °C, 2.5 h

and then warmed to
–5 to –10 °C, 2 h
~ 80%

References

1. (a) Mukaiyama, T.; Narasaka, K.; Banno, K. *Chem. Lett.* **1973,** 1011–1014. (b) Mukaiyama, T.; Narasaka, K.; Banno, K. *J. Am. Chem. Soc*. **1974,** *96*, 7503–7509.
2. Ishihara, K.; Kondo, S.; Yamamoto, H. *J. Org. Chem.* **2000,** *65*, 9125–9128.
3. Armstrong, A.; Critchley, T. J.; Gourdel-Martin, M.-E.; Kelsey, R. D.; Mortlock, A. A. *J. Chem. Soc., Perkin Trans. 1* **2002,** 1344–1350.
4. Clézio, I. L.; Escudier, J.-M.; Vigroux, A. *Org. Lett.* **2003,** *5*, 161–164.
5. Ishihara, K.; Yamamoto, H. *Boron and Silicon Lewis Acids for Mukaiyama Aldol Reactions.* In *Modern Aldol Reactions* Mahrwald, R., Ed.; **2004,** 25–68. (Review).
6. Mukaiyama, T. *Angew. Chem. Int. Ed.* **2004,** *43*, 5590–5614. (Review).
7. Adhikari, S.; Caille, S.; Hanbauer, M.; Ngo, V. X.; Overman, L. E. *Org. Lett.* **2005,** *7*, 2795–2797.
8. Acocella, M. R.; Massa, A.; Palombi, L.; Villano, R.; Scettri, A. *Tetrahedron Lett.* **2005,** *46*, 6141–6144.
9. Jiang, X.; Liu, B.; Lebreton, S.; De Brabander, J. K. *J. Am. Chem. Soc.* **2007,** *129*, 6386–6387.
10. Webb, M. R.; Addie, M. S.; Crawforth, C. M.; Dale, J. W.; Franci, X.; Pizzonero, M.; Donald, C.; Taylor, R. J. K. *Tetrahedron* **2008,** *64*, 4778–4791.
11. Frings, M.; Atodiresei, I.; Runsink, J.; Raabe, G.; Bolm, C. *Chem. Eur. J.* **2009,** *15*, 1566–1569.
12. Gao, S.; Wang, Q.; Chen, C. *J. Am. Chem. Soc.* **2009,** *131*, 1410–1412.
13. Matsuo, J.-i.; Murakami, M. *Angew. Chem. Int. Ed.* **2013,** *52*, 9109–9118. (Review).
14. Chung, J. Y. L.; Zhong, Y.-L.; Maloney, K. M.; Reamer, R.A.; Moore, J. C.; Strotman, H.; Kalinin, A.; Feng, R.; Strotman, N. A.; Xiang, B.; et al. *Org. Lett.* **2014,** *16*, 5890–5893.
15. Hosokawa, S. *Tetrahedron Lett.* **2018,** *59*, 77–88. (Review).
16. Feng, W.-D.; Zhuo, S.-M.; Zhang, F.-L. *Org. Process Res. Dev.* **2019,** *23*, 1979–1989.
17. Bressin, R. K.; Osman, S.; Pohorilets, I.; Basu, U.; Koide, K. *J. Org. Chem.* **2020,** *85*, 4637–4647.

Mukaiyama Michael Addition

Lewis acid-catalyzed Michael addition of silyl enol ether to an α,β-unsaturated system.

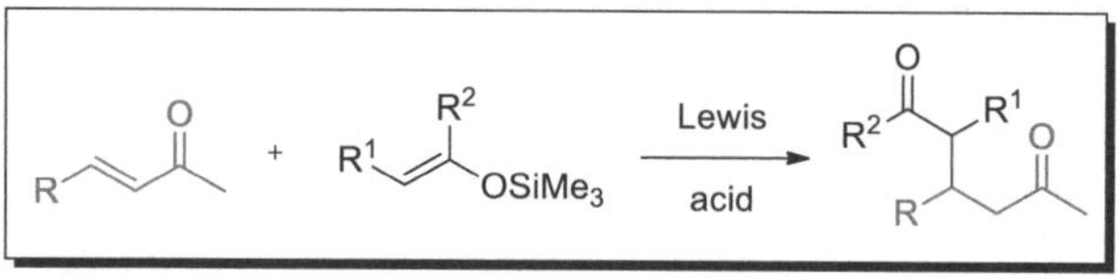

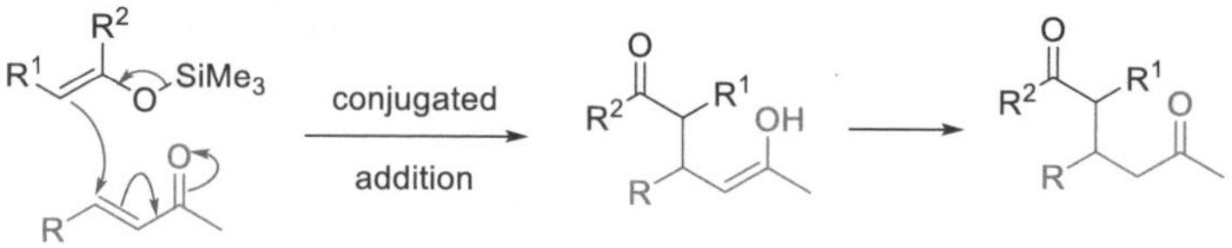

Example 1[2]

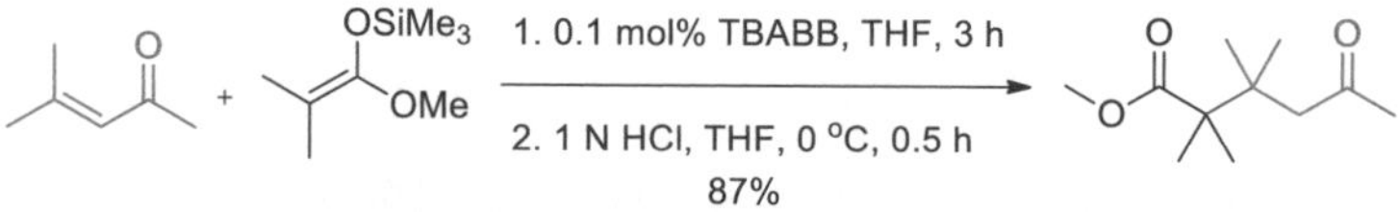

TBABB = tetra-*n*-butylammonium bibenzoate

Example 2[5]

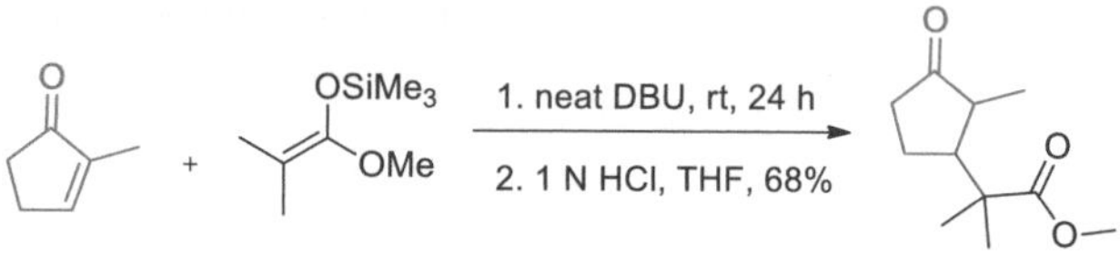

Example 3[8]

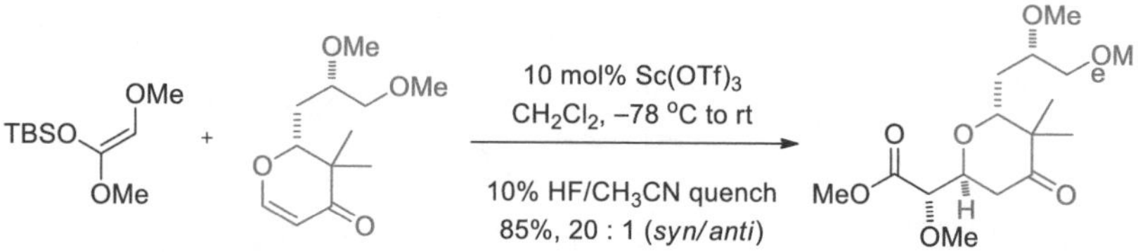

© Springer Nature Switzerland AG 2021
J. J. Li, *Name Reactions*, https://doi.org/10.1007/978-3-030-50865-4_100

Example 4, Intramolecular Mukaiyama–Michael reaction[9]

Example 5, Enantioselective Mukaiyama–Michael reaction[11]

Example 6, Mukaiyama–Michael reaction to prepare Rauhut–Currier-type products[12]

Example 7, A γ-addition[13]

Example 8[14]

References

1. (a) Mukaiyama, T.; Narasaka, K.; Banno, K. *Chem. Lett.* **1973**, 1011–1014. (b) Mukaiyama, T.; Narasaka, K.; Banno, K. *J. Am. Chem. Soc.* **1974**, *96*, 7503–7509. (c) Mukaiyama, T. *Angew. Chem. Int. Ed.* **2004**, *43*, 5590–5614. (Review).
2. Gnaneshwar, R.; Wadgaonkar, P. P.; Sivaram, S. *Tetrahedron Lett.* **2003**, *44*, 6047–6049.
3. Wang, X.; Adachi, S.; Iwai, H.; Takatsuki, H.; Fujita, K.; Kubo, M.; Oku, A.; Harada, T. *J. Org. Chem.* **2003**, *68*, 10046–10057.
4. Jaber, N.; Assie, M.; Fiaud, J.-C.; Collin, J. *Tetrahedron* **2004**, *60*, 3075–3083.
5. Shen, Z.-L.; Ji, S.-J.; Loh, T.-P. *Tetrahedron Lett.* **2005**, *46*, 507–508.
6. Wang, W.; Li, H.; Wang, J. *Org. Lett.* **2005**, *7*, 1637–1639.
7. Ishihara, K.; Fushimi, M. *Org. Lett.* **2006**, *8*, 1921–1924.
8. Jewett, J. C.; Rawal, V. H. *Angew. Chem. Int. Ed.* **2007**, *46*, 6502–6504.
9. Liu, Y.; Zhang, Y.; Jee, N.; Doyle, M. P. *Org. Lett.* **2008**, *10*, 1605–1608.
10. Takahashi, A.; Yanai, H.; Taguchi, T. *Chem. Commun.* **2008**, 2385–2387.
11. Rout, S.; Ray, S. K.; Singh, V. K. *Org. Biomol. Chem.* **2013**, *11*, 4537–4545.
12. Frias, M.; Mas-Ballesté, R.; Arias, S.; Alvarado, C.; Alemán, J. *J. Am. Chem. Soc.* **2017**, *139*, 672–679.
13. Sharma, B. M.; Shinde, D. R.; Jain, R.; Begari, E.; Sathaiya, S.; Gonnade, R. G.; Kumar, P. *Org. Lett.* **2018**, *20*, 2787–2797.
14. Gu, Q.; Wang, X.; Sun, B.; Lin, G. *Org. Lett.* **2019**, *21*, 5082–5085.
15. Kortet, S.; Claraz, A.; Pihko, P. M. *Org. Lett.* **2020**, *22*, 3010–3013.

Mukaiyama Reagent

Pyridinium halide reagent for esterification or amide formation.

General scheme:

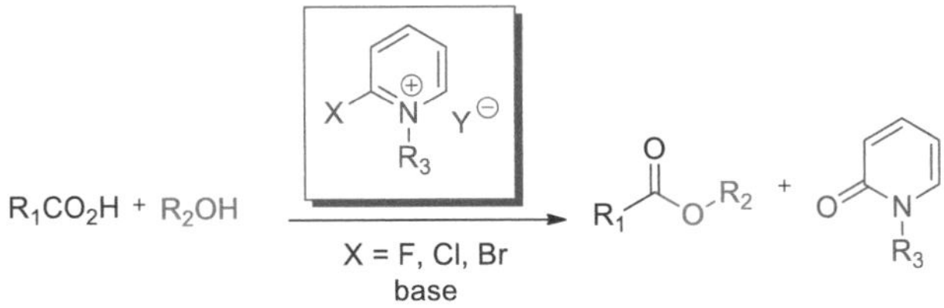

$$R_1CO_2H + R_2OH \xrightarrow[\text{base}]{\substack{X = F, Cl, Br}} R_1 \overset{O}{\underset{}{C}} O\text{-}R_2 + \text{(pyridinone)}$$

Example 1[1c]

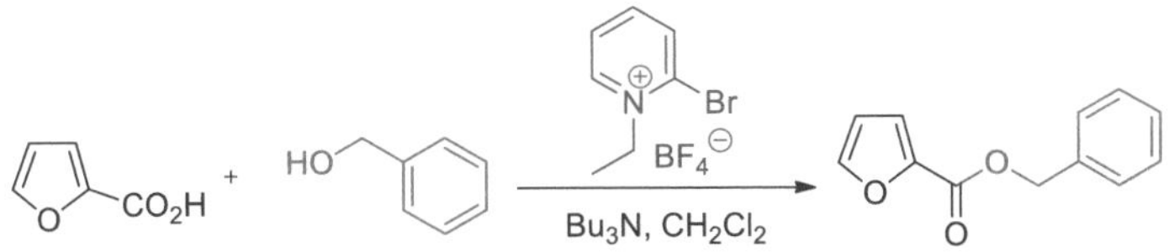

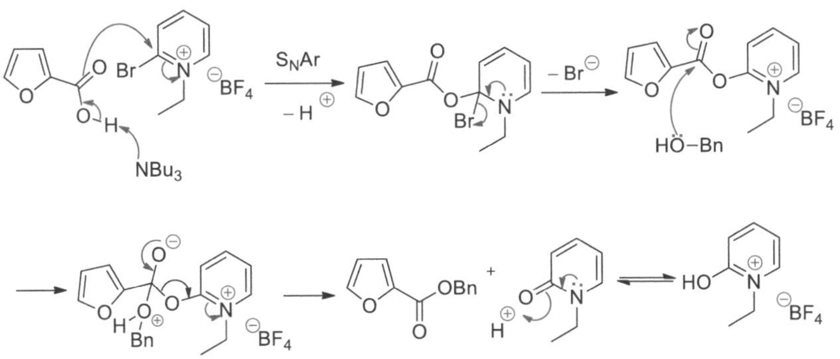

Amide formation using the Mukaiyama reagent follows a similar mechanistic pathway.[1d]

Example 2, Polymer-supported Mukaiyama reagent[5]

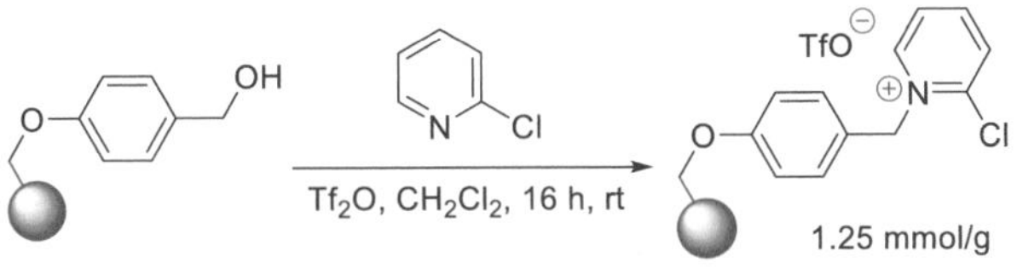

1.25 mmol/g

© Springer Nature Switzerland AG 2021
J. J. Li, *Name Reactions*, https://doi.org/10.1007/978-3-030-50865-4_101

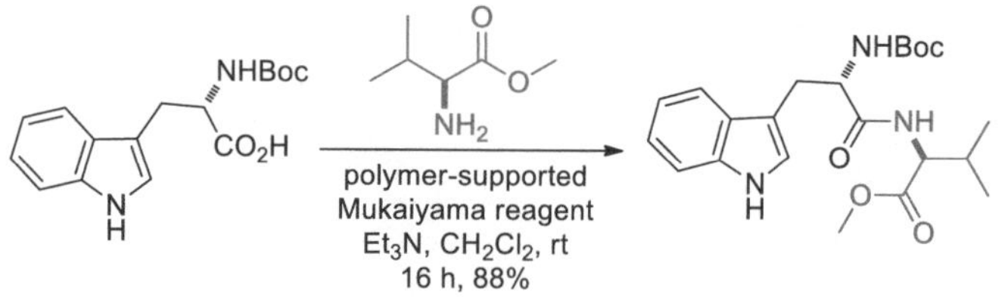

Example 3[9]

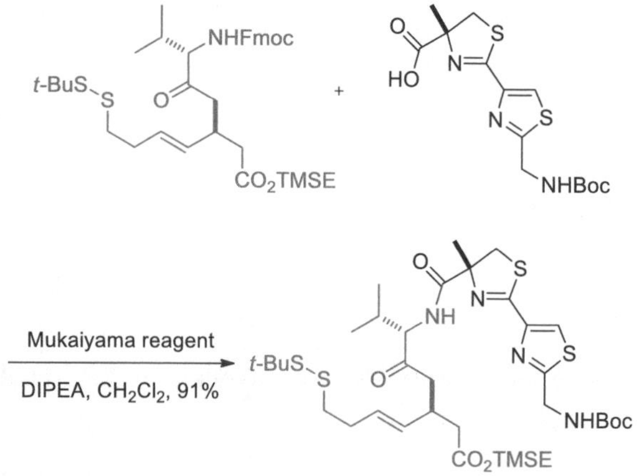

Example 4, Fluorous Mukaiyama reagent[10]

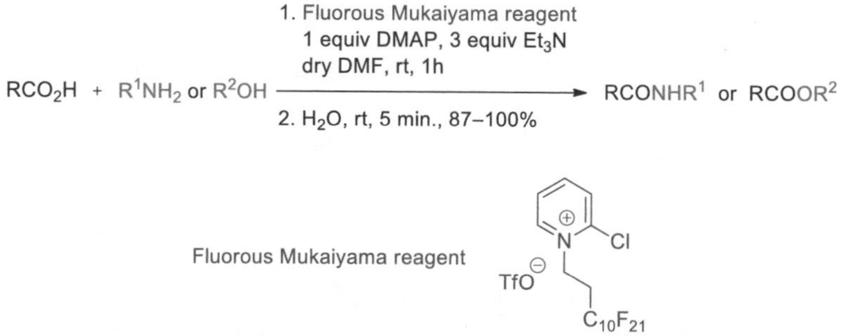

$$RCO_2H \ + \ R^1NH_2 \ or \ R^2OH \xrightarrow[\text{2. H}_2\text{O, rt, 5 min., 87--100\%}]{\substack{\text{1. Fluorous Mukaiyama reagent} \\ \text{1 equiv DMAP, 3 equiv Et}_3\text{N} \\ \text{dry DMF, rt, 1h}}} RCONHR^1 \ or \ RCOOR^2$$

Fluorous Mukaiyama reagent

Example 5, Lactamization[12]

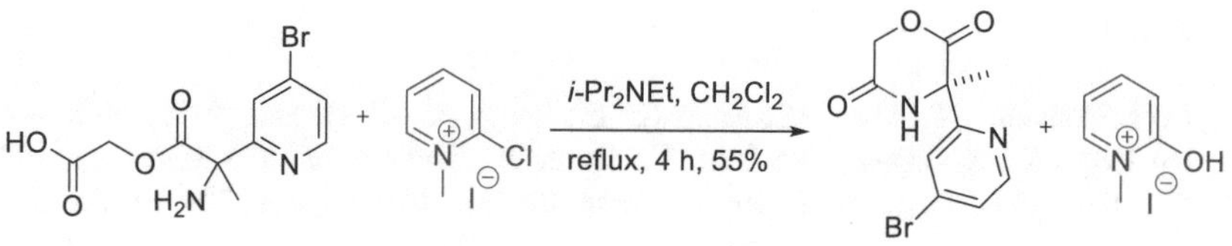

Example 6, Preparing an UDP-3-*O*-(acyl)-*N*-acetylglucosamine deacetylase (LpxC) inhibitor[13]

Example 7, Preparing selective acetyl-coa carboxylase (ACC) 1 inhibitors[14]

References

1. (a) Mukaiyama, T.; Usui, M.; Shimada, E.; Saigo, K. *Chem. Lett.* **1975,** 1045–1048. (b) Hojo, K.; Kobayashi, S.; Soai, K.; Ikeda, S.; Mukaiyama, T. *Chem. Lett.* **1977,** 635–636. (c) Mukaiyama, T. *Angew. Chem. Int. Ed.* **1979,** *18*, 707–708. (d) For amide

formation, see: Huang, H.; Iwasawa, N.; Mukaiyama, T. *Chem. Lett.* **1984,** 1465–1466.

2. Nicolaou, K. C.; Bunnage, M. E.; Koide, K. *J. Am. Chem. Soc.* **1994,** *116*, 8402–8403.
3. Yong, Y. F.; Kowalski, J. A.; Lipton, M. A. *J. Org. Chem.* **1997,** *62*, 1540–1542.
4. Folmer, J. J.; Acero, C.; Thai, D. L.; Rapoport, H. *J. Org. Chem.* **1998,** *63*, 8170–8182.
5. Crosignani, S.; Gonzalez, J.; Swinnen, D. *Org. Lett.* **2004,** *6*, 4579–4582.
6. Mashraqui, S. H.; Vashi, D.; Mistry, H. D. *Synth. Commun.* **2004,** *34*, 3129–3134.
7. Donati, D.; Morelli, C.; Taddei, M. *Tetrahedron Lett.* **2005,** *46*, 2817–2819.
8. Vandromme, L.; Monchaud, D.; Teulade-Fichou, M.-P. *Synlett* **2006,** 3423–3426.
9. Ren, Q.; Dai, L.; Zhang, H.; Tan, W.; Xu, Z.; Ye, T. *Synlett* **2008,** 2379–2383.
10. Matsugi, M.; Suganuma, M.; Yoshida, S.; Hasebe, S.; Kunda, Y.; Hagihara, K.; Oka, S. *Tetrahedron Lett.* **2008,** *49*, 6573–6574.
11. Novosjolova, I. *Synlett* **2013,** *24*, 135–136. (Review).
12. Murphy-Benenato, K. E.; Olivier, N.; Choy, A.; Ross, P. L.; Miller, M. D.; Thresher, J.; Gao, N.; Hale, M. R. *ACS Med. Chem. Lett.* **2014,** *5*, 1213–1218.
13. Rombouts, F. J. R.; Tresadern, G.; Delgado, O.; Martinez-Lamenca, C.; Van Gool, M.; Garcia-Molina, A.; Alonso de Diego, S. A.; Oehlrich, D.; Prokopcova, H.; Alonso, J. M.; et al. *J. Med. Chem.* **2015,** *58*, 8216–8235.
14. Mizojiri, R.; Asano, M.; Tomita, D.; Banno, H.; Nii, N.; Sasaki, M.; Sumi, H.; Satoh, Y.; Yamamoto, Y.; Moriya, T.; et al. *J. Med. Chem.* **2018,** *61*, 1098–1117.
15. Chen, L.; Luo, G. *Tetrahedron Lett.* **2019,** *60*, 268–271.
16. Ikeuchi, K.; Ueji, T.; Matsumoto, S.; Wakamori, S.; Yamada, H. *Eur. J. Org. Chem.* **2020,** 2077–2085.

Nazarov Cyclization

Acid-catalyzed electrocyclic formation of cyclopentenone from di-vinyl ketone.

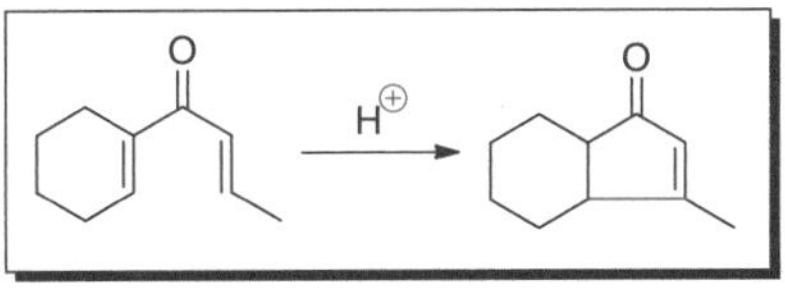

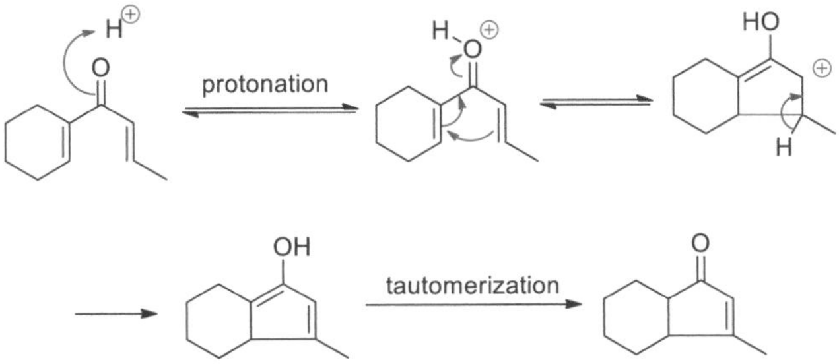

Example 1[2]

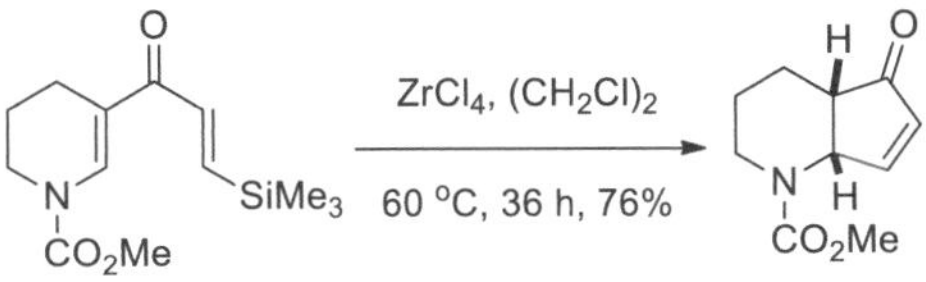

ZrCl$_4$, (CH$_2$Cl)$_2$
60 °C, 36 h, 76%

Example 2[6]

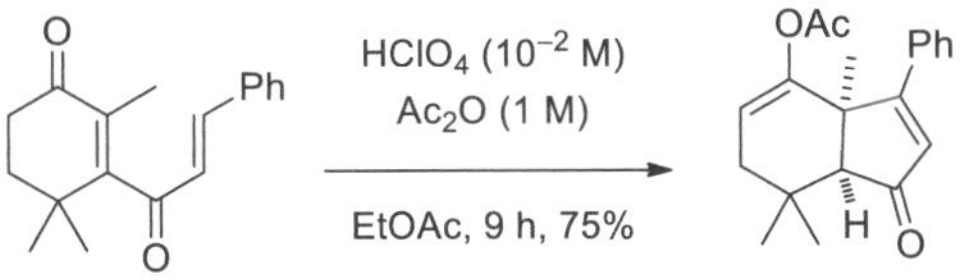

HClO$_4$ (10^{-2} M)
Ac$_2$O (1 M)

EtOAc, 9 h, 75%

Example 3[9]

5 mol% Cu(ClO$_4$)$_2$

DCE, 45 °C, 8 h, 80%

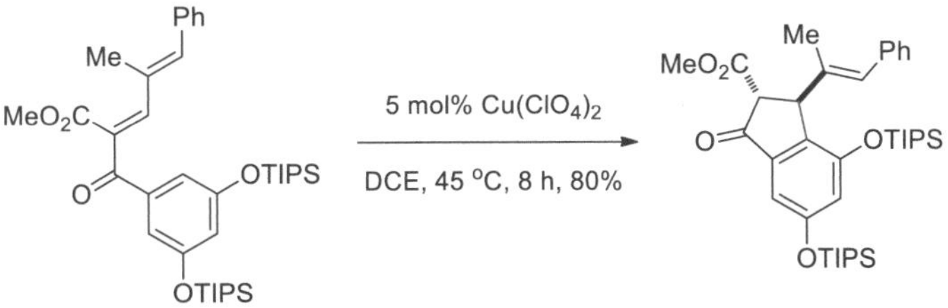

© Springer Nature Switzerland AG 2021
J. J. Li, *Name Reactions*, https://doi.org/10.1007/978-3-030-50865-4_102

Example 4[10]

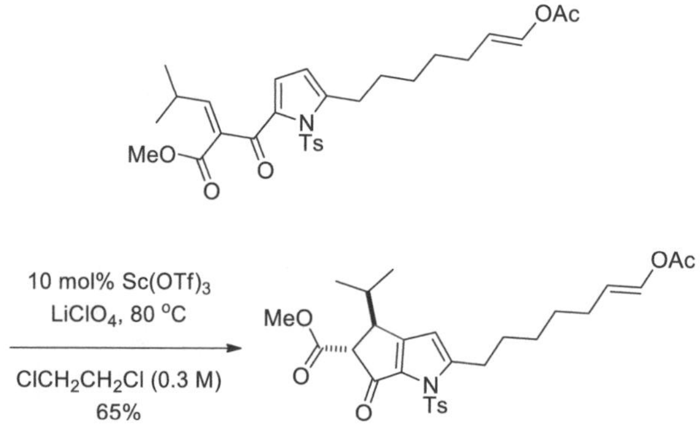

Example 5, An example with a different mechanism[11]

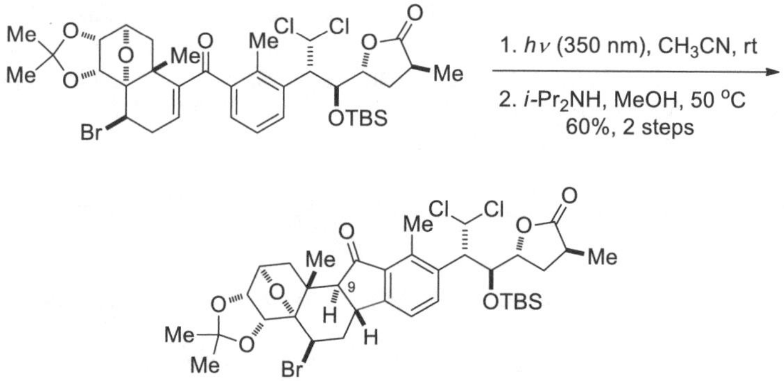

Example 6[12]

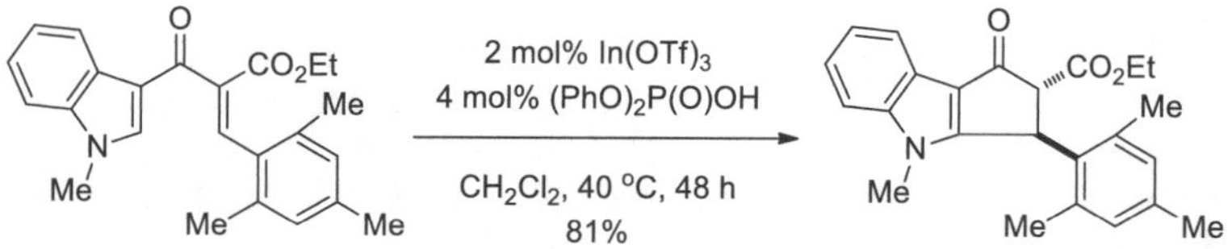

Example 7, Iron-mediated domino interrupted iso-Nazarov/dearomative (3 + 2)-cycloaddition of electrophilic indoles[14]

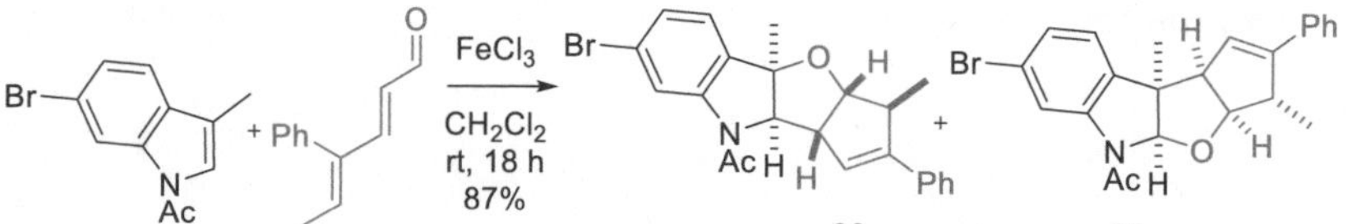

Example 8, One-pot cationic cascade to haloindene involving a halo-Nazarov cyclization[16]

References

1. Nazarov, I. N.; Torgov, I. B.; Terekhova, L. N. *Bull. Acad. Sci. (USSR)* **1942**, 200. I. N. Nazarov (1900–1957), a Soviet Union scientist, discovered this reaction in 1942. It was said that almost as many young synthetic chemists have been lost in the pursuit of an asymmetric Nazarov cyclization as of the Bayliss–Hillman reaction.
2. Denmark, S. E.; Habermas, K. L.; Hite, G. A. *Helv. Chim. Acta* **1988**, *71*, 168–194; 195–208.
3. Habermas, K. L.; Denmark, S. E.; Jones, T. K. *Org. React.* **1994**, *45*, 1–158. (Review).
4. Kim, S.-H.; Cha, J. K. *Synthesis* **2000**, 2113–2116.
5. Giese, S.; West, F. G. *Tetrahedron* **2000**, *56*, 10221–10228.
6. Mateos, A. F.; de la Nava, E. M. M.; González, R. R. *Tetrahedron* **2001**, *57*, 1049–1057.
7. Harmata, M.; Lee, D. R. *J. Am. Chem. Soc.* **2002**, *124*, 14328–14329.
8. Leclerc, E.; Tius, M. A. *Org. Lett.* **2003**, *5*, 1171–1174.
9. Marcus, A. P.; Lee, A. S.; Davis, R. L.; Tantillo, D. J.; Sarpong, R. *Angew. Chem. Int. Ed.* **2008**, *47*, 6379–6383.
10. Bitar, A. Y.; Frontier, A. J. *Org. Lett.* **2009**, *11*, 49–52.
11. Gao, S.; Wang, Q.; Chen, C. *J. Am. Chem. Soc.* **2009**, *131*, 1410–1412.
12. Xi, Z.-G.; Zhu, L.; Luo, S.; Cheng, J.-P. *J. Org. Chem.* **2013**, *78*, 606–613.
13. Di Grandi, M. J. *Org. Biomol. Chem.* **2014**, *12*, 5331–5345. (Review).
14. Marques, A.-S.; Coeffard, V.; Chataigner, I.; Vincent, G.; Moreau, X. *Org. Lett.* **2016**, *18*, 5296–5299.
15. Vinogradov, M. G.; Turova, O. V.; Zlotin, S. G. *Org. Biomol. Chem.* **2017**, *15*, 8245–8269. (Review).
16. Holt, C.; Alachouzos, G.; Frontier, A. J. *J. Am. Chem. Soc.* **2019**, *141*, 5461–5469.
17. Corbin, J. R.; Ketelboeter, D. R.; Fernandez, I.; Schomaker, J. M. *J. Am. Chem. Soc.* **2020**, *142*, 5568–5573.

Neber Rearrangement

α-Aminoketone from tosyl ketoxime and base. The net conversion of a ketone into an α-aminoketone *via* the oxime.

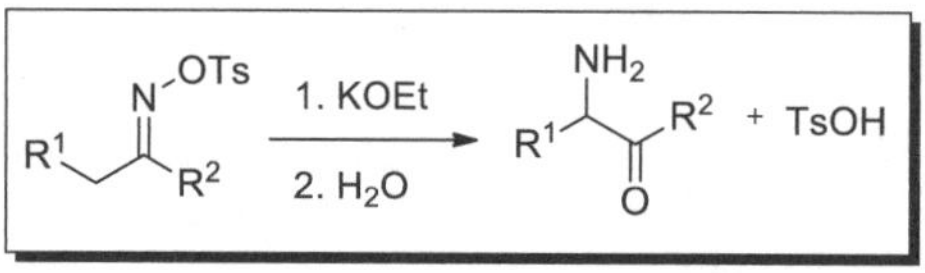

ketoxime   α-aminoketone

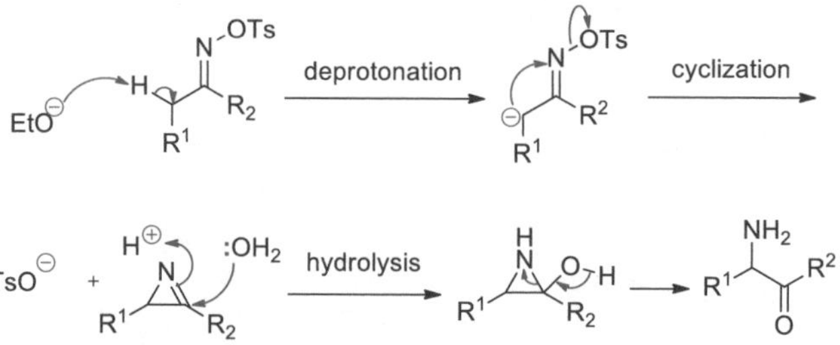

azirine intermediate

Example 1[3]

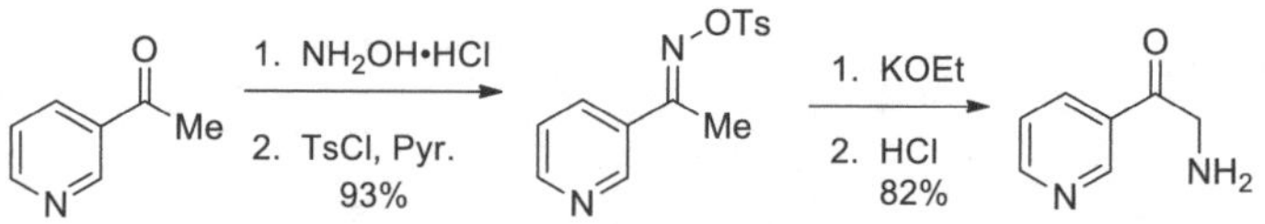

Example 2, A variant using iminochloride[5]

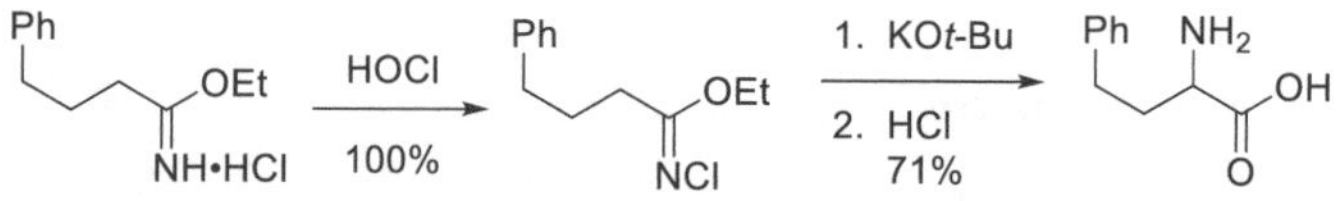

Example 3[8]

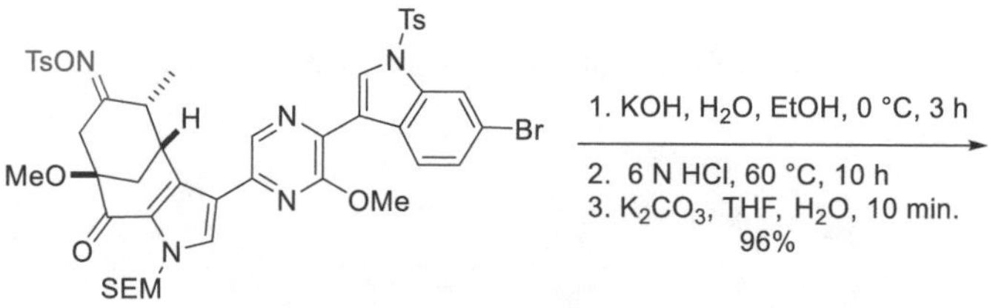

1. KOH, H₂O, EtOH, 0 °C, 3 h

2. 6 N HCl, 60 °C, 10 h
3. K₂CO₃, THF, H₂O, 10 min.
    96%

© Springer Nature Switzerland AG 2021
J. J. Li, *Name Reactions*, https://doi.org/10.1007/978-3-030-50865-4_103

Example 4[9]

Example 5, PNB = *p*-nitrobenzyl[11]

Example 6, One-pot synthesis of pyridine employing the Neber reaction[13]

Example 7, Toward total synthesis of (*R*)-(*Z*)-antazirine[14]

Example 8, *C*-Attack prevails (acidic conditions favor *O*-attack)[15]

Example 9, Trifluoromethyl-azirines via metal-free, Et₃N-mediated Neber reaction[16]

References

1. Neber, P. W.; v. Friedolsheim, A. *Ann.* **1926,** *449,* 109–134.
2. O'Brien, C. *Chem. Rev.* **1964,** *64,* 81–89. (Review).
3. LaMattina, J. L.; Suleske, R. T. *Synthesis* **1980,** 329–330.
4. Verstappen, M. M. H.; Ariaans, G. J. A.; Zwanenburg, B. *J. Am. Chem. Soc.* **1996,** *118,* 8491–8492.
5. Oldfield, M. F.; Botting, N. P. *J. Labeled Compd. Radiopharm.* **1998,** *16,* 29–36.
6. Palacios, F.; Ochoa de Retana, A. M.; Gil, J. I. *Tetrahedron Lett.* **2002,** *41,* 5363–5366.
7. Ooi, T.; Takahashi, M.; Doda, K.; Maruoka, K. *J. Am. Chem. Soc.* **2002,** *124,* 7640–7641.
8. Garg, N. K.; Caspi, D. D.; Stoltz, B. M. *J. Am. Chem. Soc.* **2005,** *127,* 5970–5978.
9. Taber, D. F.; Tian, W. *J. Am. Chem. Soc.* **2006,** *128,* 1058–1059.
10. Richter, J. M. *Neber Rearrangement.* In *Name Reactions for Homologations-Part I*; Li, J. J., Ed.; Wiley: Hoboken, NJ, **2009,** pp 464–473. (Review).
11. Cardoso, A. L.; Gimeno, L.; Lemos, A.; Palacios, F.; Teresa, M. V. D.; Melo, P. *J. Org. Chem.* **2013,** *78,* 6983–6991.
12. Khlebnikov, A. F.; Novikov, M. S. *Tetrahedron* **2013,** *69,* 3363–3401. (Review).
13. Jiang, Y.; Park, C.-M.; Loh, T.-P. *Org. Lett.* **2014,** *16,* 3432–3435.
14. Kadama, V. D.; Sudhakar, G. *Tetrahedron* **2015,** *71,* 1058–1067.
15. Ning, Y.; Otani, Y.; Ohwada, T. *J. Org. Chem.* **2018,** *83,* 203–219.
16. Huang, Y.-J.; Qiao, B.; Zhang, F.-G.; Ma, J.-A. *Tetrahedron* **2018,** *74,* 3791–3796.
17. Khlebnikov, A. F.; Novikov, M. S.; Rostovskii, N. V. *Tetrahedron* **2019,** *75,* 2555–2624. (Review).
18. Alves, C.; Grosso, C.; Barrulas, P.; Paixao, J. A.; Cardoso, A. L.; Burke, A. J.; Lemos, A.; Pinho e Melo, T. M. V. D. *Synlett* **2020,** *31,* 553–558.

Nef Reaction

Conversion of a primary or secondary nitroalkane into the corresponding carbonyl compound.

Example 1[4]

Example 2[6]

Example 3[7]

© Springer Nature Switzerland AG 2021
J. J. Li, *Name Reactions*, https://doi.org/10.1007/978-3-030-50865-4_104

Example 4[9]

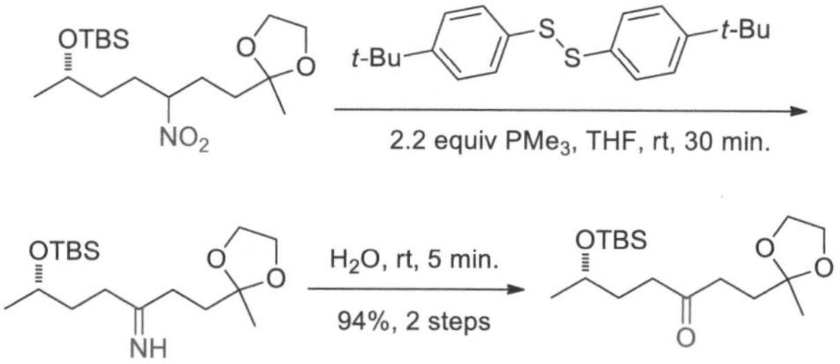

Example 5[10]

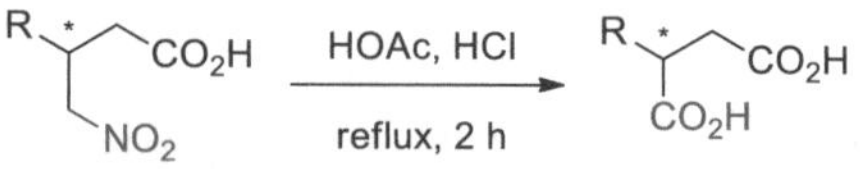

Example 6[11]

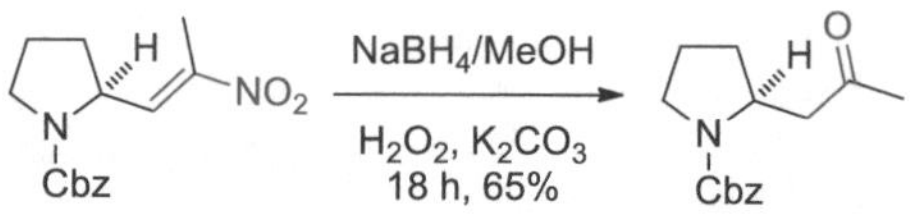

Example 7, Nef reaction was followed by a DBU-mediated alkene isomerization to afford thermodynamically more stable α,β-unsaturated ketone[13]

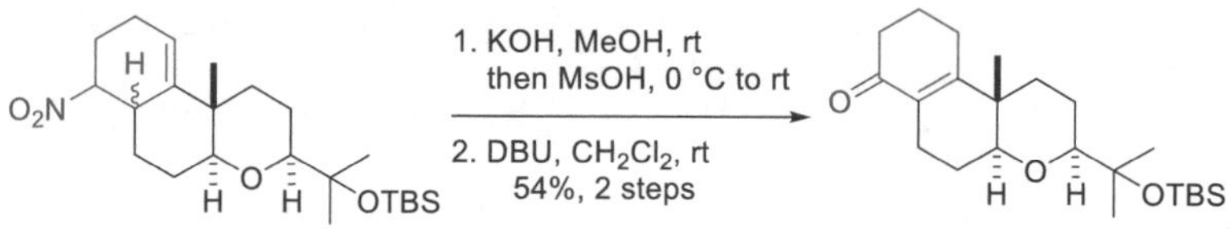

Example 8, The Nef nitro substrates prepared from palladium-catalyzed a-arylation of aryl nitromethanes[14]

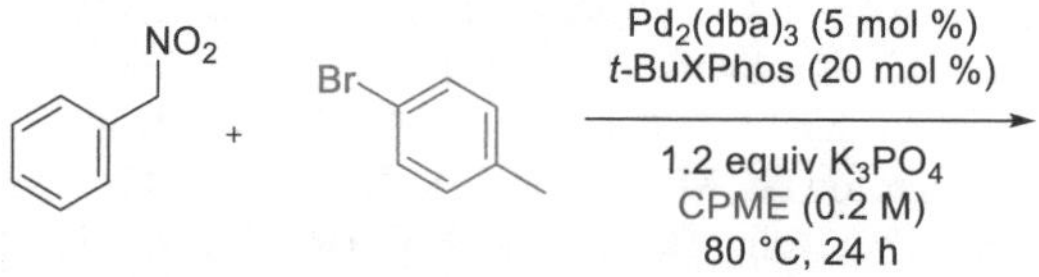

CPME = Cyclopentyl methyl ether, a solvent more resistant to auto-oxidation than THF and ethers

Example 9[15]

References

1. Nef, J. U. *Ann.* **1894**, *280*, 263–342. John Ulrich Nef (1862–1915) was born in Switzerland and immigrated to the US at the age of four with his parents. He went to Munich, Germany to study with Adolf von Baeyer, earning a Ph.D. In 1886. Back to the States, he served as a professor at Purdue University, Clark University, and the University of Chicago. The Nef reaction was discovered at Clark University in Worcester, Massachusetts. Nef was temperamental and impulsive, suffering from a couple of mental breakdowns. He was also highly individualistic, and had never published with a coworker save for three early articles.
2. Pinnick, H. W. *Org. React.* **1990**, *38*, 655–792. (Review).
3. Adam, W.; Makosza, M.; Saha-Moeller, C. R.; Zhao, C.-G. *Synlett* **1998**, 1335–1336.
4. Thominiaux, C.; Rousse, S.; Desmaele, D.; d'Angelo, J.; Riche, C. *Tetrahedron: Asymmetry* **1999**, *10*, 2015–2021.
5. Ballini, R.; Bosica, G.; Fiorini, D.; Petrini, M. *Tetrahedron Lett.* **2002**, *43*, 5233–5235.
6. Chung, W. K.; Chiu, P. *Synlett* **2005**, 55–58.
7. Tishkov, A. A.; Schmidhammer, U.; Roth, S.; Riedle, E.; Mayr, H. *Angew. Chem. Int. Ed.* **2005**, *44*, 4623–4626.
8. Wolfe, J. P. *Nef Reaction*. In *Name Reactions for Functional Group Transformations*; Li, J. J., Ed.; Wiley: Hoboken, NJ, **2007**, pp 645–652. (Review).
9. Burés, J.; Vilarrasa, J. *Tetrahedron Lett.* **2008**, *49*, 441–444.
10. Felluga, F.; Pitacco, G.; Valentin, E.; Venneri, C. D. *Tetrahedron: Asymmetry* **2008**, *19*, 945–955.
11. Chinmay Bhat, C.; Tilve, S. G. *Tetrahedron* **2013**, *69*, 6129–6143.
12. Ballini, R.; Petrini, M. *Adv. Synth. Catal.* **2015**, *357*, 2371–2402. (Review).
13. Sharpe, R. J.; Johnson, J. S. *J. Org. Chem.* **2015**, *80*, 9740–9766.
14. VanGelder, K. F.; Kozlowski, M. C. *Org. Lett.* **2015**, *17*, 5748–5751.
15. Huang, W.-L.; Raja, A.; Hong, B.-C.; Lee, G.-H. *Org. Lett.* **2017**, *19*, 3494–3497.
16. Ju, M.; Guan, W.; Schomaker, J. M.; Harper, K. C. *Org. Lett.* **2019**, *21*, 8893–8898.
17. Ferreira, J. R. M.; Nunes da Silva, R.; Rocha, J.; Silva, A. M. S.; Guieu, S. *Synlett* **2020**, *31*, 632–634.

Negishi Cross-Coupling Reaction

The Negishi cross-coupling reaction is the nickel- or palladium-catalyzed coupling of organozinc compounds with various halides or triflates (aryl, alkenyl, alkynyl, and acyl).

Example 1[3]

Example 2[4]

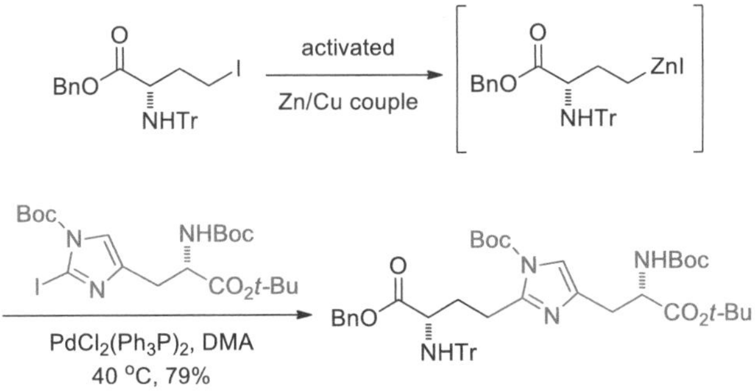

Example 3[8]

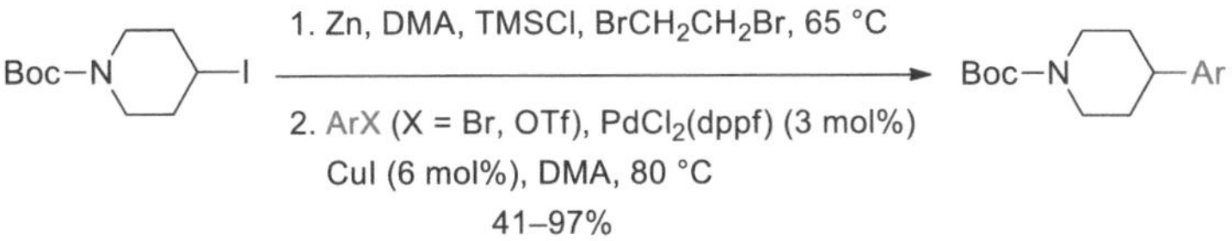

Example 4[9]

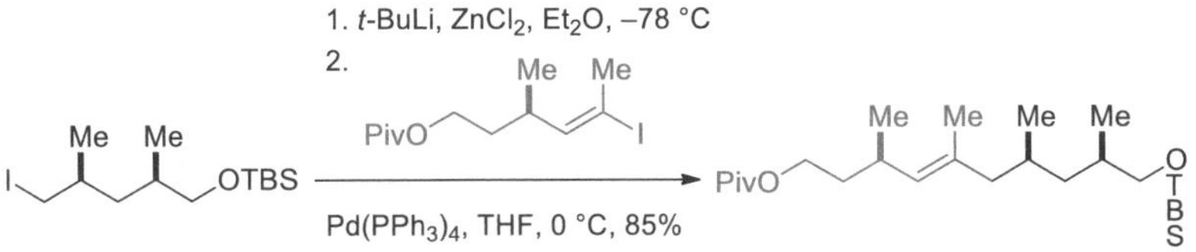

Example 5[11]

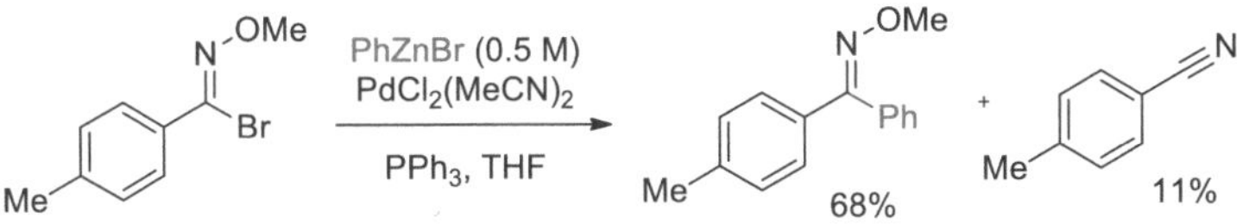

Example 6, Site-selective N,N-di-Boc activation for N–C cleavage of primary amides[12]

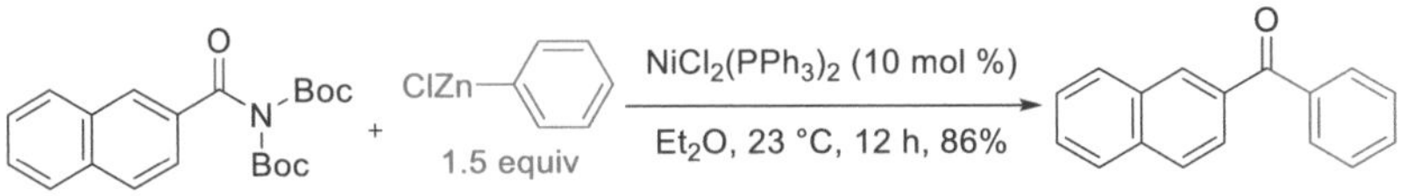

Example 7[13]

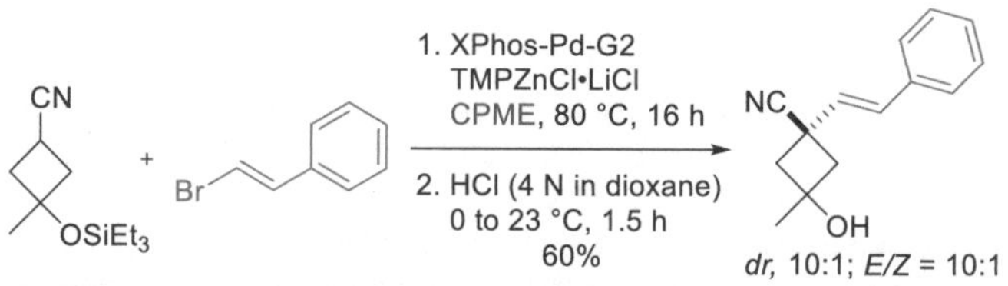

CPME = Cyclopentyl methyl ether, a solvent more resistant to auto-oxidation than THF and ethers

Example 8, Alkylpyridinium salts as electrophiles in deaminative alkyl–alkyl cross-couplings[15]

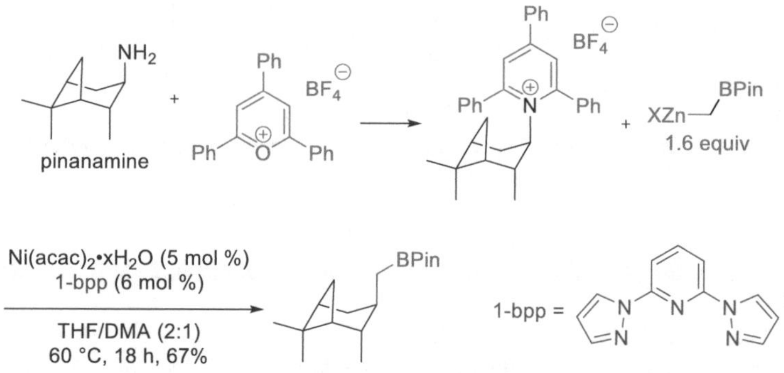

Example 9[16]

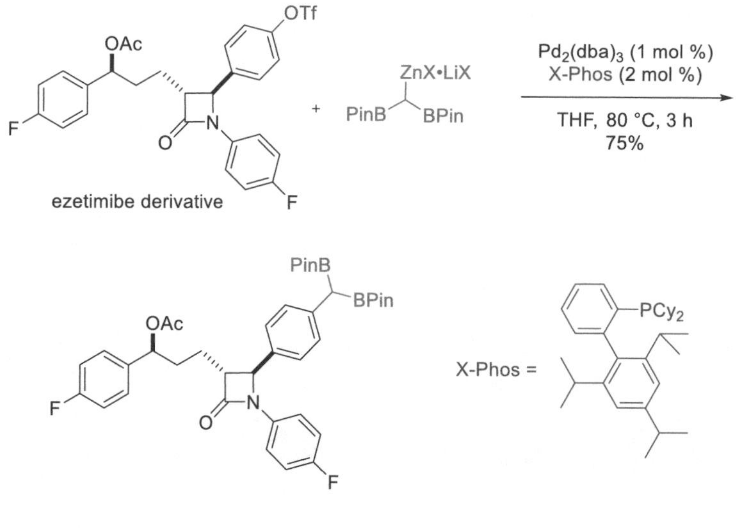

References

1. (a) Negishi, E.-I.; Baba, S. *J. Chem. Soc., Chem. Commun.* **1976**, 596–597. (b) Negishi, E.-I.; King, A. O.; Okukado, N. *J. Org. Chem.* **1977**, *42*, 1821–1823. (c) Negishi, E.-I. *Acc. Chem. Res.* **1982**, *15*, 340–348. (Review). Negishi is a professor at Purdue University. He won Nobel Prize in 2010 along with Richard F. Heck and Akira Suzuki "for palladium-catalyzed cross couplings in organic synthesis".
2. Erdik, E. *Tetrahedron* **1992**, *48*, 9577–9648. (Review).
3. De Vos, E.; Esmans, E. L.; Alderweireldt, F. C.; Balzarini, J.; De Clercq, E. *J. Heterocycl. Chem.* **1993**, *30*, 1245–1252.
4. Evans, D. A.; Bach, T. *Angew. Chem. Int. Ed.* **1993**, *32*, 1326–1327.
5. Negishi, E.-I.; Liu, F. In *Metal-Catalyzed Cross-Coupling Reactions;* Diederich, F.; Stang, P. J., Eds.; Wiley–VCH: Weinheim, Germany, **1998**, pp 1–47. (Review).
6. Arvanitis, A. G.; Arnold, C. R.; Fitzgerald, L. W.; Frietze, W. E.; Olson, R. E.; Gilligan, P. J.; Robertson, D. W. *Bioorg. Med. Chem. Lett.* **2003**, *13*, 289–291.
7. Ma, S.; Ren, H.; Wei, Q. *J. Am. Chem. Soc.* **2003**, *125*, 4817–4830.
8. Corley, E. G.; Conrad, K.; Murry, J. A.; Savarin, C.; Holko, J.; Boice, G. *J. Org. Chem.* **2004**, *69*, 5120–5123.
9. Inoue, M.; Yokota, W.; Katoh, T. *Synthesis* **2007**, 622–637.
10. Yet, L. *Negishi cross-coupling reaction.* In *Name Reactions for Homologations-Part I*; Li, J. J., Ed.; Wiley: Hoboken, NJ, **2009**, pp 70–99. (Review).
11. Dolliver, D. D.; Bhattarai, B. T.; et al. *J. Org. Chem.* **2013**, *78*, 3676–3687.
12. Shi, S.; Szostak, M. *Org. Lett.* **2016**, *18*, 5872–5875.
13. Dalziel, M. E.; Chen, P.; Carrera, D. E.; Zhang, H.; Gosselin, F. *Org. Lett.* **2017**, *19*, 3446–3449.
14. Brittain, W. D. G.; Cobb, S. L. *Org. Biomol.Chem.* **2018**, *16*, 10–20. (Review).
15. Plunkett, S.; Basch, C. H.; Santana, S. O.; Watson, M. P. *J. Am. Chem. Soc.* **2019**, *141*, 2257–2262.
16. Lee, H.; Lee, Y.; Cho, S. H. *Org. Lett.* **2019**, *21*, 5912–5916.
17. Lutter, F. H.; Grokenberger, L.; Benz, M.; Knochel, P. *Org. Lett.* **2020**, *22*, 3028–3032.

Newman–Kwart Rearrangement

Transformation of phenol to the corresponding thiophenol, a variant of the Smiles reaction.

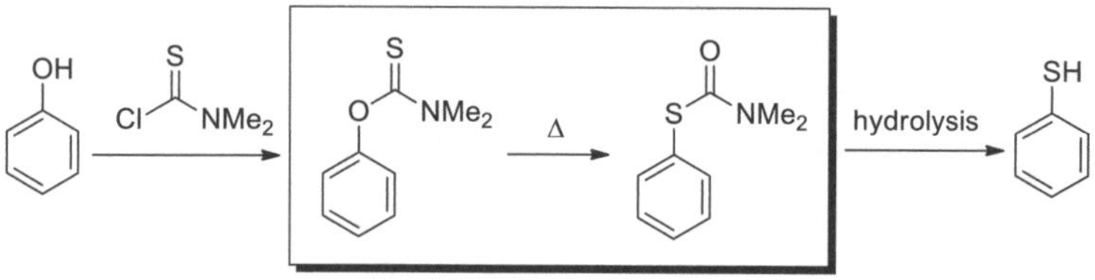

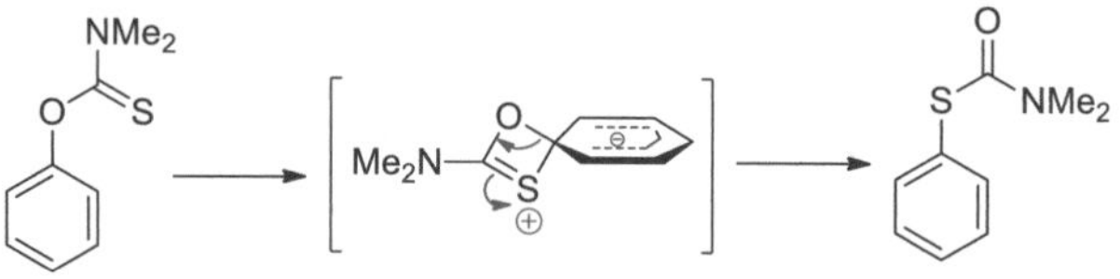

The Newman–Kwart rearrangement is a member of a series of related rearrangements, such as the **Schönberg rearrangement** and the **Chapman rearrangement**, in which aryl groups migrate intramolecularly between nonadjacent atoms. The Schönberg rearrangement is the most similar and involves the 1,3-migration of an aryl group from oxygen to sulfur in a diarylthiocarbonate. The Chapman rearrangement involves an analogous migration but to nitrogen.

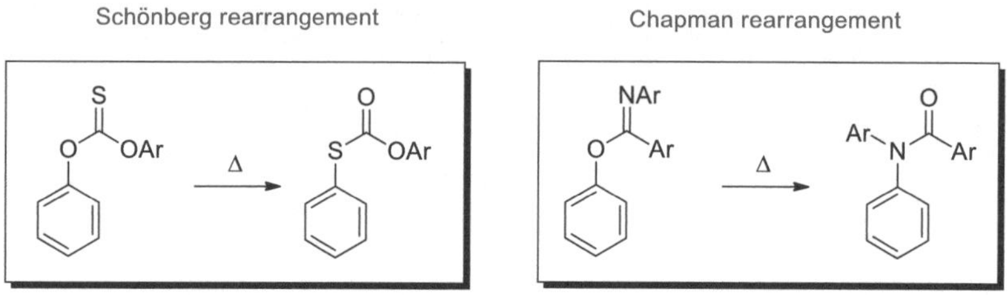

Schönberg rearrangement

Chapman rearrangement

Example 1[5]

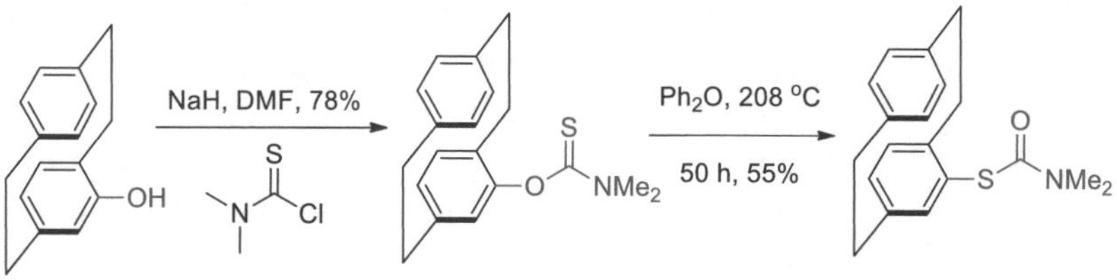

© Springer Nature Switzerland AG 2021
J. J. Li, *Name Reactions*, https://doi.org/10.1007/978-3-030-50865-4_106

Example 2[6]

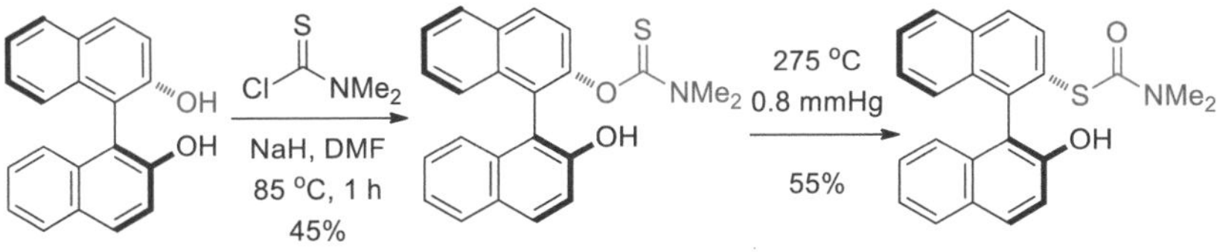

Example 3[7]

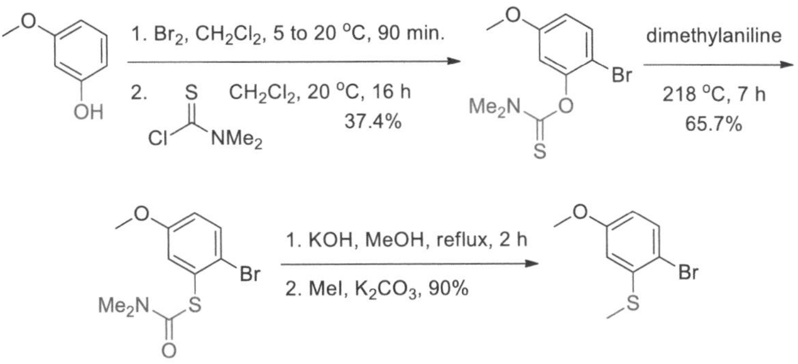

Example 4, Benzylic thio- or seleno-Newman–Kwart rearrangement[13]

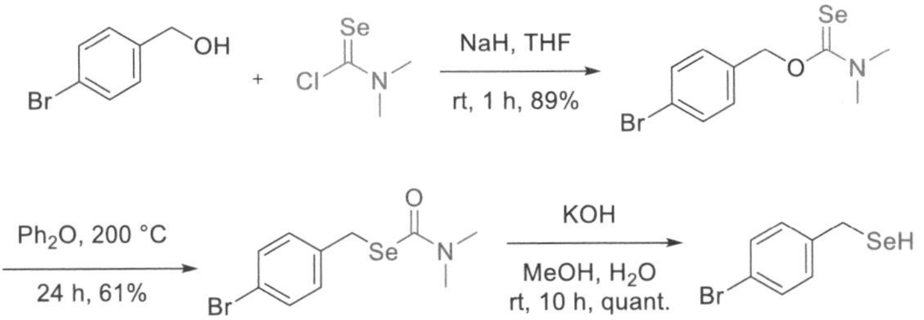

Example 5, One-electron oxidation at room temperature[14]

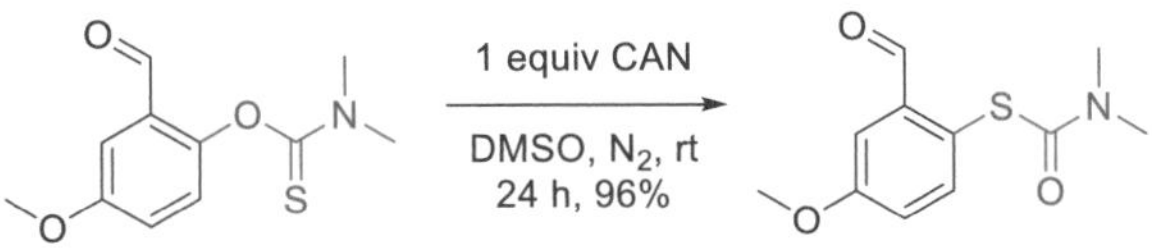

Example 6[15]

Mohr's salt (5 mol %)
1 equiv $(NH_4)_2S_2O_8$

CH_3CN/H_2O (3:1)
2 h, 86%

Mohr's salt = $(NH_4)_2Fe(SO_4)_2 \cdot 6H_2O$

References

1. (a) Kwart, H.; Evans, E. R. *J. Org. Chem.* **1966**, *31*, 410–413. (b) Newman, M. S.; Karnes, H. A. *J. Org. Chem.* **1966**, *31*, 3980–3984. (c) Newman, M. S.; Hetzel, F. W. *J. Org. Chem.* **1969**, *34*, 3604–3606.
2. Cossu, S.; De Lucchi, O.; Fabbri, D.; Valle, G.; Painter, G. F.; Smith, R. A. J. *Tetrahedron* **1997**, *53*, 6073–6084.
3. Lin, S.; Moon, B.; Porter, K. T.; Rossman, C. A.; Zennie, T.; Wemple, J. *Org. Prep. Proc. Int.* **2000**, *32*, 547–555.
4. Ponaras, A. A.; Zain, Ö. In *Encyclopedia of Reagents for Organic Synthesis,* Paquette, L. A., Ed.; Wiley: New York, **1995**, 2174–2176. (Review).
5. Kane, V. V.; Gerdes, A.; Grahn, W.; Ernst, L.; Dix, I.; Jones, P. G.; Hopf, H. *Tetrahedron Lett.* **2001**, *42*, 373–376.
6. Albrow, V.; Biswas, K.; Crane, A.; Chaplin, N.; Easun, T.; Gladiali, S.; Lygo, B.; Woodward, S. *Tetrahedron: Asymmetry* **2003**, *14*, 2813–2819.
7. Bowden, S. A.; Burke, J. N.; Gray, F.; McKown, S.; Moseley, J. D.; Moss, W. O.; Murray, P. M.; Welham, M. J.; Young, M. J. *Org. Process Res. Dev.* **2004**, *8*, 33–44.
8. Nicholson, G.; Silversides, J. D.; Archibald, S. J. *Tetrahedron Lett.* **2006**, *47*, 6541–6544.
9. Gilday, J. P.; Lenden, P.; Moseley, J. D.; Cox, B. G. *J. Org. Chem.* **2008**, *73*, 3130–3134.
10. Lloyd-Jones, G. C.; Moseley, J. D.; Renny, J. S. *Synthesis* **2008**, 661–689.
11. Tilstam, U.; Defrance, T.; Giard, T.; Johnson, M. D. *Org. Process Res. Dev.* **2009**, *13*, 321–323.
12. Das, J.; Le Cavelier, F.; Rouden, J.; Blanchet, J. *Synthesis* **2012**, *44*, 1349–1352.
18. Perkowski, A. J.; Cruz, C. L.; Nicewicz, D. A. *J. Am. Chem. Soc.* **2015**, *137*, 15684–15687.
13. Eriksen, K.; Ulfkjær, A.; Sølling, T. I.; Pittelkow, M. *J. Org. Chem.* **2018**, *83*, 10786–10797.
14. Pedersen, S. K.; Ulfkjær, A.; Newman, M. N.; Yogarasa, S.; Petersen, A. U.; Sølling, T. I.; Pittelkow, M. *J. Org. Chem.* **2018**, *83*, 12000–12006.
15. Gendron, T.; Pereira, R.; Abdi, H. Y.; Witney, T. H.; Årstad, E. *Org. Lett.* **2020**, *22*, 274–278.

Nicholas Reaction

Hexacarbonyldicobalt-stabilized propargyl cation is captured by a nucleophile. Subsequent oxidative demetallation then gives the propargylated product.

propargyl cation intermediate (stabilized by the hexacarbonyldicobalt complex).

Example 1, A chromium variant of the Nicholas reaction[3]

© Springer Nature Switzerland AG 2021
J. J. Li, *Name Reactions*, https://doi.org/10.1007/978-3-030-50865-4_107

Example 2, A Nicholas–Pauson–Khand sequence[4]

Example 3, Intramolecular Nicholas reaction using chromium[7]

Example 4[9]

Example 5, Cobalt cokmplex to boost steric hindrance[12]

Example 6[13]

Example 7[14]

References

1. Nicholas, K. M.; Pettit, R. *J. Organomet. Chem.* **1972**, *44*, C21–C24.
2. Nicholas, K. M. *Acc. Chem. Res.* **1987**, *20*, 207–214. (Review).
3. Corey, E. J.; Helal, C. J. *Tetrahedron Lett.* **1996**, *37*, 4837–4840.
4. Jamison, T. F.; Shambayati, S. *J. Am. Chem. Soc.* **1997**, *119*, 4353–4363.
5. Teobald, B. J. *Tetrahedron* **2002**, *58*, 4133–4170. (Review).
6. Takase, M.; Morikawa, T.; Abe, H.; Inouye, M. *Org. Lett.* **2003**, *5*, 625–628.
7. Ding, Y.; Green, J. R. *Synlett* **2005**, 271–274.
8. Pinacho Crisóstomo, F. R.; Carrillo, R. *Tetrahedron Lett.* **2005**, *46*, 2829–2832.
9. Hamajima, A.; Isobe, M. *Org. Lett.* **2006**, *8*, 1205–1208.
10. Shea, K. M. *Nicholas Reaction*. In *Name Reactions for Homologations-Part I*; Li, J. J., Ed.; Wiley: Hoboken, NJ, **2009**, pp 284–298. (Review).
11. Mukai, C.; Kojima, T.; Kawamura, T.; Inagaki, F. *Tetrahedron* **2013**, *69*, 7659–7669.
12. Feldman, K. S.; Folda, T. S. *J. Org. Chem.* **2016**, *81*, 4566–4575.
13. Shao, H.; Bao, W.; Jing, Z.-R.; Wang, Y.-P.; Zhang, F.-M.; Wang, S.-H.; Tu, Y.-Q. *Org. Lett.* **2017**, *19*, 4648–4651.
14. Johnson, R. E.; Ree, H.; Hartmann, M.; Lang, L.; Sawano, S.; Sarpong, R. *J. Am. Chem. Soc.* **2019**, *141*, 2233–2237.
15. Kaczmarek, R.; Korczynski, D.; Green, J. R.; Dembinski, R. *Beilst. J. Org. Chem.* **2020**, *16*, 1–8.

Noyori Asymmetric Hydrogenation

Asymmetric reduction of carbonyls and alkenes *via* hydrogenation, catalyzed by a ruthenium(II) BINAP complex.

(R)-BINAP-Ru =

$$[RuCl_2(binap)(solv)_2] \xrightarrow[-\ HCl]{H_2} [RuHCl(binap)(solv)_2]$$

The catalytic cycle:

J. J. Li, *Name Reactions*, https://doi.org/10.1007/978-3-030-50865-4_108

Example 1[1b]

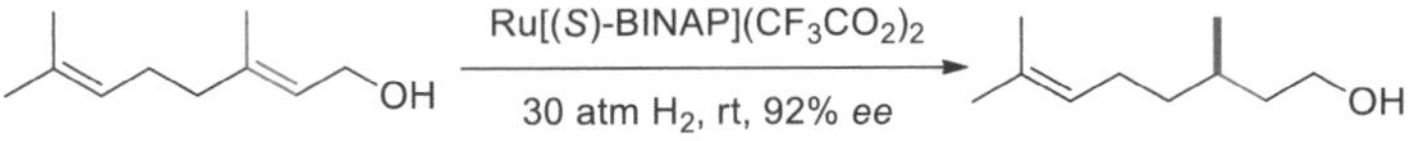
OH
Ru[(S)-BINAP](CF3CO2)2
30 atm H2, rt, 92% ee
OH

Example 2[1c]

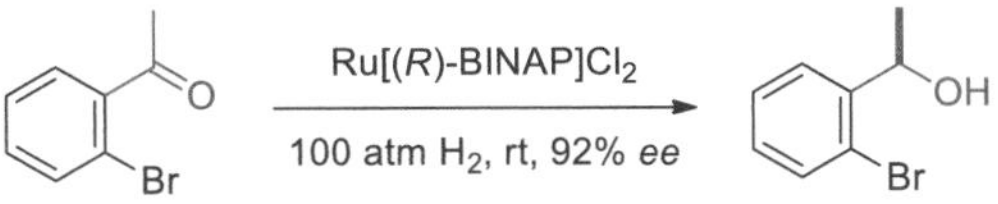
O
Br
Ru[(R)-BINAP]Cl2
100 atm H2, rt, 92% ee
OH
Br

Example 3[9]

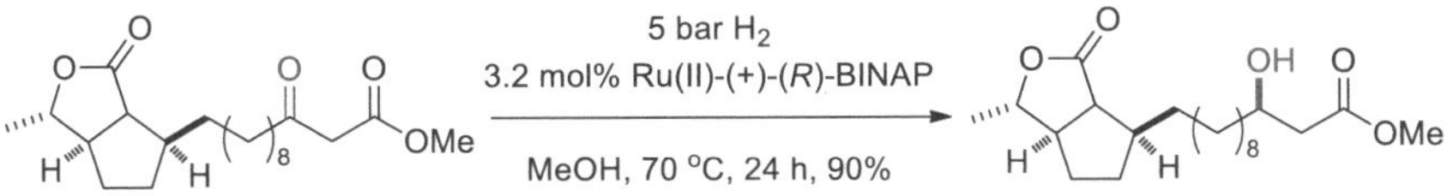
5 bar H2
3.2 mol% Ru(II)-(+)-(R)-BINAP
MeOH, 70 °C, 24 h, 90%
OMe
OH
OMe

Example 4[10]

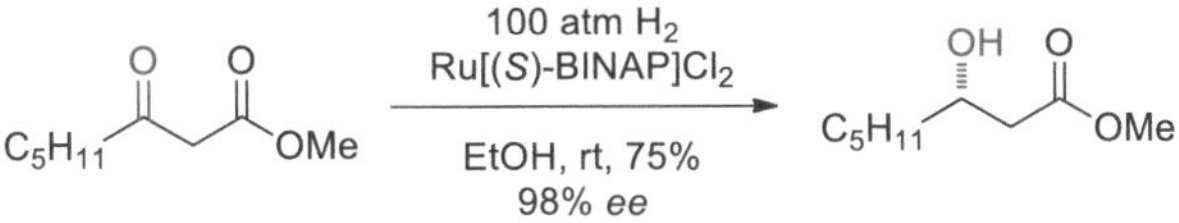
C5H11
OMe
100 atm H2
Ru[(S)-BINAP]Cl2
EtOH, rt, 75%
98% ee
OH
C5H11
OMe

Example 5[11]

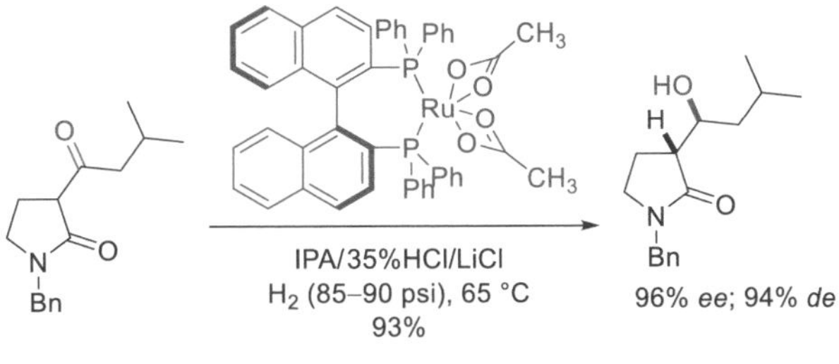
Ph Ph
CH3
Ru
CH3
Ph Ph
N
Bn
IPA/35%HCl/LiCl
H2 (85–90 psi), 65 °C
93%
HO
H
N
Bn
96% ee; 94% de

Example 6, Noyori asymmetric transfer hydrogenation (ATH) with resulted in a dynamic kinetic resolution (DKR) with spontaneous lactonization[12]

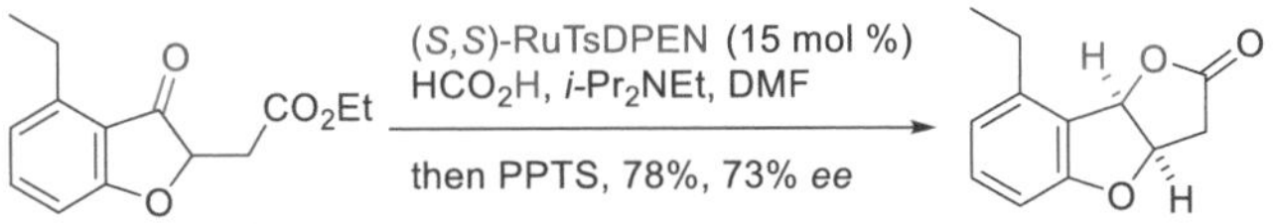
CO2Et
(S,S)-RuTsDPEN (15 mol %)
HCO2H, i-Pr2NEt, DMF
then PPTS, 78%, 73% ee
H
O
H

(S,S)-RuTsDPEN =

Example 7[13]

(S,S)-RuTsDPEN
(15 mol %)

i-PrOH, rt
94%, dr = 94:6

Example 8, Ruthenium-catalyzed asymmetric transfer hydrogenation (ATH)[14]

[RuCl(n^6-arene)(R,R)-TsDPEN
(0.5 mol %)

HCO$_2$H/Et$_3$N (5:2), CH$_2$Cl$_2$
30 °C, 7 h, 97%, > 99% ee

Example 9[15]

0.1 equiv
RuCl(p-cymene)(S,S)-TsDPEN

7.51 equiv HCO$_2$H/Et$_3$N (5:2)
65%, dr > 20:1

References

1. (a) Noyori, R.; Ohta, M.; Hsiao, Y.; Kitamura, M.; Ohta, T.; Takaya, H. *J. Am. Chem. Soc.* **1986**, *108*, 7117–7119. Ryoji Noyori (Japan, 1938–) and William S. Knowles (USA, 1917–2012) shared half of the Nobel Prize in Chemistry in 2001 for their work on chirally catalyzed hydrogenation reactions. K. Barry Sharpless (USA, 1941–) shared the other half for his work on chirally catalyzed oxidation reactions. (b) Takaya, H.; Ohta, T.; Sayo, N.; Kumobayashi, H.; Akutagawa, S.; Inoue, S.; Kasahara, I.; Noyori, R. *J. Am. Chem. Soc.* **1987**, *109*, 1596–1598. (c) Kitamura, M.; Ohkuma, T.; Inoue, S.; Sayo, N.; Kumobayashi, H.; Akutagawa, S.; Ohta, T.; Takaya, H.; Noyori, R. *J. Am. Chem. Soc.* **1988**, *110*, 629–631. (d) Noyori, R.; Ohkuma, T.; Kitamura, H.; Takaya, H.; Sayo, H.; Kumobayashi, S.; Akutagawa, S. *J. Am. Chem. Soc.* **1987**, *109*, 5856–5858. (e) Noyori, R.; Ohkuma, T. *Angew. Chem. Int. Ed.* **2001**, *40*, 40–73. (Review). (f) Noyori, R. *Angew. Chem. Int. Ed.* **2002**, *41*, 2008–2022. (Review, Nobel Prize Address).
2. Noyori, R. In *Asymmetric Catalysis in Organic Synthesis;* Ojima, I., ed.; Wiley: New York, **1994**, Chapter 2. (Review).

3. Chung, J. Y. L.; Zhao, D.; Hughes, D. L.; McNamara, J. M.; Grabowski, E. J. J.; Reider, P. J. *Tetrahedron Lett.* **1995,** *36*, 7379–7382.
4. Bayston, D. J.; Travers, C. B.; Polywka, M. E. C. *Tetrahedron: Asymmetry* **1998,** *9*, 2015–2018.
5. Berkessel, A.; Schubert, T. J. S.; Mueller, T. N. *J. Am. Chem. Soc.* **2002,** *124*, 8693–8698.
6. Fujii, K.; Maki, K.; Kanai, M.; Shibasaki, M. *Org. Lett.* **2003,** *5*, 733–736.
7. Ishibashi, Y.; Bessho, Y.; Yoshimura, M.; Tsukamoto, M.; Kitamura, M. *Angew. Chem. Int. Ed.* **2005,** *44*, 7287–7290.
8. Lall, M. S. *Noyori Asymmetric Hydrogenation*, In *Name Reactions for Functional Group Transformations*; Li, J. J., Ed.; Wiley: Hoboken, NJ, **2007,** pp 46–66. (Review).
9. Bouillon, M. E.; Meyer, H. H. *Tetrahedron* **2007,** *63*, 2712–2723.
10. Case-Green, S. C.; Davies, S. G.; Roberts, P. M.; Russell, A. J.; Thomson, J. E. *Tetrahedron: Asymmetry* **2008,** *19*, 2620–2631.
11. Magnus, N. A.; Astleford, B. A.; Laird, D. L. T.; Maloney, T. D.; McFarland, A. D.; Rizzo, J. R.; Ruble, J. C.; Stephenson, G. A.; Wepsiec, J. P. *J. Org. Chem.* **2013,** *78,* 5768–5774.
12. Alnafta, N.; Schmidt, J. P.; Nesbitt, C. L.; McErlean, C. S. P. *Org. Lett.* **2016,** *18*, 6520–6522.
13. Dias, L. C.; Polo, E. C. *J. Org. Chem.* **2017,** *82,* 4072–4112.
14. Zheng, L.-S.; Phansavath, P.; Ratovelomanana-Vidal, V. *Org. Lett.* **2018,** *20*, 5107–5111.
15. Blitz, M.; Heine, R. C.; Harms, K.; Koert, U. *Org. Lett.* **2019,** *21*, 785–788.
16. Zhao, M. M.; Zhang, H.; Iimura, S.; Bednarz, M. S.; Kanamarlapudi, R. C.; Yan, J.; Lim, N.-K.; Wu, W. *Org. Process Res. Dev.* **2020,** *24*, 261–273.

Nozaki–Hiyama–Kishi Reaction

Cr–Ni bimetallic catalyst-promoted redox addition of vinyl- or propargyl-halides to aldehydes.

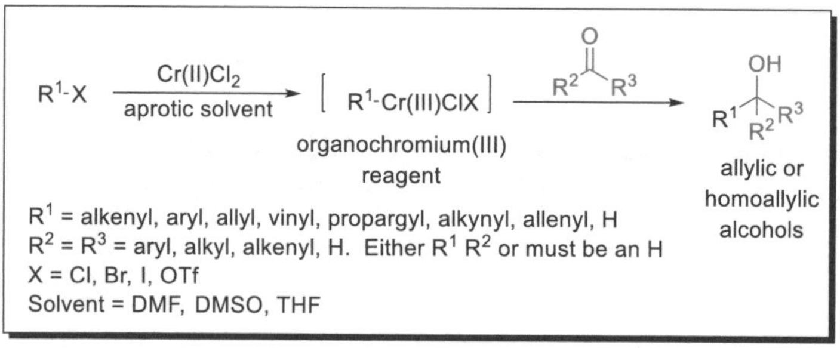
R1-X
Cr(II)Cl2
aprotic solvent
R1-Cr(III)ClX
organochromium(III) reagent
allylic or homoallylic alcohols
R1 = alkenyl, aryl, allyl, vinyl, propargyl, alkynyl, allenyl, H
R2 = R3 = aryl, alkyl, alkenyl, H. Either R1 R2 or must be an H
X = Cl, Br, I, OTf
Solvent = DMF, DMSO, THF

The catalytic cycle:[2]

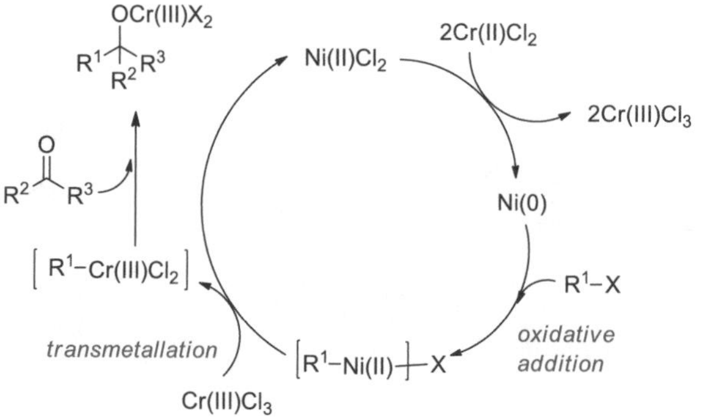
OCr(III)X2
Ni(II)Cl2
2Cr(II)Cl2
2Cr(III)Cl3
Ni(0)
R1–X
oxidative addition
R1–Ni(II)–X
transmetallation
Cr(III)Cl3
R1–Cr(III)Cl2

Example 1[3]

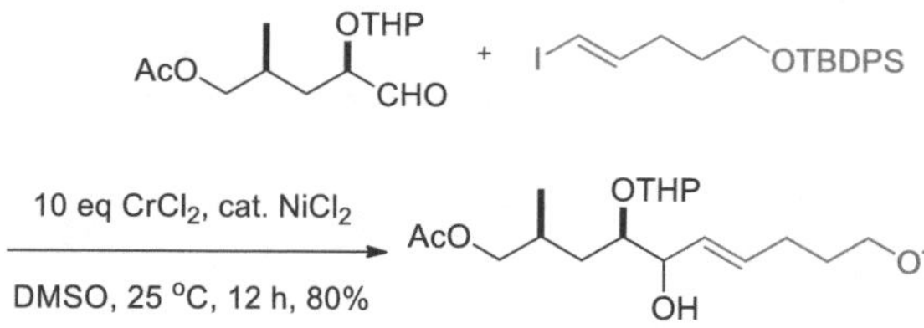
OTHP
AcO
CHO
OTBDPS
10 eq CrCl2, cat. NiCl2
DMSO, 25 °C, 12 h, 80%
OH

Example 2[5]

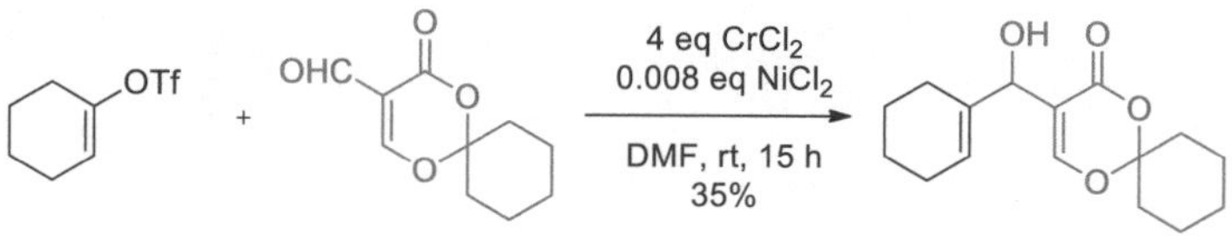
OTf
OHC
4 eq CrCl2
0.008 eq NiCl2
DMF, rt, 15 h
35%
OH

J. J. Li, *Name Reactions*, https://doi.org/10.1007/978-3-030-50865-4_109

Example 3, Intramolecular Nozaki–Hiyama–Kishi reaction[8]

Example 4, Intramolecular Nozaki–Hiyama–Kishi reaction[9]

Example 5, Asymmetric Nozaki–Hiyama–Kishi reaction[11]

Example 6, Late-stage Nozaki–Hiyama–Kishi macrolactonization[12]

Example 7[14]

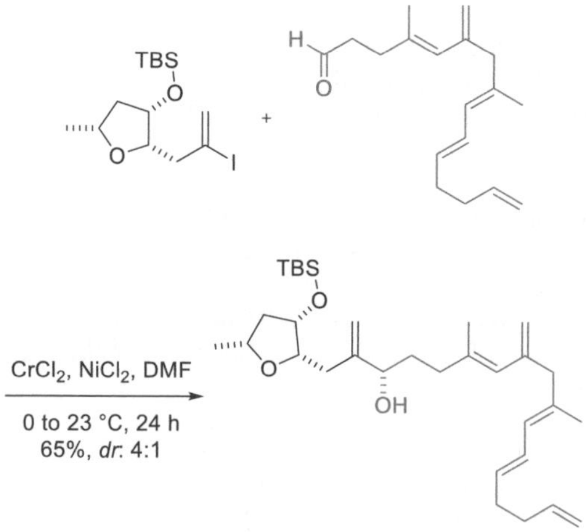
TBS
O
O
I
H
O
+
CrCl2, NiCl2, DMF
0 to 23 °C, 24 h
65%, dr: 4:1
OH

Example 8[15]

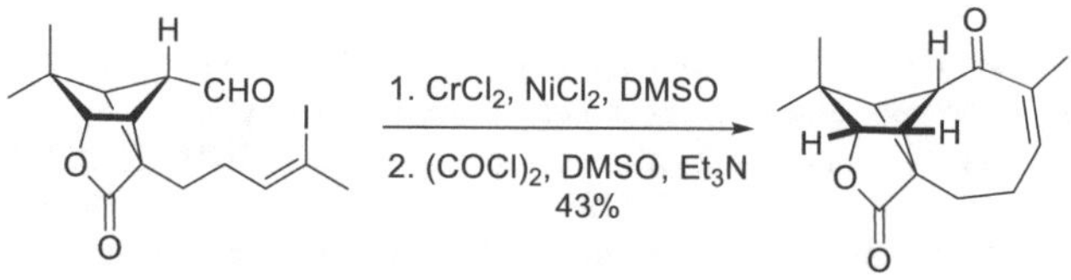
H
CHO
O
O
I
1. CrCl2, NiCl2, DMSO
2. (COCl)2, DMSO, Et3N
43%
H
H
H

Example 9[17]

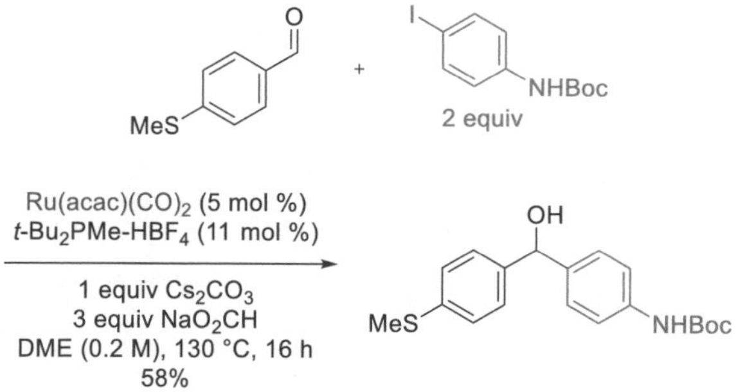
O
MeS
+
I
NHBoc
2 equiv
Ru(acac)(CO)2 (5 mol %)
t-Bu2PMe-HBF4 (11 mol %)
1 equiv Cs2CO3
3 equiv NaO2CH
DME (0.2 M), 130 °C, 16 h
58%
OH
MeS
NHBoc

References

1. (a) Okude, C. T.; Hirano, S.; Hiyama, T.; Nozaki, H. *J. Am. Chem. Soc.* **1977,** *99*, 3179–3181. Hitosi Nozaki and T. Hiyama are professors at the Japanese Academy. (b) Takai, K.; Kimura, K.; Kuroda, T.; Hiyama, T.; Nozaki, H. *Tetrahedron Lett.* **1983,** *24*, 5281–5284. Kazuhiko Takai was Prof. Nozaki's student during the discovery of

the reaction and is a professor at Okayama University. (c) Jin, H.; Uenishi, J.; Christ, W. J.; Kishi, Y. *J. Am. Chem. Soc.* **1986,** *108,* 5644–5646. Yoshito Kishi at Harvard independently discovered the catalytic effect of nickel during his total synthesis of polytoxin. (d) Takai, K.; Tagahira, M.; Kuroda, T.; Oshima, K.; Utimoto, K.; Nozaki, H. *J. Am. Chem. Soc.* **1986,** *108,* 6048–6050. (e) Kress, M. H.; Ruel, R.; Miller, L. W. H.; Kishi, Y. *Tetrahedron Lett.* **1993,** *34,* 5999–6002.

2. Fürstner, A.; Shi, N. *J. Am. Chem. Soc.* **1996,** *118,* 12349–12357. (The catalytic cycle).
3. Chakraborty, T. K.; Suresh, V. R. *Chem. Lett.* **1997,** 565–566.
4. Fürstner, A. *Chem. Rev.* **1999,** *99,* 991–1046. (Review).
5. Blaauw, R. H.; Benningshof, J. C. J.; van Ginkel, A. E.; van Maarseveen, J. H.; Hiemstra, H. *J. Chem. Soc., Perkin Trans. 1* **2001,** 2250–2256.
6. Berkessel, A.; Menche, D.; Sklorz, C. A.; Schroder, M.; Paterson, I. *Angew. Chem. Int. Ed.* **2003,** *42,* 1032–1035.
7. Takai, K. *Org. React.* **2004,** *64,* 253–612. (Review).
8. Karpov, G. V.; Popik, V. V. *J. Am. Chem. Soc.* **2007,** *129,* 3792–3793.
9. Valente, C.; Organ, M. G. *Chem. Eur. J.* **2008,** *14,* 8239–8245.
10. Yet, L. *Nozaki–Hiyama–Kishi Reaction.* In *Name Reactions for Homologations-Part I*; Li, J. J., Ed.; Wiley: Hoboken, NJ, **2009,** pp 299–318. (Review).
11. Austad, B. C.; Benayoud, F.; Calkins, T. L.; et al. *Synlett* **2013,** *17,* 327–332.
12. Bolte, B.; Basutto, J. A.; Bryan, C. S.; Garson, M. J.; Banwell, M. G.; Ward, J. S. *J. Org. Chem.* **2015,** *80,* 460–470.
13. Tian, Q.; Zhang, G. *Synthesis* **2016,** *48,* 4038–4049. (Review).
14. Ghosh, A. K.; Nyalapatla, P. R. *Org. Lett.* **2016,** *18,* 2286–2299.
15. Wang, B.; Xie, Y.; Yang, Q.; Zhang, G.; Gu, Z. *Org. Lett.* **2016,** *18,* 5388–5391.
16. Gil, A.; Albericio, F.; Alvarez, M. *Chem. Rev.* **2017,** *117,* 8420–8446. (Review).
17. Swyka, R. A.; Zhang, W.; Richardson, J.; Ruble, J. C.; Krische, M. J. *J. Am. Chem. Soc.* **2019,** *141,* 1828–1832.
18. Rafaniello, A. A.; Rizzacasa, M. A. *Org. Lett.* **2020,** *22,* 1972–1975.

Olefin Metathesis

Grubbs, Schrock, Heveyda and many others have made significant contributions to the field of olefin metathesis. Rather than giving one name reaction, it is collectively called olefin metathesis here.

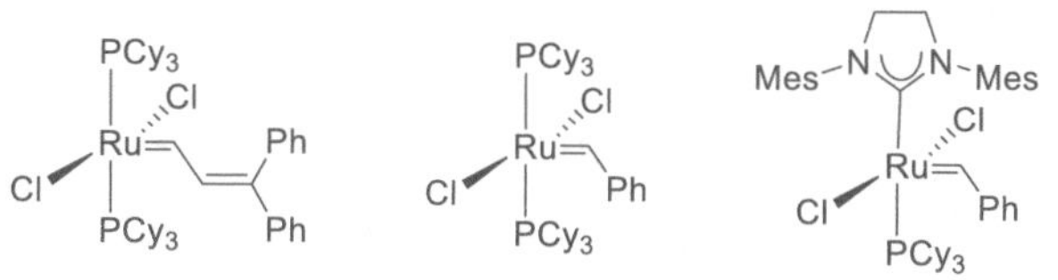

original catalyst   Grubbs I   Grubbs II

Mes = mesityl

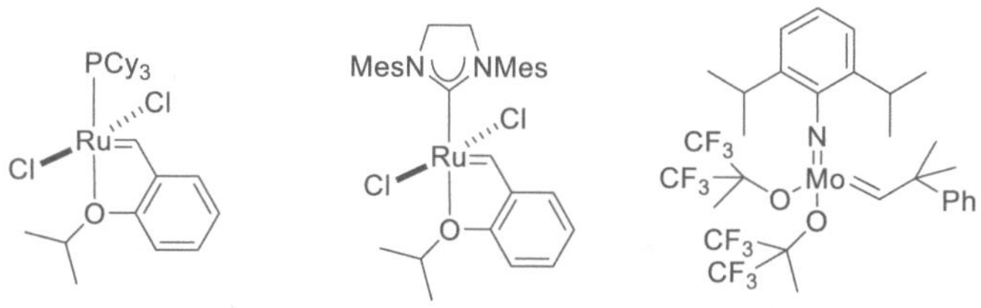

Hoveyda–Grubbs I   Hoveyda–Grubbs II   Schrock's catalyst

All three catalysts are illustrated as "$L_nM=CHR$" in the mechanism below.

Generation of the real catalyst from the precatalysts:

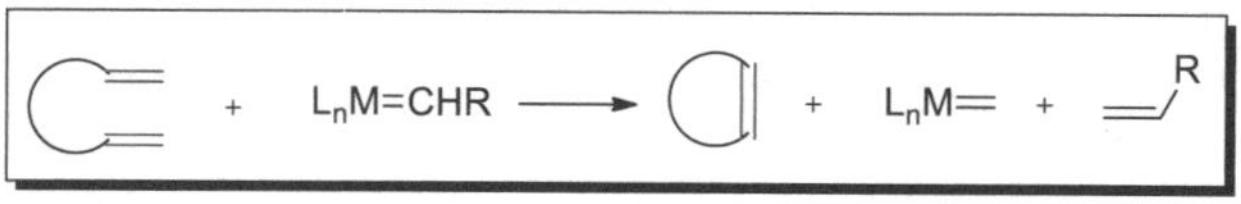

the active catalyst

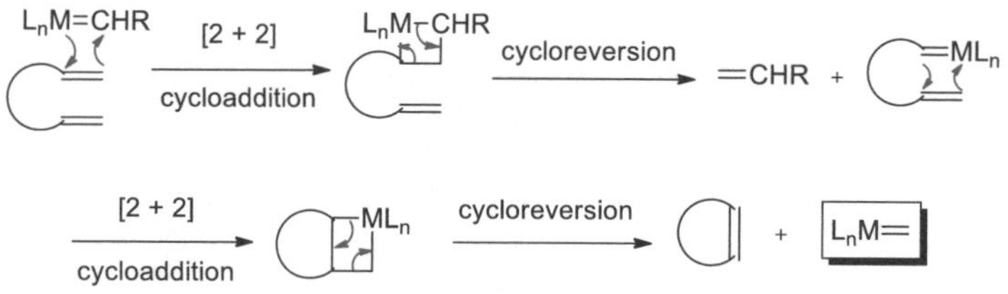

Catalytic cycle:

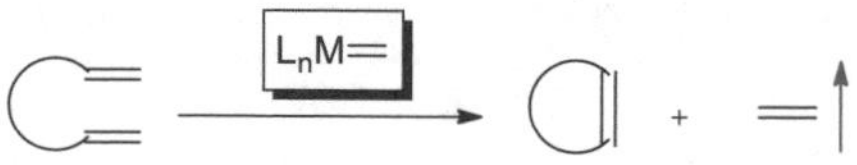

J. J. Li, *Name Reactions*, https://doi.org/10.1007/978-3-030-50865-4_110

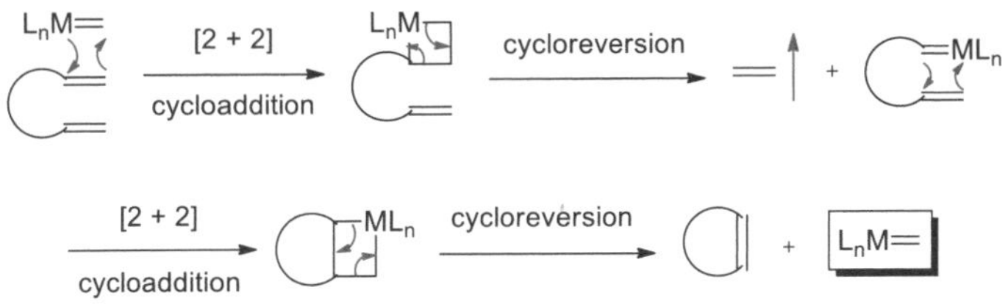

Example 1[3]

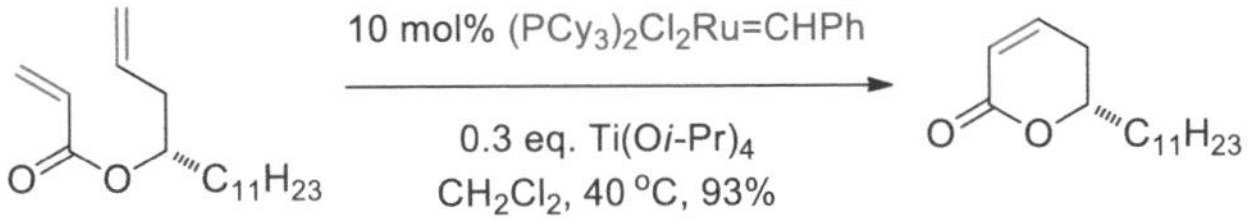

10 mol% (PCy₃)₂Cl₂Ru=CHPh

0.3 eq. Ti(O*i*-Pr)₄
CH₂Cl₂, 40 °C, 93%

Example 2[4]

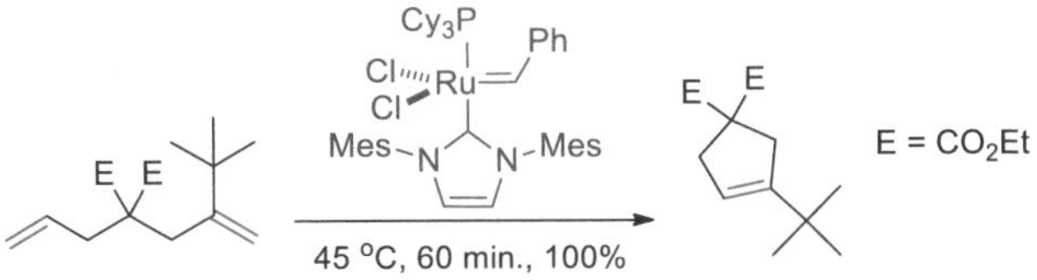

45 °C, 60 min., 100%

E = CO₂Et

Example 3[7]

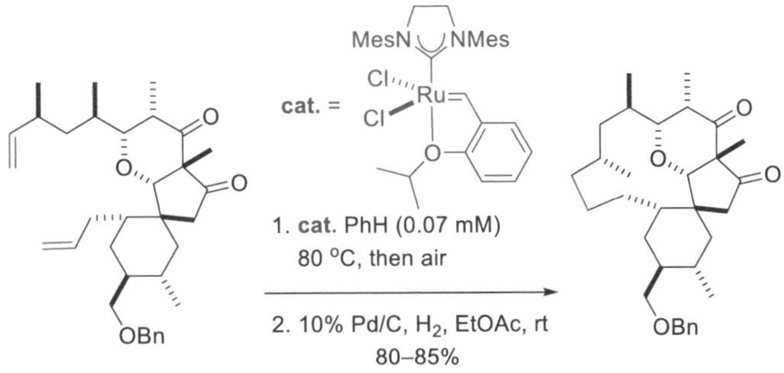

cat. =

1. cat. PhH (0.07 mM)
   80 °C, then air

2. 10% Pd/C, H₂, EtOAc, rt
   80–85%

Example 4[9]

5.4 mol%

CH₂Cl₂, rt, 73%

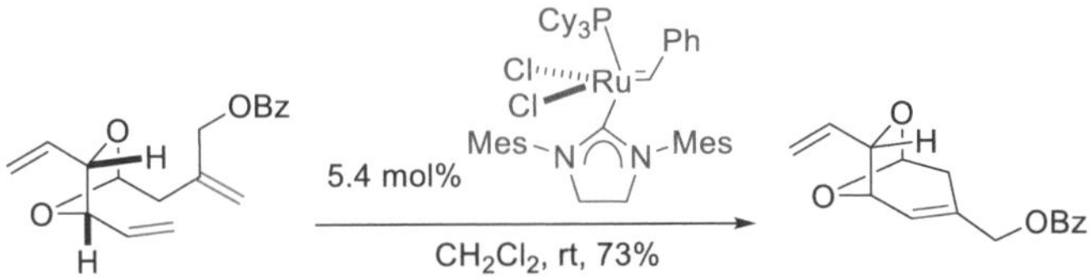

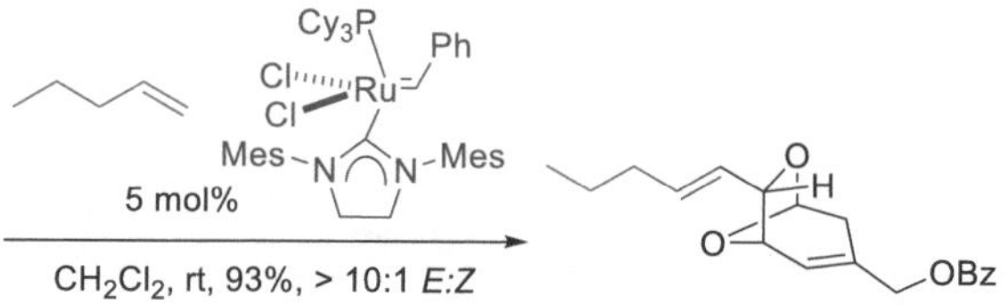

Example 5[10]

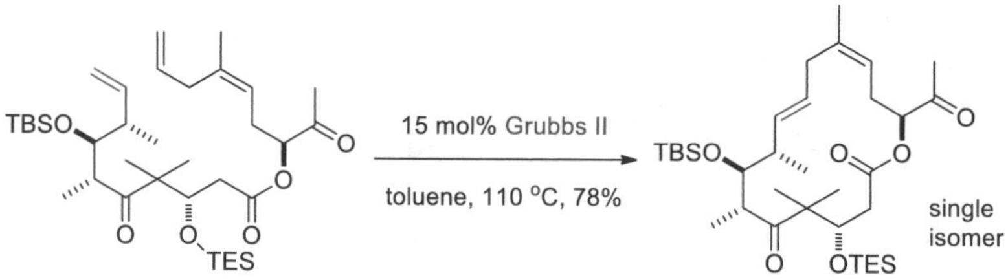

Example 6[12]

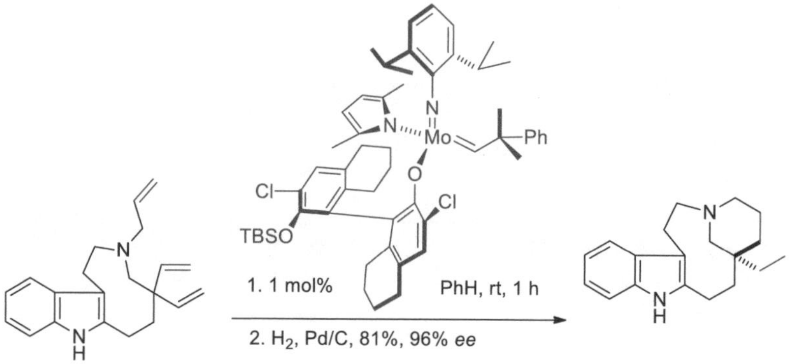

Example 7[13]

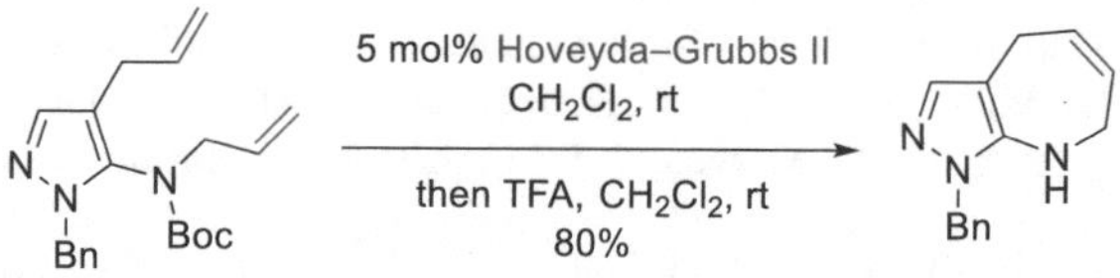

Example 8, Intermolecular olefin metathesis[14]

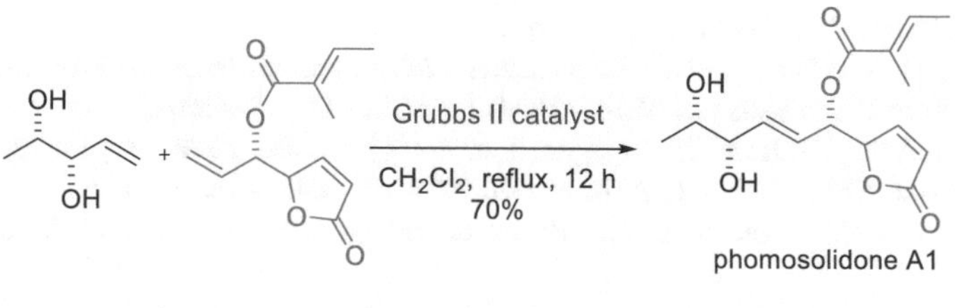

phomosolidone A1

Example 9[15]

Example 10[18]

Example 11[19]

References

1. Schrock, R. R.; Murdzek, J. S.; Bazan, G. C.; Robbins, J.; DiMare, M.; O′Regan, M. *J. Am. Chem. Soc.* **1990,** *112,* 3875–3886. Richard Schrock is a professor at MIT. He shared the 2005 Nobel Prize in Chemistry with Robert Grubbs of Caltech and Yves Chauvin of Institut Français du Pétrole in France for their contributions to metathesis.
2. Grubbs, R. H.; Miller, S. J.; Fu, G. C. *Acc. Chem. Res.* **1995,** *28,* 446–452. (Review).
3. Scholl, M.; Tunka, T. M.; Morgan, J. P.; Grubbs, R. H. *Tetrahedron Lett.* **1999,** *40,* 2247–2250.
4. Fellows, I. M.; Kaelin, D. E., Jr.; Martin, S. F. *J. Am. Chem. Soc.* **2000,** *122,* 10781–10787.
5. Timmer, M. S. M.; Ovaa, H.; Filippov, D. V.; van der Marel, G. A.; van Boom, J. H. *Tetrahedron Lett.* **2000,** *41,* 8635–8638.
6. Thiel, O. R. *Alkene and alkyne metathesis in organic synthesis.* In *Transition Metals for Organic Synthesis (2nd Edn.),* **2004,** *1,* pp 321–333. (Review).
7. Smith, A. B., III; Basu, K.; Bosanac, T. *J. Am. Chem. Soc.* **2007,** *129,* 14872–14874.
8. Hoveyda, A.H.; Zhugralin, A. R. *Nature* **2007,** *450,* 243–251. (Review).
9. Marvin, C. C.; Clemens, A. J. L.; Burke, S. D. *Org. Lett.* **2007,** *9,* 5353–5356.

10. Keck, G. E.; Giles, R. L.; Cee, V. J.; Wager, C. A.; Yu, T.; Kraft, M. B. *J. Org. Chem.* **2008**, *73*, 9675–9691.
11. Donohoe, T. J.; Fishlock, L. P.; Procopiou, P. A. *Chem. Eur. J.* **2008**, *14*, 5716–5726. (Review).
12. Sattely, E. S.; Meek, S. J.; Malcolmson, S. J.; Schrock, R. R.; Hoveyda, A. H. *J. Am. Chem. Soc.* **2009**, *131*, 943–953.
13. Moss, T. A. *Tetrahedron Lett.* **2013**, *54*, 993–997.
14. Raju, K. S.; Sabitha, G. *Tetrahedron: Asymmetry* **2016**, *27*, 639–642.
15. Burnley, J.; Wang, Z. J.; Jackson, W. R.; Robinson, A. J. *J. Org. Chem.* **2017**, *82*, 8497–8505.
16. Yu, M.; Lou, S.; Gonzalez-Bobes, F. *Org. Process Res. Dev.* **2018**, *22*, 918–946. (Review).
17. Lecourt, C.; Dhambri, S.; Allievi, L.; Sanogo, Y.; Zeghbib, N.; Othman, R. B.; Lannou, M.-I.; Sorin, G.; Ardisson, J. *Nat. Prod. Rep.* **2018**, *35*, 105–124. (Review).
18. Atkin, L.; Chen, Z.; Robertson, A.; Sturgess, D.; White, J. M.; Rizzacasa, M. A. *Org. Lett.* **2018**, *20*, 4255–4258.
19. Cheng-Sánchez, I.; Carrillo, P.; Sánchez-Ruiz, A.; Martinez-Poveda, B.; Quesada, A. R.; Medina, M. A.; López-Romero, J. M.; Sarabia, F. *J. Org. Chem.* **2018**, *83*, 5365–5383.
20. Li, J.; Stoltz, B. M.; Grubbs, R. H. *Org. Lett.* **2019**, *21*, 10139–10142.
21. Yamanushkin, P.; Smith, S. P.; Petillo, P. A.; Rubin, M. *Org. Lett.* **2020**, *22*, 3542–3546.

Oppenauer Oxidation

Alkoxide-catalyzed oxidation of secondary alcohols. Reverse of the Meerwein–Ponndorf–Verley reduction.

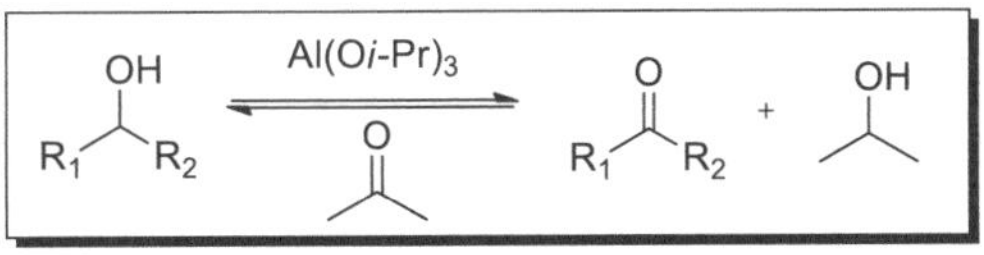

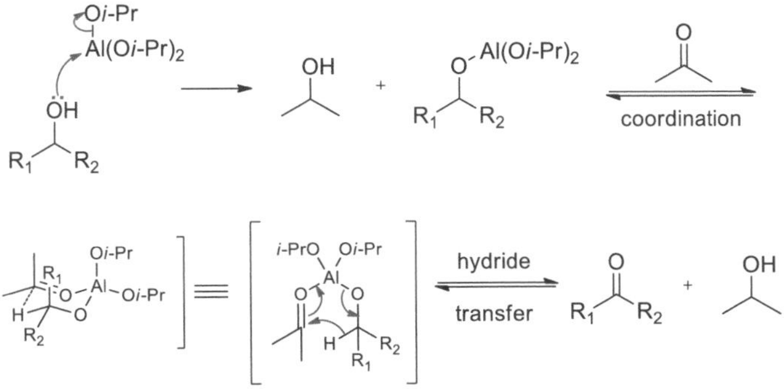

cyclic transition state

Example 1, Mg-Oppenauer oxidation[3]

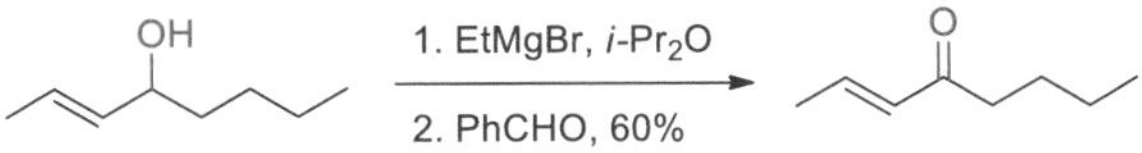

Example 2[6]

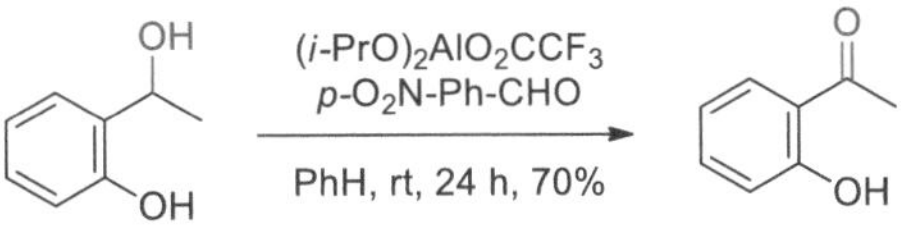

Example 3, Mg-Oppenauer oxidation[8]

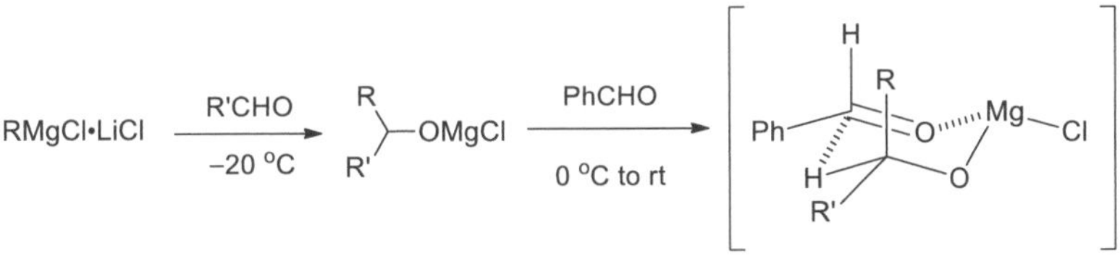

© Springer Nature Switzerland AG 2021
J. J. Li, *Name Reactions*, https://doi.org/10.1007/978-3-030-50865-4_111

Example 4[10]

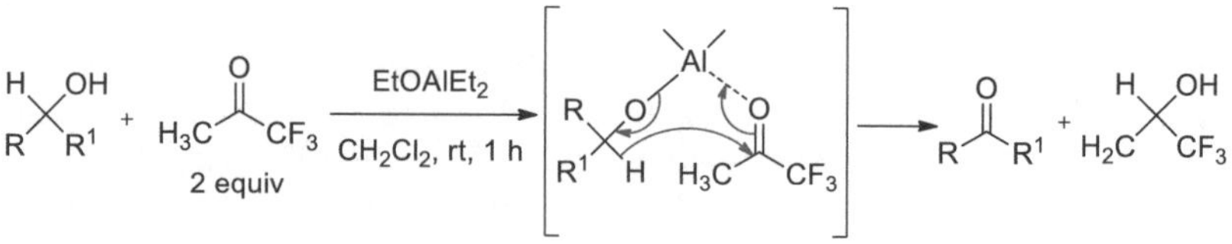

Example 5, Tandem nucleophilic addition–Oppenauer oxidation[12]

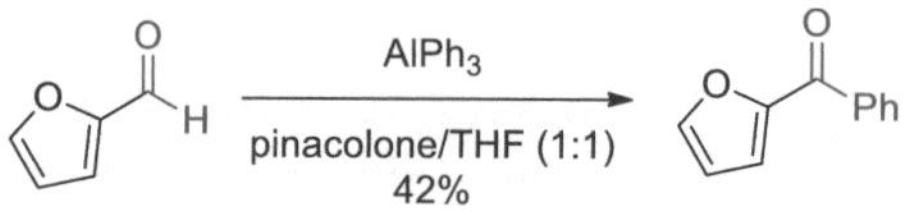

AlPh₃

pinacolone/THF (1:1)
42%

$AlPh_3$

Example 6, Ruthenium(II)-NNN complex as the catalyst[13]

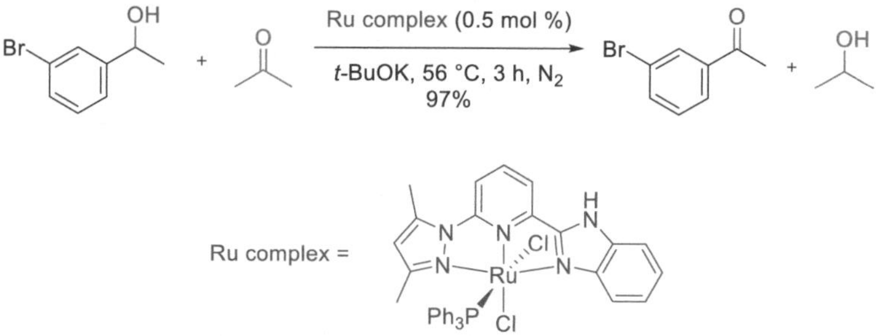

Ru complex (0.5 mol %)

t-BuOK, 56 °C, 3 h, N₂
97%

Ru complex =

Ph₃P

Example 7, Indium(III) isopropoxide as a hydrogen transfer catalyst[14]

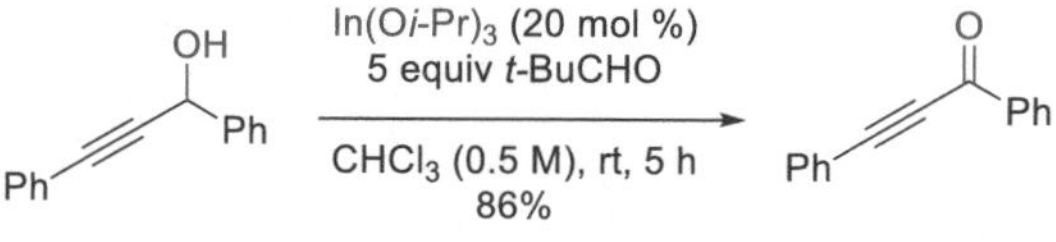

In(Oi-Pr)₃ (20 mol %)
5 equiv t-BuCHO

CHCl₃ (0.5 M), rt, 5 h
86%

References

1. Oppenauer, R. V. *Rec. Trav. Chim.* **1937**, *56*, 137–144. Rupert V. Oppenauer (1910–1969), born in Burgstall, Italy, studied at ETH in Zurich under Ruzicka and Reichstei, both Nobel laureates. After a string of academic appointments around Eu-

rope and a stint at Hoffman–La Roche, Oppenauer worked for the Ministry of Public Health in Buenos Aires, Argentina.

2. Djerassi, C. *Org. React.* **1951,** *6*, 207–235. (Review).

3. Byrne, B.; Karras, M. *Tetrahedron Lett.* **1987,** *28*, 769–772.

4. Ooi, T.; Otsuka, H.; Miura, T.; Ichikawa, H.; Maruoka, K. *Org. Lett.* **2002,** *4*, 2669–2672.

5. Suzuki, T.; Morita, K.; Tsuchida, M.; Hiroi, K. *J. Org. Chem.* **2003,** *68*, 1601–1602.

6. Auge, J.; Lubin-Germain, N.; Seghrouchni, L. *Tetrahedron Lett.* **2003,** *44*, 819–822.

7. Hon, Y.-S.; Chang, C.-P.; Wong, Y.-C. Byrne, B.; Karras, M. *Tetrahedron Lett.* **2004,** *45*, 3313–3315.

8. Kloetzing, R. J.; Krasovskiy, A.; Knochel, P. *Chem. Eur. J.* **2007,** *13*, 215–227.

9. Fuchter, M. J. *Oppenauer Oxidation.* In *Name Reactions for Functional Group Transformations*; Li, J. J., Ed.; Wiley: Hoboken, NJ, **2007,** pp 265–373. (Review).

10. Mello, R.; Martinez-Ferrer, J.; Asensio, G.; Gonzalez-Nunez, M. E. *J. Org. Chem.* **2008,** *72*, 9376–9378.

11. Borzatta, V.; Capparella, E.; Chiappino, R.; Impala, D.; Poluzzi, E.; Vaccari, A. *Cat. Today* **2009,** *140*, 112–116.

12. Fu, Y.; Yang, Y.; Hügel, H. M.; Du, Z.; Wang, K.; Huang, D.; Hu, Y. *Org. Biomol. Chem.* **2013,** *11,* 4429–4432.

13. Wang, Q.; Du, W.; liu, T.; Chai, H.; Yu, Z. *Tetrahedron Lett.* **2014,** *55*, 1585–1588.

14. Ogiwara, Y.; Ono, Y.; Sakai, N. *Synthesis* **2016,** *48*, 4143–4148.

15. Krasniqi, B.; Geerts, K.; Dehaen, W. *J. Org. Chem.* **2019,** *84*, 5027–5034.

Overman Rearrangement

Stereoselective transformation of allylic alcohol to allylic trichloroacetamide *via* trichloroacet*imidate* intermediate.

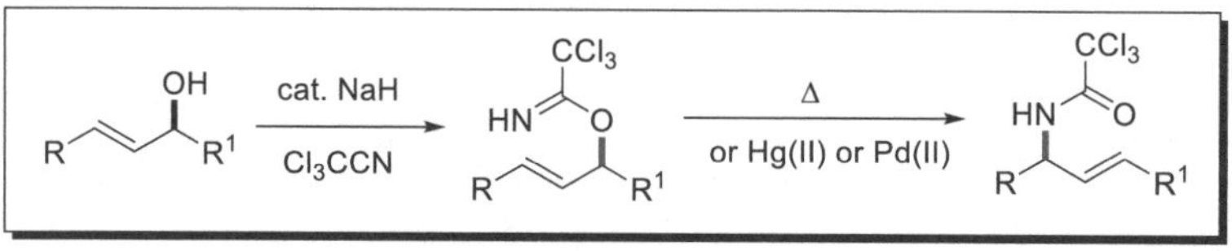

trichloroacetimidate

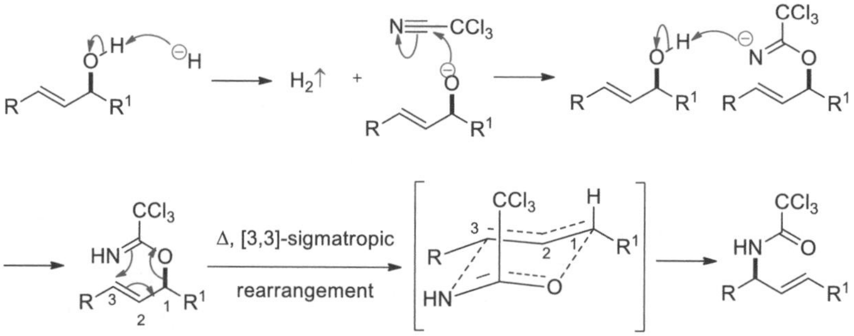

Example 1[5]

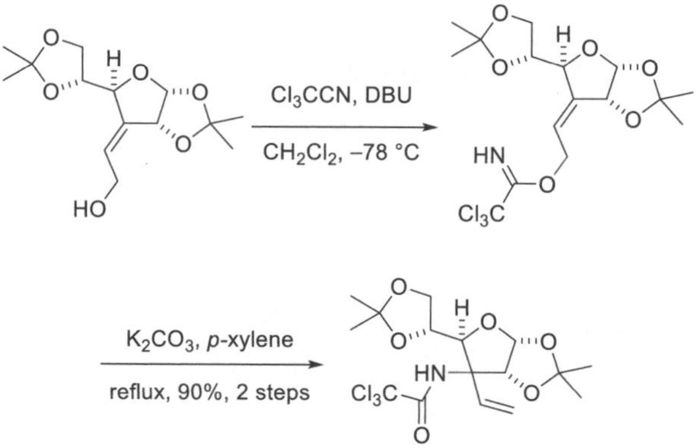

Example 2[6]

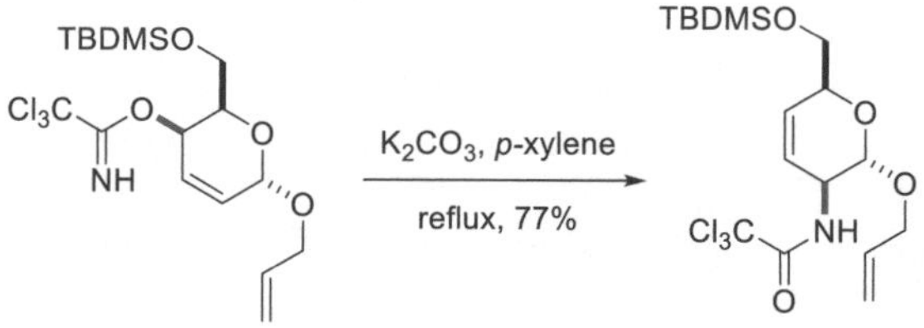

© Springer Nature Switzerland AG 2021

J. J. Li, *Name Reactions*, https://doi.org/10.1007/978-3-030-50865-4_112

Example 3[7]

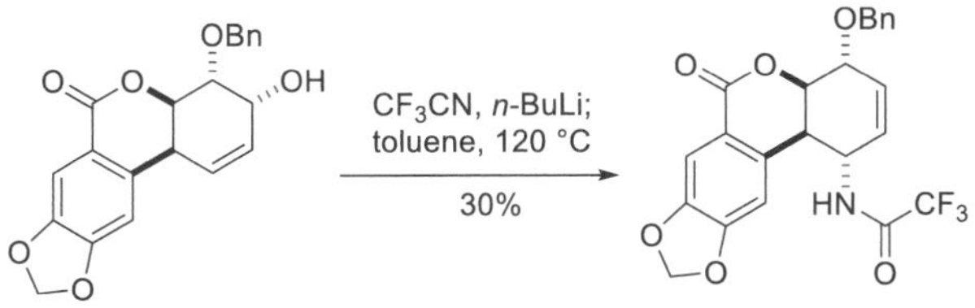

Example 4[9]

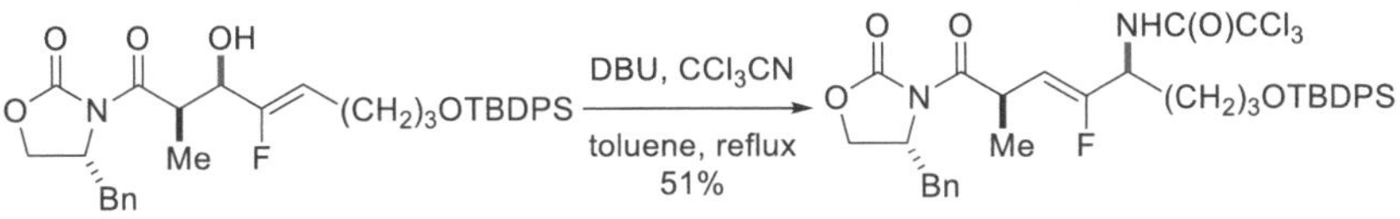

Example 5, Cascade-type Overman rearrangement[11]

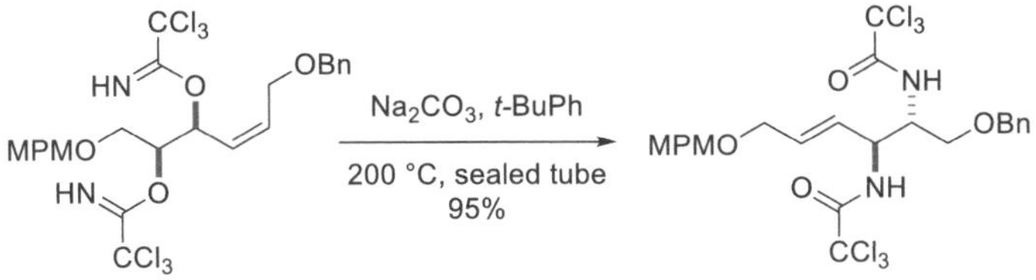

Example 6, Thermal [3,3]- and [3,5]-rearranegment[12]

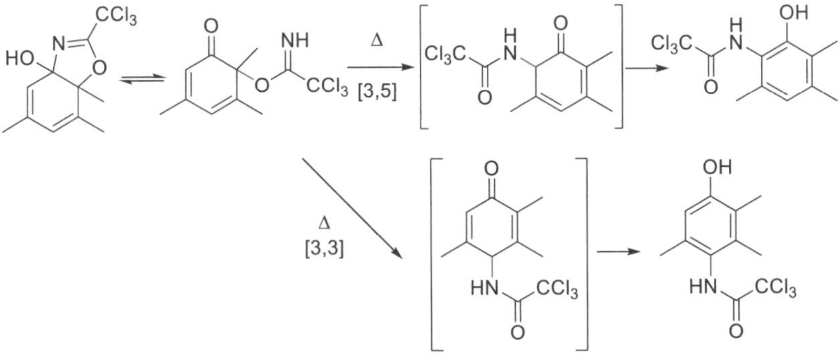

Example 7[13]

Example 8[14]

References

1. (a) Overman, L. E. *J. Am. Chem. Soc.* **1974**, *96*, 597–599. (b) Overman, L. E. *J. Am. Chem. Soc.* **1976**, *98*, 2901–2910. (c) Overman, L. E. *Acc. Chem. Res.* **1980**, *13*, 218–224. (Review).
2. Demay, S.; Kotschy, A.; Knochel, P. *Synthesis* **2001**, 863–866.
3. Oishi, T.; Ando, K.; Inomiya, K.; Sato, H.; Iida, M. *Org. Lett.* **2002**, *4*, 151–154.
4. Reilly, M.; Anthony, D. R.; Gallagher, C. *Tetrahedron Lett.* **2003**, *44*, 2927–2930.
5. Tsujimoto, T.; Nishikawa, T.; Urabe, D.; Isobe, M. *Synlett* **2005**, 433–436.
6. Montero, A.; Mann, E.; Herradon, B. *Tetrahedron Lett.* **2005**, *46*, 401–405.
7. Hakansson, A. E.; Palmelund, A.; Holm, H. *Chem. Eur. J.* **2006**, *12*, 3243–3253.
8. Bøjstrup, M.; Fanejord, M.; Lundt, I. *Org. Biomol. Chem.* **2007**, *5*, 3164–3171.
9. Lamy, C.; Hifmann, J.; Parrot-Lopez, H.; Goekjian, P. *Tetrahedron Lett.* **2007**, *48*, 6177–6180.
10. Wu, Y.-J. *Overman Rearrangement.* In *Name Reactions for Homologations-Part II*; Li, J. J., Ed.; Wiley: Hoboken, NJ, 2009, pp 210–225. (Review).
11. Nakayama, Y.; Sekiya, R.; Oishi, H.; Hama, N.; Yamazaki, M.; Sato, T.; Chida, N. *Chem. Eur. J.* **2013**, *19*, 12052–12058.
12. Sharma, S.; Rajale, T.; Unruh, D. K.; Birney, D. M. *J. Org. Chem.* **2015**, *80*, 11734–11743.
13. Martinez-Alsina, L. A.; Murray, J. C.; Buzon, L. M.; Bundesmann, M. W.; Young, J. M.; O'Neill, B. T. *J. Org. Chem.* **2017**, *82*, 12246–12256.
14. Fernandes, R. A.; Kattanguru, P.; Gholap, S. P.; Chaudhari, D. A. *Org. Biomol. Chem.* **2017**, *15*, 2672–2710. (Review).
15. Velasco-Rubio, A.; Alexy, E. J.; Yoritate, M.; Wright, A. C.; Stoltz, B. M. *Org. Let.* **2019**, *21*, 8962–8965.
16. Tjeng, A. A.; Handore, K. L.; Batey, R. A. *Org. Let.* **2020**, *22*, 3050–3055.

Paal–Knorr Pyrrole Synthesis

Reaction between 1,4-diketones and primary amines (or ammonia) to give pyrroles.
A variation of the Knorr pyrazole synthesis.

Example 1[4]

© Springer Nature Switzerland AG 2021
J. J. Li, *Name Reactions*, https://doi.org/10.1007/978-3-030-50865-4_113

Example 2[5]

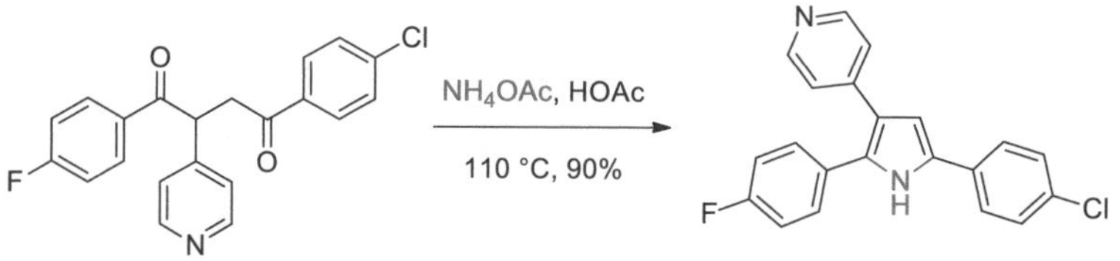

Example 3[9]

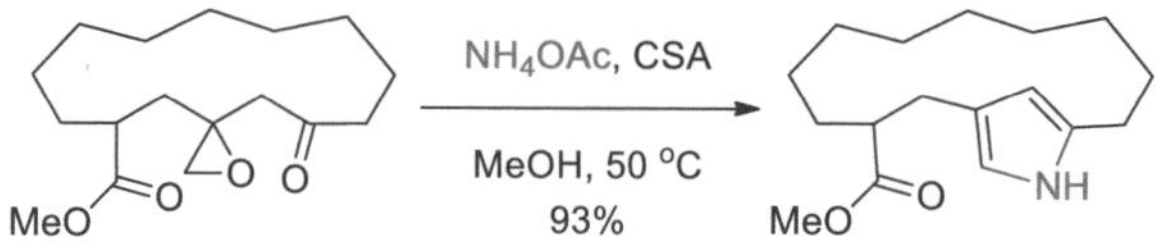

Example 4[10]

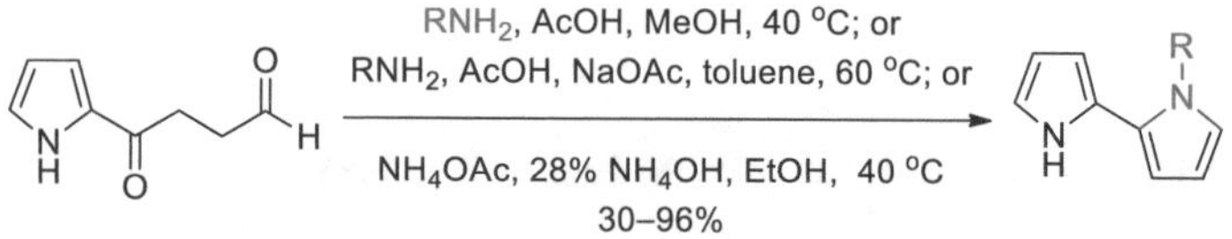

Example 5, Furan ring opening–pyrrole ring closure[10]

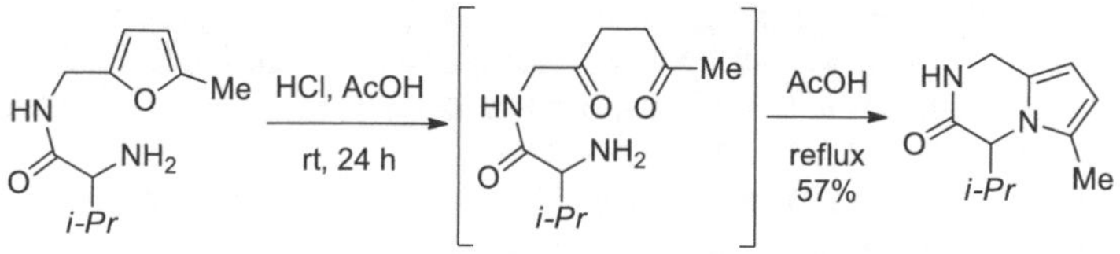

Example, A route to 2-substituted 3-cyanopyrroles[12]

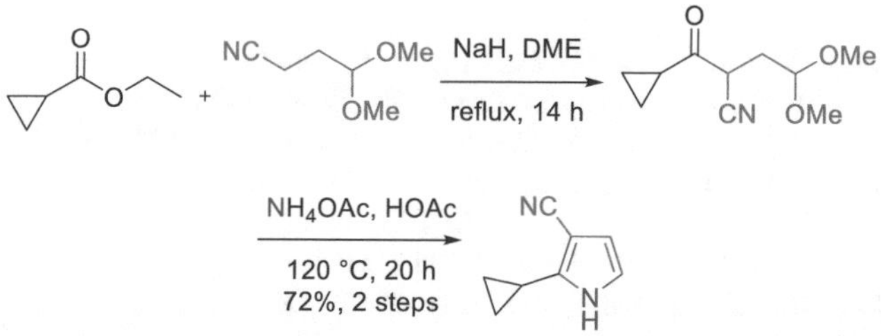

Example 6, Toward the synthesis of dihydropyrrin[13]

1. 10% aq. HCl/DMF (1:20), 30 min

2. NH₄OAc/Et₃N, 50 °C, 4 h
72%

Example 7, In total synthesis of marineosin A[14]

1. NH₄Ac, MW

2. Tf₂O, pyrrole
33%, 2 steps

References

1. (a) Paal, C. *Ber.* **1885,** *18*, 367–371. (b) Paal, C. *Ber.* **1885,** *18*, 2251–2254. (c) Knorr, L. *Ber.* **1885,** *18*, 299–311.
2. Corwin, A. H. *Heterocyclic Compounds Vol. 1*, Wiley, NY, **1950**; Chapter 6. (Review).
3. Jones, R. A.; Bean, G. P. *The Chemistry of Pyrroles*, Academic Press, London, **1977**, pp 51–57, 74–79. (Review).
4. (a) Brower, P. L.; Butler, D. E.; Deering, C. F.; Le, T. V.; Millar, A.; Nanninga, T. N.; Roth, B. D. *Tetrahedron Lett.* **1992,** *33*, 2279-2282. (b) Baumann, K. L.; Butler, D. E.; Deering, C. F.; Mennen, K. E.; Millar, A.; Nanninga, T. N.; Palmer, C. W.; Roth, B. D. *Tetrahedron Lett.* **1992,** *33*, 2279, 2283–2284.
5. de Laszlo, S. E.; Visco, D.; et al. *Bioorg. Med. Chem. Lett.* **1998,** *8*, 2689–2694.
6. Braun, R. U.; Zeitler, K.; Müller, T. J. J. *Org. Lett.* **2001,** *3*, 3297–3300.
7. Quiclet-Sire, B.; Quintero, L.; Sanchez-Jimenez, G.; Zard, Z. *Synlett* **2003**, 75–78.
8. Gribble, G. W. *Knorr and Paal–Knorr Pyrrole Syntheses*. In *Name Reactions in Heterocyclic Chemistry*; Li, J. J., Corey, E. J., Eds, Wiley: Hoboken, NJ, **2005**, 77–88. (Review).
9. Salamone, S. G.; Dudley, G. B. *Org. Lett.* **2005,** *7*, 4443–4445.
10. Fu, L.; Gribble, G. W. *Tetrahedron Lett.* **2008,** *49*, 7352–7354.
11. Trushkov, I. V.; Nevolina, T. A. *Tetrahedron Lett.* **2013,** *54*, 3974–3976.
12. Wiest, J. M.; Bach, T. *J. Org. Chem.* **2016,** *81*, 6149–6156.
13. Liu, Y.; Lindsey, J. S. *J. Org. Chem.* **2016,** *81*, 11882–11897.
14. Xu, B.; Li, G.; Li, J.; Shi, Y. *Org. Lett.* **2016,** *18*, 2028–2031.
15. Chen, J.-J.; Xu, Y.-C.; Gan, Z.-L.; Peng, X.; Yi, X.-Y. *Eur. J. Inorg. Chem.* **2019**, 1733–1739.
16. Zelina, E. Y.; Nevolina, T. A.; Sorotskaja, L. N.; Skvortsov, D. A.; Trushkov, I. V.; Uchuskin, M. G. *Tetrahedron Lett.* **2020,** *61*, 151532.

Parham Cyclization

The Parham cyclization is the generation by halogen–lithium exchange of aryllithiums and heteroaryllithiums, and their subsequent intramolecular cyclization onto an electrophilic site.

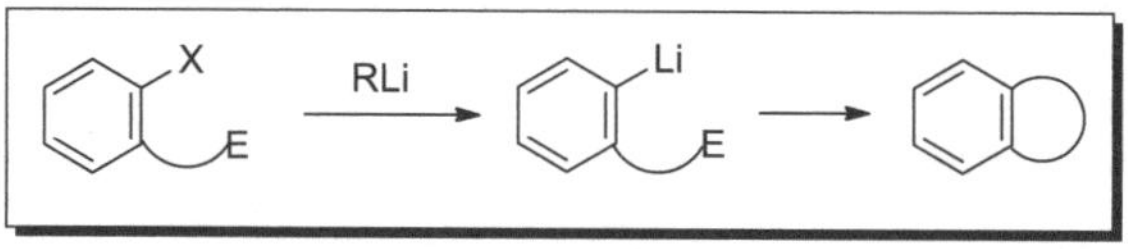

Example 1[1b]

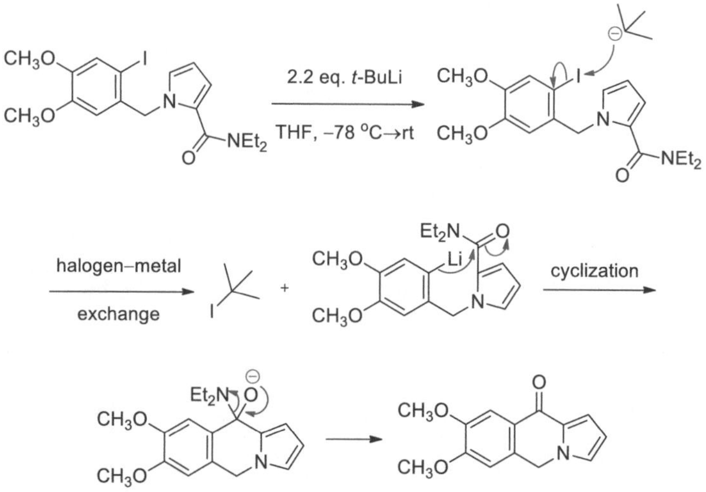

The fate of the second equivalent of t-BuLi:

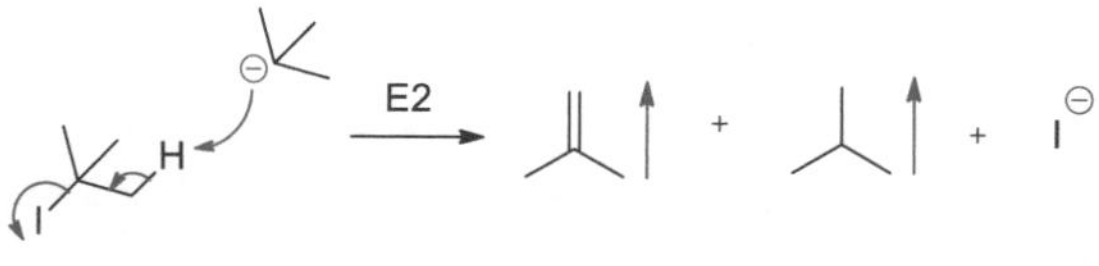

Example 2[2]

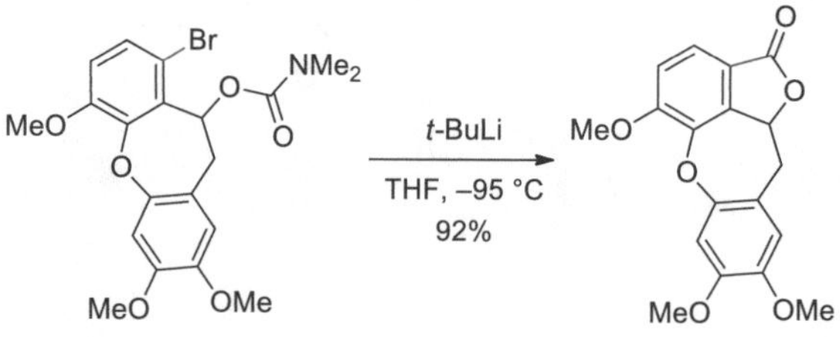

J. J. Li, *Name Reactions*, https://doi.org/10.1007/978-3-030-50865-4_114

Example 3[4]

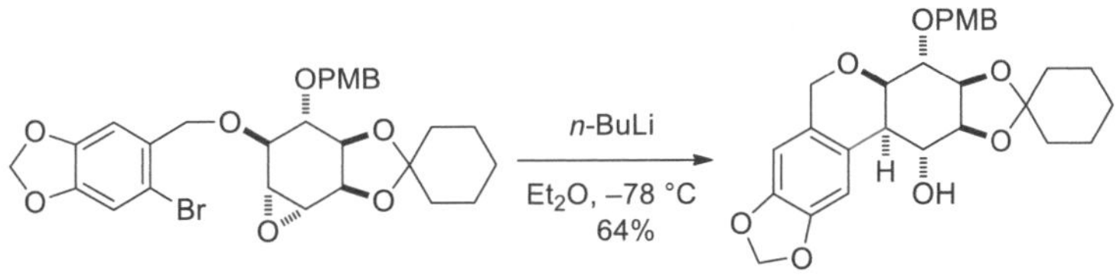

Example 4[5]

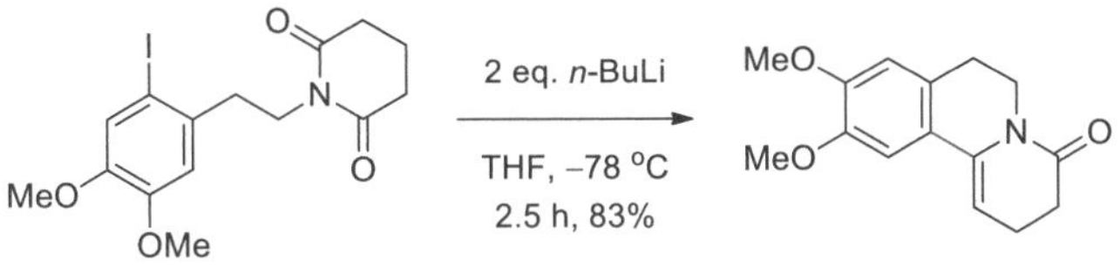

Example 5[9]

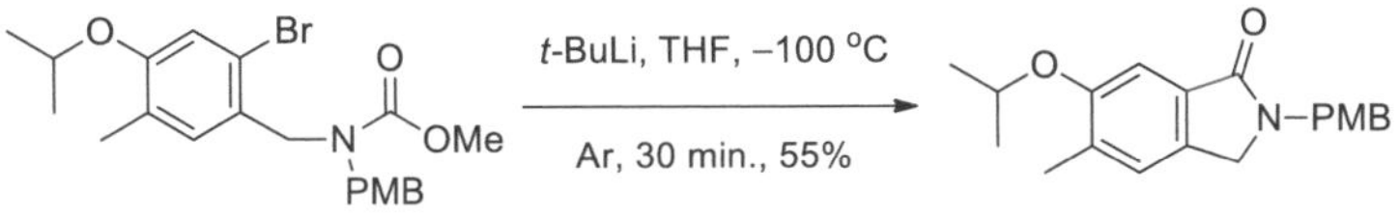

Example 6, Diaryl-fused seven-membered ring heterocyclic ketones[12]

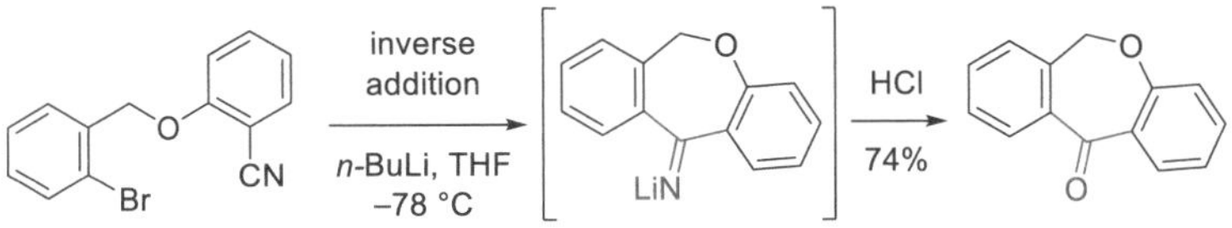

Example 7, A one-pot Parham–Aldol sequence[13]

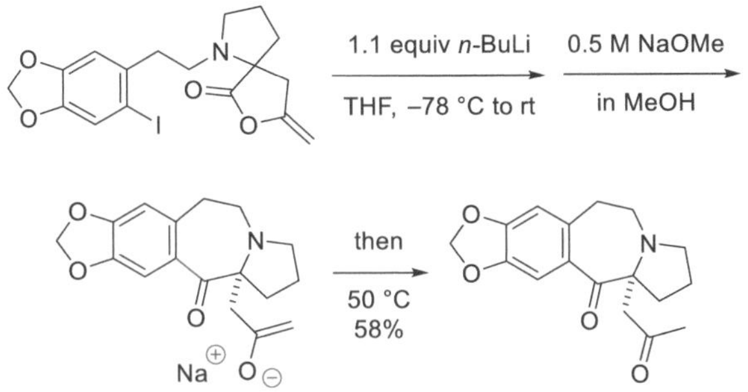

References

1. (a) Parham, W. E.; Jones, L. D.; Sayed, Y. *J. Org. Chem.* **1975**, *40*, 2394–2399. William E. Parham was a professor at Duke University. (b) Parham, W. E.; Jones, L. D.; Sayed, Y. *J. Org. Chem.* **1976**, *41*, 1184–1186. (c) Parham, W. E.; Bradsher, C. K. *Acc. Chem. Res.* **1982**, *15*, 300–305. (Review).
2. Paleo, M. R.; Lamas, C.; Castedo, L.; Domínguez, D. *J. Org. Chem.* **1992**, *57*, 2029–2033.
3. Gray, M.; Tinkl, M.; Snieckus, V. In *Comprehensive Organometallic Chemistry II*; Abel, E. W., Stone, F. G. A., Wilkinson, G., Eds.; Pergamon: Exeter, **1995**; Vol. 11; p 66. (Review).
4. Gauthier, D. R., Jr.; Bender, S. L. *Tetrahedron Lett.* **1996**, *37*, 13–16.
5. Collado, M. I.; Manteca, I.; Sotomayor, N.; Villa, M.-J.; Lete, E. *J. Org. Chem.* **1997**, *62*, 2080–2092.
6. Mealy, M. M.; Bailey, W. F. *J. Organomet. Chem.* **2002**, *646*, 59–67. (Review).
7. Sotomayor, N.; Lete, E. *Current Org. Chem.* **2003**, *7*, 275–300. (Review).
8. González-Temprano, I.; Osante, I.; Lete, E.; Sotomayor, N. *J. Org. Chem.* **2004**, *69*, 3875–3885.
9. Moreau, A.; Couture, A.; Deniau, E.; Grandclaudon, P.; Lebrun, S. *Org. Biomol. Chem.* **2005**, *3*, 2305–2309.
10. Gribble, G. W. *Parham Cyclization.* In *Name Reactions for Homologations-Part II;* Li, J. J., Ed.; Wiley: Hoboken, NJ, **2009**, pp 749–764. (Review).
11. Aranzamendi, E.; Sotomayor, N.; Lete, E. *J. Org. Chem.* **2012**, *77*, 2986–2991.
12. Farrokh, J.; Campos, C.; Hunt, D. A. *Tetrahedron Lett.* **2015**, *56*, 5245–5247.
13. Siitonen, J. H.; Yu, L.; Danielsson, J.; Di Gregorio, G.; Somfai, P. *J. Org. Chem.* **2018**, *83*, 11318–11322.
14. Melzer, B. C.; Plodek, A.; Bracher, F. *Beilst. J. Org. Chem.* **2019**, *15*, 2304–2310.

Passerini Reaction

Three-component condensation (3CC), one of the multicomponent reactions (MCRs), of carboxylic acids, C-isocyanides, and carbonyl compounds to afford α-acyloxycarboxamides. Also known as three-component reaction (3CR). *Cf.* Ugi reaction.

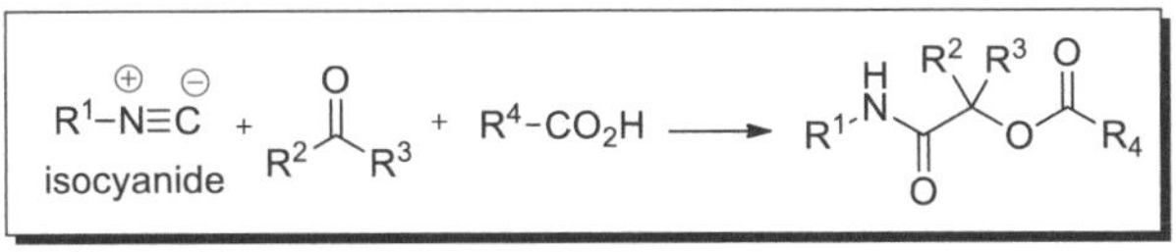

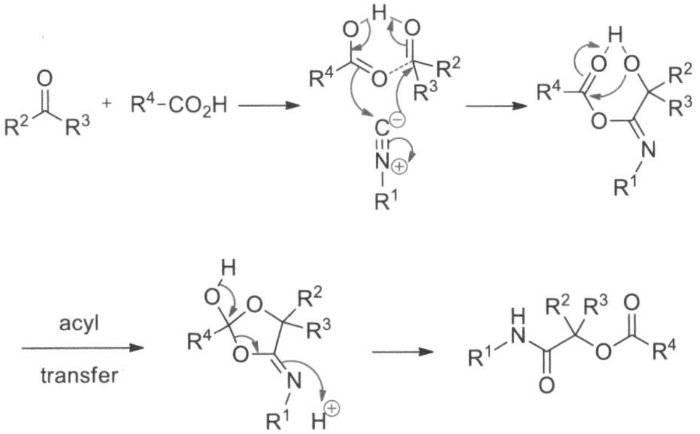

Example 1[3]

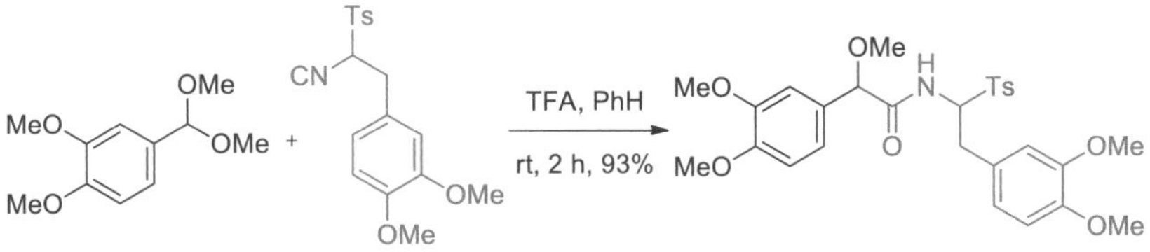

Example 2[5]

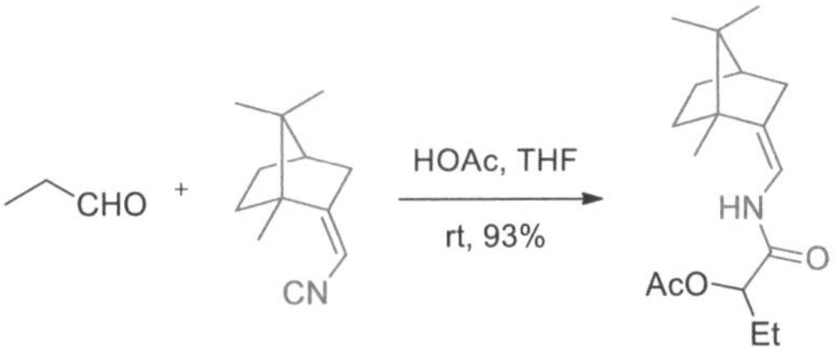

J. J. Li, *Name Reactions*, https://doi.org/10.1007/978-3-030-50865-4_115

Example 3[6]

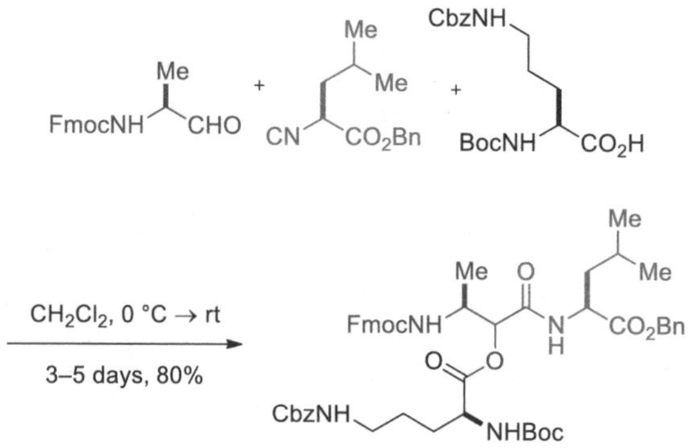

CH₂Cl₂, 0 °C → rt

3–5 days, 80%

Example 4[7]

CH₂Cl₂, 0 °C → rt

2 days, 59%

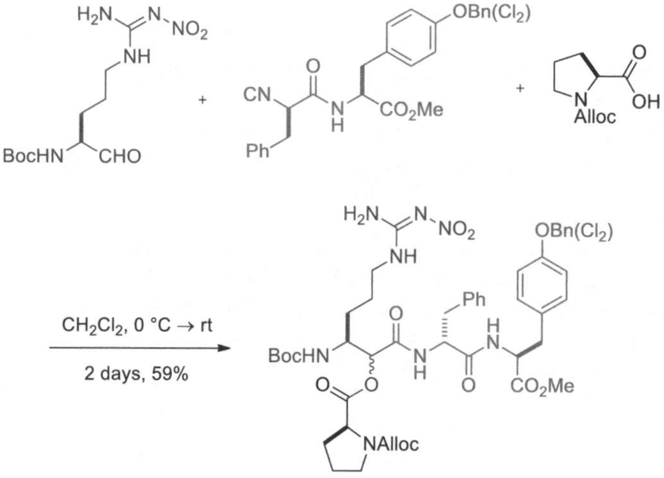

Example 5, Glycomimetics[12]

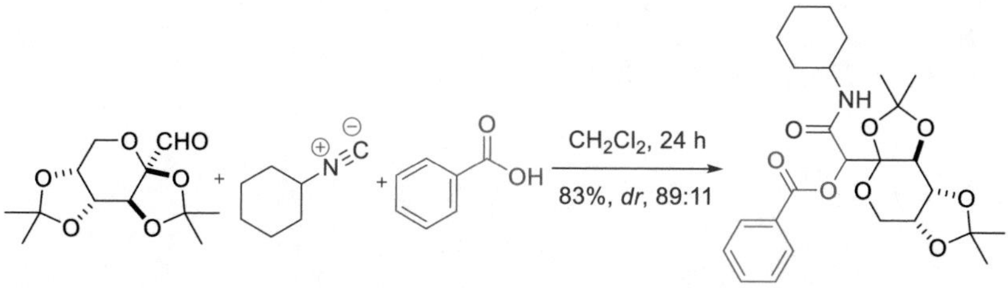

CH₂Cl₂, 24 h

83%, dr, 89:11

Example 6, Odorless isocyanides and *in situ* capture[13]

References

1. Passerini, M. *Gazz. Chim. Ital.* **1921**, *51*, 126–129. (b) Passerini, M. *Gazz. Chim. Ital.* **1921**, *51*, 181–188. Mario Passerini (1891–1962) was born in Scandicci, Italy. He obtained his Ph.D. in chemistry and pharmacy at the University of Florence, where he was a professor for most of his career.
2. Ferosie, I. *Aldrichimica Acta* **1971**, *4*, 21. (Review).
3. Barrett, A. G. M.; Barton, D. H. R.; Falck, J. R.; Papaioannou, D.; Widdowson, D. A. *J. Chem. Soc., Perkin Trans. 1* **1979**, 652–661.
4. Ugi, I.; Lohberger, S.; Karl, R. In *Comprehensive Organic Synthesis*; Trost, B. M.; Fleming, I., Eds.; Pergamon: Oxford, **1991**, *Vol. 2*, p.1083. (Review).
5. Bock, H.; Ugi, I. *J. Prakt. Chem.* **1997**, *339*, 385–389.
6. Banfi, L.; Guanti, G.; Riva, R. *Chem. Commun.* **2000**, 985–986.
7. Owens, T. D.; Semple, J. E. *Org. Lett.* **2001**, *3*, 3301–3304.
8. Xia, Q.; Ganem, B. *Org. Lett.* **2002**, *4*, 1631–1634.
9. Banfi, L.; Riva, R. *Org. React.* **2005**, *65*, 1–140. (Review).
10. Klein, J. C.; Williams, D. R. *Passerini Reaction*. In *Name Reactions for Homologations-Part II*; Li, J. J., Ed.; Wiley: Hoboken, NJ, **2009**, pp 765–785. (Review).
11. Sato, K.; Ozu, T.; Takenaga, N. *Tetrahedron Lett.* **2013**, *54*, 661–664.
12. Vlahoviček-Kahlina, K.; Vazdar, M.; Jakas, A.; Smrečki, V.; Jerić, I. *J. Org. Chem.* **2018**, *83*, 13146–13156.
13. Liu, N.; Chao, F.; Liu, M.-G.; Huang, N.-Y.; Zou, K.; Wang, L. *J. Org. Chem.* **2019**, *84*, 2366–2371.
14. So, W. H.; Xia, J. *Org. Lett.* **2020**, *22*, 214–218.

Paternó–Büchi Reaction

Photoinduced electrocyclization of a carbonyl with an alkene to form polysubstituted oxetane ring systems

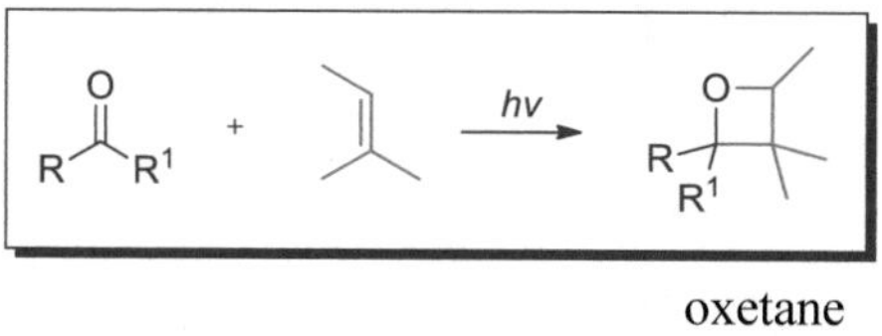
hv
oxetane

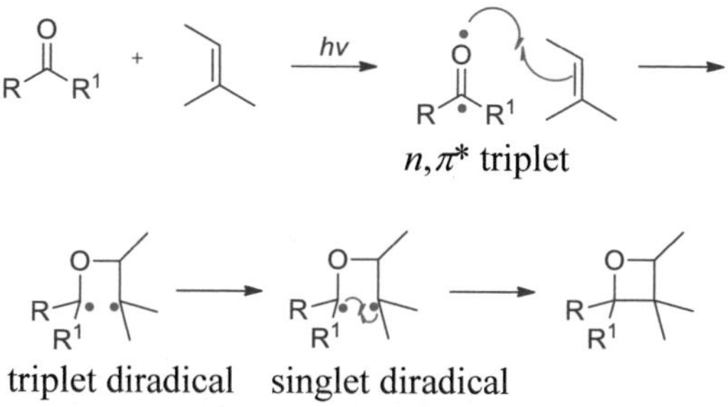
hv
n,π* triplet
triplet diradical
singlet diradical

Example 1[2]

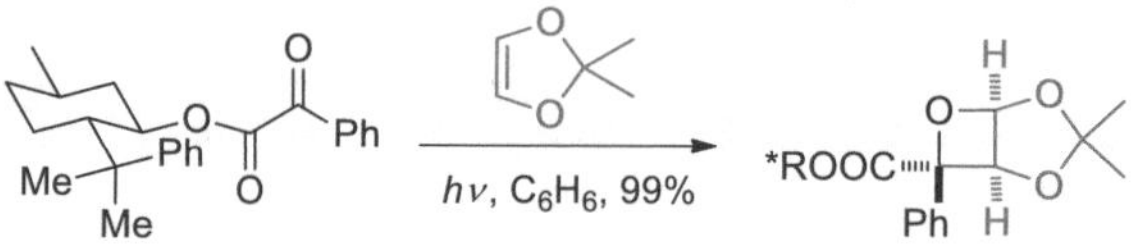
hν, C6H6, 99%
*ROOC
Ph

Example 2[4]

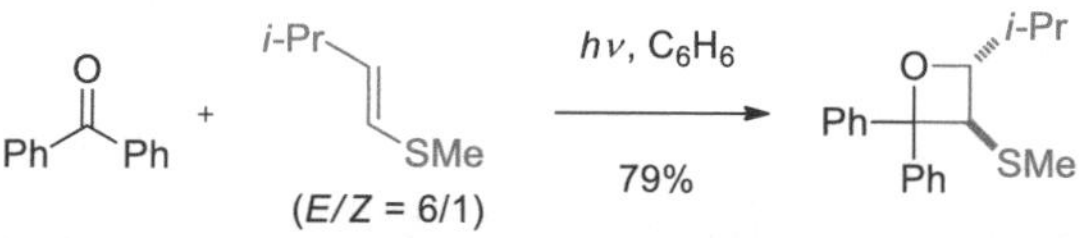
i-Pr
SMe
(E/Z = 6/1)
hν, C6H6
79%

Example 3[6]

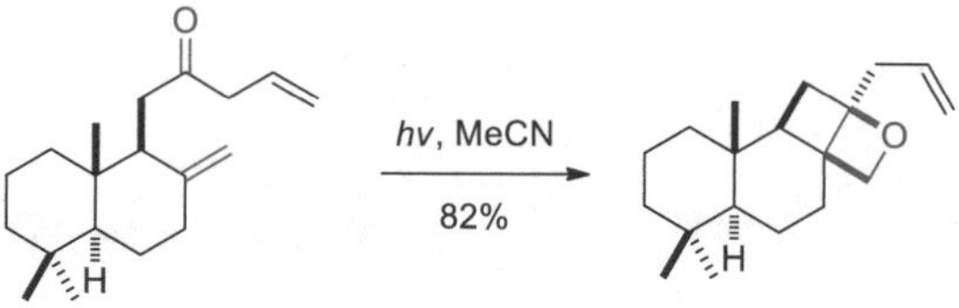
hv, MeCN
82%

J. J. Li, *Name Reactions*, https://doi.org/10.1007/978-3-030-50865-4_116

Example 4[8]

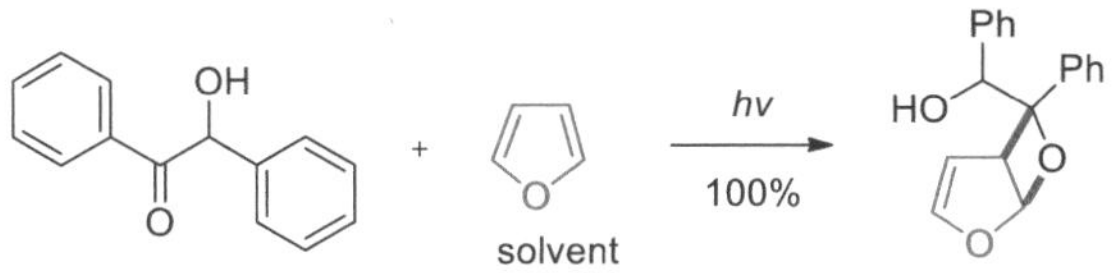

Example 5[9]

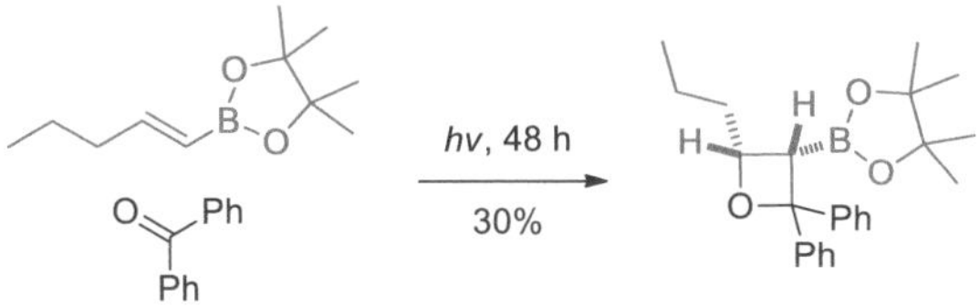

Example 6, Flow chemistry, MPL = medium pressure lamp; FEP = fluorinated ethylene propene[12]

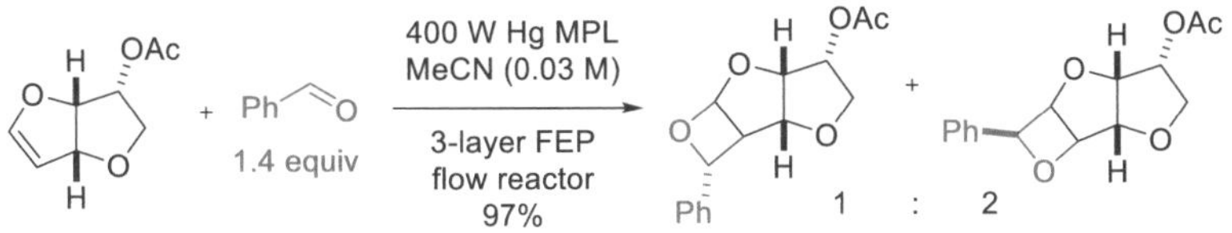

Example 7, Transposed Paternó–Büchi reaction: $\pi\pi^*$ inplace of $n\pi^*$ excited state[13]

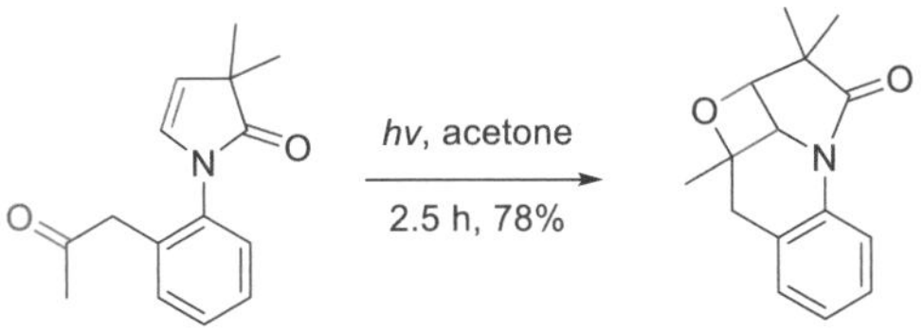

Example 8, Metal to ligand charge transfer (MLCT) as an alternative to the Paternó–Büchi reaction: Cu-catalyzed carbonyl–olefin [2 + 2] photocycloaddition (COPC)[14]

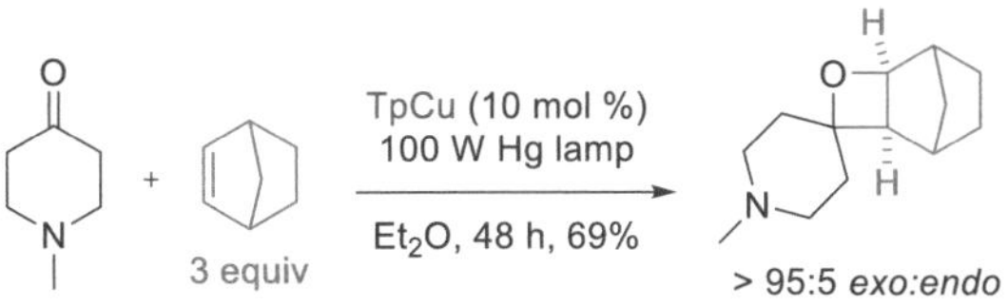

Tp = tris(pyrazolyl)borate =

References

1. (a) Paternó, E.; Chieffi, G. *Gazz. Chim. Ital.* **1909**, *39*, 341–361. Emaubuele Paternó (1847–1935) was born in Palermo, Sicily, Italy. It was 104 years ago when he first described photoinduced oxetane formation. (b) Büchi, G.; Inman, C. G.; Lipinsky, E. S. *J. Am. Chem. Soc.* **1954**, *76*, 4327–4331. George H. Büchi (1921–1998) was born in Baden, Switzerland. He was a professor at MIT when he elucidated the structure of oxetanes, the products from the light-catalyzed addition of carbonyl compounds to olefins, which had been observed by E. Paternó in 1909. Büchi died of heart failure while hiking with his wife in his native Switzerland.
2. Koch, H.; Runsink, J.; Scharf, H.-D. *Tetrahedron Lett.* **1983**, *24*, 3217–3220.
3. Carless, H. A. J. In *Synthetic Organic Photochemistry*; Horspool, W. M., Ed.; Plenum Press: New York, **1984**, 425. (Review).
4. Morris, T. H.; Smith, E. H.; Walsh, R. *J. Chem. Soc., Chem. Commun.* **1987**, 964–965.
5. Porco, J. A., Jr.; Schreiber, S. L. In *Comprehensive Organic Synthesis;* Trost, B. M.; Fleming, I., Eds.; Pergamon: Oxford, **1991**, *Vol. 5*, 151–192. (Review).
6. de la Torre, M. C.; Garcia, I.; Sierra, M. A. *J. Org. Chem.* **2003**, *68*, 6611–6618.
7. Griesbeck, A. G.; Mauder, H.; Stadtmüller, S. *Acc. Chem. Res.* **1994**, *27*, 70–75. (Review).
8. D'Auria, M.; Emanuele, L.; Racioppi, R. *Tetrahedron Lett.* **2004**, *45*, 3877–3880.
9. Liu, C. M. *Paternó–Büchi Reaction.* In *Name Reactions in Heterocyclic Chemistry*; Li, J. J., Ed.; Wiley: Hoboken, NJ, **2005**, pp 44–49. (Review).
10. Cho, D. W.; Lee, H.-Y.; Oh, S. W.; Choi, J. H.; Park, H. J.; Mariano, P. S.; Yoon, U. C. *J. Org. Chem.* **2008**, *73*, 4539–4547.
11. D'Annibale, A.; D'Auria, M.; Prati, F.; Romagnoli, C.; Stoia, S.; Racioppi, R.; Viggiani, L. *Tetrahedron* **2013**, *69*, 3782–3795.
12. Ralph, M.; Ng, S.; Booker-Milburn, K. I. *Org. Lett.* **2016**, *18*, 968–971.
13. Kumarasamy, R.; Raghunathan, R.; Kandappa, S. K.; Sreenithya, A.; Jockusch, S.; Sunoj, R. B.; Sivaguru, J. *J. Am. Chem. Soc.* **2017**, *139*, 655–662.
14. Flores, D. M.; Schmidt, V. A. *J. Am. Chem. Soc.* **2019**, *141*, 8741–8745.
15. Li, H.-F.; Cao, W.; Ma, X.; Ouyang, Z.; Xie, X.; Xia, Y. *J. Am. Chem. Soc.* **2020**, *142*, 3499–3505.

Pauson–Khand Reaction

Formal [2 + 2 + 1] cycloaddition of an alkene, alkyne, and carbon monoxide mediated by octacarbonyl dicobalt to form cyclopentenones.

hexacarbonyldicobalt complex

exo complex   sterically-favored isomer

Example 1[3]

Example 2, A catalytic version[6]

J. J. Li, *Name Reactions*, https://doi.org/10.1007/978-3-030-50865-4_117

Example 3, Intramolecular Pauson–Khand reaction[9]

Example 4, Intramolecular Pauson–Khand reaction[10]

Example 5, Intramolecular Pauson–Khand reaction[12]

Example 6, Intramolecular Pauson–Khand reaction toward the synthesis of (±)-5-epi-cyanthiwigin I, NMO = N-methylmorpholine N-oxide (NMMO)[13]

Example 7, In total synthesis[14]

Example 8, Vinyl fluorides as substrates[15]

References

1. (a) Pauson, P. L.; Khand, I. U.; Knox, G. R.; Watts, W. E. *J. Chem. Soc., Chem. Commun.* **1971**, 36. Ihsan U. Khand and Peter L. Pauson were at the University of Strathclyde, Glasgow in Scotland. (b) Khand, I. U.; Knox, G. R.; Pauson, P. L.; Watts, W. E.; Foreman, M. I. *J. Chem. Soc., Perkin Trans. 1* **1973**, 975–977. (c) Bladon, P.; Khand, I. U.; Pauson, P. L. *J. Chem. Res. (S)*, **1977**, 9. (d) Pauson, P. L. *Tetrahedron* **1985**, *41*, 5855–5860. (Review).
2. Schore, N. E. *Chem. Rev.* **1988**, *88*, 1081–1119. (Review).
3. Billington, D. C.; Kerr, W. J.; Pauson, P. L.; Farnocchi, C. F. *J. Organomet. Chem.* **1988**, *356*, 213–219.
4. Schore, N. E. In *Comprehensive Organic Synthesis*; Paquette, L. A.; Fleming, I.; Trost, B. M., Eds.; Pergamon: Oxford, **1991**, *Vol. 5*, p.1037. (Review).
5. Schore, N. E. *Org. React.* **1991**, *40*, 1–90. (Review).
6. Jeong, N.; Hwang, S. H.; Lee, Y.; Chung, J. *J. Am. Chem. Soc.* **1994**, *116*, 3159–3160.
7. Brummond, K. M.; Kent, J. L. *Tetrahedron* **2000**, *56*, 3263–3283. (Review).
8. Tsujimoto, T.; Nishikawa, T.; Urabe, D.; Isobe, M. *Synlett* **2005**, 433–436.
9. Miller, K. A.; Martin, S. F. *Org. Lett.* **2007**, *9*, 1113–1116.
10. Kaneda, K.; Honda, T. *Tetrahedron* **2008**, *64*, 11589–11593.
11. Torres, R. R. *The Pauson-Khand Reaction: Scope, Variations and Applications*, Wiley: Hoboken, NJ, 2012. (Review).
12. McCormack, M. P.; Waters, S. P. *J. Org. Chem.* **2013**, *78*, 1127–1137.
13. Chang, Y.; Shi, L.; Huang, J.; Shi, L.; Zhang, Z.; Hao, H.-D.; Gong, J.; Yang, Z. *Org. Lett.* **2018**, *20*, 2876–2879.
14. Hugelshofer, C. L.; Palani, V.; Sarpong, R. *J. Am. Chem. Soc.* **2019**, *141*, 8431–8435.
15. Román, R.; Mateu, N.; López, I.; Medio-Simon, M.; Fustero, S.; Barrio, P. *Org. Lett.* **2019**, *21*, 2569–2573.
16. Dibrell, S. E.; Maser, M. R.; Reisman, S. E. *J. Am. Chem. Soc.* **2020**, *142*, 6483–6487.

Payne Rearrangement

The isomerization of 2,3-epoxy alcohol under the influence of a base to 1,2-epoxy-3-ol is referred to as the Payne reaction. Also known as epoxide migration.

Example 1[2]

Example 2, Chemo-selective Payne rearrangement[3]

Example 3, Aza-Payne rearrangement[8]

Example 4, Aza-Payne rearrangement[9]

© Springer Nature Switzerland AG 2021
J. J. Li, *Name Reactions*, https://doi.org/10.1007/978-3-030-50865-4_118

Example 5, Lipase-mediated dynamic kinetic resolution via a *vinylogous* Payne rearrangement[11]

Example 6, Trapping the intermediates[13]

Example 7, LiAlH$_4$-induced thia-aza-Payne rearrangement[14]

Example 8, Aza-Payne/hydroamination sequence[15]

References

1. Payne, G. B. *J. Org. Chem.* **1962**, *27*, 3819–3822. George B. Payne was a chemist at Shell Development Co. in Emeryville, CA.
2. Buchanan, J. G.; Edgar, A. R. *Carbohydr. Res.* **1970**, *10*, 295–302.

3. Corey, E. J.; Clark, D. A.; Goto, G.; Marfat, A.; Mioskowski, C.; Samuelsson, B.; Hammerstrom, S. *J. Am. Chem. Soc.* **1980,** *102,* 1436–1439, and 3663–3665.
4. Ibuka, T. *Chem. Soc. Rev.* **1998,** *27,* 145–154. (Review).
5. Hanson, R. M. *Org. React.* **2002,** *60,* 1–156. (Review).
6. Yamazaki, T.; Ichige, T.; Kitazume, T. *Org. Lett.* **2004,** *6,* 4073–4076.
7. Bilke, J. L.; Dzuganova, M.; Froehlich, R.; Wuerthwein, E.-U. *Org. Lett.* **2005,** *7,* 3267–3270.
8. Feng, X.; Qiu, G.; Liang, S.; Su, J.; Teng, H.; Wu, L.; Hu, X. *Russ. J. Org. Chem.* **2006,** *42,* 514–500.
9. Feng, X.; Qiu, G.; Liang, S.; Teng, H.; Wu, L.; Hu, X. *Tetrahedron: Asymmetry* **2006,** *17,* 1394–1401.
10. Kumar, R. R.; Perumal, S. *Payne Rearrangement.* In *Name Reactions for Homologations-Part II*; Li, J. J., Ed.; Wiley: Hoboken, NJ, **2009,** pp 474–488. (Review).
11. Hoye, T. R.; Jeffrey, C. S.; Nelson, D. P. *Org. Lett.* **2010,** *12,* 52–55.
12. Kulshrestha, A.; Salehi Marzijarani, N.; Dilip Ashtekar, K.; Staples, R.; Borhan, B. *Org. Lett.* **2012,** *14,* 3592–3595.
13. Jung, M. E.; Sun, D. L. *Tetrahedron Lett.* **2015,** *56,* 3082–3085.
14. Dolfen, J.; Van Hecke, K.; D'hooghe, M. *Eur. J. Org. Chem.* **2017,** 3229–3233.
15. Gholami, H.; Kulshrestha, A.; Favor, O. K.; Staples, R. J.; Borhan, B. *Angew. Chem. Int. Ed.* **2019,** *58,* 10110–10113.

Petasis Reaction

Benzylic or allylic amine from the three-component reaction of an aryl- or a vinyl-boronic acid, a carbonyl and an amine. Also known as boron-Mannich or Petasis boronic acid-Mannich reaction. *Cf.* Mannich reaction.

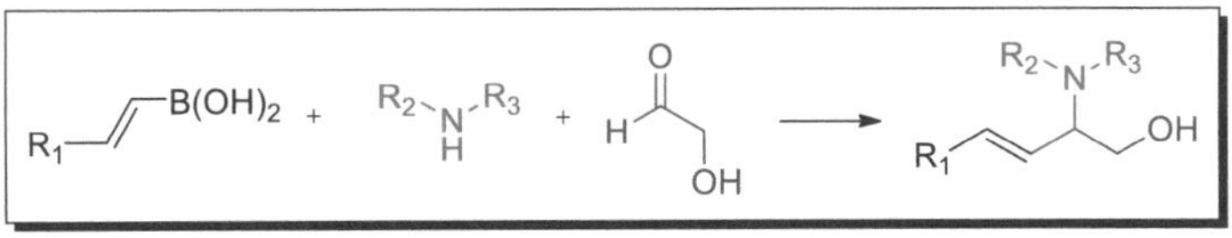

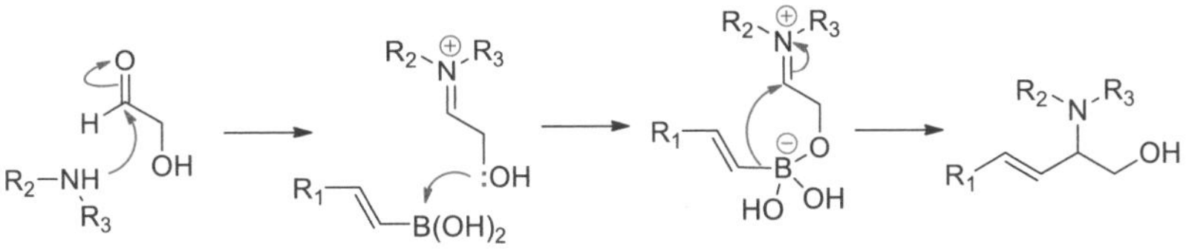

Example 1[2]

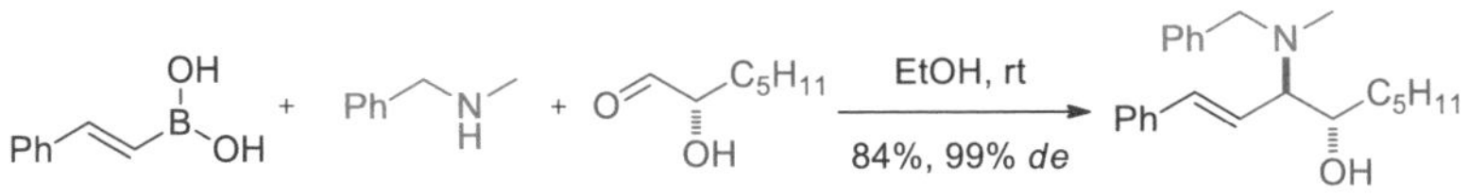

Example 2[4]

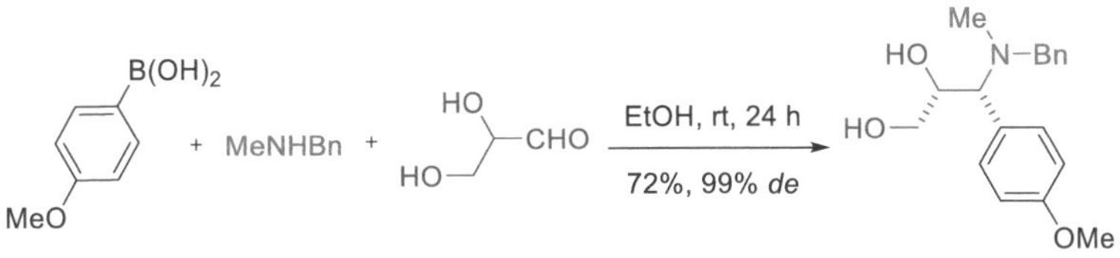

Example 3[9]

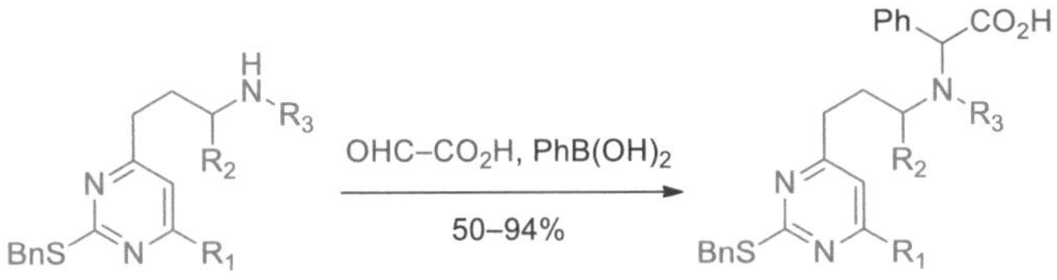

J. J. Li, *Name Reactions*, https://doi.org/10.1007/978-3-030-50865-4_119

Example 4, Asymmetric Petasis reaction, and VAPOL = 2,2'-diphenyl-(4-biphen-anthrol)[10]

Example 5, Asymmetric Petasis reaction[11]

Example 6, Amide in place of amine also works under right conditions[13]

Example 7, Enantioselective synthesis of allenes by catatlytic traceless Petasis reaction[14]

> 20:1 *dr*, 98:2 *er*

Example 8[15]

[Ir{dF(CF$_3$)$_2$ppy}$_2$(bpy)]PF$_6$ (2 mol %)

1 equiv NaHSO$_4$, 1,4-dioxane
blue LEDs, rt, 24 h, 51%

[Ir{dF(CF$_3$)$_2$ppy}$_2$(bpy)]PF$_6$

References

1. (a) Petasis, N. A.; Akritopoulou, I. *Tetrahedron Lett.* **1993**, *34*, 583–586. (b) Petasis, N. A.; Zavialov, I. A. *J. Am. Chem. Soc.* **1997**, *119*, 445–446. (c) Petasis, N. A.; Goodman, A.; Zavialov, I. A. *Tetrahedron* **1997**, *53*, 16463–16470. (d) Petasis, N. A.; Zavialov, I. A. *J. Am. Chem. Soc.* **1998**, *120*, 11798–11799. Nicos A. Petasis is a professor at the University of Southern California in Los Angeles.
2. Koolmeister, T.; Södergren, M.; Scobie, M. *Tetrahedron Lett.* **2002**, *43*, 5969–5970.
3. Orru, R. V. A.; deGreef, M. *Synthesis* **2003**, 1471–1499. (Review).
4. Sugiyama, S.; Arai, S.; Ishii, K. *Tetrahedron: Asymmetry* **2004**, *15*, 3149–3153.
5. Chang, Y. M.; Lee, S. H.; Nam, M. H.; Cho, M. Y.; Park, Y. S.; Yoon, C. M. *Tetrahedron Lett.* **2005**, *46*, 3053–3056.
6. Follmann, M.; Graul, F.; Schaefer, T.; Kopec, S.; Hamley, P. *Synlett* **2005**, 1009–1011.
7. Danieli, E.; Trabocchi, A.; Menchi, G.; Guarna, A. *Eur. J. Org. Chem.* **2007**, 1659–1668.
8. Konev, A. S.; Stas, S.; Novikov, M. S.; Khlebnikov, A. F.; Abbaspour Tehrani, K. *Tetrahedron* **2007**, *64*, 117–123.
9. Font, D.; Heras, M.; Villalgordo, J. M. *Tetrahedron* **2007**, *64*, 5226–5235.
10. Lou, S.; Schaus, S. E. *J. Am. Chem. Soc.* **2008**, *130*, 6922–6923.
11. Abbaspour Tehrani, K.; Stas, S.; Lucas, B.; De Kimpe, N. *Tetrahedron* **2009**, *65*, 1957–1966.
12. Han, W.-Y.; Zuo, J.; Zhang, X.-M.; Yuan, W.-C. *Tetrahedron* **2013**, *69*, 537–541.
13. Beisel, T.; Manolikakes, G. *Org. Lett.* **2013**, *15*, 6046–6049.

14. Jiang, Y.; Diagne, A. B.; Thomson, R. J.; Schaus, S. E. *J. Am. Chem. Soc.* **2017,** *139*, 1998–2005.
15. Yi, J.; Badir, S. O.; Alam, R.; Molander, G. A. *Org. Lett.* **2019,** *21*, 4853–4858.
16. Wu, P.; Givskov, M.; Nielsen, T. E. *Chem. Rev.* **2019,** *119*, 11245–11290. (Review).
17. Sim, Y. E.; Nwajiobi, O.; Mahesh, S.; Cohen, R. D.; Reibarkh, M. Y.; Raj, M. *Chem. Sci.* **2020,** *11*, 53–61.

Peterson Olefination

Alkenes from α-silyl carbanions and carbonyl compounds. Also known as the sila-Wittig reaction.

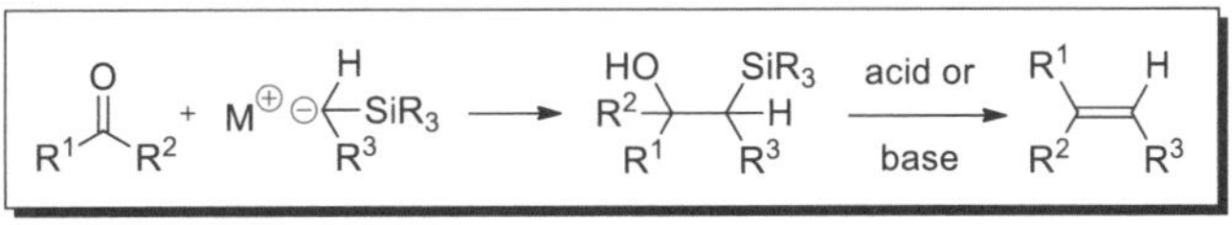

Basic conditions:

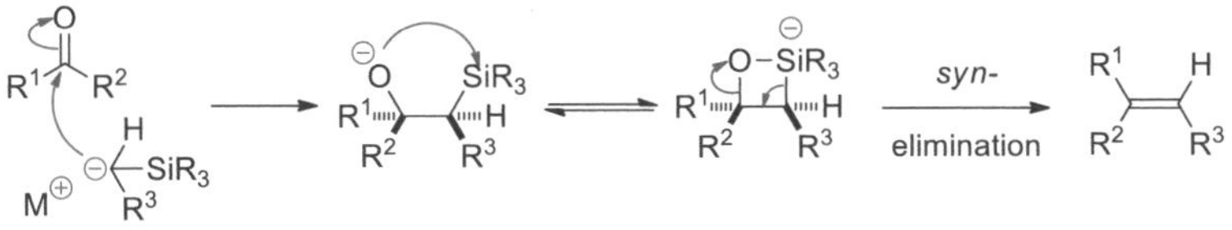

β-silylalkoxide intermediate

Acidic conditions:

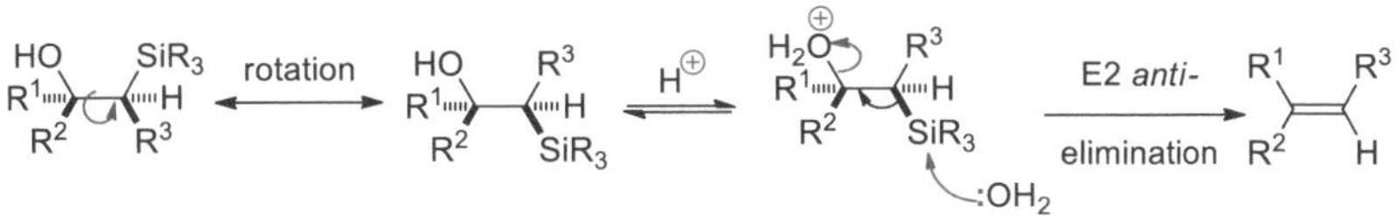

β-hydroxysilane

Example 1[6]

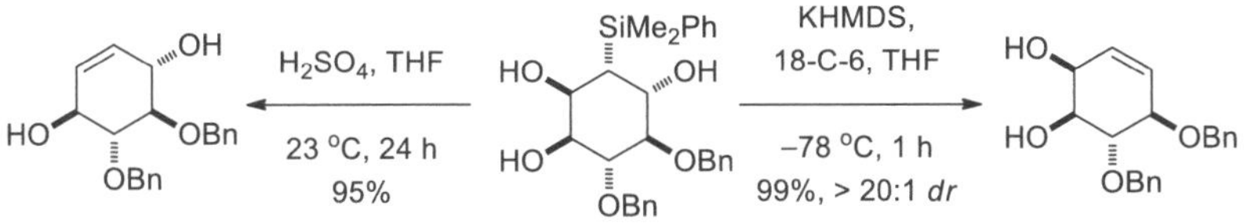

Example 2[7]

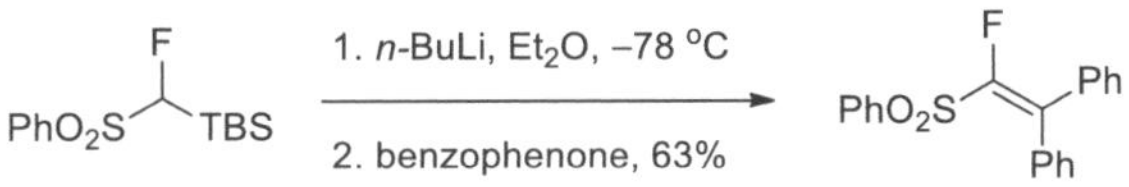

© Springer Nature Switzerland AG 2021
J. J. Li, *Name Reactions*, https://doi.org/10.1007/978-3-030-50865-4_120

Example 3[8]

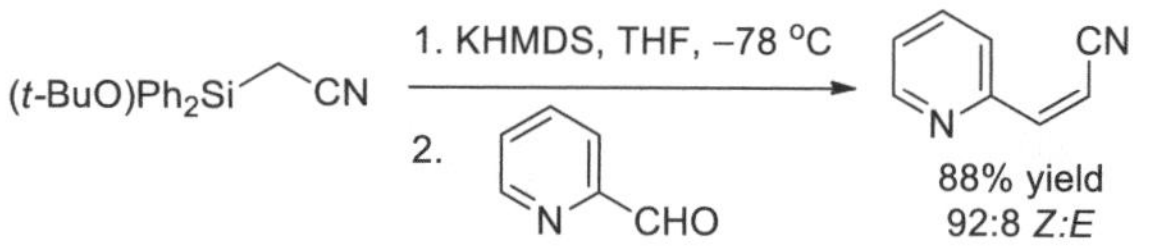
(t-BuO)Ph2Si CN
1. KHMDS, THF, −78 °C
2.
N CHO
N
CN
88% yield
92:8 Z:E

Example 4[10]

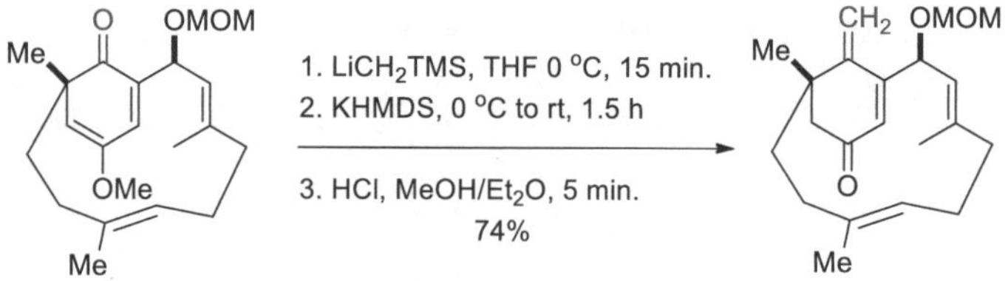
O OMOM
Me
OMe
Me
1. LiCH2TMS, THF 0 °C, 15 min.
2. KHMDS, 0 °C to rt, 1.5 h
3. HCl, MeOH/Et2O, 5 min.
74%
CH2 OMOM
Me
O
Me

Example 5[12]

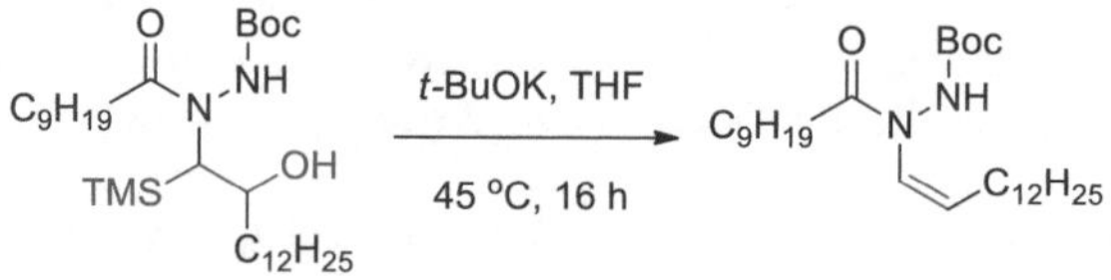
O Boc
NH
C9H19 N
TMS OH
C12H25
t-BuOK, THF
45 °C, 16 h
O Boc
NH
C9H19 N
C12H25

Example 6, Flow chemistry[13]

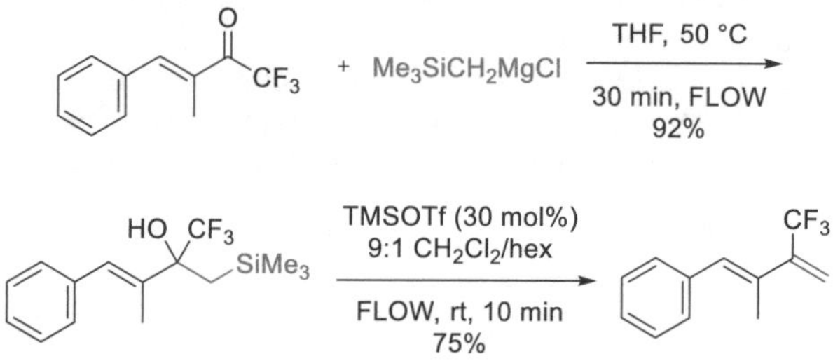
O
CF3
+ Me3SiCH2MgCl
THF, 50 °C
30 min, FLOW
92%
HO CF3
SiMe3
TMSOTf (30 mol%)
9:1 CH2Cl2/hex
FLOW, rt, 10 min
75%
CF3

Example 7, In total synthesis[14]

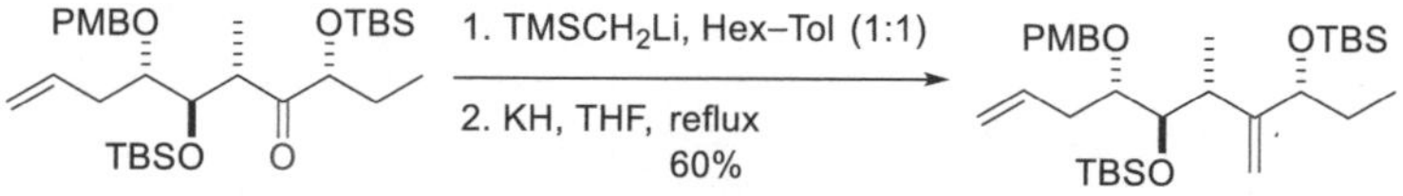
PMBO OTBS
TBSO O
1. TMSCH2Li, Hex–Tol (1:1)
2. KH, THF, reflux
60%
PMBO OTBS
TBSO

Example 8, Grignard addition followed by elimination of dimethylsilanol to prepare terminal olefin[15]

References

1. Peterson, D. J. *J. Org. Chem.* **1968**, *33*, 780–784.
2. Ager, D. J. *Org. React.* **1990**, *38*, 1–223. (Review).
3. Barrett, A. G. M.; Hill, J. M.; Wallace, E. M.; Flygare, J. A. *Synlett* **1991**, 764–770. (Review).
4. van Staden, L. F.; Gravestock, D.; Ager, D. J. *Chem. Soc. Rev.* **2002**, *31*, 195–200. (Review).
5. Ager, D. J. *Science of Synthesis* **2002**, *4*, 789–809. (Review).
6. Heo, J.-N.; Holson, E. B.; Roush, W. R. *Org. Lett.* **2003**, *5*, 1697–1700.
7. Asakura, N.; Usuki, Y.; Iio, H. *J. Fluorine Chem.* **2003**, *124*, 81–84.
8. Kojima, S.; Fukuzaki, T.; Yamakawa, A.; Murai, Y. *Org. Lett.* **2004**, *6*, 3917–3920.
9. Kano, N.; Kawashima, T. *The Peterson and Related Reactions* in *Modern Carbonyl Olefination;* Takeda, T., Ed.; Wiley-VCH: Weinheim, Germany, **2004**, 18–103. (Review).
10. Huang, J.; Wu, C.; Wulff, W. D. *J. Am. Chem. Soc.* **2007**, *129*, 13366.
11. Ahmad, N. M. *Peterson Olefination.* In *Name Reactions for Homologations-Part I*; Li, J. J., Ed., Wiley: Hoboken, NJ, **2009**, pp 521–538. (Review).
12. Beveridge, R. E.; Batey, R. A. *Org. Lett.* **2013**, *15*, 3086–3089.
13. Hamlin, T. A.; Lazarus, G. M. L.; Kelly, C. B.; Leadbeater, N. E. *Org. Process Res. Dev.* **2014**, *18*, 1253–1258.
14. Wang, L.; Wu, F.; Jia, X.; Xu, Z.; Guo, Y.; Ye, T. *Org. Lett.* **2018**, *20*, 2213–2215.
15. Tiniakos, A. F.; Wittmann, S.; Audic, A.; Prunet, J. *Org. Lett.* **2019**, *21*, 589–592.
16. Britten, T. K.; McLaughlin, M. G. *J. Org. Chem.* **2020**, *85*, 301–305.

Pictet–Spengler Tetrahydroisoquinoline Synthesis

Tetrahydroisoquinolines from condensation of β-arylethylamines and carbonyl compounds followed by cyclization.

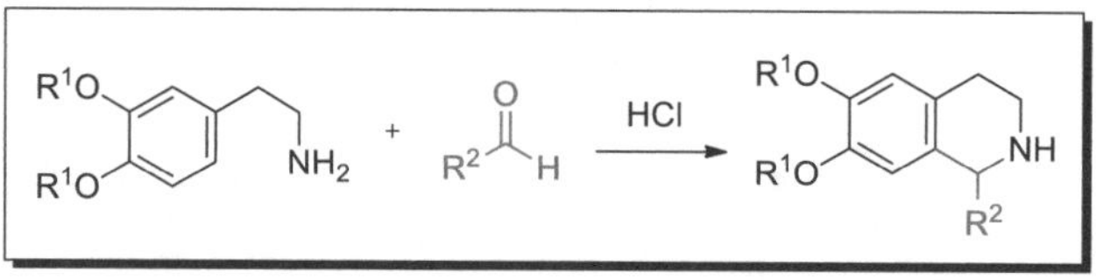

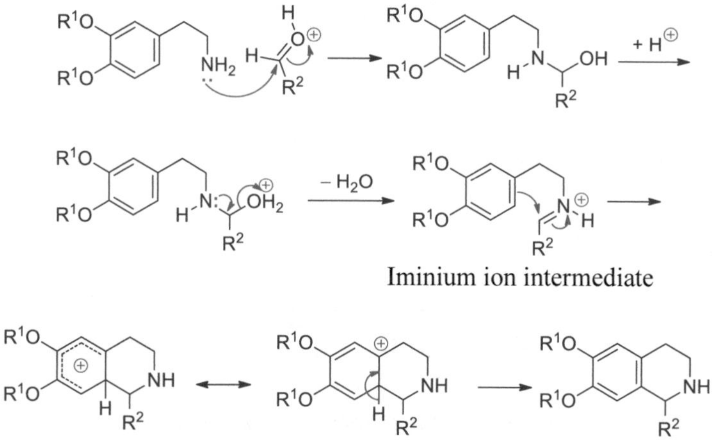

Iminium ion intermediate

Example 1[4]

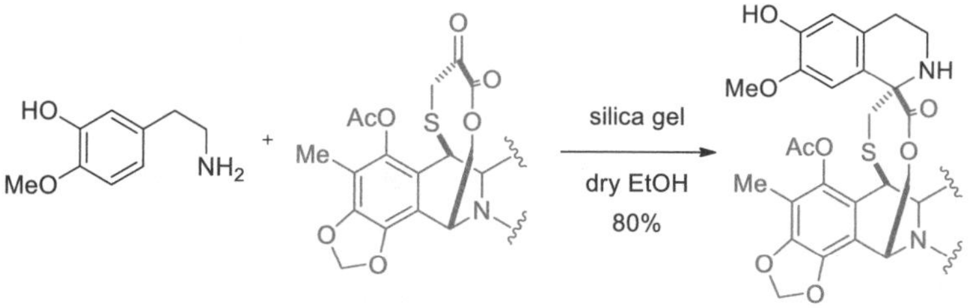

Example 2[7]

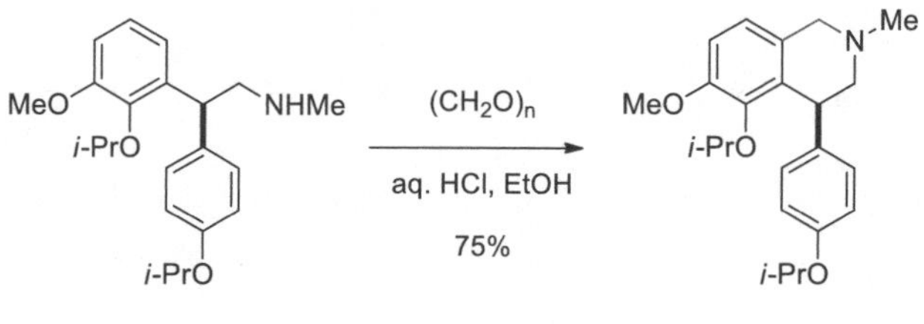

J. J. Li, *Name Reactions*, https://doi.org/10.1007/978-3-030-50865-4_121

Example 3, Asymmetric acyl Pictet–Spengler[9]

Example 4, Oxa-Pictet–Spengler[10]

Example 5, A diastereoselective Pictet–Spengler reaction to make tetrahydro-β-carboline glycosides[11]

Example 6, Formation of 7-membered heterocycles[12]

Example 7, An interrupted Pictet–Spengler reaction[13]

Example 8, *En route* to the synthesis of monoterpene indole alkaloid (–)-alstoscholarine[14]

References

1. Pictet, A.; Spengler, T. *Ber.* **1911**, *44*, 2030–2036.
2. Cox, E. D.; Cook, J. M. *Chem. Rev.* **1995**, *95*, 1797–1842. (Review).
3. Corey, E. J.; Gin, D. Y.; Kania, R. S. *J. Am. Chem. Soc.* **1996**, *118*, 9202–9203.
4. Zhou, B.; Guo, J.; Danishefsky, S. J. *Org. Lett.* **2002**, *4*, 43–46.
5. Yu, J.; Wearing, X. Z.; Cook, J. M. *Tetrahedron Lett.* **2003**, *44*, 543–547.
6. Tsuji, R.; Nakagawa, M.; Nishida, A. *Tetrahedron: Asymmetry* **2003**, *14*, 177–180.
7. Couture, A.; Deniau, E.; Grandclaudon, P.; Lebrun, S. *Tetrahedron: Asymmetry* **2003**, *14*, 1309–1320.
8. Tinsley, J. M. *Pictet–Spengler Isoquinoline Synthesis*. In *Name Reactions in Heterocyclic Chemistry*; Li, J. J., Ed.; Wiley: Hoboken, NJ, **2005**, 469–479. (Review).
9. Mergott, D. J.; Zuend, S. J.; Jacobsen, E. N. *Org. Lett.* **2008**, *10*, 745–748.
10. Eid, C. N.; Shim, J.; Bikker, J.; Lin, M. *J. Org. Chem.* **2009**, *74*, 423–426.
11. Pradhan, P.; Nandi, D.; Pradhan, S. D.; Jaisankar, P.; Giri, V. S. *Synlett* **2013**, *24*, 85–89.
12. Katte, T. A.; Reekie, T. A.; Jorgensen, W. T.; Kassiou, M. *J. Org. Chem.* **2016**, *81*, 4883–4889.
13. Gabriel, P.; Gregory, A. W.; Dixon, D. *Org. Lett.* **2019**, *21*, 6658–6662.
14. Yao, J.-N.; Liang, X.; Wei, K.; Yang, Y.-R. *Org. Lett.* **2019**, *21*, 8485–8487.
15. Zheng, C.; You, S.-L. *Acc. Chem. Res.* **2020**, *53*, 974–987. (Review).

Pinacol Rearrangement

Acid-catalyzed rearrangement of vicinal diols (pinacols) to carbonyl compounds.

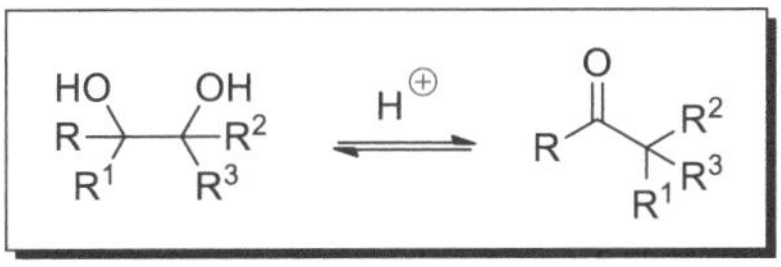

The most electron-rich alkyl group (more substituted carbon) migrates first. The general migration order:

tertiary alkyl > cyclohexyl > secondary alkyl > benzyl > phenyl > primary alkyl > methyl >> H.

For substituted aryls:
p-MeO-Ar > *p*-Me-Ar > *p*-Cl-Ar > *p*-Br-Ar > *p*-O_2N-Ar

Example 1[4]

Example 2[5]

J. J. Li, *Name Reactions*, https://doi.org/10.1007/978-3-030-50865-4_122

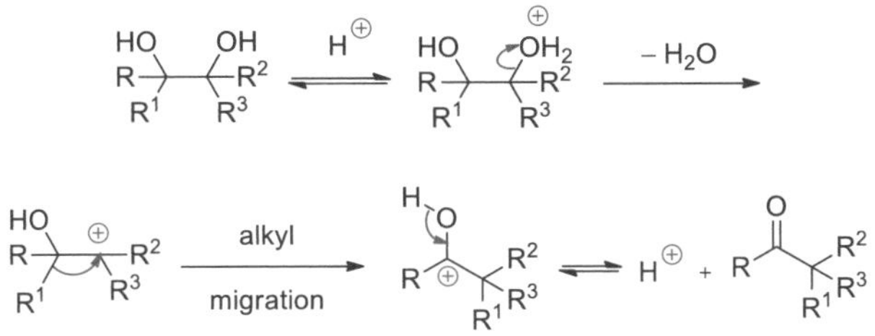

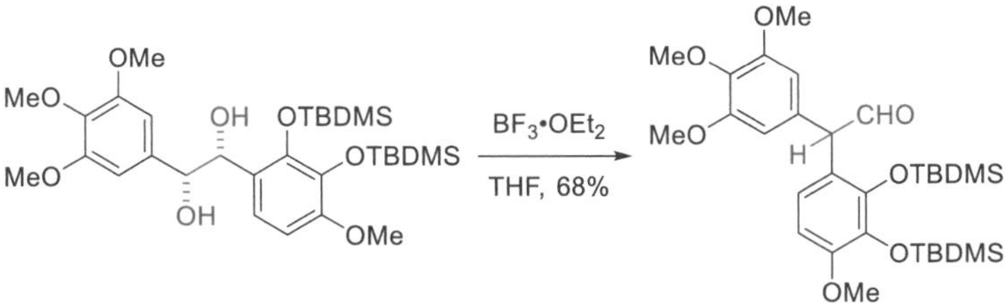

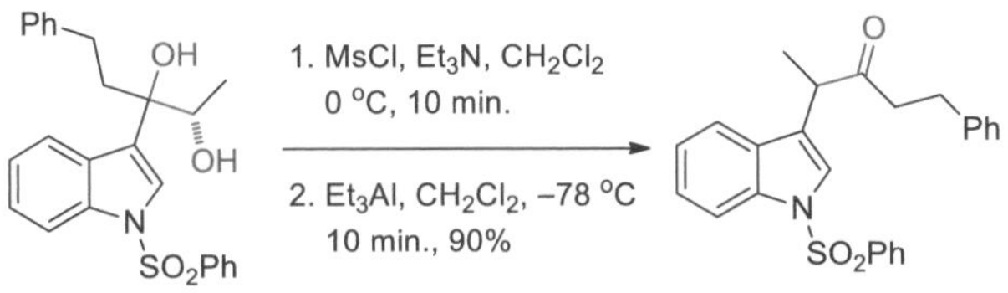

Example 3[7]

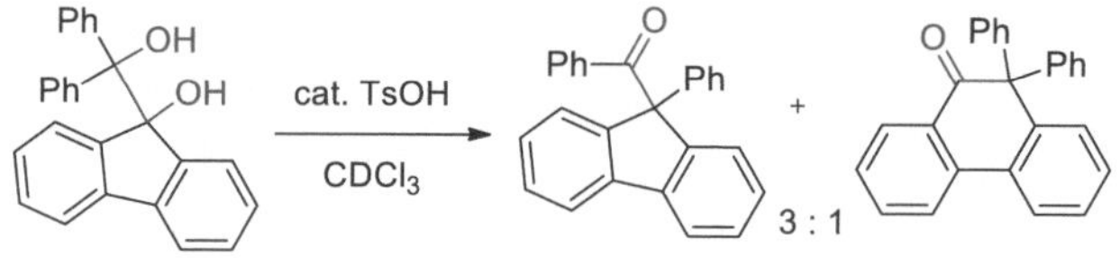
Ph
OH
Ph
OH
cat. TsOH
CDCl3
Ph
O
Ph
+
O
Ph
Ph
3 : 1

Example 4[9]

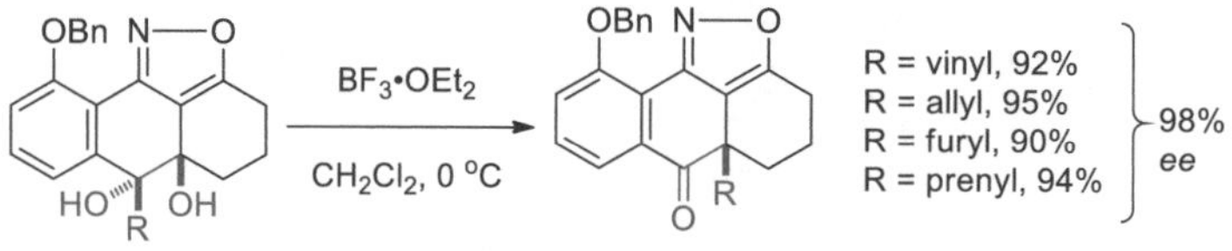
OBn
N
O
HO
R
OH
BF3•OEt2
CH2Cl2, 0 °C
OBn
N
O
O
R
R = vinyl, 92%
R = allyl, 95%
R = furyl, 90%
R = prenyl, 94%
98%
ee

Example 5, A trivalent organophosphorus reagent induced pinacol rearrangement[11]

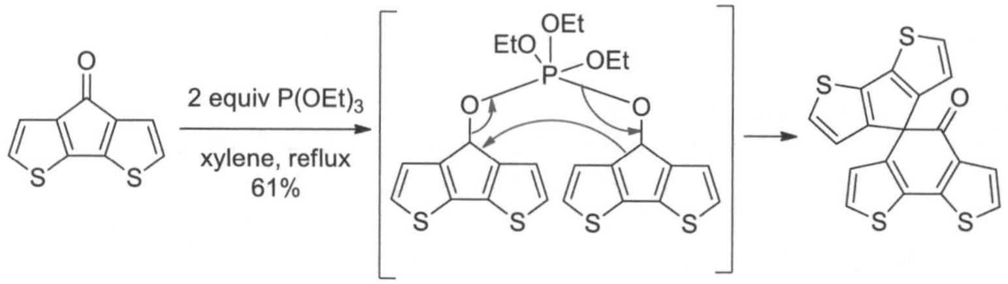
O
S
S
2 equiv P(OEt)3
xylene, reflux
61%
OEt
EtO
OEt
P
O
O
S
S
S
S
S
S
O
S
S

Example 6, Oxonium ion-induced pinacol rearrangement[13]

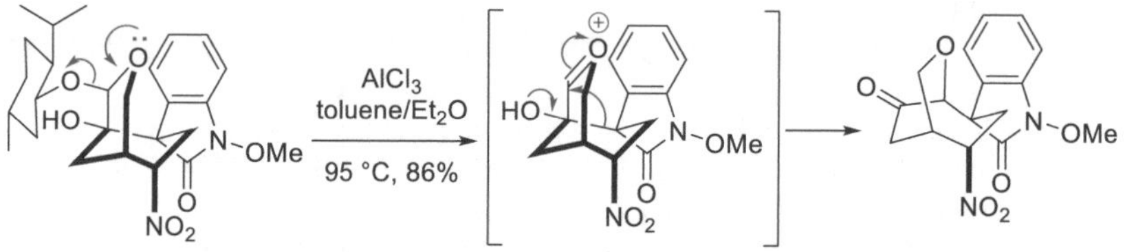
O
O
HO
N
OMe
O
NO2
AlCl3
toluene/Et2O
95 °C, 86%
O
HO
N
OMe
O
NO2
O
O
N
OMe
O
NO2

Example 7, Lewis acid-assisted electrophilic fluorine-catalyzed pinacol rearrangement of hydrobenzoin substrates (NFSI = *N*-fluorobenzenesulfonimide)[14]

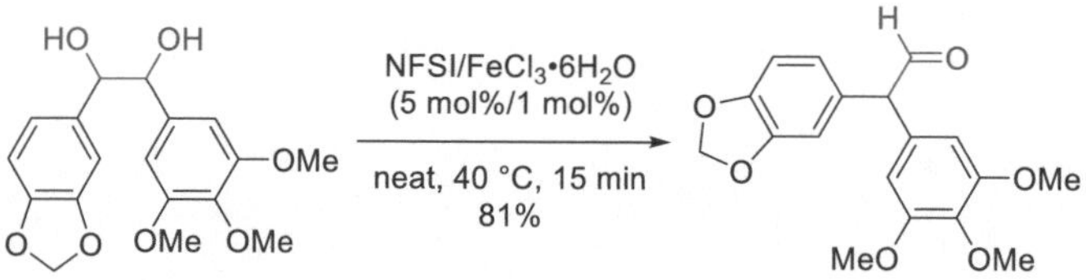
HO
OH
OMe
O
O
OMe
OMe
NFSI/FeCl3•6H2O
(5 mol%/1 mol%)
neat, 40 °C, 15 min
81%
H
O
O
O
OMe
MeO
OMe

Example 8, Catalytic enantioselective pinacol rearrangement[15]

References

1. Fittig, R. *Ann.* **1860**, *114*, 54–63.
2. Magnus, P.; Diorazio, L.; Donohoe, T. J.; Giles, M.; Pye, P.; Tarrant, J.; Thom, S. *Tetrahedron* **1996**, *52*, 14147–14176.
3. Razavi, H.; Polt, R. *J. Org. Chem.* **2000**, *65*, 5693–5706.
4. Pettit, G. R.; Lippert III, J. W.; Herald, D. L. *J. Org. Chem.* **2000**, *65*, 7438–7444.
5. Shinohara, T.; Suzuki, K. *Tetrahedron Lett.* **2002**, *43*, 6937–6940.
6. Overman, L. E.; Pennington, L. D. *J. Org. Chem.* **2003**, *68*, 7143–7157. (Review).
7. Mladenova, G.; Singh, G.; Acton, A.; Chen, L.; Rinco, O.; Johnston, L. J.; Lee-Ruff, E. *J. Org. Chem.* **2004**, *69*, 2017–2023.
8. Birsa, M. L.; Jones, P. G.; Hopf, H. *Eur. J. Org. Chem.* **2005**, 3263–3270.
9. Suzuki, K.; Takikawa, H.; Hachisu, Y.; Bode, J. W. *Angew. Chem. Int. Ed.* **2007**, *46*, 3252–3254.
10. Goes, B. *Pinacol Rearrangement. In Name Reactions for Homologations-Part I*; Li, J. J., Ed., Wiley: Hoboken, NJ, **2009**, pp 319–333. (Review).
11. Marin, L.; Zhang, Y.; Robeyns, K.; Champagne, B.; Adriaensens, P.; Lutsen, L.; Vanderzande, D.; Bevk, D.; Maes, W. *Tetrahedron Lett.* **2013**, *54*, 526–529.
12. Yu, Y.; Li, G.; Zu, L. *Synlett* **2016**, *27*, 1303–1309. (Review).
13. Wang, P.; Gao, Y.; Ma, D. *J. Am. Chem. Soc.* **2018**, *140*, 11608–11612.
14. Shi, H.; Du, C.; Zhang, X.; Xie, F.; Wang, X.; Cui, S.; Peng, X.; Cheng, M.; Lin, B.; Liu, Y. *J. Org. Chem.* **2018**, *83*, 1312–1319.
15. Wu, H.; Wang, Q.; Zhu, J. *J. Am. Chem. Soc.* **2019**, *141*, 11372–11377.
16. Liang, X.-T.; Chen, J.-H.; Yang, Z. *J. Am. Chem. Soc.* **2020**, *142*, 8116–8121.

Pinner Reaction

Transformation of a nitrile into an imino ether, which can be converted to either an ester or an amidine.

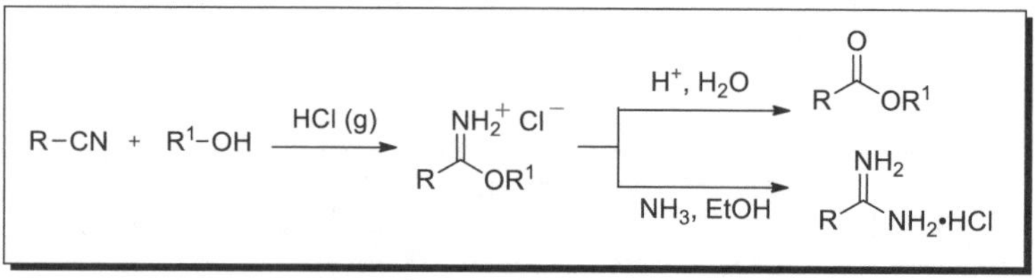

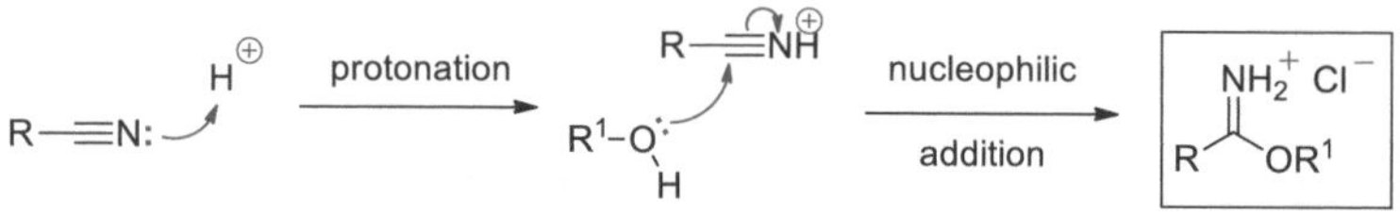

common intermediate

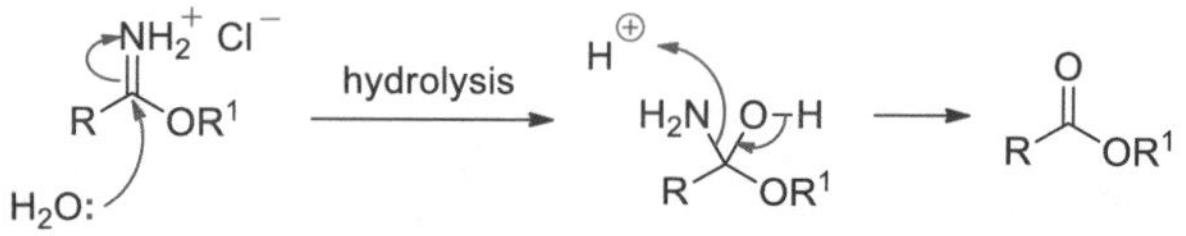

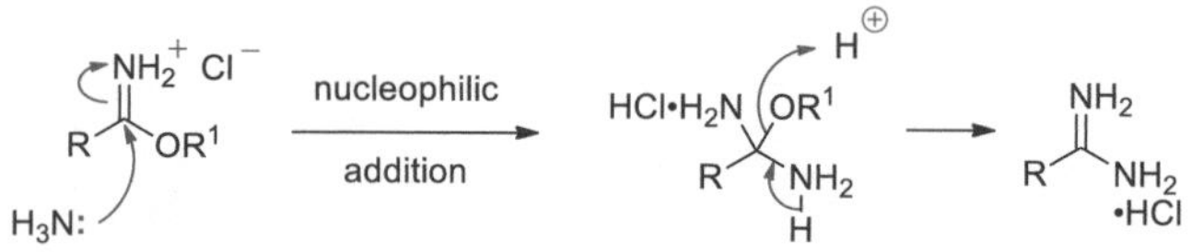

imidate hydrochloride

Example 1[2]

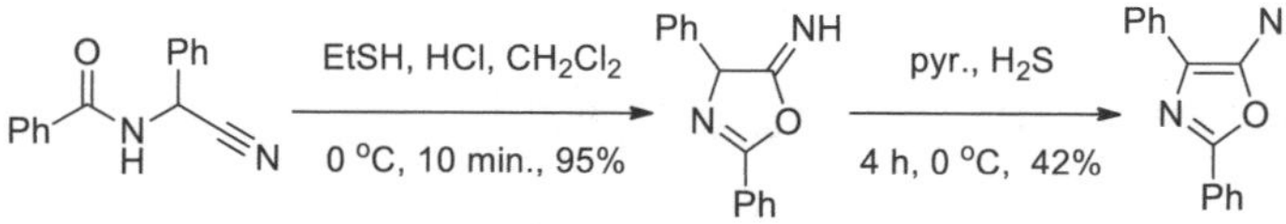

EtSH, HCl, CH$_2$Cl$_2$
0 °C, 10 min., 95%

pyr., H$_2$S
4 h, 0 °C, 42%

Example 2[2]

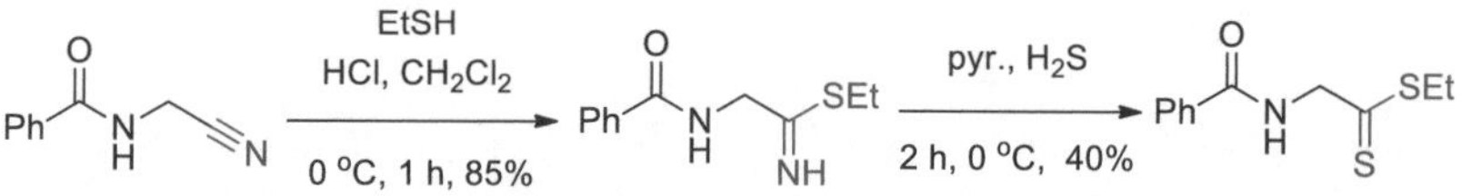

EtSH
HCl, CH$_2$Cl$_2$
0 °C, 1 h, 85%

pyr., H$_2$S
2 h, 0 °C, 40%

J. J. Li, Name Reactions, https://doi.org/10.1007/978-3-030-50865-4_123

Example 3[6]

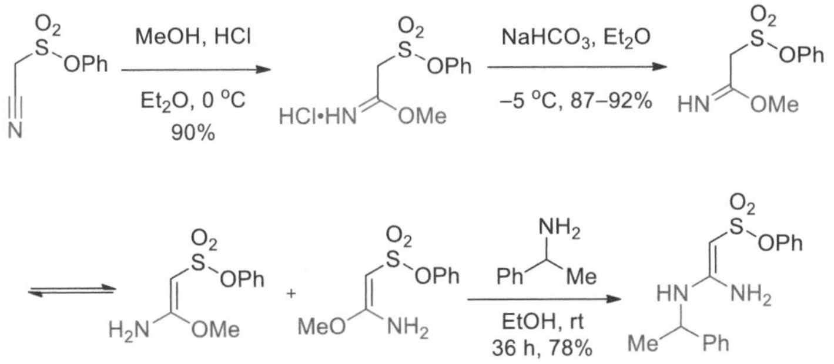

Example 4[10]

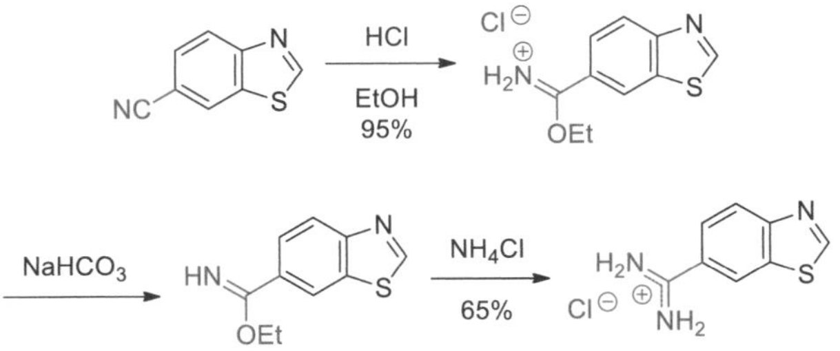

Example 5[11]

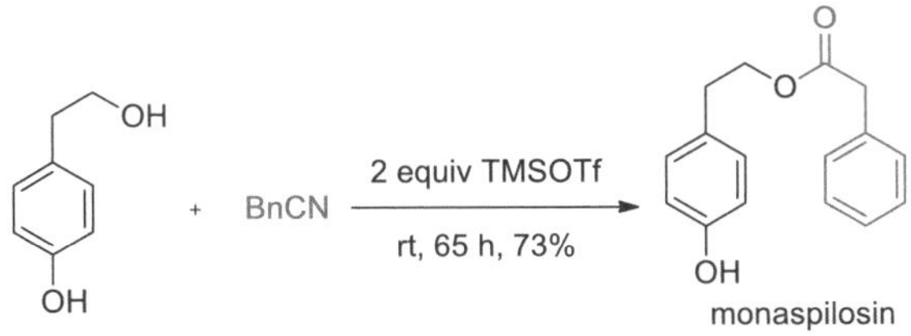

Example 6, Intramolecular 5-*oxo-dig* cyclization[12]

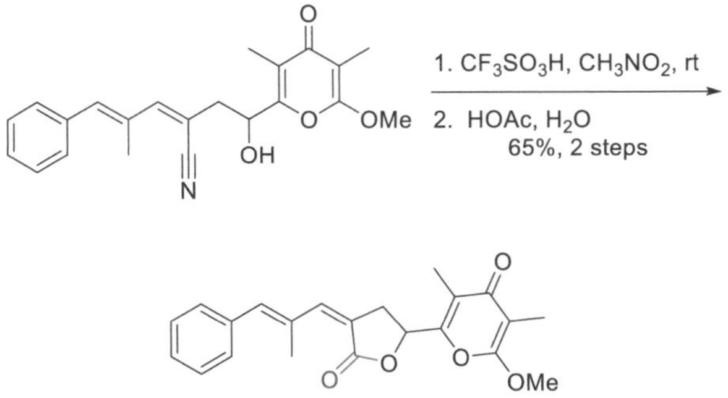

Example 7, Utility in drug discovery[13]

1. HCl(g), EtOH
2. ethylenediamine, EtOH
68%, 2 steps

Example 8, Pinner reaction followed by a Dimroth rearrangement[14]

CuCl (20 mol%)
5 equiv B$_2$(OH)$_4$
H$_2$O/MeOH (1:1)
60 oC, 3 h, 71%

References

1. (a) Pinner, A.; Klein, F. *Ber.* **1877**, *10*, 1889–1897. (b) Pinner, A.; Klein, F. *Ber.* **1878**, *11*, 1825.
2. Poupaert, J.; Bruylants, A.; Crooy, P. *Synthesis* **1972**, 622–624.
3. Lee, Y. B.; Goo, Y. M.; Lee, Y. Y.; Lee, J. K. *Tetrahedron Lett.* **1990**, *31*, 1169–1170.
4. Cheng, C. C. *Org. Prep. Proced. Int.* **1990**, *22*, 643–645.
5. Siskos, A. P.; Hill, A. M. *Tetrahedron Lett.* **2003**, *44*, 789–794.
6. Fischer, M.; Troschuetz, R. *Synthesis* **2003**, 1603–1609.
7. Fringuelli, F.; Piermatti, O.; Pizzo, F. *Synthesis* **2003**, 2331–2334.
8. Cushion, M. T.; Walzer, P. D.; Collins, M. S.; Rebholz, S.; Vanden Eynde, J. J.; Mayence, A.; Huang, T. L. *Antimicrob. Agents Chemother.* **2004**, *48*, 4209–4216.
9. Li, J.; Zhang, L.; Shi, D.; Li, Q.; Wang, D.; Wang, C.; Zhang, Q.; Zhang, L.; Fan, Y. *Synlett* **2008**, 233–236.
10. Racané, L.; Tralic-Kulenovic, V.; Mihalic, Z.; Pavlovic, G.; Karminski-Zamola, G. *Tetrahedron* **2008**, *64*, 11594–11602.
11. Pfaff, D.; Nemecek, G.; Podlech, J. *Beilst. J. Org. Chem.* **2013**, *9*, 1572–1577.
12. Henrot, M.; Jean, A.; Peixoto, P. A.; Maddaluno, J.; De Paolis, M. *J. Org. Chem.* **2016**, *81*, 5190–5201.
13. Sović, I.; Cindrić, M.; Perin, N.; Boček, I.; Novaković, I.; Damjanovic, A.; Stanojković, T.; Zlatović, M.; Hranjec, M.; Bertoša, B. *Chem. Res. Toxicol.* **2019**, *32*, 1880–1892.
14. Liu, Q.; Sui, Y.; Zhang, Y.; Zhang, K.; Chen, Y.; Zhou, H. *Synlett* **2020**, *31*, 275–279.

Polonovski Reaction

Treatment of a tertiary *N*-oxide with an activating agent such as acetic anhydride, resulting in rearrangement where an *N,N*-disubstituted acetamide and an aldehyde are generated.

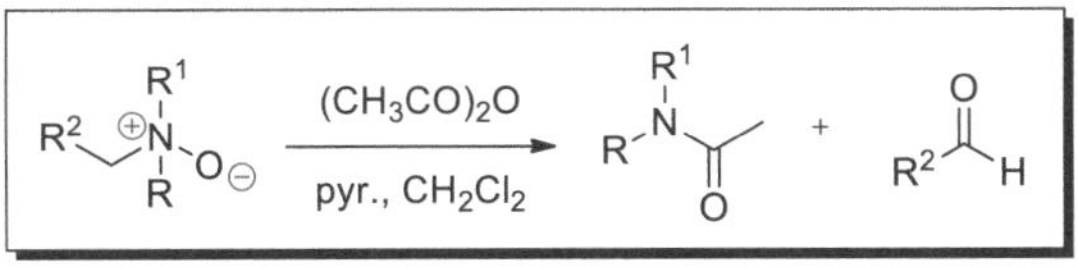

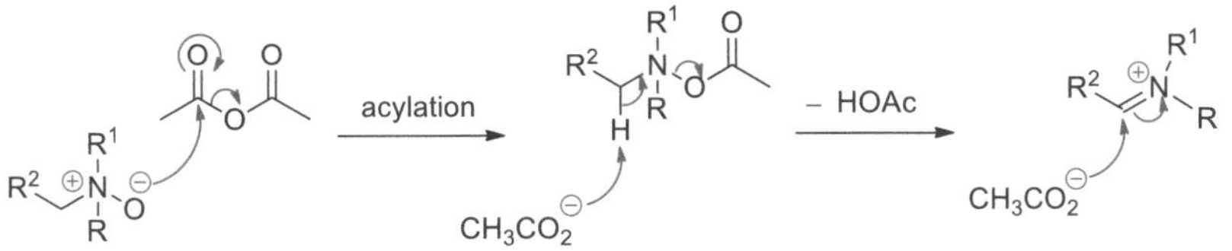

iminium ion

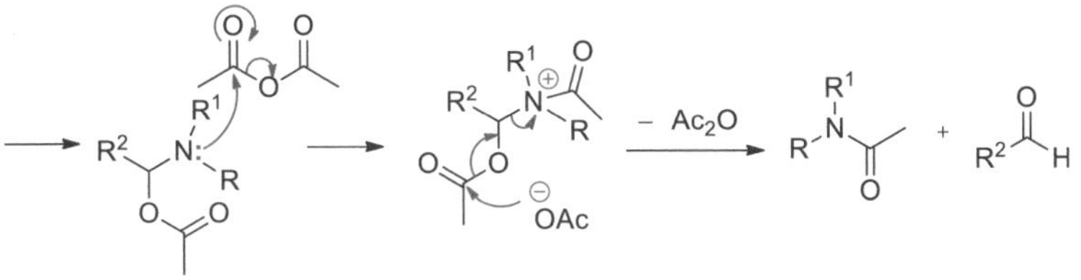

The intramolecular pathway is also operative:

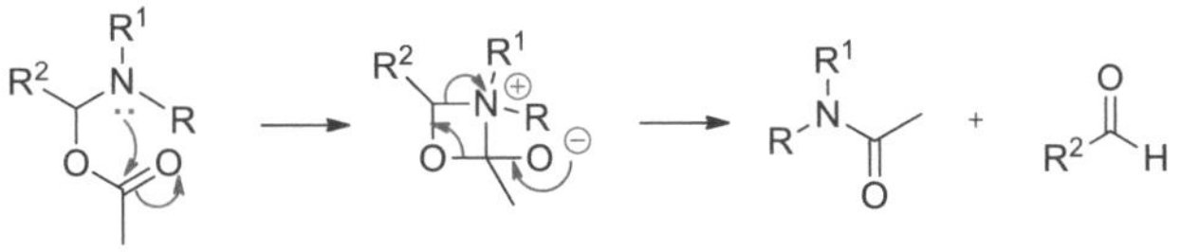

Example 1[1]

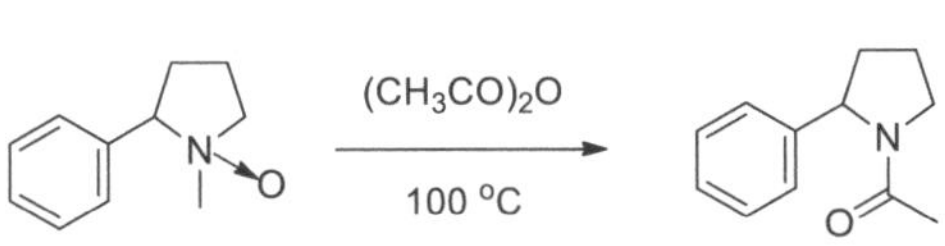

Example 2[2]

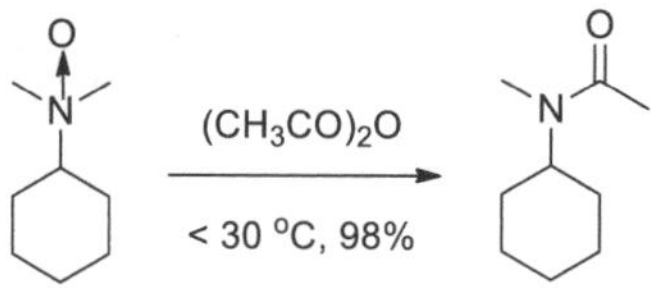

Example 3, Iron salt-mediated Polonovski reaction[9]

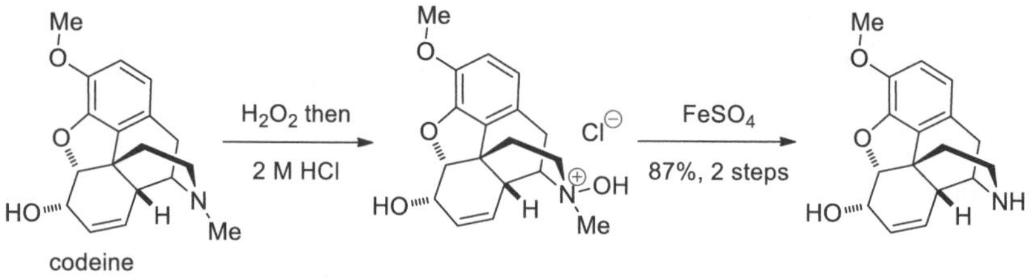

Example 4[11]

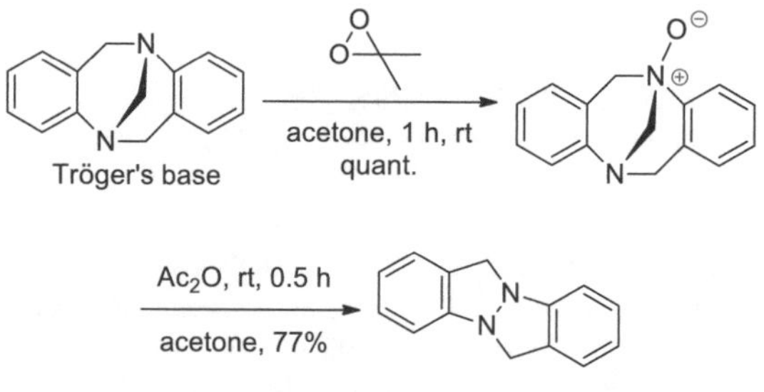

Example 5, Iron-catalyzed acylative dealkylation of *N*-alkylsulfoximines[12]

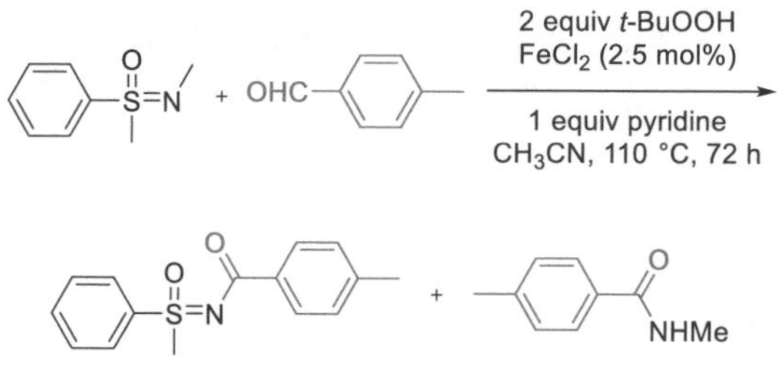

Example 6, A non-classical Polonovski reaction[13]

References

1. Polonovski, M.; Polonovski, M. *Bull. Soc. Chim. Fr.* **1927**, *41*, 1190–1208.
2. Michelot, R. *Bull. Soc. Chim. Fr.* **1969**, 4377–4385.
3. Lounasmaa, M.; Karvinen, E.; Koskinen, A.; Jokela, R. *Tetrahedron* **1987**, *43*, 2135–2146.
4. Tamminen, T.; Jokela, R.; Tirkkonen, B.; Lounasmaa, M. *Tetrahedron* **1989**, *45*, 2683–2692.
5. Grierson, D. *Org. React.* **1990**, *39*, 85–295. (Review).
6. Morita, H.; Kobayashi, J. *J. Org. Chem.* **2002**, *67*, 5378–5381.
7. McCamley, K.; Ripper, J. A.; Singer, R. D.; Scammells, P. J. *J. Org. Chem.* **2003**, *68*, 9847–9850.
8. Nakahara, S.; Kubo, A. *Heterocycles* **2004**, *63*, 1849–1854.
9. Thavaneswaran, S.; Scammells, P. J. *Bioorg. Med. Chem. Lett.* **2006**, *16*, 2868–2871.
10. Volz, H.; Gartner, H. *Eur. J. Org. Chem.* **2007**, 2791–2801.
11. Pacquelet, S.; Blache, Y.; Kimny, T.; Dubois, M.-A. L.; Desbois, N. *Synth. Commun.* **2013**, *43*, 1092–1100.
12. Lamers, P.; Priebbenow, D. L.; Bolm, C. *Eur. J. Org. Chem.* **2015**, 5594–5602.
13. Bupp, J. E.; Tanga, M. J. *J. Label. Compd. Radiopharm.* **2016**, *59*, 291–293.
14. Bush, T. S.; Yap, G. P. A.; Chain, W. *J. Org. Lett.* **2018**, *20*, 5406–5409.

Polonovski–Potier Reaction

A modification of the Polonovski reaction where trifluoroacetic anhydride is used in place of acetic anhydride. Because the reaction conditions for the Polonovski–Potier reaction are mild, it has largely replaced the Polonovski reaction.

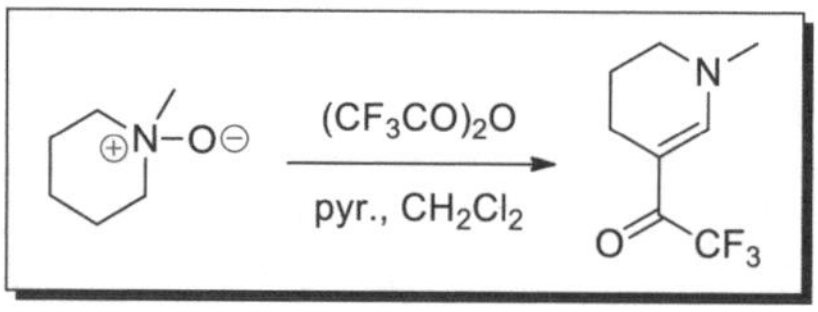

tertiary *N*-oxide

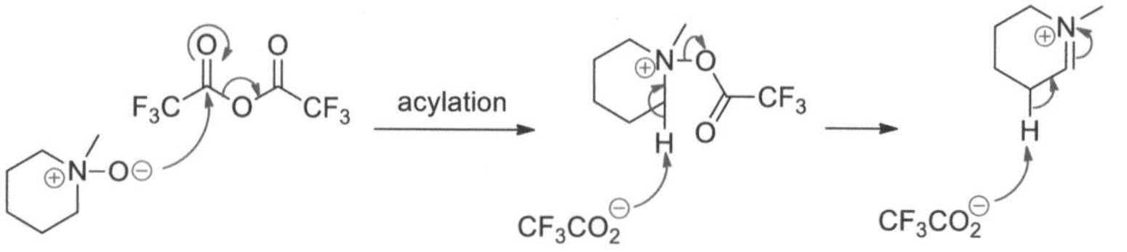

iminium ion

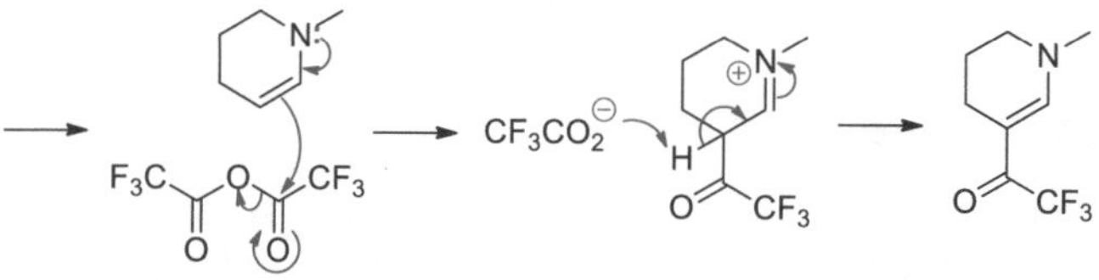

enamine

Example 1[2]

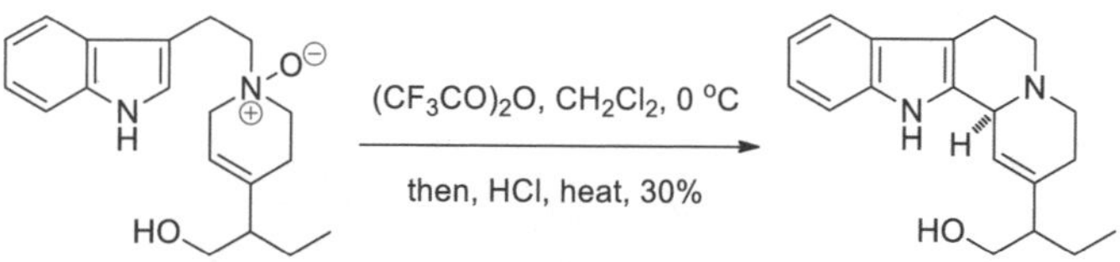

Example 2[5]

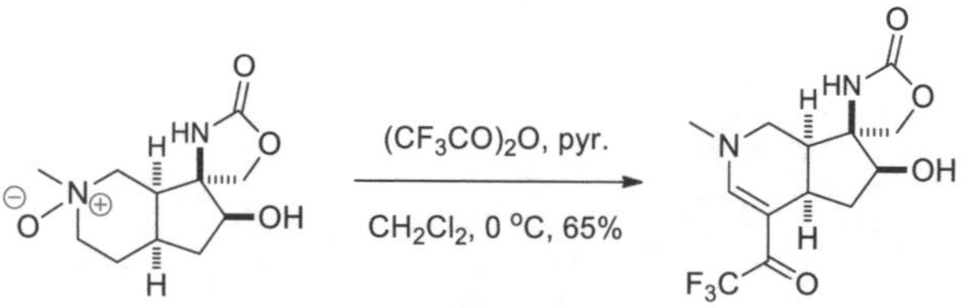

© Springer Nature Switzerland AG 2021
J. J. Li, *Name Reactions*, https://doi.org/10.1007/978-3-030-50865-4_125

Example 3[8]

Example 4, Here, *m*-CPBA also concurrently oxidized the aldehyde[10]

Example 5, Oxidative α-Cyanation[13]

Example 6, Cyclization onto the iminium intermediate[14]

Example 7, Here, *m*-CPBA also concurrently oxidized the aldehyde[15]

References

1. Ahond, A.; Cavé, A.; Kan-Fan, C.; Husson, H.-P.; de Rostolan, J.; Potier, P. *J. Am. Chem. Soc.* **1968,** *90,* 5622–5623.
2. Husson, H.-P.; Chevolot, L.; Langlois, Y.; Thal, C.; Potier, P. *J. Chem. Soc., Chem. Commun.* **1972,** 930–931.
3. Grierson, D. *Org. React.* **1990,** *39,* 85–295. (Review).
4. Sundberg, R. J.; Gadamasetti, K. G.; Hunt, P. J. *Tetrahedron* **1992,** *48,* 277–296.
5. Kende, A. S.; Liu, K.; Brands, J. K. M. *J. Am. Chem. Soc.* **1995,** *117,* 10597–10598.
6. Renko, D.; Mary, A.; Guillou, C.; Potier, P.; Thal, C. *Tetrahedron Lett.* **1998,** *39,* 4251–4254.
7. Suau, R.; Nájera, F.; Rico, R. *Tetrahedron* **2000,** *56,* 9713–9720.
8. Thomas, O. P.; Zaparucha, A.; Husson, H.-P. *Tetrahedron Lett.* **2001,** *42,* 3291–3293.
9. Lim, K.-H.; Low, Y.-Y.; Kam, T.-S. *Tetrahedron Lett.* **2006,** *47,* 5037–5039.
10. Gazak, R.; Kren, V.; Sedmera, P.; Passarella, D.; Novotna, M.; Danieli, B. *Tetrahedron* **2007,** *63,* 10466–10478.
11. Nishikawa, Y.; Kitajima, M.; Kogure, N.; Takayama, H. *Tetrahedron* **2009,** *65,* 1608–1617.
12. Han-ya, Y.; Tokuyama, H.; Fukuyama, T. *Angew. Chem. Int. Ed.* **2011,** *50,* 4884–4887.
13. Perry, M. A.; Morin, M. D.; Slafer, B. W.; Rychnovsky, S. D. *J. Org. Chem.* **2012,** *77,* 3390–3400.
14. Benimana, S. E.; Cromwell, N. E.; Meer, H. N.; Marvin, C. C. *Tetrahedron Lett.* **2016,** *57,* 5062–5064.
15. Zhang, X.; Kakde, B. N.; Guo, R.; Yadav, S.; Gu, Y.; Li, A. *Angew. Chem. Int. Ed.* **2019,** *58,* 6053–6058.
16. Lee, S.; Kang, G.; Chung, G.; Kim, D.; Lee, H.-Y.; Han, S. *Angew. Chem. Int. Ed.* **2020,** *59,* 6894–6901.

Prins Reaction

The Prins reaction is the acid-catalyzed addition of aldehydes to alkenes and gives different products depending on the reaction conditions.

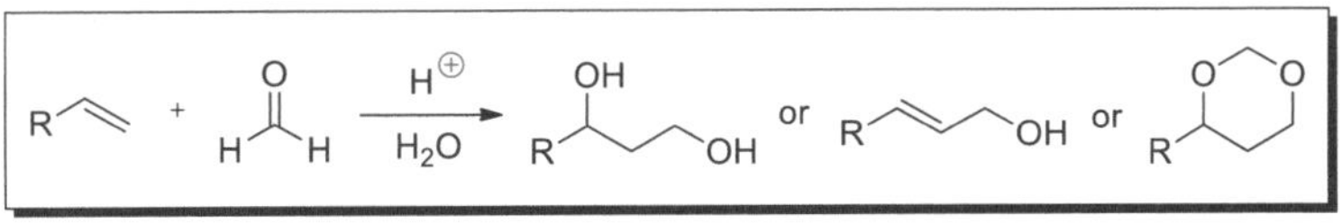

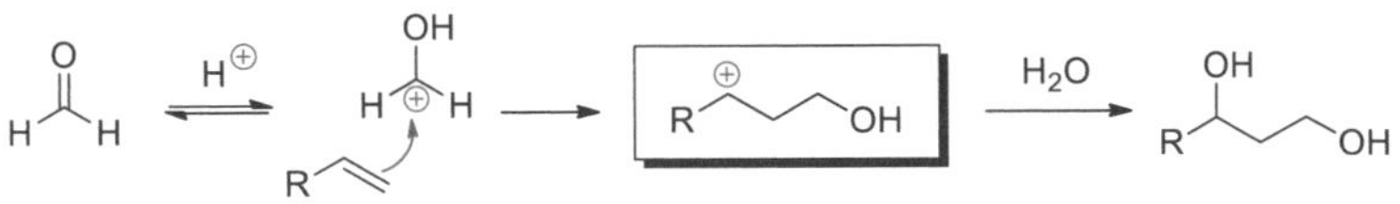

the common intermediate

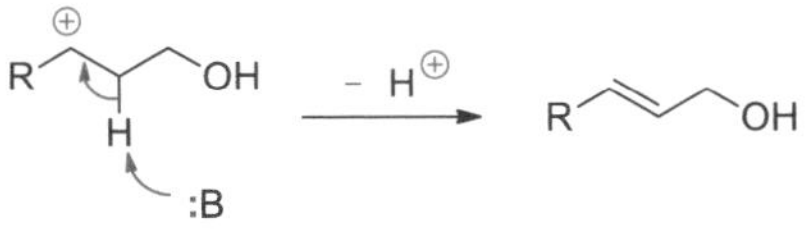

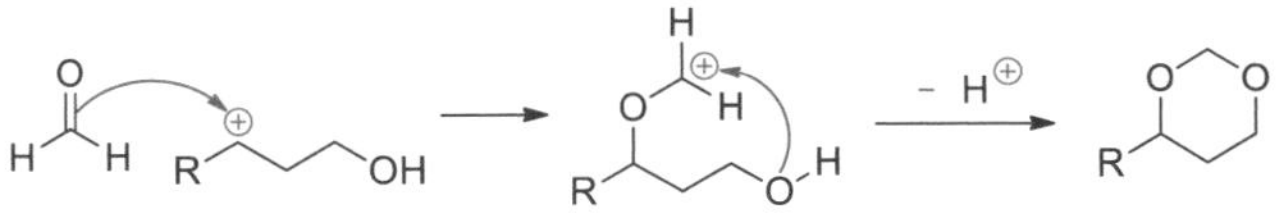

Example 1[5]

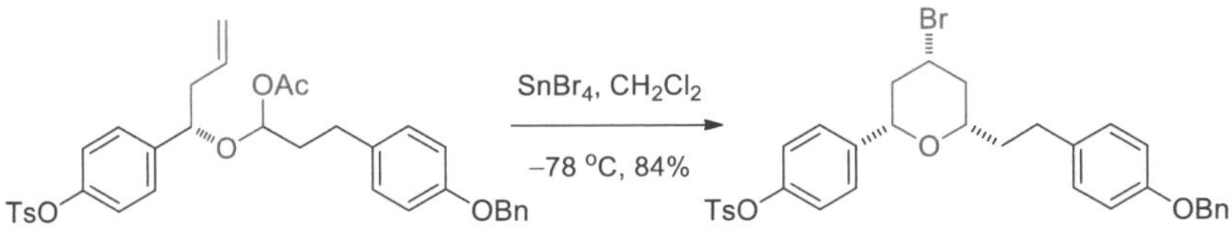

SnBr$_4$, CH$_2$Cl$_2$

−78 °C, 84%

Example 2[7]

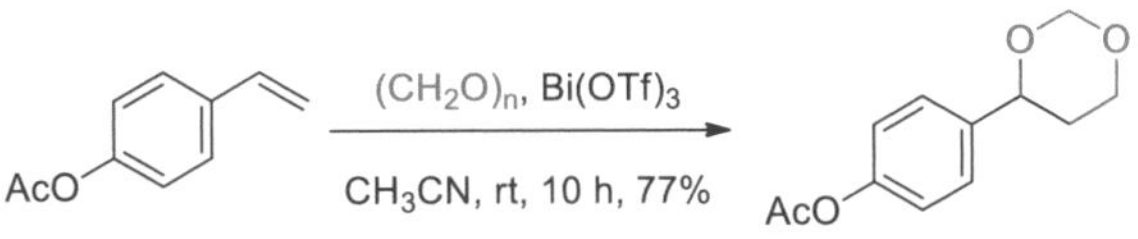

(CH$_2$O)$_n$, Bi(OTf)$_3$

CH$_3$CN, rt, 10 h, 77%

Example 3[9]

TMSEO$_2$C

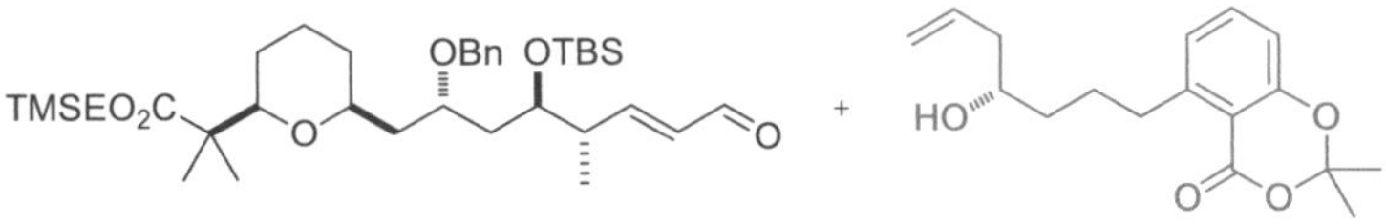

© Springer Nature Switzerland AG 2021
J. J. Li, *Name Reactions*, https://doi.org/10.1007/978-3-030-50865-4_126

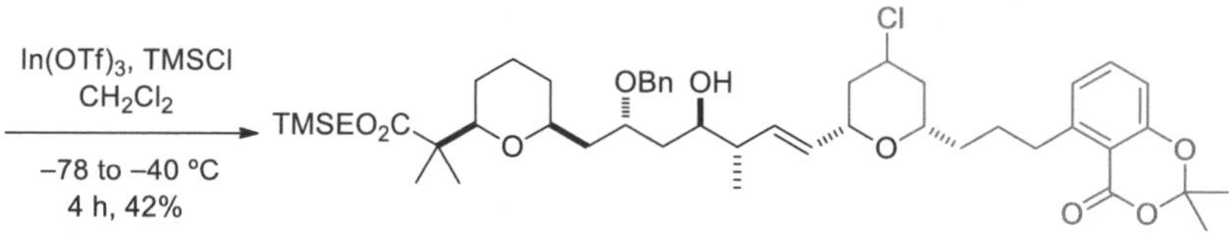

Example 4[10]

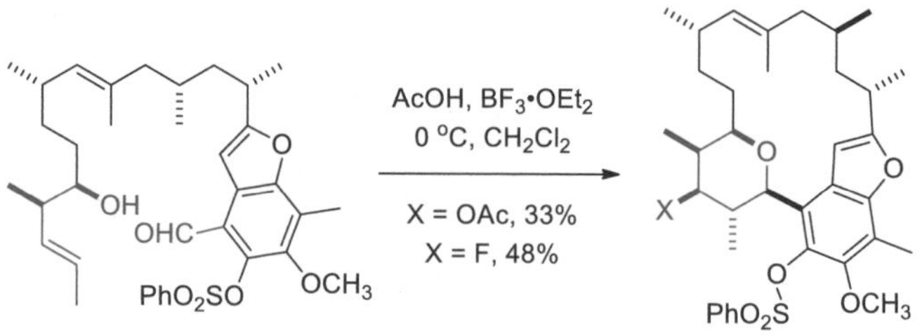

Example 5, A cascade of the Prins/Ritter amidation reaction[11]

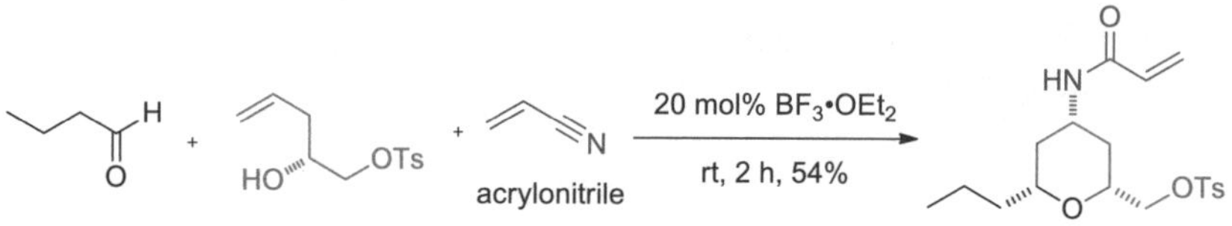

Example 6[12]

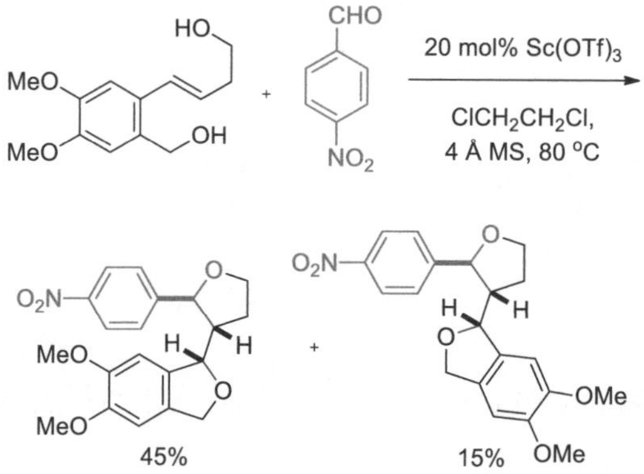

Example 7, SnCl$_4$-promoted oxonium-Prins cyclization[13]

Example 8, Prins reaction of homoallenyl alcohols: access to substituted pyrans in the halichondrin series[14]

Example 9, Aza-Prins cyclization of endocyclic N-acyliminium ions[17]

Example 10, Re_2O_7-catalyzed approach to spirocyclic ether formation from acyclic precursors[19]

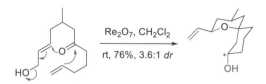

References

1. Prins, H. J. *Chem. Weekblad* **1919**, *16*, 1072–1023. Born in Zaandam, The Netherlands, Hendrik J. Prins (1889–1958) was not even an organic chemist *per se*. After obtaining a doctorate in chemical engineering, Prins worked for an essential oil company and then a company dealing with the rendering of condemned meats and carcasses. But he had a small laboratory near his house where he carried out his experiments in his spare time, which obviously was not a big distraction—for he rose to be the president-director of the firm he worked for.
2. Adam, D. R.; Bhatnagar, S. P. *Synthesis* **1977**, 661–672. (Review).
3. Hanaki, N.; Link, J. T.; MacMillan, D. W. C.; Overman, L. E.; Trankle, W. G.; Wurster, J. A. *Org. Lett.* **2000**, *2*, 223–226.
4. Davis, C. E.; Coates, R. M. *Angew. Chem. Int. Ed.* **2002**, *41*, 491–493.
5. Marumoto, S.; Jaber, J. J.; Vitale, J. P.; Rychnovsky, S. D. *Org. Lett.* **2002**, *4*, 3919–3922.
6. Braddock, D. C.; Badine, D. M.; Gottschalk, T. *Synlett* **2003**, 345–348.
7. Sreedhar, B.; Swapna, V.; Sridhar, C.; *Synth. Commun.* **2005**, *35*, 1177–1182.
8. Aubele, D. L.; Wan, S.; Floreancig, P. E. *Angew. Chem. Int. Ed.* **2005**, *44*, 3485–3488.
9. Chan, K.-P.; Ling, Y. H.; Loh, T.-P. *Chem. Commun.* **2007**, 939–941.
10. Bahnck, K. B.; Rychnovsky, S. D. *J. Am. Chem. Soc.* **2008**, *130*, 13177–13181.
11. Yadav, J. S.; Reddy, Y. J.; Reddy, P. A. N. *Org. Lett.* **2013**, *15*, 546–549.
12. Subba Reddy, B. V.; Jalal, S.; Borkar, P. *Tetrahedron Lett.* **2013**, *54*, 1519–1523.
13. Abas, H.; Linsdall, S. M.; Mamboury, M.; Rzepa, H. S.; Spivey, A. C. *Org. Lett.* **2017**, *19*, 2486–2489.
14. Choi, H.-W.; Fang, F. G.; Fang, H.; Kim, D.-S.; Mathieu, S. R.; Yu, R. T. *Org. Lett.* **2017**, *19*, 6092–6095.
15. Subba Reddy, B. V.; Nair, P. N.; Antony, A.; Lalli, C.; Gree, R. *Eur. J. Org. Chem.* **2017**, 1805–1819 (Review).
16. Subba Reddy, B. V.; Nair, P. N.; Antony, A.; Srivastava, N. *Eur. J. Org. Chem.* **2017**, 5484–5496. (Review).
17. Das, M.; Saikia, A. K. *J. Org. Chem.* **2018**, *83*, 6178–6185.
18. Doro, F.; Akeroyd, N.; Schiet, F.; Narula, A. *Angew. Chem. Int. Ed.* **2019**, *58*, 7174–7179. (Review).
19. Afeke, C.; Xie, Y.; Floreancig, P. E. *Org. Lett.* **2019**, *21*, 5064–54067.
20. Han, M.-Y.; Pan, H.; Li, P.; Wang, L. *J. Org. Chem.* **2020**, *85*, 5825–5837.

Pummerer Rearrangement

The transformation of sulfoxides into α-acyloxythioethers using acetic anhydride.

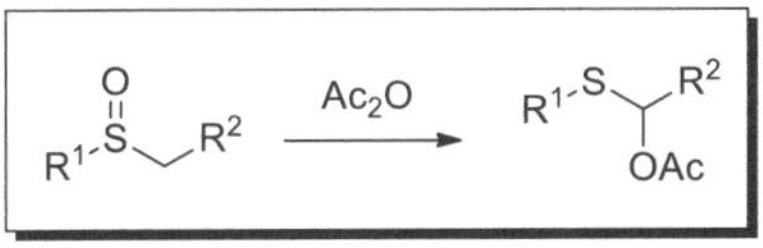

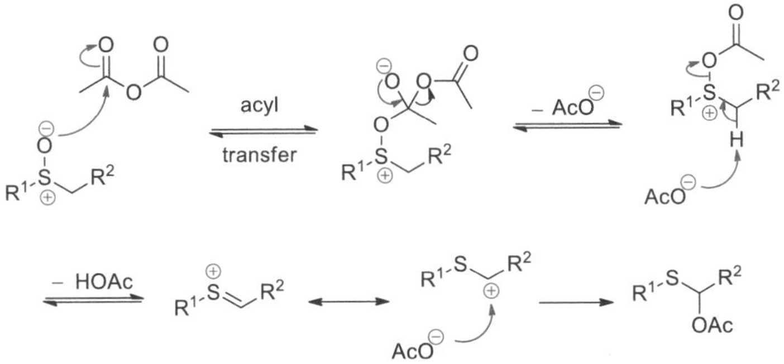

Example 1[2]

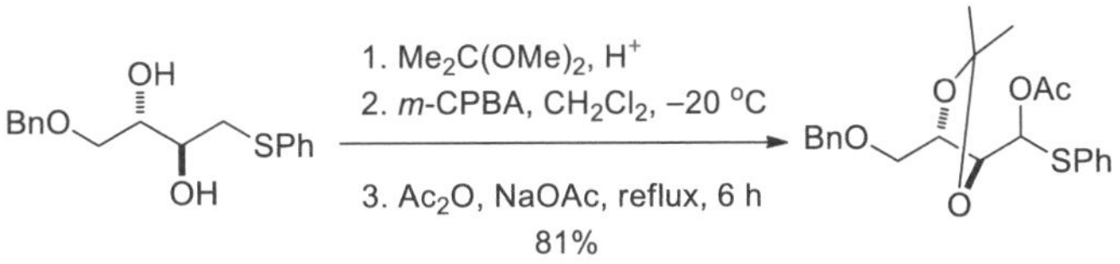

1. Me$_2$C(OMe)$_2$, H$^+$
2. m-CPBA, CH$_2$Cl$_2$, −20 °C
3. Ac$_2$O, NaOAc, reflux, 6 h
81%

Example 2[7]

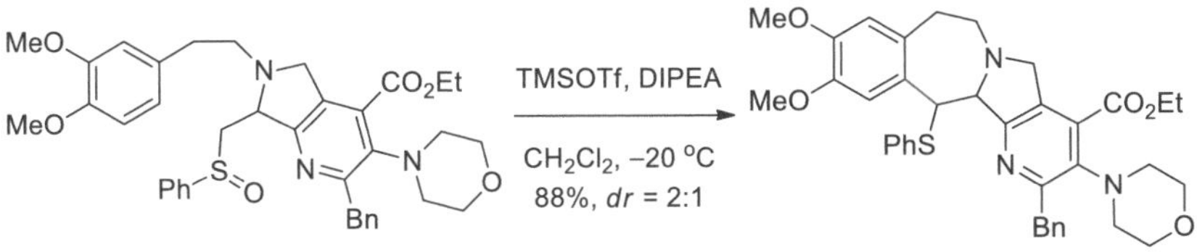

TMSOTf, DIPEA

CH$_2$Cl$_2$, −20 °C
88%, dr = 2:1

Example 3, A tandem Pummerer/Mannich cyclization sequence[8]

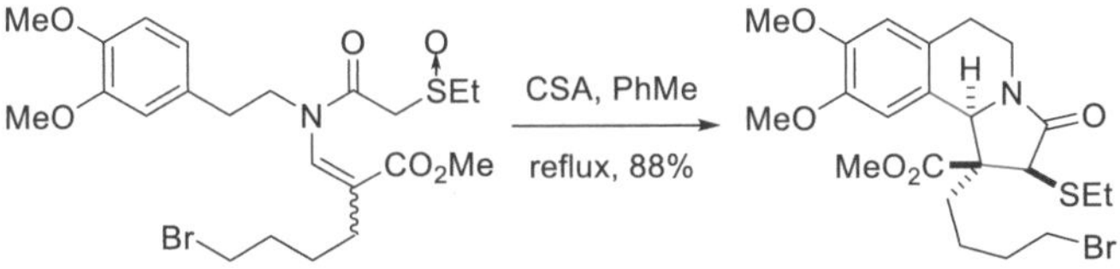

CSA, PhMe

reflux, 88%

© Springer Nature Switzerland AG 2021
J. J. Li, Name Reactions, https://doi.org/10.1007/978-3-030-50865-4_127

Example 4[9]

Example 5, Stereoselective Pummerer rearrangement[10,12]

Example 6, An aromatic Pummerer rearrangement[13]

Example 7, Stereoselective Pummerer rearrangement[14]

Example 8, Pummerer rearrangement to afford vinyl sulfide by α-hydrogen elimination[15]

References

1. Pummerer, R. *Ber.* **1910**, *43*, 1401–1412. Rudolf Pummerer, born in Austria in 1882, studied under von Baeyer, Willstätter, and Wieland. He worked for BASF for a few years and in 1921 he was appointed head of the organic division of the Munich Laboratory, fulfilling his long-desired ambition.

2. Katsuki, T.; Lee, A. W. M.; Ma, P.; Martin, V. S.; Masamune, S.; Sharpless, K. B.; Tuddenham, D.; Walker, F. J. *J. Org. Chem.* **1982,** *47*, 1373–1378.
3. De Lucchi, O.; Miotti, U.; Modena, G. *Org. React.* **1991,** *40*, 157–406. (Review).
4. Padwa, A.; Gunn, D. E., Jr.; Osterhout, M. H. *Synthesis* **1997,** 1353–1378. (Review).
5. Padwa, A.; Waterson, A. G. *Curr. Org. Chem.* **2000,** *4*, 175–203. (Review).
6. Padwa, A.; Bur, S. K.; Danca, D. M.; Ginn, J. D.; Lynch, S. M. *Synlett* **2002,** 851–862. (Review).
7. Gámez Montaño, R.; Zhu, J. *Chem. Commun.* **2002,** 2448–2449.
8. Padwa, A.; Danca, M. D.; Hardcastle, K.; McClure, M. *J. Org. Chem.* **2003,** *68*, 929–941.
9. Suzuki, T.; Honda, Y.; Izawa, K.; Williams, R. M. *J. Org. Chem.* **2005,** *70*, 7317–7323.
10. Nagao, Y.; Miyamoto, S.; Miyamoto, M.; Takeshige, H.; Hayashi, K.; Sano, S.; Shiro, M.; Yamaguchi, K.; Sei, Y. *J. Am. Chem. Soc.* **2006,** *128*, 9722–9729.
11. Ahmad, N. M. *Pummerer Rearrangement.* In *Name Reactions for Homologations-Part II*; Li, J. J., Ed.; Wiley: Hoboken, NJ, **2009,** pp 334–352. (Review).
12. Patil, M.; Loerbroks, C.; Thiel, W. *Org. Lett.* **2013,** *15*, 1682–1685.
13. Bao, X.; Yao, J.; Zhou, H.; Xu, G. *Org. Lett.* **2017,** *19*, 5780–5782.
14. Li, X.; Carter, R. G. *Org. Lett.* **2018,** *20*, 5546–5549.
15. Yan, Z.; Zhao, C.; Gong, J.; Yang, Z. *Org. Lett.* **2020,** *22*, 1644–1647.

Ramberg–Bäcklund Reaction

Olefin synthesis *via* α-halosulfone extrusion.

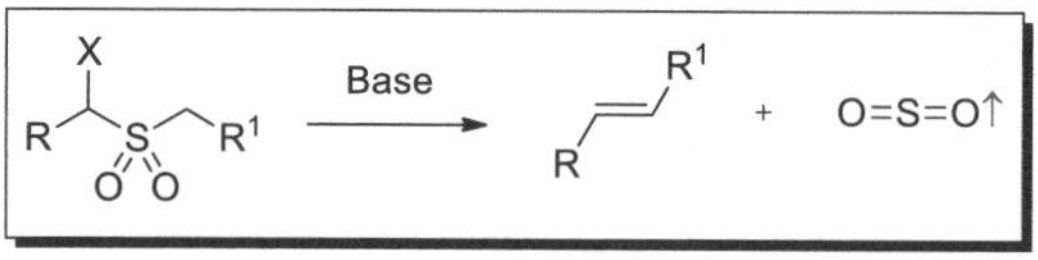

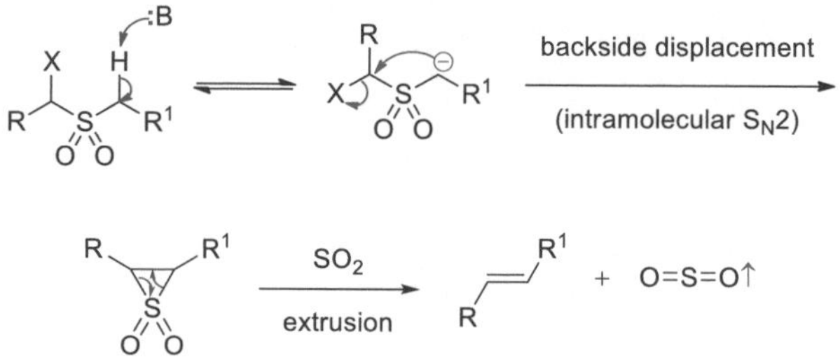

episulfone intermediate

Example 1[4]

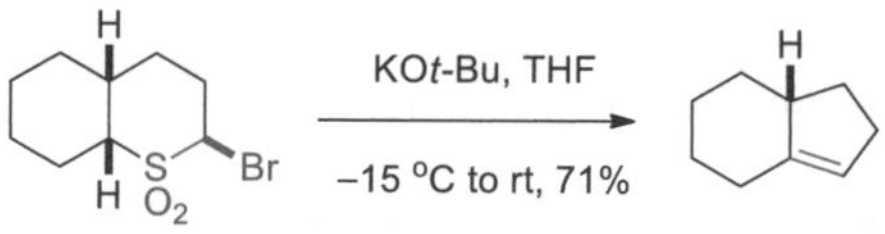

KOt-Bu, THF

−15 °C to rt, 71%

Example 2[5]

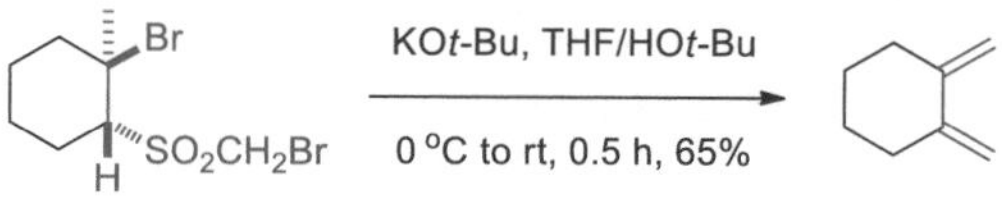

KOt-Bu, THF/HOt-Bu

0 °C to rt, 0.5 h, 65%

Example 3[6]

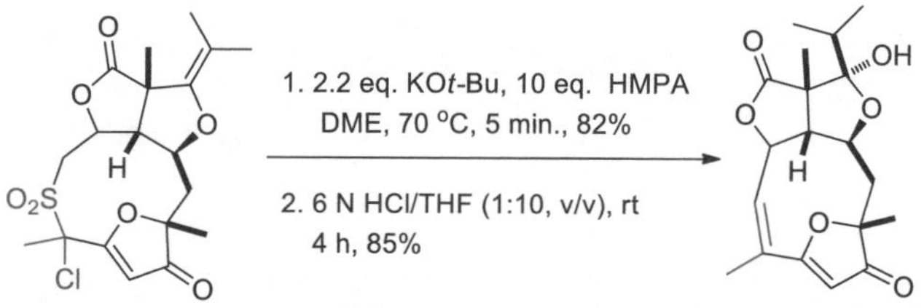

1. 2.2 eq. KOt-Bu, 10 eq. HMPA
   DME, 70 °C, 5 min., 82%

2. 6 N HCl/THF (1:10, v/v), rt
   4 h, 85%

© Springer Nature Switzerland AG 2021
J. J. Li, *Name Reactions*, https://doi.org/10.1007/978-3-030-50865-4_128

Example 4, *in situ* chlorination[7]

Example 5, Direct conversion of dipropargylic sulfones into ene-diynes by a modified one-flask Ramberg–Bäcklund reaction[8]

Example 6, Pyrroline synthesis[14]

Example 7, Towards oseltamivir (Tamiflu)[15]

Example 8, Cyclic olefin from optically pure alcohols (sulfone is th leaving group)[17]

References

1. Ramberg, L.; Bäcklund, B. *Arkiv. Kemi, Mineral Geol.* **1940**, *13A*, 1–50.
2. Paquette, L. A. *Acc. Chem. Res.* **1968**, *1*, 209–216. (Review).
3. Paquette, L. A. *Org. React.* **1977**, *25*, 1–71. (Review).
4. Becker, K. B.; Labhart, M. P. *Helv. Chim. Acta* **1983**, *66*, 1090–1100.
5. Block, E.; Aslam, M.; Eswarakrishnan, V.; Gebreyes, K.; Hutchinson, J.; Iyer, R.; Laffitte, J. A.; Wall, A. *J. Am. Chem. Soc.* **1986**, *108*, 4568–4580.
6. Boeckman, R. K., Jr.; Yoon, S. K.; Heckendorn, D. K. *J. Am. Chem. Soc.* **1991**, *113*, 9682–9684.
7. Trost, B. M.; Shi, Z. *J. Am. Chem. Soc.* **1994**, *116*, 7459–7460.
8. Cao, X.-P.; Chan, T.-L.; Chow, H.-F. *Tetrahedron Lett.* **1996**, *37*, 1049–1052.
9. Taylor, R. J. K. *Chem. Commun.* **1999**, 217–227. (Review).
10. Taylor, R. J. K.; Casy, G. *Org. React.* **2003**, *62*, 357–475. (Review).
11. Li, J. J. *Ramberg–Bäcklund olefin synthesis.* In *Name Reactions for Functional Group Transformations*; Li, J. J., Ed.; Wiley: Hoboken, NJ, **2007**, pp 386–404. (Review).
12. Pal, T. K.; Pathak, T. *Carbohydrate Res.* **2008**, *343*, 2826–2829.
13. Baird, L. J.; Timmer, M. S. M.; Teesdale-Spittle, P. H.; Harvey, J. E. *J. Org. Chem.* **2009**, *74*, 2271–2277.
14. Söderman, S. C.; Schwan, A. L. *Org. Lett.* **2013**, *15*, 4434–4437.
15. Chavan, S. P.; Chavan, P. N.; Gonnade, R. G. *RSC Adv.* **2014**, *4*, 62281–62284.
16. Lou, X. *Mini-Rev. Org. Chem.* **2015**, *412*, 449–454. (Review).
17. Pasetto, P.; Naginskaya, J. *Tetrahedron Lett.* **2018**, *59*, 2797–2899.

Reformatsky Reaction

Nucleophilic addition of organozinc reagents generated from α-haloesters to carbonyls.

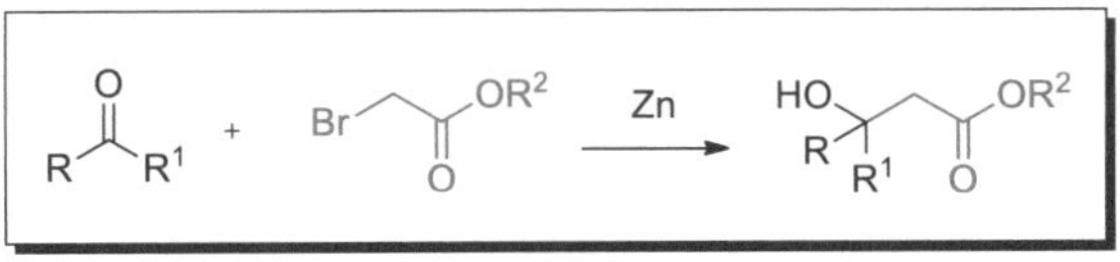

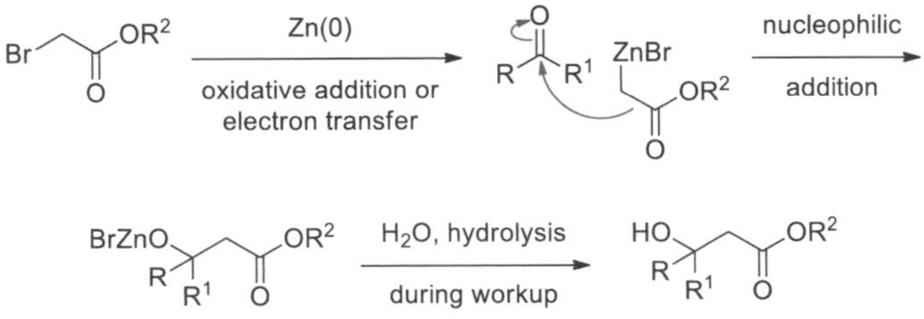

Example 1[4]

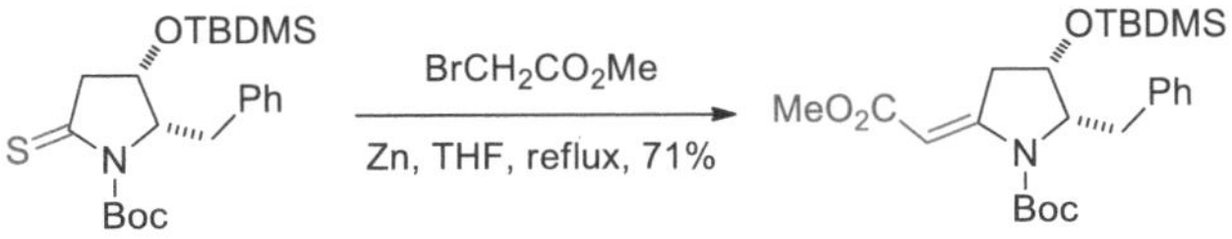

Example 2[6]

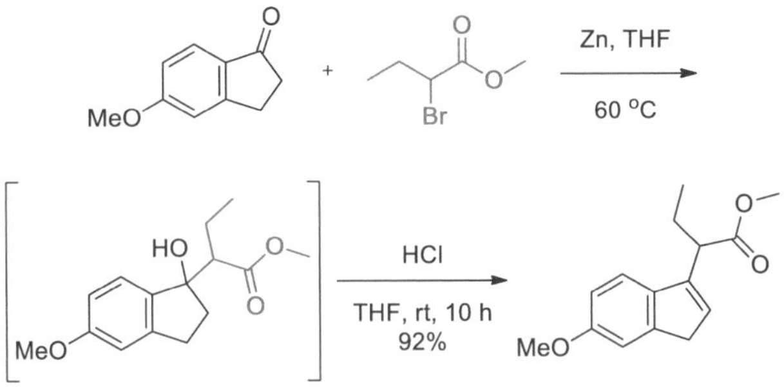

© Springer Nature Switzerland AG 2021
J. J. Li, *Name Reactions*, https://doi.org/10.1007/978-3-030-50865-4_129

Example 3, Boron-mediated Reformatsky reaction[8]

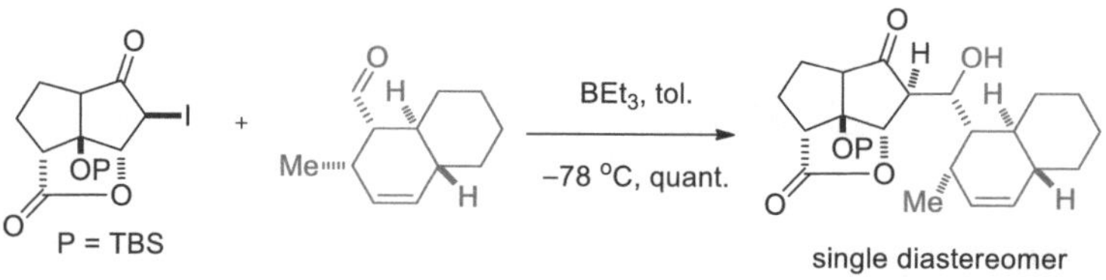
P = TBS
BEt$_3$, tol.
−78 °C, quant.
single diastereomer

Example 4, SmI_2-mediated Reformatsky reaction[9]

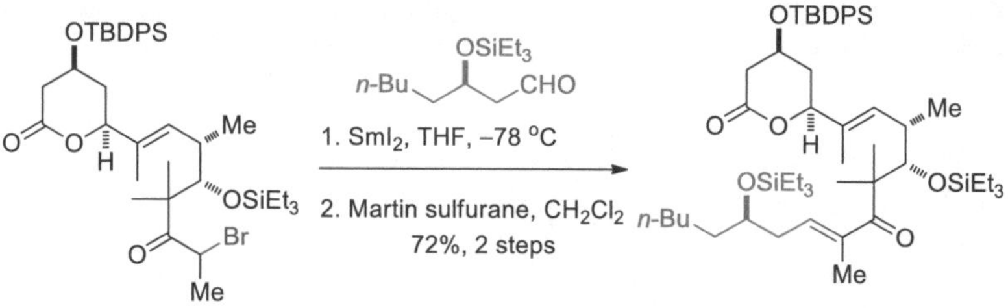
1. SmI$_2$, THF, −78 °C
2. Martin sulfurane, CH$_2$Cl$_2$
72%, 2 steps

Example 5[6]

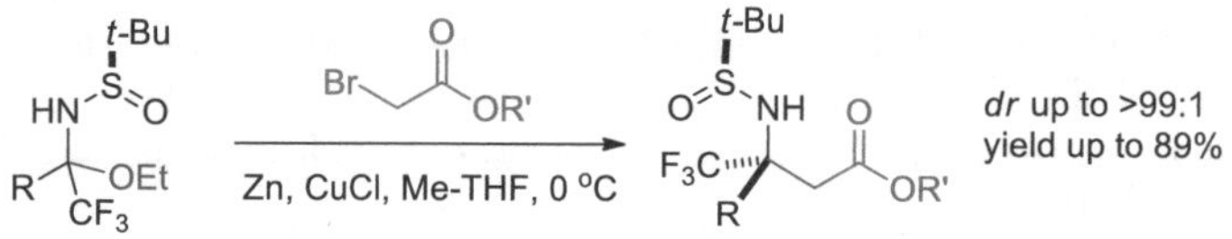
Zn, CuCl, Me-THF, 0 °C
dr up to >99:1
yield up to 89%

Example 6, SmI_2-mediated Reformatsky reaction[13]

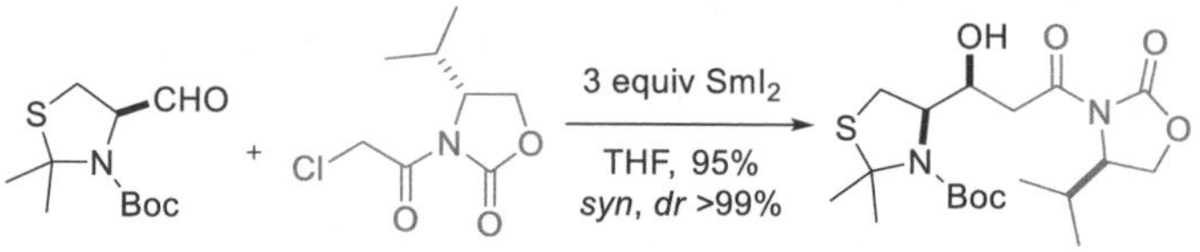
3 equiv SmI$_2$
THF, 95%
syn, dr >99%

Example 7, Diastereoselective Reformatsky reaction[14]

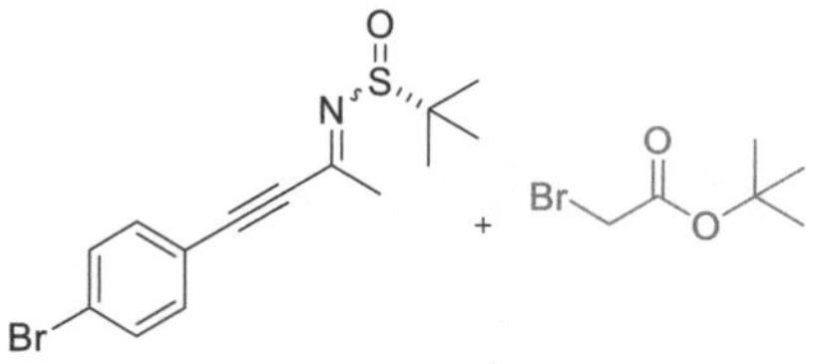

References

1. Reformatsky, S. *Ber.* **1887,** *20,* 1210–1211. Sergei Reformatsky (1860–1934) was born in Russia. He studied at the University of Kazan in Russia, the cradle of Russian chemistry professors, where he found competent guidance of a distinguished chemist, Alexander M. Zaĭtsev. Reformatsky then studied at Göttingen, Heidelberg, and Leipzig in Germany. After returning to Russia, Reformatsky became the Chair of Organic Chemistry at the University of Kiev.

2. Rathke, M. W. *Org. React.* **1975,** *22,* 423–460. (Review).

3. Fürstner, A. *Synthesis* **1989,** 571–590. (Review).

4. Lee, H. K.; Kim, J.; Pak, C. S. *Tetrahedron Lett.* **1999,** *40,* 2173–2174.

5. Fürstner, A. In *Organozinc Reagents* Knochel, P., Jones, P., Eds.; Oxford University Press: New York, **1999,** pp 287–305. (Review).

6. Zhang, M.; Zhu, L.; Ma, X. *Tetrahedron: Asymmetry* **2003,** *14,* 3447–3453.

7. Ocampo, R.; Dolbier, W. R., Jr. *Tetrahedron* **2004,** *60,* 9325–9374. (Review).

8. Lambert, T. H.; Danishefsky, S. J. *J. Am. Chem. Soc.* **2006,** *128,* 426–427.

9. Moslin, R. M.; Jamison, T. F. *J. Am. Chem. Soc.* **2006,** *128,* 15106–15107.

10. Cozzi, P. G. *Angew. Chem. Int. Ed.* **2007,** *46,* 2568–2571. (Review).

11. Ke, Y.-Y.; Li, Y.-J.; Jia, J.-H.; Sheng, W.-J.; Han, L.; Gao, J.-R. *Tetrahedron Lett.* **2009,** *50,* 1389–1391.

12. Grellepois, F. *J. Org. Chem.* **2013,** *78,* 1127–1137.

13. Segade, Y.; Montaos, M. A.; Rodríguez, J.; Jiménez, C. *Org. Lett.* **2014,** *16,* 5820–5823.

14. Fernández-Sánchez, L.; Sánchez-Salas, J. A.; Maestro, M. C.; García Runano, J. L. *J. Org. Chem.* **2018,** *83,* 12903–12910.

15. Maestro, A.; Martinez de Marigorta, E.; Palacios, F.; Vicario, J. *Org. Lett.* **2019,** *21,* 9473–9477.

Ritter Reaction

Amides from nitriles and alcohols in strong acids.
General scheme:

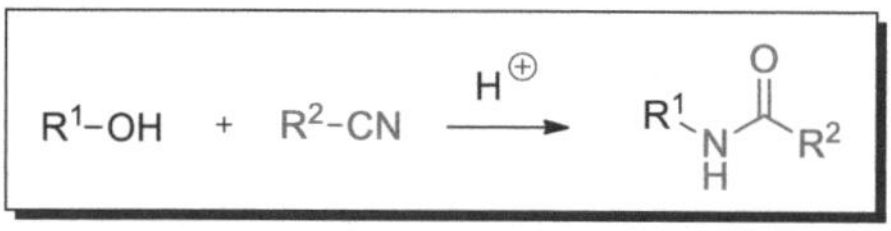

Example 1

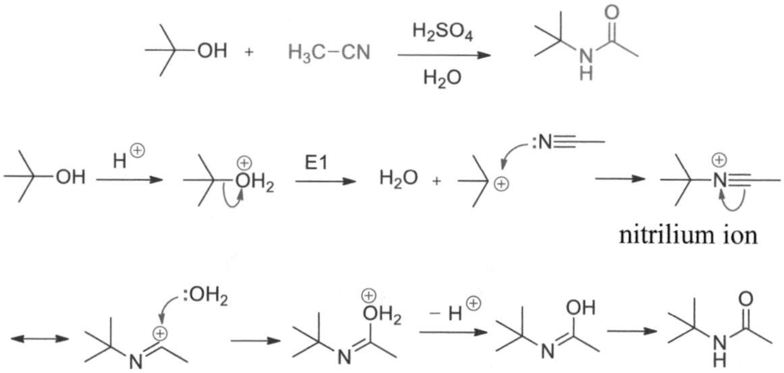

nitrilium ion

Similarly:

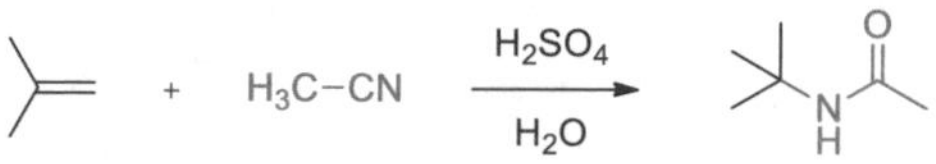

Example 2[3]

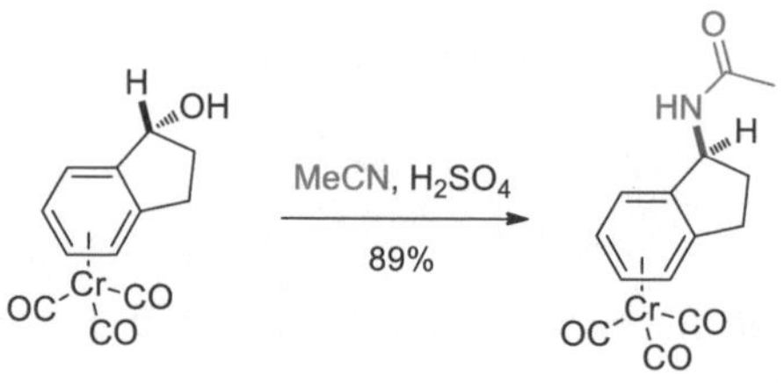

© Springer Nature Switzerland AG 2021
J. J. Li, *Name Reactions*, https://doi.org/10.1007/978-3-030-50865-4_130

Example 3[4]

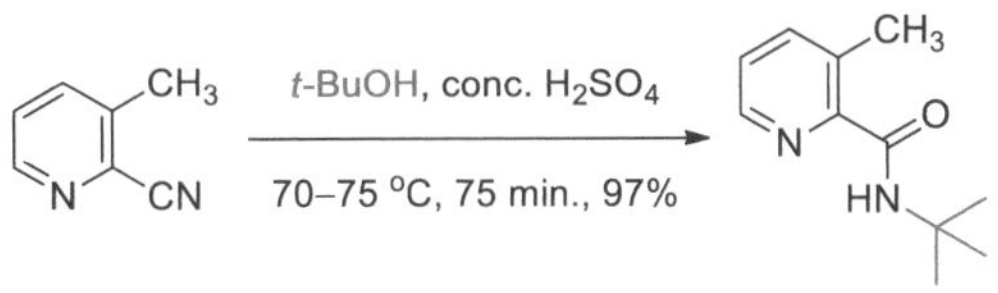

Example 4[5]

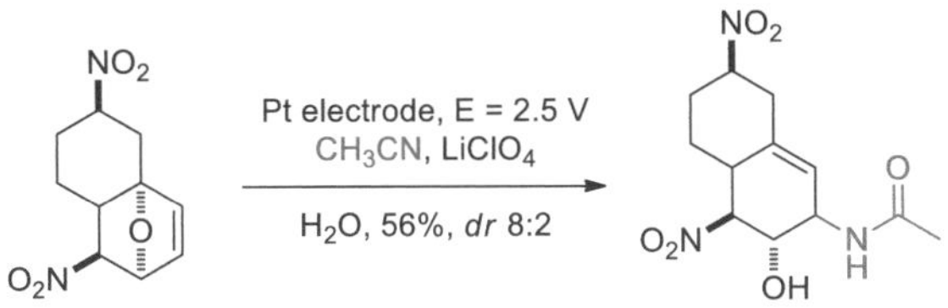

Example 5[6]

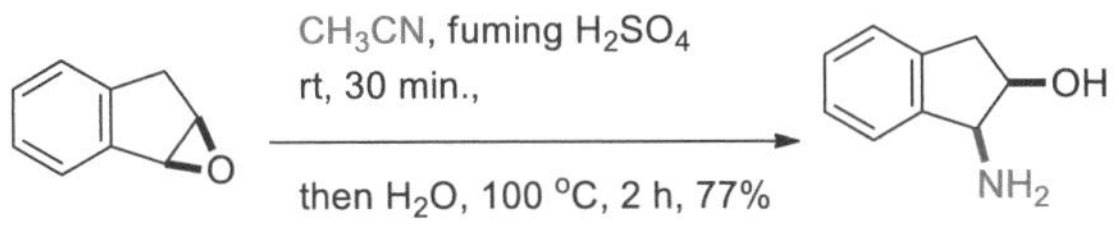

Example 6, A cascade of the Prins/Ritter amidation reaction[12]

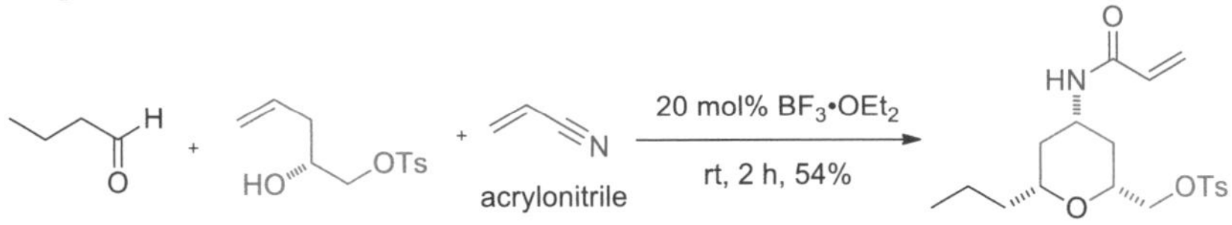

Example 7, Hypervalent iodine(III)-mediated decarboxylative Ritter-type amination[13]

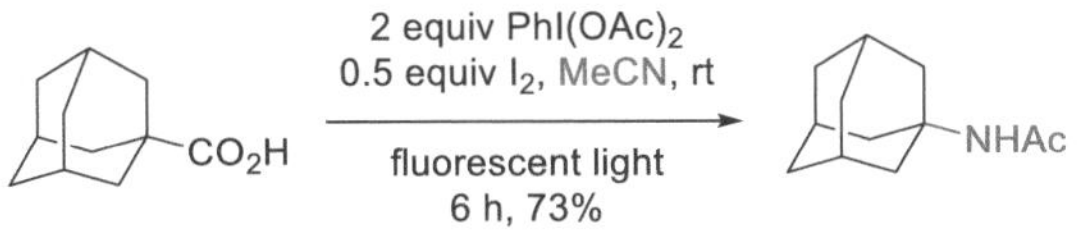

Example 8, A multicomponent reaction involving a Ritter-type route[14]

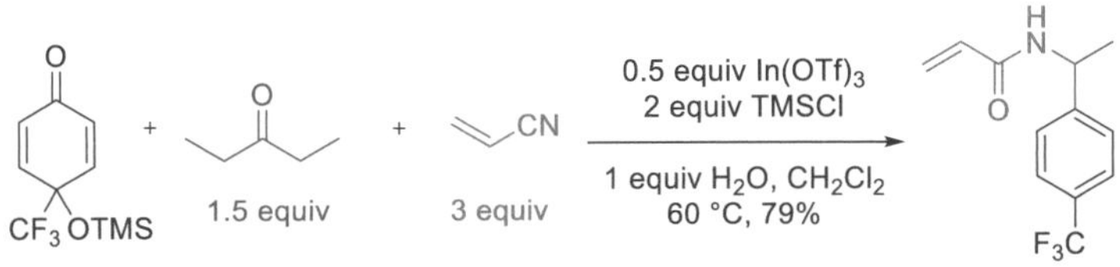

Example 9, Process-scale Ritter reaction[15]

References

1. (a) Ritter, J. J.; Minieri, P. P. *J. Am. Chem. Soc.* **1948,** *70,* 4045–4048. (b) Ritter, J. J.; Kalish, J. *J. Am. Chem. Soc.* **1948,** *70,* 4048–4050.
2. Krimen, L. I.; Cota, D. J. *Org. React.* **1969,** *17,* 213–329. (Review).
3. Top, S.; Jaouen, G. *J. Org. Chem.* **1981,** *46,* 78–82.
4. Schumacher, D. P.; Murphy, B. L.; Clark, J. E.; Tahbaz, P.; Mann, T. A. *J. Org. Chem.* **1989,** *54,* 2242–2244.
5. Le Goanvic, D; Lallemond, M.-C.; Tillequin, F.; Martens, T. *Tetrahedron Lett.* **2001,** *42,* 5175–5176.
6. Tanaka, K.; Kobayashi, T.; Mori, H.; Katsumura, S. *J. Org. Chem.* **2004,** *69,* 5906–5925.
7. Nair, V.; Rajan, R.; Rath, N. P. *Org. Lett.* **2002,** *4,* 1575–1577.
8. Concellón, J. M.; Riego, E.; Suárez, J. R.; García-Granda, S.; Díaz, M. R. *Org. Lett.* **2004,** *6,* 4499–4501.
9. Brewer, A. R. E. *Ritter reaction.* In *Name Reactions for Functional Group Transformations*; Li, J. J., Ed.; Wiley: Hoboken, NJ, **2007,** pp 471–476. (Review).
10. Baum, J. C.; Milne, J. E.; Murry, J. A.; Thiel, O. R. *J. Org. Chem.* **2009,** *74,* 2207–2209.
11. Guerinot, A.; Reymond, S.; Cossy, J. *Eur. J. Org. Chem.* **2012,** 19–28. (Review).
12. Yadav, J. S.; Reddy, Y. J.; Reddy, P. A. N.; Reddy, B. V. S. *Org. Lett.* **2013,** *15,* 546–549.
13. Kiyokawa, K.; Watanabe, T.; Fra, L.; Kojima, T.; Minakata, S. *J. Org. Chem.* **2017,** *82,* 117711–11720.
14. Feng, C.; Li, Y.; Sheng, X.; Pan, L.; Liu, Q. *Org. Lett.* **2018,** *20,* 6449–6452.
15. Zhang, Y.; Chen, S.; Liu, Y.; Wang, Q. *Org. Process Res. Dev.* **2020,** *24,* 216–227.

Robinson Annulation

Michael addition of cyclohexanones to methyl vinyl ketone followed by intramolecular aldol condensation to afford six-membered α,β-unsaturated ketones.

methyl vinyl ketone (MVK)

Example 1, Homo-Robinson[7]

Example 2[8]

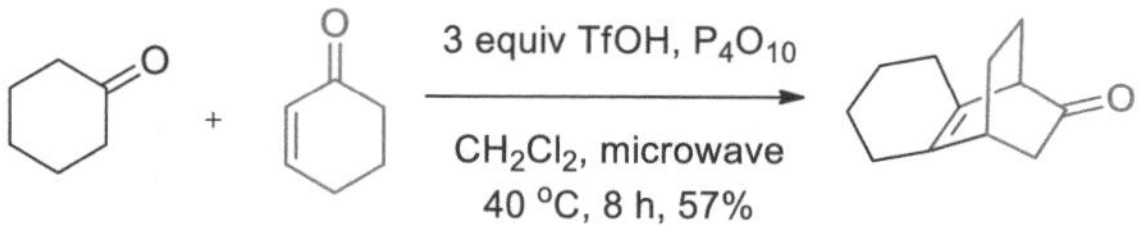

Example 3, Double Robinson-type cyclopentene annulation[9]

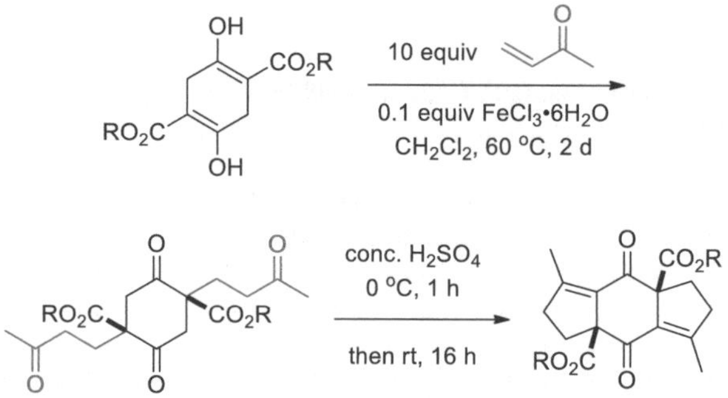

Example 4[10]

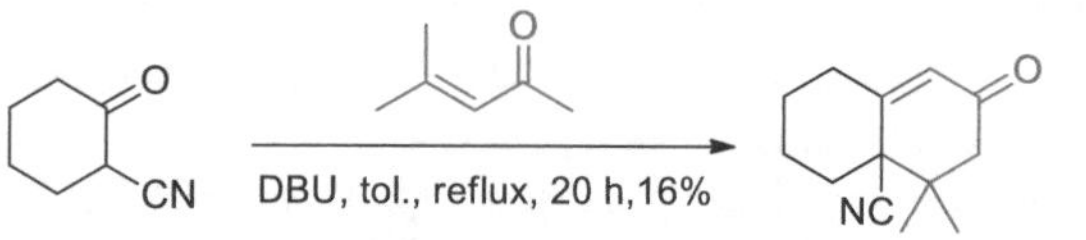

Example 5, The thermodynamically less stable Robinson annulation product was converted to the thermaodynamically more stable product under mild basic conditions[13]

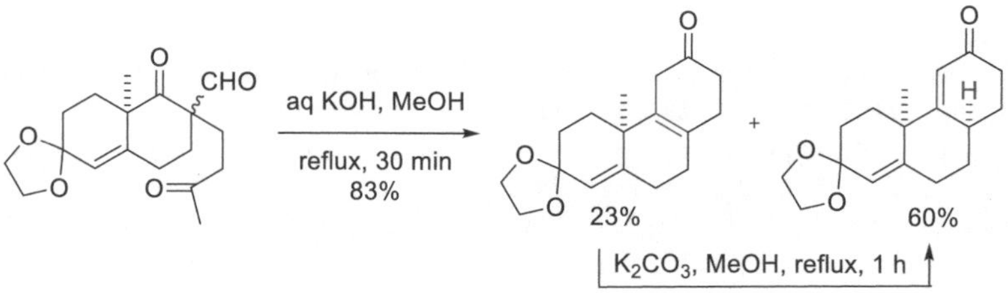

Example 6, Enantioselective Robinson annulation, PTSA = *p*-toluenesulfonic acid[14]

References

1. Rapson, W. S.; Robinson, R. *J. Chem. Soc.* **1935,** 1285–1288. Robert Robinson (1886–1975) used the Robinson annulaton in his total synthesis of cholesterol. Here is a story told by Derek Barton about Robinson and Woodward: "By pure chance, the two great men met early in a Monday morning on an Oxford train station platform in 1951. Robinson politely asked Woodward what kind of research he was doing these days; Woodward replied that he thought that Robinson would be interested in his recent total synthesis of cholesterol. Robinson, incensed and shouting 'Why do you always steal my research topic?', hit Woodward with his umbrella."—An excerpt from Barton, Derek, H. R. *Some Recollections of Gap Jumping,* American Chemical Society, Washington, D.C., **1991.**

2. Gawley, R. E. *Synthesis* **1976,** 777–794. (Review).

3. Guarna, A.; Lombardi, E.; Machetti, F.; Occhiato, E. G.; Scarpi, D. *J. Org. Chem.* **2000,** *65,* 8093–8096.

4. Tai, C.-L.; Ly, T. W.; Wu, J.-D.; Shia, K.-S.; Liu, H.-J. *Synlett* **2001,** 214–217.

5. Jung, M. E.; Piizzi, G. *Org. Lett.* **2003,** *5,* 137–140.

6. Singletary, J. A.; Lam, H.; Dudley, G. B. *J. Org. Chem.* **2005,** *70,* 739–741.

7. Yun, H.; Danishefsky, S. J. *Tetrahedron Lett.* **2005,** *46,* 3879–3882.

8. Jung, M. E.; Maderna, A. *Tetrahedron Lett.* **2005,** *46,* 5057–5061.

9. Zhang, Y.; Christoffers, J. *Synthesis* **2007,** 3061–3067.

10. Jahnke, A.; Burschka, C.; Tacke, R.; Kraft, P. *Synthesis* **2009,** 62–68.

11. Bradshaw, B.; Parra, C.; Bonjoch, J. *Org. Lett.* **2013,** *15,* 2458–2461.

12. Gallier, F.; Martel, A.; Dujardin, G. *Angew. Chem. Int. Ed.* **2017,** *56,* 12424–12458. (Review).

13. Kapras, V.; Vyklicky, V.; Budesinsky, M.; Cisarova, I.; Vyklicky, L.; Chodounska, H.; John, U. *Org. Lett.* **2018,** *20,* 946–949.

14. Zhang, Q.; Zhang, F.-M.; Zhang, C.-S.; Liu, S.-Z.; Tian, J.-M.; Wang, S.-H.; Zhang, X.-M.; Tu, Y.-Q. *Nat. Commun.* **2019,** *10,* 2507.

15. Quevedo-Acosta, Y.; Jurberg, I. D.; Gamba-Sanchez, D. *Org. Lett.* **2020,** *22,* 239–243.

16. Zhang, Y.; Chen, S.; Liu, Y.; Wang, Q. *Org. Process Res. Dev.* **2020,** *24,* 216–227.

Sandmeyer Reaction

Haloarenes from the reaction of a diazonium salt with CuX.

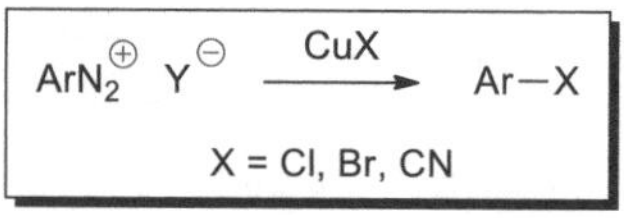

$$ArN_2^{\oplus} \ Y^{\ominus} \xrightarrow{\text{CuX}} Ar-X$$

$$X = Cl, Br, CN$$

Mechanism:

$$ArN_2^{\oplus} \ Cl^{\ominus} \xrightarrow{\text{CuCl}} N_2\uparrow \ + \ Ar\bullet \ + \ CuCl_2 \longrightarrow Ar-Cl \ + \ CuCl$$

Example 1[4]

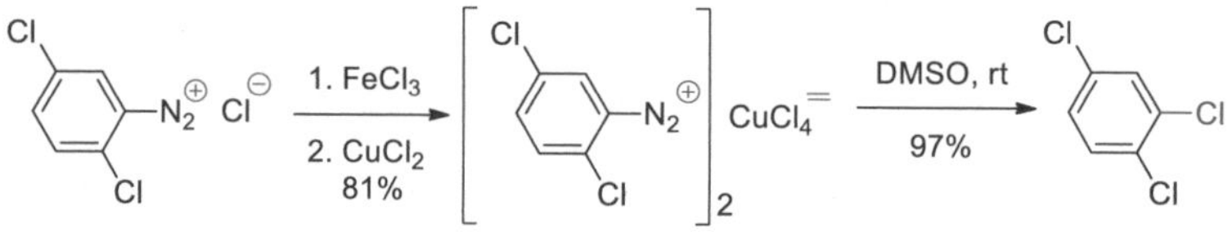

Example 2[7]

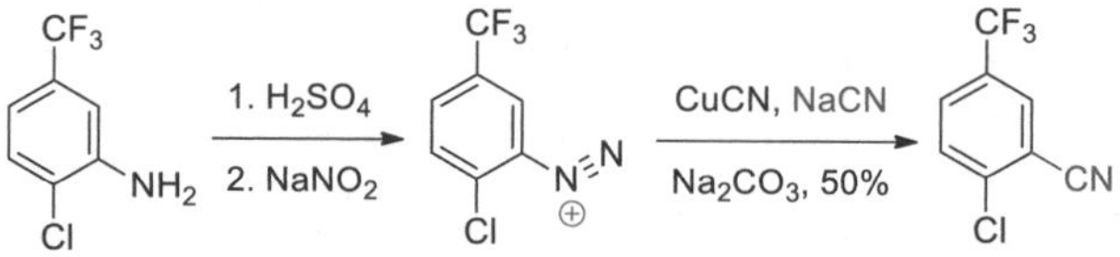

Example 3[8]

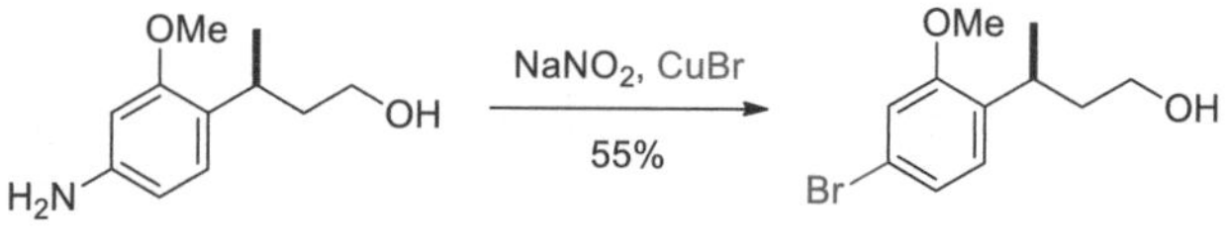

Example 4[9]

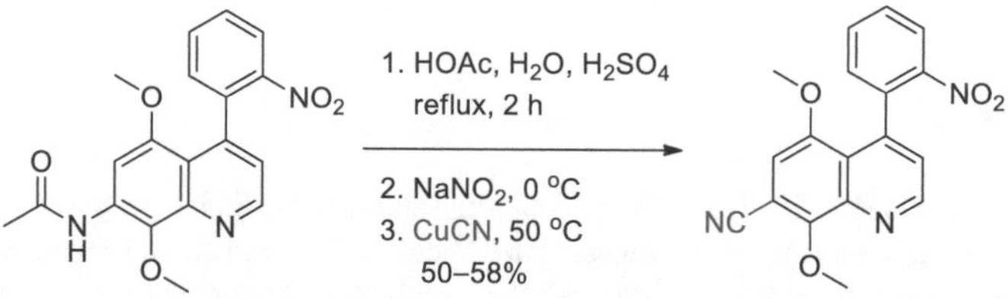

© Springer Nature Switzerland AG 2021
J. J. Li, *Name Reactions*, https://doi.org/10.1007/978-3-030-50865-4_132

Example 5, Trifluoromethylation[11]

Umemoto's reagent

Example 6, Difluoromethylation of (hetero-)arenediazonium salts[13]

Example 7, Safe continuous-flow process[14]

Example 8, Silver-mediated trifluoromethylation of (hetero-)arenediazonium tetra-fluoroborates[16]

References

1.   Sandmeyer, T. *Ber.* **1884,** *17*, 1633. Traugott Sandmeyer (1854–1922) was born in Wettingen, Switzerland. He apprenticed under Victor Meyer and Arthur Hantzsch although he never took a doctorate. He later spent 31 years at the company J. R. Geigy, which is now part of Novartis.

2. Suzuki, N.; Azuma, T.; Kaneko, Y.; Izawa, Y.; Tomioka, H.; Nomoto, T. *J. Chem. Soc., Perkin Trans. 1* **1987,** 645–647.
3. Merkushev, E. B. *Synthesis* **1988,** 923–937. (Review).
4. Obushak, M. D.; Lyakhovych, M. B.; Ganushchak, M. I. *Tetrahedron Lett.* **1998,** *39*, 9567–9570.
5. Hanson, P.; Jones, J. R.; Taylor, A. B.; Walton, P. H.; Timms, A. W. *J. Chem. Soc., Perkin Trans. 2* **2002,** 1135–1150.
6. Daab, J. C.; Bracher, F. *Monatsh. Chem.* **2003,** *134*, 573–583.
7. Nielsen, M. A.; Nielsen, M. K.; Pittelkow, T. *Org. Process Res. Dev.* **2004,** *8*, 1059–1064.
8. Kim, S.-G.; Kim, J.; Jung, H. *Tetrahedron Lett.* **2005,** *46*, 2437–2439.
9. LaBarbera, D. V.; Bugni, T. S.; Ireland, C. M. *J. Org. Chem.* **2007,** *72*, 8501–8505.
10. Gehanne, K.; Lancelot, J.-C.; Lemaitre, S.; El-Kashef, H.; Rault, S. *Heterocycles* **2008,** *75*, 3015–3024.
11. Dai, J.-J.; Fang, C.; Xiao, B.; Yi, J.; Xu, J.; Liu, Z.-J.; Lu, X.; Liu, L.; Fu, Y. *J. Am. Chem. Soc.* **2013,** *135*, 8436–8439.
12. Browne, D. L. *Angew. Chem. Int. Ed.* **2014,** *53*, 1482–1484. (Review).
13. Matheis, C.; Jouvin, K.; Goossen, L. J. *Org. Lett.* **2014,** *16*, 5984–5987.
14. D'Attoma, J.; Camara, T.; Brun, P. L.; Robin, Y.; Bostyn, S.; Buron, F.; Routier, S. *Org. Process Res. Dev.* **2017,** *21*, 44–51.
15. Mo, F.; Qiu, D.; Zhang, Y.; Wang, J. *Acc. Chem. Res.* **2018,** *53*, 496–506. (Review).
16. Yang, Y.-M.; Yao, J.-F.; Yan, W.; Luo, Z.; Tang, Z.-Y. *Org. Lett.* **2019,** *21*, 8003–8007.
17. Schafer, G.; Fleischer, T.; Ahmetovic, M.; Abele, S. *Org. Process Res. Dev.* **2020,** *24*, 228–234.

Schiemann Reaction

Fluoroarene formation from anilines. Also known as the Balz–Schiemann reaction.

$$\text{Ar-NH}_2 + \text{HNO}_2 + \text{HBF}_4 \longrightarrow \text{ArN}_2^{\oplus} \text{ BF}_4^{\ominus} \xrightarrow{\Delta} \text{Ar-F} + \text{N}_2\uparrow + \text{BF}_3$$

Example 1[4]

R = 2,3-5-tri-*O*-acetyl-β-*D*-ribofuranose

Example 2, Photo-Schiemann reaction[6]

Example 3, Photo-Schiemann reaction, bmim = 1-Butyl-3-methylimidazolium[8]

© Springer Nature Switzerland AG 2021
J. J. Li, *Name Reactions*, https://doi.org/10.1007/978-3-030-50865-4_133

Example 4, Synthesis of 3-fluorothiophene[10]

Example 5, From aminoquinoline substrate[12]

Example 6, Hypervalent iodine (III)-catalyzed fluorination under mild conditions[13]

Example 7, Organotrifluoroborates as competent sources of fluoride ion for fluoro-dediazoniation[14]

Example 8, One-step synthesis of 2-fluoroadenine using hydrogen fluoride pyridine in continuous flow operation[15]

References

1. Balz, G.; Schiemann, G. *Ber.* **1927,** *60*, 1186–1190. Günther Schiemann was born in Breslau, Germany in 1899. In 1925, he received his doctorate at Breslau, where he became an assistant professor. In 1950, he became the Chair of Technical Chemistry at Istanbul, where he extensively studied aromatic fluorine compounds.
2. Roe, A. *Org. React.* **1949,** *5*, 193–228. (Review).
3. Sharts, C. M. *J. Chem. Educ.* **1968,** *45*, 185–192. (Review).
4. Montgomery, J. A.; Hewson, K. *J. Org. Chem.* **1969,** *34*, 1396–1399.
5. Laali, K. K.; Gettwert, V. J. *J. Fluorine Chem.* **2001,** *107*, 31–34.
6. Dolensky, B.; Takeuchi, Y.; Cohen, L. A.; Kirk, K. L. *J. Fluorine Chem.* **2001,** *107*, 147–152.
7. Gronheid, R.; Lodder, G.; Okuyama, T. *J. Org. Chem.* **2002,** *67*, 693–720.
8. Heredia-Moya, J.; Kirk, K. L. *J. Fluorine Chem.* **2007,** *128*, 674–678.
9. Gribble, G. W. *Balz-Schiemann reaction.* In *Name Reactions for Functional Group Transformations*; Li, J. J., Ed.; Wiley: Hoboken, NJ, **2007,** pp 552–563. (Review).
10. Pomerantz, M.; Turkman, N. *Synthesis* **2008,** 2333–2336.
11. Cresswell, A. J.; Davies, S. G.; Roberts, P. M.; Thomson, J. E. *Chem. Rev.* **2015,** *115*, 566–611. (Review).
12. Terzić, N.; Konstantinović, J.; Tot, M.; Burojević, J.; Djurković-Djaković, O.; Srbljanović, J.; Stajner, T.; Verbić, T.; Zlatović, M.; Machado, M.; et al. *J. Med. Chem.* **2016,** *59*, 264–281.
13. Xing, B.; Ni, C.; Hu, J. *Angew. Chem. Int. Ed.* **2018,** *57*, 9896–9900.
14. Mohy El Dine, T.; Sadek, O.; Gras, E.; Perrin, D. M. *Chem. Eur. J.* **2018,** *24*, 14933–14937.
15. Salehi Marzijarani, Nastaran; Snead, David R.; McMullen, Jonathan P.; Lévesque, F.; Weisel, M.; Varsolona, R. J.; Lam, Y.-h.; Liu, Z.; Naber, J. R. *Org. Process Res. Dev.* **2019,** *23*, 1522–1528.

Schmidt Rearrangement

The Schmidt reactions refer to the acid-catalyzed reactions of hydrazoic acid with electrophiles, such as carbonyl compounds, tertiary alcohols and alkenes. These substrates undergo rearrangement and extrusion of nitrogen to furnish amines, nitriles, amides or imines.

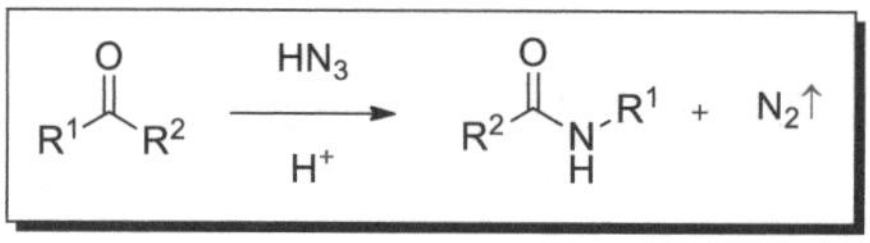

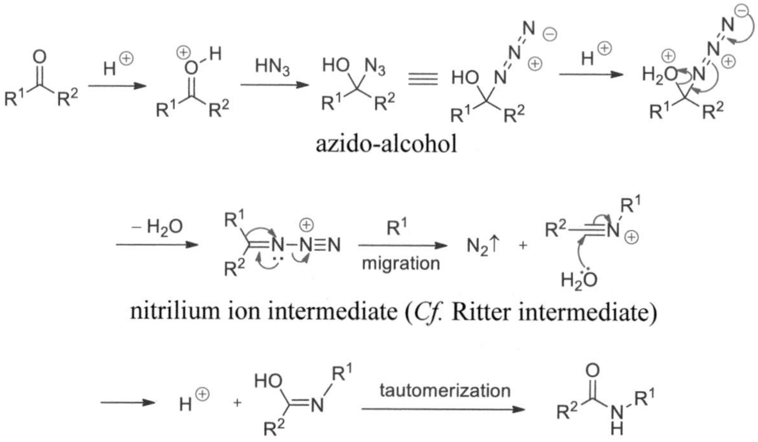

azido-alcohol

nitrilium ion intermediate (*Cf.* Ritter intermediate)

Example 1, A classic example[3]

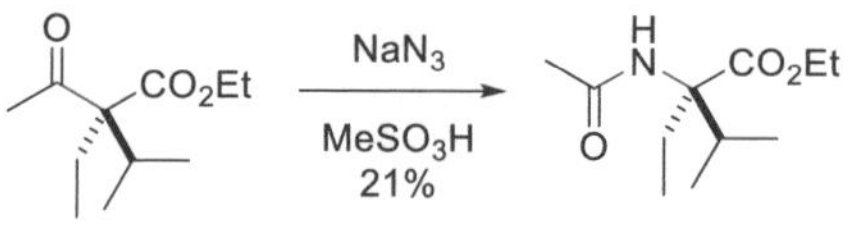

Example 2[5]

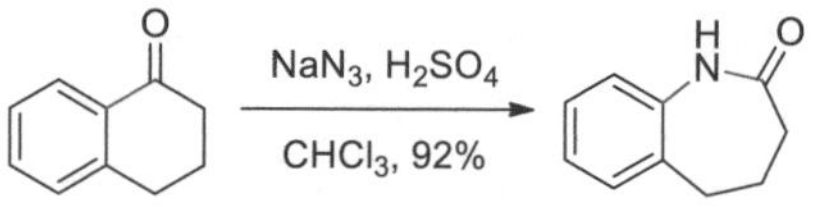

© Springer Nature Switzerland AG 2021
J. J. Li, *Name Reactions*, https://doi.org/10.1007/978-3-030-50865-4_134

Example 3, Intramolecular Schmidt rearrangement[6]

Example 4, Intramolecular Schmidt rearrangement[8]

Example 5, Intermolecular Schmidt rearrangement[9]

Example 6[11]

Example 7, Intramolecular Schmidt rearrangement[12]

Example 8, Making 2-oxoindoles[13]

Example 9, Nitromethane as a nitrogen donor in Schmidt-type formation of amides and nitriles[14]

References

1. (a) Schmidt, K. F. *Angew. Chem.* **1923**, *36*, 511. Karl Friedrich Schmidt (1887–1971) collaborated with Curtius at the University of Heidelberg, where Schmidt became a Professor of Chemistry after 1923. (b) Schmidt, K. F. *Ber.* **1924**, *57*, 704–706.
2. Wolff, H. *Org. React.* **1946**, *3*, 307–336. (Review).
3. Tanaka, M.; Oba, M.; Tamai, K.; Suemune, H. *J. Org. Chem.* **2001**, *66*, 2667–2573.
4. Golden, J. E.; Aubé, J. *Angew. Chem. Int. Ed.* **2002**, *41*, 4316–4318.
5. Johnson, P. D.; Aristoff, P. A. *Bioorg. Med. Chem. Lett.* **2003**, *13*, 4197–4200.
6. Wrobleski, A.; Sahasrabudhe, K.; Aubé, J. *J. Am. Chem. Soc.* **2004**, *126*, 5475–5481.
7. Gorin, D. J.; Davis, N. R.; Toste, F. D. *J. Am. Chem. Soc.* **2005**, *127*, 11260–11261.
8. Iyengar, R.; Schidknegt, K.; Morton, M.; Aubé, J. *J. Org. Chem.* **2005**, *70*, 10645–10652.
9. Amer, F. A.; Hammouda, M.; El-Ahl, A. A. S.; Abdel-Wahab, B. F. *Synth. Commun.* **2009**, *39*, 416–425.
10. Wu, Y.-J. *Schmidt Reactions.* In *Name Reactions for Homologations-Part II*; Li, J. J., Ed.; Wiley: Hoboken, NJ, **2009**, pp 353–372. (Review).
11. Gu, P.; Sun, J.; Kang, X.-Y.; Yi, M.; Li, X.-Q.; Xue, P.; Li, R. *Org. Lett.* **2013**, *15*, 1124–1127.
12. Kim, C.; Kang, S.; Rhee, Y. H. *J. Org. Chem.* **2014**, *79*, 11119–11124.
13. Ding, S.-L.; Ji, Y.; Su, Y.; Li, R.; Gu, P. *J. Org. Chem.* **2019**, *84*, 2012–2021.
14. Liu, J.; Zhang, C.; Zhang, Z.; Wen, X.; Dou, X.; Wei, J.; Qiu, X.; Song, S.; Jiao, N. *Sci.* **2020**, *367*, 281–285.

Shapiro Reaction

The Shapiro reaction is a variant of the Bamford–Stevens reaction. The former uses bases such as alkyl lithium and Grignard reagents whereas the latter employs bases such as Na, NaOMe, LiH, NaH, NaNH$_2$, *etc.* Consequently, *the Shapiro reaction generally affords the less-substituted olefins (the kinetic products), while the Bamford–Stevens reaction delivers the more-substituted olefins (the thermodynamic products).*

Example 1, Kinetic product is the major product[2]

Example 2[3]

© Springer Nature Switzerland AG 2021
J. J. Li, *Name Reactions*, https://doi.org/10.1007/978-3-030-50865-4_135

Example 3[7]

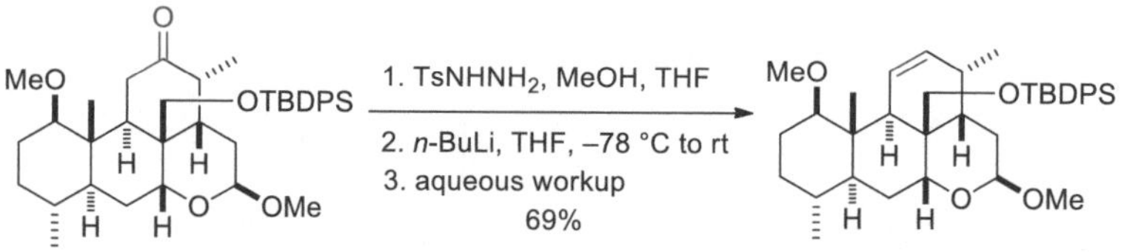

1. TsNHNH₂, MeOH, THF

2. *n*-BuLi, THF, −78 °C to rt
3. aqueous workup
69%

Example 4[8]

1. *n*-BuLi
2. MgBr₂•OEt₂
3.

55% yield
one diastereomer

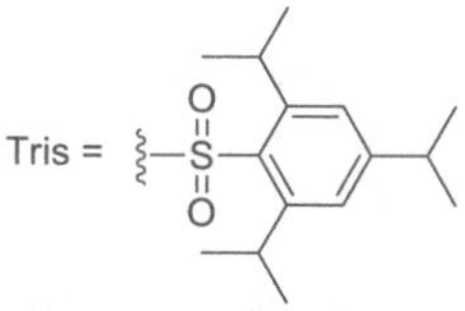

$$Tris = \text{(2,4,6-triisopropylbenzenesulfonyl)}$$

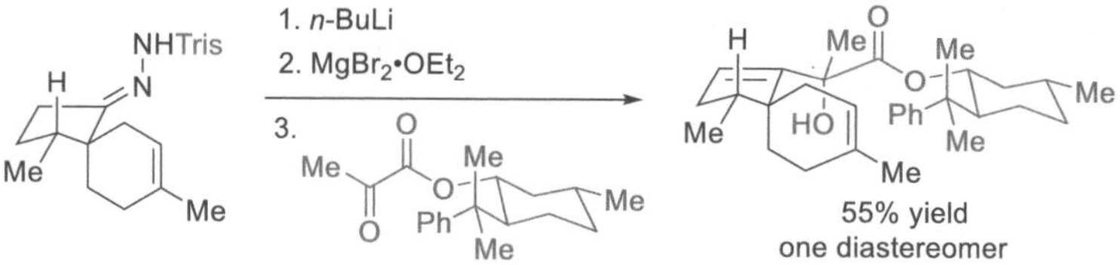

Example 5, NFSI = *N*-fluorobenzenesulfonimide[11]

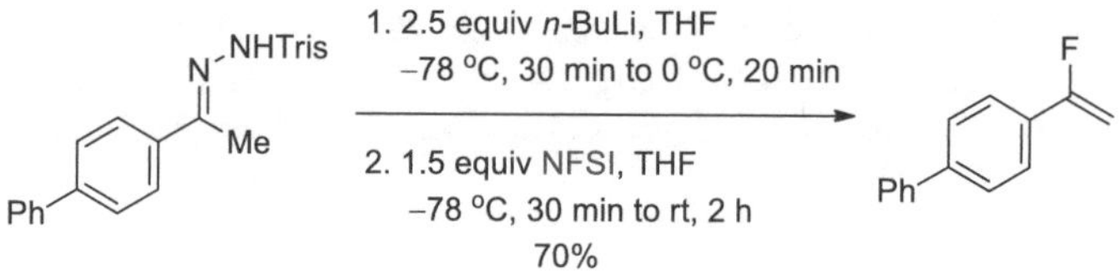

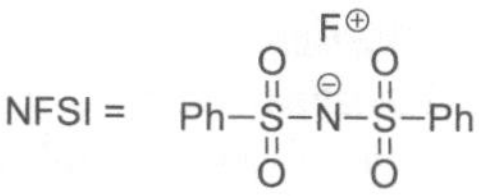

1. 2.5 equiv *n*-BuLi, THF
−78 °C, 30 min to 0 °C, 20 min

2. 1.5 equiv NFSI, THF
−78 °C, 30 min to rt, 2 h
70%

$$NFSI = \quad Ph-\overset{O}{\underset{O}{S}}-\overset{F^{\oplus}}{\underset{\ominus}{N}}-\overset{O}{\underset{O}{S}}-Ph$$

Example 6, Toward total synthesis of paspaline[12]

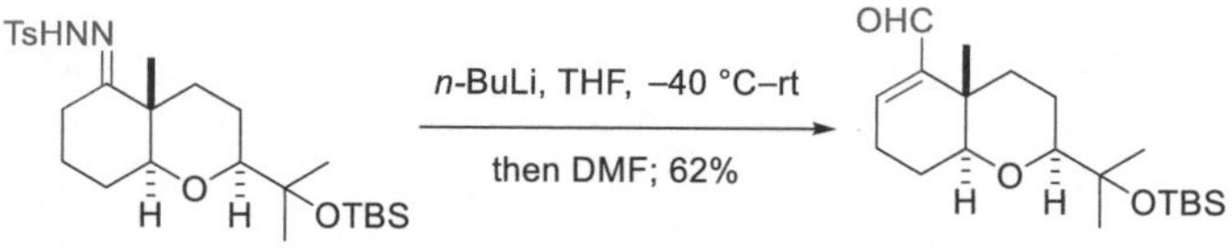

n-BuLi, THF, −40 °C–rt

then DMF; 62%

Example 7, To make ursodeoxycholic acid[13]

LiH, TMEDA

toluene, reflux
18 h, 74%

Example 8[14]

1. TsNNH$_2$, cat. p-TsOH
   MeOH

2. 2 equiv n-BuLi
   −78 to 25 °C

References

1. Shapiro, R. H.; Duncan, J. H.; Clopton, J. C. *J. Am. Chem. Soc.* **1967,** *89,* 471–472. Robert H. Shapiro published this 1967 JACS paper when he was an assistant professor at the University of Colorado. He was denied tenure despite having been immortalized with a reaction named after him.
2. Shapiro, R. H.; Heath, M. J. *J. Am. Chem. Soc.* **1967,** *89,* 5734–5735.
3. Dauben, W. G.; Lorber, M. E.; Vietmeyer, N. D.; Shapiro, R. H.; Duncan, J. H.; Tomer, K. *J. Am. Chem. Soc.* **1968,** *90,* 4762–4763.
4. Shapiro, R. H. *Org. React.* **1976,** *23,* 405–507. (Review).
5. Adlington, R. M.; Barrett, A. G. M. *Acc. Chem. Res.* **1983,** *16,* 55–59. (Review).
6. Chamberlin, A. R.; Bloom, S. H. *Org. React.* **1990,** *39,* 1–83. (Review).
7. Grieco, P. A.; Collins, J. L.; Moher, E. D.; Fleck, T. J.; Gross, R. S. *J. Am. Chem. Soc.* **1993,** *115,* 6078–6093.
8. Tamiya, J.; Sorensen, E. J. *Tetrahedron* **2003,** *59,* 6921–6932.
9. Wolfe, J. P. *Shapiro reaction.* In *Name Reactions for Functional Group Transformations;* Li, J. J., Corey, E. J., eds, Wiley: Hoboken, NJ, **2007,** pp 405–413.
10. Bettinger, H. F.; Mondal, R.; Toenshoff, C. *Org. Biomol. Chem.* **2008,** *6,* 3000–3004.
11. Yang, M.-H.; Matikonda, S. S.; Altman, R. A. *Org. Lett.* **2013,** *15,* 3894–3897.
12. Sharpe, R. J.; Johnson, J. S. *J. Org. Chem.* **2015,** *80,* 9740–9766.
13. Dou, Q.; Jiang, Z. *Synth.* **2016,** *48,* 588–594.
14. Erden, I.; Gleason, C. J. *Tetrahedron Lett.* **2018,** *59,* 284–286.
15. Pfaff, P.; Mouhib, H.; Kraft, P. *Eur. J. Org. Chem.* **2019,** 2643–2652.

Sharpless Asymmetric Amino-hydroxylation

Osmium-mediated *cis*-addition of nitrogen and oxygen to olefins. Regio-selectivity may be controlled by ligand. Nitrogen sources (X–NClNa) include:

The catalytic cycle:

Example 1[1b]

5 mol% (DHQD)$_2$PHAL
4 mol% K$_2$OsO$_2$(OH)$_4$

n-PrOH/H$_2$O (1:1)
51%, 63% *ee*

© Springer Nature Switzerland AG 2021
J. J. Li, *Name Reactions*, https://doi.org/10.1007/978-3-030-50865-4_136

$(DHQD)_2$-PHAL = 1,4-bis(9-*O*-dihydroquinidine)phthalazine:

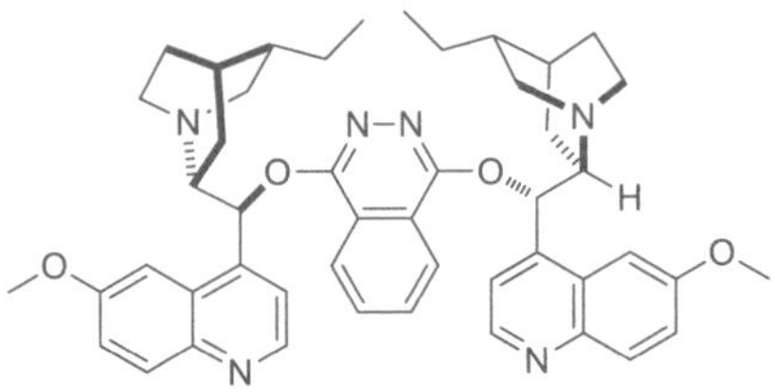

Example 2[2]

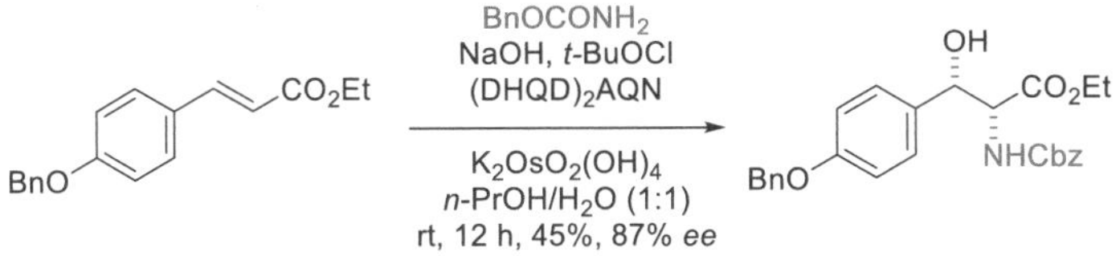

Example 3[6]

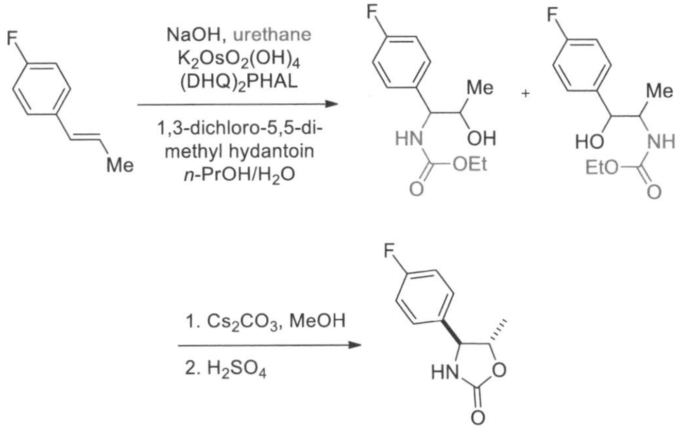

Example 4[13]

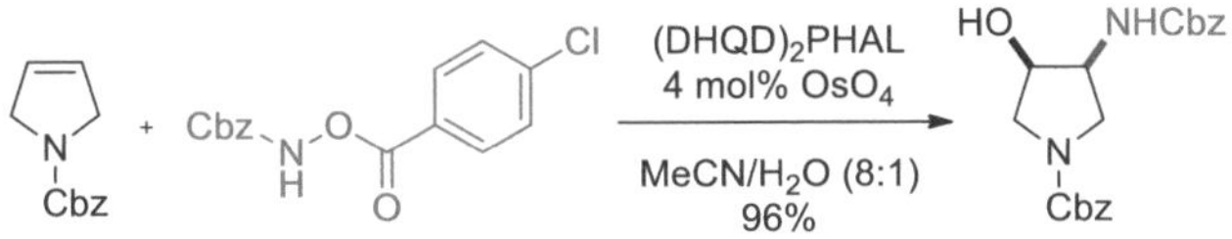

Example 5, Bromamine-T as an efficient amine source[14]

Bromamine-T   Chloramine-T

$K_2OsO_2(OH)_4$ (4 mol%)
$(DHQ)_2PHAL$ (5 mol%)

t-BuOH/H_2O (1:1)
rt, 1.5 h, 75%, 99% ee

Bromamine-T

Example 6, Anti-malarial drug[15]

3.1 equiv $BnOCONH_2$
3.05 equiv t-BuOCl
$(DHQ)_2PHAL$ (5 mol%)

$K_2OsO_2(OH)_4$ (4 mol%)
n-PrOH/H_2O (1:1), NaOH
48%, 80% ee

Example 7, Using FmocNH•HCl as the amine source[16]

$(DHQ)_2PHAL$
$K_2OsO_2(OH)_4$

NaOH, n-PrOH/H_2O
70%, 95:5 er

References

1. (a) Herranz, E.; Sharpless, K. B. *J. Org. Chem.* **1978**, *43*, 2544–2548. K. Barry Sharpless (USA, 1941–) shared the Nobel Prize in Chemistry in 2001 with Herbert William S. Knowles (USA, 1917–) and Ryoji Noyori (Japan, 1938–) for his work on chirally catalyzed oxidation reactions. (b) Li, G.; Angert, H. H.; Sharpless, K. B. *Angew. Chem. Int. Ed.* **1996**, *35*, 2813–2817. (c) Rubin, A. E.; Sharpless, K. B. *Angew. Chem. Int. Ed.* **1997**, *36*, 2637–2640. (d) Kolb, H. C.; Sharpless, K. B. *Transition Met. Org. Synth.* **1998**, *2*, 243–260. (Review). (e) Thomas, A.; Sharpless, K. B. *J. Org. Chem.* **1999**, *64*, 8379–8385. (f) Gontcharov, A. V.; Liu, H.; Sharpless, K. B. *Org. Lett.* **1999**, *1*, 783–786.
2. Nicolaou, K. C.; Boddy, C. N. C.; Li, H.; Koumbis, A. E.; Hughes, R.; Natarajan, S.; Jain, N. F.; Ramanjulu, J. M.; Braese, S.; Solomon, M. E. *Chem. Eur. J.* **1999**, *5*, 2602–2621.
3. Lohr, B.; Orlich, S.; Kunz, H. *Synlett* **1999**, 1139–1141.
4. Boger, D. L.; Lee, R. J.; Bounaud, P.-Y.; Meier, P. *J. Org. Chem.* **2000**, *65*, 6770–6772.
5. Demko, Z. P.; Bartsch, M.; Sharpless, K. B. *Org. Lett.* **2000**, *2*, 2221–2223.

6. Barta, N. S.; Sidler, D. R.; Somerville, K. B.; Weissman, S. A.; Larsen, R. D.; Reider, P. J. *Org. Lett.* **2000,** *2,* 2821–2824.
7. Bolm, C.; Hildebrand, J. P.; Muñiz, K. In *Catalytic Asymmetric Synthesis;* 2nd edn., Ojima, I., Ed.; Wiley–VCH: New York, **2000,** 399. (Review).
8. Bodkin, J. A.; McLeod, M. D. *J. Chem. Soc., Perkin 1* **2002,** 2733–2746. (Review).
9. Rahman, N. A.; Landais, Y. *Cur. Org. Chem.* **2000,** *6,* 1369–1395. (Review).
10. Nilov, D.; Reiser, O. *Recent Advances on the Sharpless Asymmetric Aminohydroxylation.* In *Organic Synthesis Highlights* Schmalz, H.-G.; Wirth, T., eds.; Wiley–VCH: Weinheim, Germany **2003,** 118–124. (Review).
11. Bodkin, J. A.; Bacskay, G. B.; McLeod, M. D. *Org. Biomol. Chem.* **2008,** *6,* 2544–2553.
12. Wong, D.; Taylor, C. M. *Tetrahedron Lett.* **2009,** *50,* 1273–1275.
13. Harris, L.; Mee, S. P. H.; Furneaux, R. H.; Gainsford, G. J.; Luxenburger, A. *J. Org. Chem.* **2011,** *76,* 358–372.
14. Kumar, J. N.; Das, B. *Tetrahedron Lett.* **2013,** *54,* 3865–3867.
15. Borah, A. J.; Phukan, P. *Tetrahedron Lett.* **2014,** *55,* 713–715.
16. Moreira, R.; Taylor, S. D. *Org. Lett.* **2018,** *20,* 7717–7720.
17. Heravi, M. M.; Lashaki, T. B.; Fattahi, B.; Zadsirjan, V. *RSC Adv.* **2018,** *8,* 6634–6659. (Review).
18. Moreira, R.; Diamandas, M.; Taylor, S. D. *J. Org. Chem.* **2019,** *84,* 15476–15485.
19. Jiang, Y.-L.; Yu, H.-X.; Li, Y.; Qu, P.; Han, Y.-X.; Chen, J.-H.; Yang, Z. *J. Am. Chem. Soc.* **2020,** *142,* 573–580.

Sharpless Asymmetric Dihydroxylation

Enantioselective *cis*-dihydroxylation of olefins using osmium catalyst in the presence of cinchona alkaloid ligands.

(DHQ)$_2$-PHAL = 1,4-bis(9-O-dihydroquinine)phthalazine:

The concerted [3 + 2] cycloaddition mechanism:[5]

Example 1[2]

© Springer Nature Switzerland AG 2021
J. J. Li, *Name Reactions*, https://doi.org/10.1007/978-3-030-50865-4_137

The catalytic cycle: (the secondary cycle is shut off by maintaining a low concentration of olefin):

NMO = N-methyl morpholine N-oxide
NMM = N-methyl morpholine

Example 2[4]

1. AD-mix-β, MeSO$_2$NH$_2$
   t-BuOH/H$_2$O (1:1), rt, 12 h

2. NosCl, Et$_3$N, CH$_2$Cl$_2$
   0 °C, 54%, 92% ee

Nos = nosylate = 4-nitrobenzenesulfonyl

Example 3[9]

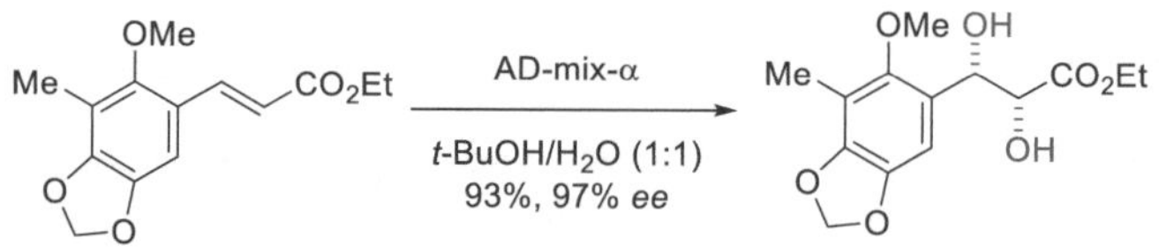

OMe
Me
CO2Et
AD-mix-α
t-BuOH/H2O (1:1)
93%, 97% ee
OMe OH
Me
CO2Et
OH

Example 4[10]

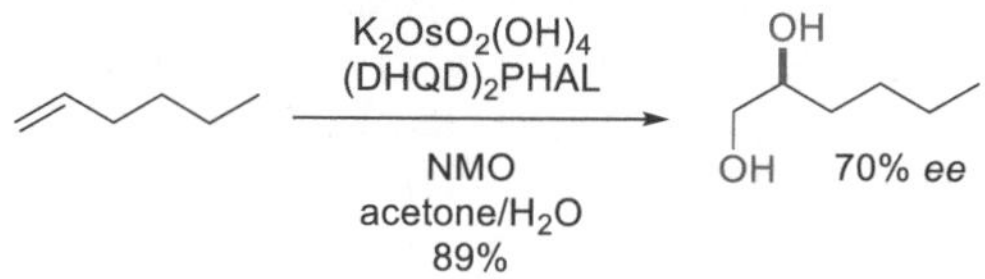

K2OsO2(OH)4
(DHQD)2PHAL
NMO
acetone/H2O
89%
OH
OH 70% ee

Example 5, A scalable synthesis[13]

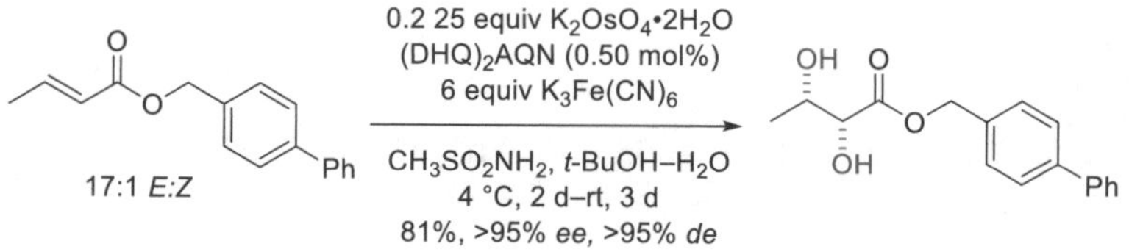

0.2 25 equiv K2OsO4•2H2O
(DHQ)2AQN (0.50 mol%)
6 equiv K3Fe(CN)6
CH3SO2NH2, t-BuOH–H2O
4 °C, 2 d–rt, 3 d
81%, >95% ee, >95% de
17:1 E:Z
Ph
OH O
OH
Ph

Example 5[14]

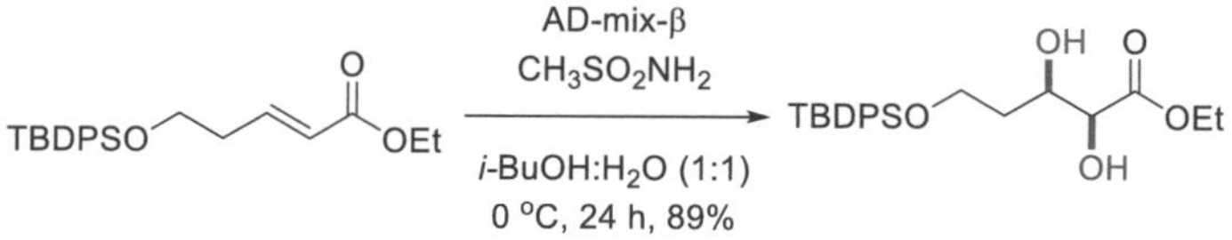

AD-mix-β
CH3SO2NH2
i-BuOH:H2O (1:1)
0 °C, 24 h, 89%
TBDPSO
OEt
TBDPSO
OH O
OEt
OH

Example 7[15]

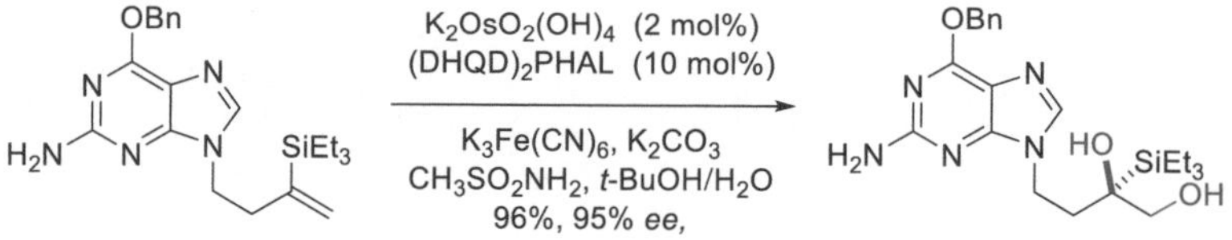

OBn
H2N
SiEt3
K2OsO2(OH)4 (2 mol%)
(DHQD)2PHAL (10 mol%)
K3Fe(CN)6, K2CO3
CH3SO2NH2, t-BuOH/H2O
96%, 95% ee,
OBn
H2N
HO
SiEt3
OH

References

1. (a) Jacobsen, E. N.; Markó, I.; Mungall, W. S.; Schröder, G.; Sharpless, K. B. *J. Am. Chem. Soc.* **1988,** *110*, 1968–1970. (b) Wai, J. S. M.; Markó, I.; Svenden, J. S.; Finn, M. G.; Jacobsen, E. N.; Sharpless, K. B. *J. Am. Chem. Soc.* **1989,** *111*, 1123–1125.

2. Kim, N.-S.; Choi, J.-R.; Cha, J. K. *J. Org. Chem.* **1993,** *58,* 7096–7699.
3. Kolb, H. C.; VanNiewenhze, M. S.; Sharpless, K. B. *Chem. Rev.* **1994,** *94*, 2483–2547. (Review).
4. Rao, A. V. R.; Chakraborty, T. K.; Reddy, K. L.; Rao, A. S. *Tetrahedron Lett.* **1994,** *35*, 5043–5046.
5. Corey, E. J.; Noe, M. C. *J. Am. Chem. Soc.* **1996,** *118*, 319–329. (Mechanism).
6. DelMonte, A. J.; Haller, J.; Houk, K. N.; Sharpless, K. B.; Singleton, D. A.; Strassner, T.; Thomas, A. A. *J. Am. Chem. Soc.* **1997,** *119*, 9907–9908. (Mechanism).
7. Sharpless, K. B. *Angew. Chem. Int. Ed.* **2002,** *41*, 2024–2032. (Review, Nobel Prize Address).
8. Zhang, Y.; O'Doherty, G. A. *Tetrahedron* **2005,** *61*, 6337–6351.
9. Chandrasekhar, S.; Reddy, N. R.; Rao, Y. S. *Tetrahedron* **2006,** *62*, 12098–12107.
10. Ferreira, F. C.; Branco, L. C.; Verma, K. K.; Crespo, J. G.; Afonso, C. A. M. *Tetrahedron Asymmetry* **2007,** *18*, 1637–1641.
11. Ramon, R.; Alonso, M.; Riera, A. *Tetrahedron: Asymmetry* **2007,** *18*, 2797–2802.
12. Krishna, P. R.; Reddy, P. S. *Synlett* **2009,** 209–212.
13. Smaltz, D. J.; Myers, A. G. *J. Org. Chem.* **2011,** *76*, 8554–8559.
14. Kamal, A.; Vangala, S. R. *Org. Biomol. Chem.* **2013,** *11*, 4442–4448.
15. Heravi, M. M.; Zadsirjan, V.; Esfandyari, M.; Lashaki, T. B. *Tetrahedron: Asym.* **2017,** *28*, 987–1043. (Review).
16. Qin, T.; Li, J.-P.; Xie, M.-S.; Qu, G.-R.; Guo, H.-M. *J. Org. Chem.* **2018,** *83*, 15512–15523.
17. Gao, D.; Li, B.; O'Doherty, G. A. *Org. Lett.* **2019,** *21*, 8334–8338.
18. Kopp, J.; Brueckner, R. *Org. Lett.* **2020,** *22*, 3607–3612.

Sharpless Asymmetric Epoxidation

Enantioselective epoxidation of allylic alcohols using *t*-butyl hydroperoxide (TBHP), titanium tetra-*iso*-propoxide, and optically pure diethyl tartrate (DET).

The catalytic cycle:

© Springer Nature Switzerland AG 2021
J. J. Li, *Name Reactions*, https://doi.org/10.1007/978-3-030-50865-4_138

The putative active catalyst:

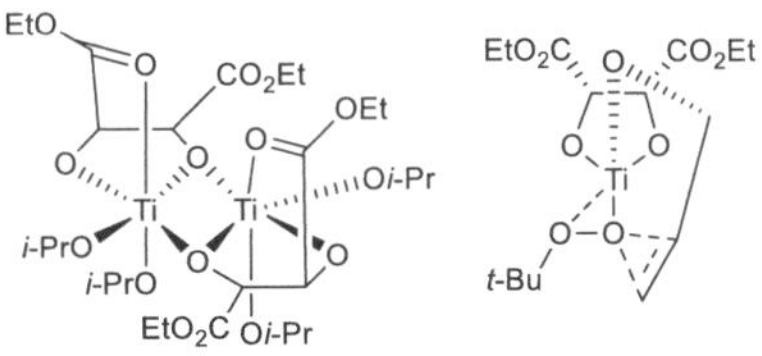
EtO
CO2Et
OEt
Oi-Pr
i-PrO
i-PrO
EtO2C Oi-Pr
EtO2C
CO2Et
t-Bu

Example 1[3]

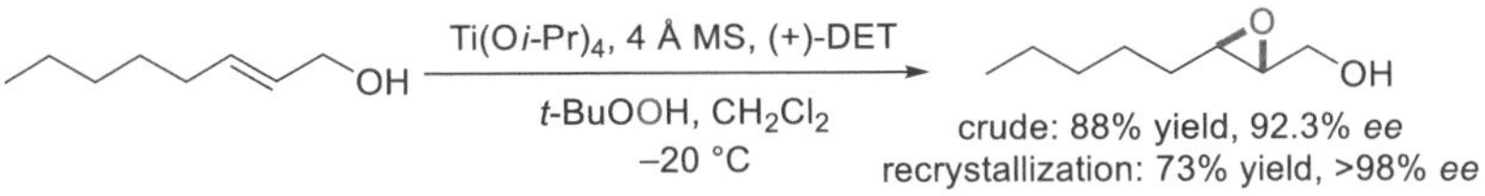
OH
Ti(Oi-Pr)4, 4 Å MS, (+)-DET
t-BuOOH, CH2Cl2
–20 °C
OH
crude: 88% yield, 92.3% ee
recrystallization: 73% yield, >98% ee

Example 2, DIPT = diisopropyl tartrate[3]

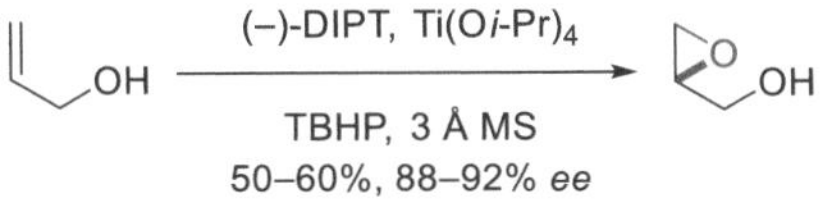
OH
(–)-DIPT, Ti(Oi-Pr)4
TBHP, 3 Å MS
50–60%, 88–92% ee
OH

Example 3[11]

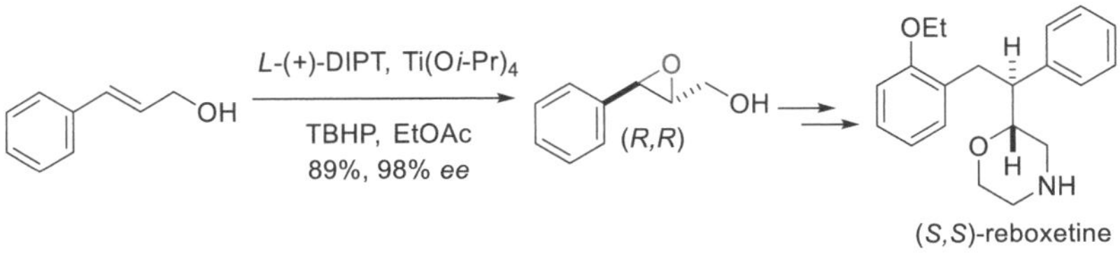
OH
L-(+)-DIPT, Ti(Oi-Pr)4
TBHP, EtOAc
89%, 98% ee
(R,R)
OH
OEt
NH
(S,S)-reboxetine

Example 4[12]

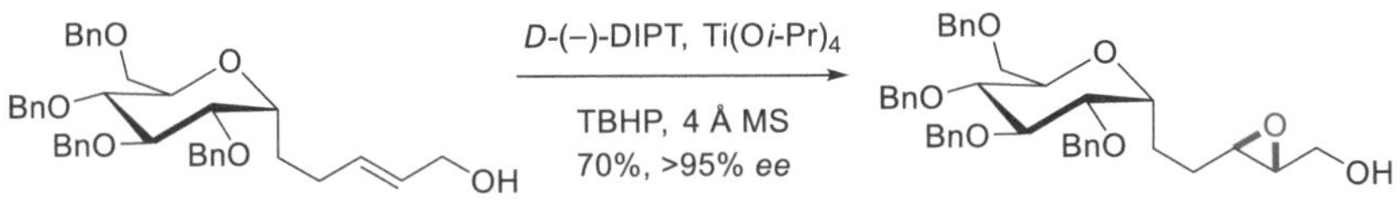
BnO
BnO
BnO
BnO
OH
D-(–)-DIPT, Ti(Oi-Pr)4
TBHP, 4 Å MS
70%, >95% ee
BnO
BnO
BnO
BnO
OH

Example 5[14]

t-BuOOH

Ti(O*i*-Pr)₄
(−)-D-DIPT

dr = 88:12

Example 6, Application in medicinal chemistry[16]

D-(−)-DET, Ti(O*i*-Pr)₄

TBHP, 4 Å MS
CH₂Cl₂, −20 °C

Example 7, Total synthesis of (+)-nivetetracyclate A[17]

(−)-DET, Ti(O*i*-Pr)₄

TBHP, 4 Å MS
82%, 93% *ee*

Example 8, Application in medicinal chemistry[18]

1. DIBAL-H, −78 °C

2. (−)-DET, Ti(O*i*-Pr)₄
TBHP, −20 °C
54%, 2 steps

References

1. (a) Katsuki, T.; Sharpless, K. B. *J. Am. Chem. Soc.* **1980,** *102*, 5974–5976. (b) Williams, I. D.; Pedersen, S. F.; Sharpless, K. B.; Lippard, S. J. *J. Am. Chem. Soc.* **1984,** *106*, 6430–6433. (c) Woodard, S. S.; Finn, M. G.; Sharpless, K. B. *J. Am. Chem. Soc.* **1991,** *113*, 106–113.
2. Pfenninger, A. *Synthesis* **1986,** 89–116. (Review).
3. Gao, Y.; Hanson, R. M.; Klunder, J. M.; Ko, S. Y.; Masamune, H.; Sharpless, K. B. *J. Am. Chem. Soc.* **1987,** *109*, 5765–5780.
4. Corey, E. J. *J. Org. Chem.* **1990,** *55*, 1693–1694. (Review).
5. Johnson, R. A.; Sharpless, K. B. In *Comprehensive Organic Synthesis*; Trost, B. M., Ed,; Pergamon Press: New York, **1991**; Vol. 7, Chapter 3.2. (Review).

6. Johnson, R. A.; Sharpless, K. B. In *Catalytic Asymmetric Synthesis*; Ojima, I., ed,; VCH: New York, **1993**; Chapter 4.1, pp 103–158. (Review).
7. Schinzer, D. *Org. Synth. Highlights II* **1995**, 3. (Review).
8. Katsuki, T.; Martin, V. S. *Org. React.* **1996**, *48*, 1–299. (Review).
9. Johnson, R. A.; Sharpless, K. B. In *Catalytic Asymmetric Synthesis;* 2nd ed., Ojima, I., ed.; Wiley-VCH: New York, **2000**, 231–285. (Review).
10. Palucki, M. *Sharpless–Katsuki Epoxidation*. In *Name Reactions in Heterocyclic Chemistry*; Li, J. J., Ed.; Wiley: Hoboken, NJ, **2005**, 50–62. (Review).
11. Henegar, K. E.; Cebula, M. *Org. Process Res. Dev.* **2007**, *11*, 354–358.
12. Pu, J.; Franck, R. W. *Tetrahedron* **2008**, *64*, 8618–8629.
13. Knight, D. W.; Morgan, I. R. *Tetrahedron Lett.* **2009**, *50*, 35–38.
14. Volchkov, I.; Lee, D. *J. Am. Chem. Soc.* **2013**, *135*, 5324–5327.
15. Heravi, M. M.; Lashaki, T. B.; Poorahmad, N. *Tetrahedron Asymmetry* **2015**, *26*, 405–495. (Review).
16. Ghosh, A. K.; Osswald, H. L.; Glauninger, K.; Agniswamy, J.; Wang, Y.-F.; Hayashi, H.; Aoki, M.; Weber, I. T.; Mitsuya, H. *J. Med. Chem.* **2016**, *59*, 6826–6837.
17. Blitz, M.; Heinze, R. C.; Harms, K.; Koert, U. *Org. Lett.* **2019**, *21*, 785–788.
18. Yoshizawa, S.-i.; Hattori, Y.; Kobayashi, K.; Akaji, K. *Bioorg. Med. Chem.* **2020**, *28*, 115273.

Simmons–Smith Reaction

Cyclopropanation of olefins using CH_2I_2 and Zn(Cu).

Example 1[2]

Example 2, An asymmetric version[3]

Example 3, Diastereoselective Simmons–Smith cyclopropanations of allylic amines and carbamates[9]

© Springer Nature Switzerland AG 2021
J. J. Li, *Name Reactions*, https://doi.org/10.1007/978-3-030-50865-4_139

Example 4[10]

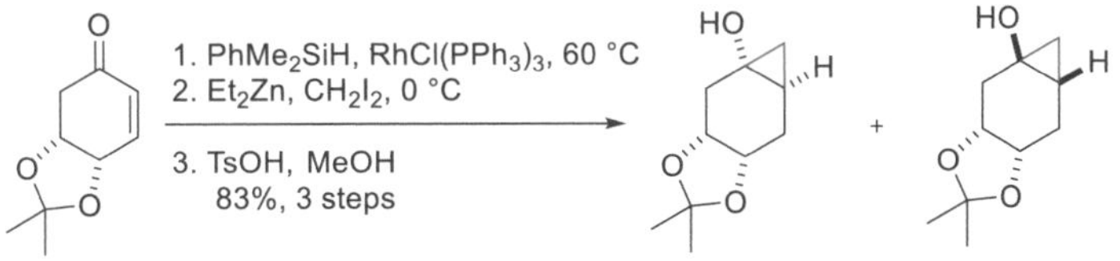

Example 5, Rearrangement after the Simmons–Smith cyclopropanation[12]

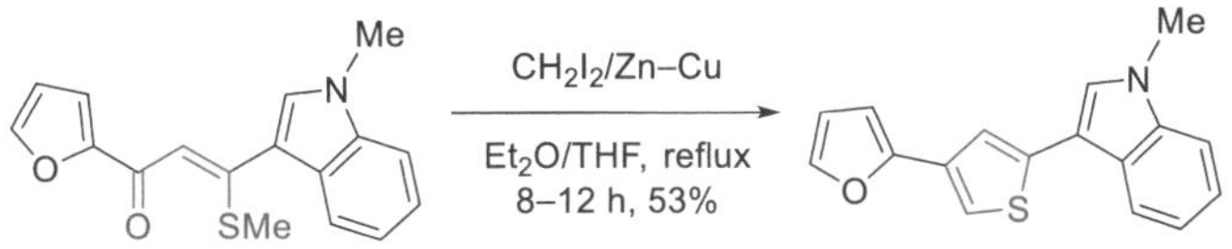

Example 6, EDTA = ethylenediaminetetraacetic acid[13]

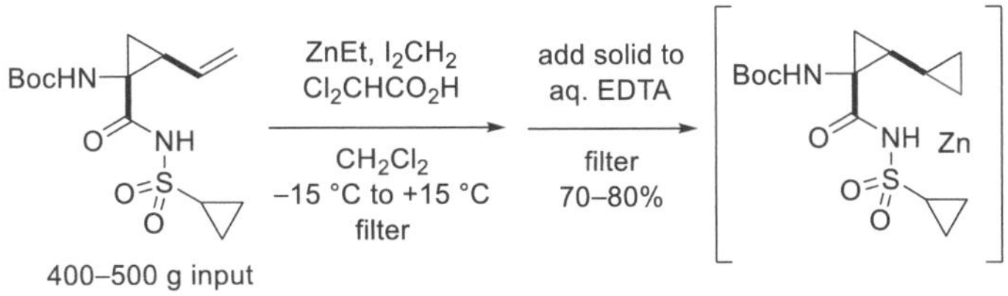

400–500 g input

Example 7, Diastereoselective borocyclopropanation of allylic ethers using a boro-methylzinc carbenoid[14]

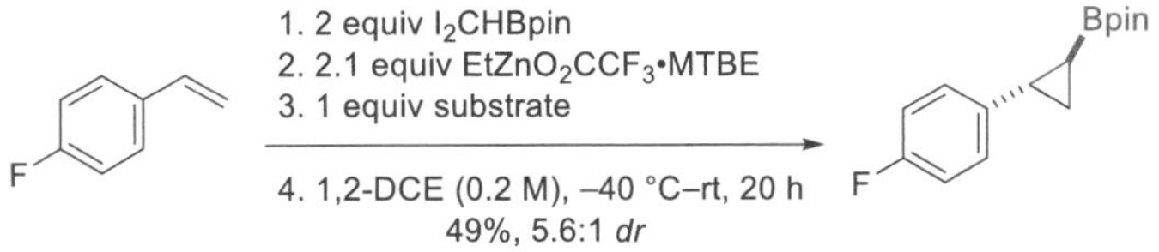

Example 8[15]

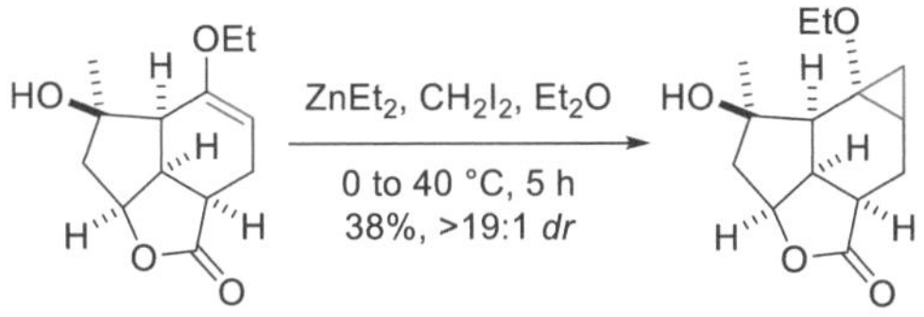

References

1. Simmons, H. E.; Smith, R. D. *J. Am. Chem. Soc.* **1958**, *80*, 5323–5324. Howard E. Sim-mons (1929–1997) was born in Norfolk, Virginia. He carried out his graduate studies at MIT under John D. Roberts and Arthur Cope. After obtaining his Ph.D. In 1954, he joined the Chemical Department of the DuPont Company, where he discovered the Sim-mons–Smith reaction with his colleague, R. D. Smith. Simmons rose to be the vice pres-ident of the Central Research at DuPont in 1979. His views on physical exercise were the same as those of Alexander Woollcot's: "If I think about exercise, I know if I wait long enough, the thought will go away."

2. Limasset, J.-C.; Amice, P.; Conia, J.-M. *Bull. Soc. Chim. Fr.* **1969**, 3981–3990.

3. Kitajima, H.; Ito, K.; Aoki, Y.; Katsuki, T. *Bull. Chem. Soc. Jpn.* **1997**, *70*, 207–217.

4. Nakamura, E.; Hirai, A.; Nakamura, M. *J. Am. Chem. Soc.* **1998**, *120*, 5844–5845.

5. Loeppky, R. N.; Elomari, S. *J. Org. Chem.* **2000**, *65*, 96–103.

6. Charette, A. B.; Beauchemin, A. *Org. React.* **2001**, *58*, 1–415. (Review).

7. Nakamura, M.; Hirai, A.; Nakamura, E. *J. Am. Chem. Soc.* **2003**, *125*, 2341–2350.

8. Long, J.; Du, H.; Li, K.; Shi, Y. *Tetrahedron Lett.* **2005**, *46*, 2737–2740.

9. Davies, S. G.; Ling, K. B.; Roberts, P. M.; Russell, A. J.; Thomson, J. E. *Chem. Com-mun.* **2007**, 4029–4031.

10. Shan, M.; O'Doherty, G. A. *Synthesis* **2008**, 3171–3179.

11. Kim, H. Y.; Salvi, L.; Carroll, P. J.; Walsh, P. J. *J. Am. Chem. Soc.* **2009**, *131*, 954–962.

12. Swaroop, T. R.; Roopashree, R.; Ila, H.; Rangappa, K. S. *Tetrahedron Lett.* **2013**, *54*, 147–150.

13. Young, I. S.; Qiu, Y.; Smith, M. J.; Hay, M. B.; Doubleday, W. W. *Org. Process Res. Dev.* **2016**, *20*, 2108–2115.

14. Benoit, G.; Charette, A. B. *J. Am. Chem. Soc.* **2017**, *139*, 1364–1367.

15. Truax, N. J.; Ayinde, S.; Van, K.; Liu, J. O.; Romo, D. *Org. Lett.* **2019**, *21*, 7394–7399.

16. Singh, U. S.; Chu, C. K. *Nucleos. Nucleot. Nucl.* **2020**, *39*, 52–68.

Smiles Rearrangement

Intramolecular nucleophilic aromatic rearrangement. General scheme:

$$X = S, SO, SO_2, O, CO_2$$
$$YH = OH, NHR, SH, CH_2R, CONHR$$
$$Z = NO_2, SO_2R$$

Mechanism:

spirocyclic anion intermediate (Meisenheimer complex)

Example 1[7]

Example 2, Microwave Smiles rearrangement[9]

© Springer Nature Switzerland AG 2021
J. J. Li, *Name Reactions*, https://doi.org/10.1007/978-3-030-50865-4_140

Example 3[10]

Example 4, The thiol group attacks the chlorine on top first, then Smiles rearrangement ensues[11]

Example 5, Mild chemotriggered generation of a fluorophore-tethered diazoalkane species via Smiles rerrangement[12]

Example 6, S–N-type Smiles rerrangement[14]

Example 7, Phenol to aniline[15]

References

1. Evans, W. J.; Smiles, S. *J. Chem. Soc.* **1935**, 181–188. Samuel Smiles began his career at King's College London as an assistant professor. He later became professor and chair there. He was elected Fellow of the Royal Society (FRS) in 1918.
2. Truce, W. E.; Kreider, E. M.; Brand, W. W. *Org. React.* **1970**, *18*, 99–215. (Review).
3. Gerasimova, T. N.; Kolchina, E. F. *J. Fluorine Chem.* **1994**, *66*, 69–74. (Review).
4. Boschi, D.; Sorba, G.; Bertinaria, M.; Fruttero, R.; Calvino, R.; Gasco, A. *J. Chem. Soc., Perkin Trans. 1* **2001**, 1751–1757.
5. Hirota, T.; Tomita, K.-I.; Sasaki, K.; Okuda, K.; Yoshida, M.; Kashino, S. *Heterocycles* **2001**, *55*, 741–752.
6. Selvakumar, N.; Srinivas, D.; Azhagan, A. M. *Synthesis* **2002**, 2421–2425.
7. Mizuno, M.; Yamano, M. *Org. Lett.* **2005**, *7*, 3629–3631.
8. Bacque, E.; El Qacemi, M.; Zard, S. Z. *Org. Lett.* **2005**, *7*, 3817–3820.
9. Bi, C. F.; Aspnes, G. E.; Guzman-Perez, A.; Walker, D. P. *Tetrahedron Lett.* **2008**, *49*, 1832–1835.
10. Jin, Y. L.; Kim, S.; Kim, Y. S.; Kim, S.-A.; Kim, H. S. *Tetrahedron Lett.* **2008**, *49*, 6835–6837.
11. Niu, X.; Yang, B.; Li, Y.; Fang, S.; Huang, Z.; Xie, C.; Ma, C. *Org. Biomol. Chem.* **2013**, *11*, 4102–4108.
12. Zhang, Z.; Li, Y.; He, H.; Qian, X.; Yang, Y. *Org. Lett.* **2016**, *18*, 4674–4677.
13. Holden, C. M.; Greaney, M. F. *Chem. Eur. J.* **2017**, *23*, 8992–9008.
14. Wang, P.; Hong, G. J.; Wilson, M. R.; Balskus, E. P. *J. Am. Chem. Soc.* **2017**, *139*, 2864–2867.
15. Chang, X.; Zhang, Q.; Guo, C. *Org. Lett.* **2019**, *21*, 4915–4918.
16. Wang, Z.-S.; Chen, Y.-B.; Zhang, H.-W.; Sun, Z.; Zhu, C.; Ye, L.-W. *J. Am. Chem. Soc.* **2020**, *142*, 3636–3644.

Truce–Smile Rearrangement

A variant of the Smiles rearrangement where Y is carbon:

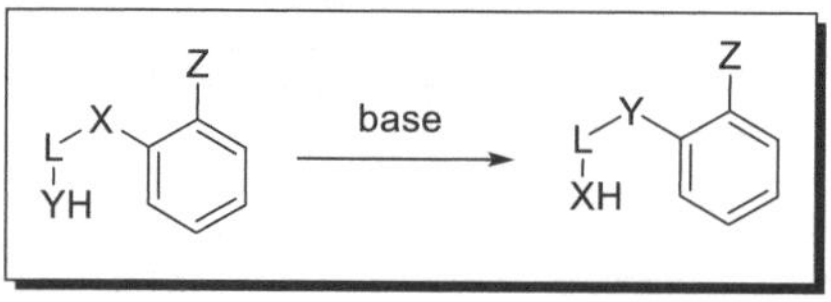

Example 1[6]

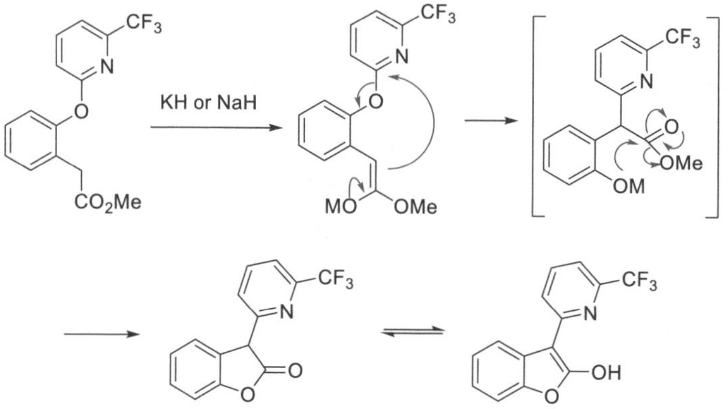

Example 2[7]

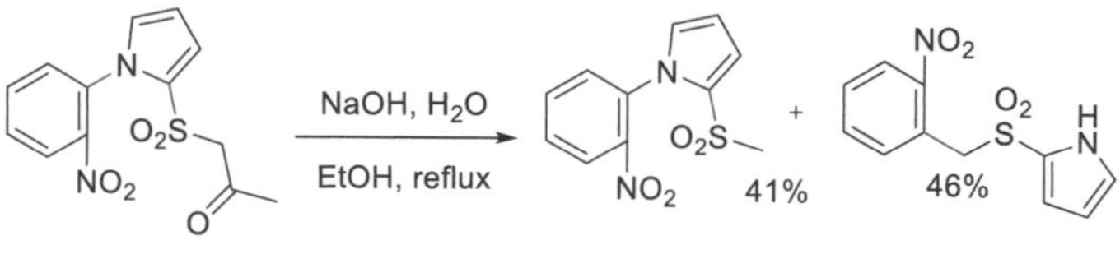

Example 3[8]

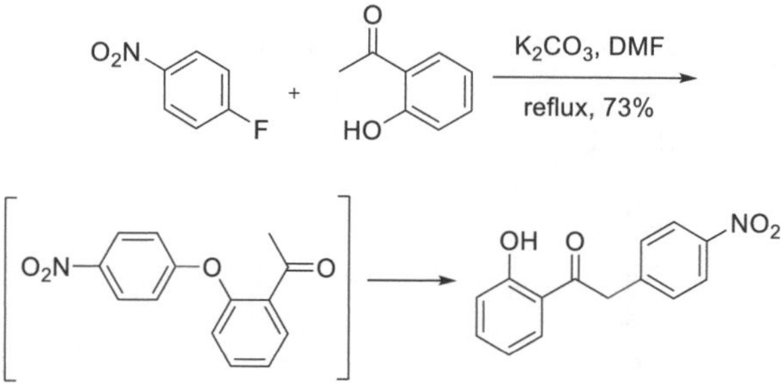

Example 4[10]

Example 4, Truce–Smile rearrangement of substituted phenylethers[11]

Example 5, A benzyne Truce–Smile rearrangement[12]

Example 6, Copper-catalyzed one-pot approach to α-aryl amidines via Truce–Smile rearrangement[13]

References

1. Truce, W. E.; Ray, W. J. Jr.; Norman, O. L.; Eickemeyer, D. B. *J. Am. Chem. Soc.* **1958,** *80,* 3625–3629. William E. Truce was a professor at Purdue University.
2. Truce, W. E.; Hampton, D. C. *J. Org. Chem.* **1963,** *28,* 2276–2279.
3. Bayne, D. W; Nicol, A. J.; Tennant, G. *J. Chem. Soc., Chem. Comm.* **1975,** *19,* 782–783.
4. Fukazawa, Y.; Kato, N.; Ito, S.; *Tetrahedron Lett.* **1982,** *23,* 437–438.
5. Hoffman, R. V.; Jankowski, B. C.; Carr, C. S.; Düsler, E. N *J. Org. Chem.* **1986,** *51,* 130–135.
6. Erickson, W. R.; McKennon, M. J. *Tetrahedron Lett.* **2000,** *41,* 4541–4544.
7. Kimbaris, A.; Cobb, J.; Tsakonas, G.; Varvounis, G. *Tetrahedron* **2004,** *60,* 8807–8815.
8. Mitchell, L. H.; Barvian, N. C. *Tetrahedron Lett.* **2004,** *45,* 5669–5672.
9. Snape, T. J. *Chem. Soc. Rev.* **2008,** *37,* 2452–2458. (Review).
10. Snape, T. J. *Synlett* **2008,** 2689–2691.
11. Kosowan, J. R.; W'Giorgis, Z.; Grewal, R.; Wood, T. E. *Org. Biomol. Chem.* **2015,** *13,* 6754–6765.
12. Holden, C. M.; Sohel, S. M. A.; Greaney, M. F. *Angew. Chem. Int. Ed.* **2016,** *55,* 2450–2453.
13. Huang, Y.; Yi, W.; Sun, Q.; Yi, F. *Adv. Synth. Catal.* **2018,** *360,* 3074–3082.

Sommelet–Hauser Rearrangement

[2,3]-Wittig rearrangement of benzylic quaternary ammonium salts upon treatment with alkali metal amides *via* the ammonium ylide intermediates. *Cf.* Stevens rearrangement.

Example 1, Intermediate investigations[3]

Example 2, Competition between Stevens and Sommelet–Hauser rearrangement[4]

© Springer Nature Switzerland AG 2021
J. J. Li, *Name Reactions*, https://doi.org/10.1007/978-3-030-50865-4_141

Example 3, Diastereoselective Sommelet–Hauser rearrangement[8]

Example 4, Quaternary carbon from a diastereoselective Sommelet–Hauser rearrangement[10]

Example 5, Potential method to make amino acids[12]

Example 6, Ring-strain effects[13]

Example 7, Cu-catalyzed Sommelet–Hauser rearrangement, resulting in de-aromatization[14]

Example 8, The aryne Sommelet–Hauser rearrangement[15]

References

1. (a) Sommelet, M. *Compt. Rend.* **1937**, *205*, 56–58. (b) Kantor, S. W.; Hauser, C. R. *J. Am. Chem. Soc.* **1951**, *73*, 4122–4131. Charles R. Hauser (1900–1970) was a professor at Duke University.
2. Shirai, N.; Sato, Y. *J. Org. Chem.* **1988**, *53*, 194–196.
3. Shirai, N.; Watanabe, Y.; Sato, Y. *J. Org. Chem.* **1990**, *55*, 2767–2770.
4. Tanaka, T.; Shirai, N.; Sugimori, J.; Sato, Y. *J. Org. Chem.* **1992**, *57*, 5034–5036.
5. Klunder, J. M. *J. Heterocycl. Chem.* **1995**, *32*, 1687–1691.
6. Maeda, Y.; Sato, Y. *J. Org. Chem.* **1996**, *61*, 5188–5190.
7. Endo, Y.; Uchida, T.; Shudo, K. *Tetrahedron Lett.* **1997**, *38*, 2113–2116.
8. Hanessian, S.; Talbot, C.; Saravanan, P. *Synthesis* **2006**, 723–734.
9. Liao, M.; Peng, L.; Wang, J. *Org. Lett.* **2008**, *10*, 693–696.
10. Tayama, E.; Orihara, K.; Kimura, H. *Org. Biomol. Chem.* **2008**, *6*, 3673–3680.
11. Zografos, A. L. In *Name Reactions in Heterocyclic Chemistry-II*, Li, J. J., Ed.; Wiley: Hoboken, NJ, 2011, pp 197–206. (Review).
12. Tayama, E.; Sato, R.; Takedachi, K.; Iwamoto, H.; Hasegawa, E. *Tetrahedron* **2012**, *68*, 4710–4718.
13. Tayama, E.; Watanabe, K.; Matano, Y. *Eur. J. Org. Chem.* **2016**, 3631–3641.
14. Pan, C.; Guo, W.; Gu, Z. *Chem. Sci.* **2018**, *9*, 5850–5854.
15. Roy, T.; Gaykar, R. N.; Bhattacharjee, S.; Biju, A. T. *Chem. Commun.* **2019**, *55*, 3004–3007.
16. Tayama, E.; Hirano, K.; Baba, S. *Tetrahedron* **2020**, *76*, 131064.

Sonogashira Reaction

Pd/Cu-catalyzed cross-coupling of organohalides with terminal alkynes. *Cf.* Cadiot–Chodkiewicz coupling and Castro–Stephens reaction. The Castro–Stephens coupling uses stoichiometric copper, whereas the Sonogashira variant uses catalytic palladium and copper.

Note that Et$_3$N may reduce Pd(II) to Pd(0) as well, where Et$_3$N is oxidized to the iminium ion at the same time:

Example 1[2]

J. J. Li, *Name Reactions*, https://doi.org/10.1007/978-3-030-50865-4_142

Example 2, Chemoselectivity for 2,5-dibromopyridine[3]

Example 3, Homocoupling in the ionic liquid[8]

Example 4, Total synthesis of the maduropeptin chromophore aglycon[9]

Example 5, Pd-catalyzed decarboxylative Sonogashira reaction via a decarboxylative bromination[14]

Example 6, Sila-Sonogashira reaction[15]

Example 7, One-pot sequential Sonogashira and Cacchi reactions[16]

References

1. (a) Sonogashira K.; Tohda, Y.; Hagihara, N. *Tetrahedron Lett.* **1975,** *50,* 4467–4470. Kenkichi Sonogashira was a professor at Fukui University. Richard Heck also discovered the same transformation using palladium but without the use of copper: *J. Organomet. Chem.* **1975,** *93,* 259–263.

2. Sakamoto, T.; Nagano, T.; Kondo, Y.; Yamanaka, H. *Chem. Pharm. Bull.* **1988,** *36,* 2248–2252.

3. Ernst, A.; Gobbi, L.; Vasella, A. *Tetrahedron Lett.* **1996,** *37,* 7959–7962.

4. Hundermark, T.; Littke, A.; Buchwald, S. L.; Fu, G. C. *Org. Lett.* **2000,** *2,* 1729–1731.

5. Batey, R. A.; Shen, M.; Lough, A. J. *Org. Lett.* **2002,** *4,* 1411–1414.

6. Sonogashira, K. In *Metal-Catalyzed Cross-Coupling Reactions*; Diederich, F.; de Meijere, A., Eds.; Wiley-VCH: Weinheim, **2004**; *Vol. 1,* 319. (Review).

7. Lemhadri, M.; Doucet, H.; Santelli, M. *Tetrahedron* **2005,** *61,* 9839–9847.

8. Li, Y.; Zhang, J.; Wang, W.; Miao, Q.; She, X.; Pan, X. *J. Org. Chem.* **2005,** *70,* 3285–3287.

9. Komano, K.; Shimamura, S.; Inoue, M.; Hirama, M. *J. Am. Chem. Soc.* **2007,** *129,* 14184–14186.

10. Nakatsuji, H.; Ueno, K.; Misaki, T.; Tanabe, Y. *Org. Lett.* **2008,** *10,* 2131–2134.

11. Gray, D. L. *Sonogashira Reaction.* In *Name Reactions for Homologations-Part II*; Li, J. J., Ed.; Wiley: Hoboken, NJ, **2009,** pp 100–133. (Review).

12. Shigeta, M.; Watanabe, J.; Konishi, G.-i. *Tetrahedron Lett.* **2013,** *54,* 1761–1764.

13. Karak, M.; Barbosa, L. C. A.; Hargaden, G. C. *RSC Adv.* **2014,** *4,* 53442–53466. (Review).

14. Jiang, Q.; Li, H.; Zhang, X.; Xu, B.; Su, W. *Org. Lett.* **2018,** *20,* 2424–2427.

15. Capani, J. S.; Cochran, J. E.; Liang, J. *J. Org. Chem.* **2019,** *84,* 9378–9384.

16. Li, J.; Smith, D.; Krishnananthan, S.; Mathur, A. *Org. Process Res. Dev.* **2020,** *24,* 454–458.

Stetter Reaction

1,4-Dicarbonyl derivatives from aldehydes and α,β-unsaturated ketones and esters. The thiazolium catalyst serves as a safe surrogate for $^-$CN. Also known as the Michael–Stetter reaction. *Cf.* Benzoin condensation.

Example 1, Intramolecular Stetter reaction[2]

© Springer Nature Switzerland AG 2021
J. J. Li, *Name Reactions*, https://doi.org/10.1007/978-3-030-50865-4_143

Example 2[3]

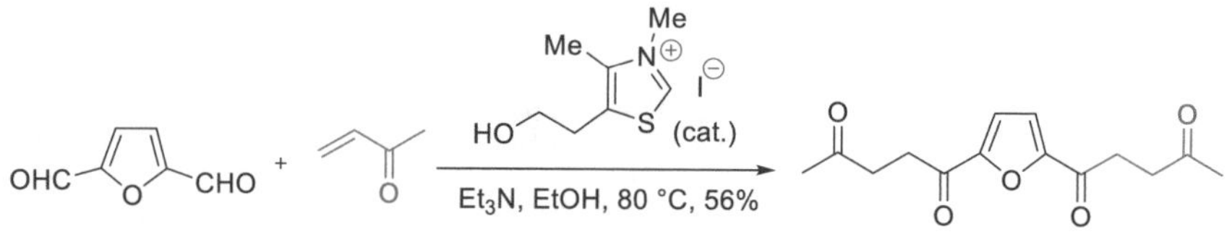

Example 3[5]

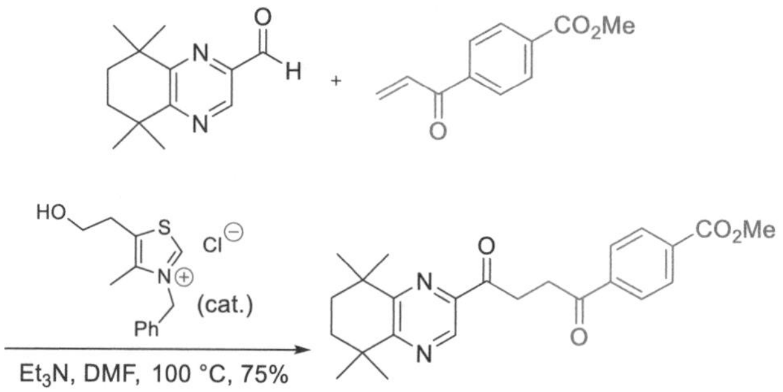

Example 4, Sila-Stetter reaction[9]

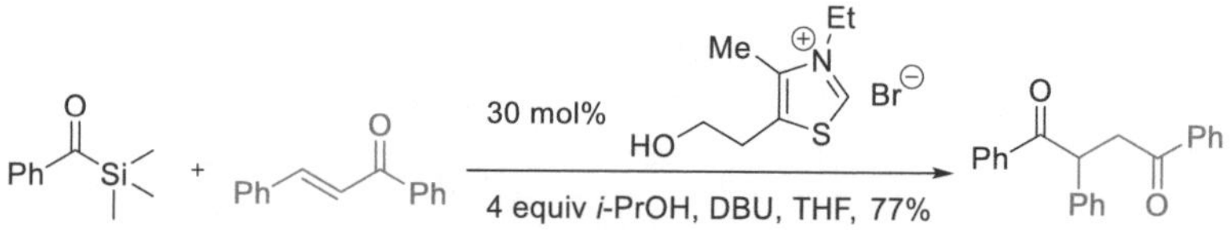

Example 5, NHC-catalyzed intramolecular asymmetric Stetter reaction under solvent-free conditions[13]

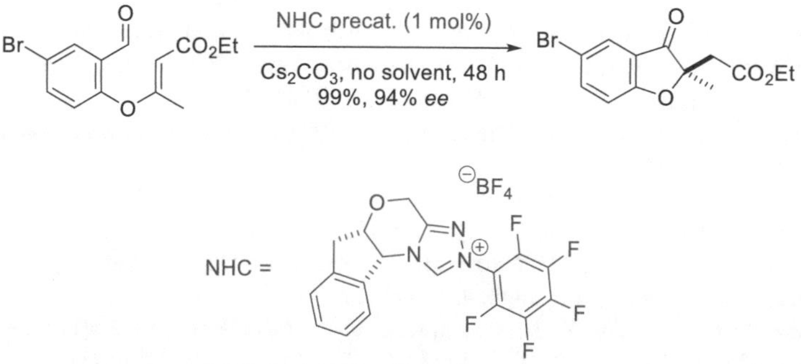

NHC = N-Heterocyclic carbene

Example 6, Construction of bisbenzopyrone via N-heterocyclic carbene-catalyzed intramolecular hydroacylation-Stetter reaction cascade[14]

Example 7, Intramolecular Stetter reaction[15]

References

1. (a) Stetter, H.; Schreckenberg, H. *Angew. Chem.* **1973**, *85*, 89. Hermann Stetter (1917–1993), born in Bonn, Germany, was a chemist at Technische Hochschule Aachen in West Germany. (b) Stetter, H. *Angew. Chem.* **1976**, *88*, 695–704. (Review). (c) Stetter, H.; Kuhlmann, H.; Haese, W. *Org. Synth.* **1987**, *65*, 26.
2. Trost, B. M.; Shuey, C. D.; DiNinno, F., Jr.; McElvain, S. S. *J. Am. Chem. Soc.* **1979**, *101*, 1284–1285.
3. El-Haji, T.; Martin, J. C.; Descotes, G. *J. Heterocycl. Chem.* **1983**, *20*, 233–235.
4. Harrington, P. E.; Tius, M. A. *Org. Lett.* **1999**, *1*, 649–651.
5. Kikuchi, K.; Hibi, S.; Yoshimura, H.; Tokuhara, N.; Tai, K.; Hida, T.; Yamauchi, T.; Nagai, M. *J. Med. Chem.* **2000**, *43*, 409–419.
6. Kobayashi, N.; Kaku, Y.; Higurashi, K. *Bioorg. Med. Chem. Lett.* **2002**, *12*, 1747–1750.
7. Read de Alaniz, J.; Rovis, T. *J. Am. Chem. Soc.* **2005**, *127*, 6284–6289.
8. Reynolds, N. T.; Rovis, T. *Tetrahedron* **2005**, *61*, 6368–6378.

9. Mattson, A. E.; Bharadwaj, A. R.; Zuhl, A. M.; Scheidt, K. A. *J. Org. Chem.* **2006,** *71*, 5715–5724.

10. Cee, V. J. *Stetter Reaction*. In *Name Reactions for Homologations-Part I*; Li, J. J., Ed.; Wiley: Hoboken, NJ, **2009,** pp 576–587. (Review).

11. Zhang, J.; Xing, C.; Tiwari, B.; Chi, Y. R. *J. Am. Chem. Soc.* **2013,** *135*, 8113–8116.

12. Yetra, S. R.; Patra, A.; Biju, A. T. *Synth.* **2015,** *47*, 1357–1378. (Review).

13. Ema, T.; Nanjo, Y.; Shiratori, S.; Terao, Y.; Kimura, R. *Org. Lett.* **2016,** *18*, 5764–5767.

14. Zhao, M.; Liu, J.-L.; Liu, H.-F.; Chen, J.; Zhou, L. *Org. Lett.* **2018,** *20*, 2676–2679.

15. Hsu, D.-S.; Cheng, C.-Y. *J. Org. Chem.* **2019,** *84*, 10832–10842.

16. Bae, C.; Park, E.; Cho, C.-G.; Cheon, C.-H. *Org. Lett.* **2020,** *22*, 2354–2358.

Stevens Rearrangement

A quaternary ammonium salt containing an electron-withdrawing group Z on one of the carbons attached to the nitrogen is treated with a strong base to give a rearranged tertiary amine. *Cf.* Sommelet–Hauser rearrangement.

The contemporary radical mechanism:

The original ionic mechanism:

Example 1, Stevens Rearrangement/Reduction Sequence[10]

J. J. Li, *Name Reactions*, https://doi.org/10.1007/978-3-030-50865-4_144

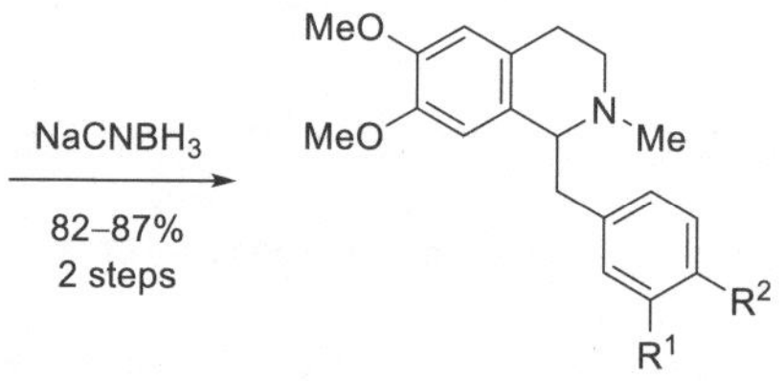

$$\xrightarrow[\substack{82\text{–}87\% \\ 2\text{ steps}}]{NaCNBH_3}$$

Example 2, The aryne-induced enantiospecific [2,3]-Stevens rearrangement[11]

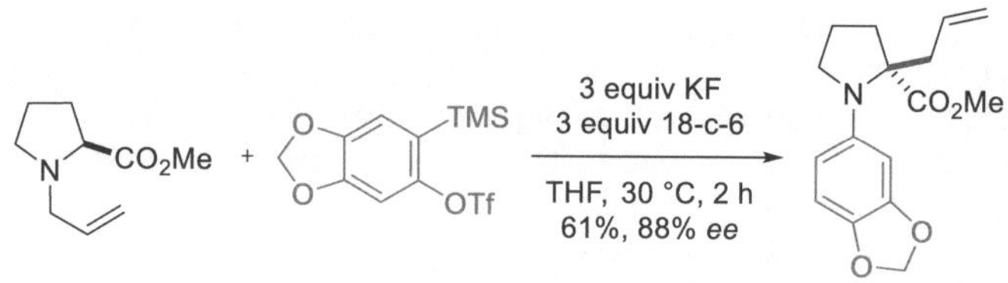

3 equiv KF
3 equiv 18-c-6

THF, 30 °C, 2 h
61%, 88% ee

Example 3, Michael addition/[1,2]-Stevens rearrangement[12]

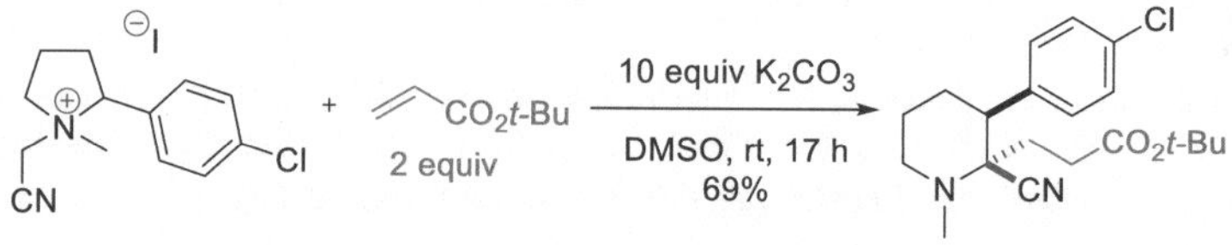

10 equiv K_2CO_3

DMSO, rt, 17 h
69%

2 equiv

Example 4, Construction of Cα-substituted prolinate via [2,3]-Stevens rearrangement[13]

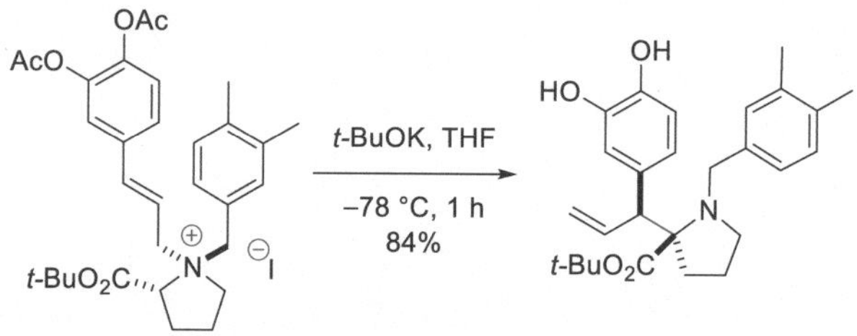

t-BuOK, THF

−78 °C, 1 h
84%

Example 5, Epoxide-mediated Stevens rearrangement of L-pipecolinic acid derivatives[14]

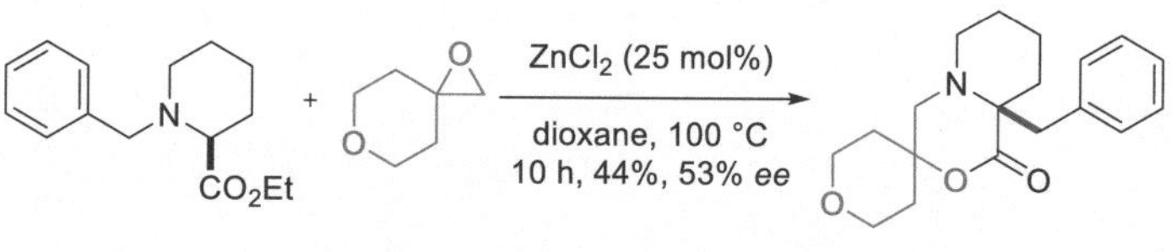

$ZnCl_2$ (25 mol%)

dioxane, 100 °C
10 h, 44%, 53% ee

References

1. Stevens, T. S.; Creighton, E. M.; Gordon, A. B.; MacNicol, M. *J. Chem. Soc.* **1928**, 3193–3197.
2. Schöllkopf, U.; Ludwig, U.; Ostermann, G.; Patsch, M. *Tetrahedron Lett.* **1969**, *10*, 3415–3418.
3. Pine, S. H.; Catto, B. A.; Yamagishi, F. G. *J. Org. Chem.* **1970**, *35*, 3663–3665. (Mechanism).
4. Doyle, M. P.; Ene, D. G.; Forbes, D. C.; Tedrow, J. S. *Tetrahedron Lett.* **1997**, *38*, 4367–4370.
5. Makita, K.; Koketsu, J.; Ando, F.; Ninomiya, Y.; Koga, N. *J. Am. Chem. Soc.* **1998**, *120*, 5764–5770.
6. Feldman, K. S.; Wrobleski, M. L. *J. Org. Chem.* **2000**, *65*, 8659-8668.
7. Kitagaki, S.; Yanamoto, Y.; Tsutsui, H.; Anada, M.; Nakajima, M.; Hashimoto, S. *Tetrahedron Lett.* **2001**, *42*, 6361–6364.
8. Knapp, S.; Morriello, G. J.; Doss, G. A. *Tetrahedron Lett.* **2002**, *43*, 5797–5800.
9. Hanessian, S.; Parthasarathy, S.; Mauduit, M.; Payza, K. *J. Med. Chem.* **2003**, *46*, 34–38.
10. Pacheco, J. C. O.; Lahm, G.; Opatz, T. *J. Org. Chem.* **2013**, *78*, 4985–4992.
11. Roy, T.; Thangaraj, M.; Kaicharla, T.; Kamath, R. V.; Gonnade, R. G.; Biju, A. T. *Org. Lett.* **2016**, *18*, 5428–5431.
12. Kowalkowska, A.; Jończyk, A.; Maurin, J. K. *J. Org. Chem.* **2018**, *83*, 4105–4110.
13. Jin, Y.-X.; Yu, B.-K.; Qin, S.-P.; Tian, S.-K. *Chem. Eur. J.* **2019**, *52*, 5169–5172.
14. Baidilov, D. *Synthesis* **2020**, *52*, 21–26. (Review on mechanism).

Stille Coupling

Palladium-catalyzed cross-coupling reaction of organostannanes with organic halides, triflates, *etc*. For the catalytic cycle, see Kumada coupling.

$$R{-}X \; + \; R^1{-}Sn(R^2)_3 \xrightarrow{\text{Pd(0)}} R{-}R^1 \; + \; X{-}Sn(R^2)_3$$

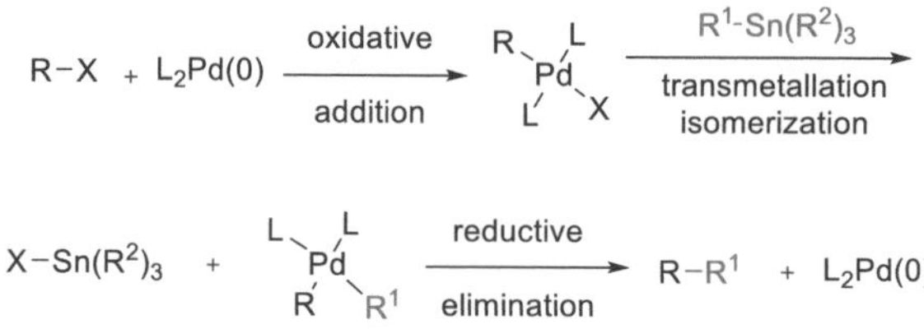

$$R{-}X \; + \; L_2Pd(0) \xrightarrow[\text{addition}]{\text{oxidative}} \underset{L'\,\diagdown\,X}{\overset{R\,\diagdown\,L}{Pd}} \xrightarrow[\substack{\text{transmetallation}\\\text{isomerization}}]{R^1{-}Sn(R^2)_3}$$

$$X{-}Sn(R^2)_3 \; + \; \underset{R'\,\diagup\,R^1}{\overset{L\,\diagup\,L}{Pd}} \xrightarrow[\text{elimination}]{\text{reductive}} R{-}R^1 \; + \; L_2Pd(0)$$

Example 1[4]

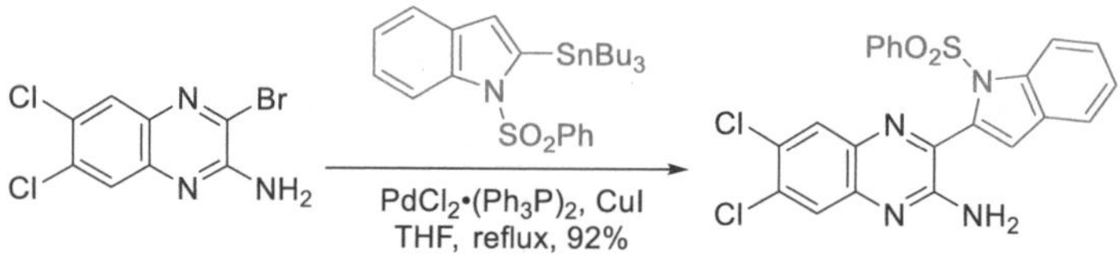

Example 2[5]

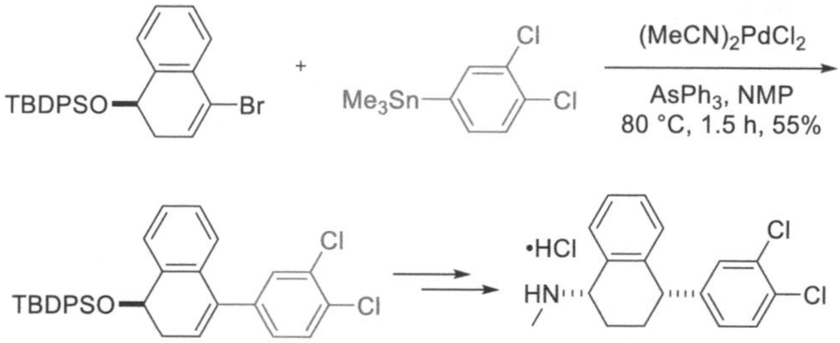

sertraline (Zoloft)

Example 3, π-Allyl Stille coupling[8]

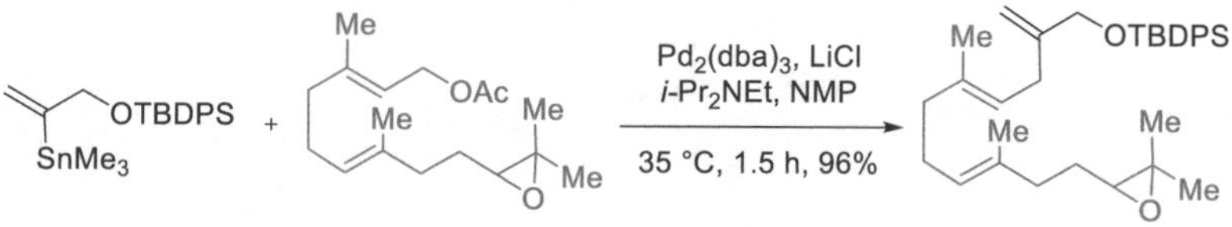

© Springer Nature Switzerland AG 2021
J. J. Li, *Name Reactions*, https://doi.org/10.1007/978-3-030-50865-4_145

Example 4, In total synthesis[9]

$$\xrightarrow[\text{DMF, 32\%}]{\text{Pd(PPh}_3)_4\text{, CuTC}}$$

Example 5, In total synthesis[11]

$$\xrightarrow[\substack{\text{MW, 150 °C, 23 min.}\\ \text{73\%}}]{\substack{\text{2.5 mol\% Pd}_2\text{(dba)}_3\\ \text{10 mol\% AsPh}_3\text{, THF}}}$$

Example 6, In total synthesis[13]

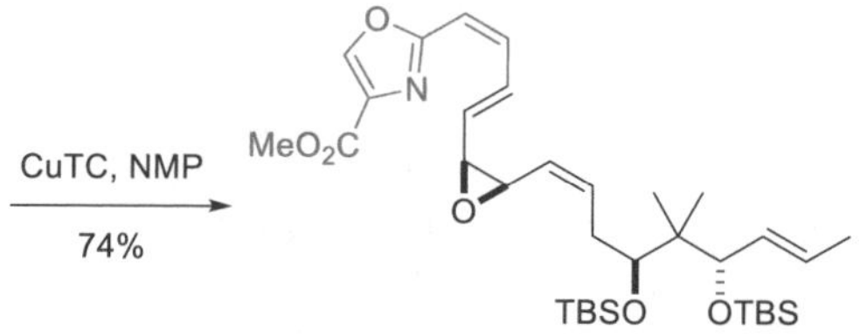

CuTC, NMP

74%

Example 7, Allylic *sp³* **stannane**[14]

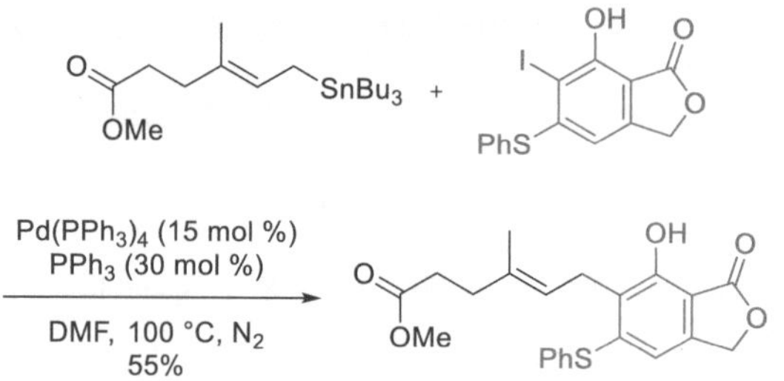

Pd(PPh₃)₄ (15 mol %)
PPh₃ (30 mol %)

DMF, 100 °C, N₂
55%

Example 8[15]

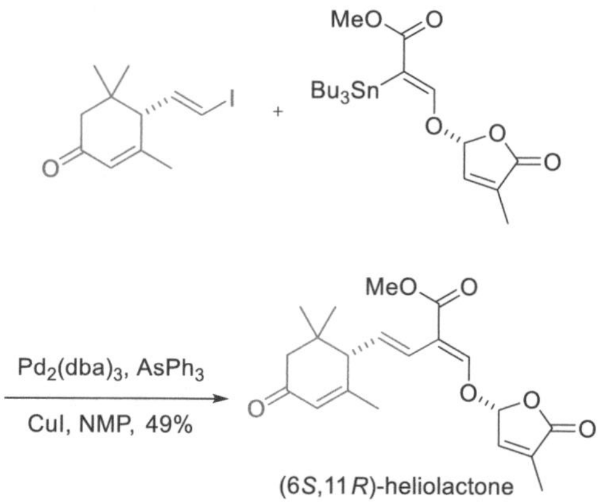

Pd₂(dba)₃, AsPh₃

CuI, NMP, 49%

(6*S*,11*R*)-heliolactone

References

1. (a) Milstein, D.; Stille, J. K. *J. Am. Chem. Soc.* **1978,** *100,* 3636–3638. John Kenneth Stille (1930–1989) was born in Tucson, Arizona. He developed the reaction bearing his name at Colorado State University. At the height of his career, Stille unfortunately died of an airplane accident returning from an ACS meeting. (b) Milstein, D.; Stille, J. K. *J. Am. Chem. Soc.* **1979,** *101,* 4992–4998. (c) Stille, J. K. *Angew. Chem. Int. Ed.* **1986,** *25,* 508–524.
2. Farina, V.; Krishnamurphy, V.; Scott, W. J. *Org. React.* **1997,** *50,* 1–652. (Review).
3. Duncton, M. A. J.; Pattenden, G. *J. Chem. Soc., Perkin Trans. 1* **1999,** 1235–1249. (Review on the intramolecular Stille reaction).

4.   Li, J. J.; Yue, W. S. *Tetrahedron Lett.* **1999,** *40,* 4507–4510.
5.   Lautens, M.; Rovis, T. *Tetrahedron,* **1999,** *55,* 8967–8976.
6.   Mitchell, T. N. *Organotin Reagents in Cross-Coupling Reactions.* In *Metal-Catalyzed Cross-Coupling Reactions* (2nd edn.) De Meijere, A.; Diederich, F. eds., **2004,** 1, 125–161. Wiley-VCH: Weinheim, Germany. (Review).
7.   Schröter, S.; Stock, C.; Bach, T. *Tetrahedron* **2005,** *61,* 2245–2267. (Review).
8.   Snyder, S. A.; Corey, E. J. *J. Am. Chem. Soc.* **2006,** *128,* 740–742.
9.   Roethle, P. A.; Chen, I. T.; Trauner, D. *J. Am. Chem. Soc.* **2007,** *129,* 8960–8961.
10. Mascitti, V. *Stille Coupling.* In *Name Reactions for Homologations-Part I*; Li, J. J., Ed.; Wiley: Hoboken, NJ, **2009,** pp 133–162. (Review).
11. Chandrasoma, N.; Brown, N.; Brassfield, A.; Nerurkar, A.; Suarez, S.; Buszek, K. R. *Tetrahedron Lett.* **2013,** *54,* 913–917.
12. Cordovilla, C.; Bartolome, C.; Martinez-Ilarduya, J. M.; Espinet, P. *ACS Catal.* **2015,** *5,* 3040–3053. (Review).
13. Nicolaou, K. C.; Bellavance, G.; Buchman, M.; Pulukuri, K. K. *J. Am. Chem. Soc.* **2017,** *139,* 15636–15639.
14. Halle, M. B.; Yudhistira, T.; Lee, W.-H.; Mulay, S. V.; Churchill, D. G. *Org. Lett.* **2018,** *20,* 3557–3561.
15. Woo, S.; McErlean, C. S. P. *Org. Lett.* **2019,** *21,* 4215–4218.
16. Drescher, C.; Keller, M.; Potterat, O.; Hamburger, M.; Brueckner, R. *Org. Lett.* **2020,** *22,* 2559–2563.

Strecker Amino Acid Synthesis

Sodium cyanide-promoted condensation of aldehyde, or ketone, with an amine to afford α-amino nitrile, which may be hydrolyzed to an α-amino acid.

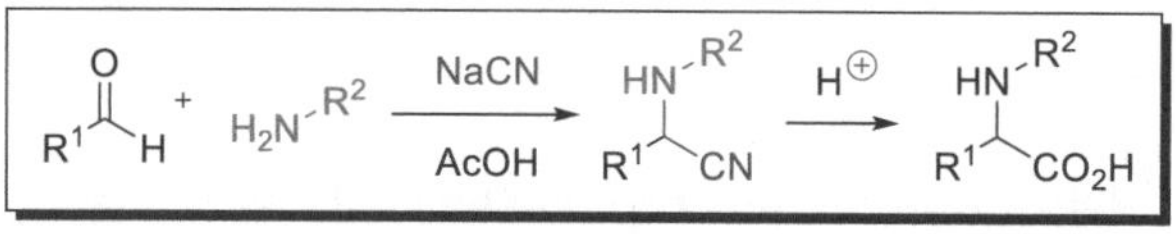

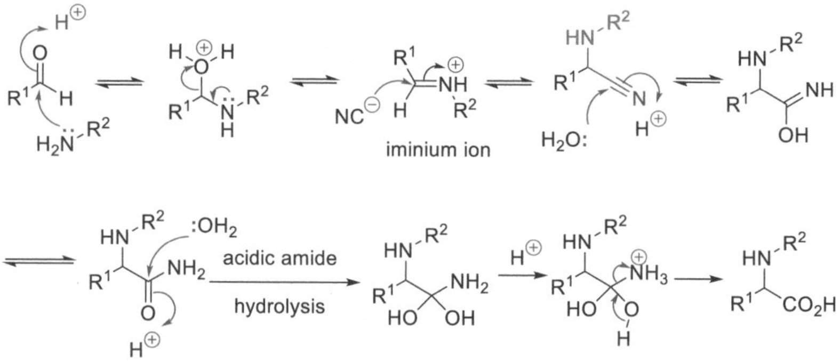

Example 1, Soluble cyanide source[2]

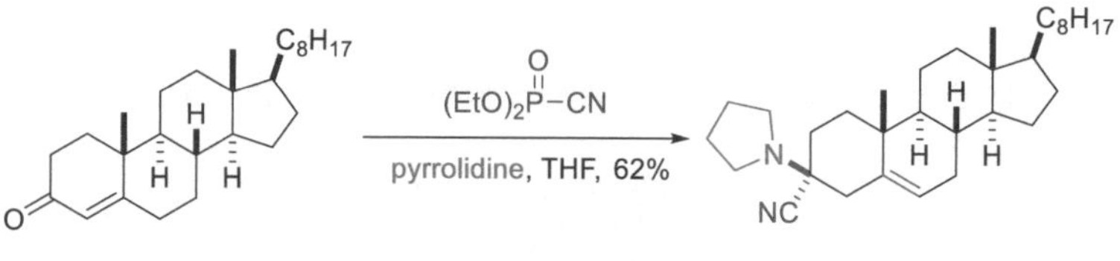

Example 2[3]

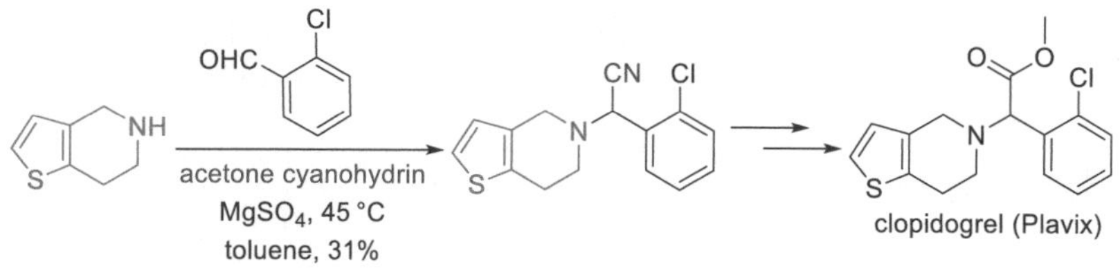

clopidogrel (Plavix)

Example 3[8]

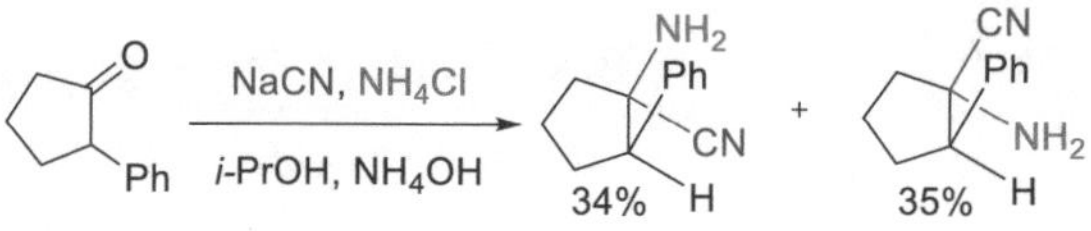

© Springer Nature Switzerland AG 2021

J. J. Li, *Name Reactions*, https://doi.org/10.1007/978-3-030-50865-4_146

Example 4[9]

Example 5, Asymmetric Strecker-type reaction of nitrones[11]

Example 6[13]

Example 7, Iridium-catalyzed reductive Strecker reaction of amides[14]

Vaska's complex = $IrCl(CO)[P(C_6H_5)_3]_2$ =

TMDS = tetramethyldisiloxane =

Example 8, Chemoselective Strecker reaction of acetals catalyzed by MgI₂-etherate[15]

Jie Jack Li   529

Example 9, Borono-Strecker reaction[16]

References

1. Strecker, A. *Ann.* **1850,** *75,* 27–45. Adolph Strecker devised this reaction over 160 years ago. In his paper he described: "The larger crystals of alanine are mother-of-pearl-shiny, hard and crunch between the teeth."
2. Harusawa, S.; Hamada, Y.; Shioiri, T. *Tetrahedron Lett.* **1979,** *20,* 4663–4666.
3. Burgos, A.; Herbert, J. M.; Simpson, I. *J. Labelled Compd. Radiopharm.* **2000,** *43,* 891–898.
4. Ishitani, H.; Komiyama, S.; Hasegawa, Y.; Kobayashi, S. *J. Am. Chem. Soc.* **2000,** *122,* 762–766.
5. Yet, L. *Recent Developments in Catalytic Asymmetric Strecker-Type Reactions,* in *Organic Synthesis Highlights V,* Schmalz, H.-G.; Wirth, T. eds.; Wiley–VCH: Weinheim, Germany, **2003,** pp 187–193. (Review).
6. Meyer, U.; Breitling, E.; Bisel, P.; Frahm, A. W. *Tetrahedron: Asymmetry* **2004,** *15,* 2029–2037.
7. Huang, J.; Corey, E. J. *Org. Lett.* **2004,** *6,* 5027–5029.
8. Cativiela, C.; Lasa, M.; Lopez, P. *Tetrahedron: Asymmetry* **2005,** *16,* 2613–2523.
9. Wrobleski, M. L.; Reichard, G. A.; Paliwal, S.; Shah, S.; Tsui, H.-C.; Duffy, R. A.; Lachowicz, J. E.; Morgan, C. A.; Varty, G. B.; Shih, N.-Y. *Bioorg. Med. Chem. Lett.* **2006,** *16,* 3859–3863.
10. Galatsis, P. *Strecker Amino Acid Synthesis.* In *Name Reactions for Functional Group Transformations*; Li, J. J., Ed.; Wiley: Hoboken, NJ, **2007,** pp 477–499. (Review).
11. Belokon, Y. N.; Hunt, J.; North, M. *Tetrahedron: Asymmetry* **2008,** *19,* 2804–2815.
12. Sakai, T.; Soeta, T.; Endo, K.; Fujinami, S.; Ukaji, Y. *Org. Lett.* **2013,** *15,* 2422–2425.
13. Netz, I.; Kucukdisli, M.; Opatz, T. *J. Org. Chem.* **2015,** *80,* 6864–6869.
14. Fuentes de Arriba, A. L.; Lenci, E.; Sonawane, M.; Formery, O.; Dixon, D. J. *Angew. Chem. Int. Ed.* **2017,** *56,* 3655–3659.
15. Li, H.; Pan, H.; Meng, X.; Zhang, X. *Synth. Commun.* **2020,** *50,* 684–691.
16. Ming, W.; Liu, X.; Friedrich, A.; Krebs, J.; Marder, T. B. *Org. Lett.* **2020,** *22,* 365–370.

Suzuki–Miyaura Coupling

Palladium-catalyzed cross-coupling reaction of organoboranes with organic halides, triflates, *etc*. In the presence of a base (transmetalation is reluctant to occur without the activating effect of a base). For the catalytic cycle, see Kumada coupling.

$$R-X \quad + \quad R^1-B(R^2)_2 \quad \xrightarrow[NaOR^3]{L_2Pd(0)} \quad R-R^1$$

$$R-X + L_2Pd(0) \xrightarrow[\text{addition}]{\text{oxidative}} \underset{\underset{X}{\overset{L}{|}}}{\overset{R}{\underset{|}{Pd}}} + R^1-B(R^2)_2 \xrightarrow[\text{base}]{NaOR^3} R^1\overset{\ominus}{\underset{|}{B}}(R^2)_2 + \underset{\underset{X}{\overset{L}{|}}}{\overset{R}{\underset{|}{Pd}}}$$

$$\xrightarrow[\text{isomerization}]{\text{transmetallation}} R^3O-B(R^2)_2 + \underset{R}{\overset{L}{Pd}}\underset{R^1}{\overset{L}{}} \xrightarrow[\text{elimination}]{\text{reductive}} R-R^1 + L_2Pd(0)$$

Example 1[2]

Example 2[4]

Example 3, Intramolecular Suzuki–Miyaura coupling[8]

J. J. Li, *Name Reactions*, https://doi.org/10.1007/978-3-030-50865-4_147

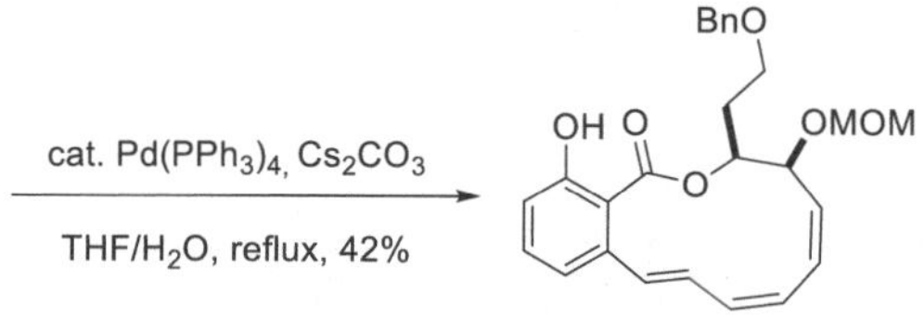
BnO
OH O
OMOM
cat. Pd(PPh3)4, Cs2CO3
THF/H2O, reflux, 42%

Example 4, Stille coupling followed by Suzuki coupling[9]

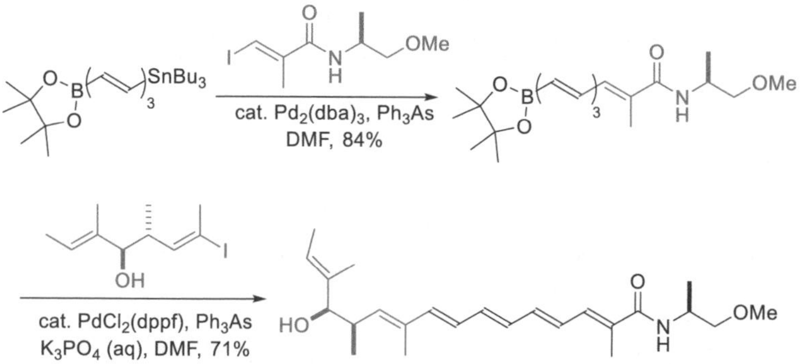
SnBu3
OMe
cat. Pd2(dba)3, Ph3As
DMF, 84%
OH
cat. PdCl2(dppf), Ph3As
K3PO4 (aq), DMF, 71%
HO

Example 5, Nickel-catalyzed Suzuki–Miyaura coupling[12]

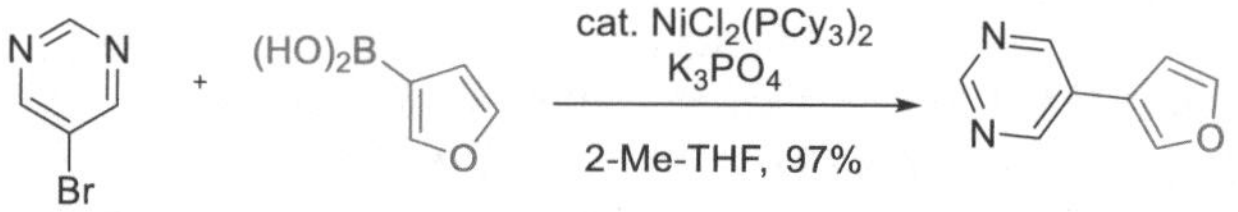
Br
(HO)2B
cat. NiCl2(PCy3)2
K3PO4
2-Me-THF, 97%

Example 6, Suzuki–Miyaura coupling of acyclic amides in catalytic carbon-nitrogen bond cleavage[15]

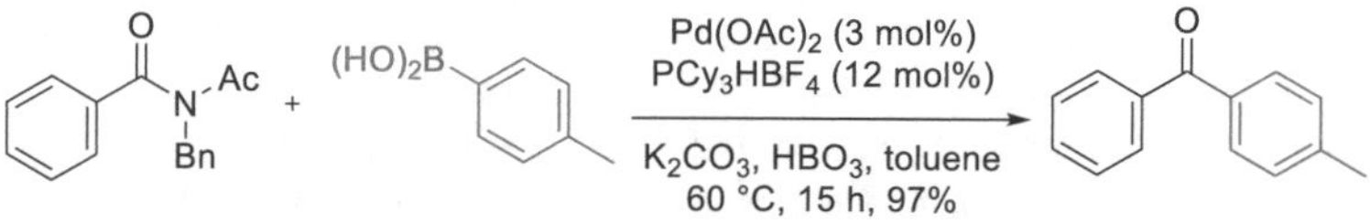
Ac
Bn
(HO)2B
Pd(OAc)2 (3 mol%)
PCy3HBF4 (12 mol%)
K2CO3, HBO3, toluene
60 °C, 15 h, 97%

Example 7, Palladium-catalyzed Suzuki–Miyaura coupling of pyrrolyl sulfonates[16]

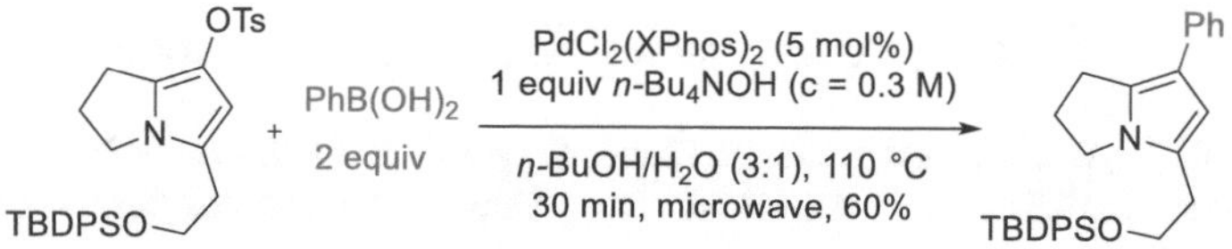
OTs
TBDPSO
PhB(OH)2
2 equiv
PdCl2(XPhos)2 (5 mol%)
1 equiv n-Bu4NOH (c = 0.3 M)
n-BuOH/H2O (3:1), 110 °C
30 min, microwave, 60%
Ph
TBDPSO

Example 8, Sequential Sandmeyer bromination and room-temperature Suzuki–Miyaura coupling (1 kg scale)[17]

References

1. (a) Miyaura, N.; Yamada, K.; Suzuki, A. *Tetrahedron Lett.* **1979**, *36*, 3437–3440. (b) Miyaura, N.; Suzuki, A. *Chem. Commun.* **1979**, 866–867. Akira Suzuki won Nobel Prize in 2010 along with Richard F. Heck and Ei-ichi Negishi "for palladium-catalyzed cross couplings in organic synthesis".
2. Tidwell, J. H.; Peat, A. J.; Buchwald, S. L. *J. Org. Chem.* **1994**, *59*, 7164–7168.
3. Miyaura, N.; Suzuki, A. *Chem. Rev.* **1995**, *95*, 2457–2483. (Review).
4. (a) Kawasaki, I.; Katsuma, H.; Nakayama, Y.; Yamashita, M.; Ohta, S. *Heterocycles* **1998**, *48*, 1887–1901. (b) Kawaski, I.; Yamashita, M.; Ohta, S. *Chem. Pharm. Bull.* **1996**, *44*, 1831–1839.
5. Suzuki, A. In *Metal-catalyzed Cross-coupling Reactions*; Diederich, F.; Stang, P. J., Eds.; Wiley–VCH: Weinhein, Germany, **1998**, 49–97. (Review).
6. Stanforth, S. P. *Tetrahedron* **1998**, *54*, 263–303. (Review).
7. Zapf, A. *Coupling of Aryl and Alkyl Halides with Organoboron Reagents (Suzuki Reaction).* In *Transition Metals for Organic Synthesis* (2nd edn.); Beller, M.; Bolm, C. eds., **2004**, 1, 211–229. Wiley–VCH: Weinheim, Germany. (Review).
8. Molander, G. A.; Dehmel, F. *J. Am. Chem. Soc.* **2004**, *126*, 10313–10318.
9. Coleman, R. S.; Lu, X.; Modolo, I. *J. Am. Chem. Soc.* **2007**, *129*, 3826–3827.
10. Wolfe, J. P.; Nakhla, J. S. *Suzuki Coupling.* In *Name Reactions for Homologations-Part I*; Li, J. J., Ed.; Wiley: Hoboken, NJ, **2009**, pp 163–184. (Review).
11. Weimar, M.; Fuchter, M. J. *Org. Biomol. Chem.* **2013**, *11*, 31–34.
12. Ramgren, S.; Hie, L.; Ye, Y.; Garg, N. K. *Org. Lett.* **2013**, *15*, 3950–3953.
13. Almond-Thynne, J.; Blakemore, D. C.; Pryde, D. C.; Spivey, A. C. *Chem. Sci.* **2017**, *8*, 40–62. (Review).
14. Taheri Kal Koshvandi, A.; Heravi, M. M.; Momeni, T. *Appl. Organomet. Chem.* **2018**, *32(3)*, 1–59. (Review).
15. Liu, C.; Li, G.; Shi, S.; Meng, G.; Lalancette, R.; Szostak, R.; Szostak, M. *ACS Catal.* **2018**, *8*, 9131–9139.
16. Sirindil, F.; Weibel, J.-M.; Pale, P.; Blanc, A. *Org. Lett.* **2019**, *21*, 5542–5546.
17. Schafer, G.; Fleischer, T.; Ahmetovic, M.; Abele, S. *Org. Process Res. Dev.* **2020**, *24*, 228–234.

Swern Oxidation

Oxidation of alcohols to the corresponding carbonyl compounds using $(COCl)_2$, DMSO, and quenching with Et_3N.

sulfur ylide

Example 1[2]

Example 2[3]

J. J. Li, *Name Reactions*, https://doi.org/10.1007/978-3-030-50865-4_148

Example 3[5]

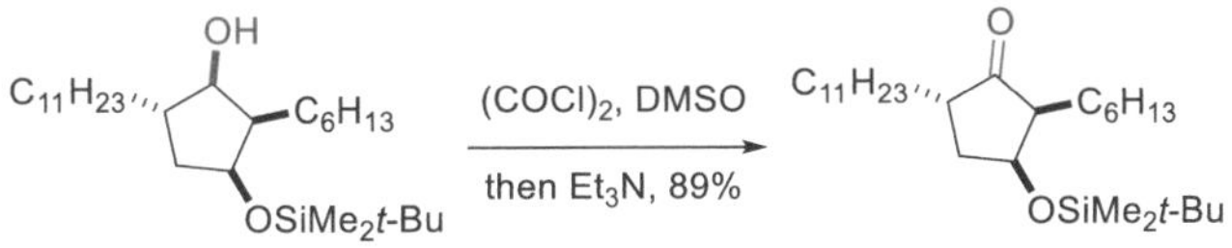

Example 4[7]

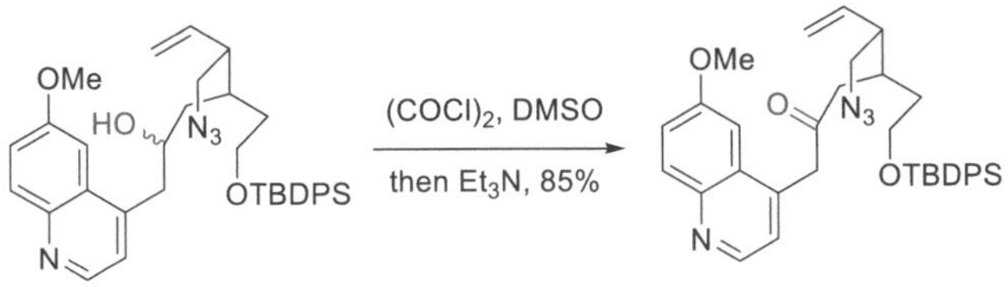

Example 5, Nitriles from primary amides under a catalytic Swern oxidation conditions[12]

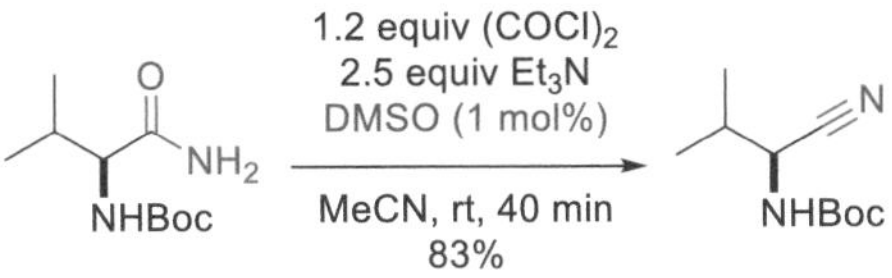

Example 6, Scale-up synthesis of tesirine[13]

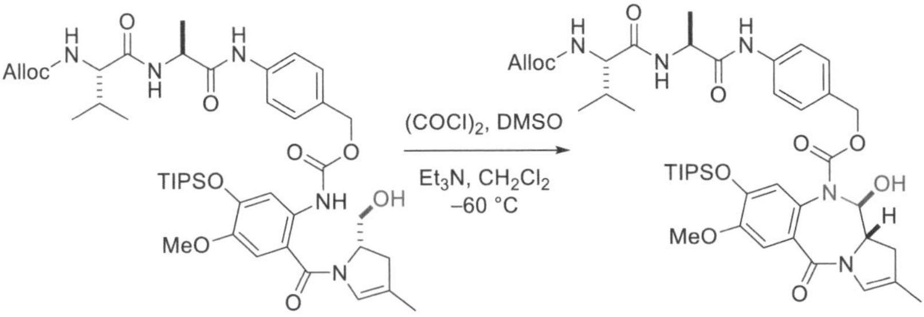

Example 7, Synthesis of resolvin E3[14]

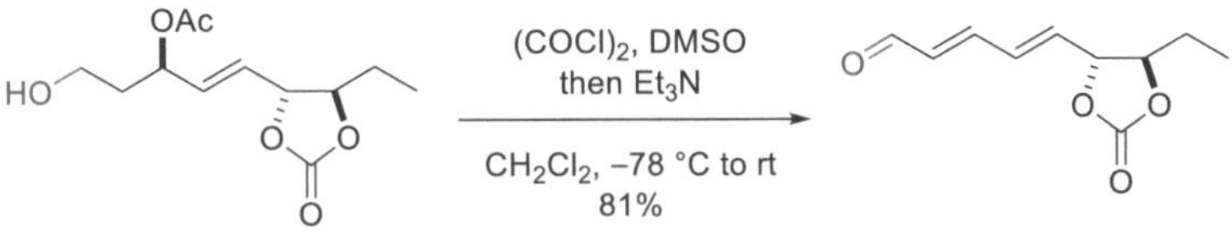

References

1. (a) Huang, S. L.; Omura, K.; Swern, D. *J. Org. Chem.* **1976,** *41,* 3329–3331. (b) Huang, S. L.; Omura, K.; Swern, D. *Synthesis* **1978,** *4,* 297–299. (c) Mancuso, A. J.; Huang, S. L.; Swern, D. *J. Org. Chem.* **1978,** *43,* 2480–2482. Daniel Swern (1916–1982) was a professor at Temple University.

2. Ghera, E.; Ben-David, Y. *J. Org. Chem.* **1988,** *53,* 2972–2979.

3. Smith, A. B., III; Leenay, T. L.; Liu, H. J.; Nelson, L. A. K.; Ball, R. G. *Tetrahedron Lett.* **1988,** *29,* 49–52.

4. Tidwell, T. T. *Org. React.* **1990,** *39,* 297–572. (Review).

5. Chadka, N. K.; Batcho, A. D.; Tang P. C.; Courtney, L. F.; Cook C. M.; Wovliulich, P. M.; Usković, M. R. *J. Org. Chem.* **1991,** *56,* 4714–4718.

6. Harris, J. M.; Liu, Y.; Chai, S.; Andrews, M. D.; Vederas, J. C. *J. Org. Chem.* **1998,** *63,* 2407–2409. (Odorless protocols).

7. Stork, G.; Niu, D.; Fujimoto, R. A.; Koft, E. R.; Bakovec, J. M.; Tata, J. R.; Dake, G. R. *J. Am. Chem. Soc.* **2001,** *123,* 3239–3242.

8. Nishide, K.; Ohsugi, S.-i.; Fudesaka, M.; Kodama, S.; Node, M. *Tetrahedron Lett.* **2002,** *43,* 5177–5179. (Another odorless protocols).

9. Ahmad, N. M. *Swern Oxidation.* In *Name Reactions for Functional Group Transformations*; Li, J. J., Ed.; Wiley: Hoboken, NJ, **2007,** pp 291–308. (Review).

10. Lopez-Alvarado, P; Steinhoff, J; Miranda, S; Avendano, C; Menendez, J. C. *Tetrahedron* **2009,** *65,* 1660–1672.

11. Zanatta, N.; Aquino, E. da C.; da Silva, F. M.; Bonacorso, H. G.; Martins, M. A. P. *Synthesis* **2012,** *44,* 3477–3482.

12. Ding, R.; Liu, Y.; Han, M.; Jiao, W.; Li, J.; Tian, H.; Sun, B. *J. Org. Chem.* **2018,** *83,* 12939–12944.

13. Tiberghien, A. C.; von Bulow, C.; Barry, C.; Ge, H.; Noti, C.; Collet Leiris, F.; McCormick, M.; Howard, P. W.; Parker, J. S. *Org. Process Res. Dev.* **2018,** *22,* 1241–1256.

14. Tanabe, S.; Kobayashi, Y. *Org. Biomol. Chem.* **2019,** *217,* 2393–2402.

15. Zhang, Z.-W.; Li, H.-B.; Li, J.; Wang, C.-C.; Feng, J.; Yang, Y.-H.; Liu, S. *J. Org. Chem.* **2020,** *85,* 537–547.

Takai Reaction

Stereoselective conversion of an aldehyde to the corresponding *E*-vinyl iodide using CHI_3 and $CrCl_2$.

A radical mechanism was proposed[10]

Example 1[2]

Example 2[3]

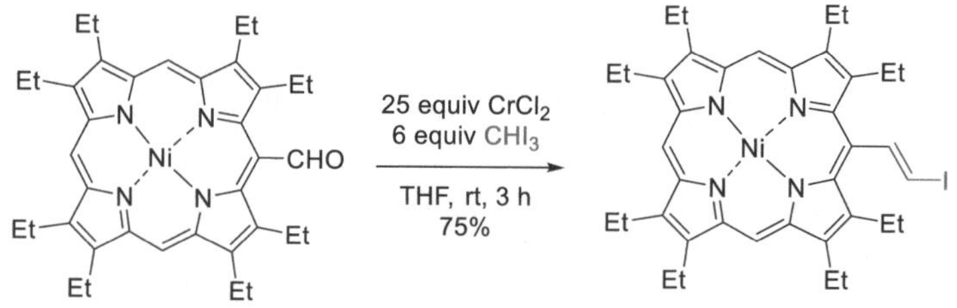

Example 3[4]

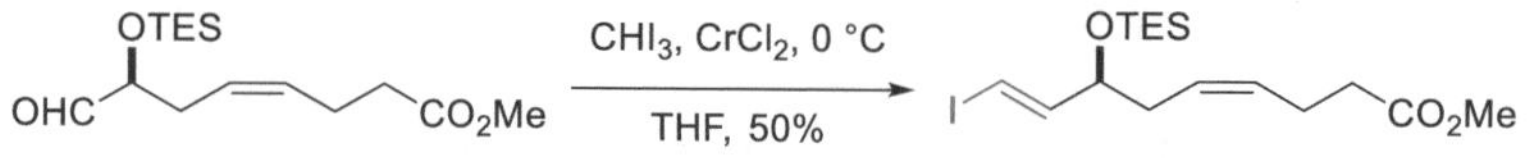

Example 4, A Br/Cl variant[9]

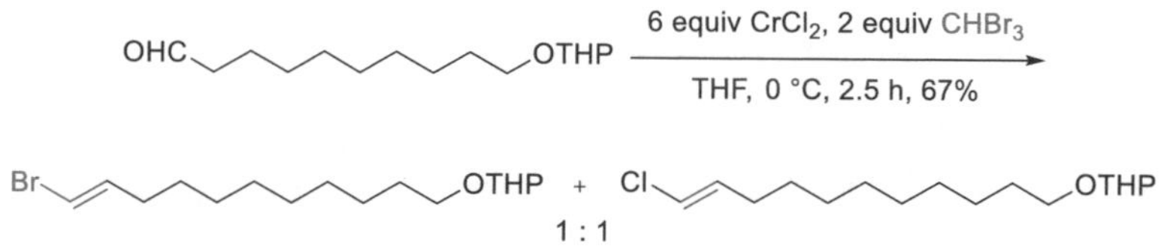

Example 5[10]

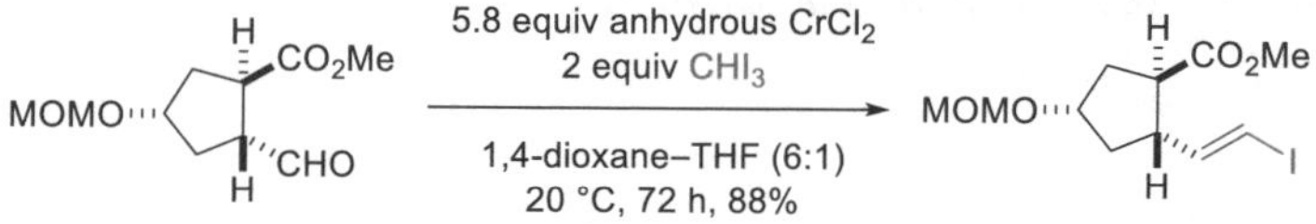

Example 5[10]

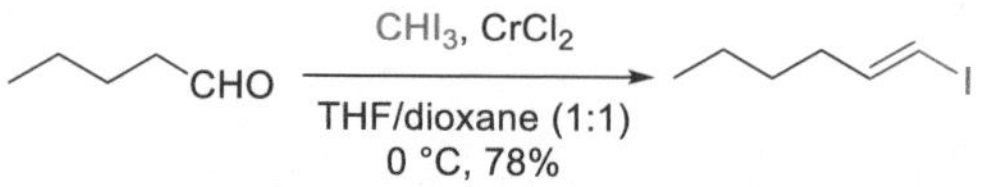

Example 6, Synthesis toward mandelalide A[12]

Example 7, Synthesis toward pestalotioprolide E[13]

Example 8, Synthesis toward raputindole A[14]

Example 9, Synthesis toward MaR2$_{\text{n-3 DPA}}$[15]

References

1.   Takai, K.; Nitta, Utimoto, K. *J. Am. Chem. Soc.* **1986,** *108,* 7408–7410. Kazuhiko Takai was a professor at Kyoto University.
2.   Andrus, M. B.; Lepore, S. D.; Turner, T. M. *J. Am. Chem. Soc.* **1997,** *119,* 12159–12169.
3.   Arnold, D. P.; Hartnell, R. D. *Tetrahedron* **2001,** *57,* 1335–1345.
4.   Rodriguez, A. R.; Spur, B. W. *Tetrahedron Lett.* **2004,** *45,* 8717–8724.
5.   Dineen, T. A.; Roush, W. R. *Org. Lett.* **2004,** *6,* 2043–2046.
6.   Lipomi, D. J.; Langille, N. F.; Panek, J. S. *Org. Lett.* **2004,** *6,* 3533–3536.
7.   Paterson, I.; Mackay, A. C. *Synlett* **2004,** 1359–1362.
8.   Concellón, J. M.; Bernad, P. L.; Méjica, C. *Tetrahedron Lett.* **2005,** *46,* 569–571.

9. Gung, B. W.; Gibeau, C.; Jones, A. *Tetrahedron: Asymmetry* **2005**, *16*, 3107–3114.

10. Legrand, F.; Archambaud, S.; Collet, S.; Aphecetche-Julienne, K.; Guingant, A.; Evain, M. *Synlett* **2008**, 389–393.

11. Saikia, B.; Joymati Devi, T.; Barua, N. C. *Org. Biomol. Chem.* **2013**, *11*, 905–913.

12. Athe, S.; Ghosh, S. *Synthesis* **2016**, *48*, 917–923.

13. Paul, D.; Saha, S.; Goswami, R. K. *Org. Lett.* **2018**, *20*, 4606–4609.

14. Kock, M.; Lindel, T. *Org. Lett.* **2018**, *6*, 5444–5447.

15. Sønderskov, J.; Tungen, J. E.; Palmas, F.; Dalli, J.; Serhan, C. N.; Stenstrøm, Y.; Vidar Hansen, T. *Tetrahedron Lett.* **2020**, *61*, 151510.

Tebbe Reagent

The Tebbe reagent, μ-chlorobis(cyclopentadienyl)(dimethylaluminium)-μ-methylenetitanium, transforms a carbonyl compound to the corresponding *exo*-olefin.

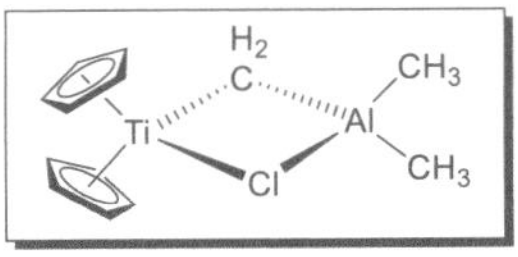

Preparation:[2,6]

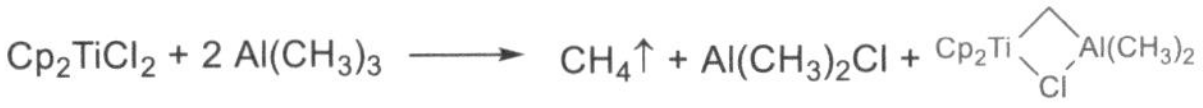
Cp2TiCl2 + 2 Al(CH3)3
CH4↑ + Al(CH3)2Cl + Cp2Ti Al(CH3)2 Cl

Mechanism:[3]

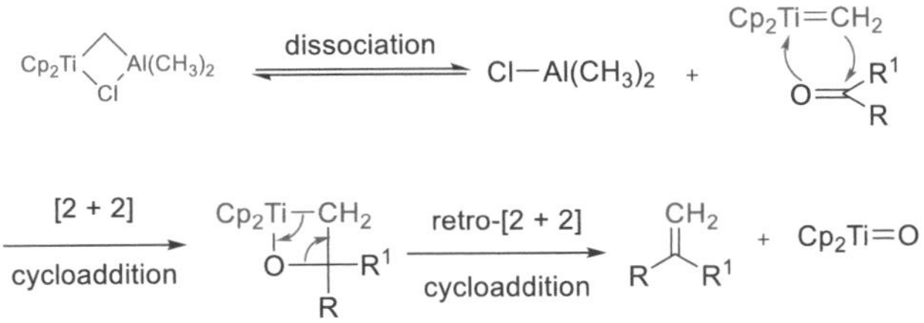
dissociation
Cl—Al(CH3)2
Cp2Ti=CH2
[2 + 2]
cycloaddition
retro-[2 + 2]
cycloaddition
Cp2Ti=O

oxatitanacyclobutane

formation of the strong Ti=O bond is the driving force.

Example 1, Chemoselective for ketone in the presence of an ester[2]

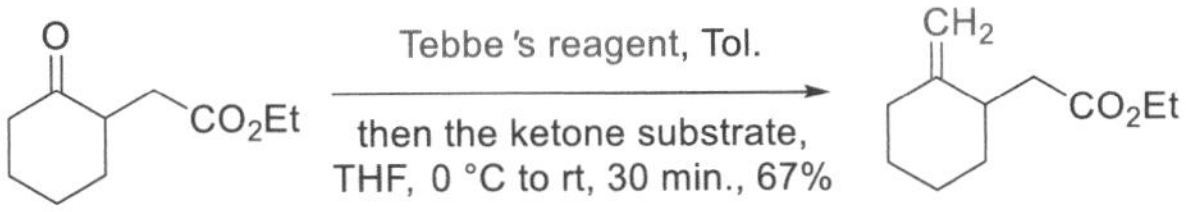
CO2Et
Tebbe 's reagent, Tol.
then the ketone substrate,
THF, 0 °C to rt, 30 min., 67%
CH2
CO2Et

Example 2, Double Tebbe[4]

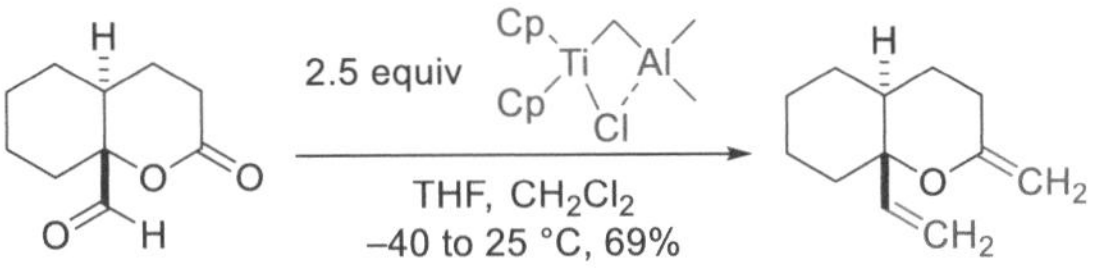
2.5 equiv
THF, CH2Cl2
–40 to 25 °C, 69%
CH2
CH2

J. J. Li, *Name Reactions*, https://doi.org/10.1007/978-3-030-50865-4_150

Example 3, Double Tebbe[5]

Example 4, N-Oxide[6]

Example 5, Amide[11]

Example 6, Olefination of methyl pyropheophorbid-α (chlorophyll derivative)[14]

MeO$_2$C methyl pyropheophorbid-α

MeO$_2$C

Example 7, Synthesis toward a leiodermatolide analog[15]

Example 8, A methylenation–Claisen–methylenation cascade[16]

1. 1.4 equiv Tebbe (0.5 M in PhMe)
   1.4 equiv pyridine, PhMe, –40 °C–rt

2. 3 equiv Tebbe (0.5 M in PhMe)
   PhMe, –40 °C–rt

3. 3 equiv pyridine, 0 °C–rt, 19 h
   72%

References

1. Tebbe, F. N.; Parshall, G. W.; Reddy, G. S. *J. Am. Chem. Soc.* **1978,** *100*, 3611–3613. Fred Tebbe worked at DuPont Central Research.
2. Pine, S. H.; Pettit, R. J.; Geib, G. D.; Cruz, S. G.; Gallego, C. H.; Tijerina, T.; Pine, R. D. *J. Org. Chem.* **1985,** *50*, 1212–1216.
3. Cannizzo, L. F.; Grubbs, R. H. *J. Org. Chem.* **1985,** *50*, 2386–2387.
4. Philippo, C. M. G.; Vo, N. H.; Paquette, L. A. *J. Am. Chem. Soc.* **1991,** *113*, 2762–2764.
5. Ikemoto, N.; Schreiber, L. S. *J. Am. Chem. Soc.* **1992,** *114*, 2524–2536.
6. Pine, S. H. *Org. React.* **1993,** *43*, 1–98. (Review).
7. Nicolaou, K. C.; Koumbis, A. E.; Snyder, S. A.; Simonsen, K. B. *Angew. Chem. Int. Ed.* **2000,** *39*, 2529–2533.
8. Straus, D. A. *Encyclopedia of Reagents for Organic Synthesis;* Wiley & Sons, **2000.** (Review).
9. Payack, J. F.; Hughes, D. L.; Cai, D.; Cottrell, I. F.; Verhoeven, T. R. *Org. Synth., Coll. Vol. 10,* **2004,** p 355.
10. Beadham, I.; Micklefield, J. *Curr. Org. Synth.* **2005,** *2*, 231–250. (Review).
11. Long, Y. O.; Higuchi, R. I.; Caferro, T.s R.; Lau, T. L. S.; Wu, M.; Cummings, M. L.; Martinborough, E. A.; Marschke, K. B.; Chang, W. Y.; Lopez, F. J.; Karanewsky, D. S.; Zhi, L. *Bioorg. Med. Chem. Lett.* **2008,** *18*, 2967–2971.
12. Zhang, J. *Tebbe reagent.* In *Name Reactions for Homolotions-Part I*; Li, J. J., Corey, E. J., Eds., Wiley: Hoboken, NJ, **2009,** pp 319–333. (Review).
13. Yamashita, S.; Suda, N.; Hayashi, Y.; Hirama, M. *Tetrahedron Lett.* **2013,** *54*, 1389–1391.
14. Tamiaki, H.; Tsuji, K.; Machida, S.; Teramura, M.; Miyatake, T. *Tetrahedron Lett.* **2016,** *57*, 788–790.
15. Reiss, A.; Maier, M. E. *Eur. J. Org. Chem.* **2018,** 4246–4255.
16. Domzalska-Pieczykolan, A. M.; Furman, B. *Synlett* **2020,** *31*, 730–736.

Tsuji–Trost Reaction

The Tsuji–Trost reaction is the palladium-catalyzed substitution of allylic leaving groups by carbon nucleophiles. These reactions proceed via π-allylpalladium intermediates.

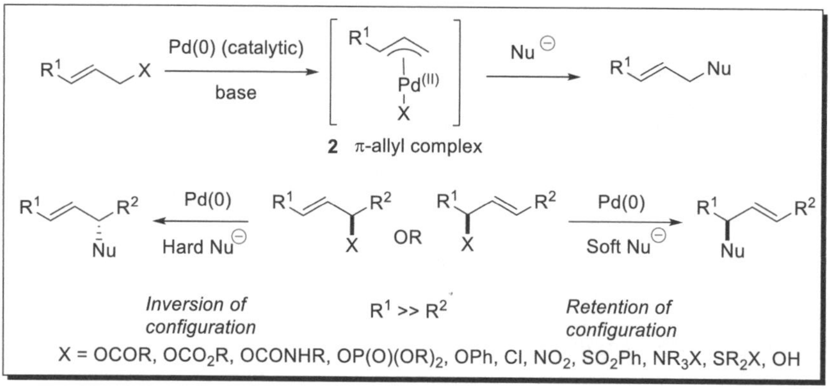

2 π-allyl complex

Inversion of configuration

$R^1 >> R^{2'}$

Retention of configuration

X = OCOR, OCO₂R, OCONHR, OP(O)(OR)₂, OPh, Cl, NO₂, SO₂Ph, NR₃X, SR₂X, OH

The catalytic cycle:

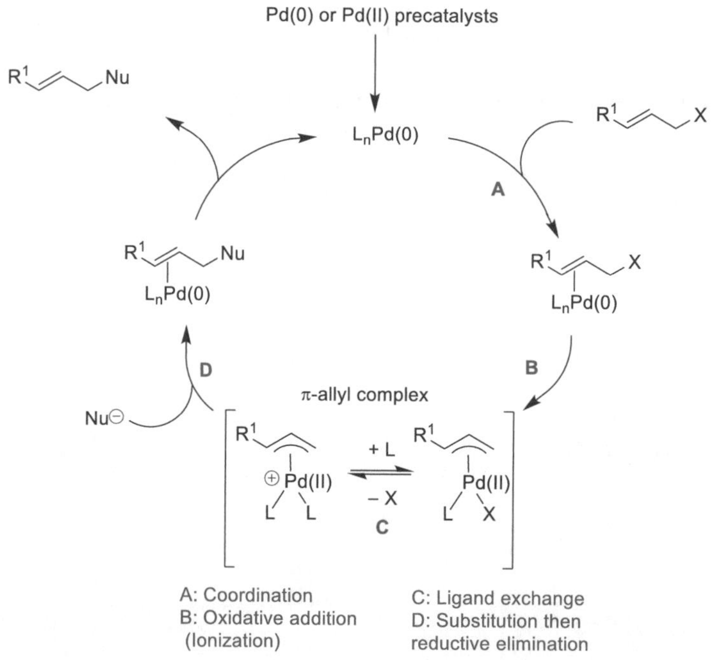

Pd(0) or Pd(II) precatalysts

π-allyl complex

A: Coordination
B: Oxidative addition
   (Ionization)

C: Ligand exchange
D: Substitution then
   reductive elimination

© Springer Nature Switzerland AG 2021
J. J. Li, *Name Reactions*, https://doi.org/10.1007/978-3-030-50865-4_151

Example 1, Allylic ether[3]

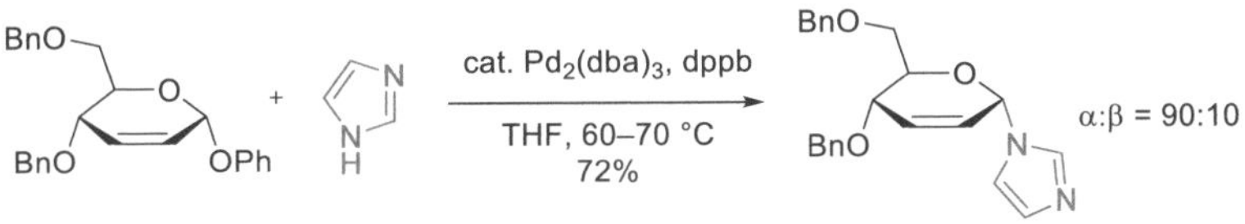

BnO
BnO
OPh
O
N
N
H
cat. $Pd_2(dba)_3$, dppb
THF, 60–70 °C
72%
α:β = 90:10

Example 2, Allylic acetate[3]

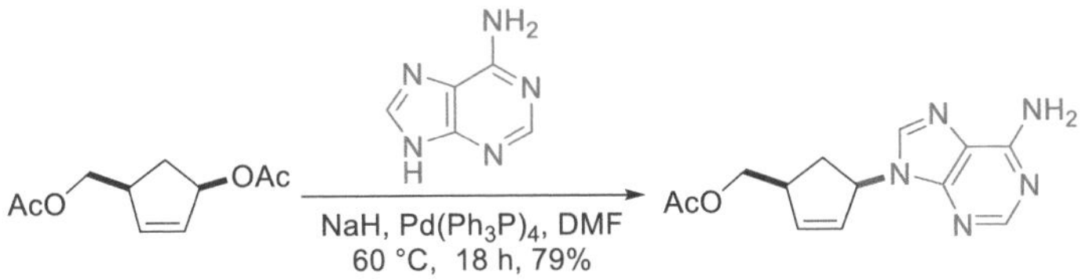

NH_2
AcO
OAc
NaH, $Pd(Ph_3P)_4$, DMF
60 °C, 18 h, 79%

Example 3, Allylic epoxide[5]

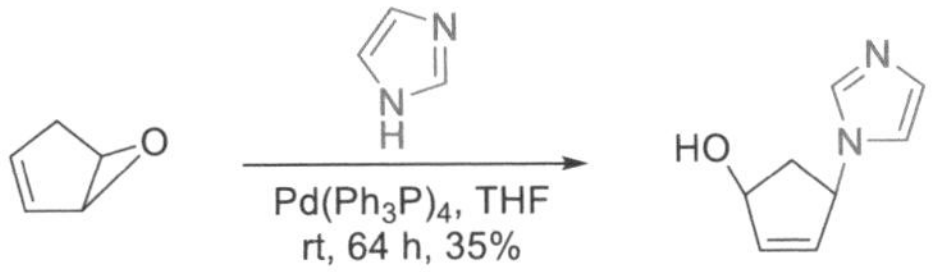

$Pd(Ph_3P)_4$, THF
rt, 64 h, 35%
HO

Example 4, Intramolecular Tsuji–Trost reaction[6]

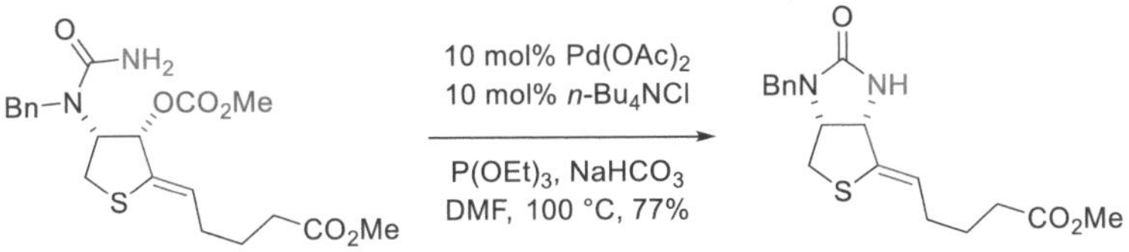

NH_2
Bn–N
OCO_2Me
S
CO_2Me
10 mol% $Pd(OAc)_2$
10 mol% n-Bu_4NCl
$P(OEt)_3$, $NaHCO_3$
DMF, 100 °C, 77%
BnN
NH

Example 5, Intramolecular Tsuji–Trost reaction[7]

Me
TESO
OTES
OMe
Ot-Bu
TMSO
$Pd_2(dba)_3$ (10 mol%)
THF (0.005 M)
40 °C, 80%
t-BuO
OTMS

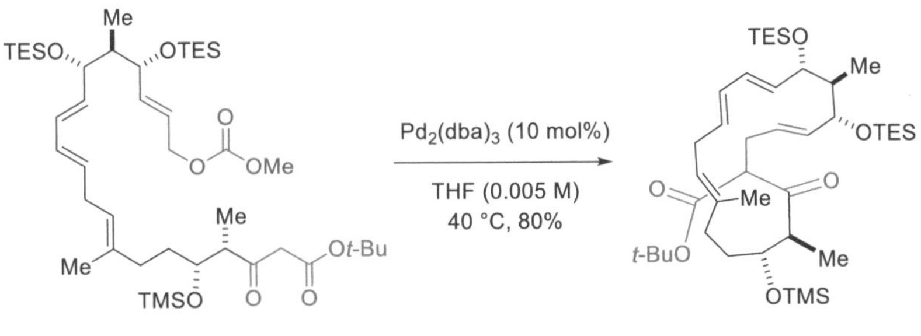

Example 6, Asymmetric Tsuji–Trost reaction[8]

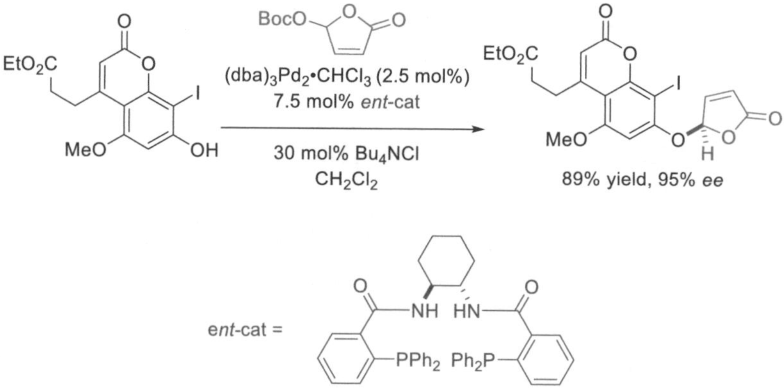
EtO2C
MeO
OH
BocO
(dba)3Pd2•CHCl3 (2.5 mol%)
7.5 mol% ent-cat
30 mol% Bu4NCl
CH2Cl2
89% yield, 95% ee
ent-cat =
NH HN
PPh2 Ph2P

Example 7, Tsuji–Trost decarboxylation–dehydrogenation sequence[12]

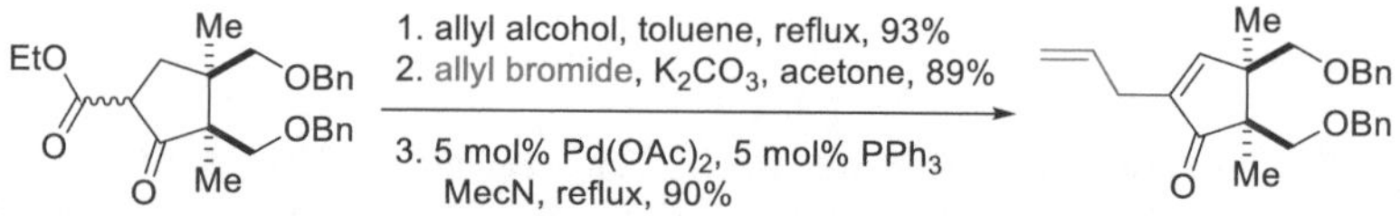
EtO
Me
OBn
OBn
Me
1. allyl alcohol, toluene, reflux, 93%
2. allyl bromide, K2CO3, acetone, 89%
3. 5 mol% Pd(OAc)2, 5 mol% PPh3
MecN, reflux, 90%

Example 8, Intramolecular Tsuji–Trost reaction[13]

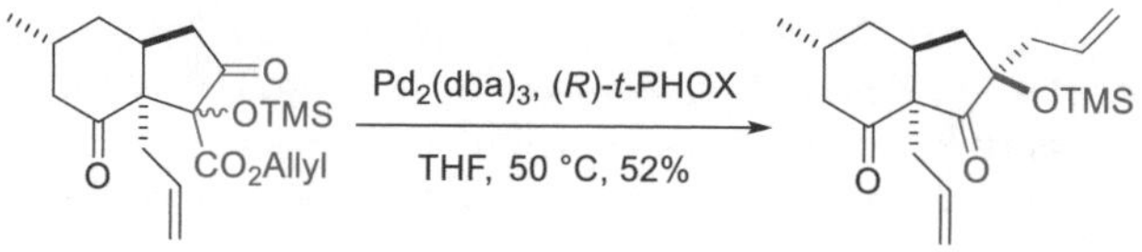
OTMS
CO2Allyl
Pd2(dba)3, (R)-t-PHOX
THF, 50 °C, 52%
OTMS

Example 9, Stereoselective Tsuji–Trost alkylation[14]

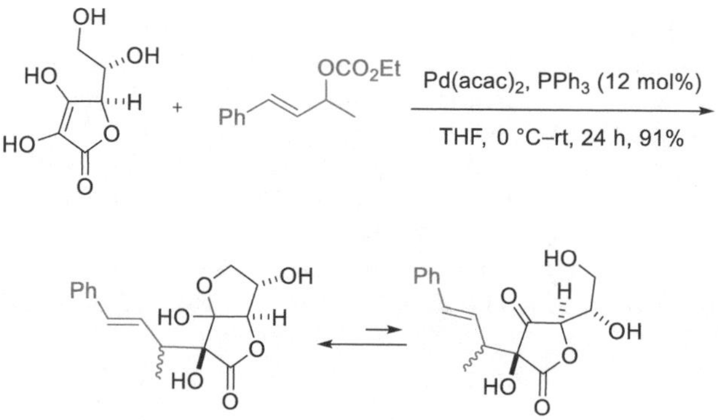
OH
OH
HO
H
HO
OCO2Et
Ph
Pd(acac)2, PPh3 (12 mol%)
THF, 0 °C–rt, 24 h, 91%
Ph
OH
HO
H
HO
Ph
HO
H
OH
HO

O₃, MeOH

then Me₂S
ca. 85%

                                    90   :   10

Example 10, Diethyl malonate as the nucleophile[15]

Pd(dppf)Cl₂ (3 mol%)

THF, rt, 1 h, 89%

References

1. (a) Tsuji, J.; Takahashi, H.; Morikawa, M. *Tetrahedron Lett.* **1965,** *6,* 4387–4388. (b) Tsuji, J. *Acc. Chem. Res.* **1969,** *2,* 144–152. (Review). Jiro Tsuji (1927–), now retired, worked at the Toyo Rayon Company in Japan.
2. Godleski, S. A. In *Comprehensive Organic Synthesis;* Trost, B. M.; Fleming, I., eds.; *Vol. 4.* Chapter 3.3. Pergamon: Oxford, 1991. (Review).
3. Bolitt, V.; Chaguir, B.; Sinou, D. *Tetrahedron Lett.* **1992,** *33,* 2481–2484.
4. Moreno-Mañas, M.; Pleixats, R. In *Advances in Heterocyclic Chemistry;* Katritzky, A. R., ed.; Academic Press: San Diego, **1996,** *66,* 73. (Review).
5. Arnau, N.; Cortes, J.; Moreno-Mañas, M.; Pleixats, R.; Villarroya, M. *J. Heterocycl. Chem.* **1997,** *34,* 233–239.
6. Seki, M.; Mori, Y.; Hatsuda, M.; Yamada, S. *J. Org. Chem.* **2002,** *67,* 5527–5536.
7. Vanderwal, C. D.; Vosburg, D. A.; Weiler, S.; Sorenson, E. J. *J. Am. Chem. Soc.* **2003,** *125,* 5393–5407.
8. Trost, B. M.; Toste, F. D. *J. Am. Chem. Soc.* **2003,** *125,* 3090–3100.
9. Behenna, D. C.; Stoltz, B. M. *J. Am. Chem. Soc.* **2004,** *126,* 15044–15045.
10. Fuchter, M. J. *Tsuji–Trost Reaction.* In *Name Reactions for Homologations-Part I*; Li, J. J., Ed.; Wiley: Hoboken, NJ, **2009,** pp 185–211. (Review).
11. Shi, L.; Meyer, K.; Greaney, M. F. *Angew. Chem. Int. Ed.* **2010,** *49,* 9250–9253.
12. Brehm, E.; Breinbauer, R. *Org. Biomol. Chem.* **2013,** *11,* 4750–4756.
13. Meng, L. *J. Org. Chem.* **2016,** *81,* 7784–7789.
14. Burtea, A.; Rychnovsky, S. D. *Org. Lett.* **2018,** *20,* 5849–5852.
15. Kučera, R.; Goetzke, F. W.; Fletcher, S. P. *Org. Lett.* **2020,** *22,* 2991–2994.

Ugi Reaction

Four-component condensation (4CC) of carboxylic acids, *C*-isocyanides, amines, and carbonyl compounds to afford diamides. Also known as four-component reaction (4CR). *Cf.* Passerini reaction.

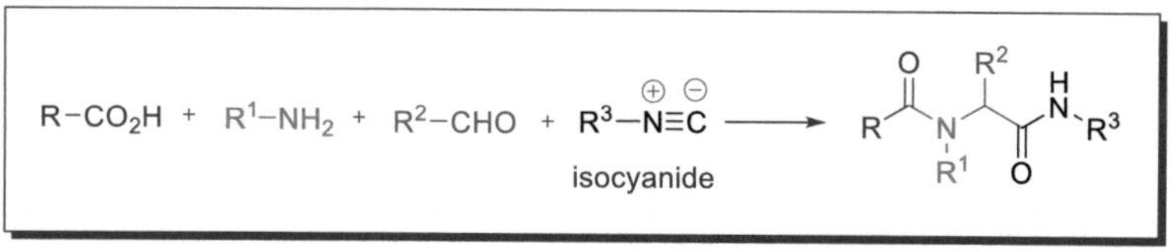

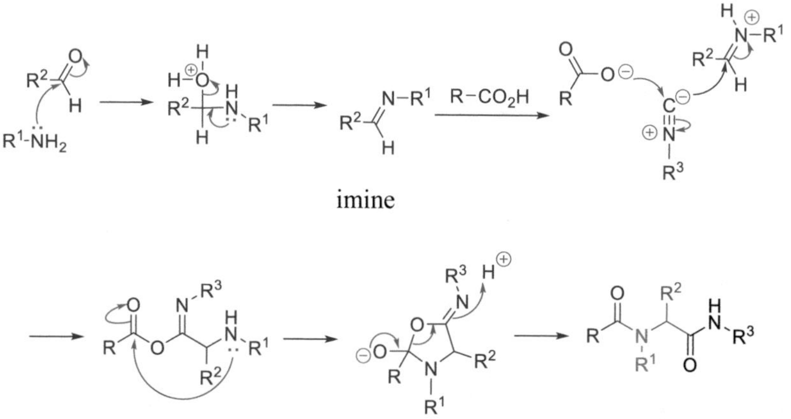

imine

Example 1[2]

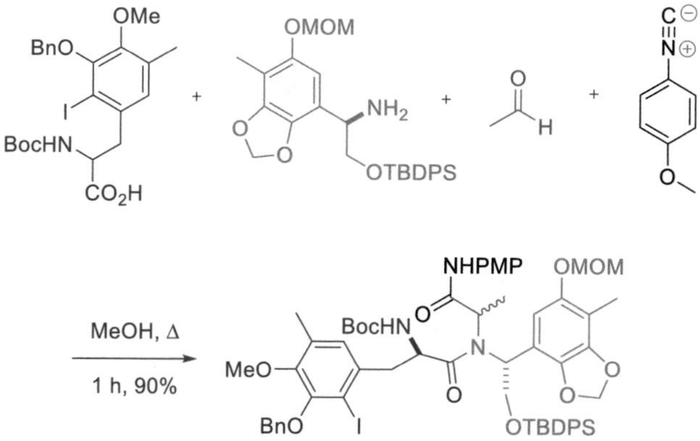

© Springer Nature Switzerland AG 2021
J. J. Li, *Name Reactions*, https://doi.org/10.1007/978-3-030-50865-4_152

Example 2[5]

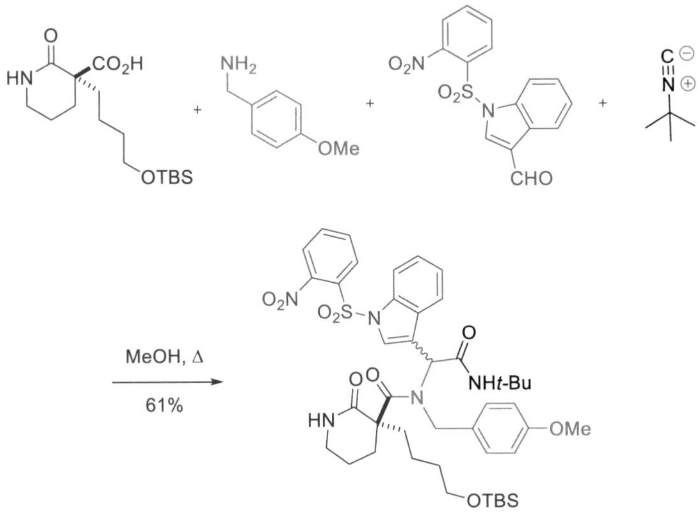

MeOH, Δ
⟶
61%

Example 3[7]

TFE, rt
⟶
78%

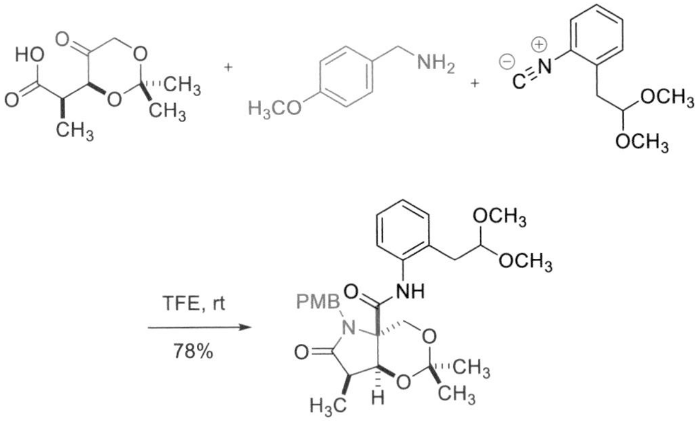

Example 4, "Double whammy" Ugi reaction[8]

Example 5[11]

Example 6, Synthesis of ivosidenib (Tibsovo), an isocitrate dehydrogenase 1 (IDH1) inhibitor[12]

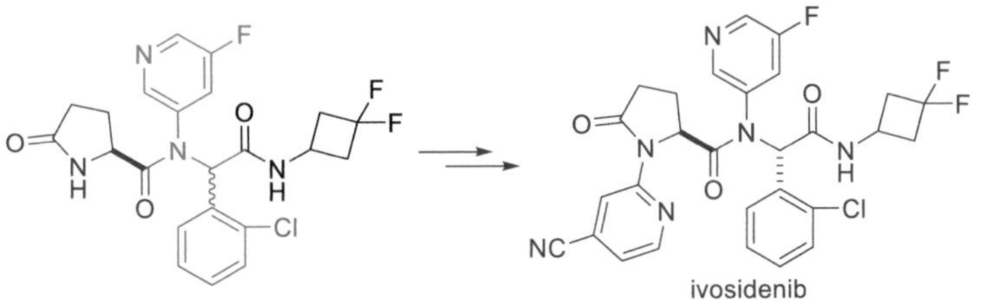

ivosidenib

Example 7, Boron-based peptidomimetics as potent inhibitors of human caseino-lytic protease P(ClpP)[13]

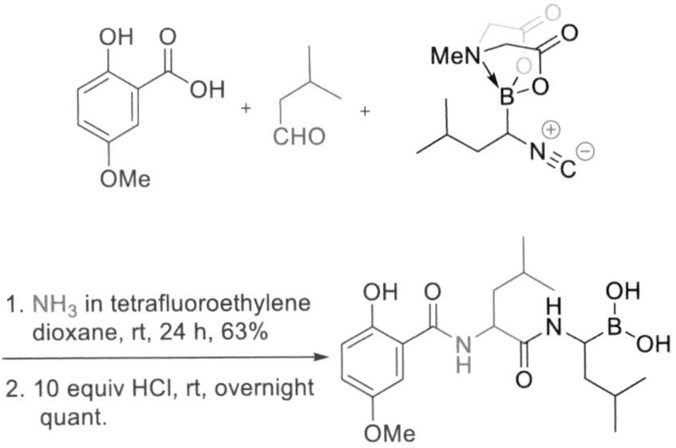

1. NH₃ in tetrafluoroethylene dioxane, rt, 24 h, 63%

2. 10 equiv HCl, rt, overnight quant.

Example 8, A lead thioredoxin reductase (TrxR) inhibitor[14]

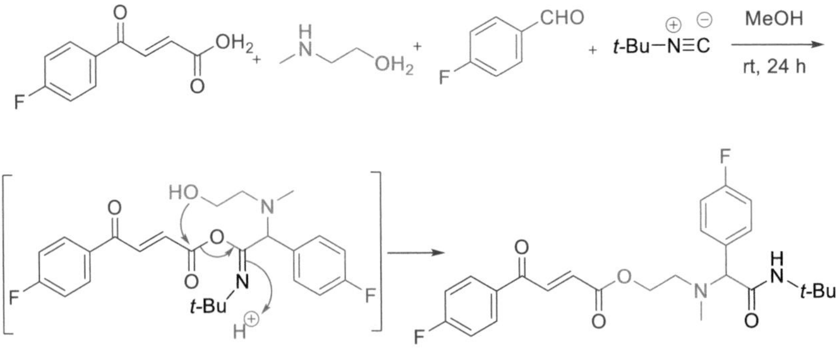

References

1.   (a) Ugi, I. *Angew. Chem. Int. Ed.* **1962**, *1*, 8–21. (b) Ugi, I.; Offermann, K.; Herlinger, H.; Marquarding, D. *Liebigs Ann. Chem.* **1967**, *709*, 1–10. (c) Ugi, I.; Kaufhold, G. *Ann.* **1967**, *709*, 11–28. (d) Ugi, I.; Lohberger, S.; Karl, R. In *Comprehensive Organic Synthesis*; Trost, B. M.; Fleming, I., Eds.; Pergamon: Oxford, **1991**, *Vol. 2*, 1083. (Review). (e) Dömling, A.; Ugi, I. *Angew. Chem. Int. Ed.* **2000**, *39*, 3168. (Review). (f) Ugi, I.

Pure Appl. Chem. **2001**, *73*, 187–191. (Review). Ivar Karl Ugi (1930–2005) earned his Ph.D. under the guidance of Prof. Rolf Huisgen. Since 1962, he worked at Bayer AG, rising through the ranks to director. But he left Bayer in 1969 to pursue his indendent academic career at the University of Southern California (USC). In 1973, he moved to the Technische Universität München, where stayed until his retirement in 1999. Ugi was one of the pioneers in multi-component reactions (MCRs).

2.   Endo, A.; Yanagisawa, A.; Abe, M.; Tohma, S.; Kan, T.; Fukuyama, T. *J. Am. Chem. Soc.* **2002**, *124*, 6552–6554.

3.   Hebach, C.; Kazmaier, U. *Chem. Commun.* **2003**, 596–597.

4.   *Multicomponent Reactions* J. Zhu, H. Bienaymé, Eds.; Wiley-VCH, Weinheim, **2005**.

5.   Oguri, H.; Schreiber, S. L. *Org. Lett.* **2005**, *7*, 47–50.

6.   Dömling, A. *Chem. Rev.* **2006**, *106*, 17–89.

7.   Gilley, C. B.; Buller, M. J.; Kobayashi, Y. *Org. Lett.* **2007**, *9*, 3631–3634.

8.   Rivera, D. G.; Pando, O.; Bosch, R.; Wessjohann, L. A. *J. Org. Chem.* **2008**, *73*, 6229–6238.

9.   Bonger, K. M.; Wennekes, T.; Filippov, D. V.; Lodder, G.; van der Marel, G. A.; Overkleeft, H. S. *Eur. J. Org. Chem.* **2008**, 3678–3688.

10. Williams, D. R.; Walsh, M. J. *Ugi Reaction*. In *Name Reactions for Homologations-Part II*; Li, J. J., Ed.; Wiley: Hoboken, NJ, **2009**, pp 786–805. (Review).

11. Tyagi, V.; Shahnawaz Khan, S.; Chauhan, P. M. S. *Tetrahedron Lett.* **2013**, *54*, 1279–1284.

12. Popovici-Muller, J.; Lemieux, R. M.; Artin, E.; Saunders, J. O.; Salituro, F. G.; Travins, J.; Cianchetta, G.; Cai, Z.; Zhou, D.; Cui, D.; et al. *ACS Med. Chem. Lett.* **2018**, *9*, 300–305.

13. Wang, Q.; Wang, D.-X.; Wang, M.-X.; Zhu, J. *Acc. Chem. Res.* **2018**, *51*, 1290–1300. (Review).

14. Reguera, L.; Rivera, D. G. *Chem. Rev.* **2019**, *119*, 9836–9860. (Review).

15. Tan, J.; Grouleff, J. J.; Jitkova, Y.; Diaz, D. B.; Griffith, E. C.; Shao, W.; Bogdanchikova, A. F.; Poda, G.; Schimmer, A. D.; Lee, R. E.; et al. *J. Med. Chem.* **2019**, *62*, 6377–6390.

16. Jovanović, M.; Zhukovsky, D.; Podolski-Renić, A.; Žalubovskis, R.; Dar'in, D.; Sharoyko, V.; Tennikova, T.; Pešić, M.; Krasavin, M. *Eur. J. Med. Chem.* **2020**, *191*, 112119.

Ullmann Coupling

Homocoupling of aryl iodides in the presence of Cu or Ni or Pd to afford biaryls.

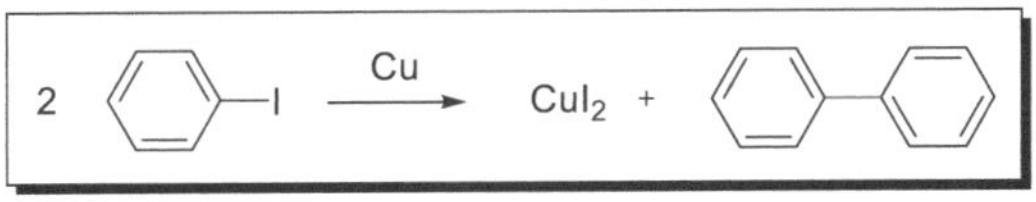

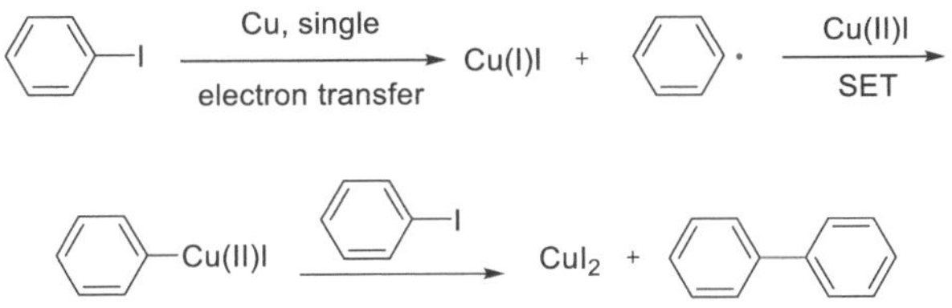

The overall transformation of PhI to PhCuI is an oxidative addition process.

Example 1[3]

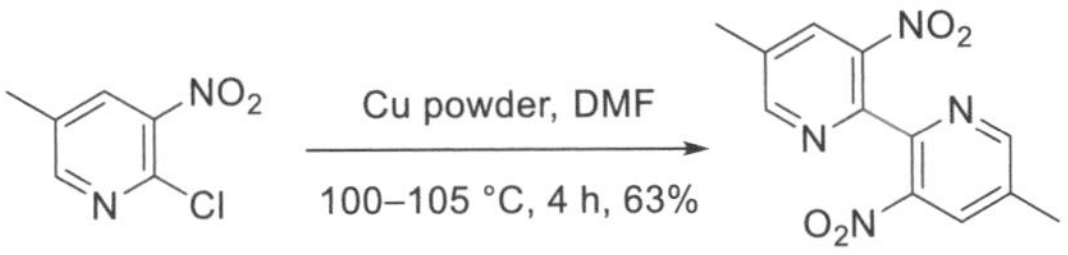

Example 2, CuTC-catalyzed Ullmann coupling, CuTC = Copper(I)-thiophene-2-carboxylate[4]

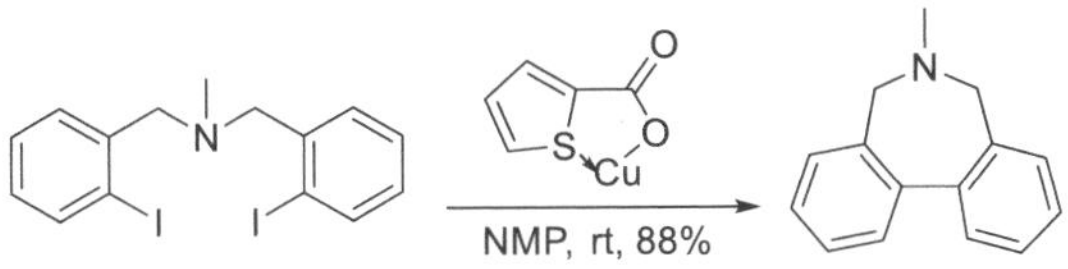

Example 3[5]

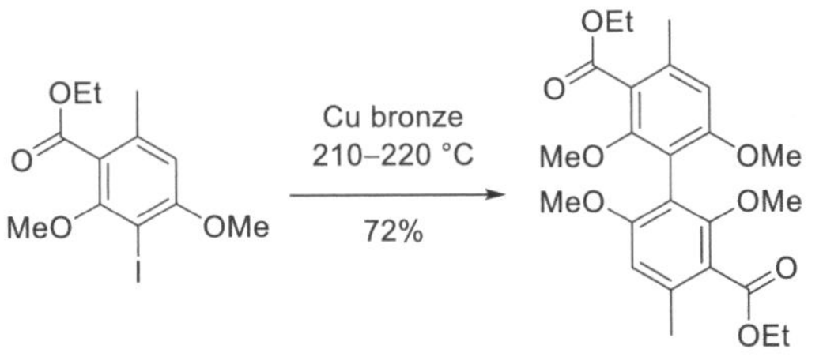

© Springer Nature Switzerland AG 2021
J. J. Li, Name Reactions, https://doi.org/10.1007/978-3-030-50865-4_153

Example 4[8]

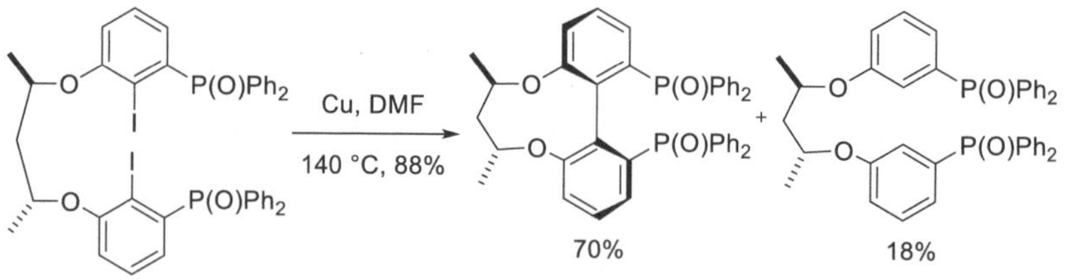

70%   18%

Example 5[9]

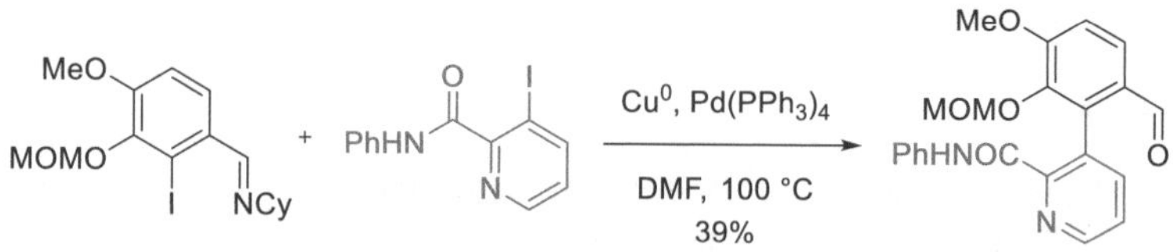

Example 6, Ullman-type C–N coupling[11]

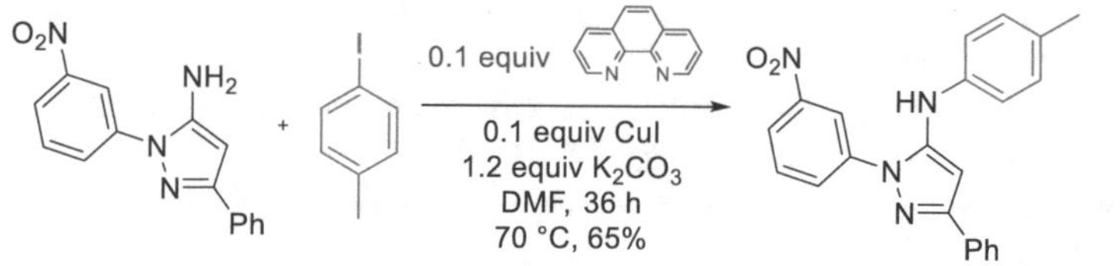

Example 7, Ullman-type C–N coupling[12]

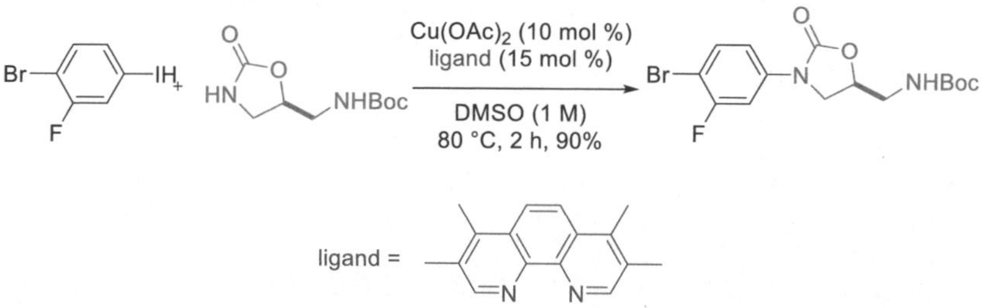

ligand =

Example 8, Palladium-catalyzed Ullman coupling[13]

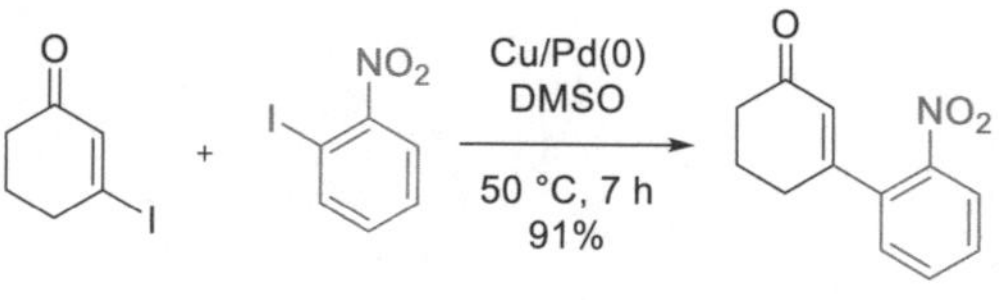

Example 8, Palladium-catalyzed Ullman-type coupling[13]

References

1. (a) Ullmann, F.; Bielecki, J. *Ber.* **1901**, *34*, 2174–2185. Fritz Ullmann (1875–1939), born in Fürth, Bavaria, studied under Graebe at Geneva. He taught at the Technische Hochschule in Berlin and the University of Geneva. (b) Ullmann, F. *Ann.* **1904**, *332*, 38–81.
2. Fanta, P. E. *Synthesis* **1974**, 9–21. (Review).
3. Kaczmarek, L.; Nowak, B.; Zukowski, J.; Borowicz, P.; Sepiol, J.; Grabowska, A. *J. Mol. Struct.* **1991**, *248*, 189–200.
4. Zhang, S.; Zhang, D.; Liebskind, L. S. *J. Org. Chem.* **1997**, *62*, 2312–2313.
5. Hauser, F. M.; Gauuan, P. J. F. *Org. Lett.* **1999**, *1*, 671–672.
6. Buck, E.; Song, Z. J.; Tschaen, D.; Dormer, P. G.; Volante, R. P.; Reider, P. J. *Org. Lett.* **2002**, *4*, 1623–1626.
7. Nelson, T. D.; Crouch, R. D. *Org. React.* **2004**, *63*, 265–556. (Review).
8. Qui, L.; Kwong, F. Y.; Wu, J.; Wai, H. L.; Chan, S.; Yu, W.-Y.; Li, Y.-M.; Guo, R.; Zhou, Z.; Chan, A. S. C. *J Am. Chem. Soc.* **2006**, *128*, 5955–5965.
9. Markey, M. D.; Fu, Y.; Kelly, T. R. *Org. Lett.* **2007**, *9*, 3255–3257.
10. Ahmad, N. M. *Ullman Coupling.* In *Name Reactions for Homologations-Part I*; Li, J. J., Ed.; Wiley: Hoboken, NJ, 2009; pp 255–267. (Review).
11. Chang, E. C.; Chen, C.-Y.; Wang, L.-Y.; Huang, Y.-Y.; Yeh, M.-Y.; Wong, F. F. *Tetrahedron* **2013**, *69*, 570–576.
12. Kelly, S. M.; Han, C.; Tung, L.; Gosselin, F. *Org. Lett.* **2017**, *19*, 3021–3024.
13. Khan, F.; Dlugosch, M.; Liu, X.; Khan, M.; Banwell, M. G.; Ward, J. S.; Carr, P. D. *Org. Lett.* **2018**, *20*, 2770–2773.
14. Waters, G. D.; Carrick, J. D. *RSC Adv.* **2020**, *10*, 10807–10815.

Vilsmeier–Haack Reaction

The Vilsmeier–Haack reagent, a chloroiminium salt, is a weak electrophile. Therefore, the Vilsmeier–Haack reaction works better with electron-rich carbocycles and heterocycles.

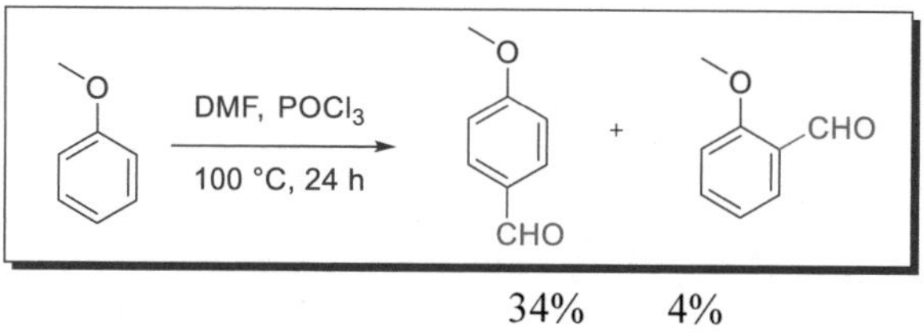

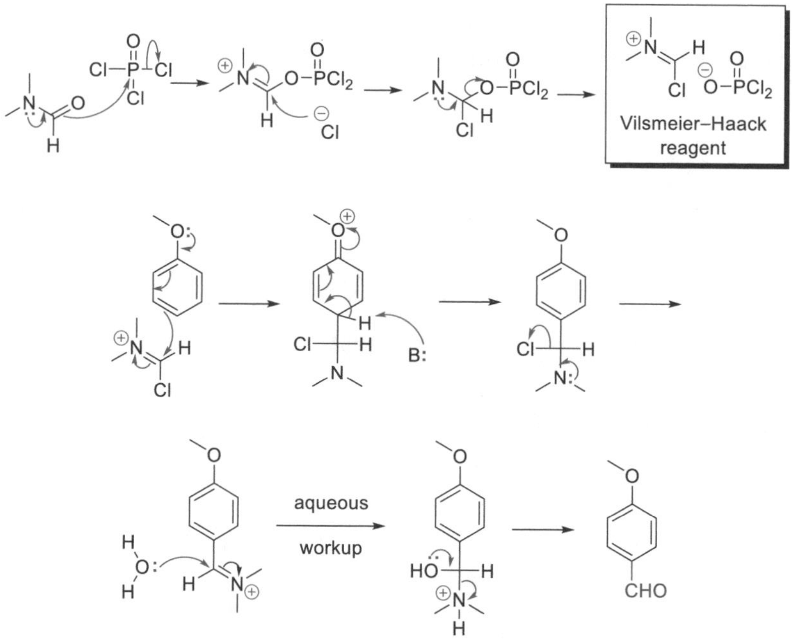

DMF, POCl3
100 °C, 24 h
CHO
CHO
34%
4%

Vilsmeier–Haack reagent
aqueous
workup
B:
CHO

Example 1[2]

MeO
OH
OMe
DMF, POCl3
100 °C, 14 h, 98%
OHC
MeO
OMe

J. J. Li, *Name Reactions*, https://doi.org/10.1007/978-3-030-50865-4_154

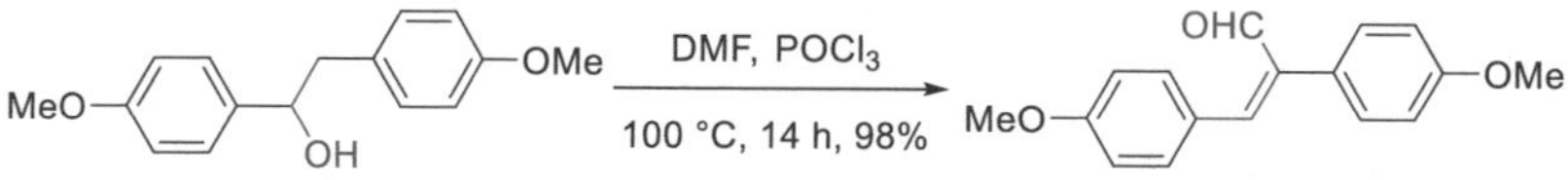

Example 2[3]

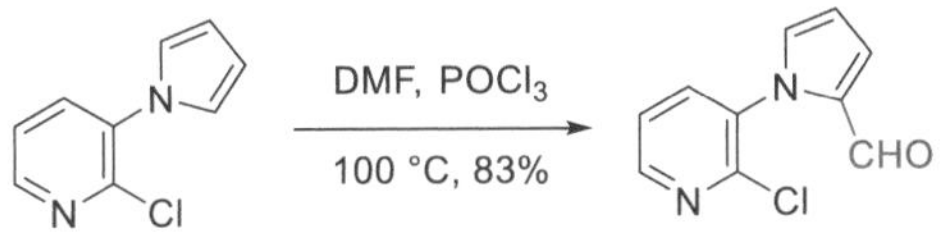

Example 3[9]

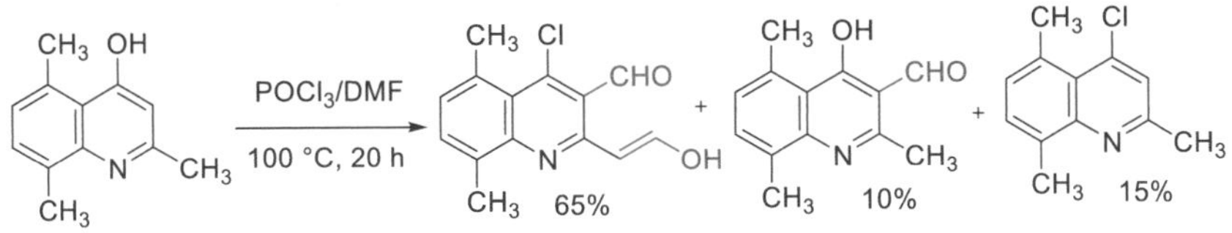

Example 4, Reaction outcomes differ as temperature differs[10]

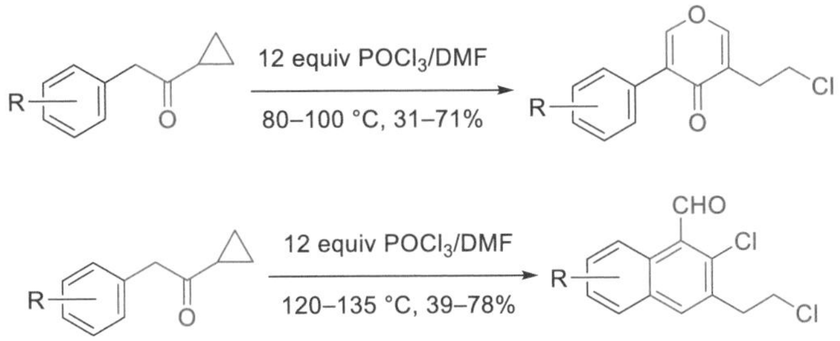

Example 5, An interesting mechanism[11]

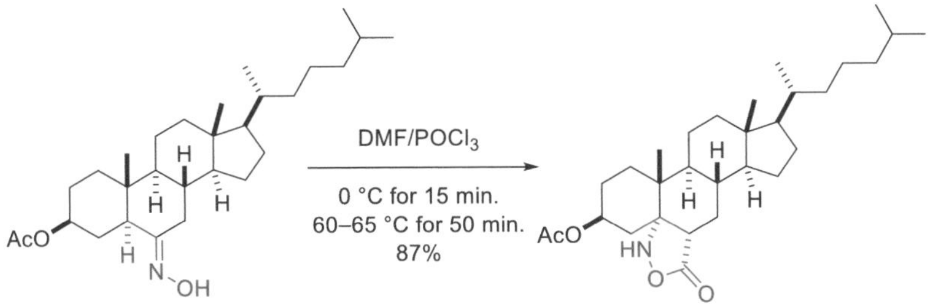

Example 6, One-pot Vilsmeier–Haack cyclization and azomethine ylide cycloaddition[12]

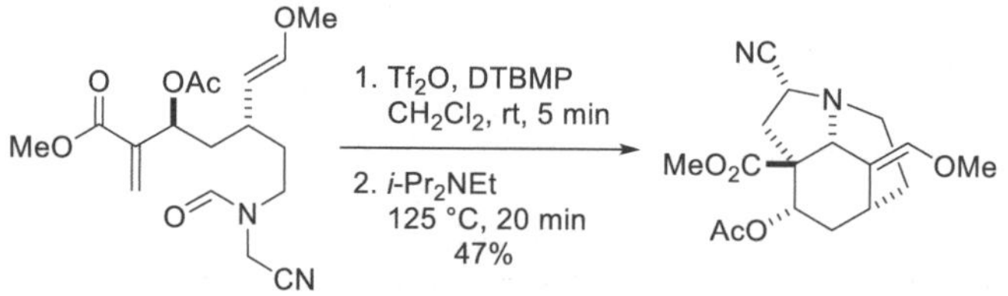

DTBMP = 2,6-di-*tert*-butyl-4-methylpyridine

Example 7, XtalFluor-E as an alternative to POCl₃ in the formylation of C2-glycal[13]

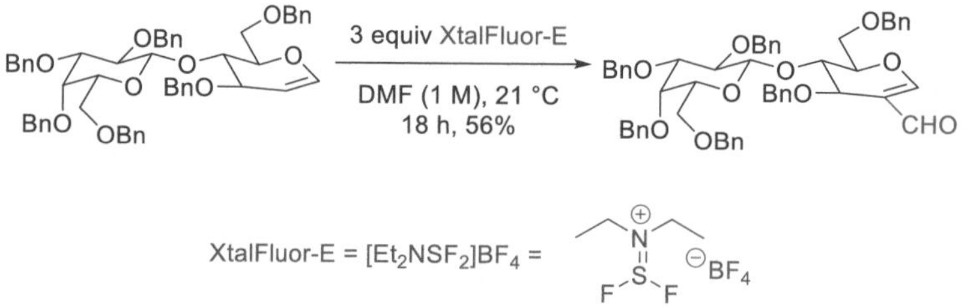

XtalFluor-E = [Et₂NSF₂]BF₄ =

Example 8, Sequential one-pot Vilsmeier–Haack and organocatalyzed Mannich cyclization[14]

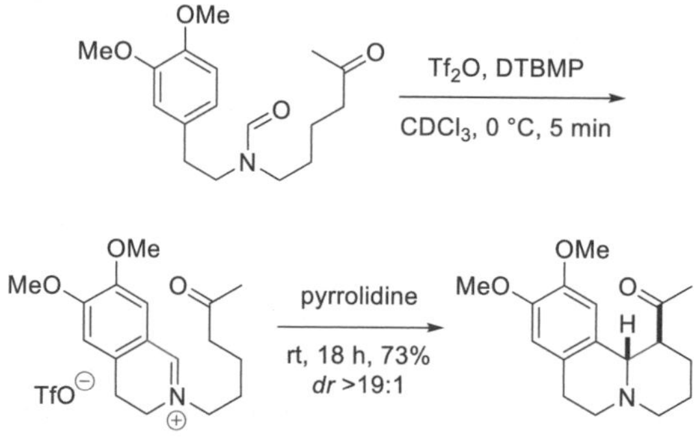

References

1. Vilsmeier, A.; Haack, A. *Ber.* **1927**, *60*, 119–122. German chemists Anton Vilsmerier and Albrecht Haack discovered this recation in 1927.
2. Reddy, M. P.; Rao, G. S. K. *J. Chem. Soc., Perkin Trans. 1* **1981**, 2662–2665.
3. Lancelot, J.-C.; Ladureé, D.; Robba, M. *Chem. Pharm. Bull.* **1985**, *33*, 3122–3128.
4. Marson, C. M.; Giles, P. R. *Synthesis Using Vilsmeier Reagents* CRC Press, **1994**. (Book).
5. Seybold, G. *J. Prakt. Chem.* **1996**, *338,* 392–396 (Review).

6. Jones, G.; Stanforth, S. P. *Org. React.* **1997,** *49*, 1–330. (Review).
7. Jones, G.; Stanforth, S. P. *Org. React.* **2000,** *56*, 355–659. (Review).
8. Tasneem, *Synlett* **2003,** 138–139. (Review of the Vilsmeier–Haack reagent).
9. Nandhakumar, R.; Suresh, T.; Jude, A. L. C.; Kannan, V. R.; Mohan, P. S. *Eur. J. Med. Chem.* **2007,** *42*, 1128–1136.
10. Tang, X.-Y.; Shi, M. *J. Org. Chem.* **2008,** *73*, 8317–8320.
11. Shamsuzzaman, Hena Khanam, H.; Mashrai, A.; Siddiqui, N. *Tetrahedron Lett.* **2013,** *54,* 874–877.
12. Hauduc, C.; Bélanger, G. *J. Org. Chem.* **2017,** *82*, 4703–4712.
13. Roudias, M.; Vallée, F.; Martel, J.; Paquin, J.-F. *J. Org. Chem.* **2018,** *83*, 8731–8738.
14. Outin, J.; Quellier, P.; Bélanger, G. *J. Org. Chem.* **2020,** *85*, 4712–4729.

von Braun Reaction

Different from the von Braun degradation reaction (amide to nitrile), the von Braun reaction refers to the treatment of tertiary amines with cyanogen bromide, resulting in a substituted cyanamide.

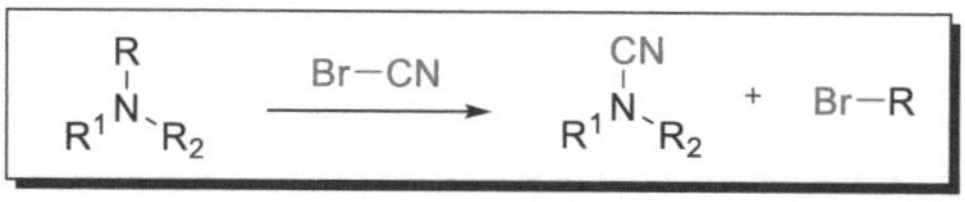

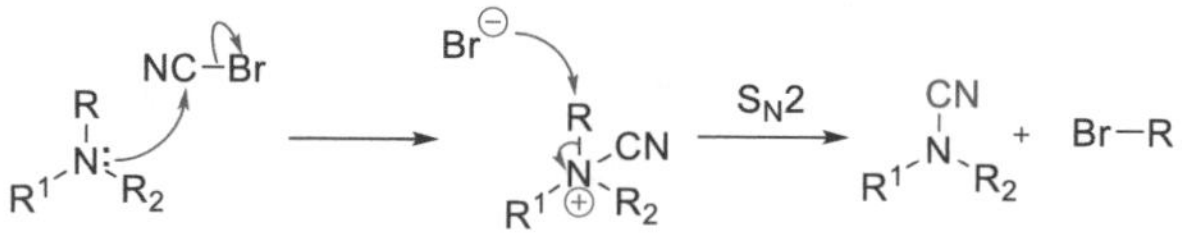

Example 1[4]

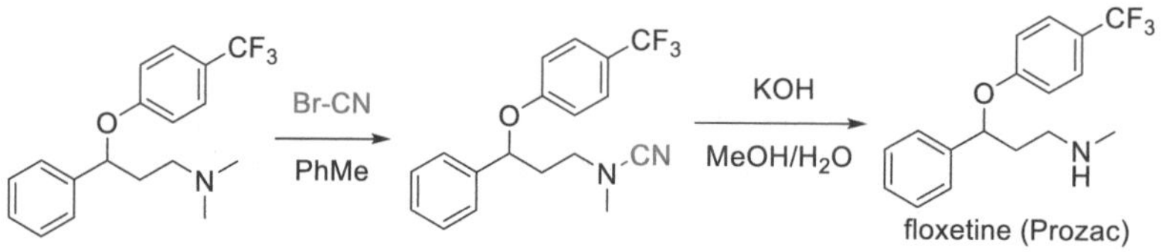

floxetine (Prozac)

Example 2[5]

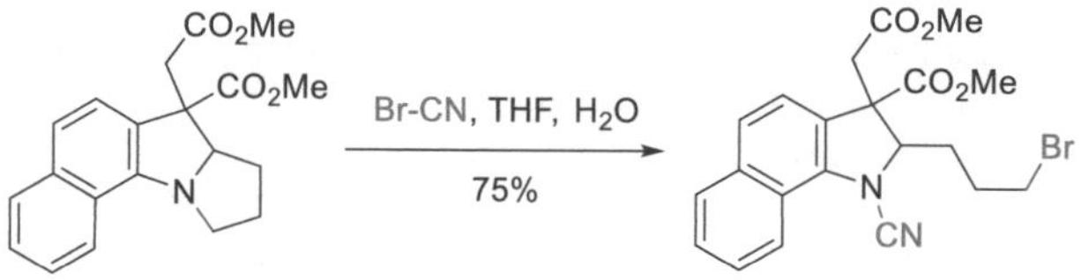

Example 3[9]

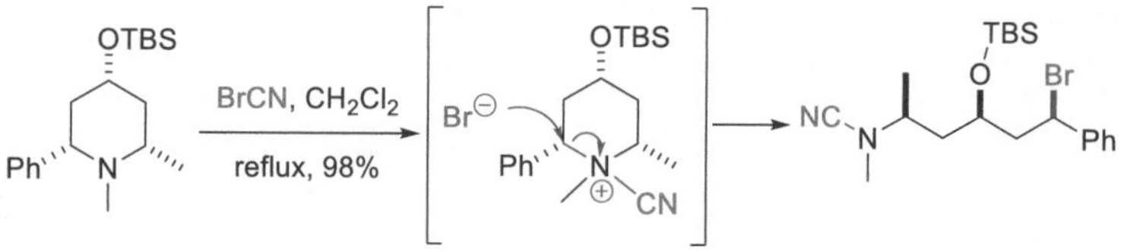

© Springer Nature Switzerland AG 2021
J. J. Li, *Name Reactions*, https://doi.org/10.1007/978-3-030-50865-4_155

Example 4, A vinylic Rosenmund–von Braun reaction[11]

Example 5, Ring-opening of azetidine by the von Braun reaction[12]

Example 6, Cleavage of C–N bond via ring opening[13]

References

1. von Braun, J. *Ber.* **1907,** *40,* 3914–3933. Julius von Braun (1875–1940), born in Warsaw, Poland, was a Professor of Chemistry at Frankfurt.
2. Hageman, H. A. *Org. React.* **1953,** *7,* 198–262. (Review).
3. Fodor, G.; Nagubandi, S. *Tetrahedron* **1980,** *36,* 1279–1300. (Review).
4. Mody, S. B.; Mehta, B. P.; Udani, K. L.; Patel, M. V.; Mahajan, Rajendra N.. Indian Patent IN177159 (1996).
5. McLean, S.; Reynolds, W. F.; Zhu, X. *Can. J. Chem.* **1987,** *65,* 200–204.
6. Chambert, S.; Thomasson, F.; Décout, J.-L. *J. Org. Chem.* **2002,** *67,* 1898–1904.
7. Hatsuda, M.; Seki, M. *Tetrahedron* **2005,** *61,* 9908–9917.
8. Thavaneswaran, S.; McCamley, K.; Scammells, P. J. *Nat. Prod. Commun.* **2006,** *1,* 885–897. (Review).
9. McCall, W. S.; Abad Grillo, T.; Comins, D. L. *Org. Lett.* **2008,** *10,* 3255–3257.
10. Tayama, E.; Sato, R.; Ito, M.; Iwamoto, H.; Hasegawa, E. *Heterocycles* **2013,** *87,* 381–388.
11. Pradal, A.; Evano, G. *Chem. Commun.* **2014,** *50,* 11907–11910.
12. Wright, K.; Drouillat, B.; Menguy, L.; Marrot, J.; Couty, F. *Eur. J. Org. Chem.* **2017,** 7195–7201.
13. Wahl, M. H.; Jandl, C.; Bach, T. *Org. Lett.* **2018,** *20,* 7674–7678.

Wacker Oxidation

Palladium-catalyzed oxidation of olefins to ketones, and aldehydes in certain cases.

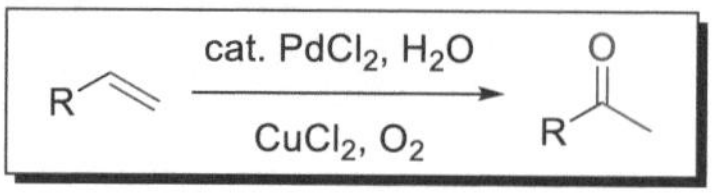

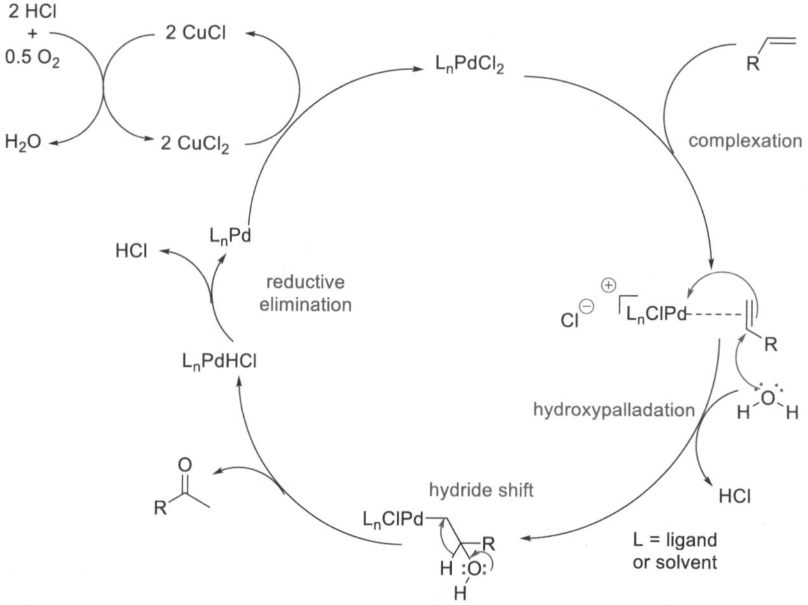

Example 1[5]

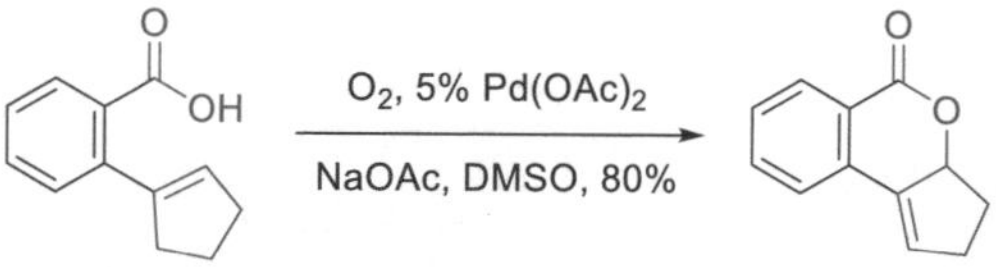

Example 2[7]

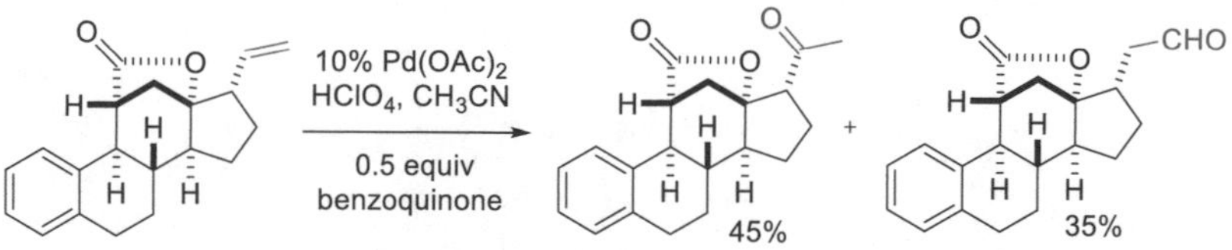

© Springer Nature Switzerland AG 2021

J. J. Li, *Name Reactions*, https://doi.org/10.1007/978-3-030-50865-4_156

Example 3[9]

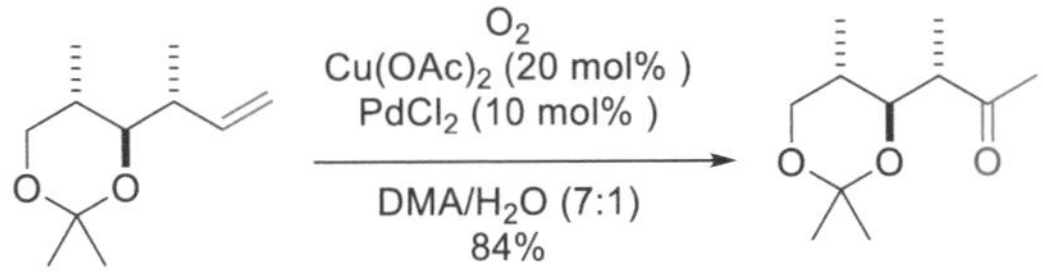

$$O_2$$
Cu(OAc)$_2$ (20 mol%)
PdCl$_2$ (10 mol%)

DMA/H$_2$O (7:1)
84%

Example 4[10]

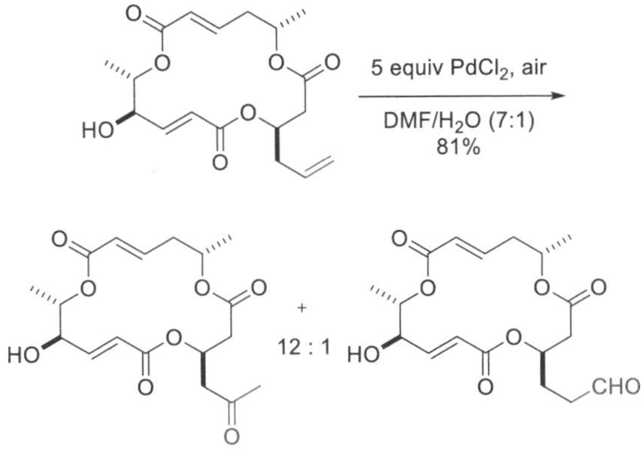

5 equiv PdCl$_2$, air

DMF/H$_2$O (7:1)
81%

+

12 : 1

CHO

Example 5[10]

5 mol% Pd(Quinox)Cl$_2$
12.5% AgSbF$_6$

12 equiv t-BuOOH (aq)
CH$_2$Cl$_2$, 0.1 M, rt, 66%

Quinox =

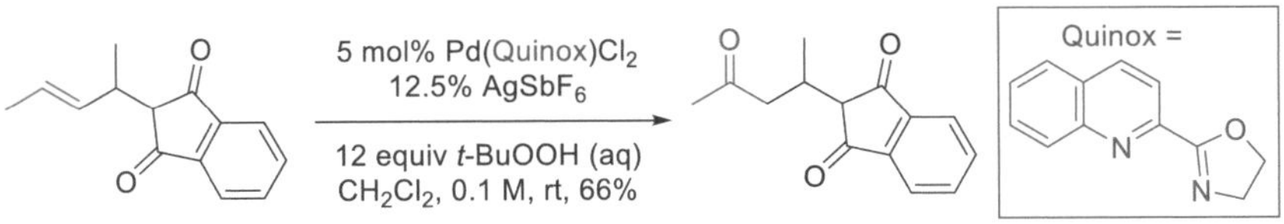

Example 6, Aldehyde-selective Wacker oxidation[16]

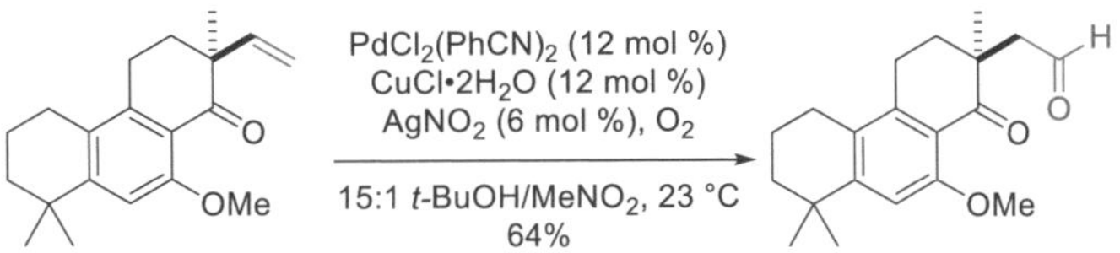

PdCl$_2$(PhCN)$_2$ (12 mol %)
CuCl•2H$_2$O (12 mol %)
AgNO$_2$ (6 mol %), O$_2$

15:1 t-BuOH/MeNO$_2$, 23 °C
64%

Example 7, Asymmetric intermolecular aza-Wacker-type reaction[17]

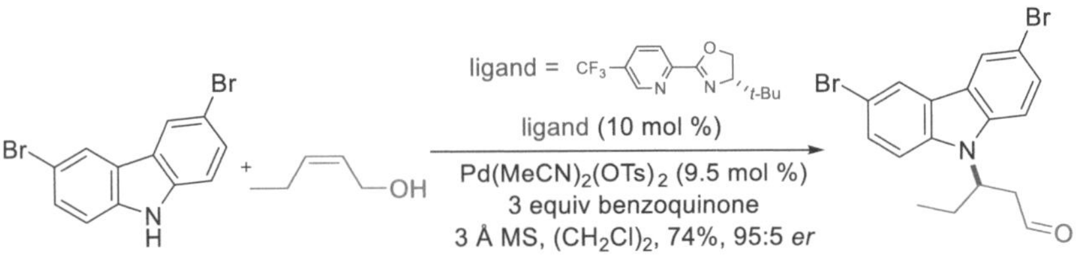

ligand =   CF$_3$ t-Bu

ligand (10 mol %)

Pd(MeCN)$_2$(OTs)$_2$ (9.5 mol %)
3 equiv benzoquinone
3 Å MS, (CH$_2$Cl)$_2$, 74%, 95:5 er

Example 8, Reacting unprotected carbohydrate-based terminal olefins through the "Uemura system" to hemiketals and α,β-unsaturated diketones[18]

References

1. Smidt, J.; Sieber, R. *Angew. Chem. Int. Ed.* **1962,** *1,* 80–88. Wacker is not a person, but a place in Germany where Wacker Chemie developed this process. Since Hoechst AG later refined the reaction, this is sometimes called Hoechst–Wacker process.
2. Tsuji, J. *Synthesis* **1984,** 369–384. (Review).
3. Hegedus, L. S. In *Comp. Org. Syn.* Trost, B. M.; Fleming, I., Eds.; Pergamon, **1991,** *Vol. 4,* 552. (Review).
4. Tsuji, J. In *Comp. Org. Syn.* Trost, B. M.; Fleming, I., Eds.; Pergamon, **1991,** *Vol. 7,* 449. (Review).
5. Larock, R. C.; Hightower, T. R. *J. Org. Chem.* **1993,** *58,* 5298–5300.
6. Hegedus, L. S. *Transition Metals in the Synthesis of Complex Organic Molecule* **1994,** University Science Books: Mill Valley, CA, pp 199–208. (Review).
7. Pellissier, H.; Michellys, P.-Y.; Santelli, M. *Tetrahedron* **1997,** *53,* 10733–10742.
8. Feringa, B. L. *Wacker Oxidation.* In *Transition Met. Org. Synth.* Beller, M.; Bolm, C., eds.; Wiley–VCH: Weinheim, Germany. **1998,** *2,* 307–315. (Review).
9. Smith, A. B.; Friestad, G. K.; Barbosa, J.; Bertounesque, E.; Hull, K. G.; Iwashima, M.; Qiu, Y.; Salvatore, B. A.; Spoors, P. G.; Duan, J. J.-W. *J. Am. Chem. Soc.* **1999,** *121,* 10468–10477.
10. Kobayashi, Y.; Wang, Y.-G. *Tetrahedron Lett.* **2002,** *43,* 4381–4384.
11. Hintermann, L. *Wacker-type Oxidations* in *Transition Met. Org. Synth. (2nd edn.)* Beller, M.; Bolm, C., eds., Wiley–VCH: Weinheim, Germany. **2004,** *2,* pp 379–388. (Review).
12. Li, J. J. *Wacker–Tsuji oxidation.* In *Name Reactions for Functional Group Transformations*; Li, J. J., Ed.; Wiley: Hoboken, NJ, **2007,** pp 309–326. (Review).
13. Okamoto, M.; Taniguchi, Y. *J. Catal.* **2009,** *261,* 195–200.
14. DeLuca, R. J.; Edwards, J. L.; Steffens, L. D.; Michel, B. W.; Qiao, X.; Zhu, C.; Cook, S. P.; Sigman, M. S. *J. Org. Chem.* **2013,** *78,* 1682–1686.
15. Baiju, T. V.; Gravel, E.; Doris, E.; Namboothiri, I. N. N. *Tetrahedron Lett.* **2016,** *57,* 3993–4000. (Review).
16. Kim, K. E.; Li, J.; Grubbs, R. H.; Stoltz, B. M. *J. Am. Chem. Soc.* **2016,** *138,* 13179–12182.
17. Allen, J. R.; Bahamonde, A.; Furukawa, Y.; Sigman, M. S. *J. Am. Chem. Soc.* **2019,** *141,* 8670–8674.
18. Runeberg, P. A.; Eklund, P. C. *Org. Lett.* **2019,** *21,* 8145–8148.
19. Tang, S.; Ben-David, Y.; Milstein, D. *J. Am. Chem. Soc.* **2020,** *142,* 5980–5984.

Wagner–Meerwein Rearrangement

Acid-catalyzed alkyl group migration of alcohols to give more substituted olefins.

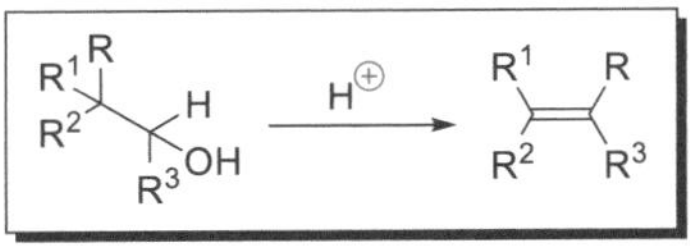

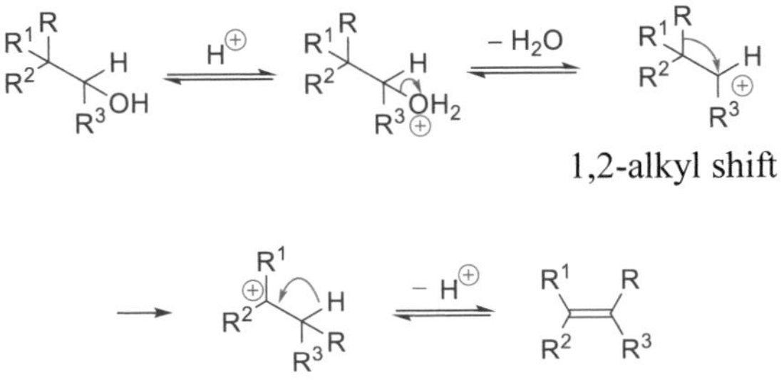

1,2-alkyl shift

Example 1[3]

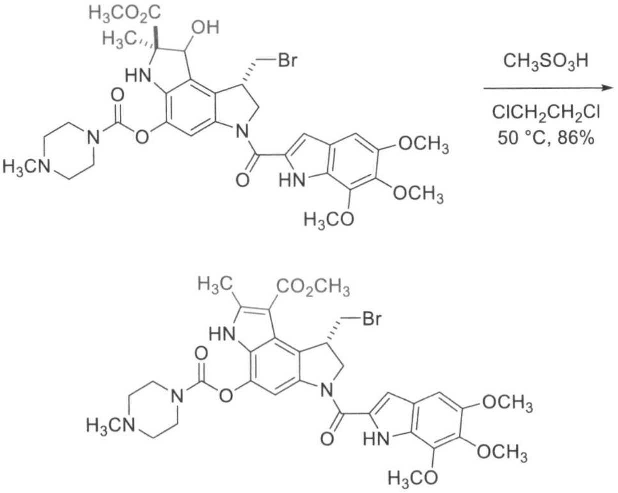

Example 2[6]

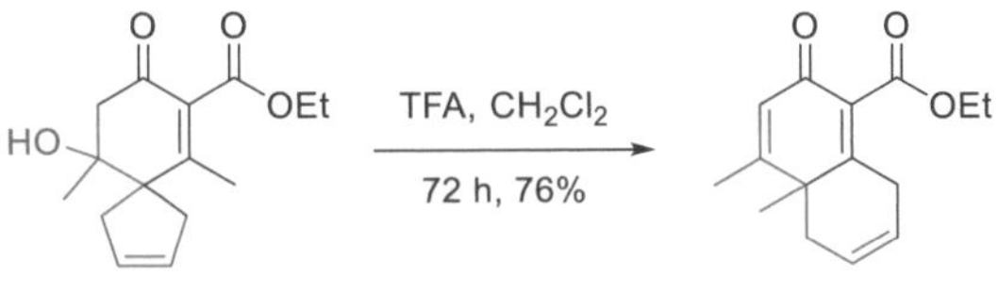

J. J. Li, *Name Reactions*, https://doi.org/10.1007/978-3-030-50865-4_157

Example 3[7]

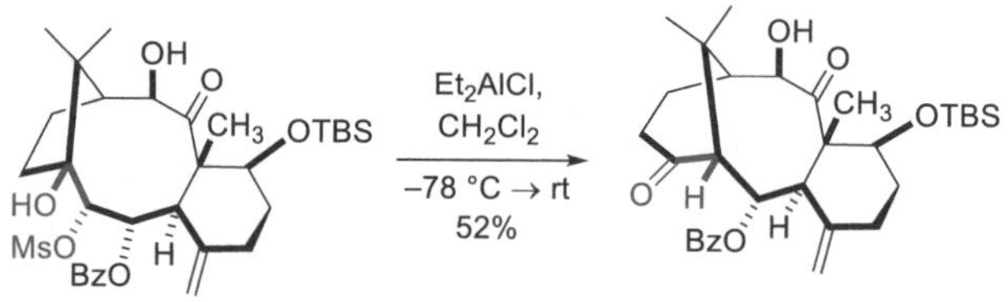

Example 4[9]

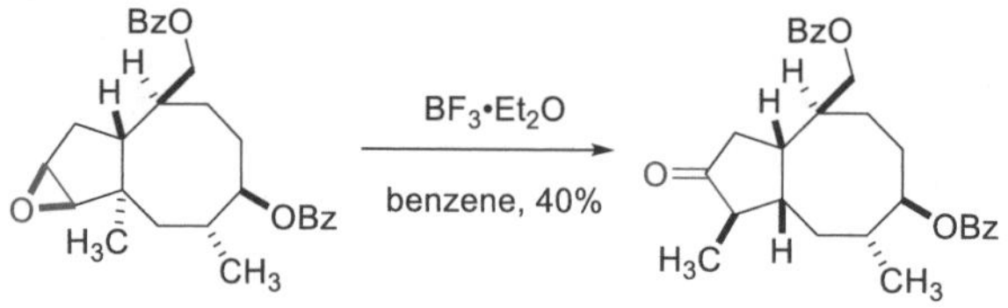

Example 5, Vinyl migration in preference to aryl migration[12]

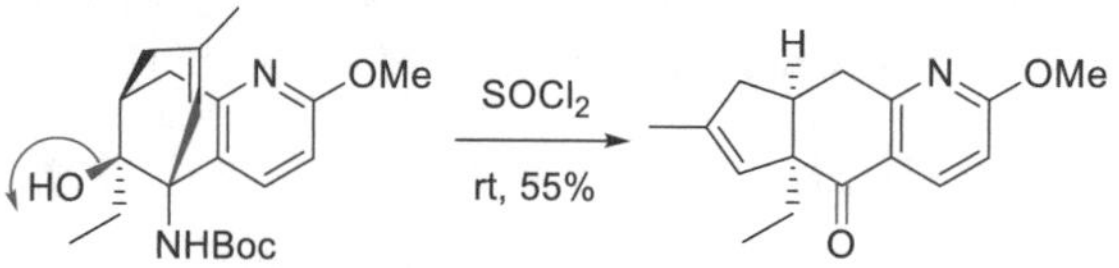

Example 6, Synthesis of cardiopetaline, Wagner–Meerwein rearrangement without preactivation of the pivotal hydroxy group[13]

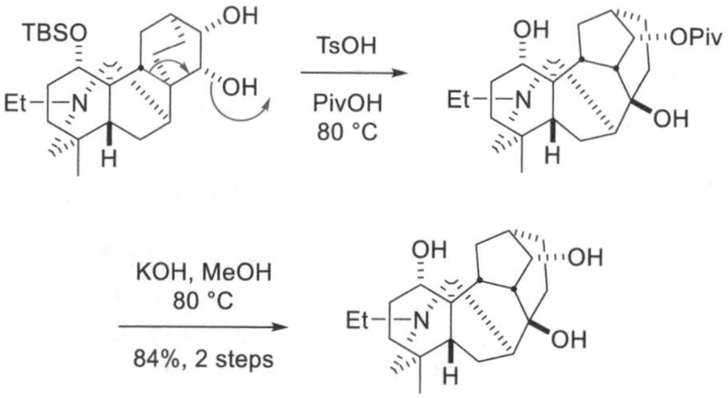

cardiopetaline

Example 7, Prins/Wagner–Meerwein rearrangement cascade[14]

SnCl$_4$
−15 °C

1 h, 64%

Example 8, Synthesis of nortriterpenoid propindilactone G[15]

1. MsCl, Et$_3$N

2. t-BuOH/H$_2$O
80%, 2 steps

References

1. Wagner, G. *J. Russ. Phys. Chem. Soc.* **1899,** *31,* 690. Wagner first observed this rearrangement in 1899 and German chemist Hans Meerwein unveiled the mechanism in 1914.
2. Hogeveen, H.; Van Kruchten, E. M. G. A. *Top. Curr. Chem.* **1979,** *80,* 89–124. (Review).
3. Kinugawa, M.; Nagamura, S.; Sakaguchi, A.; Masuda, Y.; Saito, H.; Ogasa, T.; Kasai, M. *Org. Process Res. Dev.* **1998,** *2,* 344–350.
4. Trost, B. M.; Yasukata, T. *J. Am. Chem. Soc.* **2001,** *123,* 7162–7163.
5. Guizzardi, B.; Mella, M.; Fagnoni, M.; Albini, A. *J. Org. Chem.* **2003,** *68,* 1067–1074.
6. Bose, G.; Ullah, E.; Langer, P. *Chem. Eur. J.* **2004,** *10,* 6015–6028.
7. Guo, X.; Paquette, L. A. *J. Org. Chem.* **2005,** *70,* 315–320.
8. Li, W.-D. Z.; Yang, Y.-R. *Org. Lett.* **2005,** *7,* 3107–3110.
9. Michalak, K.; Michalak, M.; Wicha, J. *Molecules* **2005,** *10,* 1084–1100.
10. Mullins, R. J.; Grote, A. L. *Wagner–Meerwein Rearrangement.* In *Name Reactions for Homologations-Part II*; Li, J. J., Ed.; Wiley: Hoboken, NJ, **2009,** pp 373–394. (Review).
11. Ghorpade, S.; Su, M.-D.; Liu, R.-S. *Angew. Chem. Int. Ed.* **2013,** *52,* 4229–4234.
12. Fu, J.-G.; Ding, R.; Sun, B.-F.; Lin, G.-Q. *Tetrahedron* **2014,** *70,* 8374–8379.
13. Nishiyama, Y.; Yokoshima, S.; Fukuyama, T. *Org. Lett.* **2017,** *19,* 5833–5835.
14. Zhou, S.; Xia, K.; Leng, X.; Li, A. *J. Am. Chem. Soc.* **2019,** *141,* 13718–13723.
15. Wang, Y.; Chen, B.; He, X.; Gui, J. *J. Am. Chem. Soc.* **2020,** *142,* 5007–5012.

Williamson Ether Synthesis

Ether from the alkylation of alkoxides by alkyl halides. In order for reaction to go smoothly, the alkyl halides are preferred to be primary. Secondary halides work as well sometimes, but tertiary halides do not work at all because E_2 elimination will be the predominant reaction pathway.

Example 1, Diastereoselective intermolecular S_N2 reaction[9]

Example 2, Phenolate revealed by $Pd(PPh_3)_4$ and K_2CO_3[10]

© Springer Nature Switzerland AG 2021
J. J. Li, *Name Reactions*, https://doi.org/10.1007/978-3-030-50865-4_158

Example 3, Cyclic intramolecular etherification[11]

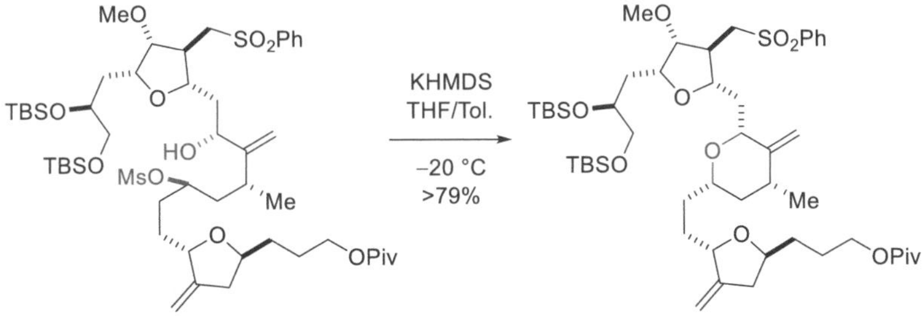

MeO
SO2Ph
TBSO
TBSO
HO
MsO
Me
OPiv
KHMDS
THF/Tol.
−20 °C
>79%
MeO
SO2Ph
TBSO
TBSO
Me
OPiv

Example 4, Williamson ether synthesis with phenols at a pertiary stereogenic carbon: formal enantioselective phenoxylation of β-keto esters[12]

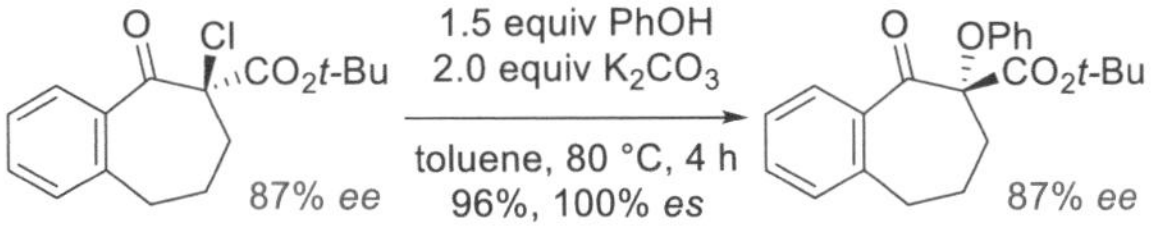

O
Cl
CO2t-Bu
87% ee
1.5 equiv PhOH
2.0 equiv K2CO3
toluene, 80 °C, 4 h
96%, 100% es
O
OPh
CO2t-Bu
87% ee

Example 5, Intramolecular S_N2 displacement[13]

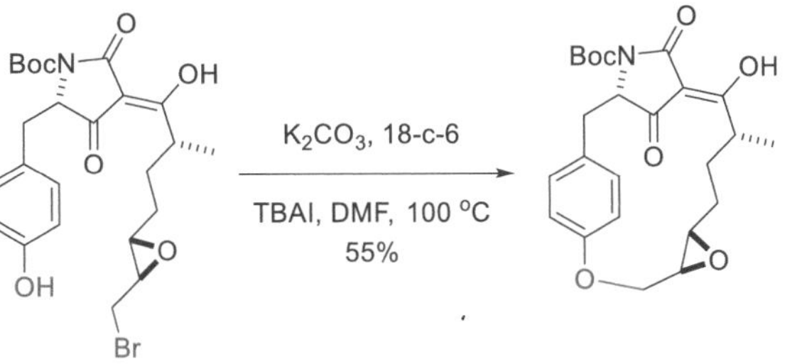

BocN
OH
OH
Br
K2CO3, 18-c-6
TBAI, DMF, 100 °C
55%
BocN
OH

Example 6, A domino oxa-Michael/aza-Michael/Williamson cycloetherification sequence[15]

EtO−P
EtO
EtO
Ph
+
Br
HN
HO
Cs2CO3
CH3CN, rt
90%
dr > 98:2
Ph
EtO−P
EtO

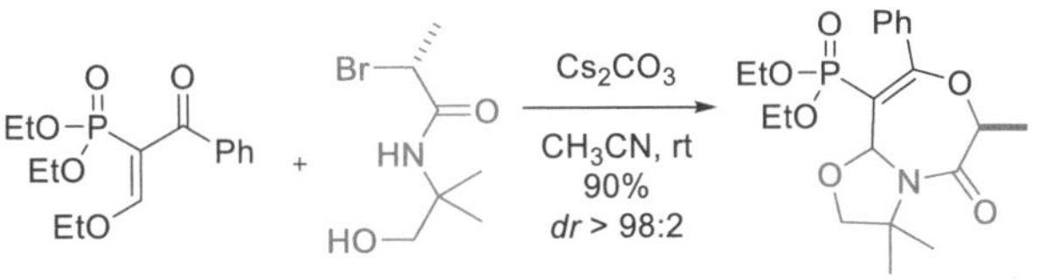

Example 7, One-pot synthesis of benzyl ethers bearing a nitrogen-containing bicycles[16]

References

1. Williamson, A. W. *J. Chem. Soc.* **1852**, *4,* 229–239. Alexander William Williamson (1824–1904) discovered this reaction in 1850 at University College, London.
2. Dermer, O. C. *Chem. Rev.* **1934,** *14,* 385–430. (Review).
3. Freedman, H. H.; Dubois, R. A. *Tetrahedron Lett.* **1975,** *16,* 3251–3254.
4. Jursic, B. *Tetrahedron* **1988,** *44,* 6677–6680.
5. Tan, S. N.; Dryfe, R. A.; Girault, H. H. *Helv. Chim. Acta* **1994,** *77,* 231–242.
6. Silva, A. L.; Quiroz, B.; Maldonado, L. A. *Tetrahedron Lett.* **1998,** *39,* 2055–2058.
7. Peng, Y.; Song, G. *Green Chem.* **2002,** *4,* 349–351.
8. Stabile, R. G.; Dicks, A. P. *J. Chem. Educ.* **2003,** *80,* 313–315.
9. Aikins, J. A.; Haurez, M.; Rizzo, J. R.; Van Hoeck, J.-P.; Brione, W.; Kestemont, J.-P.; Stevens, C.; Lemair, X.; Stephenson, G. A.; Marlot, E.; et al. *J. Org. Chem.* **2005,** *70,* 4695–4705.
10. Barnickel, B.; Schobert, R. *J. Org. Chem.* **2010,** *75,* 6716–6719.
11. Austad, B. C.; Benayoud, F.; Calkins, T. L.; et al. *Synlett* **2013,** *17,* 327–332.
12. Shibatomi, K.; Kotozaki, M.; Sasaki, N.; Fujisawa, I.; Iwasa, S. *Chem. Eur. J.* **2015,** *21,* 14095–14098.
13. Haase, R. G.; Schobert, R. *Org. Lett.* **2016,** *18,* 6352–6355.
14. Mandal, S.; Mandal, S.; Ghosh, S. K.; Sar, P.; Ghosh, A.; Saha, R.; Saha, B. *RSC Adv.* **2016,** *6,* 69605–69614. (Review).
15. El Bouakher, A.; Tasserie, J.; Le Goff, R.; Lhoste, J.; Martel, A.; Comesse, S. *J. Org. Chem.* **2017,** *82,* 5798–5809.
16. López, J. J.; Pérez, E. G. *Synth. Commun.* **2019,** *49,* 715–723.
17. Yearty, K. L.; Maynard, R. K.; Cortes, C. N.; Morrison, R. W. *J. Chem. Educ.* **2020,** *97,* 578–581.

Wittig Reaction

Olefination of carbonyls using phosphorus ylides, typically the Z-olefin is the major isomer obtained.

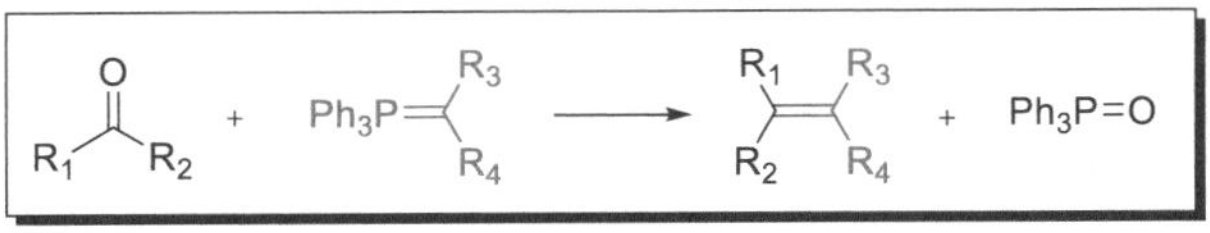

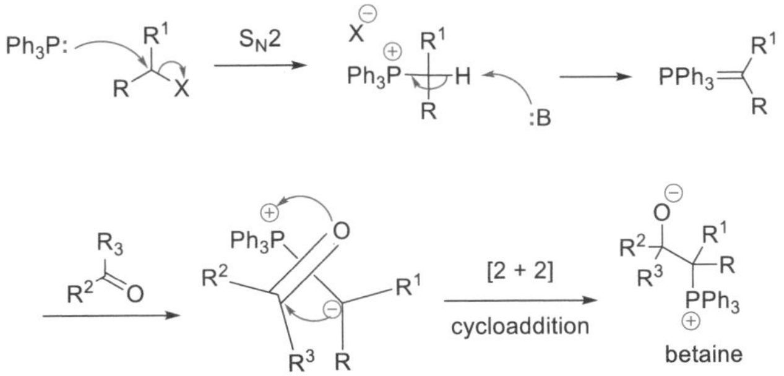

The "puckered" transition state, irreversible and concerted

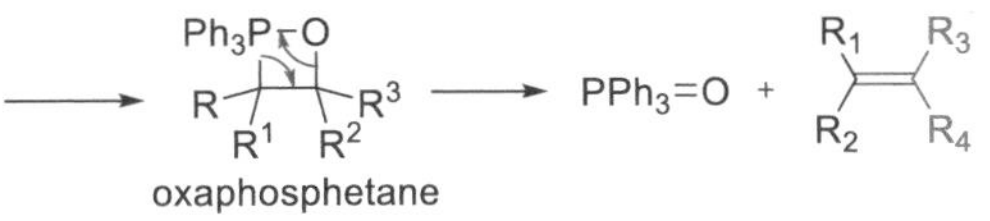

oxaphosphetane

Example 1[3]

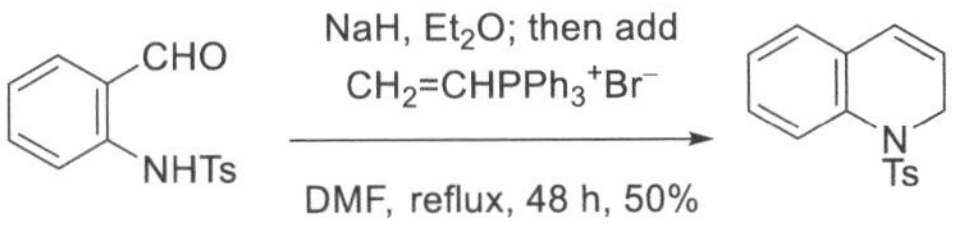

Example 2[4]

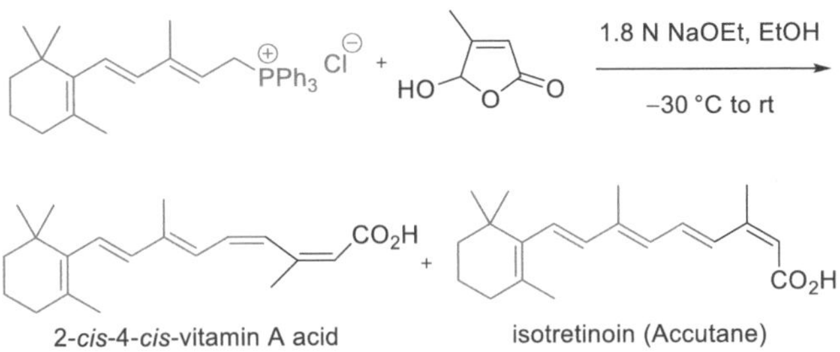

2-cis-4-cis-vitamin A acid   isotretinoin (Accutane)

J. J. Li, *Name Reactions*, https://doi.org/10.1007/978-3-030-50865-4_159

Example 3, Intramolecular Wittig reaction[5]

Example 4[9]

Example 5[11]

Z/E = 60:40

Example 6, Process-scale[14]

Example 7, Manufacturing route for Janus kinase (JAK) inhibitor ASP3627[16]

Example 8, Aza-Wittig reaction[17]

References

1. Wittig, G.; Schöllkopf, U. *Ber.* **1954**, *87*, 1318–1330. Georg Wittig (Germany, 1897–1987), born in Berlin, Germany, received his Ph.D. from K. von Auwers. He shared the Nobel Prize in Chemistry in 1981 with Herbert C. Brown (USA, 1912–2004) for their development of organic boron and phosphorous compounds.
2. Maercker, A. *Org. React.* **1965**, *14*, 270–490. (Review).
3. Schweizer, E. E.; Smucker, L. D. *J. Org. Chem.* **1966**, *31*, 3146–3149.
4. Garbers, C. F.; Schneider, D. F.; van der Merwe, J. P. *J. Chem. Soc. (C)* **1968**, 1982–1983.
5. Ernest, I.; Gosteli, J.; Greengrass, C. W.; Holick, W.; Jackman, D. E.; Pfaendler, H. R.; Woodward, R. B. *J. Am. Chem. Soc.* **1978**, *100*, 8214–8222.
6. Murphy, P. J.; Brennan, J. *Chem. Soc. Rev.* **1988**, *17*, 1–30. (Review).
7. Maryanoff, B. E.; Reitz, A. B. *Chem. Rev.* **1988**, *89*, 863–927. (Review).
8. Vedejs, E.; Peterson, M. J. *Top. Stereochem.* **1994**, *21*, 1–157. (Review).
9. Nicolaou, K. C. *Angew. Chem. Int. Ed.* **1996**, *35*, 589–607. (Review).
10. Rong, F. *Wittig reaction* in. In *Name Reactions for Homologations-Part I*; Li, J. J., Ed.; Wiley: Hoboken, NJ, **2009**, pp 588–612. (Review).
11. Kajjout, M.; Smietana, M.; Leroy, J.; Rolando, C. *Tetrahedron Lett.* **2013**, *38*, 1658–1660.
12. Rocha, D. H. A.; Pinto, D. C. G. A.; Silva, A. M. S. *Eur. J. Org. Chem.* **2018**, 2443–2457. (Review).
13. Karanam, P.; Reddy, G. M.; Lin, W. *Synlett* **2018**, *29*, 2608–2622. (Review).

14. Zhu, F.; Aisa, H. A.; Zhang, J.; Hu, T.; Sun, C.; He, Y.; Xie, Y.; Shen, J. *Org. Process Res. Dev.* **2018,** *22,* 91–96.
15. Longwitz, L.; Werner, T. *Pure Appl. Chem.* **2019,** *91,* 95–102. (Review).
16. Hirasawa, S.; Kikuchi, T.; Kawazoe, S. *Org. Process Res. Dev.* **2019,** *23,* 2378–2387.
17. Luo, J.; Kang, Q.; Huang, W.; Zhu, J.; Wang, T. *Synth. Commun.* **2020,** *50,* 692–699.

[1,2]-Wittig Rearrangement

Treatment of ethers with bases such as alkyl lithium results in alcohols.

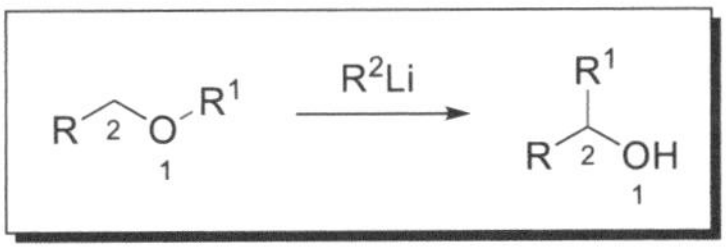

The [1,2]-Wittig rearrangement is believed to proceed via a radical mechanism:

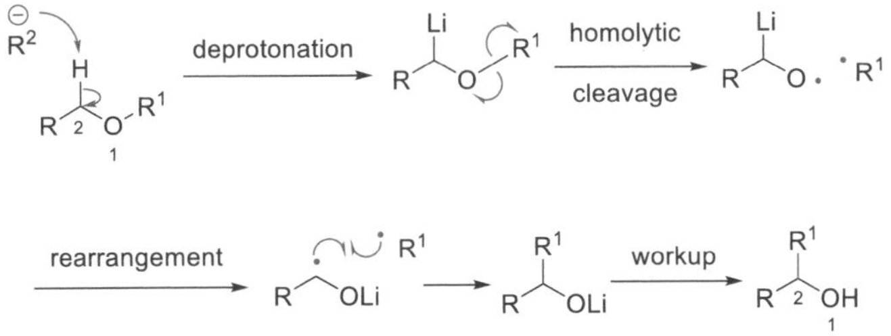

Example 1, Aza [1,2]-Wittig rearrangement[2]

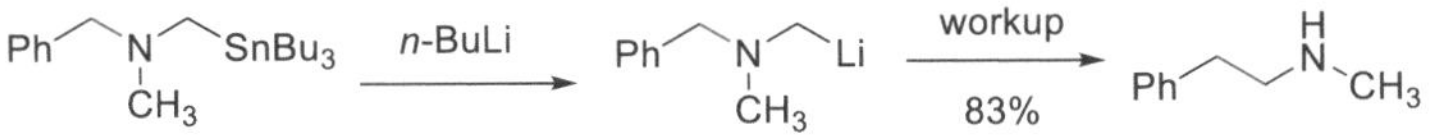

Example 2[3]

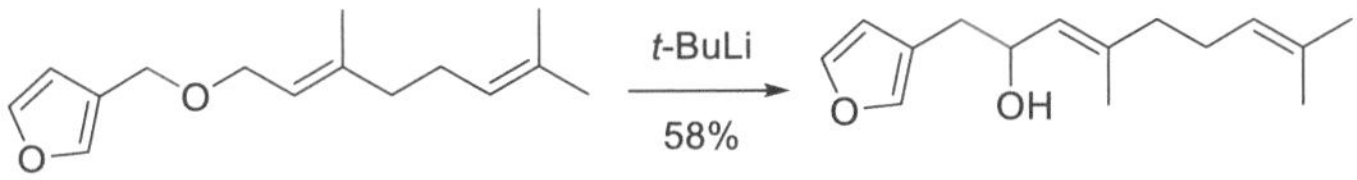

Example 3[4]

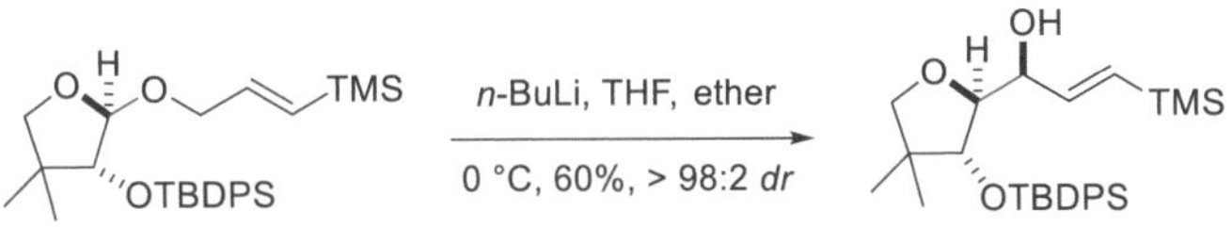

Example 4[6]

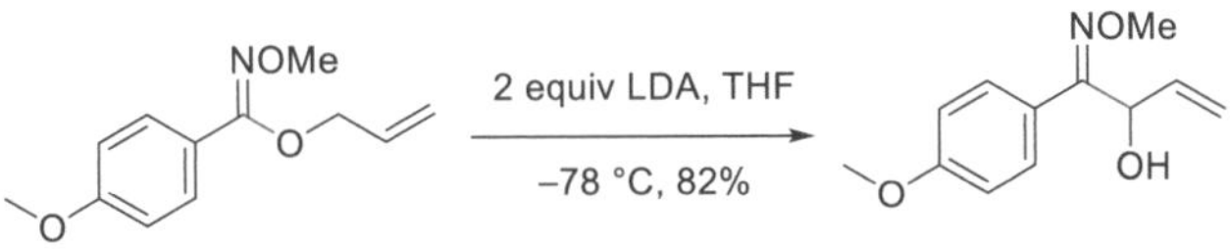

Example 5[8]

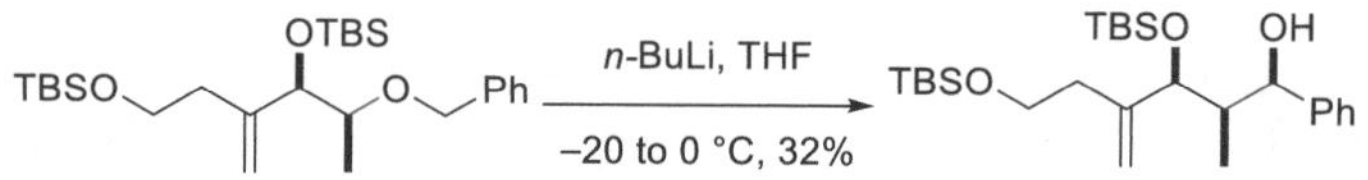

Example 6[9]

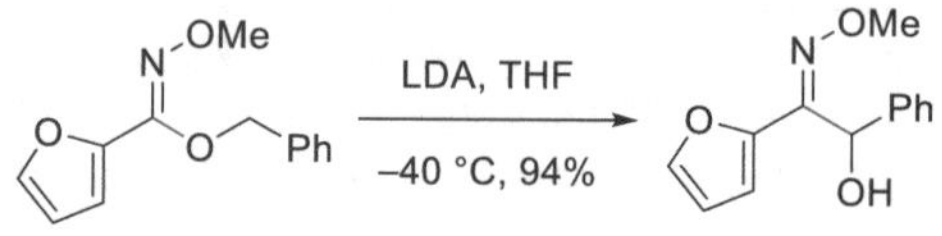

Example 7[11]

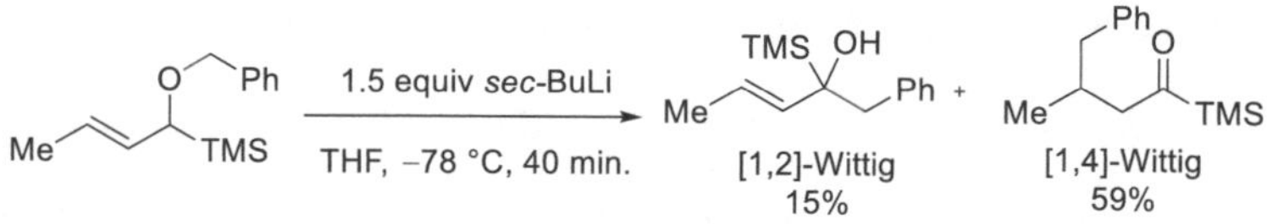

Example 8, Synthesis of (Z)-allylic alcohol[12]

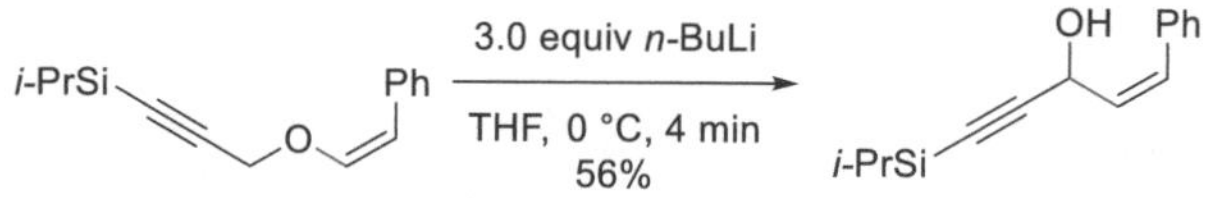

Example 9, [1,2]-Wittig rearrangement of 6H-benzo[c]chromene[13]

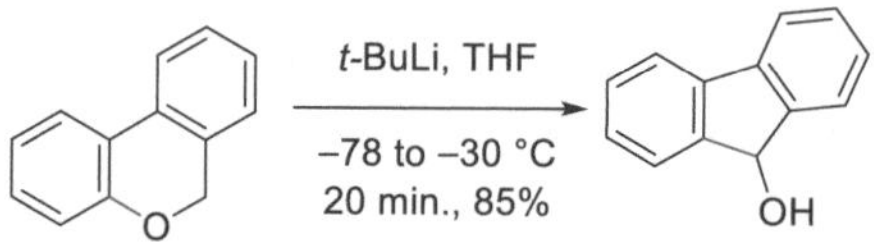

Example 10, CPME = cyclopentyl methyl ether[14]

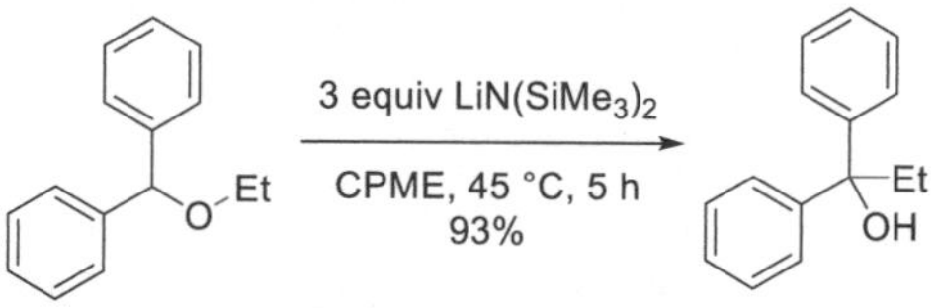

References

1   Wittig, G.; Löhmann, L. *Ann.* **1942,** *550*, 260–268.
2   Peterson, D. J.; Ward, J. F. *J. Organomet. Chem.* **1974,** *66*, 209–217.
3   Tsubuki, M.; Okita, H.; Honda, T. *J. Chem. Soc., Chem. Commun.* **1995,** 2135–2136.
4   Tomooka, K.; Yamamoto, H.; Nakai, T. *J. Am. Chem. Soc.* **1996,** *118*, 3317–3318.
5   Maleczka, R. E., Jr.; Geng, F. *J. Am. Chem. Soc*. **1998,** *120*, 8551–8552.
6   Miyata, O.; Asai, H.; Naito, T. *Synlett* **1999,** 1915–1916.
7   Katritzky, A. R.; Fang, Y. *Heterocycles* **2000,** *53*, 1783–1788.
8   Tomooka, K.; Kikuchi, M.; Igawa, K.; Suzuki, M.; Keong, P.-H.; Nakai, T. *Angew. Chem. Int. Ed.* **2000,** *39*, 4502–4505.
9   Miyata, O.; Asai, H.; Naito, T. *Chem. Pharm. Bull.* **2005,** *53*, 355–360.
10 Wolfe, J. P.; Guthrie, N. J. *[1,2]-Wittig Rearrangement.* In *Name Reactions for Homologations-Part II*; Li, J. J., Ed.; Wiley: Hoboken, NJ, **2009,** pp 226–240. (Review).
11 Onyeozili, E. N.; Mori-Quiroz, L. M.; Maleczka, R. E., Jr. *Tetrahedron* **2013,** *69*, 849–860.
12 Kurosawa, F.; Nakano, T.; Soeta, T.; Endo, K.; Ukaji, Y. *J. Org. Chem.* **2015,** *80*, 5696–5703.
13 Velasco, R.; Silva López, C.; Nieto Faza, O.; Sanz, R. *Chem. Eur. J.* **2016,** *22*, 15058–15068.
14 Liu, Z.; Li, M.; Wang, B.; Deng, G.; Chen, W.; Kim, B.-S.; Zhang, H.; Yang, X.; Walsh, P. *J. Org. Chem. Front.* **2018,** *5*, 1870–1876.

[2,3]-Wittig Rearrangement

Transformation of allyl ethers into homoallylic alcohols by treatment with base. Also known as the Wittig–Still rearrangement, or Still [2,3]-Wittig rearrangement. *Cf.* Sommelet–Hauser rearrangement.

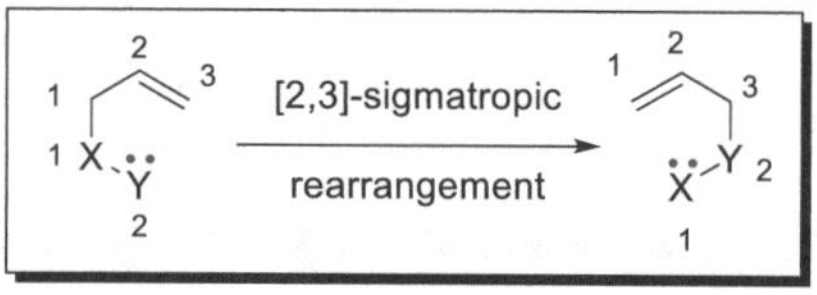

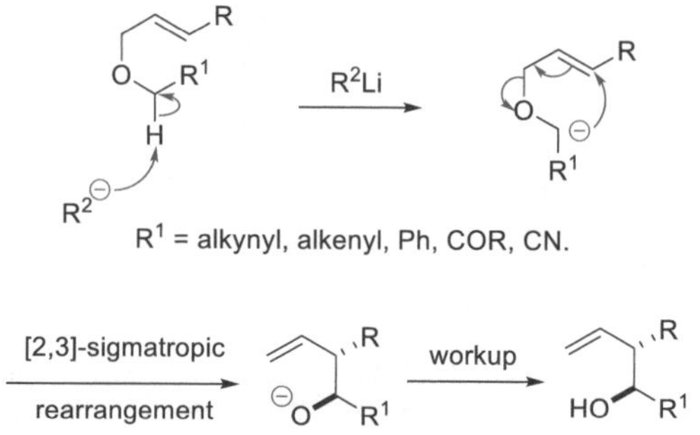

R^1 = alkynyl, alkenyl, Ph, COR, CN.

Example 1[4]

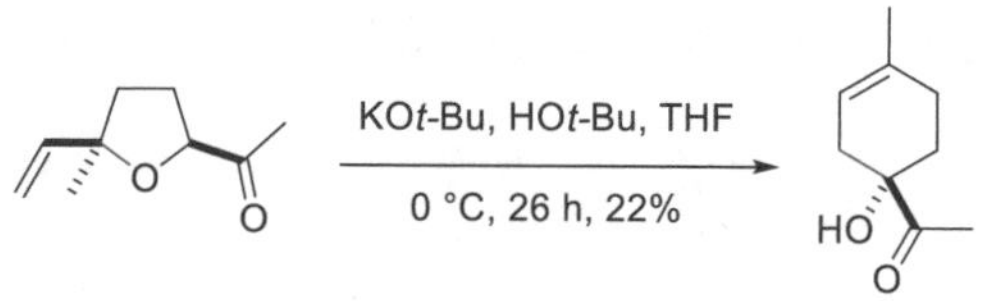

Example 2[5]

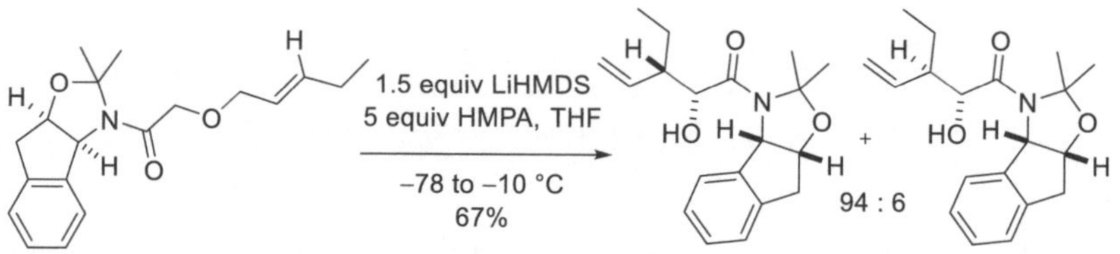

Example 3, Total synthesis of the pseudopterolide kalllolide A[6]

Example 4, Tandem Wittig rearrangement/alkylative cyclization reactions[11]

Example 5, Aza-[2,3]-Wittig rearrangement[12]

Example 6, Fidelity of the chirality conserved[14]

Example 7, The homoallylic alcohol was obtained as a single stereoisomer[15]

Example 8, Making mechanism-based inhibitors (MBIs) of a-L-arabino-furanosidases[16]

References

1. Cast, J.; Stevens, T. S.; Holmes, J. *J. Chem. Soc.* **1960**, 3521–3527.
2. Thomas, A. F.; Dubini, R. *Helv. Chim. Acta* **1974**, *57*, 2084–2087.
3. Nakai, T.; Mikami, K.; Taya, S.; Kimura, Y.; Mimura, T. *Tetrahedron Lett.* **1981**, *22*, 69–72.
4. Nakai, T.; Mikami, K. *Org. React.* **1994**, *46*, 105–209. (Review).
5. Kress, M. H.; Yang, C.; Yasuda, N.; Grabowski, E. J. J. *Tetrahedron Lett.* **1997**, *38*, 2633–2636.
6. Marshall, J. A.; Liao, J. *J. Org. Chem.* **1998**, *63*, 5962–5970.
7. Maleczka, R. E., Jr.; Geng, F. *Org. Lett.* **1999**, *1*, 1111–1113.
8. Tsubuki, M.; Kamata, T.; Nakatani, M.; Yamazaki, K.; Matsui, T.; Honda, T. *Tetrahedron: Asymmetry* **2000**, *11*, 4725–4736.
9. Schaudt, M.; Blechert, S. *J. Org. Chem.* **2003**, *68*, 2913–2920.
10. Ahmad, N. M. *[2,3]-Wittig Rearrangement*. In *Name Reactions for Homologations-Part II*; Li, J. J., Ed.; Wiley: Hoboken, NJ, 2009, pp 241–256. (Review).
11. Everett, R. K.; Wolfe, J. P. *Org. Lett.* **2013**, *15*, 2926–2929.
12. Everett, R. K.; Wolfe, J. P. *J. Org. Chem.* **2015**, *80*, 9041–9056.
13. Rycek, L.; Hudlicky, T. *Angew. Chem. Int. Ed.* **2017**, *56*, 6022–6066. (Review).
14. Han, P.; Zhou, Z.; Si, C.-M.; Sha, X.-Y.; Gu, Z.-Y.; Wei, B.-G.; Lin, G.-Q. *Org. Lett.* **2017**, *19*, 6732–6735.
15. Leon, R. M.; Ravi, D.; An, J. S.; del Genio, C. L.; Rheingold, A. L.; Gaur, A. B.; Micalizio, G. C. *Org. Lett.* **2019**, *21*, 3193–3197.
16. McGregor, N. G. S.; Artola, M.; Nin-Hill, A.; Linzel, D.; Haon, M.; Reijngoud, J.; Ram, A.; Rosso, M.-N.; van der Marel, G. A.; Codee, J. D. C.; et al. *J. Am. Chem. Soc.* **2020**, *142*, 4648–4662.

Wolff Rearrangement

Conversion of an α-diazoketone into a ketene. Often employed to conduct ring con-straction.

Step-wise mechanism:

Treatment of the ketene with water would give the corresponding homologated car-boxylic acid.

Concerted mechanism:

Example 1, Hydride migrates with higher migratory priority[2]

Example 2, Oxindoles[3]

© Springer Nature Switzerland AG 2021
J. J. Li, *Name Reactions*, https://doi.org/10.1007/978-3-030-50865-4_160

Example 3, TFEA = trifluoroethyl trifluoroacetate[4]

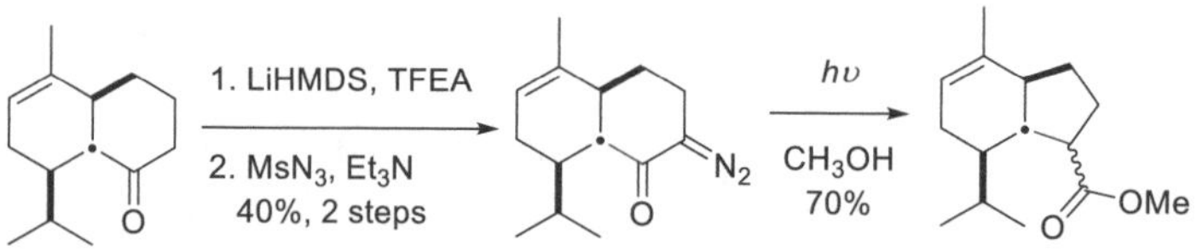

Example 4, Indole ring migrates[9]

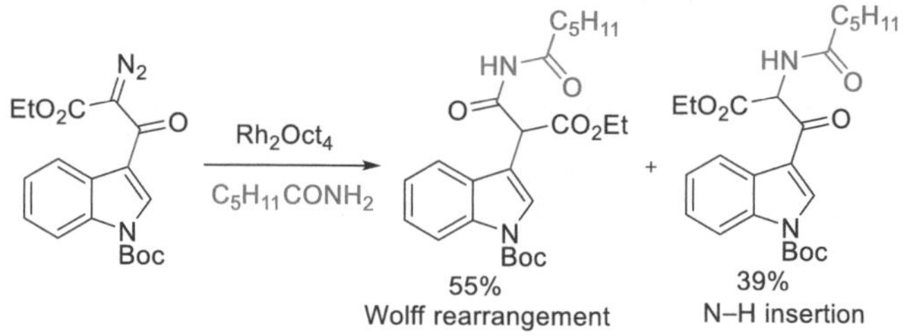

55%
Wolff rearrangement

39%
N–H insertion

Example 5[11]

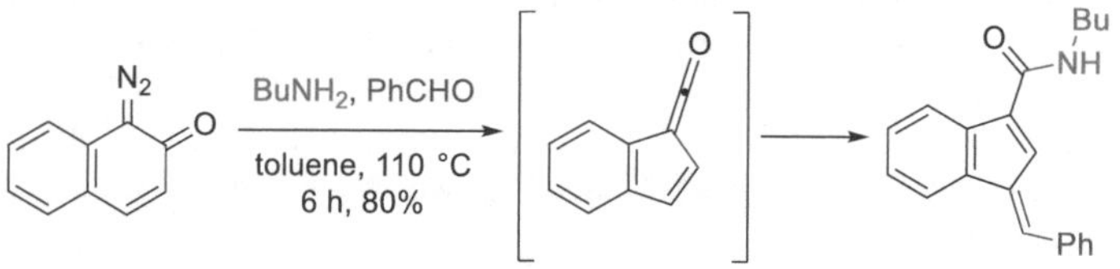

Example 6, The first example of a tandem Wolff rearrangement/catalytic ketene addition[12]

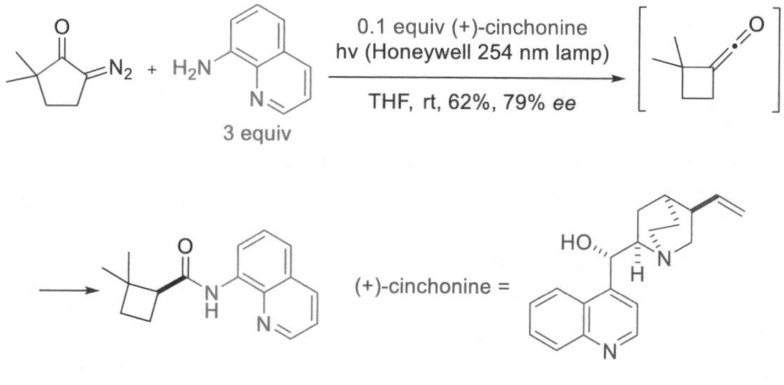

(+)-cinchonine =

Example 7, Indoles as nucleophiles adding to the ketene intermediates of the Wolff rearrangement[13]

Example 8, Ring contraction[14]

References

1. Wolff, L. *Ann.* **1912**, *394*, 23–108. Johann Ludwig Wolff (1857–1919) earned his doctorate in 1882 under Fittig at Strasbourg, where he later became an instructor. In 1891, Wolff joined the faculty of Jena, where he collaborated with Knorr for 27 years.
2. Zeller, K.-P.; Meier, H.; Müller, E. *Tetrahedron* **1972**, *28*, 5831–5838.
3. Kappe, C.; Fäber, G.; Wentrup, C.; Kappe, T. *Ber.* **1993**, *126*, 2357–2360.
4. Taber, D. F.; Kong, S.; Malcolm, S. C. *J. Org. Chem.* **1998**, *63*, 7953–7956.
5. Yang, H.; Foster, K.; Stephenson, C. R. J.; Brown, W.; Roberts, E. *Org. Lett.* **2000**, *2*, 2177–2179.
6. Kirmse, W. "100 years of the Wolff Rearrangement" *Eur. J. Org. Chem.* **2002**, 2193–2256. (Review).
7. Julian, R. R.; May, J. A.; Stoltz, B. M.; Beauchamp, J. L. *J. Am. Chem. Soc.* **2003**, *125*, 4478–4486.
8. Zeller, K.-P.; Blocher, A.; Haiss, P. *Minirev. Org. Chem.* **2004**, *1*, 291–308. (Review).
9. Davies, J. R.; Kane, P. D.; Moody, C. J.; Slawin, A. M. Z. *J. Org. Chem.* **2005**, *70*, 5840–5851.
10. Kumar, R. R.; Balasubramanian, M. *Wolff Rearrangement.* In *Name Reactions for Homologations-Part II*; Li, J. J., Ed.; Wiley: Hoboken, NJ, **2009**, pp 257–273. (Review).
11. Somai Magar, K. B.; Lee, Y. R. *Org. Lett.* **2013**, *15*, 4288–4291.
12. Chapman, L. M.; Beck, J. C.; Wu, L.; Reisman, S. E. *J. Am. Chem. Soc.* **2016**, *138*, 9803–9806.
13. Hu, X.; Chen, F.; Deng, Y.; Jiang, H.; Zeng, W. *Org. Lett.* **2018**, *20*, 6140–6143.
14. Hancock, E. N.; Kuker, E. L.; Tantillo, D. J.; Brown, M. K. *Angew. Chem. Int. Ed.* **2020**, *59*, 436–441.

Wolff–Kishner Reaction

Carbonyl reduction to methylene using basic hydrazine.

Example 1, The Huang Minlon modification, with loss of ethylene[5]

Example 2[7]

© Springer Nature Switzerland AG 2021
J. J. Li, *Name Reactions*, https://doi.org/10.1007/978-3-030-50865-4_161

Example 3[8]

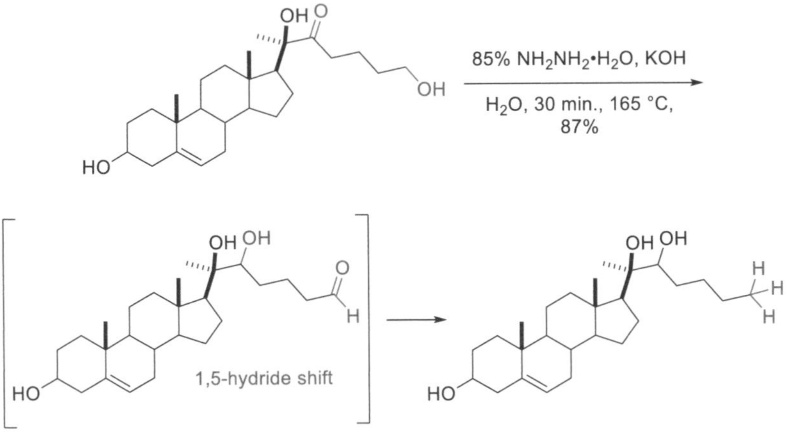

Example 4, Huang Minlon modification[10]

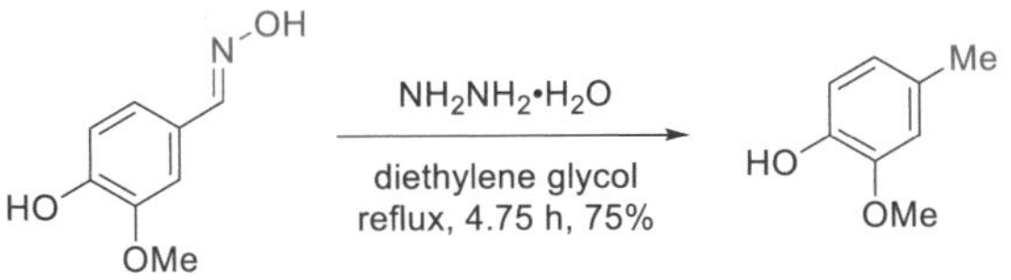

Example 3, Large-scale Wolff–Kishner reaction[13]

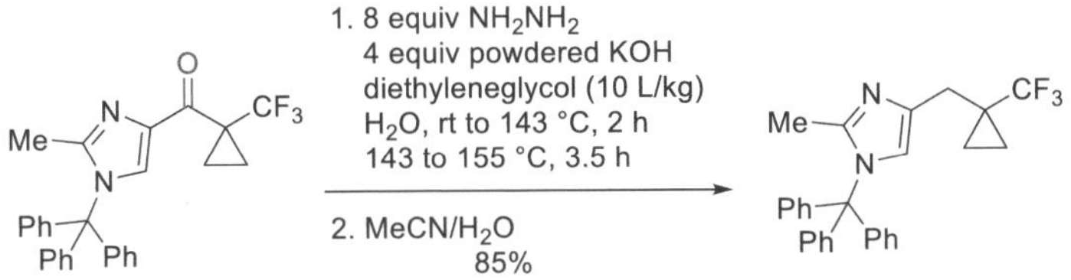

Example 4, Oxydative deoxygenation is followed by Wolff–Kishner reaction[14]

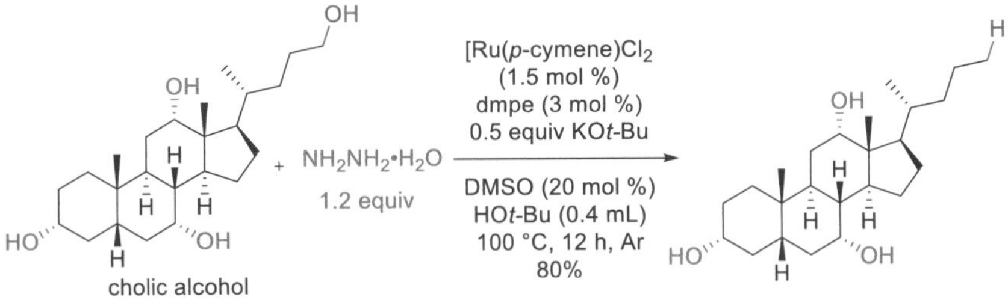

cholic alcohol

Example 5, Wolff–Kishner reaction works better than the Baron–McCombie reaction for deoxygenation here:[15,16]

References

1. (a) Kishner, N. *J. Russ. Phys. Chem. Soc.* **1911**, *43*, 582–595. Nicolai Kishner was a Russian chemist. (b) Wolff, L. *Ann.* **1912**, *394*, 86. (c) Huang, Minlon *J. Am. Chem. Soc.* **1946**, *68*, 2487–2488. (d) Huang, Minlon *J. Am. Chem. Soc.* **1949**, *71*, 3301–3303. (The Huang Minlon modification).
2. Cram, D. J.; Sahyun, M. R. V.; Knox, G. R. *J. Am. Chem. Soc.* **1962**, *84*, 1734–1735.
3. Szmant, H. H. *Angew. Chem. Int. Ed.* **1968**, *7*, 120–128. (Review).
4. Murray, R. K., Jr.; Babiak, K. A. *J. Org. Chem.* **1973**, *38*, 2556–2557.
5. Lemieux, R. P.; Beak, P. *Tetrahedron Lett.* **1989**, *30*, 1353–1356.
6. Taber, D. F.; Stachel, S. J. *Tetrahedron Lett.* **1992**, *33*, 903–906.
7. Gadhwal, S.; Baruah, M.; Sandhu, J. S. *Synlett* **1999**, 1573–1592.
8. Szendi, Z.; Forgó, P.; Tasi, G.; Böcskei, Z.; Nyerges, L.; Sweet, F. *Steroids* **2002**, *67*, 31–38.
9. Bashore, C. G.; Samardjiev, I. J.; Bordner, J.; Coe, J. W. *J. Am. Chem. Soc.* **2003**, *125*, 3268–3272.
10. Pasha, M. A. *Synth. Commun.* **2006**, *36*, 2183–2187.
11. Song, Y.-H.; Seo, J. *J. Heterocycl. Chem.* **2007**, *44*, 1439–1443.
12. Shibahara, M.; Watanabe, M.; Aso, K.; Shinmyozu, T. *Synthesis* **2008**, 3749–3754.
13. Kuethe, J. T.; Childers, K. G.; Peng, Z.; Journet, M.; Humphrey, G. R.; Vickery, T.; Bachert, D.; Lam, T. T. *Org. Process Res. Dev.* **2009**, *13*, 576–580.
14. Dai, X.-J.; Li, C.-J. *J. Am. Chem. Soc.* **2016**, *138*, 5433–5440.
15. Wu, G.-J.; Zhang, Y.-H.; Tan, D.-X.; Han, F.-S. *Nat. Commun.* **2018**, *9*, 1–8.
16. Wu, G.-J.; Zhang, Y.-H.; Tan, D.-X.; He, L.; Cao, B.-C.; He, Y.-P.; Han, F.-S. *J. Org. Chem.* **2019**, *84*, 3223–3238.
17. Li, C.-J.; Huang, J.; Dai, X.-J.; Wang, H.; Chen, N.; Wei, W.; Zeng, H.; Tang, J.; Li, C.; Zhu, D.; et al. *Synlett* **2019**, *30*, 1508–1524. (Review).
18. Wang, S.; Cheng, B.-Y.; Srsen, M.; Koenig, B. *J. Am. Chem. Soc.* **2020**, *142*, 7524–7531.

Index

© Springer Nature Switzerland AG 2021
J. J. Li, *Name Reactions*, https://doi.org/10.1007/978-3-030-50865-4

Printed in the United States
by Baker & Taylor Publisher Services